Prealgebra

Prealgebra

Fifth Edition

Elayn Martin-Gay

University of New Orleans

PEARSON

Prentice
Hall

Upper Saddle River, New Jersey 07458

Library of Congress Cataloging-in-Publication Data

Martin-Gay, K. Elayn
  Prealgebra/K. Elayn Martin-Gay.—5th ed.
    p. cm.
  Includes index.
  ISBN 0-13-231951-9 (pbk.: alk. paper) — ISBN 0-13-157642-9 (alk. paper)
  1. Algebra.   I. Title.

QA152.3.M37 2008
512.9—dc21

Executive Editor: *Paul Murphy*
Editorial Director: *Christine Hoag*
Project Manager: *Mary Beckwith*
Production Management: *Elm Street Publishing Services, Inc.*
Senior Managing Editor: *Linda Mihatov Behrens*
Executive Managing Editor: *Kathleen Schiaparelli*
Media Project Manager, Developmental Math: *Audra J. Walsh*
Media Production Editor: *Ashley Booth*
Assistant Managing Editor, Science and Math Supplements: *Karen Bosch*
Manufacturing Buyer: *Maura Zaldivar*
Manufacturing Manager: *Alexis Heydt-Long*
Director of Marketing: *Patrice Jones*
Senior Marketing Manager: *Kate Valentine*
Marketing Assistant: *Jennifer de Leeuwerk*
Development Editor: *Elaine Page*
Editor in Chief, Development: *Carol Trueheart*
Editorial Assistant/Print Supplements Editor: *Abigail Rethore*
Art Director: *Maureen Eide*
Interior/Cover Designer: *Suzanne Behnke*
Art Editor: *Thomas Benfatti*
Director of Creative Services: *Paul Belfanti*
Cover Photo: *Getty Images, Inc./Altrendo Images*
Manager, Cover Visual Research & Permissions: *Karen Sanatar*
Director, Image Resource Center: *Melinda Patelli*
Manager, Rights and Permissions: *Zina Arabia*
Manager, Visual Research: *Beth Brenzel*
Image Permission Coordinator: *Craig Jones*
Photo Researcher: *Julie Tesser*
Composition: *ICC McMillan Inc.*
Art Studios: *Scientific Illustrators and Laserwords*

Printed in the United States of America
10 9 8 7 6 5 4 3 2 1

ISBN: (paperback) 0-13-231951-9; (case bound) 0-13-157642-9

Pearson Education LTD., *London*
Pearson Education Australia PTY., Limited, *Sydney*
Pearson Education Singapore, Pte. Ltd.
Pearson Education North Asia Ltd., *Hong Kong*
Pearson Education Canada, Ltd., *Canada*
Pearson Educacíon de Mexico, S.A. de C.V.
Pearson Education—Japan, *Tokyo*
Pearson Education Malaysia, Pte. Ltd.

To Jewett B. Gay,
and to the memory of
her husband, Jack Gay

Contents

1   The Whole Numbers   1

2   Integers and Introduction to Solving Equations   95

Tools to Help Students Succeed

Your textbook includes a number of features designed to help you succeed in this math course—as well as the next math course you take. These features include:

Feature	Benefit	Page
Well-crafted Exercise Sets: We learn math by doing math	The exercise sets in your text offer an ample number of exercises carefully ordered so you can master basic mathematical skills and concepts while developing all-important problem solving skills. Exercise sets include Mixed Practice exercises to help you master multiple key concepts, as well as Vocabulary and Readiness Check, Writing, Applications, Concept Check, Concept Extension, and Review exercises.	179–181
Solutions-to-Selected Exercises: Built-in solutions at the back of the text	If you need to review problems you find difficult, this built-in solutions manual at the back of the text provides the step-by-step solutions to every other odd-numbered exercise in the exercise sets.	A25
Study Skills Builders: Maximize your chances for success	Study Skills Builders reinforce the material in *Section 1.1—Tips for Success in Mathematics.* Study Skills Builders are a great resource for study ideas and self-assessment to maximize your opportunity for success in this course. Take your new study skills with you to help you succeed in your next math course.	140
The Bigger Picture: Succeed in this math course and the next one you take	The Bigger Picture focuses on the key concept of this course—operations on sets of numbers and solving equations—and asks you to keep an ongoing outline so you can recognize and perform operations on different types of numbers and solve equations. A strong foundation in these topics will help you succeed in this prealgebra course, as well as the next math course you take.	191
Examples: Step-by-step instruction for you	Examples in the text provide you with clear, concise step-by-step instructions to help you learn. Annotations in the examples provide additional instruction.	176
Helpful Hints: Help where you'll need it most	Helpful Hints provide tips and advice at exact locations where students need it most. Strategically placed where you might have the most difficulty, Helpful Hints will help you work through common trouble spots.	176
Practice Problems: Immediate reinforcement	Practice Problems offer immediate reinforcement after every example. Try each Practice Problem after studying the corresponding example to make sure you have a good working knowledge of the concept.	176
Integrated Review: Mid-chapter progress check	To ensure that you understand the key concepts covered in the first sections of the chapter, work the exercises in the Integrated Review before you continue with the rest of the chapter.	182
Vocabulary Check: Key terms and vocabulary **Vocabulary and Readiness Check**	Make sure you understand key terms and vocabulary in each chapter with the Vocabulary Check at the end of the chapter. Use the Vocabulary and Readiness Checks to build your skills in the sections.	200, 188
Chapter Highlights: Study smart	Chapter Highlights outline the key concepts of the chapter along with examples to help you focus your studying efforts as you prepare for your test.	201
Chapter Test: Take a practice test	In preparation for your classroom test, take this practice test to make sure you understand the key topics in the chapter. Be sure to use the **Chapter Test Prep Video CD** included with this text to see the author present a fully worked-out solution to each exercise in the Chapter Test.	208

Martin-Gay's CD VIDEO RESOURCES Help Students Succeed

Martin-Gay's Chapter Test Prep Video CD (available with this text)

- Provides students with help during their most "teachable moment"—while they are studying for a test.
- Text author Elayn Martin-Gay presents step-by-step solutions to the exact exercises found in each Chapter Test in the book.
- Easy video navigation allows students to instantly access the worked-out solutions to the exercises they want to review.
- Captioned in English and Spanish.

Martin-Gay's CD Lecture Series (with Tips for Success in Mathematics)

- Text author Elayn Martin-Gay presents the key concepts from every section of the text in 10–15 minute mini-lectures.
- Students can easily review a section or a specific topic before a homework assignment, quiz, or test.
- Includes fully worked-out solutions to exercises marked with a CD Video icon ( ) in each section.
- Includes *Section 1.1, Tips for Success in Mathematics.*
- Captioned in English and Spanish.
- Ask your bookstore for information about Martin-Gay's *Prealgebra,* Fifth Edition, CD Lecture Series, or visit www.prenhall.com.

Additional Resources to Help You Succeed

Student Study Pack
A single, easy-to-use package—available bundled with your textbook or by itself—for purchase through your bookstore. This package contains the following resources to help you succeed:

Student Solutions Manual
- Contains worked-out solutions to odd-numbered exercises from each section exercise set, Practice Problems, Mental Math exercises, and all exercises found in the Chapter Review and Chapter Tests.

Prentice Hall Math Tutor Center
- Staffed by qualified math instructors who provide students with tutoring on examples and odd-numbered exercises from the textbook. Tutoring is available via toll-free telephone, toll-free fax, email, or the Internet.

Martin-Gay's CD Lecture Series
- Text author Elayn Martin-Gay presents the key concepts from every section of the text with 10–15 minute mini-lectures. Students can easily review a section or a specific topic before a homework assignment, quiz, or test.
- Includes fully worked-out solutions to exercises marked with a CD Video icon ( ) in each section. Also includes *Section 1.1, Tips for Success in Mathematics*.

Online Homework and Tutorial Resources

MyMathLab *MyMathLab*
MyMathLab

MyMathLab is a series of text specific, easily customizable, online courses for Prentice Hall textbooks in mathematics and statistics. MyMathLab is powered by Course Compass™—Pearson Education's online teaching and learning environment—and by MathXL®—our online homework, tutorial, and assessment system. MyMathLab gives instructors the tools they need to deliver all or a portion of their course online, whether students are in a lab setting or working from home. MyMathLab provides a rich and flexible set of course materials, featuring free-response exercises that are algorithmically generated for unlimited practice and mastery. Students can also use online tools, such as video lectures, animations, and a multimedia textbook, to independently improve their understanding and performance. MyMathLab is available to qualified adopters. For more information, visit our Web site at www.mymathlab.com or contact your Prentice Hall sales representative. (MyMathLab must be set up and assigned by your instructor.)

MathXL® www.mathxl.com *Math XL*
MathXL®

MathXL is a powerful online homework, tutorial, and assessment system that accompanies the text. With MathXL, instructors can create, edit, and assign online homework and tests using algorithmically generated exercises correlated to your textbook. All student work is tracked in MathXL's online gradebook. Students can take chapter tests in MathXL and receive personalized study plans based on their test results. The study plan diagnoses weaknesses and links students directly to tutorial exercises for the objectives they need to study and retest. Students can also access supplemental animations and video clips directly from selected exercises. MathXL is available to qualified adopters. For more information, visit our Web site at www.mathxl.com, or contact your Prentice Hall sales representative for a product demonstration. (MathXL must be set up and assigned by your instructor.)

Preface

***Prealgebra,* Fifth Edition** was written to help students make the transition from arithmetic to algebra. To reach this goal, I introduce algebraic concepts early and repeat them as I treat traditional arithmetic topics, thus laying the groundwork for the next algebra course your students will take. A second goal was to show students the relevancy of the mathematics in everyday life and in the workplace.

The many factors that contributed to the success of the previous editions have been retained. In preparing the Fifth Edition, I considered comments and suggestions of colleagues, students, and many users of the prior edition throughout the country.

What's New in the Fifth Edition?

Enhanced Exercise Sets

- **NEW!** Three forms of mixed sections of exercises have been added to the Fifth Edition.
  - **Mixed Practice** exercises combining objectives within a section
  - **Mixed Practice** exercises combining previous sections
  - **Mixed Review** exercises included at the end of the Chapter Review

  These exercises require students to determine the problem type and strategy needed in order to solve it. In doing so, students need to think about key concepts to proceed with a correct method of solving—just as they would need to do on a test.

- **NEW! Vocabulary and Readiness Check** exercises appear at the beginning of most exercise sets. These exercises quickly check a student's understanding of new vocabulary words so that forthcoming instructions in the problem sets will be clear. The **readiness** exercises center on a student's understanding of a concept that is necessary in order to continue with the exercise set. For example, such a readiness exercise might be for students to write out both $\frac{2^3}{7}$ and $\left(\frac{2}{7}\right)^3$.

- **NEW! Concept Check exercises** have been added to the section exercise sets. These exercises are related to the Concept Check(s) found within the section. They help students measure their understanding of key concepts by focusing on common trouble areas. These exercises may ask students to identify a common error, and/or provide an explanation.

- **NEW! Concept Extensions** (formerly Combining Concepts) have been revised. These exercises extend the concepts and require students to combine several skills or concepts to solve the exercises in this section.

Increased Emphasis on Study Skills and Student Success

- **NEW! Study Skills Builders** (formerly Study Skill Reminders) Found at the end of many exercise sets, Study Skills Builders allow instructors to assign exercises that will help students improve their study skills and take responsibility for their part of the learning process. Study Skills Builders reinforce the material found in Section 1.1, "Tips for Success in Mathematics," and serve as an excellent tool for self-assessment.

- **NEW! The Bigger Picture** is a recurring feature that focuses on the key concepts of the course—1. operations on sets of numbers, and 2. solving equations—and helps students develop an outline to recognize and perform operations on different sets of numbers or to solve different types of equations. By working the exercises and developing this outline throughout the text, students can begin to transition from thinking "section by section" to thinking about how the mathematics in this course is part of the "bigger picture" of mathematics in general. A completed outline is provided in Appendix B so students have a model for their work.

- **Chapter Test Prep Video CD** New captioning options available in both English and Spanish provide students with help during their most "teachable moment"—while they are studying for a test. Included with every copy of the student edition of the text, this video CD provides fully worked-out solutions by the author to every exercise from each Chapter Test in the text. The easy video navigation allows students to instantly access the solutions to the exercises they want to review. The problems are solved by the author in the same manner as in the text.

- **NEW! Chapter Test files in TestGen** provide algorithms specific to each exercise from each Chapter Test in the text. Allows for easy replication of Chapter Tests with consistent, algorithmically generated problem types for additional assignments or assessment purposes.

Content Changes in the Fifth Edition

- **Enhanced Treatment of Equation Solving** to help students better prepare for their work in Chapter 3, Solving Equations and Problem Solving, and to expose them to prealgebra topics earlier in the course, the idea of equation solving is now introduced in Section 1.8, and revisited in Section 2.6, with dedicated coverage following in Chapter 3.

  This progression helps students gain greater exposure to and experience with equation solving—an essential tool for the prealgebra course as well as future courses. They are better prepared to work with equation solving in dedicated Chapter 3 and throughout the text. The earlier exposure helps reduce overlap with basic math topics and improve student motivation.

  In Section 1.8 (Introduction to Variables, Algebraic Expressions, and Equations) students determine whether a number is indeed a solution of a given equation. In Section 2.6, students are introduced to the multiplication and addition properties of equality and solve equations requiring one application of a property. The properties are reviewed in Section 3.2, and students solve equations requiring more than one property.

- **The global concepts in prealgebra that students struggle with are revisited in the text in the Review exercises, Mixed Practice exercises and Concept Extensions.** These concepts include:
  - Understanding the difference between an equation and an expression and between solving equations and simplifying expressions
  - Translating word phrases and statements to expressions and equations
  - Remembering operations on fractions

- **Adding and Subtracting Whole Numbers are now covered in Section 1.3** versus separate sections, to improve the pace of the chapter.

- To help students remember and retain concepts and operations on fractions, often a side-by-side illustration is used to help those who are visual learners. For example in Section 4.5, Adding and Subtracting Unlike Fractions, this side-by-side approach is used. On one side, figures illustrating unlike fractions are added. On the other side, the same fractions are added algebraically.

- Two new appendices have been added. They cover negative exponents and scientific notation.

- All exercise sets have been reviewed and updated to ensure that even- and odd-numbered exercises are paired.

Key Pedagogical Features

The following key features have been retained and/or updated for the Fifth Edition of the text:

Problem Solving Process   This is formally introduced in Chapter 3 with a four-step process that is integrated throughout the text. The four steps are **Understand, Translate, Solve,** and **Interpret.** The repeated use of these steps in a variety of examples shows their wide applicability. Reinforcing the steps can increase students' comfort level and confidence in tackling problems.

Exercise Sets Revised and Updated   The exercise sets have been carefully examined and extensively revised. Special focus was placed on making sure that even- and odd-numbered exercises are paired.

Examples   Detailed step-by-step examples were added, deleted, replaced, or updated as needed. Many of these reflect real life. Additional instructional support is provided in the annotated examples.

Practice Problems   Throughout the text, each worked-out example has a parallel Practice Problem. These invite students to be actively involved in the learning process. Students should try each Practice Problem after finishing the corresponding example. Learning by doing will help students grasp ideas before moving on to other concepts. Answers to the Practice Problems are provided at the bottom of each page.

Helpful Hints   Helpful Hints contain practical advice on applying mathematical concepts. Strategically placed where students are most likely to need immediate reinforcement, Helpful Hints help students avoid common trouble areas and mistakes.

Concept Checks   This feature allows students to gauge their grasp of an idea as it is being presented in the text. Concept Checks stress conceptual understanding at the point-of-use and help suppress misconceived notions before they start. Answers appear at the bottom of the page. Exercises related to Concept Checks are now included in the exercise sets.

Selected Solutions   Solutions to every other odd exercise are included in the back of the text. This built-in solutions manual allows students to check their work.

Integrated Reviews   A unique, mid-chapter exercise set that helps students assimilate new skills and concepts that they have learned separately over several sections. These reviews provide yet another opportunity for students to work with "mixed" exercises as they master the topics.

Vocabulary Check   Provides an opportunity for students to become more familiar with the use of mathematical terms as they strengthen their verbal skills. These appear at the end of each chapter before the Chapter Highlights.

Chapter Highlights   Found at the end of every chapter, these contain key definitions and concepts with examples to help students understand and retain what they have learned and help them organize their notes and study for tests.

Chapter Review   The end of every chapter contains a comprehensive review of topics introduced in the chapter. The Chapter Review offers exercises keyed to every section in the chapter, as well as Mixed Review **(NEW!)** exercises that are not keyed to sections.

Chapter Test and Chapter Test Prep Video CD   The Chapter Test is structured to include those problems that involve common student errors. The **Chapter Test Prep Video CD** gives students instant author access to a step-by-step video solution of each exercise in the Chapter Test. New captioning options are available in English and Spanish.

Cumulative Review   Follows every chapter in the text (except Chapter 1). Each odd-numbered exercise contained in the Cumulative Review is an earlier worked example in the text that is referenced in the back of the book along with the answer.

Writing Exercises ✎   These exercises occur in almost every exercise set and require students to provide a written response to explain concepts or justify their thinking.

Applications   Real-world and real-data applications have been thoroughly updated and many new applications are included. These exercises occur in almost every exercise set and show the relevance of mathematics and help students gradually, and continuously develop their problem solving skills.

Review Exercises (formerly Review and Preview exercises)   These exercises occur in each exercise set (except in Chapter 1) and are keyed to earlier sections. They review concepts learned earlier in the text that will be needed in the next section or chapter.

Exercise Set Resource Icons   Located at the opening of each exercise set, these icons remind students of the resources available for extra practice and support:

See Student Resource descriptions pages xxi–xxii for details on the individual resources available.

Exercise Icons   These icons facilitate the assignment of specialized exercises and let students know what resources can support them.

  • CD Video icon: exercise worked on Martin-Gay's CD Lecture Series.
  △ Triangle icon: identifies exercises involving geometric concepts.
  ✎ Pencil icon: indicates a written response is needed.
  ▦ Calculator icon: optional exercises intended to be solved using a scientific or graphing calculator.

Group Activities   Found at the end of each chapter, these activities are for individual or group completion, and are usually hands-on or data-based activities that extend the concepts found in the chapter allowing students to make decisions and interpretations and to think and write about algebra.

Optional: Calculator Exploration Boxes and Calculator Exercises   The optional Calculator Explorations provide key strokes and exercises at appropriate points to provide an opportunity for students to become familiar with these tools. Section exercises that are best completed by using a calculator are identified by ▦ for ease of assignment.

A Word about Textbook Design and Student Success

The design of developmental mathematics textbooks has become increasingly important. As students and instructors have told Prentice Hall in focus groups and market research surveys, these textbooks cannot look "cluttered" or "busy." A "busy" design can distract a student from what is most important in the text. It can also heighten math anxiety.

As a result of the conversations and meetings we have had with students and instructors, we concluded the design of this text should be understated and focused on the most important pedagogical elements. Students and instructors helped us to identify the primary elements that are central to student success. These primary elements include:

- Exercise Sets
- Examples and Practice Problems
- Helpful Hints
- Rules, Property, and Definition boxes

As you will notice in this text, these primary features are the most prominent elements in the design. We have made every attempt to make sure these elements are the features the eye is drawn to. The remaining features, the secondary elements in the design, blend into the "fabric" or "grain" of the overall design. These secondary elements complement the primary elements without becoming distractions.

Prentice Hall's thanks goes to all of the students and instructors (as noted by the author in Acknowledgments) who helped us develop the design of this text. At every step in the design process, their feedback proved valuable in helping us to make the right decisions. Thanks to your input, we're confident the design of this text will be both practical and engaging as it serves its educational and learning purposes.

Sincerely,

Paul Murphy

Executive Editor
Developmental Mathematics
Prentice Hall

Instructor and Student Resources

The following resources are available to help instructors and students use this text more effectively.

Instructor Resources

Annotated Instructor's Edition (0-13-157643-7)

- Answers to all exercises printed on the same text page
- Teaching Tips throughout the text placed at key points
- Includes Vocabulary Check Tips at the beginning of relevant sections
- General tips and suggestions for classroom or group activities

Instructor Solutions Manual (0-13-157640-2)

- Solutions to the even-numbered exercises
- Solutions to every Practice Problem
- Solutions to every exercise in the Integrated Reviews, Chapter Reviews, Chapter Tests, and Cumulative Reviews

Instructor's Resource Manual with Tests (0-13-157641-0)

- **NEW!** Includes Mini-Lectures for every section from the text
- Group Activities
- Free Response Test Forms, Multiple Choice Test Forms, Cumulative Tests, and Additional Exercises
- Answers to all items

TestGen (0-13-157647-X)

- Enables instructors to build, edit, print, and administer tests
- Features a computerized bank of questions developed to cover all text objectives
- Available on dual-platform Windows/Macintosh CD-ROM

Instructor Adjunct Resource Kit (0-13-157637-2)

The Martin-Gay Instructor/Adjunct Resource Kit (IARK) contains tools and resources to help adjuncts and instructors succeed in the classroom. The IARK includes:

- Instructor-to-Instructor CD Videos that offer tips, suggestions, and strategies for engaging students and presenting key topics
- PDF files of the Instructor Solutions Manual and the Instructor's Resource Manual
- TestGen

MyMathLab Instructor Version (0-13-147898-2)
MyMathLab www.mymathlab.com

MyMathLab is a series of text specific, easily customizable, online courses for Prentice Hall textbooks in mathematics and statistics. MyMathLab is powered by Course Compass™—Pearson Education's online teaching and learning environment—and by MathXL®—our online homework, tutorial, and assessment system. MyMathLab gives instructors the tools they need to deliver all or a portion of their course online, whether students are in a lab setting or working from home. MyMathLab provides a rich and flexible set of course materials, featuring free-response exercises that are algorithmically generated for unlimited practice and mastery. Students can also use online tools, such as video lectures, animations, and a multimedia textbook, to independently improve their understanding and performance. Instructors can use MyMathLab's homework and test managers to select and assign online exercises correlated directly to the text, and they can import TestGen tests into MyMathLab for added flexibility. MyMathLab's online gradebook—designed specifically for mathematics and statistics—automatically tracks students' homework and test results and gives the instructor control over how to calculate final grades. Instructors can also add offline (paper-and-pencil) grades to the gradebook. MyMathLab is available to qualified adopters. For more information, visit our website at www.mymathlab.com or contact your Prentice Hall sales representative.

MathXL Instructor Version (0-13-147895-8)
MathXL® www.mathxl.com

MathXL is a powerful online homework, tutorial, and assessment system that accompanies the text. With MathXL, instructors can create, edit, and assign online homework and tests using algorithmically generated exercises correlated to your textbook. All student work is tracked in MathXL's online gradebook. Students can take chapter tests in MathXL and receive personalized study plans based on their test results. The study plan diagnoses weaknesses and links students directly to tutorial exercises for the objectives they need to study and retest. Students can also access supplemental animations and video clips directly from selected exercises. MathXL is available to qualified adopters. For more information, visit our Web site at www.mathxl.com, or contact your Prentice Hall sales representative for a product demonstration.

Interact Math® Tutorial Web site www.interactmath.com

Get practice and tutorial help online! This interactive tutorial Web site provides algorithmically generated practice exercises that correlate directly to the exercises in your textbook. You can retry an exercise as many times as you like with new values each time for unlimited practice and mastery. Every exercise is accompanied by an interactive guided solution that gives you helpful feedback if you enter an incorrect answer, and you can also view a worked-out sample problem that steps you through an exercise similar to the one you're working on.

Student Resources

Student Solutions Manual (0-13-157638-0)

- Solutions to the odd-numbered section exercises
- Solutions to the Practice Problems
- Solutions to every exercise found in the Chapter Reviews and Chapter Tests

Martin-Gay's CD Lecture Series (0-13-157634-8)

- Perfect for review of a section or a specific topic, these mini-lectures by Elayn Martin-Gay cover the key concepts from each section of the text in approximately 10–15 minutes
- Includes fully worked-out solutions to exercises in each section marked with a 💿
- Includes coverage of Section 1.1, "Tips for Success Mathematics"
- New! Captioning options available in both English and Spanish

New—Worksheets for Classroom and Lab Practice

- Convenient, ready to use format, with ample work space
- Helps students stay organized and show their work step-by-step. Students work sample examples in class with their instructor. Then solve examples independently on the same page.

Prentice Hall Math Tutor Center (0-13-064604-0)

- Staffed by qualified math instructors who provide students with tutoring on examples and odd-numbered exercises from the textbook
- Tutoring is available via toll-free telephone, toll-free fax, e-mail, or the Internet
- Whiteboard technology allows tutors and students to see problems worked while they "talk" in real time over the Internet during tutoring sessions

Prealgebra, Fifth Edition Student Study Pack (0-13-134902-3)

The Student Study Pack includes:

- Martin-Gay's CD Lecture Series
- Student Solutions Manual
- Prentice Hall Math Tutor Center access code

Chapter Test Prep Video CD—Standalone (0-13-157644-5)

- Includes fully worked-out solutions to every problem from each Chapter Test in the text.

MathXL Tutorials on CD—Standalone (0-13-157648-8)

- Provides algorithmically generated practice exercises that correlate to exercises at the end of sections.
- Every exercise is accompanied by an example and a guided solution, selected exercises include a video clip.
- The software recognizes student errors and provides feedback. It can also generate printed summaries of students progress.

Interact Math® Tutorial Web Site www.interactmath.com

Get practice and tutorial help online! This interactive tutorial Web site provides algorithmically generated practice exercises that correlate directly to the exercises in your textbook. You can retry an exercise as many times as you like with new values each time for unlimited practice and mastery. Every exercise is accompanied by an interactive guided solution that gives you helpful feedback if you enter an incorrect answer, and you can also view a worked-out sample problem that steps you through an exercise similar to the one you're working on.

Acknowledgments

There are many people who helped me develop this text, and I will attempt to thank some of them here. Carrie Green and Cindy Trimble were *invaluable* for contributing to the overall accuracy of the text. Elaine Page and Suellen Robinson were *invaluable* for their many suggestions and contributions during the development and writing of this Fifth Edition. Ingrid Benson provided guidance throughout the production process.

A special thanks to my editor, Paul Murphy, for all of his assistance, support, and contributions to this project. A very special thank you goes to my project manager, Mary Beckwith, for being there 24/7/365, as my students say. Last, my thanks to the staff at Prentice Hall for all their support: Linda Behrens, Maura Zaldivar, Patty Burns, Tom Benfatti, Paul Belfanti, Maureen Eide, Suzanne Behnke, Kate Valentine, Patrice Jones, Chris Hoag, Paul Corey, and Tim Bozik.

I would like to thank the following reviewers for their input and suggestions:

Laurel Berry, *Bryant & Stratton*

Anita Collins, *Mesa Community College*

Robert Diaz, *Fullerton College*

Kathryn Gunderson, *Three Rivers Community College*

Elizabeth Hamman, *Cypress College*

Lloyd Harris, *Gulf Coast Community College*

Jeff Koleno, *Lorain County Community College*

Judy Langer, *Westchester Community College*

Sandy Lofstock, *St. Petersburg College*

Stan Mattoon, *Merced College*

Dr. Kris Mudunuri, *Long Beach City College*

Greg Nguyen, *Fullerton College*

Darlene Ornelas, *Fullerton College*

Warren Powell, *Tyler Junior College*

Jeanette Shea, *Central Texas College*

Edward Wagner, *Central Texas College*

Jenny Wilson, *Tyler Junior College*

I would also like to thank the following dedicated group of instructors who participated in our focus groups, Martin-Gay Summits, and our design review for the series. Their feedback and insights have helped to strengthen this edition of the text. These instructors include:

Cedric Atkins, *Mott Community College*

Laurel Berry, *Bryant & Stratton*

John Beyers, *University of Maryland*

Bob Brown, *Community College of Baltimore County–Essex*

Lisa Brown, *Community College of Baltimore County–Essex*

Gail Burkett, *Palm Beach Community College*

Cheryl Cantwell, *Seminole Community College*

Jackie Cohen, *Augusta State College*

Julie Dewan, *Mohawk Valley Community College*

Janice Ervin, *Central Piedmont Community College*

Pauline Hall, *Iowa State College*

Sonya Johnson, *Central Piedmont Community College*

Irene Jones, *Fullerton College*

Paul Jones, *University of Cincinnati*

Nancy Lange, *Inver Hills Community College*

Sandy Lofstock, *St. Petersburg College*

Jean McArthur, *Joliet Junior College*

Marica Molle, *Metropolitan Community College*

Greg Nguyen, *Fullerton College*

Linda Padilla, *Joliet Junior College*

Ena Salter, *Manatee Community College*

Carole Shapero, *Oakton Community College*

Anne Smallen, *Mohawk Valley Community College*

Jennifer Strehler, *Oakton Community College*

Tanomo Taguchi, *Fullerton College*

Leigh Ann Wheeler, *Greenville Technical Community College*

Valerie Wright, *Central Piedmont Community College*

A special thank you to those students who participated in our design review: Katherine Browne, Mike Bulfin, Nancy Canipe, Ashley Carpenter, Jeff Chojnachi, Roxanne Davis, Mike Dieter, Amy Dombrowski, Kay Herring, Todd Jaycox, Kaleena Levan, Matt Montgomery, Tony Plese, Abigail Polkinghorn, Harley Price, Eli Robinson, Avery Rosen, Robyn Schott, Cynthia Thomas, and Sherry Ward.

Additional Acknowledgments

As usual, I would like to thank my husband, Clayton, for his constant encouragement. A special thanks to my children, Eric and Bryan, for providing most of the cooking and humor in our household. I would also like to thank my extended family for their help and wonderful sense of humor. Their contributions are too numerous to list. They are Rod and Karen Pasch; Peter, Michael, Christopher, Matthew, and Jessica Callac; Stuart and Earline Martin; Josh, Mandy, Bailey, Ethan, and Avery Barnes; Mark, Sabrina, and Madison Martin; Leo and Barbara Miller; and Jewett Gay.

Elayn Martin-Gay

About the Author

Elayn Martin-Gay has taught mathematics at the University of New Orleans for more than 25 years. Her numerous teaching awards include the local University Alumni Association's Award for Excellence in Teaching, and Outstanding Developmental Educator at University of New Orleans, presented by the Louisiana Association of Developmental Educators.

Prior to writing textbooks, Elayn Martin-Gay developed an acclaimed series of lecture videos to support developmental mathematics students in their quest for success. These highly successful videos originally served as the foundation material for her texts. Today, the videos are specific to each book in the Martin-Gay series. The author has also created Chapter Test Prep Videos to help students during their most "teachable moment"—as they prepare for a test, along with Instructor-to-Instructor videos that provide teaching tips, hints, and suggestions for each developmental mathematics course, including basic mathematics, prealgebra, beginning algebra, and intermediate algebra.

Elayn is the author of 10 published textbooks as well as multimedia interactive mathematics, all specializing in developmental mathematics courses. She has participated as an author across the broadest range of educational materials: textbooks, videos, tutorial software, and courseware. All of these components are designed to work together. This offers an opportunity of various combinations for an integrated teaching and learning package offering great consistency for the student.

Applications Index

Prealgebra

1

The Whole Numbers

Whole numbers are the basic building blocks of mathematics. The whole numbers answer the question "How many?"

This chapter covers basic operations on whole numbers. Knowledge of these operations provides a good foundation on which to build further mathematical skills.

A river basin is the geographic area drained by a river and its tributaries. The Mississippi River Basin, drawn to the right, is the third largest in the world. It is divided into the six sub-basins shown. In Section 1.3, Exercises 75 through 78, page 27, we will see how whole numbers can be used to measure the area of the Mississippi River Basin.

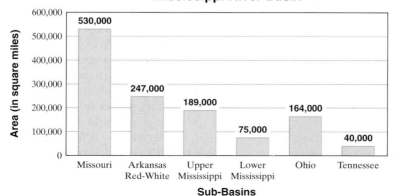

Mississippi River Basin

(Area in square miles, by Sub-Basins)

- Missouri: 530,000
- Arkansas Red-White: 247,000
- Upper Mississippi: 189,000
- Lower Mississippi: 75,000
- Ohio: 164,000
- Tennessee: 40,000

Before reading this section, remember that your instructor is your best source of information. Please see your instructor for any additional help or information.

Objective **A** Getting Ready for This Course

Now that you have decided to take this course, remember that a *positive attitude* will make all the difference in the world. Your belief that you can succeed is just as important as your commitment to this course. Make sure you are ready for this course by having the time and positive attitude that it takes to succeed.

Next, make sure that you have scheduled your math course at a time that will give you the best chance for success. For example, if you are also working, you may want to check with your employer to make sure that your work hours will not conflict with your course schedule.

On the day of your first class period, double-check your schedule and allow yourself extra time to arrive on time in case of traffic problems or difficulty locating your classroom. Make sure that you bring at least your textbook, paper, and a writing instrument. Are you required to have a lab manual, graph paper, calculator, or some other supplies besides this text? If so, also bring this material with you.

Objective **B** General Tips for Success

Below are some general tips that will increase your chance for success in a mathematics class. Many of these tips will also help you in other courses you may be taking.

Exchange names and phone numbers or e-mail addresses with at least one other person in class. This contact person can be a great help if you miss an assignment or want to discuss math concepts or exercises that you find difficult.

Choose to attend all class periods. If possible, sit near the front of the classroom. This way, you will see and hear the presentation better. It may also be easier for you to participate in classroom activities.

Do your homework. You've probably heard the phrase "practice makes perfect" in relation to music and sports. It also applies to mathematics. You will find that the more time you spend solving mathematics exercises, the easier the process becomes. Be sure to schedule enough time to complete your assignments before the next class period.

Check your work. Review the steps you made while working a problem. Learn to check your answers in the original problems. You may also compare your answers with the "Answers to Selected Exercises" section in the back of the book. If you have made a mistake, try to figure out what went wrong. Then correct your mistake. If you can't find what went wrong, don't erase your work or throw it away. Bring your work to your instructor, a tutor in a math lab, or a classmate. It is easier for someone to find where you had trouble if he or she looks at your original work.

Learn from your mistakes. Everyone, even your instructor, makes mistakes. Use your errors to learn and to become a better math student. The key is finding and understanding your errors. Was your mistake a careless one, or did you make it because you can't read your own math writing? If so, try to work more slowly or write more neatly and make a conscious effort to carefully check your work. Did you make a mistake because you don't understand a concept? If so, take the time to review the concept or ask questions to better understand it.

Know how to get help if you need it. It's all right to ask for help. In fact, it's a good idea to ask for help whenever there is something that you don't understand. Make sure you know when your instructor has office hours and how to find his or her office. Find out whether math tutoring services are available on your campus. Check on the hours, location, and requirements of the tutoring service. Know whether software is available and how to access this resource.

Organize your class materials, including homework assignments, graded quizzes and tests, and notes from your class or lab. All of these items will make valuable references throughout your course and when studying for upcoming tests and the final exam. Make sure that you can locate these materials when you need them.

Read your textbook before class. Reading a mathematics textbook is unlike reading a novel or a newspaper. Your pace will be much slower. It is helpful to have paper and a pencil with you when you read. Try to work out examples on your own as you encounter them in your text. You should also write down any questions that you want to ask in class. When you read a mathematics textbook, sometimes some of the information in a section will be unclear. But after you hear a lecture or watch a lecture video on that section, you will understand it much more easily than if you had not read your text beforehand.

Don't be afraid to ask questions. You are not the only person in class with questions. Other students are normally grateful that someone has spoken up.

Hand in assignments on time. This way you can be sure that you will not lose points for being late. Show every step of a problem and be neat and organized. Also be sure that you understand which problems are assigned for homework. If allowed, you can always double-check the assignment with another student in your class.

Objective C Using This Text

There are many helpful resources that are available to you in this text. It is important that you become familiar with and use these resources. They should increase your chances for success in this course.

- *Practice Problems*. Each example in every section has a parallel Practice Problem. As you read a section, try each Practice Problem after you've finished the corresponding example. This "learn-by-doing" approach will help you grasp ideas before you move on to other concepts.

- *Chapter Test Prep Video CD*. This book contains a CD. This CD contains all of the Chapter Test exercises worked out by the author. This supplement is very helpful before a classroom chapter test.

- *Lecture Video CDs*. Exercises marked with a ⊚ are fully worked out by the author on video CDs. Check with your instructor for the availability of these video CDs.

- *Symbols at the beginning of an exercise set*. If you need help with a particular section, the symbols listed at the beginning of each exercise set will remind you of the numerous supplements available.

- *Objectives*. The main section of exercises in each exercise set is referenced by an objective, such as **A** or **B**, and also an example(s). There is also often a section of exercises entitled "Mixed Practice," which is referenced by two or more objectives or sections. These are mixed exercises written to prepare you for your next exam. Use all of this referencing if you have trouble completing an assignment from the exercise set.

- *Icons (Symbols)*. Make sure that you understand the meaning of the icons that are beside many exercises. ⊚ tells you that the corresponding exercise may be viewed on the video segment that corresponds to that section. ✎ tells you that this exercise is a writing exercise in which you should answer in complete sentences. △ icon tells you that the exercise involves geometry.

- *Integrated Reviews*. Found in the middle of each chapter, these reviews offer you a chance to practice—in one place—the many concepts that you have learned separately over several sections.

- *End of Chapter Opportunities*. There are many opportunities at the end of each chapter to help you understand the concepts of the chapter.

  Vocabulary Checks contain key vocabulary terms introduced in the chapter.

  Chapter Highlights contain chapter summaries and examples.

  Chapter Reviews contain review problems. The first part is organized section by section and the second part contains a set of mixed exercises.

Chapter Tests are sample tests to help you prepare for an exam. The Chapter Test Prep Video CD, found in this text, contains all the Chapter Test exercises worked by the author.

Cumulative Reviews are reviews consisting of material from the beginning of the book to the end of that particular chapter.

- *The Bigger Picture.* This feature contains the directions for building an outline to be used throughout the course. The purpose of this outline is to help you make the transition from thinking "section by section" to thinking about how the mathematics in this course is part of a bigger picture.

- *Study Skills Builder.* This feature is found at the end of many exercise sets. In order to increase your chance of success in this course, please read and answer the questions in the Study Skills Builder. For your convenience, the table below contains selected Study Skills Builder titles and their location.

Study Skills Builder Title	Page of First Occurence
Learning New Terms	Page 29
What To Do the Day of an Exam?	Page 74
Have You Decided to Complete This Course Successfully?	Page 83
Organizing a Notebook	Page 121
How Are Your Homework Assignments Going?	Page 140
Are You Familiar with Your Textbook Supplements?	Page 172
How Well Do You Know Your Textbook?	Page 199
Are You Organized?	Page 226
Are You Satisfied with Your Performance on a Particular Quiz or Exam?	Page 276
Tips for Studying for an Exam	Page 313
Are You Getting All the Mathematics Help That You Need?	Page 342
How Are You Doing?	Page 442
Are You Preparing for Your Final Exam?	Page 550

See the Preface at the beginning of this text for a more thorough explanation of the features of this text.

Objective D Getting Help

If you have trouble completing assignments or understanding the mathematics, get help as soon as you need it! This tip is presented as an objective on its own because it is so important. In mathematics, usually the material presented in one section builds on your understanding of the previous section. This means that if you don't understand the concepts covered during a class period, there is a good chance that you will not understand the concepts covered during the next class period. If this happens to you, get help as soon as you can.

Where can you get help? Many suggestions have been made in this section on where to get help, and now it is up to you to do it. Try your instructor, a tutoring center, or a math lab, or you may want to form a study group with fellow classmates. If you do decide to see your instructor or go to a tutoring center, make sure that you have a neat notebook and are ready with your questions.

Objective E Preparing for and Taking an Exam

Make sure that you allow yourself plenty of time to prepare for a test. If you think that you are a little "math anxious," it may be that you are not preparing for a test in a way that will ensure success. The way that you prepare for a test in mathematics is important. To prepare for a test:

1. Review your previous homework assignments.
2. Review any notes from class and section-level quizzes you have taken. (If this is a final exam, also review chapter tests you have taken.)
3. Review concepts and definitions by reading the Chapter Highlights at the end of each chapter.
4. Practice working out exercises by completing the Chapter Review found at the end of each chapter. (If this is a final exam, go through a Cumulative Review. There is one found at the end of each chapter except Chapter 1. Choose the review found at the end of the latest chapter that you have covered in your course.) *Don't stop here!*
5. It is important that you place yourself in conditions similar to test conditions to find out how you will perform. In other words, as soon as you feel that you know the material, get a few blank sheets of paper and take a sample test. There is a Chapter Test available at the end of each chapter, or you can work selected problems from the Chapter Review. Your instructor may also provide you with a review sheet. During this sample test, do not use your notes or your textbook. Then check your sample test. If you are not satisfied with the results, study the areas that you are weak in and try again.
6. On the day of the test, allow yourself plenty of time to arrive at where you will be taking your exam.

When taking your test:

1. Read the directions on the test carefully.
2. Read each problem carefully as you take the test. Make sure that you answer the question asked.
3. Watch your time and pace yourself so that you can attempt each problem on your test.
4. If you have time, check your work and answers.
5. Do not turn your test in early. If you have extra time, spend it double-checking your work.

Objective F Managing Your Time

As a college student, you know the demands that classes, homework, work, and family place on your time. Some days you probably wonder how you'll ever get everything done. One key to managing your time is developing a schedule. Here are some hints for making a schedule:

1. Make a list of all of your weekly commitments for the term. Include classes, work, regular meetings, extracurricular activities, etc. You may also find it helpful to list such things as laundry, regular workouts, grocery shopping, etc.
2. Next, estimate the time needed for each item on the list. Also make a note of how often you will need to do each item. Don't forget to include time estimates for the reading, studying, and homework you do outside of your classes. You may want to ask your instructor for help estimating the time needed.
3. In the exercise set that follows, you are asked to block out a typical week on the schedule grid given. Start with items with fixed time slots like classes and work.
4. Next, include the items on your list with flexible time slots. Think carefully about how best to schedule items such as study time.
5. Don't fill up every time slot on the schedule. Remember that you need to allow time for eating, sleeping, and relaxing! You should also allow a little extra time in case some items take longer than planned.
6. If you find that your weekly schedule is too full for you to handle, you may need to make some changes in your workload, classload, or in other areas of your life. You may want to talk to your advisor, manager or supervisor at work, or someone in your college's academic counseling center for help with such decisions.

1. What is your instructor's name?

2. What are your instructor's office location and office hours?

3. What is the best way to contact your instructor?

4. Do you have the name and contact information of at least one other student in class?

5. Will your instructor allow you to use a calculator in this class?

6. Is tutorial software available to you? If so, what type and where?

7. Is there a tutoring service available on campus? If so, what are its hours? What services are available?

8. Have you attempted this course before? If so, write down ways that you might improve your chances of success during this second attempt.

9. List some steps that you can take if you begin having trouble understanding the material or completing an assignment.

10. How many hours of studying does your instructor advise for each hour of instruction?

11. What does the ✎ icon in this text mean?

12. What does the 🔘 icon in this text mean?

13. What does the △ icon in this text mean?

14. Search the minor columns in your text. What are Practice Problems?

15. When might be the best time to work a Practice Problem?

16. Where are the answers to Practice Problems?

17. What answers are contained in this text and where are they?

18. What solutions are contained in this text and where are they?

19. What and where are Integrated Reviews?

20. What video CD is contained in this book, where is it, and what material is on it?

21. Chapter Highlights are found at the end of each chapter. Find the Chapter 1 Highlights and explain how you might use it and how it might be helpful.

22. Chapter Reviews are found at the end of each chapter. Find the Chapter 1 Review and explain how you might use it and how it might be useful.

23. Chapter Tests are found at the end of each chapter. Find the Chapter 1 Test and explain how you might use it and how it might be helpful when preparing for an exam on Chapter 1. Include how the Chapter Test Prep Video in this book may help.

24. Read or reread objective **F** and fill out the schedule grid below.

	Monday	Tuesday	Wednesday	Thursday	Friday	Saturday	Sunday
7:00 a.m.							
8:00 a.m.							
9:00 a.m.							
10:00 a.m.							
11:00 a.m.							
12:00 noon							
1:00 p.m.							
2:00 p.m.							
3:00 p.m.							
4:00 p.m.							
5:00 p.m.							
6:00 p.m.							
7:00 p.m.							
8:00 p.m.							
9:00 p.m.							

1.2 PLACE VALUE AND NAMES FOR NUMBERS

Objectives

A Find the Place Value of a Digit in a Whole Number.

B Write a Whole Number in Words and in Standard Form.

C Write a Whole Number in Expanded Form.

D Read Tables.

The **digits** 0, 1, 2, 3, 4, 5, 6, 7, 8, and 9 can be used to write numbers. For example, the **whole numbers** are

0, 1, 2, 3, 4, 5, 6, 7, 8, 9, 10, 11, . . .

and the **natural numbers** are 1, 2, 3, 4, 5, 6, 7, 8, 9, 10, 11, . . .

The three dots (. . .) after the 11 mean that this list continues indefinitely. That is, there is no largest whole number. The smallest whole number is 0.

Objective **A** Finding the Place Value of a Digit in a Whole Number

The position of each digit in a number determines its **place value.** For example, the distance (in miles) between the planet Mercury and the planet Earth can be represented by the whole number 48,337,000. Below is a place-value chart for this whole number.

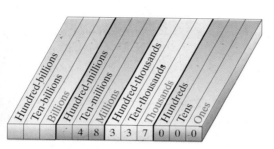

The two 3s in 48,337,000 represent different amounts because of their different placements. The place value of the 3 on the left is hundred-thousands. The place value of the 3 on the right is ten-thousands.

EXAMPLES Find the place value of the digit 3 in each whole number.

1. 396,418
↑
hundred-thousands

2. 93,192
↑
thousands

3. 534,275,866
↑
ten-millions

▨ **Work Practice Problems 1–3**

Objective **B** Writing a Whole Number in Words and in Standard Form

A whole number such as 1,083,664,500 is written in **standard form.** Notice that commas separate the digits into groups of three, starting from the right. Each group of three digits is called a **period.** The names of the first four periods are shown in red.

PRACTICE PROBLEMS 1–3

Find the place value of the digit 8 in each whole number.

1. 38,760,005

2. 67,890

3. 481,922

Answers

1. millions   **2.** hundreds

3. ten-thousands

Periods			
Billions	Millions	Thousands	Ones

Hundred-billions	Ten-billions	Billions	Hundred-millions	Ten-millions	Millions	Hundred-thousands	Ten-thousands	Thousands	Hundreds	Tens	Ones
		1	0	8	3	6	6	4	5	0	0

Writing a Whole Number in Words

To write a whole number in words, write the number in each period followed by the name of the period. (The ones period is usually not written.) This same procedure can be used to read a whole number.

For example, we write 1,083,664,500 as

one **billion**,

eighty-three **million**,

six hundred sixty-four **thousand**,

five **hundred**

> **Helpful Hint** Notice the commas after the name of each period.

> **Helpful Hint** The name of the ones period is not used when reading and writing whole numbers. For example,
>
> 9,265
>
> is read as
>
> "nine **thousand**, two **hundred** sixty-five."

PRACTICE PROBLEMS 4–6

Write each number in words.

4. 54

5. 678

6. 93,205

EXAMPLES Write each number in words.

4. 72   seventy-two

5. 546   five hundred forty-six

6. 27,034   twenty-seven thousand, thirty-four

▢ **Work Practice Problems 4–6**

> **Helpful Hint** The word "and" is *not* used when reading and writing whole numbers. It is used when reading and writing mixed numbers and some decimal values, as shown later in this text.

PRACTICE PROBLEM 7

Write 679,430,105 in words.

EXAMPLE 7 Write 308,063,557 in words.

Solution: 308,063,557 is written as

three hundred eight **million**, sixty-three **thousand**, five **hundred** fifty-seven

▢ **Work Practice Problem 7**

Answers

4. fifty-four   **5.** six hundred seventy-eight   **6.** ninety-three thousand, two hundred five   **7.** six hundred seventy-nine million, four hundred thirty thousand, one hundred five

✔ Concept Check Answer

false

✔ **Concept Check** True or false? When writing a check for $2600, the word name we write for the dollar amount of the check is "two thousand sixty." Explain your answer.

Writing a Whole Number in Standard Form

To write a whole number in standard form, write the number in each period, followed by a comma.

EXAMPLES Write each number in standard form.

8. forty-one   41   9. seven hundred eight   708
10. six thousand, four hundred ninety-three

6,493   or   6493

11. three million, seven hundred forty-six thousand, five hundred twenty-two

3,746,522

 Work Practice Problems 8–11

Helpful Hint

A comma may or may not be inserted in a four-digit number. For example, both

6,493   and   6493

are acceptable ways of writing six thousand, four hundred ninety-three.

PRACTICE PROBLEMS 8–11

Write each number in standard form.

8. thirty-seven
9. two hundred twelve
10. eight thousand, two hundred seventy-four
11. five million, fifty-seven thousand, twenty-six

Objective **C** Writing a Whole Number in Expanded Form

The place value of a digit can be used to write a number in expanded form. The **expanded form** of a number shows each digit of the number with its place value. For example, 5672 is written in expanded form as

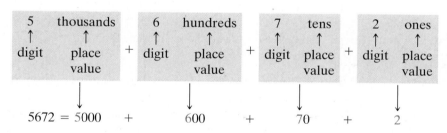

$$5672 = 5000 \quad + \quad 600 \quad + \quad 70 \quad + \quad 2$$

EXAMPLE 12 Write 5,207,034 in expanded form.

Solution: 5,000,000 + 200,000 + 7,000 + 30 + 4

 Work Practice Problem 12

PRACTICE PROBLEM 12

Write 4,026,301 in expanded form.

We can visualize whole numbers by points on a line. The line below is called a **number line.** This number line has equally spaced marks for each whole number. The arrow to the right simply means that the whole numbers continue indefinitely. In other words, there is no largest whole number.

Number Line

We will study number lines further in Section 1.4.

Answers

8. 37 9. 212 10. 8,274 11. 5,057,026
12. 4,000,000 + 20,000 + 6000 + 300 + 1

Objective D Reading Tables

Now that we know about place value and names for whole numbers, we introduce one way that whole number data may be presented. **Tables** are often used to organize and display facts that contain numbers. The following table shows the countries that have won the most medals during the Olympic Winter Games for the years 1924 through 2006. (Although the medals are truly won by athletes from the various countries, for simplicity we will state that countries have won the medals.)

Most Medals—Olympic Winter (1924–2006) Games										
Country	**Gold**	**Silver**	**Bronze**	**Total**	**Country**	**Gold**	**Silver**	**Bronze**	**Total**	
Germany*	118	116	92	326	Finland	41	57	52	150	
Russia	121	89	86	296	Sweden	46	32	44	122	
Norway	96	100	83	279	Canada	37	38	43	118	
United States	78	80	58	216	Switzerland	37	37	43	117	
Austria	50	64	71	185						

(*Source:* International Olympic Committee)
*Includes West Germany 1952, 1968–1988; East Germany 1968–1988

For example, by reading from left to right along the row marked "United States," we find that the United States won 78 gold, 80 silver, and 58 bronze medals during the Olympic Winter Games from the years 1924 to 2006.

PRACTICE PROBLEM 13

Use the Winter Games table to answer each question.

a. How many gold medals did Norway win during the Olympic Winter Games?

b. Which countries shown have won more than 100 gold medals?

EXAMPLE 13 Use the Winter Games table to answer each question.

a. How many silver medals did Sweden win during the Winter Games of the Olympics?

b. Which countries shown have won fewer bronze medals than Austria?

Solution:

a. Find "Sweden" in the left-hand column. Then read from left to right until the "silver" column is reached. We find that Sweden has won 32 silver medals.

b. Austria has won 71 bronze medals. United States, Finland, Sweden, Canada, and Switzerland have each won fewer than 71 bronze medals.

■ **Work Practice Problem 13**

Answers

13. a. 96   **b.** Germany, Russia

Vocabulary and Readiness Check

Use the choices below to fill in each blank.

~~standard form~~   ~~period~~   whole

expanded form   ~~place value~~   ~~words~~

1. The numbers 0, 1, 2, 3, 4, 5, 6, 7, 8, 9, 10, 11, 12, … are called *whole* numbers.
2. The number 1,286 is written in *standard*
3. The number "twenty-one" is written in *words*.
4. The number 900 + 60 + 5 is written in *expanded*
5. In a whole number, each group of three digits is called a(n) *period*.
6. The *place value* of the digit 4 in the whole number 264 is ones.

1.2 EXERCISE SET

FOR EXTRA HELP

Student Solutions Manual   PH Math/Tutor Center   CD/Video for Review   Math XL   MathXL®   MyMathLab MyMathLab

Objective A *Determine the place value of the digit 5 in each whole number. See Examples 1 through 3.*

1. 657   *tens* / *Hundred thousand*
2. 905   *ones*
3. 5423   *thousand*
4. 6527   *hundred*

5. 43,526,000   *Ten thousand*
6. 79,050,000
7. 5,408,092   *million*
8. 51,682,700   *ten million*

Objective B *Write each whole number in words. See Examples 4 through 7.*

9. 354
10. 316
11. 8279
12. 5445

13. 26,990
14. 42,009
15. 2,388,000
16. 3,204,000

17. 24,350,185
18. 47,033,107

Write each number in the sentence in words. See Examples 4 through 7.

19. As of this writing, the population of Bermuda is 65,773. (*Source: The World Factbook*)

20. Each Home Depot store in the United States and Canada stocks at least 40,000 different kinds of building materials, home improvement supplies, and lawn and garden products. (*Source:* The Home Depot, Inc.)

21. The world's tallest building, the Taipei 101 building in Taiwan is 1679 feet tall. (*Source:* Council on Tall Buildings and Urban Habitat)

22. In a recent year, there were 6912 patients in the United States waiting for an organ transplant. (*Source:* United Network for Organ Sharing)

23. Each day, FedEx Ground delivers 2,800,000 packages throughout the United States, Canada, and Puerto Rico. (*Source:* FedEx Corporation)

24. In a recent year, the approximate attendance figure for Disney World's Magic Kingdom was 15,200,000 people. (*Source: Amusement Business*)

25. The highest point in Oregon is Mount Hood, at an elevation of 11,239 feet. (*Source:* U.S. Geological Survey)

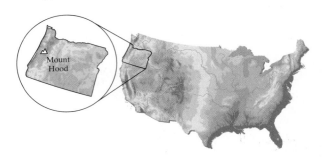

26. The highest point in Colorado is Mount Elbert, at an elevation of 14,433 feet. (*Source:* U.S. Geological Survey)

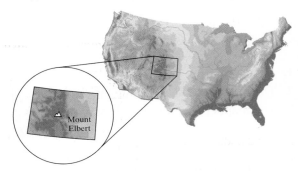

27. The Goodyear blimp *Eagle* holds 202,700 cubic feet of helium. (*Source:* The Goodyear Tire & Rubber Company)

28. In a recent year, zinc mines in the United States mined 799,000 metric tons of zinc. (*Source:* U.S. Dept. of Interior)

Write each whole number in standard form. See Examples 8 through 11.

29. Six thousand, five hundred eighty-seven

30. Four thousand, four hundred sixty-eight

31. Fifty-nine thousand, eight hundred

32. Seventy-three thousand, two

33. Thirteen million, six hundred one thousand, eleven

34. Sixteen million, four hundred five thousand, sixteen

35. Seven million, seventeen

36. Two million, twelve

37. Two hundred sixty thousand, nine hundred ninety-seven

38. Six hundred forty thousand, eight hundred eighty-one

Write the whole number in each sentence in standard form. See Examples 8 through 11.

39. The Hubble Space Telescope orbits above Earth at an altitude of three hundred fifty-three miles. (*Source:* Hubblesite.org)

40. The average distance between the surfaces of the Earth and the Moon is about two hundred thirty-four thousand miles.

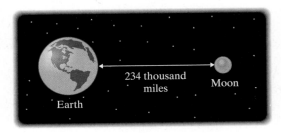

41. The price for a 2006 Porsche Carrera GT is four hundred eighty-four thousand, two hundred thirty-five dollars. (*Source:* Porsche Cars North America)

42. The world's tallest self-supporting structure is the CN Tower in Toronto, Canada. It is one thousand, eight hundred fifteen feet tall. (*Source:* The World Almanac, 2006)

43. The Buena Vista film *Pirates of the Caribbean: Dead Man's Chest* set the world record for opening day income when it took in fifty-four million, five hundred thousand dollars on July 7, 2006. (*Source: Chicago Tribune,* 2006)

44. The Twentieth Century Fox film *StarWars: Episode III—Revenge of the Sith* holds the record for second highest opening day income; it took in fifty million, thirteen thousand, eight hundred fifty-nine dollars on May 19, 2005. (*Source: Guinness Book,* 2006)

45. In 2006, Barry Bonds surpassed Babe Ruth's home run record in Major League baseball of seven hundred fourteen home runs.

46. As of this writing, Hank Aaron holds the career record for home runs in Major League baseball since 1974, with a total of seven hundred fifty-five home runs. (*Source:* Major League Baseball)

Objective **C** *Write each whole number in expanded form. See Example 12.*

47. 209

48. 789

49. 3470

50. 6040

51. 80,774

52. 20,215

53. 66,049

54. 99,032

55. 39,680,000

56. 47,703,029

Objectives **B** **C** **D** **Mixed Practice** *The table shows the six tallest mountains in New England and their elevations. Use this table to answer Exercises 57 through 62. See Example 13.*

Mountain (State)	Elevation (in feet)
Boott Spur (NH)	5492
Mt. Adams (NH)	5774
Mt. Clay (NH)	5532
Mt. Jefferson (NH)	5712
Mt. Sam Adams (NH)	5584
Mt. Washington (NH)	6288
Source: U.S. Geological Survey	

Elevation in feet

57. Write the elevation of Mt. Clay in standard form and then in words.

58. Write the elevation of Mt. Washington in standard form and then in words.

59. Write the height of Boott Spur in expanded form.

60. Write the height of Mt. Jefferson in expanded form.

61. Which mountain is the tallest in New England?

62. Which mountain is the second tallest in New England?

The table shows the top ten popular breeds of dogs in 2004 according to the American Kennel Club. Use this table to answer Exercises 63 through 68. See Example 13.

Top Ten American Kennel Club Registrations in 2004			
Breed	**Number of Registered Dogs**	**Average Dog Maximum Height (in inches)**	**Average Dog Maximum Weight (in pounds)**
Beagle	44,557	15	30
Boxer	37,744	25	70
Chihuahua	24,853	9	6
Dachshund	40,774	9	25
German shepherd dog	46,054	26	95
Golden retriever	52,560	24	80
Labrador retriever	146,714	25	75
Poodle (standard, miniature, and toy)	32,673	standard: 26	standard: 70
Shih Tzu	28,960	11	16
Yorkshire terrier	43,527	9	7
(*Source:* American Kennel Club)			

63. Which breed has fewer dogs registered, Chihuahua or Golden retriever?

64. Which breed has more dogs registered, Beagle or Yorkshire terrier?

65. Which breed has the most American Kennel Club registrations? Write the number of registrations for this breed in words.

66. Which of the listed breeds has the fewest registrations? Write the number of registered dogs for this breed in words.

67. What is the maximum weight of an average-size Dachshund?

68. What is the maximum height of an average-size Yorkshire terrier?

Concept Extensions

69. Write the largest four-digit number that can be made from the digits 1, 9, 8, and 6 if each digit must be used once.

_____ _____ _____ _____

70. Write the largest five-digit number that can be made using the digits 5, 3, and 7 if each digit must be used at least once.

_____ _____ _____ _____

Check to see whether each number written in standard form matches the number written in words. If not, correct the number in words. See the Concept Check in this section.

71.

72.

73. If a number is given in words, describe the process used to write this number in standard form.

74. If a number is written in standard form, describe the process used to write this number in expanded form.

75. The world's fastest super computer is Blue Gene/L, a joint project of IBM and the Department of Energy. It can perform 135.5 trillion calculations in a second. Look up "trillion" in a dictionary and use the definition to write this number in standard form (*Source:* IBM).

1.3 ADDING AND SUBTRACTING WHOLE NUMBERS, AND PERIMETER

Objectives

A Add Whole Numbers.

B Subtract Whole Numbers.

C Find the Perimeter of a Polygon.

D Solve Problems by Adding or Subtracting Whole Numbers.

Objective A Adding Whole Numbers

The iPod is a hard drive–based portable audio player. As of 2006, it is the most popular digital music player in the United States.

Suppose that an electronics store received a shipment of two boxes of iPods one day and an additional four boxes of iPods the next day. The **total** shipment in the two days can be found by adding 2 and 4.

$$2 \text{ boxes of iPods} + 4 \text{ boxes of iPods} = 6 \text{ boxes of iPods}$$

The **sum** (or total) is 6 boxes of iPods. Each of the numbers 2 and 4 is called an **addend,** and the process of finding the sum is called **addition.**

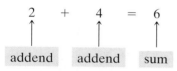

To add whole numbers, we add the digits in the ones place, then the tens place, then the hundreds place, and so on. For example, let's add 2236 + 160.

```
  2236        Line up numbers vertically so that the place values correspond. Then
+  160        add digits in corresponding place values, starting with the ones place.
  2396
```
sum of ones
sum of tens
sum of hundreds
sum of thousands

EXAMPLE 1 Add: 46 + 713

Solution:
```
   46
+ 713
  759
```

◻ **Work Practice Problem 1**

Adding by Carrying

When the sum of digits in corresponding place values is more than 9, **carrying** is necessary. For example, to add 365 + 89, add the ones-place digits first.

Carrying
```
    1
   365
+   89     5 ones + 9 ones = 14 ones or 1 ten + 4 ones
     4     Write the 4 ones in the ones place and carry the 1 ten to the tens place.
```

Next, add the tens-place digits.
```
   1 1
   365
+   89     1 ten + 6 tens + 8 tens = 15 tens or 1 hundred + 5 tens
    54     Write the 5 tens in the tens place and carry the 1 hundred to the hundreds place.
```

Next, add the hundreds-place digits.
```
   1 1
   365
+   89     1 hundred + 3 hundreds = 4 hundreds
   454     Write the 4 hundreds in the hundreds place.
```

PRACTICE PROBLEM 1
Add: 4135 + 252

Answer
1. 4387

15

PRACTICE PROBLEM 2

Add: 47,364 + 135,898

EXAMPLE 2 Add: 46,278 + 124,931

Solution:

$$
\begin{array}{r}
\overset{1\ 1\ \ 1}{46{,}278} \\
+\ 124{,}931 \\
\hline
171{,}209
\end{array}
$$

▣ **Work Practice Problem 2**

✔**Concept Check** What is wrong with the following computation?

$$
\begin{array}{r}
394 \\
+\ 283 \\
\hline
577
\end{array}
$$

Before we continue adding whole numbers, let's review some properties of addition that you may have already discovered. The first property that we will review is the **addition property of 0.** This property reminds us that the sum of 0 and any number is that same number.

Addition Property of 0

The sum of 0 and any number is that number. For example,

$$7 + 0 = 7$$
$$0 + 7 = 7$$

Next, notice that we can add any two whole numbers in any order and the sum is the same. For example,

$$4 + 5 = 9 \quad \text{and} \quad 5 + 4 = 9$$

We call this special property of addition the **commutative property of addition.**

Commutative Property of Addition

Changing the **order** of two addends does not change their sum. For example,

$$2 + 3 = 5 \quad \text{and} \quad 3 + 2 = 5$$

Another property that can help us when adding numbers is the **associative property of addition.** This property states that when adding numbers, the grouping of the numbers can be changed without changing the sum. We use parentheses to group numbers. They indicate which numbers to add first. For example, let's use two different groupings to find the sum of 2 + 1 + 5.

$$(2 + 1) + 5 = 3 + 5 = 8$$

Also,

$$2 + (1 + 5) = 2 + 6 = 8$$

Both groupings give a sum of 8.

Answer

2. 183,262

✔ **Concept Check Answer**

forgot to carry 1 hundred to the hundreds place

Associative Property of Addition

Changing the **grouping** of addends does not change their sum. For example,

$$3 + \underbrace{(5 + 7)} = 3 + 12 = 15 \quad \text{and} \quad \underbrace{(3 + 5)} + 7 = 8 + 7 = 15$$

The commutative and associative properties tell us that we can add whole numbers using any order and grouping that we want.

When adding several numbers, it is often helpful to look for two or three numbers whose sum is 10, 20, and so on. Why? Adding multiples of 10 such as 10 and 20 is easier.

EXAMPLE 3 Add: $13 + 2 + 7 + 8 + 9$

Solution: $13 + 2 + 7 + 8 + 9 = 39$

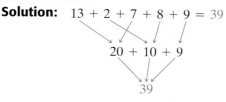

$$20 + 10 + 9$$

$$39$$

▣ **Work Practice Problem 3**

Feel free to use the process of Example 3 anytime when adding.

EXAMPLE 4 Add: $1647 + 246 + 32 + 85$

Solution:

$$\begin{array}{r} \overset{1\,2\,2}{1647} \\ 246 \\ 32 \\ +85 \\ \hline 2010 \end{array}$$

▣ **Work Practice Problem 4**

Objective **B** Subtracting Whole Numbers

If you have $5 and someone gives you $3, you have a total of $8, since $5 + 3 = 8$. Similarly, if you have $8 and then someone borrows $3, you have $5 left. **Subtraction** is finding the **difference** of two numbers.

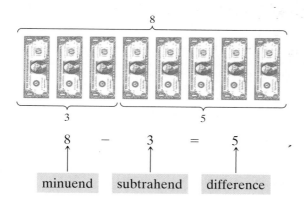

minuend   subtrahend   difference

PRACTICE PROBLEM 3

Add: $12 + 4 + 8 + 6 + 5$

PRACTICE PROBLEM 4

Add: $6432 + 789 + 54 + 28$

Answers

3. 35 **4.** 7303

In this example, 8 is the **minuend,** and 3 is the **subtrahend.** The **difference** between these two numbers, 8 and 3, is 5.

Notice that addition and subtraction are very closely related. In fact, subtraction is defined in terms of addition.

$$8 - 3 = 5 \text{ because } 5 + 3 = 8$$

This means that subtraction can be *checked* by addition, and we say that addition and subtraction are reverse operations.

PRACTICE PROBLEM 5

Subtract. Check each answer by adding.
a. $14 - 6$
b. $20 - 8$
c. $93 - 93$
d. $42 - 0$

EXAMPLE 5 Subtract. Check each answer by adding.

a. $12 - 9$ b. $22 - 7$ c. $35 - 35$ d. $70 - 0$

Solution:

a. $12 - 9 = 3$ because $3 + 9 = 12$
b. $22 - 7 = 15$ because $15 + 7 = 22$
c. $35 - 35 = 0$ because $0 + 35 = 35$
d. $70 - 0 = 70$ because $70 + 0 = 70$

◼ **Work Practice Problem 5**

Look again at Examples 5(c) and 5(d).

5(c) $35 - 35 = 0$ 5(d) $70 - 0 = 70$

| same | difference | a number | difference is the |
| number | is 0 | minus 0 | same number |

These two examples illustrate the subtraction properties of 0.

Subtraction Properties of 0

The difference of any number and that same number is 0. For example,

$$11 - 11 = 0$$

The difference of any number and 0 is that same number. For example,

$$45 - 0 = 45$$

To subtract whole numbers we subtract the digits in the ones place, then the tens place, then the hundreds place, and so on. When subtraction involves numbers of two or more digits, it is more convenient to subtract vertically. For example, to subtract $893 - 52$,

$$
\begin{array}{r}
8\,9\,3 \quad \longleftarrow \text{minuend} \\
-\,5\,2 \quad \longleftarrow \text{subtrahend} \\
\hline
8\,4\,1 \quad \longleftarrow \text{difference}
\end{array}
$$

Line up the numbers vertically so that the minuend is on top and the place values correspond. Subtract in corresponding place values, starting with the ones place.

$3 - 2$
$9 - 5$
$8 - 0$

To check, add.

$$
\begin{array}{r}
\text{difference} \\
+\,\text{subtrahend} \\
\hline
\text{minuend}
\end{array}
\quad \text{or} \quad
\begin{array}{r}
841 \\
+\ 52 \\
\hline
893
\end{array}
$$

← Since this is the original minuend, the problem checks.

Answers

5. a. 8 **b.** 12 **c.** 0 **d.** 42

Subtract: 7826 − 505. Check by adding.

Solution: 7826 ←
             −505
             ———
             7321

Check: 7321
            +505
            ———
            7826

🖳 **Work Practice Problem 6**

Subtracting by Borrowing

When subtracting vertically, if a digit in the second number (subtrahend) is larger than the corresponding digit in the first number (minuend), **borrowing** is necessary. For example, consider

    8|1
  − 6|3

Since the 3 in the ones place of 63 is larger than the 1 in the ones place of 81, borrowing is necessary. We borrow 1 ten from the tens place and add it to the ones place.

Borrowing

    8 − 1 = 7 →   7 11 ← 1 ten + 1 one = 11 ones
  tens ten tens   8̸ 1̸
                   − 6 3

Now we subtract the ones-place digits and then the tens-place digits.

      7 11
      8̸ 1̸ ←
    − 6 3
    ——————
      1 8 ← 11 − 3 = 8
         └─ 7 − 6 = 1

Check: 18
             +63
             ———
              81   The original minuend.

EXAMPLE 7 **Subtract:** 543 − 29. Check by adding.

Solution: 3 13
               5 4̸ 3̸ ←
               − 2 9
               ——————
                5 1 4

Check: 514
            +29
            ———
            543

🖳 **Work Practice Problem 7**

Sometimes we may have to borrow from more than one place. For example, to subtract 7631 − 152, we first borrow from the tens place.

           2 11
      7 6 3̸ 1̸
    −   1 5 2
    —————————
            9   ← 11 − 2 = 9

In the tens place, 5 is greater than 2, so we borrow again. This time we borrow from the hundreds place.

      ┌─ 6 hundreds − **1 hundred** = 5 hundreds
    ↓ 12 ←
      5 2̸ 11
    7 6̸ 3̸ 1̸ ←   **1 hundred** + 2 tens
    −   1 5 2   or
    —————————   10 tens + 2 tens = 12 tens
    7 4 7 9

Check: 7479
            +152
            ———
            7631   The original minuend.

PRACTICE PROBLEM 6

Subtract. Check by adding.
a. 9123 − 142
b. 978 − 851

PRACTICE PROBLEM 7

Subtract. Check by adding.
a.   697
       − 49

b.   326
       − 245

c.   1234
       − 822

Answers
6. a. 8981 **b.** 127
7. a. 648 **b.** 81 **c.** 412

PRACTICE PROBLEM 8

Subtract. Check by adding.

a.   400
      − 164

b.   1000
      − 762

EXAMPLE 8 Subtract: 900 − 174. Check by adding.

Solution: In the ones place, 4 is larger than 0, so we borrow from the tens place. But the tens place of 900 is 0, so to borrow from the tens place we must first borrow from the hundreds place.

$$\begin{array}{r} \overset{8\ \ 10}{9\ \cancel{0}\ 0} \\ -\ 1\ 7\ 4 \\ \end{array}$$

Now borrow from the tens place.

$$\begin{array}{r} \overset{9}{\underset{8\ \cancel{10}\ 10}{\cancel{9}\ \cancel{0}\ \cancel{0}}} \\ -1\ 7\ 4 \\ \hline 7\ 2\ 6 \\ \end{array}$$

Check: 726
        +174
        ─────
         900

⊡ **Work Practice Problem 8**

Objective C Finding the Perimeter of a Polygon

In geometry, addition is used to find the perimeter of a polygon. A **polygon** can be described as a flat figure formed by line segments connected at their ends. (For more review, see Appendix A.1.) Geometric figures such as triangles, squares, and rectangles are called polygons.

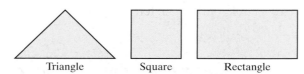

Triangle   Square   Rectangle

The **perimeter** of a polygon is the *distance around* the polygon. This means that the perimeter of a polygon is the sum of the lengths of its sides.

PRACTICE PROBLEM 9

Find the perimeter of the polygon shown. (A centimeter is a unit of length in the metric system.)

EXAMPLE 9 Find the perimeter of the polygon shown.

Solution: To find the perimeter (distance around), we add the lengths of the sides.

2 in. + 3 in. + 1 in. + 3 in. + 4 in. = 13 in.

The perimeter is 13 inches.

⊡ **Work Practice Problem 9**

To make the addition appear simpler, we will often not include units with the addends. If you do this, make sure units are included in the final answer.

Answers

8. a. 236   **b.** 238
9. 30 cm

EXAMPLE 10 **Calculating the Perimeter of a Building**

The largest commercial building in the world under one roof is the flower auction building of the cooperative VBA in Aalsmeer, Netherlands. The floor plan is a rectangle that measures 776 meters by 639 meters. Find the perimeter of this building. (A meter is a unit of length in the metric system.) (*Source: The Handy Science Answer Book,* Visible Ink Press)

Solution: Recall that opposite sides of a rectangle have the same length. To find the perimeter of this building, we add the lengths of the sides. The sum of the lengths of its sides is

$$
\begin{array}{r}
639 \\
639 \\
776 \\
+\ 776 \\
\hline
2830
\end{array}
$$

The perimeter of the building is 2830 meters.

■ **Work Practice Problem 10**

PRACTICE PROBLEM 10

A park is in the shape of a triangle. Each of the park's three sides is 647 feet. Find the perimeter of the park.

Objective D **Solving Problems by Adding or Subtracting**

Often, real-life problems occur that can be solved by adding or subtracting. The first step in solving any word problem is to *understand* the problem by reading it carefully.

Descriptions of problems solved through addition or subtraction *may* include any of these key words or phrases:

Addition		
Key Words or Phrases	Examples	Symbols
added to	5 added to 7	7 + 5
plus	0 plus 78	0 + 78
increased by	12 increased by 6	12 + 6
more than	11 more than 25	25 + 11
total	the total of 8 and 1	8 + 1
sum	the sum of 4 and 133	4 + 133

Subtraction		
Key Words or Phrases	Examples	Symbols
subtract	subtract 5 from 8	8 − 5
difference	the difference of 10 and 2	10 − 2
less	17 less 3	17 − 3
less than	2 less than 20	20 − 2
take away	14 take away 9	14 − 9
decreased by	7 decreased by 5	7 − 5
subtracted from	9 subtracted from 12	12 − 9

Helpful Hint

Be careful when solving applications that suggest subtraction. Although order *does not* matter when adding, order *does* matter when subtracting. For example, 20 − 15 and 15 − 20 do not simplify to the same number.

Answer

10. 1941 ft

✔**Concept Check**   In each of the following problems, identify which number is the minuend and which number is the subtrahend.

a. What is the result when 6 is subtracted from 40?

b. What is the difference of 15 and 8?

c. Find a number that is 15 fewer than 23.

    To solve a word problem that involves addition or subtraction, we first use the facts given to write an addition or subtraction statement. Then we write the corresponding solution of the real-life problem. It is sometimes helpful to write the statement in words (brief phrases) and then translate to numbers.

PRACTICE PROBLEM 11

The radius of Uranus is 15,759 miles. The radius of Neptune is 458 miles less than the radius of Uranus. What is the radius of Neptune? (*Source:* National Space Science Data Center)

EXAMPLE 11   **Finding the Radius of a Planet**

The radius of Jupiter is 43,441 miles. The radius of Saturn is 7257 miles less than the radius of Jupiter. Find the radius of Saturn. (*Source:* National Space Science Data Center)

Solution:

In Words		**Translate to Numbers**
radius of Jupiter	⟶	43,441
− 7257	⟶	− 7257
radius of Saturn	⟶	36,184

The radius of Saturn is 36,184 miles.

▣ **Work Practice Problem 11**

Helpful Hint
    Since subtraction and addition are reverse operations, don't forget that a subtraction problem can be checked by adding.

    Graphs can be used to visualize data. The graph shown next is called a **bar graph**. For this bar graph, the height of each bar is labeled above the bar. To check this height, follow the top of each bar to the vertical line to the left. For example, the first bar is labeled 370. Follow the top of that bar to the left until the vertical line is reached, not quite halfway between 350 and 400, or 370.

Answer

11. 15,301 miles

✔ **Concept Check Answers**

a. minuend: 40; subtrahend: 6
b. minuend: 15; subtrahend: 8
c. minuend: 23; subtrahend: 15

EXAMPLE 12 **Reading a Bar Graph**

Airline executives are studying selected aircraft models and their seating capacity for possible equipment expansion and replacement. In the following graph, each bar represents an aircraft model and the height of each bar represents the corresponding seating capacity.

Selected Aircraft Seating Capacity

Source: Air Transport Association of America, Inc.

a. Which model shown has the greatest seating capacity?

b. Find the total number of seats for the DC-9, the B757-300, and the MD-80.

Solution:

a. The model with the greatest capacity corresponds to the highest bar, which is the B747-400.

b. The key word here is "total." To find the total number of seats in the DC-9, the B757-300, and the MD-80, we add.

In Words		Translate to Numbers
DC-9	→	101
B757-300	→	235
MD-80	→	+ 134
	Total	470

The total number of seats in the DC-9, the B757-300, and the MD-80 is 470.

■ **Work Practice Problem 12**

PRACTICE PROBLEM 12

Use the graph in Example 12 to answer the following:

a. Which aircraft model shown contains the fewest seats?

b. Find the total number of seats for the B747-400, the F-100, and the B737-400.

Answers

12. a. F-100   **b.** 598

▤ CALCULATOR EXPLORATIONS Adding and Subtracting Numbers

Adding Numbers

To add numbers on a calculator, find the keys marked ⊞ and ⌷=⌷ or ⌷ENTER⌷.

For example, to add 5 and 7 on a calculator, press the keys ⌷5⌷ ⌷+⌷ ⌷7⌷ ⌷=⌷ or ⌷ENTER⌷.
The display will read ⌷ 12 ⌷.
Thus, $5 + 7 = 12$.

To add 687, 981, and 49 on a calculator, press the keys
⌷687⌷ ⌷+⌷ ⌷981⌷ ⌷+⌷ ⌷49⌷ ⌷=⌷ or ⌷ENTER⌷.
The display will read ⌷ 1717 ⌷.

Thus, $687 + 981 + 49 = 1717$. (Although entering 687, for example, requires pressing more than one key, here numbers are grouped together for easier reading.)

Use a calculator to add.

1. $89 + 45$ **2.** $76 + 97$
3. $285 + 55$ **4.** $8773 + 652$
5.   985   **6.**   465
        1210   9888
         562   620
       + 77   + 1550
      ———   ———

Subtracting Numbers

To subtract numbers on a calculator, find the keys marked ⌷−⌷ and ⌷=⌷ or ⌷ENTER⌷.

For example, to find $83 − 49$ on a calculator, press the keys ⌷83⌷ ⌷−⌷ ⌷49⌷ ⌷=⌷ or ⌷ENTER⌷.
The display will read ⌷ 34 ⌷. Thus, $83 − 49 = 34$.

Use a calculator to subtract.

7. $865 − 95$
8. $76 − 27$
9. $147 − 38$
10. $366 − 87$
11. $9625 − 647$
12. $10,711 − 8925$

Vocabulary and Readiness Check

Use the choices below to fill in each blank.

0	order	addend	associative
sum	number	grouping	commutative
perimeter	minuend	subtrahend	difference

1. The sum of 0 and any number is the same _____.
2. In $35 + 20 = 55$, the number 55 is called the _____ and 35 and 20 are each called a(n) _____.
3. The difference of any number and that same number is _____.
4. The difference of any number and 0 is the same _____.
5. In $37 − 19 = 18$, the number 37 is the _____, the 19 is the _____, and the 18 is _____.
6. The distance around a polygon is called its _____.
7. Since $7 + 10 = 10 + 7$, we say that changing the _____ in addition does not change the sum. This property is called the _____ property of addition.
8. Since $(3 + 1) + 20 = 3 + (1 + 20)$, we say that changing the _____ in addition does not change the sum. This property is called the _____ property of addition.

Objective A *Add. See Examples 1 through 4.*

1. 14 + 22

2. 27 + 31

3. 62
   + 230

4. 37
   + 542

5. 12
  13
  + 24

6. 23
  45
  + 30

7. 5267
   + 132

8. 236
  + 6243

9. 22,781 + 186,297

10. 17,427 + 821,059

11. 8
  9
  2
  5
  + 1

12. 3
  5
  8
  5
  + 7

13. 81
  17
  23
  79
  + 12

14. 64
  28
  56
  25
  + 32

15. 24 + 9006 + 489 + 2407

16. 16 + 1056 + 748 + 7770

17. 6820
  4271
  + 5626

18. 6789
  4321
  + 5555

19. 49
  628
  5 762
  + 29,462

20. 26
  582
  4 763
  + 62,511

21. 121,742
  57,279
  26,586
  + 426,782

22. 504,218
  321,920
  38,507
  + 594,687

Objective B *Subtract. Check by adding. See Examples 5 through 8.*

23. 62
  − 37

24. 55
  − 29

25. 749
  − 149

26. 957
  − 257

27. 922
  − 634

28. 674
  − 299

29. 600
  − 432

30. 300
  − 149

31. 6283
  − 560

32. 5349
  − 720

33. 533
  − 29

34. 724
  − 16

35. 1983 − 1904

36. 1983 − 1914

37. 50,000 − 17,289

38. 40,000 − 23,582

39. 7020 − 1979

40. 6050 − 1878

41. 51,111 − 19,898

42. 62,222 − 39,898

Objectives A B Mixed Practice *Add or subtract as indicated. See Examples 1 through 8.*

43. 986
  + 48

44. 986
  − 48

45. 76 − 67

46. 80 + 93 + 17 + 9 + 2

47.   9000
          − 482

48.   10,000
          − 1786

49.   10,962
            4851
          + 7063

50.   12,468
            3211
          + 1988

Objective **C** *Find the perimeter of each figure. See Examples 9 and 10.*

51.

7 feet   8 feet
10 feet

52.
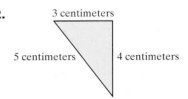
3 centimeters
5 centimeters   4 centimeters

 53.
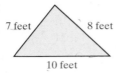
4 inches
Rectangle 8 inches

54.

8 miles
Rectangle   4 miles

55.

8 inches
1 inch   3 inches
5 inches
5 inches
7 inches

56.
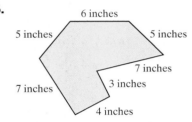
6 inches
5 inches   5 inches
7 inches
7 inches   3 inches
4 inches

57.

10 meters
5 meters
5 meters   12 meters
?
? 5

58.
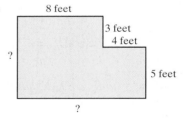
8 feet
3 feet
4 feet
?
5 feet
?

Objectives **C D Mixed Practice—Translating** *Solve. See Examples 9 through 12.*

59. Find the sum of 297 and 1796.

60. Find the sum of 802 and 6487.

61. Find the total of 76, 39, 8, 17, and 126.

62. Find the total of 89, 45, 2, 19, and 341.

63. Find the difference of 41 and 21.

64. Find the difference of 16 and 5.

65. What is 452 increased by 92?

66. What is 712 increased by 38?

67. Find 108 less 36.

68. Find 25 less 12.

69. Find 12 subtracted from 100.

70. Find 86 subtracted from 90.

71. The population of California is projected to grow from 36,039 thousand in 2005 to 38,067 thousand in 2010. What is California's projected population increase over this time period?

72. The population of Illinois is projected to grow from 12,699 thousand in 2005 to 12,917 thousand in 2010. What is Illinois's projected population increase over this time period?

73. A new DVD player with remote control costs $295. A college student has $914 in her savings account. How much will she have left in her savings account after she buys the DVD player?

74. A stereo that regularly sells for $547 is discounted by $99 in a sale. What is the sale price?

A river basin is the geographic area drained by a river and its tributaries. The Mississippi River Basin is the third largest in the world and is divided into six sub-basins, whose areas are shown in the following bar graph. Use this graph for Exercises 75 through 78.

75. Find the total U.S. land area drained by the Upper Mississippi and Lower Mississippi sub-basins.

76. Find the total U.S. land area drained by the Ohio and Tennessee sub-basins.

77. How much more square feet of land is drained by the Missouri sub-basin than the Arkansas Red-White sub-basin?

78. How much more square feet of land is drained by the Upper Mississippi sub-basin than the Lower Mississippi sub-basin?

79. A homeowner is installing a fence in his backyard. How many feet of fencing are needed to enclose the yard below?

80. A homeowner is considering installing gutters around her home. Find the perimeter of her rectangular home.

81. Professor Graham is reading a 503-page book. If she has just finished reading page 239, how many more pages must she read to finish the book?

82. When a couple began a trip, the odometer read 55,492. When the trip was over, the odometer read 59,320. How many miles did they drive on their trip?

83. The two top-selling automobiles in the United States are the Honda Accord, with sales of 386,119 cars and the Toyota Camry with sales of 426,990 cars in 2005. What is the total number of Accords and Camrys sold in the United States in 2005? (*Source:* Toyota Corp. and Honda Corp.)

84. In 2005, the country of New Zealand had 35,758,801 more sheep than people. If the human population of New Zealand in 2005 was 4,141,199, what was the sheep population? (*Source: Statistics: New Zealand*)

The decibel (dB) is a unit of measurement for sound. Every increase of 10 dB is a tenfold increase in sound intensity. The following bar graph below shows the decibel levels for some common sounds. Use this graph for Exercises 85 through 88.

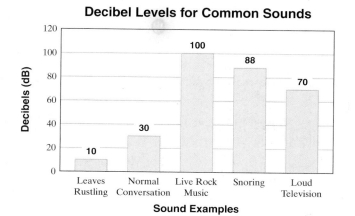

85. What is the dB rating for live rock music?

86. Which is the quietest of all the sounds shown in the graph?

87. How much louder is the sound of snoring than normal conversation?

88. What is the difference is sound intensity between live rock music and loud television?

89. In 2006, there were 5696 Blockbuster video rental stores located in the United States and 3346 located outside the United States. How many Blockbuster video rental stores were located worldwide? (*Source: Blockbuster Inc.*)

90. Automobile classes are defined by the amount of interior room. A subcompact car is defined as a car with a maximum interior space of 99 cubic feet. A midsize car is defined as a car with a maximum interior space of 119 cubic feet. What is the difference in volume between a midsize and a subcompact car?

91. The largest permanent Monopoly board is made of granite and is located in San Jose, California. It is in the shape of a square with side lengths of 31 ft. Find the perimeter of the square playing board.

92. The smallest commercially available jigsaw puzzle is a 1000-piece puzzle manufactured in Spain. It is in the shape of a rectangle with length of 18 inches and width of 12 inches. Find the perimeter of this rectangular-shaped puzzle.

The table shows the number of Target stores in ten states. Use this table to answer Exercises 93 through 98.

The Top States for Target Stores in 2005	
State	**Number of Stores**
Arizona	41
California	205
Florida	95
Georgia	44
Illinois	75
New York	49
Michigan	53
Minnesota	64
Ohio	49
Texas	121
(*Source:* Target Corporation)	

93. Which state has the most Target stores?

94. Which of the states listed in the table has the fewest number of Target stores?

95. What is the total number of Target stores located in the three states with the most Target stores?

96. How many Target stores are located in the ten states listed in the table?

97. Which pair of neighboring states have more Target stores combined, Florida and Georgia or Michigan and Ohio?

98. Target operates stores in 47 states. There are 601 Target stores located in the states not listed in the table. How many Target stores are in the United States?

99. The State of Delaware has 2029 miles of urban highways and 3865 miles of rural highways. Find the total highway mileage in Delaware. (*Source:* U.S. Federal Highway Administration)

100. The state of Rhode Island has 5193 miles of urban highways and 1222 miles of rural highways. Find the total highway mileage in Rhode Island. (*Source:* U.S. Federal Highway Administration)

Concept Extensions

For each exercise, identify which number is the minuend and which number is the subtrahend. See the Concept Check in this section.

101.
$$\begin{array}{r} 48 \\ -\ 1 \\ \hline \end{array}$$

102.
$$\begin{array}{r} 2863 \\ -1904 \\ \hline \end{array}$$

103. Subtract 7 from 70.

104. Find 86 decreased by 25.

105. In your own words, explain the commutative property of addition.

106. In your own words, explain the associative property of addition.

Check each addition below. If it is incorrect, find the correct answer. See the Concept Check in this section.

107.
```
   566
   932
 + 871
  2369
```

108.
```
   773
   659
 + 481
  1913
```

109.
```
    14
   173
    86
 + 257
   520
```

110.
```
    19
   214
    49
 + 651
   923
```

Identify each answer as correct or incorrect. Use addition to check. If the answer is incorrect, then write the correct answer.

111.
```
   741
 −  56
   675
```

112.
```
   478
 −  89
   389
```

113.
```
  1029
 −  888
   141
```

114.
```
  7615
 −  547
  7168
```

Fill in the missing digits in each problem.

115.
```
   526_
 − 2_85
   28_4
```

116.
```
  10,_4_
 − 8 5_4
    _710
```

117. Is there a commutative property of subtraction? In other words, does order matter when subtracting? Why or why not?

118. Explain why the phrase "Subtract 7 from 10" translates to "10 − 7."

 119. The local college library is having a Million Pages of Reading promotion. The freshmen have read a total of 289,462 pages; the sophomores have read a total of 369,477 pages; the juniors have read a total of 218,287 pages; and the seniors have read a total of 121,685 pages. Have they reached a goal of one million pages? If not, how many more pages need to be read?

STUDY SKILLS BUILDER

Learning New Terms

Many of the terms used in this text may be new to you. It will be helpful to make a list of new mathematical terms and symbols as you encounter them and to review them frequently. Placing these new terms (including page references) on 3 × 5 index cards might help you later when you're preparing for a quiz.

Answer the following.

1. Name one way you might place a word and its definition on a 3 × 5 card.

2. How do new terms stand out in this text so that they can be found?

1.4 ROUNDING AND ESTIMATING

Objective **A** Rounding Whole Numbers

Rounding a whole number means approximating it. A rounded whole number is often easier to use, understand, and remember than the precise whole number. For example, instead of trying to remember the South Carolina state population as 4,198,068, it is much easier to remember it rounded to the nearest million: 4,000,000, or 4 million people.

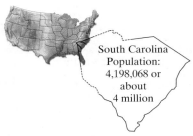

South Carolina Population: 4,198,068 or about 4 million

Recall from Section 1.2 that the line below is called a number line. To **graph** a whole number on this number line, we darken the point representing the location of the whole number. For example, the number 4 is graphed below.

On the number line, the whole number 36 is closer to 40 than 30, so 36 rounded to the nearest ten is 40.

The whole number 52 is closer to 50 than 60, so 52 rounded to the nearest ten is 50.

In trying to round 25 to the nearest ten, we see that 25 is halfway between 20 and 30. It is not closer to either number. In such a case, we round to the larger ten, that is, to 30.

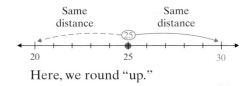

Here, we round "up."

To round a whole number without using a number line, follow these steps:

Rounding Whole Numbers to a Given Place Value

Step 1: Locate the digit to the right of the given place value.

Step 2: If this digit is 5 or greater, add 1 to the digit in the given place value and replace each digit to its right by 0.

Step 3: If this digit is less than 5, replace it and each digit to its right by 0.

EXAMPLE 1   Round 568 to the nearest ten.

Solution: 5 6 ⑧   The digit to the right of the tens place is the ones place,
            ↑   which is circled.
        tens place

        5 6 ⑧   Since the circled digit is 5 or greater, add 1 to the 6 in
        ↑   ↖   the tens place and replace the digit to the right by 0.
      Add 1.   Replace
            with 0.

We find that 568 rounded to the nearest ten is 570.

■ **Work Practice Problem 1**

PRACTICE PROBLEM 1
Round to the nearest ten.
a. 57
b. 641
c. 325

EXAMPLE 2   Round 278,362 to the nearest thousand.

Solution:

        Thousands place
            ↓   ┌─ 3 is less than 5.
        278, ③ 62
            ↑   ↑
    Do not add 1.   Replace with zeros.

The number 278,362 rounded to the nearest thousand is 278,000.

■ **Work Practice Problem 2**

PRACTICE PROBLEM 2
Round to the nearest thousand.
a. 72,304
b. 9222
c. 671,800

EXAMPLE 3   Round 248,982 to the nearest hundred.

Solution:

        Hundreds place
            ↓ ┌─ 8 is greater than or equal to 5.
        248,9 ⑧ 2
            ↑
        Add 1.   9 + 1 = 10, so replace the digit 9 by 0 and carry 1 to the
                   place value to the left.

              8 + 1   0
        2 4 8, 9   8 2
              ↑   ↑
           Add 1.   Replace with zeros.

The number 248,982 rounded to the nearest hundred is 249,000.

■ **Work Practice Problem 3**

PRACTICE PROBLEM 3
Round to the nearest hundred.
a. 3474
b. 76,243
c. 978,865

✔**Concept Check** Round each of the following numbers to the nearest
hundred. Explain your reasoning.

a. 59   **b.** 29

Answers
1. a. 60   **b.** 640   **c.** 330
2. a. 72,000   **b.** 9000   **c.** 672,000
3. a. 3500   **b.** 76,200   **c.** 978,900

✔ **Concept Check Answers**
a. 100   **b.** 0

Objective B Estimating Sums and Differences

By rounding addends, minuends, and subtrahends, we can estimate sums and differences. An estimated sum or difference is appropriate when the exact number is not necessary. Also, an estimated sum or difference can help us determine if we made a mistake in calculating an exact amount. To estimate the sum below, round each number to the nearest hundred and then add.

768	rounds to	800
1952	rounds to	2000
225	rounds to	200
+ 149	rounds to	+ 100
		3100

The estimated sum is 3100, which is close to the **exact** sum of 3094.

EXAMPLE 4

Round each number to the nearest hundred to find an estimated sum.

    294
    625
    1071
    + 349

Solution:

Exact:		Estimate:
294	rounds to	300
625	rounds to	600
1071	rounds to	1100
+ 349	rounds to	+ 300
		2300

The estimated sum is 2300. (The exact sum is 2339.)

◻ **Work Practice Problem 4**

EXAMPLE 5

Round each number to the nearest hundred to find an estimated difference.

    4725
    − 2879

Solution:

Exact:		Estimate:
4725	rounds to	4700
− 2879	rounds to	− 2900
		1800

The estimated difference is 1800. (The exact difference is 1846.)

◻ **Work Practice Problem 5**

Objective C Solving Problems by Estimating

Making estimates is often the quickest way to solve real-life problems when solutions do not need to be exact.

PRACTICE PROBLEM 4

Round each number to the nearest ten to find an estimated sum.

    49
    25
    32
    51
    + 98

PRACTICE PROBLEM 5

Round each number to the nearest thousand to find an estimated difference.

    3785
    − 2479

Answers
4. 260   **5.** 2000

EXAMPLE 6 **Estimating Distances**

Jose Guillermo is trying to estimate quickly the distance from Temple, Texas, to Brenham, Texas. Round each distance given on the map to the nearest ten to estimate the total distance.

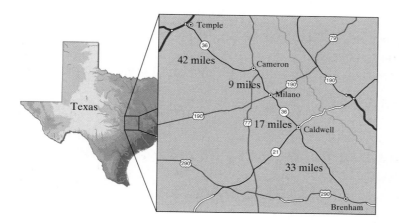

Solution:

Exact Distance:		Estimate:
42	rounds to	40
9	rounds to	10
17	rounds to	20
+33	rounds to	+ 30
		100

It is approximately 100 miles from Temple to Brenham. (The exact distance is 101 miles.)

◻ **Work Practice Problem 6**

EXAMPLE 7 **Estimating Data**

In three recent years the numbers of reported cases of mumps in the United States were 906, 1537, and 1692. Round each number to the nearest hundred to estimate the total number of cases reported over this period. (*Source:* Centers for Disease Control and Prevention)

Solution:

Exact Number of Cases:		Estimate:
906	rounds to	900
1537	rounds to	1500
+ 1692	rounds to	+ 1700
		4100

The approximate number of cases reported over this period was 4100. (The exact number of cases was 4135.)

◻ **Work Practice Problem 7**

PRACTICE PROBLEM 6

Tasha Kilbey is trying to estimate how far it is from Grove, Kansas, to Hays, Kansas. Round each given distance on the map to the nearest ten to estimate the total distance.

PRACTICE PROBLEM 7

In a recent year, there were 120,710 reported cases of chicken pox, 22,878 reported cases of tuberculosis, and 45,974 reported cases of salmonellosis in the United States. Round each number to the nearest ten-thousand to estimate the total number of cases reported for these diseases. (*Source:* Centers for Disease Control and Prevention)

Answers

6. 80 mi **7.** 190,000

Vocabulary and Readiness Check

Use the choices below to fill in each blank.

60	rounding	exact
70	estimate	graph

1. To _____ a number on a number line, darken the point representing the location of the number.

2. Another word for approximating a whole number is _____.

3. The number 65 rounded to the nearest ten is _____, but the number 61 rounded to the nearest ten is _____.

4. A(n) _____ number of products is 1265, but a(n) _____ is 1000.

FOR EXTRA HELP

1.4 EXERCISE SET

 Student Solutions Manual   PH Math/Tutor Center   CD/Video for Review   Math XL MathXL®   MyMathLab MyMathLab

Objective A *Round each whole number to the given place. See Examples 1 through 3.*

1. 423 to the nearest ten

2. 273 to the nearest ten

3. 635 to the nearest ten

4. 846 to the nearest ten

5. 2791 to the nearest hundred

6. 8494 to the nearest hundred

7. 495 to the nearest ten

8. 898 to the nearest ten

9. 21,094 to the nearest thousand

10. 82,198 to the nearest thousand

11. 33,762 to the nearest thousand

12. 42,682 to the nearest ten-thousand

13. 328,495 to the nearest hundred

14. 179,406 to the nearest hundred

15. 36,499 to the nearest thousand

16. 96,501 to the nearest thousand

17. 39,994 to the nearest ten

18. 99,995 to the nearest ten

19. 29,834,235 to the nearest ten-million

20. 39,523,698 to the nearest million

Complete the table by estimating the given number to the given place value.

		Tens	Hundreds	Thousands
21.	5281			
22.	7619			
23.	9444			
24.	7777			
25.	14,876			
26.	85,049			

Round each number to the indicated place.

27. In Fall 2005, the University of Texas at El Paso enrolled a student body of 19,264. Round this number to the nearest thousand. (*Source:* University of Texas at El Paso)

28. The number of passengers handled in 2006 by the Hartsfield Atlanta International Airport was 85,907,423. Round this number to the nearest hundred-thousand. (*Source:* Airports Council International)

29. Kareem Abdul-Jabbar holds the NBA record for points scored, a total of 38,387 over his NBA career. Round this number to the nearest thousand. (*Source:* National Basketball Association)

30. It takes 60,149 days for Neptune to make a complete orbit around the Sun. Round this number to the nearest hundred. (*Source:* National Space Science Data Center)

31. The most valuable brand in the world in 2006 was Coca-Cola, with an estimated brand value of $67,520,000,000. Round this to the nearest billion. (*Source:* BBC news)

32. According to the 2000 U.S. Census, the population of the United States was 281,421,906. Round this population figure to the nearest million. (*Source:* U.S. Census Bureau)

33. The average salary for a Boston Red Sox baseball player during the 2006 season was $3,023,894. Round this average salary to the nearest hundred-thousand. (*Source: USA Today*)

34. In 2005, the Procter & Gamble Company had $56,720,000,000 in sales. Round this sales figure to the nearest billion. (*Source:* Procter & Gamble)

35. The United States currently has 207,897,000 cellular mobile phone users, while India has 69,193,300 users. Round each of the user numbers to the nearest million. (*Source*: Cellular Telecommunications Industry Association)

36. In 2005, U.S. farms produced 114,878,000 bushels of oats. Round the oat production figure to the nearest ten-million. (*Source:* U.S. Department of Agriculture)

Objective **B** *Estimate the sum or difference by rounding each number to the nearest ten. See Examples 4 and 5.*

37.	**38.**	**39.**	**40.**
39	52	449	555
45	33	− 373	− 235
22	15		
+ 17	+ 29		

Estimate the sum or difference by rounding each number to the nearest hundred. See Examples 4 and 5.

41.	**42.**	**43.**	**44.**	**45.**	**46.**
1913	4050	1774	1989	3995	799
1886	3133	− 1492	− 1870	2549	1655
+ 1925	+ 1220			+ 4944	+ 271

Three of the given calculator answers below are incorrect. Find them by estimating each sum.

47. 463 + 219   680

48. 522 + 785   1307

49. 229 + 443 + 606   1278

50. 542 + 789 + 198   2139

51. 7806 + 5150   12,956

52. 5233 + 4988   9011

> **Helpful Hint**
>
> Estimation is useful to check for incorrect answers when using a calculator. For example, pressing a key too hard may result in a double digit, while pressing a key too softly may result in the digit not appearing in the display.

Objective **C** *Solve each problem by estimating. See Examples 6 and 7.*

53. An appliance store advertises three refrigerators on sale at $899, $1499, and $999. Round each cost to the nearest hundred to estimate the total cost.

54. Suppose you scored 89, 97, 100, 79, 75, and 82 on your biology tests. Round each score to the nearest ten to estimate your total score.

55. The distance from Kansas City to Boston is 1429 miles and from Kansas City to Chicago is 530 miles. Round each distance to the nearest hundred to estimate how much farther Boston is from Kansas City than Chicago is.

56. The Gonzales family took a trip and traveled 588, 689, 277, 143, 59, and 802 miles on six consecutive days. Round each distance to the nearest hundred to estimate the distance they traveled.

57. The peak of Mt. McKinley, in Alaska, is 20,320 feet above sea level. The top of Mt. Rainier, in Washington, is 14,410 feet above sea level. Round each height to the nearest thousand to estimate the difference in elevation of these two peaks. (*Source:* U.S. Geological Survey)

58. A student is pricing new car stereo systems. One system sells for $1895 and another system sells for $1524. Round each price to the nearest hundred dollars to estimate the difference in price of these systems.

59. In 2005, the population of Jacksonville, Florida, was 782,623, and the population of Miami was 382,894. Round each population to the nearest ten-thousand to estimate how much larger Jacksonville was than Miami. (*Source:* Internet research)

60. Round each distance given on the map to the nearest ten to estimate the total distance from North Platte, Nebraska, to Lincoln, Nebraska.

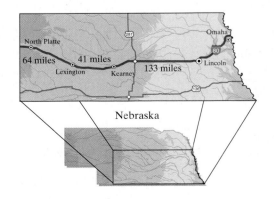

61. Head Start is a national program that provides developmental and social services for America's low-income preschool children ages three to five. Enrollment figures in Head Start programs showed a decrease from 909,608 children in 2003 to 906,993 children in 2005. Round each number of children to the nearest thousand to estimate this decrease. (*Source:* Head Start Bureau)

62. Enrollment figures at a local community college showed an increase from 49,713 credit hours in 2005 to 51,746 credit hours in 2006. Round each number to the nearest thousand to estimate the increase.

Mixed Practice (*Sections 1.3 and 1.4*) *The following table shows a few of the top leading television advertisers in the United States for 2005 and the amount of money spent that year on local and national television advertising. Complete this table. The first line is completed for you.*

Advertiser	Amount Spent on Television Advertising in 2005 (in millions of dollars)	Amount Written in Standard Form	Standard Form Rounded to Nearest Ten-Million	Standard Form Rounded to Nearest Hundred-Million
Daimler Chrysler AG	528	$528,000,000	$530,000,000	$500,000,000
63. Honda	339			
64. Yum Brands	214			
65. Federated Department Stores	179			
66. General Mills	155			
(*Source:* Television Bureau of Advertising, Inc.)				

Concept Extensions

67. Find one number that when rounded to the nearest hundred is 5700.

68. Find one number that when rounded to the nearest ten is 5700.

69. A number rounded to the nearest hundred is 8600.
  a. Determine the smallest possible number.
  b. Determine the largest possible number.

70. On August 23, 1989, it was estimated that 1,500,000 people joined hands in a human chain stretching 370 miles to protest the fiftieth anniversary of the pact that allowed what was then the Soviet Union to annex the Baltic nations in 1939. If the estimate of the number of people is to the nearest hundred-thousand, determine the largest possible number of people in the chain.

71. In your own words, explain how to round a number to the nearest thousand.

72. Estimate the perimeter of the triangle by first rounding the length of each side to the nearest hundred.

5950 miles   7693 miles
8203 miles

73. Estimate the perimeter of the rectangle by first rounding the length of each side to the nearest ten.

54 meters

Rectangle   17 meters

1.5 MULTIPLYING WHOLE NUMBERS AND AREA

Multiplication Shown as Repeated Addition Suppose that we wish to count the number of laptops provided in a computer class. The laptops are arranged in 5 rows, and each row has 6 laptops.

6 laptops in each row

Adding 5 sixes gives the total number of laptops. We can write this as $6 + 6 + 6 + 6 + 6 = 30$ laptops. When each addend is the same, we refer to this as **repeated addition.**

Multiplication is repeated addition but with different notation.

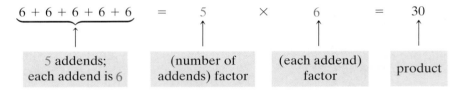

The $\times$ is called a **multiplication sign.** The numbers 5 and 6 are called **factors.** The number 30 is called the **product.** The notation 5×6 is read as "five times six." The symbols $\cdot$ and ( ) can also be used to indicate multiplication.

$$5 \times 6 = 30, \quad 5 \cdot 6 = 30, \quad (5)(6) = 30, \quad \text{and} \quad 5(6) = 30$$

✔ Concept Check

a. Rewrite $5 + 5 + 5 + 5 + 5 + 5 + 5$ using multiplication.

b. Rewrite 3×16 as repeated addition. Is there more than one way to do this? If so, show all ways.

Objective **A** Using the Properties of Multiplication

As with addition, we memorize products of one-digit whole numbers and then use certain properties of multiplication to multiply larger numbers. (If necessary, review the multiplication of one-digit numbers.)

Notice that when any number is multiplied by 0, the result is always 0. This is called the **multiplication property of 0.**

✔ **Concept Check Answers**

a. $7 \times 5 = 35$

b. $16 + 16 + 16 = 48$; yes,
$3 + 3 + 3 + 3 + 3 + 3 + 3 + 3 + 3 + 3 + 3 + 3 + 3 + 3 + 3 + 3 = 48$

Multiplication Property of 0

The product of 0 and any number is 0. For example,

$$5 \cdot 0 = 0 \quad \text{and} \quad 0 \cdot 8 = 0$$

Also notice that when any number is multiplied by 1, the result is always the original number. We call this result the **multiplication property of 1.**

Multiplication Property of 1

The product of 1 and any number is that same number. For example,

$$1 \cdot 9 = 9 \quad \text{and} \quad 6 \cdot 1 = 6$$

EXAMPLE 1 Multiply.

a. 4×1 **b.** $0(3)$ **c.** $1 \cdot 64$ **d.** $(48)(0)$

Solution:

a. $4 \times 1 = 4$ **b.** $0(3) = 0$

c. $1 \cdot 64 = 64$ **d.** $(48)(0) = 0$

◼ **Work Practice Problem 1**

PRACTICE PROBLEM 1

Multiply.

a. 6×0

b. $(1)8$

c. $(50)(0)$

d. $75 \cdot 1$

Like addition, multiplication is commutative and associative. Notice that when multiplying two numbers, the order of these numbers can be changed without changing the product. For example,

$$3 \cdot 5 = 15 \quad \text{and} \quad 5 \cdot 3 = 15$$

This property is called the **commutative property of multiplication.**

Commutative Property of Multiplication

Changing the **order** of two factors does not change their product. For example,

$$9 \cdot 2 = 18 \quad \text{and} \quad 2 \cdot 9 = 18$$

Another property that can help us when multiplying is the **associative property of multiplication.** This property states that when multiplying numbers, the grouping of the numbers can be changed without changing the product. For example,

$$(2 \cdot 3) \cdot 4 = 6 \cdot 4 = 24$$

Also,

$$2 \cdot (3 \cdot 4) = 2 \cdot 12 = 24$$

Both groupings give a product of 24.

Answers

1. a. 0 **b.** 8 **c.** 0 **d.** 75

Associative Property of Multiplication

Changing the **grouping** of factors does not change their product. From the previous work, we know that for example,

$$(2 \cdot 3) \cdot 4 = 2 \cdot (3 \cdot 4)$$

With these properties, along with the **distributive property,** we can find the product of any whole numbers. The distributive property says that multiplication **distributes** over addition. For example, notice that $3(2 + 5)$ simplifies to the same number as $3 \cdot 2 + 3 \cdot 5$.

$$3(2 + 5) = 3(7) = 21$$

$$3 \cdot 2 + 3 \cdot 5 = 6 + 15 = 21$$

Since $3(2 + 5)$ and $3 \cdot 2 + 3 \cdot 5$ both simplify to 21, then

$$3(2 + 5) = 3 \cdot 2 + 3 \cdot 5$$

Notice in $3(2 + 5) = 3 \cdot 2 + 3 \cdot 5$ that each number inside the parentheses is multiplied by 3.

Distributive Property

Multiplication distributes over addition. For example,

$$2(3 + 4) = 2 \cdot 3 + 2 \cdot 4$$

PRACTICE PROBLEM 2

Rewrite each using the distributive property.

a. $6(4 + 5)$
b. $30(2 + 3)$
c. $7(2 + 8)$

EXAMPLE 2 Rewrite each using the distributive property.

a. $5(6 + 5)$ **b.** $20(4 + 7)$ **c.** $2(7 + 9)$

Solution: Using the distributive property, we have

a. $5(6 + 5) = 5 \cdot 6 + 5 \cdot 5$
b. $20(4 + 7) = 20 \cdot 4 + 20 \cdot 7$
c. $2(7 + 9) = 2 \cdot 7 + 2 \cdot 9$

▣ **Work Practice Problem 2**

Objective B Multiplying Whole Numbers

Let's use the distributive property to multiply $7(48)$. To do so, we begin by writing the expanded form of 48 (see Section 1.2) and then applying the distributive property.

$$7(48) = 7(40 + 8) \quad \text{Write 48 in expanded form.}$$
$$= 7 \cdot 40 + 7 \cdot 8 \quad \text{Apply the distributive property.}$$
$$= 280 + 56 \quad \text{Multiply.}$$
$$= 336 \quad \text{Add.}$$

Answers

2. a. $6(4 + 5) = 6 \cdot 4 + 6 \cdot 5$
   b. $30(2 + 3) = 30 \cdot 2 + 30 \cdot 3$
   c. $7(2 + 8) = 7 \cdot 2 + 7 \cdot 8$

This is how we multiply whole numbers. When multiplying whole numbers, we will use the following notation.

First:

$$\begin{array}{r} \overset{5}{4}\,8 \\ \times\ 7 \\ \hline 3\,3\,6 \end{array}$$ ← $7 \cdot 8 = 56$

Write 6 in the ones place and carry 5 to the tens place.

Next:

$$\begin{array}{r} \overset{5}{4}\,8 \\ \times\ 7 \\ \hline 3\,3\,6 \end{array}$$ ↑ $7 \cdot 4 + 5 = 28 + 5 = 33$

The product of 48 and 7 is 336.

EXAMPLE 3 Multiply:

a. $\begin{array}{r} 25 \\ \times\ 8 \end{array}$ b. $\begin{array}{r} 246 \\ \times\ 5 \end{array}$

Solution:

a. $\begin{array}{r} \overset{4}{2}5 \\ \times\ 8 \\ \hline 200 \end{array}$ b. $\begin{array}{r} \overset{2\,3}{2}46 \\ \times\ 5 \\ \hline 1230 \end{array}$

🔲 **Work Practice Problem 3**

To multiply larger whole numbers, use the following similar notation. Multiply 89×52.

Step 1

$$\begin{array}{r} \overset{1}{8}9 \\ \times\ 52 \\ \hline 178 \end{array}$$ ← Multiply 89×2.

Step 2

$$\begin{array}{r} \overset{4}{8}9 \\ \times\ 52 \\ \hline 178 \\ 4450 \end{array}$$ ← Multiply 89×50.

Step 3

$$\begin{array}{r} 89 \\ \times\ 52 \\ \hline 178 \\ 4450 \\ \hline 4628 \end{array}$$ Add.

The numbers 178 and 4450 are called **partial products.** The sum of the partial products, 4628, is the product of 89 and 52.

EXAMPLE 4 Multiply: 236×86

Solution:
$$\begin{array}{r} 236 \\ \times\ 86 \\ \hline 1\,416 \\ 18\,880 \\ \hline 20{,}296 \end{array}$$ ← 6(236)
← 80(236)
Add.

🔲 **Work Practice Problem 4**

EXAMPLE 5 Multiply: 631×125

Solution:
$$\begin{array}{r} 631 \\ \times\ 125 \\ \hline 3\,155 \\ 12\,620 \\ 63\,100 \\ \hline 78{,}875 \end{array}$$ ← 5(631)
← 20(631)
← 100(631)
Add.

🔲 **Work Practice Problem 5**

PRACTICE PROBLEM 3
Multiply.

a. $\begin{array}{r} 29 \\ \times\ 6 \end{array}$ b. $\begin{array}{r} 648 \\ \times\ 5 \end{array}$

PRACTICE PROBLEM 4

Multiply.
$$\begin{array}{r} 306 \\ \times\ 81 \end{array}$$

PRACTICE PROBLEM 5

Multiply.
$$\begin{array}{r} 726 \\ \times 142 \end{array}$$

Answers

3. a. 174 b. 3240
4. 24,786
5. 103,092

✔ **Concept Check** Find and explain the error in the following multiplication problem.

$$
\begin{array}{r}
102 \\
\times\ 33 \\
\hline
306 \\
306 \\
\hline
612
\end{array}
$$

Objective C Finding the Area of a Rectangle

A special application of multiplication is finding the **area** of a region. Area measures the amount of surface of a region. For example, we measure a plot of land or the living space of a home by its area. The figures below show two examples of units of area measure. (A centimeter is a unit of length in the metric system.)

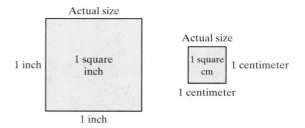

For example, to measure the area of a geometric figure such as the rectangle below, count the number of square units that cover the region.

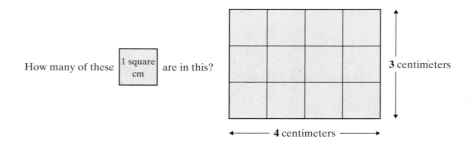

This rectangular region contains 12 square units, each 1 square centimeter. Thus, the area is 12 square centimeters. This total number of squares can be found by counting or by multiplying **4 · 3** (length · width).

$$
\begin{aligned}
\text{Area of a rectangle} &= \text{length} \cdot \text{width} \\
&= (4 \text{ centimeters})(3 \text{ centimeters}) \\
&= 12 \text{ square centimeters}
\end{aligned}
$$

In this section, we find the areas of rectangles only. In later sections, we will find the areas of other geometric regions.

✔ **Concept Check Answer**

$$
\begin{array}{r}
102 \\
\times\ 33 \\
\hline
306 \\
3060 \\
\hline
3366
\end{array}
$$

EXAMPLE 6 Finding the Area of a State

The state of Colorado is in the shape of a rectangle whose length is 380 miles and whose width is 280 miles. Find its area.

Solution:

The area of a rectangle is the product of its length and its width.

$$\text{Area} = \text{length} \cdot \text{width}$$
$$= (380 \text{ miles})(280 \text{ miles})$$
$$= 106{,}400 \text{ square miles}$$

The area of Colorado is 106,400 square miles.

Work Practice Problem 6

PRACTICE PROBLEM 6

The state of Wyoming is in the shape of a rectangle whose length is 360 miles and whose width is 280 miles. Find its area.

Objective D Solving Problems by Multiplying

There are several words or phrases that indicate the operation of multiplication. Some of these are as follows:

Multiplication		
Key Words or Phrases	**Examples**	**Symbols**
multiply	multiply 5 by 7	$5 \cdot 7$
product	the product of 3 and 2	$3 \cdot 2$
times	10 times 13	$10 \cdot 13$

Many key words or phrases describing real-life problems that suggest addition might be better solved by multiplication instead. For example, to find the **total** cost of 8 shirts, each selling for $27, we can either add

$$27 + 27 + 27 + 27 + 27 + 27 + 27 + 27$$

or we can multiply 8(27).

EXAMPLE 7 Finding DVD Space

A digital video disc (DVD) can hold about 4800 megabytes (MB) of information. How many megabytes can 12 DVDs hold?

Solution:

Twelve DVDs will hold 12×4800 megabytes.

In Words		Translate to Numbers
megabytes per disk	→	4800
$\times$ DVDs	→	$\times$ 12
		9600
		48000
total megabytes		57,600

Twelve DVDs will hold 57,600 megabytes.

Work Practice Problem 7

PRACTICE PROBLEM 7

A particular computer printer can print 16 pages per minute in color. How many pages can it print in 45 minutes?

Answers

6. 100,800 sq mi
7. 720 pages

PRACTICE PROBLEM 8

Ken Shimura purchased DVDs and CDs through a club. Each DVD was marked at $11 and each CD cost $9. Ken bought eight DVDs and five CDs. Find the total cost of the order.

EXAMPLE 8 Budgeting Money

Ann Sheridan and two friends agree to take their children to the San Antonio Zoo. The ticket price for each child is $7 and for each adult, $9. If 8 children and 3 adults plan to go, how much money is needed for admission? (*Source:* The San Antonio Zoo)

Solution: If the price of one child's ticket is $7, the price for 8 children is $8 \cdot 7 = \$56$. The price of one adult ticket is $9, so the price for 3 adults is $9 \cdot 3 = \$27$. The total cost is:

In Words		Translate to Numbers
price of 8 children	→	56
+ price of 3 adults	→	+ 27
total cost		83

The total cost is $83.

☐ **Work Practice Problem 8**

PRACTICE PROBLEM 9

If an average page in a book contains 163 words, estimate, rounding each number to the nearest hundred, the total number of words contained on 391 pages.

EXAMPLE 9 Estimating Word Count

The average page of a book contains 259 words. Estimate, rounding each number to the nearest hundred, the total number of words contained on 212 pages.

Solution: The exact number of words is 259×212. Estimate this product by rounding each factor to the nearest hundred.

$$259 \quad \text{rounds to} \quad 300$$
$$\times 212 \quad \text{rounds to} \quad \times 200, \qquad 300 \times 200 = 60,000$$
$$3 \cdot 2 = 6$$

There are approximately 60,000 words contained on 212 pages.

☐ **Work Practice Problem 9**

Answers

8. $133
9. 80,000 words

🖩 **CALCULATOR EXPLORATIONS** Multiplying Numbers

To multiply numbers on a calculator, find the keys marked $\boxed{\times}$ and $\boxed{=}$ or $\boxed{\text{ENTER}}$. For example, to find $31 \cdot 66$ on a calculator, press the keys $\boxed{31}$ $\boxed{\times}$ $\boxed{66}$ $\boxed{=}$ or $\boxed{\text{ENTER}}$. The display will read $\boxed{2046}$. Thus, $31 \cdot 66 = 2046$.

Use a calculator to multiply.

1. 72×48 **2.** 81×92

3. $163 \cdot 94$ **4.** $285 \cdot 144$

5. $983(277)$ **6.** $1562(843)$

Vocabulary and Readiness Check

Use the choices below to fill in each blank.

area	grouping	commutative	1	product	length
factor	order	associative	0	distributive	number

1. The product of 0 and any number is _____.
2. The product of 1 and any number is the _____.
3. In $8 \cdot 12 = 96$, the 96 is called the _____ and 8 and 12 are each called a(n) _____.
4. Since $9 \cdot 10 = 10 \cdot 9$, we say that changing the _____ in multiplication does not change the product. This property is called the _____ property of multiplication.
5. Since $(3 \cdot 4) \cdot 6 = 3 \cdot (4 \cdot 6)$, we say that changing the _____ in multiplication does not change the product. This property is called the _____ property of multiplication.
6. _____ measures the amount of surface of a region.
7. Area of a rectangle = _____ $\cdot$ width.
8. We know $9(10 + 8) = 9 \cdot 10 + 9 \cdot 8$ by the _____ property.

1.5 EXERCISE SET

FOR EXTRA HELP

Student Solutions Manual   PH Math/Tutor Center   CD/Video for Review   MathXL®   MyMathLab

Objective A *Multiply. See Example 1.*

1. $1 \cdot 24$
2. $55 \cdot 1$
3. $0 \cdot 19$
4. $27 \cdot 0$

5. $8 \cdot 0 \cdot 9$
6. $7 \cdot 6 \cdot 0$
7. $87 \cdot 1$
8. $1 \cdot 41$

Use the distributive property to rewrite each expression. See Example 2.

9. $6(3 + 8)$
10. $5(8 + 2)$
11. $4(3 + 9)$

12. $6(1 + 4)$
13. $20(14 + 6)$
14. $12(12 + 3)$

Objective B *Multiply. See Example 3.*

15. $\begin{array}{r} 64 \\ \times\ 8 \\ \hline \end{array}$
16. $\begin{array}{r} 79 \\ \times\ 3 \\ \hline \end{array}$
17. $\begin{array}{r} 613 \\ \times\ 6 \\ \hline \end{array}$
18. $\begin{array}{r} 638 \\ \times\ 5 \\ \hline \end{array}$

19. 277×6
20. 882×2
21. 1074×6
22. 9021×3

Objectives **Ⓐ Ⓑ** **Mixed Practice** *Multiply. See Examples 1 through 5.*

23. 89
     × 13

24. 91
     × 72

25. 421
     × 58

26. 526
     × 23

27. 306
     × 81

28. 708
     × 21

29. (780)(20)

30. (720)(80)

31. (495)(13)(0)

32. (593)(47)(0)

33. (640)(1)(10)

34. (240)(1)(20)

35. 1234×39

36. 1357×79

37. 609×234

38. 807×127

39. 8649
     × 274

40. 1234
     × 567

41. 589
     × 110

42. 426
     × 110

43. 1941
     × 2035

44. 1876
     × 1407

Objective **Ⓒ** *Find the area of each rectangle. See Example 6.*

45.

9 meters
7 meters

46.
3 inches
13 inches

47.

17 feet
40 feet

48.
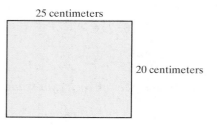
25 centimeters
20 centimeters

Objectives **Ⓒ Ⓓ** **Mixed Practice** *Estimate the products by rounding each factor to the nearest hundred. See Example 9.*

49. 576×354

50. 982×650

51. 604×451

52. 111×999

Without actually calculating, mentally round, multiply, and choose the best estimate.

53. $38 \times 42 =$
 a. 16
 b. 160
 c. 1600
 d. 16,000

54. $2872 \times 12 =$
 a. 2872
 b. 28,720
 c. 287,200
 d. 2,872,000

55. $612 \times 29 =$
 a. 180
 b. 1800
 c. 18,000
 d. 180,000

56. $706 \times 409 =$
 a. 280
 b. 2800
 c. 28,000
 d. 280,000

Translating *Solve. See Examples 6 through 9.*

57. Multiply 80 by 11.

58. Multiply 70 by 12.

59. Find the product of 6 and 700.

60. Find the product of 9 and 900.

61. Find 2 times 2240.

62. Find 3 times 3310.

63. One tablespoon of olive oil contains 125 calories. How many calories are in 3 tablespoons of olive oil? (*Source: Home and Garden Bulletin No. 72*, U.S. Department of Agriculture).

64. One ounce of hulled sunflower seeds contains 14 grams of fat. How many grams of fat are in 8 ounces of hulled sunflower seeds? (*Source: Home and Garden Bulletin No. 72*, U.S. Department of Agriculture).

65. The textbook for a course in biology costs $94. There are 35 students in the class. Find the total cost of the biology books for the class.

66. The seats in a large lecture hall are arranged in 14 rows with 34 seats in each row. Find how many seats are in this room.

67. Cabot Creamery is packing a palette of 20-lb boxes of cheddar cheese to send to a local restaurant. There are five layers of boxes on the pallet, and each layer is four boxes wide by five boxes deep.

   a. How many boxes are in one layer?
   b. How many boxes are on the pallet?
   c. What is the weight of the cheese on the pallet?

68. An apartment building has *three floors*. Each floor has five rows of apartments with four apartments in each row.

   a. How many apartments are on 1 floor?
   b. How many apartments are in the building?

69. A plot of land measures 80 feet by 110 feet. Find its area.

70. A house measures 45 feet by 60 feet. Find the floor area of the house.

71. The largest hotel lobby can be found at the Hyatt Regency in San Francisco, CA. It is in the shape of a rectangle that measures 350 feet by 160 feet. Find its area.

72. Recall from an earlier section that the largest commercial building in the world under one roof is the flower auction building of the cooperative VBA in Aalsmeer, Netherlands. The floor plan is a rectangle that measures 776 meters by 639 meters. Find the area of this building. (A meter is a unit of length in the metric system.) (*Source: The Handy Science Answer Book*, Visible Ink Press)

73. A pixel is a rectangular dot on a graphing calculator screen. If a graphing calculator screen contains 62 pixels in a row and 94 pixels in a column, find the total number of pixels on a screen.

74. A compact disc (CD) can hold 700 megabytes (MB) of information. How many MBs can 17 discs hold?

75. A line of print on a computer contains 60 characters (letters, spaces, punctuation marks). Find how many characters there are in 35 lines.

76. An average cow eats 3 pounds of grain per day. Find how much grain a cow eats in a year. (Assume 365 days in 1 year.)

77. One ounce of Planters® Dry Roasted Peanuts has 160 calories. How many calories are in 8 ounces? (*Source:* RJR Nabisco, Inc.)

78. One ounce of Planters® Dry Roasted Peanuts has 13 grams of fat. How many grams of fat are in 16 ounces? (*Source:* RJR Nabisco, Inc.)

79. The Thespian club at a local community college is ordering T-shirts. T-shirts size S, M, or L cost $10 each and T-shirts size XL or XXL cost $12 each. Use the table below to find the total cost. (The first row is filled in for you.)

T-Shirt Size	Number of Shirts Ordered	Cost per Shirt	Cost per Size Ordered
S	4	$10	$40
M	6		
L	20		
XL	3		
XXL	3		

Total Cost $372

80. The student activities group at North Shore Community College is planning a trip to see the local minor league baseball team. Tickets cost $5 for students, $7 for nonstudents, and $2 for children under 12. Use the following table to find the total cost.

Person	Number of Persons	Cost per Person	Cost per Category
Student	24	$5	$120
Nonstudent	4		
Children under 12	5		

Total Cost $158

81. Hershey's main chocolate factory in Hershey, Pennsylvania, uses 700,000 quarts of milk each day. How many quarts of milk would be used during the month of March, assuming that chocolate is made at the factory every day of the month? (*Source:* Hershey Foods Corp.)

82. Among older Americans (ages 65 and over) there are about 3 times as many widows as widowers. There were about 2,641,000 widowers in 2005. How many widows were there in 2005? (*Source:* Administration on Aging, U.S. Census Bureau)

Mixed Practice (*Sections 1.3, 1.5*) *Perform each indicated operation.*

83. $\begin{array}{r} 128 \\ +\ \ 7 \\ \hline \end{array}$

84. $\begin{array}{r} 126 \\ -\ \ 8 \\ \hline \end{array}$

85. $\begin{array}{r} 134 \\ \times\ 16 \\ \hline \end{array}$

86. $47 + 26 + 10 + 231 + 50$

87. Find the sum of 19 and 4.

88. Find the product of 19 and 4.

89. Find the difference of 19 and 4.

90. Find the total of 19 and 4.

Concept Extensions

Solve. See the first Concept Check in this section.

91. Rewrite $6 + 6 + 6 + 6 + 6$ using multiplication.

92. Rewrite $11 + 11 + 11 + 11 + 11 + 11$ using multiplication.

93. a. Rewrite $3 \cdot 5$ as repeated addition.
   b. Explain why there is more than one way to do this.

94. a. Rewrite $4 \cdot 5$ as repeated addition.
   b. Explain why there is more than one way to do this.

Find and explain the error in each multiplication problem. See the second Concept Check in this section.

95.
```
    203
 ×   14
    812
    203
   1015
```

96.
```
     31
 ×   50
    155
```

Fill in the missing digits in each problem.

97.
```
     4_
 ×   _3
    126
   3780
   3906
```

98.
```
     _7
 ×   6_
    171
   3420
   3591
```

99. Explain how to multiply two 2-digit numbers using partial products.

100. During the NBA's 2005–2006 regular season, Dirk Nowitzki of the Dallas Mavericks scored 110 three-point field goals, 641 two-point field goals, and 539 free throws (worth one point each). How many points did Nowitzki score during the 2005–2006 regular season? (*Source:* NBA)

101. A window washer in New York City is bidding for a contract to wash the windows of a 23-story building. To write a bid, the number of windows in the building is needed. If there are 7 windows in each row of windows on 2 sides of the building and 4 windows per row on the other 2 sides of the building, find the total number of windows.

1.6 DIVIDING WHOLE NUMBERS

Suppose three people pooled their money and bought a raffle ticket at a local fund-raiser. Their ticket was the winner and they won a $75 cash prize. They then divided the prize into three equal parts so that each person received $25.

Objective **A** Dividing Whole Numbers

The process of separating a quantity into equal parts is called **division.** The division above can be symbolized by several notations.

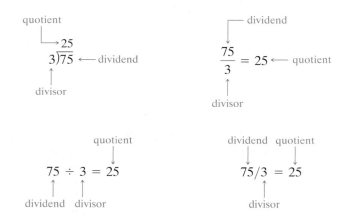

(In the notation $\dfrac{75}{3}$, the bar separating 75 and 3 is called a **fraction bar.**) Just as subtraction is the reverse of addition, division is the reverse of multiplication. This means that division can be checked by multiplication.

$$25 \atop 3\overline{)75}$$ because $25 \cdot 3 = 75$

$$\boxed{\text{Quotient}} \cdot \boxed{\text{Divisor}} = \boxed{\text{Dividend}}$$

Since multiplication and division are related in this way, you can use your knowledge of multiplication facts to review quotients of one-digit divisors if necessary.

EXAMPLE 1 Find each quotient. Check by multiplying.

a. $42 \div 7$ **b.** $\dfrac{64}{8}$ **c.** $3\overline{)21}$

Solution:

a. $42 \div 7 = 6$ because $6 \cdot 7 = 42$

b. $\dfrac{64}{8} = 8$ because $8 \cdot 8 = 64$

c. $3\overline{)21}^{\,7}$ because $7 \cdot 3 = 21$

■ **Work Practice Problem 1**

EXAMPLE 2 Find each quotient. Check by multiplying.

a. $1\overline{)7}$ **b.** $12 \div 1$ **c.** $\dfrac{6}{6}$ **d.** $9 \div 9$ **e.** $\dfrac{20}{1}$ **f.** $18\overline{)18}$

Solution:

a. $1\overline{)7}^{\,7}$ because $7 \cdot 1 = 7$

b. $12 \div 1 = 12$ because $12 \cdot 1 = 12$

c. $\dfrac{6}{6} = 1$ because $1 \cdot 6 = 6$

d. $9 \div 9 = 1$ because $1 \cdot 9 = 9$

e. $\dfrac{20}{1} = 20$ because $20 \cdot 1 = 20$

f. $18\overline{)18}^{\,1}$ because $1 \cdot 18 = 18$

■ **Work Practice Problem 2**

Example 2 illustrates the important properties of division described next:

Division Properties of 1

The quotient of any number (except 0) and that same number is 1. For example,

$$8 \div 8 = 1 \qquad \frac{5}{5} = 1 \qquad 4\overline{)4}^{\,1}$$

The quotient of any number and 1 is that same number. For example,

$$9 \div 1 = 9 \qquad \frac{6}{1} = 6 \qquad 1\overline{)3}^{\,3} \qquad \frac{0}{1} = 0$$

EXAMPLE 3 Find each quotient. Check by multiplying.

a. $9\overline{)0}$ **b.** $0 \div 12$ **c.** $\dfrac{0}{5}$ **d.** $\dfrac{3}{0}$

Solution:

a. $9\overline{)0}^{\,0}$ because $0 \cdot 9 = 0$ **b.** $0 \div 12 = 0$ because $0 \cdot 12 = 0$

c. $\dfrac{0}{5} = 0$ because $0 \cdot 5 = 0$

Continued on next page

PRACTICE PROBLEM 1

Find each quotient. Check by multiplying.

a. $9\overline{)72}$
b. $40 \div 5$
c. $\dfrac{24}{6}$

PRACTICE PROBLEM 2

Find each quotient. Check by multiplying.

a. $\dfrac{7}{7}$ **b.** $5 \div 1$
c. $1\overline{)11}$ **d.** $4 \div 1$
e. $\dfrac{10}{1}$ **f.** $21 \div 21$

PRACTICE PROBLEM 3

Find each quotient. Check by multiplying.

a. $\dfrac{0}{7}$ **b.** $8\overline{)0}$
c. $7 \div 0$ **d.** $0 \div 14$

Answers

1. a. 8 **b.** 8 **c.** 4 **2. a.** 1 **b.** 5
c. 11 **d.** 4 **e.** 10 **f.** 1 **3. a.** 0
b. 0 **c.** undefined **d.** 0

d. If $\frac{3}{0} = $ a *number,* then the *number* times $0 = 3$. Recall from Section 1.5 that any number multiplied by 0 is 0 and not 3. We say, then, that $\frac{3}{0}$ is **undefined.**

⬛ **Work Practice Problem 3**

Example 3 illustrates important division properties of 0.

Division Properties of 0

The quotient of 0 and any number (except 0) is 0. For example,

$$0 \div 9 = 0 \qquad \frac{0}{5} = 0 \qquad 14\overline{)0}^{\,0}$$

The quotient of any number and 0 is not a number. We say that

$$\frac{3}{0}, \quad 0\overline{)3}, \quad \text{and} \quad 3 \div 0$$

are **undefined.**

Objective **B** Performing Long Division

When dividends are larger, the quotient can be found by a process called **long division.** For example, let's divide 2541 by 3.

$$\text{divisor} \longrightarrow 3\overline{)2541}$$
$$\underset{\text{dividend}}{\uparrow}$$

We can't divide 3 into 2, so we try dividing 3 into the first two digits.

$$\begin{array}{r} 8 \\ 3\overline{)2541} \end{array} \qquad 25 \div 3 = 8 \text{ with 1 left, so our best estimate is 8. We place 8 over the 5 in 25.}$$

Next, multiply 8 and 3 and subtract this product from 25. Make sure that this difference is less than the divisor.

$$\begin{array}{r} 8 \\ 3\overline{)2541} \\ -24 \\ \hline 1 \end{array} \qquad \begin{array}{l} 8(3) = 24 \\ 25 - 24 = 1, \text{ and 1 is less than the divisor 3.} \end{array}$$

Bring down the next digit and go through the process again.

$$\begin{array}{r} 84 \\ 3\overline{)2541} \\ -24\downarrow \\ \hline 14 \\ -12 \\ \hline 2 \end{array} \qquad \begin{array}{l} 14 \div 3 = 4 \text{ with 2 left} \\ \\ \\ 4(3) = 12 \\ 14 - 12 = 2 \end{array}$$

Once more, bring down the next digit and go through the process.

$$\begin{array}{r} 847 \\ 3\overline{)2541} \\ -24 \\ \hline 14 \\ -12\downarrow \\ \hline 21 \\ -21 \\ \hline 0 \end{array} \qquad \begin{array}{l} 21 \div 3 = 7 \\ \\ \\ \\ \\ 7(3) = 21 \\ 21 - 21 = 0 \end{array}$$

The quotient is 847. To check, see that $847 \times 3 = 2541$.

EXAMPLE 4 Divide: $3705 \div 5$. Check by multiplying.

Solution:

$$
\begin{array}{r}
7 \\
5)\overline{3705} \\
-35\downarrow \\
\hline
20
\end{array}
$$

$37 \div 5 = 7$ with 2 left. Place this estimate, 7, over the 7 in 37.

$7(5) = 35$

$37 - 35 = 2$, and 2 is less than the divisor 5.

Bring down the 0.

$$
\begin{array}{r}
74 \\
5)\overline{3705} \\
-35 \\
\hline
20 \\
-20\downarrow \\
\hline
05
\end{array}
$$

$20 \div 5 = 4$

$4(5) = 20$

$20 - 20 = 0$, and 0 is less than the divisor 5.

Bring down the 5.

$$
\begin{array}{r}
741 \\
5)\overline{3705} \\
-35 \\
\hline
20 \\
-20\downarrow \\
\hline
5 \\
-5 \\
\hline
0
\end{array}
$$

$5 \div 5 = 1$

$1(5) = 5$

$5 - 5 = 0$

Check:

$$
\begin{array}{r}
741 \\
\times \quad 5 \\
\hline
3705
\end{array}
$$

🔲 **Work Practice Problem 4**

Helpful Hint

Since division and multiplication are reverse operations, don't forget that a division problem can be checked by multiplying.

EXAMPLE 5 Divide and check: $1872 \div 9$

Solution:

$$
\begin{array}{r}
208 \\
9)\overline{1872} \\
-18\downarrow\downarrow \\
\hline
07 \\
-0\downarrow \\
\hline
72 \\
-72 \\
\hline
0
\end{array}
$$

$2(9) = 18$

$18 - 18 = 0$; bring down the 7.

$0(9) = 0$

$7 - 0 = 7$; bring down the 2.

$8(9) = 72$

$72 - 72 = 0$

Check: $208 \cdot 9 = 1872$

🔲 **Work Practice Problem 5**

Naturally, quotients don't always "come out even." Making 4 rows out of 26 chairs, for example, isn't possible if each row is supposed to have exactly the same number of chairs. Each of 4 rows can have 6 chairs, but 2 chairs are still left over.

4 rows   6 chairs in each row

2 chairs left over

We signify "leftovers" or **remainders** in this way:

$$\begin{array}{r} 6 \quad \text{R } 2 \\ 4\overline{)26} \end{array}$$

The **whole number part of the quotient** is 6; the **remainder part of the quotient** is 2. Checking by multiplying,

whole number part · divisor + remainder part = dividend

6	·	4 +	2	
		24 +	2	= 26

PRACTICE PROBLEM 6

Divide and check.

a. $4\overline{)939}$

b. $5\overline{)3287}$

EXAMPLE 6 Divide and check: $2557 \div 7$

Solution:

$$\begin{array}{r} 365 \quad \text{R } 2 \\ 7\overline{)2557} \\ -21 \\ \hline 45 \\ -42 \\ \hline 37 \\ -35 \\ \hline 2 \end{array}$$

$3(7) = 21$
$25 - 21 = 4$; bring down the 5.
$6(7) = 42$
$45 - 42 = 3$; bring down the 7.
$5(7) = 35$
$37 - 35 = 2$; the remainder is 2.

Check: 365   ·   7   +   2   =   2557

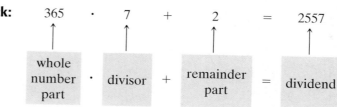

whole number part · divisor + remainder part = dividend

■ **Work Practice Problem 6**

Answers

6. a. 234 R 3   **b.** 657 R 2

EXAMPLE 7 Divide and check: 56,717 ÷ 8

Solution:

```
        7089   R 5
    8)56717
     -56↓||
       07||        7(8) = 56
      - 0↓|         Subtract and bring down the 7.
        71|         0(8) = 0
      - 64↓         Subtract and bring down the 1.
        77          8(8) = 64
      - 72          Subtract and bring down the 7.
         5          9(8) = 72
                    Subtract. The remainder is 5.
```

Check: 7089   ·   8   +   5   =   56,717

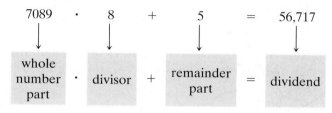

whole number part · divisor + remainder part = dividend

🔲 **Work Practice Problem 7**

When the divisor has more than one digit, the same pattern applies. For example, let's find 1358 ÷ 23.

```
         5          135 ÷ 23 = 5 with 20 left over. Our estimate is 5.
    23)1358
     -115↓          5(23) = 115
       208          135 - 115 = 20. Bring down the 8.
```

Now we continue estimating.

```
        59   R 1    208 ÷ 23 = 9 with 1 left over.
    23)1358
     -115
       208
      -207          9(23) = 207
         1          208 - 207 = 1. The remainder is 1.
```

To check, see that 59 · 23 + 1 = 1358.

EXAMPLE 8 Divide: 6819 ÷ 17

Solution:

```
        401   R 2
    17)6819
     -68↓|
       01|          4(17) = 68
      - 0↓           Subtract and bring down the 1.
        19           0(17) = 0·
      - 17           Subtract and bring down the 9.
         2           1(17) = 17
                     Subtract. The reminder is 2.
```

To check, see that 401 · 17 + 2 = 6819.

🔲 **Work Practice Problem 8**

PRACTICE PROBLEM 9

Divide: $33,282 \div 678$

EXAMPLE 9 Divide: $51,600 \div 403$

Solution:

$$
\begin{array}{r}
128 \text{ R } 16 \\
403\overline{)51600} \\
-403 \\
\hline
1130 \\
-806 \\
\hline
3240 \\
-3224 \\
\hline
16
\end{array}
$$

$1(403) = 403$
Subtract and bring down the 0.
$2(403) = 806$
Subtract and bring down the 0.
$8(403) = 3224$
Subtract. The remainder is 16.

To check, see that $128 \cdot 403 + 16 = 51,600$.

Work Practice Problem 9

Division Shown as Repeated Subtraction To further understand division, recall from Section 1.5 that addition and multiplication are related in the following manner:

$$\underbrace{3 + 3 + 3 + 3}_{\text{4 addends; each addend is 3}} = 4 \times 3 = 12$$

In other words, multiplication is repeated addition. Likewise, division is repeated subtraction.

For example, let's find

$$35 \div 8$$

by repeated subtraction. Keep track of the number of times 8 is subtracted from 35. We are through when we can subtract no more because the difference is less than 8.

$35 \div 8$: Repeated Subtraction

$$
\begin{array}{r}
35 \\
-8
\end{array} \Big\} \text{ 1 time}
$$
$$
\begin{array}{r}
27 \\
-8
\end{array} \Big\} \text{ 2 times}
$$
$$
\begin{array}{r}
19 \\
-8
\end{array} \Big\} \text{ 3 times}
$$
$$
\begin{array}{r}
11 \\
-8
\end{array} \Big\} \text{ 4 times}
$$
$$3 \longleftarrow \text{ Remainder}$$
(We cannot subtract 8 again.)

35 dollars

8 dollars 1 time
8 dollars 2 times
8 dollars 3 times
8 dollars 4 times
3 dollars left over

Thus, $35 \div 8 = 4$ R 3.

To check, perform the same multiplication as usual, but finish by adding in the remainder.

whole number part of quotient	·	divisor	+	remainder	=	dividend
4	·	8	+	3	=	35

Answer
9. 49 R 60

Objective C Solving Problems by Dividing

Below are some key words and phrases that may indicate the operation of division:

Division		
Key Words or Phrases	**Examples**	**Symbols**
divide	divide 10 by 5	$10 \div 5$ or $\dfrac{10}{5}$
quotient	the quotient of 64 and 4	$64 \div 4$ or $\dfrac{64}{4}$
divided by	9 divided by 3	$9 \div 3$ or $\dfrac{9}{3}$
divided or shared equally among	$100 divided equally among five people	$100 \div 5$ or $\dfrac{100}{5}$
per	100 miles per 2 hours	$\dfrac{100 \text{ miles}}{2 \text{ hours}}$

✔ **Concept Check** Which of the following is the correct way to represent "the quotient of 60 and 12"? Or are both correct? Explain your answer.

a. $12 \div 60$
b. $60 \div 12$

EXAMPLE 10 **Finding Shared Earnings**

Three college students share a paper route to earn money for expenses. The total in their fund after expenses was $2895. How much is each person's equal share?

Solution:

In words: | Each person's share | = | total money | ÷ | number of persons |

Translate: Each person's share   =   2895   ÷   3

Then

$$
\begin{array}{r}
965 \\
3\overline{)2895} \\
-27 \\
\hline
19 \\
-18 \\
\hline
15 \\
-15 \\
\hline
0
\end{array}
$$

Each person's share is $965.

⬛ **Work Practice Problem 10**

PRACTICE PROBLEM 10

Three students bought 171 blank CDs to share equally. How many CDs did each person get?

Answer
10. 57 CDs

✔ **Concept Check Answers**
a. incorrect **b.** correct

PRACTICE PROBLEM 11

Printers can be packed 12 to a box. If 532 printers are to be packed but only full boxes are shipped, how many full boxes will be shipped? How many printers are left over and not shipped?

EXAMPLE 11 Dividing Number of Downloads

As part of a promotion, Becky Foster receives 238 cards, each good for one free song download from MSN music. If Becky wants to share them evenly with 19 friends, how many download cards will each friend receive? How many will be left over?

Solution:

In words: $\boxed{\text{Number of cards for each person}} = \boxed{\text{number of cards}} \div \boxed{\text{number of friends}}$

Translate: $\boxed{\text{Number of cards for each person}} = 238 \div 19$

$$
\begin{array}{r}
12 \text{ R } 10 \\
19\overline{)238} \\
-19 \\
\hline
48 \\
-38 \\
\hline
10
\end{array}
$$

Each friend will receive 12 download cards. The cards cannot be divided equally among her friends since there is a nonzero remainder. There will be 10 download cards left over.

■ **Work Practice Problem 11**

Objective **D** **Finding Averages**

A special application of division (and addition) is finding the average of a list of numbers. The **average** of a list of numbers is the sum of the numbers divided by the *number* of numbers.

$$\text{average} = \frac{\text{sum of numbers}}{\textit{number} \text{ of numbers}}$$

PRACTICE PROBLEM 12

To compute a safe time to wait for reactions to occur after allergy shots are administered, a lab technician is given a list of elapsed times between administered shots and reactions. Find the average of the times 4 minutes, 7 minutes, 35 minutes, 16 minutes, 9 minutes, 3 minutes, and 52 minutes.

EXAMPLE 12 Averaging Scores

A mathematics instructor is checking a simple program she wrote for averaging the scores of her students. To do so, she averages a student's scores of 75, 96, 81, and 88 by hand. Find this average score.

Solution: To find the average score, we find the sum of his scores and divide by 4, the number of scores.

$$
\begin{array}{r}
75 \\
96 \\
81 \\
+88 \\
\hline
340 \text{ sum}
\end{array}
\qquad
\text{average} = \frac{340}{4} = 85
\qquad
\begin{array}{r}
85 \\
4\overline{)340} \\
-32 \\
\hline
20 \\
-20 \\
\hline
0
\end{array}
$$

The average score is 85.

■ **Work Practice Problem 12**

▣ CALCULATOR EXPLORATIONS Dividing Numbers

To divide numbers on a calculator, find the keys marked [÷] and [=] or [ENTER]. For example, to find 435 ÷ 5 on a calculator, press the keys [435] [÷] [5] [=] or [ENTER]. The display will read [87]. Thus, 435 ÷ 5 = 87.

Use a calculator to divide.

1. 848 ÷ 16

2. 564 ÷ 12

3. 95)‾5890‾

4. 27)‾1053‾

5. $\dfrac{32,886}{126}$

6. $\dfrac{143,088}{264}$

7. 0 ÷ 315

8. 315 ÷ 0

Vocabulary and Readiness Check

Use the choices below to fill in each blank.

1	number	divisor	dividend
0	undefined	average	quotient

1. In 90 ÷ 2 = 45, the answer 45 is called the _____, 90 is called the _____, and 2 is called the _____.

2. The quotient of any number and 1 is the same _____.

3. The quotient of any number (except 0) and the same number is _____.

4. The quotient of 0 and any number (except 0) is _____.

5. The quotient of any number and 0 is _____.

6. The _____ of a list of numbers is the sum of the numbers divided by the _____ of numbers.

1.6 EXERCISE SET

FOR EXTRA HELP

 Student Solutions Manual   PH Math/Tutor Center   CD/Video for Review   Math XL · MathXL®   MyMathLab · MyMathLab

Objective A *Find each quotient. See Examples 1 through 3.*

1. 54 ÷ 9

2. 72 ÷ 9

3. 36 ÷ 3

4. 24 ÷ 3

5. 0 ÷ 8

6. 0 ÷ 4

7. 31 ÷ 1

8. 38 ÷ 1

9. $\dfrac{18}{18}$

10. $\dfrac{49}{49}$

11. $\dfrac{24}{3}$

12. $\dfrac{45}{9}$

13. 26 ÷ 0

14. $\dfrac{12}{0}$

15. 26 ÷ 26

16. 6 ÷ 6

17. 0 ÷ 14

18. 7 ÷ 0

19. 18 ÷ 2

20. 18 ÷ 3

Objectives **A** **B** **Mixed Practice** *Divide and then check by multiplying. See Examples 1 through 5.*

21. $3\overline{)87}$ **22.** $5\overline{)85}$ **23.** $3\overline{)222}$ **24.** $8\overline{)640}$ **25.** $3\overline{)1014}$ **26.** $4\overline{)2104}$

27. $\dfrac{30}{0}$ **28.** $\dfrac{0}{30}$ **29.** $63 \div 7$ **30.** $56 \div 8$ **31.** $150 \div 6$ **32.** $121 \div 11$

Divide and then check by multiplying. See Examples 6 and 7.

33. $7\overline{)479}$ **34.** $7\overline{)426}$ **35.** $6\overline{)1421}$ **36.** $3\overline{)1240}$

37. $305 \div 8$ **38.** $167 \div 3$ **39.** $2286 \div 7$ **40.** $3333 \div 4$

Divide and then check by multiplying. See Examples 8 and 9.

41. $55\overline{)715}$ **42.** $23\overline{)736}$ **43.** $23\overline{)1127}$ **44.** $42\overline{)2016}$ **45.** $97\overline{)9417}$

46. $1938 \div 44$ **47.** $3146 \div 15$ **48.** $7354 \div 12$ **49.** $6578 \div 13$ **50.** $5670 \div 14$

51. $9299 \div 46$ **52.** $2505 \div 64$ **53.** $\dfrac{12{,}744}{236}$ **54.** $\dfrac{5781}{123}$ **55.** $\dfrac{10{,}297}{103}$

56. $\dfrac{23{,}092}{240}$ **57.** $20{,}619 \div 102$ **58.** $40{,}853 \div 203$ **59.** $244{,}989 \div 423$ **60.** $164{,}592 \div 543$

Divide. See Examples 1 through 9.

61. $7\overline{)119}$ **62.** $8\overline{)104}$ **63.** $7\overline{)3580}$ **64.** $5\overline{)3017}$

65. $40\overline{)85{,}312}$ **66.** $50\overline{)85{,}747}$ **67.** $142\overline{)863{,}360}$ **68.** $214\overline{)650{,}560}$

Objective **C** **Translating** *Solve. See Examples 10 through 11.*

69. Find the quotient of 117 and 5.   **70.** Find the quotient of 94 and 7.

71. Find 200 divided by 35.   **72.** Find 116 divided by 32.

73. Find the quotient of 62 and 3.

74. Find the quotient of 78 and 5.

75. Martin Thieme teaches American Sign Language classes for $65 per student for a 7-week session. He collects $2145 from the group of students. Find how many students are in the group.

76. Kathy Gomez teaches Spanish lessons for $85 per student for a 5-week session. From one group of students, she collects $4930. Find how many students are in the group.

77. The gravity of Jupiter is 318 times as strong as the gravity of Earth, so objects on Jupiter weigh 318 times as much as they weigh on Earth. If a person would weigh 52,470 pounds on Jupiter, find how much the person weighs on Earth.

78. Twenty-one people pooled their money and bought lottery tickets. One ticket won a prize of $5,292,000. Find how many dollars each person received.

79. An 18-hole golf course is 5580 yards long. If the distance to each hole is the same, find the distance between holes.

80. A truck hauls wheat to a storage granary. It carries a total of 5768 bushels of wheat in 14 trips. How much does the truck haul each trip if each trip it hauls the same amount?

81. There is a bridge over highway I-35 every three miles. The first bridge is at the beginning of a 265-mile stretch of highway. Find how many bridges there are over 265 miles of I-35.

82. The white stripes dividing the lanes on a highway are 25 feet long, and the spaces between them are 25 feet long. Let's call a "lane divider" a stripe followed by a space. Find how many whole "lane dividers" there are in 1 mile of highway. (A mile is 5280 feet.)

83. Ari Trainor is in the requisitions department of Central Electric Lighting Company. Light poles along a highway are placed 492 feet apart. The first light pole is at the beginning of a 1-mile strip. Find how many poles he should order for the 1-mile strip of highway. (A mile is 5280 feet.)

84. Professor Lopez has a piece of rope 176 feet long that she wants to cut into pieces for an experiment in her physics class. Each piece of rope is to be 8 feet long. Determine whether she has enough rope for her 22-student class. Determine the amount extra or the amount short.

85. Broad Peak in Pakistan is the twelfth-tallest mountain in the world. Its elevation is 26,400 feet. A mile is 5280 feet. How many miles tall is Broad Peak? (*Source:* National Geographic Society)

86. Shaun Alexander of the Seattle Seahawks led the NFL in touchdowns during the 2005 regular football season, scoring a total of 168 points from touchdowns. If a touchdown is worth 6 points, how many touchdowns did Alexander make during the 2005 season? (*Source:* National Football League)

87. Find how many yards are in 1 mile. (A mile is 5280 feet; a yard is 3 feet.)

88. Find how many whole feet are in 1 rod. (A mile is 5280 feet; 1 mile is 320 rods.)

Objective D *Find the average of each list of numbers. See Example 12.*

89. 10, 24, 35, 22, 17, 12

90. 37, 26, 15, 29, 51, 22

91. 205, 972, 210, 161

92. 121, 200, 185, 176, 163

93. 86, 79, 81, 69, 80

94. 92, 96, 90, 85, 92, 79

The normal monthly temperature in degrees Fahrenheit for Salt Lake City, Utah, is given in the graph. Use this graph to answer Exercises 95 and 96. (Source: National Climatic Data Center)

95. Find the average temperature for June, July, and August.

96. Find the average temperature for October, November, and December.

Mixed Practice (Sections 1.3, 1.5, 1.6) *Perform each indicated operation. Watch the operation symbol.*

97. 82 + 463 + 29 + 8704

98. 23 + 407 + 92 + 7011

99. $\begin{array}{r} 546 \\ \times\ 28 \end{array}$

100. $\begin{array}{r} 712 \\ \times\ 54 \end{array}$

101. $\begin{array}{r} 722 \\ -\ 43 \end{array}$

102. $\begin{array}{r} 712 \\ -\ 54 \end{array}$

103. $\dfrac{45}{0}$

104. $\dfrac{0}{23}$

105. 228 ÷ 24

106. 304 ÷ 31

Concept Extensions

Match each word phrase to the correct translation. (Not all letter choices will be used.) See the Concept Check in this section.

107. The quotient of 40 and 8

a. 20 ÷ 200

108. The quotient of 200 and 20

b. 200 ÷ 20

109. 200 divided by 20

c. 40 ÷ 8

110. 40 divided by 8

d. 8 ÷ 40

The following table shows the top five leading U.S. television advertisers in 2005 and the amount of money spent that year on television advertising. Use this table to answer Exercises 111 and 112.

Advertiser	Amount Spent on Television Advertising in 2005
General Motors, Dealers and corporate	$797,000,000
Ford Motors, Dealers and corporate	$667,000,000
Toyota, Dealers and corporate	$544,000,000
DaimlerChrysler AG	$528,000,000
Honda	$339,000,000

111. Find the average amount of money spent on television ads for the year by the top two advertisers.

112. Find the average amount of money spent on television ads by the top four advertisers.

In Example 12 in this section, we found that the average of 75, 96, 88, and 81 is 85. Use this information to answer Exercises 113 and 114.

113. If the number 75 is removed from the list of numbers, does the average increase or decrease? Explain why.

114. If the number 96 is removed from the list of numbers, does the average increase or decrease? Explain why.

115. Without computing it, tell whether the average of 126, 135, 198, 113 is 86. Explain why it is or why it is not.

116. If the area of a rectangle is 60 square feet and its width is 5 feet, what is its length?

117. Write down any two numbers whose quotient is 25.

118. Find 26 ÷ 5 using the process of repeated subtraction.

 THE BIGGER PICTURE Operations on Sets of Numbers

This is a special feature that we begin in this section. Among other concepts introduced later in the text, it is very important for you to be able to perform operations on different sets of numbers. To help you remember these operations, we begin an outline below and continually expand this outline throughout this text. Although suggestions are given, this outline should be in your own words. Once you complete the new portion of your outline, try the exercises below. Remember: Study your outline often as you proceed through this text.

I. Some Operations on Sets of Numbers

  A. Whole Numbers

    1. Add or Subtract:

$$
\begin{array}{r} 13 \\ +29 \\ \hline 42 \end{array}
\qquad
\begin{array}{r} 400 \\ -38 \\ \hline 362 \end{array}
$$

2. Multiply or Divide:

$$
\begin{array}{r} 238 \\ \times 47 \\ \hline 1666 \\ 9520 \\ \hline 11186 \end{array}
\qquad
\begin{array}{r} 127\ \text{R}\,2 \\ 7\overline{)891} \\ -7 \\ \hline 19 \\ -14 \\ \hline 51 \\ -49 \\ \hline 2 \end{array}
$$

Perform indicated operations.

1. 82 + 39   **2.** 82 − 39

3. 82 × 39   **4.** 2592 ÷ 29

5. 0 · 15   **6.** 0 ÷ 11

7. 26 · 1   **8.** 36 ÷ 0

9. 82 ÷ 1   **10.** 2000 − 156

Operations on Whole Numbers

Answers

1. _____
2. _____
3. _____
4. _____
5. _____
6. _____
7. _____
8. _____
9. _____
10. _____
11. _____
12. _____
13. _____
14. _____
15. _____
16. _____
17. _____
18. _____
19. _____
20. _____
21. _____
22. _____
23. _____
24. _____
25. _____
26. _____
27. _____
28. _____
29. _____
30. _____

Perform each indicated operation.

1.
$$\begin{array}{r} 42 \\ 63 \\ +\,89 \\ \hline \end{array}$$

2.
$$\begin{array}{r} 7006 \\ -\,451 \\ \hline \end{array}$$

3.
$$\begin{array}{r} 87 \\ \times\,52 \\ \hline \end{array}$$

4. $8\overline{)4496}$

5. $1 \cdot 67$

6. $\dfrac{36}{0}$

7. $16 \div 16$

8. $5 \div 1$

9. $0 \cdot 21$

10. $7 \cdot 0 \cdot 8$

11. $0 \div 7$

12. $12 \div 4$

13. $9 \cdot 7$

14. $45 \div 5$

15.
$$\begin{array}{r} 207 \\ -\,69 \\ \hline \end{array}$$

16.
$$\begin{array}{r} 207 \\ +\,69 \\ \hline \end{array}$$

17. $3718 - 2549$

18. $1861 + 7965$

19. $7\overline{)1278}$

20.
$$\begin{array}{r} 1259 \\ \times\,63 \\ \hline \end{array}$$

21. $7\overline{)7695}$

22. $9\overline{)1000}$

23. $32\overline{)21,240}$

24. $65\overline{)70,000}$

25. $4000 - 2963$

26. $10,000 - 101$

27.
$$\begin{array}{r} 303 \\ \times\,101 \\ \hline \end{array}$$

28. $(475)(100)$

29. Find the total of 62 and 9.

30. Find the product of 62 and 9.

31. Find the quotient of 62 and 9.

32. Find the difference of 62 and 9.

33. Subtract 17 from 200.

34. Find the difference of 432 and 201.

Complete the table by rounding the given number to the given place value.

		Tens	Hundreds	Thousands
35.	9735			
36.	1429			
37.	20,801			
38.	432,198			

Find the perimeter and area of each figure.

△ **39.**

Square   6 feet

△ **40.**

14 inches

Rectangle   7 inches

Find the perimeter of each figure.

△ **41.**

13 miles   9 miles

6 miles

△ **42.**

3 meters

4 meters

3 meters

3 meters

Find the average of each list of numbers.

43. 19, 15, 25, 37, 24

44. 108, 131, 98, 159

45. The Mackinac Bridge is a suspension bridge that connects the lower and upper peninsulas of Michigan across the Straits of Mackinac. Its total length is 26,372 feet. The Lake Pontchartrain Bridge is a twin concrete trestle bridge in Slidell, Louisiana. Its total length is 28,547 feet. Which bridge is longer and by how much? (*Sources:* Mackinac Bridge Authority and Federal Highway Administration, Bridge Division)

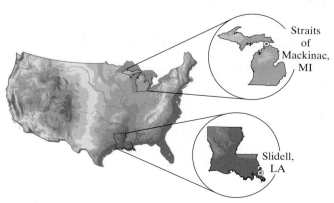

Straits
of
Mackinac,
MI

Slidell,
LA

46. In North America, the average toy expenditure per child is $347 per year. On average, how much is spent on toys for a child by the time he or she reaches age 18? (*Source:* The NPD Group Worldwide)

31. _____

32. _____

33. _____

34. _____

35. see table _____

36. see table _____

37. see table _____

38. see table _____

39. _____

40. _____

41. _____

42. _____

43. _____

44. _____

45. _____

46. _____

A Write Repeated Factors Using Exponential Notation.

B Evaluate Expressions Containing Exponents.

C Use the Order of Operations.

D Find the Area of a Square.

Objective A Using Exponential Notation

In the product $3 \cdot 3 \cdot 3 \cdot 3 \cdot 3$, notice that 3 is a factor several times. When this happens, we can use a shorthand notation, called an **exponent**, to write the repeated multiplication.

$\underbrace{3 \cdot 3 \cdot 3 \cdot 3 \cdot 3}$ can be written as

3 is a factor 5 times

exponent

3^5 Read as "three to the fifth power."

base

This is called **exponential notation.** The **exponent**, 5, indicates how many times the **base**, 3, is a factor.

The table below shows examples of reading exponential notation in words.

Expression	In Words
5^2	"five to the second power" or "five squared"
5^3	"five to the third power" or "five cubed"
5^4	"five to the fourth power"

Usually, an exponent of 1 is not written, so when no exponent appears, we assume that the exponent is 1. For example, $2 = 2^1$ and $7 = 7^1$.

PRACTICE PROBLEMS 1–4

Write using exponential notation.

1. $8 \cdot 8 \cdot 8 \cdot 8$

2. $3 \cdot 3 \cdot 3$

3. $10 \cdot 10 \cdot 10 \cdot 10 \cdot 10$

4. $5 \cdot 5 \cdot 4 \cdot 4 \cdot 4 \cdot 4 \cdot 4 \cdot 4$

EXAMPLES Write using exponential notation.

1. $7 \cdot 7 \cdot 7 = 7^3$

2. $3 \cdot 3 = 3^2$

3. $6 \cdot 6 \cdot 6 \cdot 6 \cdot 6 = 6^5$

4. $3 \cdot 3 \cdot 3 \cdot 3 \cdot 9 \cdot 9 \cdot 9 = 3^4 \cdot 9^3$

Work Practice Problems 1–4

Objective B Evaluating Exponential Expressions

To **evaluate** an exponential expression, we write the expression as a product and then find the value of the product.

PRACTICE PROBLEMS 5–8

Evaluate.

5. 4^2 **6.** 8^3

7. 11^1 **8.** $2 \cdot 3^2$

EXAMPLES Evaluate.

5. $9^2 = 9 \cdot 9 = 81$

6. $6^1 = 6$

7. $3^4 = 3 \cdot 3 \cdot 3 \cdot 3 = 81$

8. $5 \cdot 6^2 = 5 \cdot 6 \cdot 6 = 180$

Work Practice Problems 5–8

Answers

1. 8^4 **2.** 3^3 **3.** 10^5 **4.** $5^2 \cdot 4^6$
5. 16 **6.** 512 **7.** 11 **8.** 18

Example 8 illustrates an important property: An exponent applies only to its base. The exponent 2, in $5 \cdot 6^2$, applies only to its base, 6.

Helpful Hint

An exponent applies only to its base. For example, $4 \cdot 2^3$ means $4 \cdot 2 \cdot 2 \cdot 2$.

Helpful Hint

Don't forget that 2^4, for example, is *not* $2 \cdot 4$. The expression 2^4 means repeated multiplication of the same factor.

$$2^4 = 2 \cdot 2 \cdot 2 \cdot 2 = 16, \quad \text{whereas } 2 \cdot 4 = 8$$

✔ **Concept Check**   Which of the following statements is correct?

a. 3^5 is the same as $5 \cdot 5 \cdot 5$.
b. "Ten cubed" is the same as 10^2.
c. "Six to the fourth power" is the same as 6^4.
d. 12^2 is the same as $12 \cdot 2$.

Objective C Using the Order of Operations

Suppose that you are in charge of taking inventory at a local cell phone store. An employee has given you the number of a certain cell phone in stock as the expression

$$6 + 2 \cdot 30$$

To calculate the value of this expression, do you add first or multiply first? If you add first, the answer is 240. If you multiply first, the answer is 66.

Contents: 30 cell phones

Contents: 30 cell phones

Mathematical symbols wouldn't be very useful if two values were possible for one expression. Thus, mathematicians have agreed that, given a choice, we multiply first.

$$6 + 2 \cdot 30 = 6 + 60 \quad \text{Multiply.}$$
$$= 66 \quad \text{Add.}$$

This agreement is one of several **order of operations** agreements.

Order of Operations

1. Perform all operations within parentheses ( ), brackets [ ], or other grouping symbols such as fraction bars, starting with the innermost set.
2. Evaluate any expressions with exponents.
3. Multiply or divide in order from left to right.
4. Add or subtract in order from left to right.

Below we practice using order of operations to simplify expressions.

PRACTICE PROBLEM 9

Simplify: $9 \cdot 3 - 8 \div 4$

EXAMPLE 9 Simplify: $2 \cdot 4 - 3 \div 3$

Solution: There are no parentheses and no exponents, so we start by multiplying and dividing, from left to right.

$$2 \cdot 4 - 3 \div 3 = 8 - 3 \div 3 \quad \text{Multiply.}$$
$$= 8 - 1 \quad \text{Divide.}$$
$$= 7 \quad \text{Subtract.}$$

Work Practice Problem 9

PRACTICE PROBLEM 10

Simplify: $48 \div 3 \cdot 2^2$

EXAMPLE 10 Simplify: $4^2 \div 2 \cdot 4$

Solution: We start by evaluating 4^2.

$$4^2 \div 2 \cdot 4 = 16 \div 2 \cdot 4 \quad \text{Write } 4^2 \text{ as 16.}$$

Next we multiply or divide *in order* from left to right. Since division appears before multiplication from left to right, we divide first, then multiply.

$$16 \div 2 \cdot 4 = 8 \cdot 4 \quad \text{Divide.}$$
$$= 32 \quad \text{Multiply.}$$

Work Practice Problem 10

PRACTICE PROBLEM 11

Simplify: $(10 - 7)^4 + 2 \cdot 3^2$

EXAMPLE 11 Simplify: $(8 - 6)^2 + 2^3 \cdot 3$

Solution: $(8 - 6)^2 + 2^3 \cdot 3 = 2^2 + 2^3 \cdot 3 \quad \text{Simplify inside parentheses.}$
$$= 4 + 8 \cdot 3 \quad \text{Write } 2^2 \text{ as 4 and } 2^3 \text{ as 8.}$$
$$= 4 + 24 \quad \text{Multiply.}$$
$$= 28 \quad \text{Add.}$$

Work Practice Problem 11

Answers
9. 25   **10.** 64   **11.** 99

EXAMPLE 12 Simplify: $4^3 + [3^2 - (10 \div 2)] - 7 \cdot 3$

Solution: Here we begin with the innermost set of parentheses.

$4^3 + [3^2 - (10 \div 2)] - 7 \cdot 3 = 4^3 + [3^2 - 5] - 7 \cdot 3$ Simplify inside parentheses.

$\qquad = 4^3 + [9 - 5] - 7 \cdot 3$ Write 3^2 as 9.

$\qquad = 4^3 + 4 - 7 \cdot 3$ Simplify inside brackets.

$\qquad = 64 + 4 - 7 \cdot 3$ Write 4^3 as 64.

$\qquad = 64 + 4 - 21$ Multiply.

$\qquad = 47$ Add and subtract from left to right.

■ **Work Practice Problem 12**

EXAMPLE 13 Simplify: $\dfrac{7 - 2 \cdot 3 + 3^2}{5(2 - 1)}$

Solution: Here, the fraction bar is like a grouping symbol. We simplify above and below the fraction bar separately.

$\dfrac{7 - 2 \cdot 3 + 3^2}{5(2 - 1)} = \dfrac{7 - 2 \cdot 3 + 9}{5(1)}$ Evaluate 3^2 and $(2 - 1)$.

$\qquad = \dfrac{7 - 6 + 9}{5}$ Multiply $2 \cdot 3$ in the numerator and multiply 5 and 1 in the denominator.

$\qquad = \dfrac{10}{5}$ Add and subtract from left to right.

$\qquad = 2$ Divide.

■ **Work Practice Problem 13**

EXAMPLE 14 Simplify: $64 \div 8 \cdot 2 + 4$

Solution: $64 \div 8 \cdot 2 + 4 = 8 \cdot 2 + 4$ Divide.

$\qquad = 16 + 4$ Multiply.

$\qquad = 20$ Add.

■ **Work Practice Problem 14**

Objective D Finding the Area of a Square

Since a square is a special rectangle, we can find its area by finding the product of its length and its width.

$\qquad$ Area of a rectangle = length · width

By recalling that each side of a square has the same measurement, we can use the following procedure to find its area:

$\qquad$ Area of a square = length · width

$\qquad\qquad\qquad\qquad = $ side · side

$\qquad\qquad\qquad\qquad = (\text{side})^2$

Square | Side

Side

PRACTICE PROBLEM 12
Simplify:
$36 \div [20 - (4 \cdot 2)] + 4^3 - 6$

PRACTICE PROBLEM 13
Simplify: $\dfrac{25 + 8 \cdot 2 - 3^3}{2(3 - 2)}$

PRACTICE PROBLEM 14
Simplify: $36 \div 6 \cdot 3 + 5$

Answers
12. 61 **13.** 7 **14.** 23

PRACTICE PROBLEM 15

Find the area of a square whose side measures 12 centimeters.

EXAMPLE 15 Find the area of a square whose side measures 4 inches.

Solution: Area of a square $= (\text{side})^2$
$$= (4 \text{ inches})^2$$
$$= 16 \text{ square inches}$$

The area of the square is 16 square inches.

4 inches

◼ **Work Practice Problem 15**

Answer

15. 144 sq cm

 CALCULATOR EXPLORATIONS   Exponents

To evaluate an exponent such as 4^7 on a calculator, find the keys marked $\boxed{y^x}$ or $\boxed{\wedge}$ and $\boxed{=}$ or $\boxed{\text{ENTER}}$. To evaluate 4^7, press the keys $\boxed{4}$ $\boxed{y^x}$ (or $\boxed{\wedge}$) $\boxed{7}$ $\boxed{=}$ or $\boxed{\text{ENTER}}$. The display will read $\boxed{16384}$. Thus, $4^7 = 16,384$.

Use a calculator to evaluate.

1. 3^6 **2.** 5^6 **3.** 4^5

4. 7^6 **5.** 2^{11} **6.** 6^8

Order of Operations

To see whether your calculator has the order of operations built in, evaluate $5 + 2 \cdot 3$ by pressing the keys $\boxed{5}$ $\boxed{+}$ $\boxed{2}$ $\boxed{\times}$ $\boxed{3}$ $\boxed{=}$ or $\boxed{\text{ENTER}}$. If the display reads $\boxed{11}$, your calculator does have the order of operations built in. This means that most of the time you can key in a

problem exactly as it is written and the calculator will perform operations in the proper order. When evaluating an expression containing parentheses, key in the parentheses. (If an expression contains brackets, key in parentheses.) For example, to evaluate $2[25 - (8 + 4)] - 11$, press the keys $\boxed{2}$ $\boxed{\times}$ $\boxed{(}$ $\boxed{25}$ $\boxed{-}$ $\boxed{(}$ $\boxed{8}$ $\boxed{+}$ $\boxed{4}$ $\boxed{)}$ $\boxed{)}$ $\boxed{-}$ $\boxed{11}$ $\boxed{=}$ or $\boxed{\text{ENTER}}$.
The display will read $\boxed{15}$.

Use a calculator to evaluate.

7. $7^4 + 5^3$

8. $12^4 - 8^4$

9. $63 \cdot 75 - 43 \cdot 10$

10. $8 \cdot 22 + 7 \cdot 16$

11. $4(15 \div 3 + 2) - 10 \cdot 2$

12. $155 - 2(17 + 3) + 185$

Vocabulary and Readiness Check

Use the choices below to fill in each blank.

addition   multiplication   exponent

subtraction   division   base

1. In $2^5 = 32$, the 2 is called the _____ and the 5 is called the _____.

2. To simplify $8 + 2 \cdot 6$, which operation should be performed first? _____

3. To simplify $(8 + 2) \cdot 6$, which operation should be performed first? _____

4. To simplify $9(3 - 2) \div 3 + 6$, which operation should be performed first? _____

5. To simplify $8 \div 2 \cdot 6$, which operation should be performed first? _____

1.7 EXERCISE SET

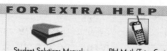

FOR EXTRA HELP

Student Solutions Manual   PH Math/Tutor Center   CD/Video for Review   MathXL MathXL®   MyMathLab MyMathLab

Objective A *Write using exponential notation. See Examples 1 through 4.*

1. $4 \cdot 4 \cdot 4$

2. $5 \cdot 5 \cdot 5 \cdot 5$

3. $7 \cdot 7 \cdot 7 \cdot 7 \cdot 7 \cdot 7$

4. $6 \cdot 6 \cdot 6 \cdot 6 \cdot 6 \cdot 6 \cdot 6$

5. $12 \cdot 12 \cdot 12$

6. $10 \cdot 10 \cdot 10$

7. $6 \cdot 6 \cdot 5 \cdot 5 \cdot 5$

8. $4 \cdot 4 \cdot 3 \cdot 3 \cdot 3$

9. $9 \cdot 8 \cdot 8$

10. $7 \cdot 4 \cdot 4 \cdot 4$

11. $3 \cdot 2 \cdot 2 \cdot 2 \cdot 2$

12. $4 \cdot 6 \cdot 6 \cdot 6 \cdot 6$

13. $3 \cdot 2 \cdot 2 \cdot 2 \cdot 2 \cdot 5 \cdot 5 \cdot 5 \cdot 5 \cdot 5$

14. $6 \cdot 6 \cdot 2 \cdot 9 \cdot 9 \cdot 9 \cdot 9$

Objective B *Evaluate. See Examples 5 through 8.*

15. 8^2

16. 6^2

17. 5^3

18. 6^3

19. 2^5

20. 3^5

21. 1^{10}

22. 1^{12}

23. 7^1

24. 8^1

25. 2^7

26. 5^4

27. 2^8

28. 3^3

29. 4^4

30. 4^3

31. 9^3

32. 8^3

33. 12^2

34. 11^2

35. 10^2

36. 10^3

37. 20^1

38. 14^1

39. 3^6

40. 4^5

41. $3 \cdot 2^6$

42. $5 \cdot 3^2$

43. $2 \cdot 3^4$

44. $2 \cdot 7^2$

Objective C *Simplify. See Examples 9 through 14.*

45. $15 + 3 \cdot 2$

46. $24 + 6 \cdot 3$

47. $28 \div 7 \cdot 1 + 3$

48. $100 \div 10 \cdot 5 + 4$

49. $32 \div 4 - 3$

50. $42 \div 7 - 6$

51. $13 + \dfrac{24}{8}$

52. $32 + \dfrac{8}{2}$

53. $6 \cdot 5 + 8 \cdot 2$

54. $3 \cdot 4 + 9 \cdot 1$

55. $\dfrac{5 + 12 \div 4}{1^7}$

56. $\dfrac{6 + 9 \div 3}{3^2}$

57. $(7 + 5^2) \div 4 \cdot 2^3$

58. $6^2 \cdot (10 - 8)$

59. $5^2 \cdot (10 - 8) + 2^3 + 5^2$

60. $5^3 \div (10 + 15) + 9^2 + 3^3$

61. $\dfrac{18 + 6}{2^4 - 2^2}$

62. $\dfrac{40 + 8}{5^2 - 3^2}$

63. $(3 + 5) \cdot (9 - 3)$

64. $(9 - 7) \cdot (12 + 18)$

65. $\dfrac{7(9 - 6) + 3}{3^2 - 3}$

66. $\dfrac{5(12 - 7) - 4}{5^2 - 18}$

67. $8 \div 0 + 37$

68. $18 - 7 \div 0$

69. $2^4 \cdot 4 - (25 \div 5)$

70. $2^3 \cdot 3 - (100 \div 10)$

71. $3^4 - [35 - (12 - 6)]$

72. $[40 - (8 - 2)] - 2^5$

73. $(7 \cdot 5) + [9 \div (3 \div 3)]$

74. $(18 \div 6) + [(3 + 5) \cdot 2]$

75. $8 \cdot [2^2 + (6 - 1) \cdot 2] - 50 \cdot 2$

76. $35 \div [3^2 + (9 - 7) - 2^2] + 10 \cdot 3$

77. $\dfrac{9^2 + 2^2 - 1^2}{8 \div 2 \cdot 3 \cdot 1 \div 3}$

78. $\dfrac{5^2 - 2^3 + 1^4}{10 \div 5 \cdot 4 \cdot 1 \div 4}$

79. $\dfrac{2 + 4^2}{5(20 - 16) - 3^2 - 5}$

80. $\dfrac{3 + 9^2}{3(10 - 6) - 2^2 - 1}$

81. $9 \div 3 + 5^2 \cdot 2 - 10$

82. $10 \div 2 + 3^3 \cdot 2 - 20$

83. $[13 \div (20 - 7) + 2^5] - (2 + 3)^2$

84. $[15 \div (11 - 6) + 2^2] + (5 - 1)^2$

85. $7^2 - \{18 - [40 \div (5 \cdot 1) + 2] + 5^2\}$

86. $29 - \{5 + 3[8 \cdot (10 - 8)] - 50\}$

Objective **D** *Find the area of each square. See Example 15.*

△ **87.**

7 meters

△ **88.**

9 centimeters

△ **89.**

23 miles

△ **90.**

41 feet

Concept Extensions

Answer the following true or false. See the Concept Check in this section.

91. "Six to the fifth power" is the same as 6^5.

92. "Seven squared" is the same as 7^2.

93. 2^5 is the same as $5 \cdot 5$.

94. 4^9 is the same as $4 \cdot 9$.

Insert grouping symbols (parentheses) so that each given expression evaluates to the given number.

95. $2 + 3 \cdot 6 - 2$; evaluate to 28

96. $2 + 3 \cdot 6 - 2$; evaluate to 20

97. $24 \div 3 \cdot 2 + 2 \cdot 5$; evaluate to 14

98. $24 \div 3 \cdot 2 + 2 \cdot 5$; evaluate to 15

△ **99.** A building contractor is bidding on a contract to install gutters on seven homes in a retirement community, all in the shape shown. To estimate the cost of materials, she needs to know the total perimeter of all seven homes. Find the total perimeter.

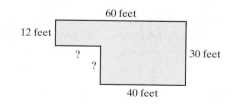

Simplify.

▦ **100.** $25^3 \cdot (45 - 7 \cdot 5) \cdot 5$

▦ **101.** $(7 + 2^4)^5 - (3^5 - 2^4)^2$

102. Explain why $2 \cdot 3^2$ is not the same as $(2 \cdot 3)^2$.

103. Write an expression that simplifies to 5. Use multiplication, division, addition, subtraction, and at least one set of parentheses. Explain the process you would use to simplify the expression.

THE BIGGER PICTURE   Operations on Sets of Numbers

Continue your outline started in Section 1.6. Suggestions are once again written to help you complete this part of your outline.

I. **Some Operations on Sets of Numbers**
    A. **Whole Numbers**
        1. **Add or Subtract** (Section 1.3)
        2. **Multiply or Divide** (Sections 1.5, 1.6)
        3. **Exponent:** $3^4 = \overbrace{3 \cdot 3 \cdot 3 \cdot 3}^{\text{4 factors of 3}} = 81$
        4. **Order of Operations:**

$$24 \div 3 \cdot 2 - (2 + 8)$$
$$= 24 \div 3 \cdot 2 - (10) \quad \text{Parentheses.}$$
$$= 8 \cdot 2 - 10 \quad \text{Multiply or divide from left to right.}$$
$$= 16 - 10 \quad \text{Multiply or divide from left to right.}$$
$$= 6 \quad \text{Add or subtract from left to right.}$$

Perform the indicated operations.

1. 6^3
2. $2^3 \cdot 6^1$
3. $8 \cdot 5^2$
4. $1 + 2(3 + 4)$
5. $2 + 5(10 - 3)$
6. $200 \div 2 \cdot 2$
7. $978 - 179$
8. $\begin{array}{r} 72 \\ \times\, 30 \\ \hline \end{array}$
9. $614 \div 58$
10. $3[(7 - 3)^2 - (25 - 22)^2] + 6$

STUDY SKILLS BUILDER

What to Do the Day of an Exam?

Your first exam may be soon. On the day of an exam, don't forget to try the following:

- Allow yourself plenty of time to arrive.
- Read the directions on the test carefully.
- Read each problem carefully as you take your test. Make sure that you answer the question asked.
- Watch your time and pace yourself so that you may attempt each problem on your test.
- Check your work and answers.
- ***Do not turn your test in early.*** If you have extra time, spend it double-checking your work.

Good luck!

Answer the following questions based on your most recent mathematics exam, whenever that was.

1. How soon before class did you arrive?
2. Did you read the directions on the test carefully?
3. Did you make sure you answered the question asked for each problem on the exam?
4. Were you able to attempt each problem on your exam?
5. If your answer to question 4 is no, list reasons why.
6. Did you have extra time on your exam?
7. If your answer to question 6 is yes, describe how you spent that extra time.

1.8 INTRODUCTION TO VARIABLES, ALGEBRAIC EXPRESSIONS, AND EQUATIONS

Objectives

A Evaluate Algebraic Expressions Given Replacement Values.

B Identify Solutions of Equations.

C Translate Phrases into Variable Expressions.

Objective **A** Evaluating Algebraic Expressions

Perhaps the most important quality of mathematics is that it is a science of patterns. Communicating about patterns is often made easier by using a letter to represent all the numbers fitting a pattern. We call such a letter a **variable.** For example, in Section 1.2 we presented the addition property of 0, which states that the sum of 0 and any number is that number. We might write

$$0 + 1 = 1$$
$$0 + 2 = 2$$
$$0 + 3 = 3$$
$$0 + 4 = 4$$
$$0 + 5 = 5$$
$$0 + 6 = 6$$
$$\vdots$$

continuing indefinitely. This is a pattern, and all whole numbers fit the pattern. We can communicate this pattern for all whole numbers by letting a letter, such as a, represent all whole numbers. We can then write

$$0 + a = a$$

Using variable notation is a primary goal of learning **algebra.** We now take some important first steps in beginning to use variable notation.

A combination of operations on letters (variables) and numbers is called an **algebraic expression** or simply an **expression.**

Algebraic Expressions

$$3 + x \qquad 5 \cdot y \qquad 2 \cdot z - 1 + x$$

If two variables or a number and a variable are next to each other, with no operation sign between them, the operation is multiplication. For example,

$$2x \quad \text{means} \quad 2 \cdot x$$

and

$$xy \text{ or } x(y) \quad \text{means} \quad x \cdot y$$

Also, the meaning of an exponent remains the same when the base is a variable. For example,

$$x^2 = \underbrace{x \cdot x}_{2 \text{ factors of } x} \quad \text{and} \quad y^5 = \underbrace{y \cdot y \cdot y \cdot y \cdot y}_{5 \text{ factors of } y}$$

Algebraic expressions such as $3x$ have different values depending on replacement values for x. For example, if x is 2, then $3x$ becomes

$$3x = 3 \cdot 2$$
$$= 6$$

If x is 7, then $3x$ becomes

$$3x = 3 \cdot 7$$
$$= 21$$

Replacing a variable in an expression by a number and then finding the value of the expression is called **evaluating the expression** for the variable. When finding the value of an expression, remember to follow the order of operations given in Section 1.7.

Copyright 2008 Pearson Education, Inc.

PRACTICE PROBLEM 1

Evaluate $x - 2$ if x is 7.

EXAMPLE 1 Evaluate $x + 6$ if x is 8.

Solution: Replace x with 8 in the expression $x + 6$.

$$x + 6 = 8 + 6 \qquad \text{Replace } x \text{ with 8.}$$
$$= 14 \qquad \text{Add.}$$

■ **Work Practice Problem 1**

When we write a statement such as "x is 5," we can use an equals symbol ($=$) to represent "is" so that

x is 5   can be written as $x = 5$.

PRACTICE PROBLEM 2

Evaluate $y(x - 3)$ for $x = 8$ and $y = 4$.

EXAMPLE 2 Evaluate $2(x - y)$ for $x = 6$ and $y = 3$.

Solution: $2(x - y) = 2(6 - 3)$ Replace x with 6 and y with 3.

$$= 2(3) \qquad \text{Subtract.}$$
$$= 6 \qquad \text{Multiply.}$$

■ **Work Practice Problem 2**

PRACTICE PROBLEM 3

Evaluate $\dfrac{y + 6}{x}$ for $x = 6$ and $y = 18$.

EXAMPLE 3 Evaluate $\dfrac{x - 5y}{y}$ for $x = 35$ and $y = 5$.

Solution: $\dfrac{x - 5y}{y} = \dfrac{35 - 5(5)}{5}$ Replace x with 35 and y with 5.

$$= \dfrac{35 - 25}{5} \qquad \text{Multiply.}$$

$$= \dfrac{10}{5} \qquad \text{Subtract.}$$

$$= 2 \qquad \text{Divide.}$$

■ **Work Practice Problem 3**

PRACTICE PROBLEM 4

Evaluate $25 - z^3 + x$ for $z = 2$ and $x = 1$.

EXAMPLE 4 Evaluate $x^2 + z - 3$ for $x = 5$ and $z = 4$.

Solution: $x^2 + z - 3 = 5^2 + 4 - 3$ Replace x with 5 and z with 4.

$$= 25 + 4 - 3 \qquad \text{Evaluate } 5^2.$$
$$= 26 \qquad \text{Add and subtract from left to right.}$$

■ **Work Practice Problem 4**

Answers
1. 5 **2.** 20 **3.** 4 **4.** 18

Helpful Hint
If you are having difficulty replacing variables with numbers, first replace each variable with a set of parentheses, then insert the replacement number within the parentheses.

Example:

$$x^2 + z - 3 = (\ )^2 + (\ ) - 3$$
$$= (5)^2 + (4) - 3$$
$$= 25 + 4 - 3$$
$$= 26$$

✔ **Concept Check** What's wrong with the solution to the following problem?
Evaluate $3x + 2y$ for $x = 2$ and $y = 3$.

Solution: $3x + 2y = 3(3) + 2(2)$
$$= 9 + 4$$
$$= 13$$

EXAMPLE 5 The expression $\dfrac{5(F - 32)}{9}$ can be used to write degrees Fahrenheit F as degrees Celsius C. Find the value of this expression for $F = 86$.

Solution: $\dfrac{5(F - 32)}{9} = \dfrac{5(86 - 32)}{9}$

$$= \dfrac{5(54)}{9}$$

$$= \dfrac{270}{9}$$

$$= 30$$

Thus 86°F is the same temperature as 30°C.

▣ **Work Practice Problem 5**

Objective Ⓑ Identifying Solutions of Equations

In Objective Ⓐ, we learned that a combination of operations on variables and numbers is called an algebraic expression or simply an expression. Frequently in this book, we have written statements like $7 + 4 = 11$ or area = length · width. Each of these statements is called an **equation.** An equation is of the form

Helpful Hint
An equation contains "=", while an expression does not.

expression = expression

An equation can be labeled as

equal sign
↓
$x + 7 = 10$
↑   ↑
left side   right side

PRACTICE PROBLEM 5

Evaluate $\dfrac{5(F - 32)}{9}$ for $F = 41$.

✔ **Concept Check Answer**
$3x + 2y = 3(2) + 2(3)$
$$= 6 + 6$$
$$= 12$$

When an equation contains a variable, deciding which values of the variable make an equation a true statement is called **solving** an equation for the variable. A **solution** of an equation is a value for the variable that makes an equation a true statement. For example, 2 is a solution of the equation $x + 5 = 7$, since replacing x with 2 results in the *true* statement $2 + 5 = 7$. Similarly, 3 is not a solution of $x + 5 = 7$, since replacing x with 3 results in the *false* statement $3 + 5 = 7$.

PRACTICE PROBLEM 6

Determine whether 8 is a solution of the equation $3(y - 6) = 6$.

EXAMPLE 6 Determine whether 6 is a solution of the equation $4(x - 3) = 12$.

Solution: We replace x with 6 in the equation.

$$4(x - 3) = 12$$
$$\downarrow$$
$$4(6 - 3) \stackrel{?}{=} 12 \quad \text{Replace } x \text{ with 6.}$$
$$4(3) \stackrel{?}{=} 12$$
$$12 = 12 \quad \text{True}$$

Since $12 = 12$ is a true statement, 6 *is* a solution of the equation.

☐ **Work Practice Problem 6**

A collection of numbers enclosed by braces is called a set. For example,

$$\{0, 1, 2, 3, \dots\}$$

is the set of whole numbers that we are studying about in this chapter. The three dots after the number 3 in the set mean that this list of numbers continues in the same manner indefinitely.

The next example contains set notation.

PRACTICE PROBLEM 7

Determine which numbers in the set $\{10, 6, 8\}$ are solutions of the equation $5n + 4 = 34$.

EXAMPLE 7 Determine which numbers in the set $\{26, 40, 20\}$ are solutions of the equation $2n - 30 = 10$.

Solution: Replace n with each number from the set to see if a true statement results.

Let n be 26.

$$2n - 30 = 10$$
$$2 \cdot 26 - 30 \stackrel{?}{=} 10$$
$$52 - 30 \stackrel{?}{=} 10$$
$$22 = 10 \quad \text{False}$$

Let n be 40.

$$2n - 30 = 10$$
$$2 \cdot 40 - 30 \stackrel{?}{=} 10$$
$$80 - 30 \stackrel{?}{=} 10$$
$$50 = 10 \quad \text{False}$$

Let n be 20.

$$2n - 30 = 10$$
$$2 \cdot 20 - 30 \stackrel{?}{=} 10$$
$$40 - 30 \stackrel{?}{=} 10$$
$$10 = 10 \quad \text{True} \checkmark$$

Thus, 20 is a solution while 26 and 40 are not solutions.

☐ **Work Practice Problem 7**

Objective **C** **Translating Phrases into Variable Expressions**

To aid us in solving problems later, we practice translating verbal phrases into algebraic expressions. Certain key words and phrases suggesting addition, subtraction, multiplication, or division are reviewed next.

Answers

6. yes
7. 6 is a solution.

Addition (+)	Subtraction (−)	Multiplication (·)	Division (÷)
sum	difference	product	quotient
plus	minus	times	divide
added to	subtract	multiply	shared equally among
more than	less than	multiply by	per
increased by	decreased by	of	divided by
total	less	double/triple	divided into

EXAMPLE 8 Write as an algebraic expression. Use x to represent "a number."

a. 7 increased by a number
b. 15 decreased by a number
c. The product of 2 and a number
d. The quotient of a number and 5
e. 2 subtracted from a number

Solution:

a. In words: 7   increased by   a number
   Translate: 7   + x

b. In words: 15   decreased by   a number
   Translate: 15   − x

c. In words: The product of
                2   and   a number
   Translate: 2   · x or $2x$

d. In words: The quotient of
              a number   and   5
   Translate: x ÷   5   or $\dfrac{x}{5}$

e. In words: 2   subtracted from   a number
   Translate: x −   2

Work Practice Problem 8

PRACTICE PROBLEM 8

Write as an algebraic expression. Use x to represent "a number."

a. Twice a number.
b. 8 increased by a number.
c. 10 minus a number.
d. 10 subtracted from a number.
e. The quotient of 6 and a number.

Helpful Hint
Remember that order is important when subtracting. Study the order of numbers and variables below.

Phrase	Translation
a number *decreased by* 5	$x - 5$
a number *subtracted from* 5	$5 - x$

Answers
8. a. $2x$ **b.** $8 + x$ **c.** $10 - x$
d. $x - 10$ **e.** $6 \div x$ or $\dfrac{6}{x}$

Vocabulary and Readiness Check

Use the choices below to fill in each blank. You may use each choice more than once.

evaluating an expression   variable(s)   expression   equation   solution

1. A combination of operations on letters (variables) and numbers is a(n) _____.
2. A letter that represents a number is a(n) _____.
3. $3x - 2y$ is called a(n) _____ and the letters x and y are _____.
4. Replacing a variable in an expression by a number and then finding the value of the expression is called _____.
5. A statement of the form "expression = expression" is called a(n) _____.
6. A value for the variable that makes the equation a true statement is called a(n) _____.

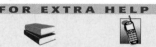 **1.8 EXERCISE SET**

FOR EXTRA HELP

Student Solutions Manual   PH Math/Tutor Center   CD/Video for Review   MathXL®   MyMathLab

Objective Ⓐ *Complete the table. The first row has been done for you. See Examples 1 through 5.*

	a	b	$a + b$	$a - b$	$a \cdot b$	$a \div b$
	45	9	54	36	405	5
1.	21	7				
2.	24	6				
3.	152	0				
4.	298	0				
5.	56	1				
6.	82	1				

Evaluate each following expression for $x = 2$, $y = 5$, and $z = 3$. See Examples 1 through 5.

7. $3 + 2z$ **8.** $7 + 3z$ **9.** $3xz - 5x$ **10.** $4yz + 2x$

11. $z - x + y$ **12.** $x + 5y - z$ **13.** $4x - z$ **14.** $2y + 5z$

15. $y^3 - 4x$ **16.** $y^3 - z$ **17.** $2xy^2 - 6$ **18.** $3yz^2 + 1$

19. $8 - (y - x)$ **20.** $3 + (2y - 4)$ **21.** $x^5 + (y - z)$ **22.** $x^4 - (y - z)$

23. $\dfrac{6xy}{z}$ **24.** $\dfrac{8yz}{15}$ **25.** $\dfrac{2y - 2}{x}$ **26.** $\dfrac{6 + 3x}{z}$

27. $\dfrac{x + 2y}{z}$ **28.** $\dfrac{2z + 6}{3}$ **29.** $\dfrac{5x}{y} - \dfrac{10}{y}$ **30.** $\dfrac{70}{2y} - \dfrac{15}{z}$

31. $2y^2 - 4y + 3$ **32.** $3x^2 + 2x - 5$ **33.** $(4y - 5z)^3$ **34.** $(4y + 3z)^2$

35. $(xy + 1)^2$ **36.** $(xz - 5)^4$ **37.** $2y(4z - x)$ **38.** $3x(y + z)$

39. $xy(5 + z - x)$ **40.** $xz(2y + x - z)$ **41.** $\dfrac{7x + 2y}{3x}$ **42.** $\dfrac{6z + 2y}{4}$

43. The expression $16t^2$ gives the distance in feet that an object falls after t seconds. Complete the table by evaluating $16t^2$ for each given value of t.

t	1	2	3	4
$16t^2$				

44. The expression $\dfrac{5(F - 32)}{9}$ gives the equivalent degrees Celsius for F degrees Fahrenheit. Complete the table by evaluating this expression for each given value of F.

F	50	59	68	77
$\dfrac{5(F - 32)}{9}$				

Objective B *Decide whether the given number is a solution of the given equation. See Example 6.*

45. Is 10 a solution of $n - 8 = 2$?

46. Is 9 a solution of $n - 2 = 7$?

47. Is 3 a solution of $24 = 80n$?

48. Is 50 a solution of $250 = 5n$?

49. Is 7 a solution of $3n - 5 = 10$?

50. Is 8 a solution of $11n + 3 = 91$?

51. Is 20 a solution of $2(n - 17) = 6$?

52. Is 0 a solution of $5(n + 9) = 40$?

53. Is 0 a solution of $5x + 3 = 4x + 13$?

54. Is 2 a solution of $3x - 6 = 5x - 10$?

55. Is 8 a solution of $7f = 64 - f$?

56. Is 5 a solution of $8x - 30 = 2x$?

Determine which numbers in each set are solutions to the corresponding equations. See Example 7.

57. $n - 2 = 10$; {10, 12, 14}

58. $n + 3 = 16$; {9, 11, 13}

59. $5n = 30$; {6, 25, 30}

60. $3n = 45$; {15, 30, 45}

61. $6n + 2 = 26$; {0, 2, 4}

62. $4n - 14 = 6$; {0, 5, 10}

63. $3(n - 4) = 10$; {5, 7, 10}

64. $6(n + 2) = 23$; {1, 3, 5}

65. $7x - 9 = 5x + 13$; {3, 7, 11}

66. $9x - 15 = 5x + 1$; {2, 4, 11}

Objective C **Translating** *Write each phrase as a variable expression. Use x to represent "a number." See Example 8.*

67. Eight more than a number

68. The sum of three and a number

69. The total of a number and eight

70. The difference of a number and five hundred

71. Twenty decreased by a number

72. A number less thirty

73. The product of 512 and a number

74. A number times twenty

75. The quotient of six and a number

76. A number divided by 11

77. The sum of seventeen and a number added to the product of five and the number

78. The difference of twice a number, and four

79. The product of five and a number

80. The quotient of twenty and a number, decreased by three

81. A number subtracted from 11

82. Twelve subtracted from a number

83. A number less 5

84. The product of a number and 7

85. 6 divided by a number

86. The sum of a number and 7

87. Fifty decreased by eight times a number

88. Twenty decreased by twice a number

Concept Extensions

For Exercises 89 through 92, use a calculator to evaluate each expression for $x = 23$ and $y = 72$.

89. $x^4 - y^2$

90. $2(x + y)^2$

91. $x^2 + 5y - 112$

92. $16y - 20x + x^3$

93. If x is a whole number, which expression is the largest: $2x$, $5x$, or $\dfrac{x}{3}$?

Explain your answer.

94. If x is a whole number, which expression is the smallest: $2x$, $5x$, or $\dfrac{x}{3}$?

Explain your answer.

95. In Exercise 43, what do you notice about the value of $16t^2$ as t gets larger?

96. In Exercise 44, what do you notice about the value of $\dfrac{5(F - 32)}{9}$ as F gets larger?

STUDY SKILLS BUILDER

Have You Decided to Complete This Course Successfully?

Ask yourself if one of your current goals is to complete this course successfully.

If it is not a goal of yours, ask yourself why? One common reason is fear of failure. Amazingly enough, fear of failure alone can be strong enough to keep many of us from doing our best in any endeavor.

Another common reason is that you simply haven't taken the time to make successfully completing this course one of your goals. How do you do this? Start by writing this goal in your mathematics notebook. Then list steps you will take to ensure success. A great first step is to read or reread Section 1.1 and make a commitment to try the suggestions in that section.

Good luck, and don't forget that a positive attitude will make a big difference.

Let's see how you are doing.

1. Have you decided to make "successfully completing this course" a goal of yours? If no, please list reasons why this has not happened. Study your list and talk to your instructor about this.

2. If your answer to question 1 is yes, take a moment and list in your notebook further specific goals that will help you achieve this major goal of successfully completing this course. (For example, "My goal this semester is not to miss any of my mathematics classes.")

3. Rate your commitment to this course with a number between 1 and 5. Use the diagram below to help.

High Commitment		Average Commitment	Not Committed at All	
5	4	3	2	1

5. If you have rated your personal commitment level (from the exercise above) as a 1, 2, or 3, list the reasons why this is so. Then determine whether it is possible to increase your commitment level to a 4 or 5.

CHAPTER 1   Group Activity

Investigating Endangered and Threatened Species

An **endangered** species is one that is thought to be in danger of becoming extinct throughout all or a major part of its habitat. A **threatened** species is one that may become endangered. The Division of Endangered Species at the U.S. Fish and Wildlife Service keeps close tabs on the state of threatened and endangered wildlife in the United States and around the world. The table on the next page was compiled from 2006 data in the Division of Endangered Species' box score. The "Total Species" column gives the total number of endangered and threatened species for each group.

1. Round each number of *endangered animal species* to the nearest ten to estimate the Animal Total.

2. Round each number of *endangered plant species* to the nearest ten to estimate the Plant Total.

3. Add the exact numbers of endangered animal species to find the exact Animal Total and record it in the table in the Endangered Species column. Add the exact numbers of endangered plant species to find the Plant Total and record it in the table in the Endangered Species column. Then find the total number of endangered species (animals and plants combined) and record this number in the table as the Grand Total in the Endangered Species column.

4. Find the Animal Total, Plant Total, and Grand Total for the Total Species column. Record these values in the table.

5. Use the data in the table to complete the Threatened Species column.

6. Write a paragraph discussing the conclusions that can be drawn from the table.

Continued on next page

Endangered and Threatened Species Worldwide			
Group	**Endangered Species**	**Threatened Species**	**Total Species**
Mammals	323		351
Birds	251		270
Reptiles	79		115
Amphibians	20		30
Fishes	84		128
Clams	64		72
Snails	25		37
Insects	51		61
Arachnids	12		12
Crustaceans	19		22
Animal Total			
Flowering Plants	571		714
Conifers	2		5
Ferns and Allies	24		26
Lichens	2		2
Plant Total			
Grand Total			

(*Source:* U.S. Fish and Wildlife Service, Division of Endangered Species)

(Animals label spans Mammals through Crustaceans; Plants label spans Flowering Plants through Lichens.)

Chapter 1 Vocabulary Check

Fill in each blank with one of the words or phrases listed below.

difference   factor   perimeter   dividend   minuend

place value   whole numbers   equation   divisor   variable

sum   set   addend   exponent   expression

solution   quotient   subtrahend   product   digits

area

1. The _____ are 0, 1, 2, 3, . . .
2. The _____ of a polygon is its distance around or the sum of the lengths of its sides.
3. The position of each digit in a number determines its _____.
4. An _____ is a shorthand notation for repeated multiplication of the same factor.
5. To find the _____ of a rectangle, multiply length times width.
6. The _____ used to write numbers are 0, 1, 2, 3, 4, 5, 6, 7, 8, and 9.
7. A letter used to represent a number is called a _____.
8. An _____ can be written in the form "expression = expression."
9. A combination of operations on variables and numbers is called an _____.
10. A _____ of an equation is a value of the variable that makes the equation a true statement.
11. A collection of numbers (or objects) enclosed by braces is called a _____.

Use the facts below for Exercises 12 through 21.

$$2 \cdot 3 = 6 \qquad 4 + 17 = 21 \qquad 20 - 9 = 11 \qquad 5\overline{)35}^{\,7}$$

12. The 21 above is called the _____.

13. The 5 above is called the _____.

14. The 35 above is called the _____.

15. The 7 above is called the _____.

16. The 3 above is called a _____.

17. The 6 above is called the _____.

18. The 20 above is called the _____.

19. The 9 above is called the _____.

20. The 11 above is called the _____.

21. The 4 above is called an _____.

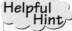

Are you preparing for your test? Don't forget to take the Chapter 1 Test on page 93. Then check your answers at the back of the text and use the Chapter Test Prep Video CD to see the fully worked-out solutions to any of the exercises you want to review.

1 Chapter Highlights

DEFINITIONS AND CONCEPTS	EXAMPLES
Section 1.2 Place Value and Names for Numbers	

The **whole numbers** are 0, 1, 2, 3, 4, 5, ...

The position of each digit in a number determines its **place value.** A place-value chart is shown next with the names of the periods given.

0, 14, 968, 5,268,619

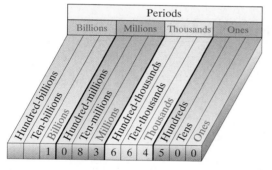

To write a whole number in words, write the number in each period followed by the name of the period. (The name of the ones period is not included.)

9,078,651,002 is written as nine billion, seventy-eight million, six hundred fifty-one thousand, two.

To write a whole number in standard form, write the number in each period, followed by a comma.

Four million, seven hundred six thousand, twenty-eight is written as 4,706,028.

Section 1.3 Adding and Subtracting Whole Numbers, and Perimeter	

To add whole numbers, add the digits in the ones place, then the tens place, then the hundreds place, and so on, carrying when necessary.

Find the sum:
$$\begin{array}{r} {\scriptstyle 2\,1\,1} \\ 2689 \leftarrow \text{addend} \\ 1735 \leftarrow \text{addend} \\ +\ \ 662 \leftarrow \text{addend} \\ \hline 5086 \leftarrow \text{sum} \end{array}$$

DEFINITIONS AND CONCEPTS	**EXAMPLES**

Section 1.3   Adding and Subtracting Whole Numbers, and Perimeter (*continued*)

To subtract whole numbers, subtract the digits in the ones place, then the tens place, then the hundreds place, and so on, borrowing when necessary.

Subtract:

$$\begin{array}{r} \overset{8\ 15}{79\cancel{5}4} \leftarrow \quad \text{minuend} \\ -5673 \leftarrow \quad \text{subtrahend} \\ \hline 2281 \leftarrow \quad \text{difference} \end{array}$$

The **perimeter** of a polygon is its distance around or the sum of the lengths of its sides.

Find the perimeter of the polygon shown.

5 feet

2 feet   3 feet

9 feet

The perimeter is 5 feet + 3 feet + 9 feet + 2 feet = 19 feet.

Section 1.4   Rounding and Estimating

ROUNDING WHOLE NUMBERS TO A GIVEN PLACE VALUE

Step 1. Locate the digit to the right of the given place value.

Step 2. If this digit is 5 or greater, add 1 to the digit in the given place value and replace each digit to its right with 0.

Step 3. If this digit is less than 5, replace it and each digit to its right with 0.

Round 15,721 to the nearest thousand.

15,⑦21

Add 1 ⤴   Since the circled digit is 5 or greater, add 1 to the
Replace   given place value and replace digits to its right with
with   zeros.
zeros.

15,721 rounded to the nearest thousand is 16,000.

Section 1.5   Multiplying Whole Numbers and Area

To multiply 73 and 58, for example, multiply 73 and 8, then 73 and 50. The sum of these partial products is the product of 73 and 58. Use the notation to the right.

$$\begin{array}{r} 73 \leftarrow \quad \text{factor} \\ \times 58 \leftarrow \quad \text{factor} \\ \hline 584 \leftarrow \quad 73 \times 8 \\ 3650 \leftarrow \quad 73 \times 50 \\ \hline 4234 \leftarrow \quad \text{product} \end{array}$$

To find the **area** of a rectangle, multiply length times width.

Find the area of the rectangle shown.

11 meters

7 meters

area of rectangle = length · width
= (11 meters)(7 meters)
= 77 square meters

DEFINITIONS AND CONCEPTS	**EXAMPLES**

Section 1.6   Dividing Whole Numbers

DIVISION PROPERTIES OF 0

The quotient of 0 and any number (except 0) is 0.

The quotient of any number and 0 is not a number. We say that this quotient is undefined.

$\dfrac{0}{5} = 0$

$\dfrac{7}{0}$ is undefined

To divide larger whole numbers, use the process called **long division** as shown to the right.

$$\begin{array}{r} 507 \ \ \text{R } 2 \leftarrow \text{quotient} \\ \text{divisor} \rightarrow 14\overline{)7100} \leftarrow \text{dividend} \\ -70\downarrow| \qquad\qquad 5(14) = 70 \\ \hline 10| \qquad\quad \text{Subtract and bring down the 0.} \\ -0\downarrow \qquad\quad 0(14) = 0 \\ \hline 100 \qquad\quad \text{Subtract and bring down the 0.} \\ -98 \qquad\quad 7(14) = 98 \\ \hline 2 \qquad\quad \text{Subtract. The remainder is 2.} \end{array}$$

To check, see that $507 \cdot 14 + 2 = 7100$.

The **average** of a list of numbers is

$$\text{average} = \dfrac{\text{sum of numbers}}{\text{number of numbers}}$$

Find the average of 23, 35, and 38.

$$\text{average} = \dfrac{23 + 35 + 38}{3} = \dfrac{96}{3} = 32$$

Section 1.7   Exponents and Order of Operations

An **exponent** is a shorthand notation for repeated multiplication of the same factor.

$$3^{4} = \underbrace{3 \cdot 3 \cdot 3 \cdot 3}_{4 \text{ factors of } 3} = 81$$

base

ORDER OF OPERATIONS

1. Perform all operations within parentheses ( ), brackets [ ], or other grouping symbols such as fraction bars, starting with the innermost set.
2. Evaluate any expressions with exponents.
3. Multiply or divide in order from left to right.
4. Add or subtract in order from left to right.

Simplify: $\dfrac{5 + 3^2}{2(7 - 6)}$

Simplify above and below the fraction bar separately.

$$\dfrac{5 + 3^2}{2(7 - 6)} = \dfrac{5 + 9}{2(1)} \quad \begin{array}{l}\text{Evaluate } 3^2 \text{ above the fraction bar.}\\ \text{Subtract: } 7 - 6 \text{ below the fraction bar.}\end{array}$$

$$= \dfrac{14}{2} \quad \begin{array}{l}\text{Add.}\\ \text{Multiply.}\end{array}$$

$$= 7 \quad \text{Divide.}$$

The **area of a square** is $(\text{side})^2$.

Find the area of a square with side length 9 inches.

$$\begin{aligned}\text{Area of the square} &= (\text{side})^2 \\ &= (9 \text{ inches})^2 \\ &= 81 \text{ square inches}\end{aligned}$$

DEFINITIONS AND CONCEPTS	EXAMPLES
Section 1.8 Introduction to Variables, Algebraic Expressions, and Equations	

A letter used to represent a number is called a **variable.**

A combination of operations on variables and numbers is called an **algebraic expression.**

Replacing a variable in an expression by a number, and then finding the value of the expression is called **evaluating the expression** for the variable.

A statement written in the form "expression = expression" is an **equation.**

A **solution** of an equation is a value for the variable that makes the equation a true statement.

Variables:

$$x, \quad y, \quad z, \quad a, \quad b$$

Algebraic expressions:

$$3 + x, \quad 7y, \quad x^3 + y - 10$$

Evaluate $2x + y$ for $x = 22$ and $y = 4$.

$$
\begin{aligned}
2x + y &= 2 \cdot 22 + 4 && \text{Replace } x \text{ with 22 and } y \text{ with 4.} \\
&= 44 + 4 && \text{Multiply.} \\
&= 48 && \text{Add.}
\end{aligned}
$$

Equations:

$$
\begin{aligned}
n - 8 &= 12 \\
2(20 - 7n) &= 32 \\
\text{Area} &= \text{length} \cdot \text{width}
\end{aligned}
$$

Determine whether 2 is a solution of the equation $4(x - 1) = 7$.

$$
\begin{aligned}
4(2 - 1) &\overset{?}{=} 7 && \text{Replace } x \text{ with 2.} \\
4(1) &\overset{?}{=} 7 && \text{Subtract.} \\
4 &\overset{?}{=} 7 && \text{False}
\end{aligned}
$$

No, 2 is not a solution.

1   CHAPTER REVIEW

(1.2) *Determine the place value of the digit 4 in each whole number.*

1. 7640

2. 46,200,120

Write each whole number in words.

3. 7640

4. 46,200,120

Write each whole number in expanded form.

5. 3158

6. 403,225,000

Write each whole number in standard form.

7. Eighty-one thousand, nine hundred

8. Six billion, three hundred four million

The following table shows the populations of the ten largest cities in the United States. Use this table to answer Exercises 9 through 12 and other exercises throughout this review.

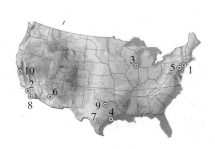

Rank	City	2005	2000	1990
1	New York, NY	8,143,197	8,008,278	7,322,564
2	Los Angeles, CA	3,844,829	3,694,820	3,485,398
3	Chicago, IL	2,842,518	2,896,016	2,783,726
4	Houston, TX	2,016,582	1,953,631	1,630,553
5	Philadelphia, PA	1,463,281	1,517,550	1,585,577
6	Phoenix, AZ	1,461,575	1,321,045	983,403
7	San Antonio, TX	1,256,509	1,144,646	935,933
8	San Diego, CA	1,255,540	1,223,400	1,110,549
9	Dallas, TX	1,213,825	1,188,580	1,006,877
10	San Jose, CA	912,332	894,943	782,248

(*Source:* U.S. Census Bureau)

9. Find the population of Houston, Texas, in 2000.

10. Find the population of Los Angeles, California, in 2005.

11. Which city in the table had the smallest population in 2005?

12. Which city had the largest population in 2005?

(1.3) *Add or subtract as indicated.*

13. $18 + 49$

14. $28 + 39$

15. $462 - 397$

16. $583 - 279$

17. $428 + 21$

18. $819 + 21$

19. $4000 - 86$

20. $8000 - 92$

21. $91 + 3623 + 497$

22. $82 + 1647 + 238$

Translating *Solve.*

23. Find the sum of 74, 342, and 918.

24. Find the sum of 49, 529, and 308.

25. Subtract 7965 from 25,862.

26. Subtract 4349 from 39,007.

27. The distance from Washington, DC, to New York City is 205 miles. The distance from New York City to New Delhi, India, is 7318 miles. Find the total distance from Washington, DC, to New Delhi if traveling by air through New York City.

28. Susan Summerline earned salaries of $62,589, $65,340, and $69,770 during the years 2004, 2005, and 2006, respectively. Find her total earnings during those three years.

Find the perimeter of each figure.

△ **29.**

52 feet
40 feet
52 feet
72 feet

△ **30.**

11 kilometers   20 kilometers
35 kilometers

Use the city population table for Exercises 31 and 32.

31. Find the increase in population for San Antonio, Texas, from 2000 to 2005.

32. Find the decrease in population for Philadelphia, Pennsylvania, from 2000 to 2005.

The following bar graph shows the monthly savings account balance for a freshman attending a local community college. Use this graph to answer Exercises 33 through 36.

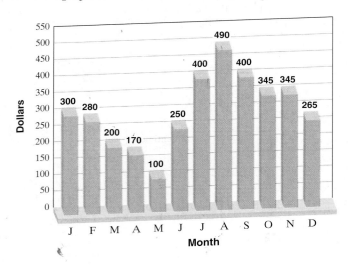

33. During what month was the balance the least?

34. During what month was the balance the greatest?

35. By how much did the balance decrease from February to April?

36. By how much did the balance increase from June to August?

(1.4) *Round to the given place.*

37. 43 to the nearest ten

38. 45 to the nearest ten

39. 876 to the nearest ten

40. 493 to the nearest hundred

41. 3829 to the nearest hundred

42. 57,534 to the nearest thousand

43. 39,583,819 to the nearest million

44. 768,542 to the nearest hundred-thousand

Estimate the sum or difference by rounding each number to the nearest hundred.

45. 3785 + 648 + 2866

46. 5925 − 1787

47. A group of students took a week-long driving trip and traveled 630, 192, 271, 56, 703, 454, and 329 miles on seven consecutive days. Round each distance to the nearest hundred to estimate the distance they traveled.

48. According to the city population table, the 2005 population of Phoenix was 1,461,575, and for Dallas it was 1,213,825. Round each number to the nearest hundred-thousand and estimate how much larger Phoenix is than Dallas.

(1.5) *Multiply.*

49. 276
   × 8

50. 349
   × 4

51. 57
   × 40

52. 69
   × 42

53. 20(7)(4)

54. 25(9)(4)

55. 26 · 34 · 0

56. 62 · 88 · 0

57. 586
   × 29

58. 242
   × 37

59. 642
   × 177

60. 347
   × 129

61. 1026
   × 401

62. 2107
   × 302

Translating *Solve.*

63. Find the product of 6 and 250.

64. Find the product of 6 and 820.

65. A golf pro orders shirts for the company sponsoring a local charity golfing event. Shirts size large cost $32 while shirts size extra-large cost $38. If 15 large shirts and 11 extra-large shirts are ordered, find the cost.

66. The cost to attend Black Hills State University full-time is $4,820 per semester. Determine the cost for 20 students to attend full-time.
(*Source: World Almanac*, 2006)

Find the area of each rectangle.

 67.

13 miles
7 miles

△ **68.**

20 centimeters
25 centimeters

(1.6) *Divide and then check.*

69. $\dfrac{49}{7}$

70. $\dfrac{36}{9}$

71. $27 \div 5$

72. $18 \div 4$

73. $918 \div 0$

74. $0 \div 668$

75. $5\overline{)167}$

76. $8\overline{)159}$

77. $26\overline{)626}$

78. $19\overline{)680}$

79. $47\overline{)23{,}792}$

80. $53\overline{)48{,}111}$

81. $207\overline{)578{,}291}$

82. $306\overline{)615{,}732}$

Translating *Solve.*

83. Find the quotient of 92 and 5.

84. Find the quotient of 86 and 4.

85. A box can hold 24 cans of corn. How many boxes can be filled with 648 cans of corn?

86. One mile is 1760 yards. Find how many miles there are in 22,880 yards.

87. Find the average of the numbers 76, 49, 32, and 47.

88. Find the average of the numbers 23, 85, 62, and 66.

(1.7) *Simplify.*

89. 8^2

90. 5^3

91. $5 \cdot 9^2$

92. $4 \cdot 10^2$

93. $18 \div 2 + 7$

94. $12 - 8 \div 4$

95. $\dfrac{5(6^2 \div 3)}{3^2 + 2}$

96. $\dfrac{7(16 - 8)}{2^3}$

97. $48 \div 8 \cdot 2$

98. $27 \div 9 \cdot 3$

99. $2 + 3[1^5 + (20 - 17) \cdot 3] + 5 \cdot 2$

100. $21 - [2^4 - (7 - 5) - 10] + 8 \cdot 2$

101. $19 - 2(3^2 - 2^2)$

102. $16 - 2(4^2 - 3^2)$

103. $4 \cdot 5 - 2 \cdot 7$

104. $8 \cdot 7 - 3 \cdot 9$

105. $(6 - 4)^3 \cdot [10^2 \div (3 + 17)]$

106. $(7 - 5)^3 \cdot [9^2 \div (2 + 7)]$

107. $\dfrac{5 \cdot 7 - 3 \cdot 5}{2(11 - 3^2)}$

108. $\dfrac{4 \cdot 8 - 1 \cdot 11}{3(9 - 2^3)}$

Find the area of each square.

△ **109.** A square with side length of 7 meters.   △ **110.**

3 inches

(1.8) *Evaluate each expression for $x = 5$, $y = 0$, and $z = 2$.*

111. $\dfrac{2x}{z}$ **112.** $4x - 3$ **113.** $\dfrac{x + 7}{y}$ **114.** $\dfrac{y}{5x}$

115. $x^3 - 2z$ **116.** $\dfrac{7 + x}{3z}$ **117.** $(y + z)^2$ **118.** $\dfrac{100}{x} + \dfrac{y}{3}$

Translating *Translate each phrase into a variable expression.*

119. Five subtracted from a number   **120.** Seven more than a number

121. Ten divided by a number   **122.** The product of 5 and a number

Decide whether the given number is a solution of the given equation.

123. Is 5 a solution of $n + 12 = 20 - 3$?   **124.** Is 23 a solution of $n - 8 = 10 + 6$?

125. Is 14 a solution of $30 = 3(n - 3)$?   **126.** Is 20 a solution of $5(n - 7) = 65$?

Determine which numbers in each set are solutions to the corresponding equations.

127. $7n = 77$; $\{6, 11, 20\}$ **128.** $n - 25 = 150$; $\{125, 145, 175\}$

129. $5(n + 4) = 90$; $\{14, 16, 26\}$ **130.** $3n - 8 = 28$; $\{3, 7, 15\}$

Mixed Review

Perform the indicated operations.

131. $485 - 68$ **132.** $729 - 47$ **133.** 732×3 **134.** 629×4

135. $374 + 29 + 698$ **136.** $593 + 52 + 766$ **137.** $13\overline{)5962}$ **138.** $18\overline{)4267}$

139. 1968×36 **140.** 5324×18 **141.** $2000 - 356$ **142.** $9000 - 519$

Round to the given place.

143. 842 to the nearest ten   **144.** 258,371 to the nearest hundred-thousand

Simplify.

145. $24 \div 4 \cdot 2$ **146.** $\dfrac{(15 + 3) \cdot (8 - 5)}{2^3 + 1}$

Solve.

147. Is 9 a solution of $5n - 6 = 40$?   **148.** Is 3 a solution of $2n - 6 = 5n - 15$?

149. A manufacturer of drinking glasses ships his delicate stock in special boxes that can hold 32 glasses. If 1714 glasses are manufactured, how many full boxes are filled? Are there any glasses left over?

150. A teacher orders 2 small white boards for $27 each and 8 boxes of dry erase pens for $4 each. What is her total bill before taxes?

1 CHAPTER TEST

 Use the Chapter Test Prep Video CD to see the fully worked-out solutions to any of the exercises you want to review.

Simplify.

1. Write 82,426 in words.

2. Write "four hundred two thousand, five hundred fifty" in standard form.

3. $59 + 82$

4. $600 - 487$

5. $\begin{array}{r} 496 \\ \times\ \ 30 \\ \hline \end{array}$

6. $52{,}896 \div 69$

7. $2^3 \cdot 5^2$

8. $98 \div 1$

9. $0 \div 49$

10. $62 \div 0$

11. $(2^4 - 5) \cdot 3$

12. $16 + 9 \div 3 \cdot 4 - 7$

13. $6^1 \cdot 2^3$

14. $2[(6 - 4)^2 + (22 - 19)^2] + 10$

15. $5698 \cdot 1000$

16. Find the average of 62, 79, 84, 90, and 95.

17. Round 52,369 to the nearest thousand.

Estimate each sum or difference by rounding each number to the nearest hundred.

18. $6289 + 5403 + 1957$

19. $4267 - 2738$

Solve.

20. Subtract 15 from 107.

21. Find the sum of 15 and 107.

22. Find the product of 15 and 107.

23. Find the quotient of 107 and 15.

24. Twenty-nine cans of Sherwin-Williams paint cost $493. How much was each can?

25. Jo McElory is looking at two new refrigerators for her apartment. One costs $599 and the other costs $725. How much more expensive is the higher-priced one?

Answers

1. _____

2. _____

3. _____

4. _____

5. _____

6. _____

7. _____

8. _____

9. _____

10. _____

11. _____

12. _____

13. _____

14. _____

15. _____

16. _____

17. _____

18. _____

19. _____

20. _____

21. _____

22. _____

23. _____

24. _____

25. _____

26. _____

27. _____

28. _____

29. _____

30. _____

31. _____

32. a. _____

b. _____

33. _____

34. _____

26. One tablespoon of white granulated sugar contains 45 calories. How many calories are in 8 tablespoons of white granulated sugar? (*Source: Home and Garden Bulletin No. 72*, U.S. Department of Agriculture)

Find the perimeter and the area of each figure.

△ **28.**

Square | 5 centimeters

30. Evaluate $5(x^3 - 2)$ for $x = 2$.

32. Translate the following phrases into mathematical expressions. Use x to represent "a number."
   a. The quotient of a number and 17
   b. Twice a number, decreased by 20

27. A small business owner recently ordered 16 digital cameras that cost $430 each and 5 printers that cost $205 each. Find the total cost for these items.

△ **29.**

20 yards

Rectangle | 10 yards

31. Evaluate $\dfrac{3x - 5}{2y}$ for $x = 7$ and $y = 8$.

33. Is 6 a solution of the equation $5n - 11 = 19$?

34. Determine which (if any) number in the set is a solution to the given equation.
   $n + 20 = 4n - 10$; {0, 10, 20}

2

Integers and Introduction to Solving Equations

Thus far, we have studied whole numbers, but these numbers are not sufficient for representing many situations in real life. For example, to express 5 degrees below zero or $100 in debt, numbers less than 0 are needed. This chapter is devoted to integers, which include numbers less than 0, and operations on these numbers.

Members of the International Astronomical Union (IAU) met in August 2006 and approved a resolution that there be three distinct categories of bodies in our Solar System: "Planet," "Dwarf Planet" (completely distinct from a planet), and "Small Solar-System Bodies." See the following chart.

In Section 2.1, Example 8 and Exercises 91 through 94, and Section 2.3, Exercises 81 and 82, we will see how integers can be used to compare the average surface temperatures of several planets in our Solar System.

Planets	Mercury, Venus, Earth, Mars, Jupiter, Saturn, Uranus, and Neptune
Dwarf Planets	Pluto, Ceres (was an asteroid), and 2003 UB313 (newly named Eris)*
Small Solar-System Bodies	Asteroids, comets, and other small bodies

*More "dwarf planets" are expected to be announced by the IAU in the near future.

Average Surface Temperature of Planets*

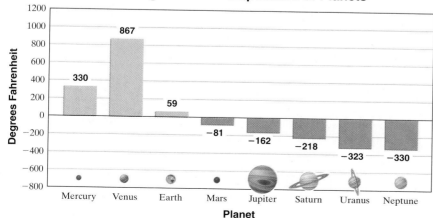

Source: The World Almanac, 2006

* For some planets, the temperature given is the temperature where the atmosphere pressure equals 1 Earth atmosphere.

Objectives

A Represent Real-Life Situations with Integers.

B Graph Integers on a Number Line.

C Compare Integers.

D Find the Absolute Value of a Number.

E Find the Opposite of a Number.

F Read Bar Graphs Containing Integers.

2.1 INTRODUCTION TO INTEGERS

Objective A Representing Real-Life Situations

Thus far in this text, all numbers have been 0 or greater than 0. Numbers greater than 0 are called **positive numbers.** However, sometimes situations exist that cannot be represented by a number greater than 0. For example,

5 degrees below 0°

Sea level

20 feet below sea level

To represent these situations, we need numbers less than 0.

Extending the number line to the left of 0 allows us to picture **negative numbers,** which are numbers that are less than 0.

Negative numbers   Zero   Positive numbers

When a single + sign or no sign is in front of a number, the number is a positive number. When a single − sign is in front of a number, the number is a negative number. Together, we call positive numbers, negative numbers, and zero the **signed numbers.**

−5 indicates "negative five."

5 and +5 both indicate "positive five."

The number 0 is neither positive nor negative.

Some signed numbers are integers. The **integers** consist of the numbers labeled on the number line above. The integers are

. . . , −3, −2, −1, 0, 1, 2, 3, . . .

Now we have numbers to represent the situations previously mentioned.

5 degrees below 0   −5°
20 feet below sea level   −20 feet

Helpful Hint 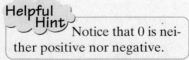 Notice that 0 is neither positive nor negative.

Helpful Hint
A − sign, such as the one in −1, tells us that the number is to the left of 0 on the number line. −1 is read "negative one."
A + sign or no sign tells us that a number lies to the right of 0 on the number line. For example, 3 and +3 both mean "positive three."

EXAMPLE 1 Representing Depth with an Integer

The world's deepest cave is Krubera (or Voronja), in the country of Georgia, located by the Black Sea in Asia. It has been explored to a depth of 6824 feet below the surface of the Earth. Represent this position using an integer. (*Source: Guinness World Records* 2006)

Solution: If 0 represents the surface of the Earth, then 6824 feet below the surface can be represented by -6824.

▢ **Work Practice Problem 1**

Objective B Graphing Integers

EXAMPLE 2 Graph 0, -3, 2, and -2 on the number line.

Solution:

▢ **Work Practice Problem 2**

Objective C Comparing Integers

We can compare integers by using a number line. For any two numbers graphed on a number line, the number to the **right** is the **greater number** and the number to the **left** is the **smaller number.** Also, the symbols $<$ and $>$ are called **inequality symbols.**

The inequality symbol $>$ means "**is greater than**" and

the inequality symbol $<$ means "**is less than.**"

For example, both -5 and -7 are graphed on the number line below.

On the graph, -7 is **to the left of** -5, so -7 **is less than** -5, written as

$$-7 < -5$$

We can also write

$$-5 > -7$$

since -5 is **to the right** of -7, so -5 **is greater than** -7.

✔ Concept Check Is there a largest positive number? Is there a smallest negative number? Explain.

EXAMPLE 3 Insert $<$ or $>$ between each pair of numbers to make a true statement.

a. $-7 \; < \; 7$ **b.** $0 \; > \; -4$ **c.** $-9 \; > \; -11$

Solution:

a. -7 is to the left of 7 on a number line, so $-7 < 7$.
b. 0 is to the right of -4 on a number line, so $0 > -4$.
c. -9 is to the right of -11 on a number line, so $-9 > -11$.

▢ **Work Practice Problem 3**

Helpful Hint

If you think of $<$ and $>$ as arrowheads, notice that in a true statement the arrow always points to the smaller number.

$$5 > -4 \qquad -3 < -1$$

smaller number   smaller number

Objective D Finding the Absolute Value of a Number

The **absolute value** of a number is the number's distance from 0 on the number line. The symbol for absolute value is | |. For example, |3| is read as "the absolute value of 3."

|3| = 3 because 3 is 3 units from 0.

|−3| = 3 because −3 is 3 units from 0.

PRACTICE PROBLEM 4

Simplify.

a. |−6|

b. |4|

c. |−12|

EXAMPLE 4 Simplify.

a. |−9| **b.** |3| **c.** |0|

Solution:

a. |−9| = 9 because −9 is 9 units from 0.

b. |3| = 3 because 3 is 3 units from 0.

c. |0| = 0 because 0 is 0 units from 0.

🔲 **Work Practice Problem 4**

Helpful Hint

Since the absolute value of a number is that number's *distance* from 0, the absolute value of a number is always 0 or positive. It is never negative.

$$|0| = 0 \qquad |-6| = 6$$

zero   a positive number

Objective E Finding Opposites

Two numbers that are the same distance from 0 on the number line but are on opposite sides of 0 are called **opposites**.

4 and −4 are opposites.

When two numbers are opposites, we say that each is the opposite of the other. Thus **4 is the opposite of −4 and −4 is the opposite of 4.**

Answers

4. a. 6   **b.** 4   **c.** 12

The phrase "the opposite of" is written in symbols as "−." For example,

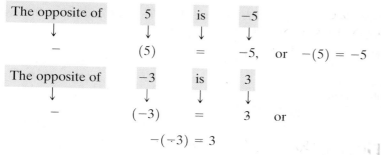

The opposite of   5   is   −5

−   (5)   =   −5,   or   −(5) = −5

The opposite of   −3   is   3

−   (−3)   =   3   or

−(−3) = 3

In general, we have the following:

Opposites

If a is a number, then $-(-a) = a$.

Notice that because "the opposite of" is written as "−", to find the opposite of a number we place a "−" sign in front of the number.

EXAMPLE 5 Find the opposite of each number.

a. 13   **b.** −2   **c.** 0

Solution:

a. The opposite of 13 is −13.
b. The opposite of −2 is −(−2) or 2.
c. The opposite of 0 is 0.

> **Helpful Hint**
> Remember that 0 is neither positive nor negative.

Work Practice Problem 5

✔ **Concept Check** True or false? The number 0 is the only number that is its own opposite.

EXAMPLE 6 Simplify.

a. −(−4)   **b.** −|−5| **c.** −|6|

Solution:

a. −(−4) = 4   The opposite of negative 4 is 4.

b. −|−5| = −5   The opposite of the absolute value of −5 is the opposite of 5, or −5.

c. −|6| = −6   The opposite of the absolute value of 6 is the opposite of 6, or −6.

Work Practice Problem 6

EXAMPLE 7 Evaluate −|−x| if $x = -2$.

Solution: Carefully replace x with −2; then simplify.

−|−x| = −|−(−2)| Replace x with −2.

Then −|−(−)2| = −|2| = −2.

Work Practice Problem 7

PRACTICE PROBLEM 5

Find the opposite of each number.

a. 14
b. −9

PRACTICE PROBLEM 6

Simplify.

a. −|−7|
b. −|4|
c. −(−12)

PRACTICE PROBLEM 7

Evaluate −|x| if $x = -6$.

Answers
5. **a.** −14   **b.** 9
6. **a.** −7   **b.** −4   **c.** 12   7. −6

✔ **Concept Check Answer**
True

Objective F Reading Bar Graphs Containing Integers

The bar graph below shows the average temperature (in Fahrenheit) of the eight planets, excluding the newly classified "dwarf planet," Pluto. Notice that a negative temperature is illustrated by a bar below the horizontal line representing 0°F, and a positive temperature is illustrated by a bar above the horizontal line representing 0°F.

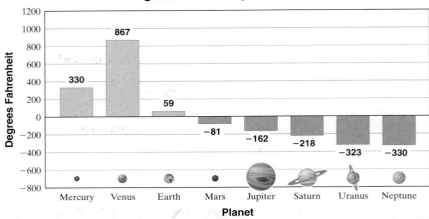

Source: The World Almanac, 2006
* For some planets, the temperature given is the temperature where the atmosphere pressure equals 1 Earth atmosphere.

PRACTICE PROBLEM 8

Which planet has the highest average temperature?

EXAMPLE 8   Which planet has the lowest average temperature?

Solution   The planet with the lowest average temperature is the one that corresponds to the bar that extends the furthest in the negative direction (downward.) Neptune has the lowest average temperature of −330°F.

■ **Work Practice Problem 8**

Answer

8. Venus; 867°F

Copyright © 2009 Pearson Education, Inc.

Vocabulary and Readiness Check

Use the choices below to fill in each blank. Not all choices will be used.

opposites	absolute value	right	is less than
inequality symbols	negative	positive	left
signed	integers	is greater than	

1. The numbers . . . $-3, -2, -1, 0, 1, 2, 3, . . .$ are called _____.

2. Positive numbers, negative numbers, and zero, together, are called _____ numbers.

3. The symbols "$<$" and "$>$" are called _____.

4. Numbers greater than 0 are called _____ numbers while numbers less than 0 are called _____ numbers.

5. The sign "$<$" means _____ and "$>$" means _____.

6. On a number line, the greater number is to the _____ of the lesser number.

7. A number's distance from 0 on the number line is the number's _____.

8. The numbers -5 and 5 are called _____.

2.1 EXERCISE SET

Objective A *Represent each quantity by an integer. See Example 1.*

1. A worker in a silver mine in Nevada works 1235 feet underground.

2. A scuba diver is swimming 25 feet below the surface of the water in the Gulf of Mexico.

3. The peak of Mount Elbert in Colorado is 14,433 feet above sea level. (*Source:* U.S. Geological Survey)

4. The lowest elevation in the United States is found at Death Valley, California, at an elevation of 282 feet below sea level. (*Source:* U.S. Geological Survey)

5. The record high temperature in Nevada is 118 degrees above zero Fahrenheit. (*Source:* National Climatic Data Center)

6. The Minnesota Viking football team gained 28 yards on a play.

7. The average depth of the Pacific Ocean is 13,000 feet below the surface of the ocean. (*Source:* U.S. Navy)

8. The Dow Jones stock market average fell 213 points in one day.

9. General Motors Corporation posted a net income loss of $10,458 million for the fiscal year 2005. (*Source:* General Motors)

10. For the fiscal year 2005, General Motors reported a loss of about $19 per share of stock. (*Source:* General Motors)

11. Two divers are exploring the wreck of the *Andrea Doria*, south of Nantucket Island, Massachusetts. Guillermo is 160 feet below the surface of the ocean and Luigi is 147 feet below the surface. Represent each quantity by an integer and determine who is deeper.

12. The temperature on one January day in Chicago was 10° below 0° Celsius. Represent this quantity by an integer and tell whether this temperature is cooler or warmer than 5° below 0° Celsius.

13. In a recent year, the number of music cassette singles shipped to retailers reflected an 81 percent loss from the previous year. Write an integer to represent the percent loss in cassette singles shipped. (*Source:* Recording Industry Association of America)

14. In a recent year, the number of music CDs shipped to retailers reflected a 49 percent loss from the previous year. Write an integer to represent the percent loss in CDs shipped. (*Source:* Recording Industry Association of America)

Objective Ⓑ *Graph each integer in the list on the same number line. See Example 2.*

15. 0, 3, 4, 6

16. 7, 5, 2, 0

17. 1, −1, 2, −2, −4,

18. 3, −3, 5, −5, 6

19. 0, 1, 9, 14

20. 0, 3, 10, 11

21. 0, −2, −7, −5

22. 0, −7, 3, −6

Objective Ⓒ *Insert < or > between each pair of integers to make a true statement. See Example 3.*

23. 0   −7   **24.** −8   0   **25.** −7   −5   **26.** −12   −10

27. −30   −35   **28.** −27   −29   **29.** −26   26   **30.** 13   −13

Objective Ⓓ *Simplify. See Example 4.*

31. |5| **32.** |7| **33.** |−8| **34.** |−19|

35. |0| **36.** |100| **37.** |−55| **38.** |−10|

Objective Ⓔ *Find the opposite of each integer. See Example 5.*

39. 5   **40.** 8   **41.** −4   **42.** −6

43. 23   **44.** 123   **45.** −85   **46.** −13

Objectives Ⓒ Ⓓ Ⓔ **Mixed Practice** *Simplify. See Example 6.*

47. |−7| **48.** |−11| **49.** −|20| **50.** −|43|

51. −|−3| **52.** −|−18| **53.** −(−43)   **54.** −(−27)

55. |−15| **56.** −(−14)   **57.** −(−33)   **58.** −|−29|

Evaluate. See Example 7.

59. $|-x|$ if $x = -6$ **60.** $-|x|$ if $x = -8$ **61.** $-|-x|$ if $x = 2$ **62.** $-|-x|$ if $x = 10$

63. $|x|$ if $x = -32$ **64.** $|x|$ if $x = 32$ **65.** $-|x|$ if $x = 7$ **66.** $|-x|$ if $x = 1$

Insert $<, >,$ or $=$ between each pair of numbers to make a true statement. See Examples 3 through 6.

67. $-12 \quad -6$ **68.** $-4 \quad -17$ **69.** $|-8| \quad |-11|$ **70.** $|-8| \quad |-4|$

71. $|-47| \quad -(-47)$ **72.** $-|17| \quad -(-17)$ **73.** $-|-12| \quad -(-12)$ **74.** $|-24| \quad -(-24)$

75. $0 \quad -9$ **76.** $-45 \quad 0$ **77.** $|0| \quad |-9|$ **78.** $|-45| \quad |0|$

79. $-|-2| \quad -|-10|$ **80.** $-|-8| \quad -|-4|$ **81.** $-(-12) \quad -(-18)$ **82.** $-22 \quad -(-38)$

Objectives **D** **E** **Mixed Practice** *Fill in the chart. See Examples 4 through 7.*

	Number	Absolute Value of Number	Opposite of Number
83.	31	31	-31
85.	28	28	-28

	Number	Absolute Value of Number	Opposite of Number
84.	-13	13	-13
86.	90	-90	90

Objective **F** *The bar graph shows the elevations of selected lakes. Use this graph For Exercises 87 through 90* (*Source:* U.S. Geological Survey). *See Example 8.*

Elevations of Selected Lakes

87. Which lake shown has the lowest elevation?

88. Which lake has an elevation at sea level?

89. Which lake shown has the highest elevation?

90. Which lake shown has the second lowest elevation?

Use the bar graph from Example 8 to answer Exercises 91 through 94.

91. Which planet has a negative average temperature closest to 0°F?

92. Which planet has an average temperature closest to 0°F?

93. Which planet has an average temperature closest to −300°F?

94. Which planet has an average temperature closest to −200°F?

Review

Add. See Section 1.3.

95. $0 + 13$

96. $9 + 0$

97. $15 + 20$

98. $20 + 15$

99. $47 + 236 + 77$

100. $362 + 37 + 90$

Concept Extensions

Write the given numbers in order from least to greatest.

101. $2^2, -|3|, -(-5), -|-8|$

102. $|10|, 2^3, -|-5|, -(-4)$

103. $|-1|, -|-6|, -(-6), -|1|$

104. $1^4, -(-3), -|7|, |-20|$

105. $-(-2), 5^2, -10, -|-9|, |-12|$

106. $3^3, -|-11|, -(-10), -4, -|2|$

Choose all numbers for x from each given list that make each statement true.

107. $|x| > 8$

    a. -9 **b.** -5 **c.** 8 **d.** -12

108. $|x| > 4$

    a. 0 **b.** -4 **c.** 5 **d.** -100

109. Evaluate: $-(-|-8|)$

110. Evaluate: $(-|-(-7)|)$

Answer true or false for Exercises 111 through 115.

111. If $a > b$, then a must be a positive number.

112. The absolute value of a number is *always* a positive number.

113. A positive number is always greater than a negative number.

114. Zero is always less than a positive number.

115. The number $-a$ is always a negative number. (*Hint:* Read "$-$" as "the opposite of.")

116. Given the number line , is it true that $b < a$?

117. Write in your own words how to find the absolute value of a signed number.

118. Explain how to determine which of two signed numbers is larger.

For Exercises 119 and 120, see the first Concept Check in this section.

119. Is there a largest negative number? If so, what is it?

120. Is there a smallest positive number? If so, what is it?

STUDY SKILLS BUILDER

Have You Decided to Complete This Course Successfully?*

Ask yourself if one of your current goals is to complete this course successfully.

If it is not a goal of yours, ask yourself why? One common reason is fear of failure. Amazingly enough, fear of failure alone can be strong enough to keep many of us from doing our best in any endeavor.

Another common reason is that you simply haven't taken the time to make successfully completing this course one of your goals. How do you do this? Start by writing this goal in your mathematics notebook. Then list steps you will take to ensure success. A great first step is to read or reread Section 1.1 and make a commitment to try the suggestions in that section.

Good luck, and don't forget that a positive attitude will make a big difference.

Let's see how you are doing.

1. Have you decided to make "successfully completing this course" a goal of yours? If no, please list reasons

why this has not happened. Study your list and talk to your instructor about this.

2. If your answer to question 1 is yes, take a moment and list in your notebook further specific goals that will help you achieve this major goal of successfully completing this course. (For example, "My goal this semester is not to miss any of my mathematics classes.")

3. Rate your commitment to this course with a number between 1 and 5. Use the diagram below to help.

High Commitment		Average Commitment		Not committed at all
5	4	3	2	1

4. If you have rated your personal commitment level (from the exercise above) as a 1, 2, or 3, list the reasons why this is so. Then determine whether it is possible to increase your commitment level to a 4 or 5.

*Because of its importance, this is a repeat of the Study Skills Builder on page 83.

A Add Integers.

B Evaluate an Algebraic Expression by Adding.

C Solve Problems by Adding Integers.

2.2 ADDING INTEGERS

Objective **A** Adding Integers

Adding integers can be visualized using a number line. A positive number can be represented on the number line by an arrow of appropriate length pointing to the right, and a negative number by an arrow of appropriate length pointing to the left.

Both arrows represent 2 or +2. They both point to the right and they are both 2 units long.

Both arrows represent −3. They both point to the left and they are both 3 units long.

EXAMPLE 1 Add using a number line: $5 + (-2)$

Solution: To add integers on a number line, such as $5 + (-2)$, we start at 0 on the number line and draw an arrow representing 5. From the tip of this arrow, we draw another arrow representing −2. The tip of the second arrow ends at their sum, 3.

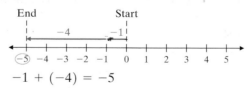

$$5 + (-2) = 3$$

🔲 **Work Practice Problem 1**

PRACTICE PROBLEM 1

Add using a number line:
$5 + (-1)$

EXAMPLE 2 Add using a number line: $-1 + (-4)$

Start at 0 and draw an arrow representing −1. From the tip of this arrow, we draw another arrow representing −4. The tip of the second arrow ends at their sum, −5.

Solution:

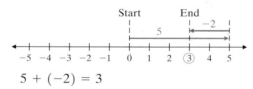

$$-1 + (-4) = -5$$

🔲 **Work Practice Problem 2**

PRACTICE PROBLEM 2

Add using a number line:
$-6 + (-2)$

EXAMPLE 3 Add using a number line: $-7 + 3$

Solution:

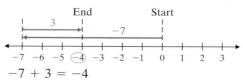

$$-7 + 3 = -4$$

🔲 **Work Practice Problem 3**

PRACTICE PROBLEM 3

Add using a number line:
$-8 + 3$

Answers

1.

$$5 + (-1) = 4$$

2.

$$-6 + (-2) = -8$$

3.

$$-8 + 3 = -5$$

Copyright 2008 Pearson Education, Inc.

Using a number line each time we add two numbers can be time consuming. Instead, we can notice patterns in the previous examples and write rules for adding signed numbers.

Rules for adding signed numbers depend on whether we are adding numbers with the same sign or different signs. When adding two numbers with the same sign, as in Example 2, notice that the sign of the sum is the same as the sign of the addends.

Adding Two Numbers with the Same Sign

Step 1: Add their absolute values.

Step 2: Use their common sign as the sign of the sum.

EXAMPLE 4 Add: $-2 + (-21)$

Solution:

Step 1: $|-2| = 2, |-21| = 21$, and $2 + 21 = 23$.
Step 2: Their common sign is negative, so the sum is negative:

$$-2 + (-21) = -23$$

Work Practice Problem 4

EXAMPLES Add.

5. $-15 + (-10) = -25$
6. $2 + 6 = 8$

Work Practice Problems 5–6

When adding two numbers with different signs, as in Examples 1 and 3, the sign of the result may be positive, negative, or the result may be 0.

Adding Two Numbers with Different Signs

Step 1: Find the larger absolute value minus the smaller absolute value.

Step 2: Use the sign of the number with the larger absolute value as the sign of the sum.

EXAMPLE 7 Add: $-2 + 5$

Solution:

Step 1: $|-2| = 2, |5| = 5$, and $5 - 2 = 3$.
Step 2: 5 has the larger absolute value and its sign is an understood +:

$$-2 + 5 = +3 \text{ or } 3$$

Work Practice Problem 7

EXAMPLE 8 Add: $3 + (-7)$

Solution:

Step 1: $|3| = 3, |-7| = 7$, and $7 - 3 = 4$.
Step 2: -7 has the larger absolute value and its sign is $-$:

$$3 + (-7) = -4$$

Work Practice Problem 8

PRACTICE PROBLEM 4
Add: $(-3) + (-19)$

PRACTICE PROBLEMS 5–6
Add.
5. $-12 + (-30)$
6. $9 + 4$

PRACTICE PROBLEM 7
Add: $-1 + 6$

PRACTICE PROBLEM 8
Add: $2 + (-8)$

Answers
4. -22 **5.** -42 **6.** 13   **7.** 5   **8.** -6

PRACTICE PROBLEMS 9–11

Add.

9. $-54 + 20$
10. $7 + (-2)$
11. $-3 + 0$

EXAMPLES Add.

9. $-18 + 10 = -8$
10. $12 + (-8) = 4$
11. $0 + (-5) = -5$ The sum of 0 and any number is the number.

Work Practice Problems 9–11

Recall that numbers such as 7 and -7 are called opposites. In general, the sum of a number and its opposite is always 0.

$$7 + (-7) = 0 \qquad -26 + 26 = 0 \qquad 1008 + (-1008) = 0$$

opposites   opposites   opposites

If a is a number, then

$-a$ is its opposite. Also,

$$\left.\begin{array}{c} a + (-a) = 0 \\ -a + a = 0 \end{array}\right\}$$ The sum of a number and its opposite is 0.

PRACTICE PROBLEMS 12–13

Add.

12. $18 + (-18)$
13. $-64 + 64$

EXAMPLES Add.

12. $-21 + 21 = 0$
13. $36 + (-36) = 0$

Work Practice Problems 12–13

✔**Concept Check**   What is wrong with the following calculation?

$$5 + (-22) = 17$$

In the following examples, we add three or more integers. Remember that by the associative and commutative properties for addition, we may add numbers in any order that we wish. In Examples 14 and 15, let's add the numbers from left to right.

PRACTICE PROBLEM 14

Add: $6 + (-2) + (-15)$

EXAMPLE 14 Add: $(-3) + 4 + (-11)$

Solution: $(-3) + 4 + (-11) = 1 + (-11)$
$$= -10$$

Work Practice Problem 14

PRACTICE PROBLEM 15

Add: $5 + (-3) + 12 + (-14)$

EXAMPLE 15 Add: $1 + (-10) + (-8) + 9$

Solution: $1 + (-10) + (-8) + 9 = -9 + (-8) + 9$
$$= -17 + 9$$
$$= -8$$

> **Helpful Hint**
> Don't forget that addition is commutative and associative. In other words, numbers may be added in any order.

The sum will be the same if we add the numbers in any order. To see this, let's add the positive numbers together and then the negative numbers together first.

$$\begin{array}{ll} 1 + 9 = 10 & \text{Add the positive numbers.} \\ (-10) + (-8) = -18 & \text{Add the negative numbers.} \\ 10 + (-18) = -8 & \text{Add these results.} \end{array}$$

The sum is -8.

Work Practice Problem 15

Answers

9. -34 **10.** 5 **11.** -3 **12.** 0 **13.** 0
14. -11 **15.** 0

✔ **Concept Check Answer**

$5 + (-22) = -17$

Objective B Evaluating Algebraic Expressions

We can continue our work with algebraic expressions by evaluating expressions given integer replacement values.

EXAMPLE 16 Evaluate $2x + y$ for $x = 3$ and $y = -5$.

Solution: Replace x with 3 and y with -5 in $2x + y$.

$$2x + y = 2 \cdot 3 + (-5)$$
$$= 6 + (-5)$$
$$= 1$$

🔲 **Work Practice Problem 16**

PRACTICE PROBLEM 16

Evaluate $x + 3y$ for $x = -6$ and $y = 2$.

EXAMPLE 17 Evaluate $x + y$ for $x = -2$ and $y = -10$.

Solution: $x + y = (-2) + (-10)$ Replace x with -2 and y with -10.
$$= -12$$

🔲 **Work Practice Problem 17**

PRACTICE PROBLEM 17

Evaluate $x + y$ for $x = -13$ and $y = -9$.

Objective C Solving Problems by Adding Integers

Next, we practice solving problems that require adding integers.

EXAMPLE 18 **Calculating Temperature**

In Barrow, Alaska, the monthly average temperature for February is $-16°$ Fahrenheit. In March, this average temperature rises 2 degrees, and in April, it rises another 13 degrees. What is the average temperature in April? (*Source:* National Climatic Data Center)

Solution:

In words:

April temperature	=	February temperature	+	rise of 2°	+	rise of 13°

Translate:

$$\text{April temperature} = -16 + (+2) + (+13)$$
$$= -14 + (+13)$$
$$= -1$$

The average temperature in Barrow, Alaska, during the month of April is $-1°$F.

🔲 **Work Practice Problem 18**

PRACTICE PROBLEM 18

If the temperature was $-7°$ Fahrenheit at 6 a.m., and it rose 4 degrees by 7 a.m. and then rose another 7 degrees in the hour from 7 a.m. to 8 a.m., what was the temperature at 8 a.m.?

Answers
16. 0   **17.** -22 **18.** $4°$F

▦ CALCULATOR EXPLORATIONS Entering Negative Numbers

To enter a negative number on a calculator, find the key marked $\boxed{+/-}$. (Some calculators have a key marked $\boxed{\text{CHS}}$ and some calculators have a special key $\boxed{(-)}$ for entering a negative sign.) To enter the number -2, for example, press the keys $\boxed{2}$ $\boxed{+/-}$. The display will read $\boxed{\quad -2}$.

To find $-32 + (-131)$, press the keys

$\boxed{32}$ $\boxed{+/-}$ $\boxed{+}$ $\boxed{131}$ $\boxed{+/-}$ $\boxed{=}$ or

$\boxed{(-)}$ $\boxed{32}$ $\boxed{+}$ $\boxed{(-)}$ $\boxed{131}$ $\boxed{\text{ENTER}}$

The display will read $\boxed{\quad -163}$. Thus $-32 + (-131) = -163$.

Use a calculator to perform each indicated operation.

1. $-256 + 97$

2. $811 + (-1058)$

3. $6(15) + (-46)$

4. $-129 + 10(48)$

5. $-108,650 + (-786,205)$

6. $-196,662 + (-129,856)$

Vocabulary and Readiness Check

Use the choices below to fill in each blank. Not all choices will be used.

$-a$ a 0 commutative   associative

1. If n is a number, then $-n + n =$ _____.

2. Since $x + n = n + x$, we say that addition is _____.

3. If a is a number, then $-(-a) =$ _____.

4. Since $n + (x + a) = (n + x) + a$, we say that addition is _____.

2.2 EXERCISE SET

Objective [A] *Add using a number line. See Examples 1 through 3.*

1. $-1 + (-6)$

2. $9 + (-4)$

3. $-4 + 7$

4. $10 + (-3)$

5. $-13 + 7$

6. $-6 + (-5)$

Add. See Examples 4 through 13.

7. $46 + 21$

8. $15 + 42$

9. $-8 + (-2)$

10. $-5 + (-4)$

11. $-43 + 43$

12. $-62 + 62$

13. $6 + (-2)$

14. $8 + (-3)$

15. $-6 + 0$

16. $-8 + 0$

17. $3 + (-5)$

18. $5 + (-9)$

19. $-2 + (-7)$

20. $-6 + (-1)$

21. $-12 + (-12)$

22. $-23 + (-23)$

23. $-640 + (-200)$

24. $-400 + (-256)$

25. $12 + (-5)$

26. $24 + (-10)$

27. $-6 + 3$

28. $-8 + 4$

29. $-56 + 26$

30. $-89 + 37$

31. $-45 + 85$

32. $-32 + 62$

33. $124 + (-144)$

34. $325 + (-375)$

35. $-82 + (-43)$

36. $-56 + (-33)$

Add. See Examples 14 and 15.

37. $-4 + 2 + (-5)$

38. $-1 + 5 + (-8)$

39. $-52 + (-77) + (-117)$

40. $-103 + (-32) + (-27)$

41. $12 + (-4) + (-4) + 12$

42. $18 + (-9) + 5 + (-2)$

43. $(-10) + 14 + 25 + (-16)$

44. $34 + (-12) + (-11) + 213$

Objective **A** **Mixed Practice** *Add. See Examples 1 through 15.*

45. $-6 + (-15) + (-7)$

46. $-12 + (-3) + (-5)$

47. $-26 + 15$

48. $-35 + (-12)$

49. $5 + (-2) + 17$

50. $3 + (-23) + 6$

51. $-13 + (-21)$

52. $-100 + 70$

53. $3 + 14 + (-18)$

54. $(-45) + 22 + 20$

55. $-92 + 92$

56. $-87 + 0$

57. $-13 + 8 + (-10) + (-27)$

58. $-16 + 6 + (-14) + (-20)$

Objective **B** *Evaluate $x + y$ for the given replacement values. See Examples 16 and 17.*

59. $x = -20$ and $y = -50$

60. $x = -1$ and $y = -29$

Evaluate $3x + y$ for the given replacement values. See Examples 16 and 17.

61. $x = 2$ and $y = -3$

62. $x = 7$ and $y = -11$

63. $x = 3$ and $y = -30$

64. $x = 13$ and $y = -17$

Objective **C** **Translating** *Translate each phrase; then simplify. See Example 18.*

65. Find the sum of -6 and 25.

66. Find the sum of -30 and 15.

67. Find the sum of -31, -9, and 30.

68. Find the sum of -49, -2, and 40.

Solve. See Example 18.

69. Suppose a deep-sea diver dives from the surface to 215 feet below the surface. He then dives down 16 more feet. Use positive and negative numbers to represent this situation. Then find the diver's present depth.

70. Suppose a diver dives from the surface to 248 meters below the surface and then swims up 8 meters, down 16 meters, down another 28 meters, and then up 32 meters. Use positive and negative numbers to represent this situation. Then find the diver's depth after these movements.

In golf, it is possible to have positive and negative scores. The following table shows the results of the eighteen-hole playoff between Pat Hurst and Annika Sorenstam at the 2006 U.S. Open. Use the table to answer Exercises 71 and 72.

Player/Hole	1	2	3	4	5	6	7	8	9	10	11	12	13	14	15	16	17	18
Sorenstam	−1	0	−1	0	0	+1	0	0	0	0	0	−1	+1	0	0	0	0	0
Hurst	+1	0	0	0	0	+2	0	0	+1	0	0	0	0	0	0	0	0	−1

(*Source:* Ladies' Professional Golf Association)

71. Find the total score for each of the athletes in the playoff.

72. In golf, the lower score is the winner. Use the result of Exercise 71 to determine who won the 2006 U.S. Open.

The following bar graph shows the yearly net income for Apple, Inc. Net income is one indication of a company's health. It measures revenue (money taken in) minus cost (money spent). Use this graph to answer Exercises 73 through 76. (Source: Apple, Inc.)

73. What was the net income (in dollars) for Apple, Inc. in 2002?

74. What was the net income (in dollars) for Apple, Inc in 2001?

75. Find the total net income for the years 2001 and 2002.

76. Find the total net income for all the years shown.

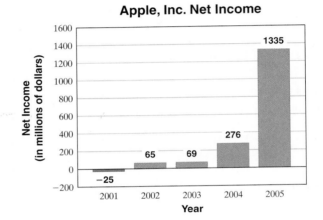

Apple, Inc. Net Income

77. The temperature at 4 p.m. on February 2 was −10° Celsius. By 11 p.m. the temperature had risen 12 degrees. Find the temperature at 11 p.m.

78. In some card games, it is possible to have both positive and negative scores. After four rounds of play, Michelle had scores of 14, −5, −8, and 7. What was her total score for the game?

A small business company reports the following net incomes. Use this table to answer Exercises 79 and 80.

Year	Net Income (in Dollars)
2003	− $10,412
2004	− $1,786
2005	$15,395
2006	$31,418

79. Find the sum of the net incomes for 2004 and 2005.

80. Find the sum of the net incomes for all four years shown.

81. The all-time record low temperature for Wisconsin is −55°F. West Virginia's all-time record low temperature is 8°F higher than Wisconsin record low. What is West Virginia's record low temperature? (*Source:* National Climatic Data Center)

82. The all-time record low temperature for Oklahoma is −27°F. In Georgia, the lowest temperature ever recorded is 10°F higher than Oklahoma's all-time low temperature. What is the all-time record low temperature for Georgia? (*Source:* National Climatic Data Center)

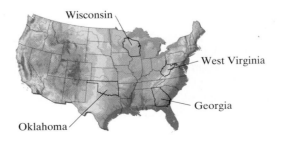

83. The deepest spot in the Pacific Ocean is the Mariana Trench, which has an elevation of 10,924 meters below sea level. The bottom of the Pacific's Aleutian Trench has an elevation 3245 meters higher than that of the Mariana Trench. Use a negative number to represent the depth of the Aleutian Trench. (*Source:* Defense Mapping Agency)

84. The deepest spot in the Atlantic Ocean is the Puerto Rico Trench, which has an elevation of 8605 meters below sea level. The bottom of the Atlantic's Cayman Trench has an elevation 1070 meters above the level of the Puerto Rico Trench. Use a negative number to represent the depth of the Cayman Trench. (*Source:* Defense Mapping Agency)

Review

Subtract. See Section 1.3.

85. $44 - 0$ **86.** $91 - 0$ **87.** $200 - 59$ **88.** $400 - 18$

Concept Extensions

89. Name 2 numbers whose sum is -17.

90. Name 2 numbers whose sum is -30.

Each calculation below is incorrect. Find the error and correct. See the Concept Check in this section.

91. $7 + (-10) \overset{?}{=} 17$

92. $-10 + (-12) \overset{?}{=} -120$

93. $-4 + 14 \overset{?}{=} -18$

94. $-15 + (-17) \overset{?}{=} 32$

For Exercises 95 through 98, determine whether each statement is true or false.

95. The sum of two negative numbers is always a negative number.

96. The sum of two positive numbers is always a positive number.

97. The sum of a positive number and a negative number is always a negative number.

98. The sum of zero and a negative number is always a negative number.

99. In your own words, explain how to add two negative numbers.

100. In your own words, explain how to add a positive number and a negative number.

2.3 SUBTRACTING INTEGERS

In Section 2.1, we discussed the opposite of an integer.

The opposite of 3 is -3.

The opposite of -6 is 6.

In this section, we use opposites to subtract integers.

Objective **A** Subtracting Integers

To subtract integers, we will write the subtraction problem as an addition problem. To see how to do this, study the examples below.

$$10 - 4 = 6$$
$$10 + (-4) = 6$$

Since both expressions simplify to 6, this means that

$$10 - 4 = 10 + (-4) = 6$$

Also,

$$3 - 2 = 3 + (-2) = 1$$
$$15 - 1 = 15 + (-1) = 14$$

Thus, to subtract two numbers, we add the first number to the opposite of the second number. (The opposite of a number is also known as its **additive inverse**.)

Subtracting Two Numbers

If a and b are numbers, then $a - b = a + (-b)$.

PRACTICE PROBLEMS 1–4

Subtract.

1. $13 - 4$

2. $-8 - 2$

3. $11 - (-15)$

4. $-9 - (-1)$

EXAMPLES Subtract.

subtraction	=	first number	+	opposite of the second number	
1. $8 - 5$	=	8	+	(-5)	= 3
2. $-4 - 10$	=	-4	+	(-10)	= -14
3. $6 - (-5)$	=	6	+	5	= 11
4. $-11 - (-7)$	=	-11	+	7	= -4

◻ Work Practice Problems 1–4

PRACTICE PROBLEMS 5–7

Subtract.

5. $6 - 9$ **6.** $-14 - 5$

7. $-3 - (-4)$

EXAMPLES Subtract.

5. $-10 - 5 = -10 + (-5) = -15$

6. $8 - 15 = 8 + (-15) = -7$

7. $-4 - (-5) = -4 + 5 = 1$

◻ Work Practice Problems 5–7

Answers

1. 9 **2.** -10 **3.** 26 **4.** -8 **5.** -3
6. -19 **7.** 1

Helpful Hint

To visualize subtraction, try the following:
The difference between $5°F$ and $-2°F$ can be found by subtracting. That is,

$$5 - (-2) = 5 + 2 = 7$$

Can you visually see from the thermometer on the right that there are actually 7 degrees between $5°F$ and $-2°F$?

$5° F$

$0° F$

$-2° F$

7 degrees

✔ **Concept Check**   What is wrong with the following calculation?

$$-9 - (-5) = -14$$

EXAMPLE 8 Subtract 7 from -3.

Solution: To subtract 7 *from* -3, we find

$$-3 - 7 = -3 + (-7) = -10$$

▢ **Work Practice Problem 8**

PRACTICE PROBLEM 8

Subtract 6 from -15.

Objective B Adding and Subtracting Integers

If a problem involves adding or subtracting more than two integers, we rewrite differences as sums and add. Recall that by associative and commutative properties, we may add numbers in any order. In Examples 9 and 10, we will add from left to right.

EXAMPLE 9 Simplify: $7 - 8 - (-5) - 1$

Solution:

$$
\begin{aligned}
7 - 8 - (-5) - 1 &= 7 + (-8) + 5 + (-1) \\
&= -1 + 5 + (-1) \\
&= 4 + (-1) \\
&= 3
\end{aligned}
$$

▢ **Work Practice Problem 9**

PRACTICE PROBLEM 9

Simplify: $-6 - 5 - 2 - (-3)$

EXAMPLE 10 Simplify: $7 + (-12) - 3 - (-8)$

Solution:

$$
\begin{aligned}
7 + (-12) - 3 - (-8) &= 7 + (-12) + (-3) + 8 \\
&= -5 + (-3) + 8 \\
&= -8 + 8 \\
&= 0
\end{aligned}
$$

▢ **Work Practice Problem 10**

PRACTICE PROBLEM 10

Simplify:
$8 + (-2) - 9 - (-7)$

Objective C Evaluating Expressions

Now let's practice evaluating expressions when the replacement values are integers.

EXAMPLE 11 Evaluate $x - y$ for $x = -3$ and $y = 9$.

Solution: Replace x with -3 and y with 9 in $x - y$.

$$
\begin{aligned}
x &\quad - \quad y \\
\downarrow &\quad \downarrow \quad \downarrow \\
= (-3) &\quad - \quad 9 \\
= (-3) &+ (-9) \\
= -12
\end{aligned}
$$

▢ **Work Practice Problem 11**

PRACTICE PROBLEM 11

Evaluate $x - y$ for $x = -5$ and $y = 13$.

Answers
8. -21 **9.** -10 **10.** 4 **11.** -18

✔ **Concept Check Answer**
$-9 - (-5) = -4$

PRACTICE PROBLEM 12

Evaluate $3y - z$ for $y = 9$ and $z = -4$.

EXAMPLE 12 Evaluate $2a - b$ for $a = 8$ and $b = -6$.

Solution: Watch your signs carefully!

$$
\begin{array}{ccc}
2a & - & b \\
\downarrow\downarrow & & \downarrow \\
= 2 \cdot 8 & - & (-6) \\
= 16 & + & 6 \\
= 22 &
\end{array}
$$

Replace a with 8 and b with -6.

Multiply.

Add.

> **Helpful Hint**
> Watch carefully when replacing variables in the expression $a - b$. Make sure that all symbols are inserted and accounted for.

◻ **Work Practice Problem 12**

Objective D Solving Problems by Subtracting Integers

Solving problems often requires subtraction of integers.

PRACTICE PROBLEM 13

The highest point in Asia is the top of Mount Everest, at a height of 29,028 feet above sea level. The lowest point is the Dead Sea, which is 1312 feet below sea level. How much higher is Mount Everest than the Dead Sea? (*Source:* National Geographic Society)

EXAMPLE 13 Finding a Change in Elevation

The highest point in the United States is the top of Mount McKinley, at a height of 20,320 feet above sea level. The lowest point is Death Valley, California, which is 282 feet below sea level. How much higher is Mount McKinley than Death Valley? (*Source:* U.S. Geological Survey)

Solution:

1. UNDERSTAND. Read and reread the problem. To find "how much higher," we subtract. Don't forget that since Death Valley is 282 feet *below* sea level, we represent its height by -282. Draw a diagram to help visualize the problem.

2. TRANSLATE.

In words:	how much higher is Mt. McKinley	=	height of Mt. McKinley	minus	height of Death Valley
	↓	↓	↓	↓	↓
Translate:	how much higher is Mt. McKinley	=	20,320	−	(−282)

3. SOLVE:

$$20{,}320 - (-282) = 20{,}320 + 282 = 20{,}602$$

4. INTERPRET. Check and state your conclusion: Mount McKinley is 20,602 feet higher than Death Valley.

◻ **Work Practice Problem 13**

Answers
12. 31   **13.** 30,340 ft

Vocabulary and Readiness Check

Multiple choice: Select the correct lettered response following each exercise.

1. It is true that $a - b =$ _____.
   a. $b - a$ **b.** $a + (-b)$ **c.** $a + b$

2. The opposite of n is _____.
   a. $-n$ **b.** $-(-n)$ **c.** n

3. To evaluate $x - y$ for $x = -10$ and $y = -14$, we replace x with -10 and y with -14 and evaluate _____.
   a. $10 - 14$ **b.** $-10 - 14$ **c.** $-14 - 10$ **d.** $-10 - (-14)$

4. The expression $-5 - 10$ equals _____.
   a. $5 - 10$ **b.** $5 + 10$ **c.** $-5 + (-10)$ **d.** $10 - 5$

2.3 EXERCISE SET

FOR EXTRA HELP

 Student Solutions Manual   PH Math/Tutor Center   CD/Video for Review   MathXL®   MyMathLab

Objective A *Subtract. See Examples 1 through 7.*

1. $-8 - (-8)$ **2.** $-6 - (-6)$ **3.** $19 - 16$ **4.** $15 - 12$

5. $3 - 8$ **6.** $2 - 5$ **7.** $11 - (-11)$ **8.** $12 - (-12)$

9. $-4 - (-7)$ **10.** $-25 - (-25)$ **11.** $-16 - 4$ **12.** $-2 - 42$

13. $3 - 15$ **14.** $8 - 9$ **15.** $42 - 55$ **16.** $17 - 63$

17. $478 - (-30)$ **18.** $844 - (-20)$ **19.** $-4 - 10$ **20.** $-5 - 8$

21. $-7 - (-3)$ **22.** $-12 - (-5)$ **23.** $17 - 29$ **24.** $16 - 45$

Translating *Translate each phrase; then simplify. See Example 8.*

25. Subtract 17 from -25.   **26.** Subtract 10 from -22.   **27.** Find the difference of -22 and -3.

28. Find the difference of -8 and -13.   **29.** Subtract -12 from 2.   **30.** Subtract -50 from -50.

Mixed Practice (*Sections 2.2, 2.3*) *Add or subtract as indicated.*

31. $-37 + (-19)$ **32.** $-35 + (-11)$ **33.** $8 - 13$ **34.** $4 - 21$

35. $-56 - 89$ **36.** $-105 - 68$ **37.** $30 - 67$ **38.** $86 - 98$

Objective **B** *Simplify. See Examples 9 and 10.*

39. $8 - 3 - 2$

40. $8 - 4 - 1$

41. $13 - 5 - 7$

42. $30 - 18 - 12$

43. $-5 - 8 - (-12)$

44. $-10 - 6 - (-9)$

45. $-11 + (-6) - 14$

46. $-15 + (-8) - 4$

47. $18 - (-32) + (-6)$

48. $23 - (-17) + (-9)$

49. $-(-5) - 21 + (-16)$

50. $-(-9) - 14 + (-23)$

51. $-10 - (-12) + (-7) - 4$

52. $-6 - (-8) + (-12) - 7$

53. $-3 + 4 - (-23) - 10$

54. $5 + (-18) - (-21) - 2$

Objective **C** *Evaluate $x - y$ for the given replacement values. See Examples 11 and 12.*

55. $x = -4$ and $y = 7$

56. $x = -7$ and $y = 1$

57. $x = 8$ and $y = -23$

58. $x = 9$ and $y = -2$

Evaluate $2x - y$ for the given replacement values. See Examples 11 and 12.

59. $x = 4$ and $y = -4$

60. $x = 8$ and $y = -10$

61. $x = 1$ and $y = -18$

62. $x = 14$ and $y = -12$

Objective **D** *Solve. See Example 13.*

The bar graph below shows the monthly average temperature in Fairbanks, Alaska. Notice that a negative temperature is illustrated by a bar below the horizontal line representing 0°F. Use this graph to answer Exercises 63 through 66.

63. Find the difference in temperature between the months of March and February.

64. Find the difference in temperature between the months of November and December.

65. Find the difference in temperature between the two months with the lowest temperatures.

66. Find the difference in temperature between the month with the warmest temperature and the month with the coldest temperature.

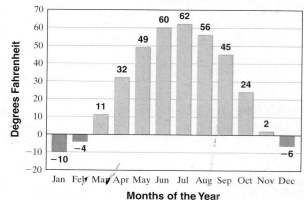

Monthly Average Temperatures in Fairbanks, AK

Source: National Climatic Data Center

67. The coldest temperature ever recorded on Earth was $-129°F$ in Antarctica. The warmest temperature ever recorded was $136°F$ in the Sahara Desert. How many degrees warmer is $136°F$ than $-129°F$? (*Source: Questions Kids Ask*, Grolier Limited, 1991, and *The World Almanac*, 2005)

68. The coldest temperature ever recorded in the United States was $-80°F$ in Alaska. The warmest temperature ever recorded was $134°F$ in California. How many degrees warmer is $134°F$ than $-80°F$? (*Source: The World Almanac*, 2005)

Solve.

69. Marta Saarens received a statement of her charge account at Old Navy. She spent $93 on purchases last month. She returned an $18 top because she didn't like the color. She also returned a $26 night shirt because it was damaged. What does she actually owe on her account?

70. Tiger Woods finished the Cialis Western Open golf tournament in 2006 in second place, with a score of −11, or 11 under par. In 82nd place was Nick Watney, with a score of +8, or eight over par. What was the difference in scores between Watney and Woods?

71. The temperature on a February morning is −4° Celsius at 6 a.m. If the temperature drops 3 degrees by 7 a.m., rises 4 degrees between 7 a.m. and 8 a.m., and then drops 7 degrees between 8 a.m. and 9 a.m., find the temperature at 9 a.m.

72. Mauna Kea in Hawaii has an elevation of 13,796 feet above sea level. The Mid-America Trench in the Pacific Ocean has an elevation of 21,857 feet below sea level. Find the difference in elevation between those two points. (*Source:* National Geographic Society and Defense Mapping Agency)

Some places on Earth lie below sea level, which is the average level of the surface of the oceans. Use this diagram to answer Exercises 73 through 76. (Source: Fantastic Book of Comparisons, Russell Ash)

73. Find the difference in elevation between Death Valley and Quattâra Depression.

74. Find the difference in elevation between Danakil and Turfan Depressions.

75. Find the difference in elevation between the two lowest elevations shown.

76. Find the difference in elevation between the highest elevation shown and the lowest elevation shown.

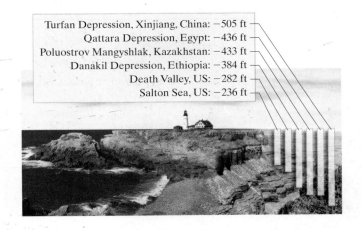

Turfan Depression, Xinjiang, China: −505 ft
Qattara Depression, Egypt: −436 ft
Poluostrov Mangyshlak, Kazakhstan: −433 ft
Danakil Depression, Ethiopia: −384 ft
Death Valley, US: −282 ft
Salton Sea, US: −236 ft

The bar graph shows heights of selected lakes. For Exercises 77 through 80, find the difference in elevation for the lakes listed. (Source: U.S. Geological Survey)

77. Lake Superior and Lake Eyre

78. Great Bear Lake and Caspian Sea

79. Lake Maracaibo and Lake Vanern

80. Lake Eyre and Caspian Sea

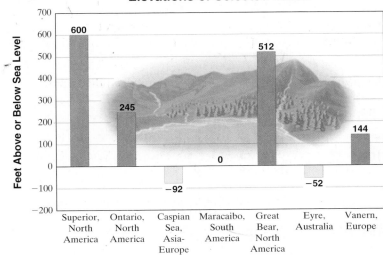

Elevations of Selected Lakes

Solve.

81. The surface temperature of the hottest planet, Venus, is 867°F, while the surface temperature of the coldest planet, Neptune, is −330°F. Find the difference in temperatures.

82. The surface temperature of Mercury is 330°F, while the surface temperature of Jupiter is −162°F Find the difference in temperatures.

83. The difference between a country's exports and imports is called the country's *trade balance*. In 2005, the United States had $895 billion in exports and $1677 billion in imports. What was the U.S. trade balance in 2005? (*Source:* U.S. Dept. of Commerce)

84. In 2005, the United States exported 369 million barrels of petroleum products and imported 1267 million barrels of petroleum products. What was the U.S. trade balance for petroleum products in 2005? (*Source:* U.S. Energy Information Administration)

Mixed Practice–Translating (*Sections 2.2, 2.3*) *Translate each phrase to an algebraic expression. Use "x" to represent "a number."*

85. The sum of -5 and a number.

86. The difference of -3 and a number.

87. Subtract a number from -20.

88. Add a number and -36.

Review

Multiply or divide as indicated. See Sections 1.5 and 1.6.

89. $\dfrac{100}{20}$

90. $\dfrac{96}{3}$

91. $\begin{array}{r} 23 \\ \times\, 46 \\ \hline \end{array}$

92. $\begin{array}{r} 51 \\ \times\, 89 \\ \hline \end{array}$

Concept Extensions

93. Name two numbers whose difference is -3.

94. Name two numbers whose difference is -10.

*Each calculation below is **incorrect**. Find the error and correct. See the Concept Check in this section.*

95. $9 - (-7) \stackrel{?}{=} 2$

96. $-4 - 8 \stackrel{?}{=} 4$

97. $10 - 30 \stackrel{?}{=} 20$

98. $-3 - (-10) \stackrel{?}{=} -13$

Simplify. (Hint: Find the absolute values first.)

99. $|-3| - |-7|$

100. $|-12| - |-5|$

101. $|-5| - |5|$

102. $|-8| - |8|$

103. $|-15| - |-29|$

104. $|-23| - |-42|$

For Exercises 103 and 104, determine whether each statement is true or false.

105. $|-8 - 3| = 8 - 3$

106. $|-2 - (-6)| = |-2| - |-6|$

107. In your own words, explain how to subtract one signed number from another.

108. A student explains to you that the first step to simplify $8 + 12 \cdot 5 - 100$ is to add 8 and 12. Is the student correct? Explain why or why not.

STUDY SKILLS BUILDER

Organizing a Notebook

It's never too late to get organized. If you need ideas about organizing a notebook for your mathematics course, try some of these:

- Use a spiral or ring binder notebook with pockets and use it for mathematics only.

- Start each page by writing the book's section number you are working on at the top.

- When your instructor is lecturing, take notes. **Always** include any examples your instructor works for you.

- Place your worked-out homework exercises in your notebook immediately after the lecture notes from that section. This way, a section's worth of material is together.

- Homework exercises: Attempt all assigned homework. For odd-numbered exercises, you are not through until you check your answers against the back of the book. Correct any exercises with incorrect answers. You may want to place a "?" by any homework exercises or notes that you need to ask questions about. Also, consider placing a "!" by any notes or exercises you feel are important.

- Place graded quizzes in the pockets of your notebook. If you are using a binder, you can place your quizzes in a special section of your binder.

Let's check your notebook organization by answering the following questions.

1. Do you have a spiral or ring binder notebook for your mathematics course only?

2. Have you ever had to flip through several sheets of notes and work in your mathematics notebook to determine what section's work you are in?

3. Are you now writing the textbook's section number at the top of each notebook page?

4. Have you ever lost or had trouble finding a graded quiz or test?

5. Are you now placing all your graded work in a dedicated place in your notebook?

6. Are you attempting all of your homework and placing all of your work in your notebook?

7. Are you checking and correcting your homework in your notebook? If not, why not?

8. Are you writing in your notebook the examples your instructor works for you in class?

2.4 MULTIPLYING AND DIVIDING INTEGERS

Multiplying and dividing integers is similar to multiplying and dividing whole numbers. One difference is that we need to determine whether the result is a positive number or a negative number.

Objective **A** Multiplying Integers

Consider the following pattern of products.

First factor decreases by 1 each time.

$$3 \cdot 2 = 6$$
$$2 \cdot 2 = 4$$
$$1 \cdot 2 = 2$$
$$0 \cdot 2 = 0$$

Product decreases by 2 each time.

This pattern can be continued, as follows.

$$-1 \cdot 2 = -2$$
$$-2 \cdot 2 = -4$$
$$-3 \cdot 2 = -6$$

This suggests that the product of a negative number and a positive number is a negative number.

What is the sign of the product of two negative numbers? To find out, we form another pattern of products. Again, we decrease the first factor by 1 each time, but this time the second factor is negative.

$$2 \cdot (-3) = -6$$
$$1 \cdot (-3) = -3$$
$$0 \cdot (-3) = 0$$

Product increases by 3 each time.

This pattern continues as:

$$-1 \cdot (-3) = 3$$
$$-2 \cdot (-3) = 6$$
$$-3 \cdot (-3) = 9$$

This suggests that the product of two negative numbers is a positive number. Thus we can determine the sign of a product when we know the signs of the factors.

Multiplying Numbers

The product of two numbers having the same sign is a positive number.

The product of two numbers having different signs is a negative number.

Product of Like Signs

$$(+)(+) = +$$
$$(-)(-) = +$$

Product of Different Signs

$$(-)(+) = -$$
$$(+)(-) = -$$

EXAMPLES Multiply.

1. $-7 \cdot 3 = -21$
2. $-2(-5) = 10$
3. $0 \cdot (-4) = 0$
4. $10(-8) = -80$

🔲 **Work Practice Problems 1–4**

Recall that by the associative and commutative properties for multiplication, we may multiply numbers in any order that we wish. In Example 5, we multiply from left to right.

EXAMPLES Multiply.

5. $7(-6)(-2) = -42(-2)$
$$= 84$$

6. $(-2)(-3)(-4) = 6(-4)$
$$= -24$$

7. $(-1)(-2)(-3)(-4) = -1(-24)$ We have -24 from Example 6.
$$= 24$$

🔲 **Work Practice Problems 5–7**

✔**Concept Check** What is the sign of the product of five negative numbers? Explain.

Recall from our study of exponents that $2^3 = 2 \cdot 2 \cdot 2 = 8$. We can now work with bases that are negative numbers. For example,

$$(-2)^3 = (-2)(-2)(-2) = -8$$

EXAMPLE 8 Evaluate: $(-5)^2$

Solution: Remember that $(-5)^2$ means 2 factors of -5.

$$(-5)^2 = (-5)(-5) = 25$$

🔲 **Work Practice Problem 8**

Helpful Hint

Have you noticed a pattern when multiplying signed numbers?
If we let $(-)$ represent a negative number and $(+)$ represent a positive number, then

The product of an even number of negative numbers is a positive result.
$$(-)(-) = (+)$$
$$(-)(-)(-) = (-)$$ ⟵ The product of an odd number of negative numbers is a negative result.
$$(-)(-)(-)(-) = (+)$$
$$(-)(-)(-)(-)(-) = (-)$$

Notice in Example 8 the parentheses around -5 in $(-5)^2$. With these parentheses, -5 is the base that is squared. Without parentheses, such as -5^2, only the 5 is squared. In other words, $-5^2 = -(5 \cdot 5) = -25$.

PRACTICE PROBLEM 9

Evaluate: -8^2

EXAMPLE 9 Evaluate: -7^2

Solution: Remember that without parentheses, only the 7 is squared.

$$-7^2 = -(7 \cdot 7) = -49$$

■ **Work Practice Problem 9**

Helpful Hint

Make sure you understand the difference between Examples 8 and 9.

parentheses, so -5 is squared

$$(-5)^2 = (-5)(-5) = 25$$

no parentheses, so only the 7 is squared

$$-7^2 = -(7 \cdot 7) = -49$$

Objective **B** **Dividing Integers**

Division of integers is related to multiplication of integers. The sign rules for division can be discovered by writing a related multiplication problem. For example,

$$\frac{6}{2} = 3 \qquad \text{because } 3 \cdot 2 = 6$$

$$\frac{-6}{2} = -3 \qquad \text{because } -3 \cdot 2 = -6$$

$$\frac{6}{-2} = -3 \qquad \text{because } -3 \cdot (-2) = 6$$

$$\frac{-6}{-2} = 3 \qquad \text{because } 3 \cdot (-2) = -6$$

Helpful Hint

Just as for whole numbers, division can be checked by multiplication.

Dividing Numbers

The quotient of two numbers having the same sign is a positive number.

The quotient of two numbers having different signs is a negative number.

Quotient of Like Signs

$$\frac{(+)}{(+)} = + \qquad \frac{(-)}{(-)} = +$$

Quotient of Different Signs

$$\frac{(+)}{(-)} = - \qquad \frac{(-)}{(+)} = -$$

PRACTICE PROBLEMS 10–12

Divide.

10. $\dfrac{42}{-7}$ 11. $-16 \div (-2)$

12. $\dfrac{-80}{10}$

EXAMPLES Divide.

10. $\dfrac{-12}{6} = -2$

11. $-20 \div (-4) = 5$

12. $\dfrac{48}{-3} = -16$

■ **Work Practice Problems 10–12**

Answers

9. -64 10. -6 11. 8 12. -8

✔ **Concept Check Answer**

$\dfrac{-36}{-9} = 4$

✔ **Concept Check** What is wrong with the following calculation?

$$\frac{-36}{-9} = -4$$

EXAMPLES Divide, if possible.

13. $\dfrac{0}{-5} = 0$ because $0 \cdot -5 = 0$

14. $\dfrac{-7}{0}$ is undefined because there is no number that gives a product of -7 when multiplied by 0.

▣ **Work Practice Problems 13–14**

PRACTICE PROBLEMS 13–14

Divide, if possible.

13. $\dfrac{-6}{0}$ **14.** $\dfrac{0}{-7}$

Objective C Evaluating Expressions

Next, we practice evaluating expressions given integer replacement values.

EXAMPLE 15 Evaluate xy for $x = -2$ and $y = 7$.

Solution: Recall that xy means $x \cdot y$.

Replace x with -2 and y with 7.

$$xy = -2 \cdot 7$$
$$= -14$$

▣ **Work Practice Problem 15**

PRACTICE PROBLEM 15

Evaluate xy for $x = 5$ and $y = -8$.

EXAMPLE 16 Evaluate $\dfrac{x}{y}$ for $x = -24$ and $y = 6$.

Solution: $\dfrac{x}{y} = \dfrac{-24}{6}$ *Replace x with -24 and y with 6.*

$$= -4$$

▣ **Work Practice Problem 16**

PRACTICE PROBLEM 16

Evaluate $\dfrac{x}{y}$ for $x = -12$ and $y = -3$.

Objective D Solving Problems by Multiplying and Dividing Integers

Many real-life problems involve multiplication and division of signed numbers.

EXAMPLE 17 Calculating Total Golf Score

A professional golfer finished seven strokes under par (-7) for each of three days of a tournament. What was her total score for the tournament?

Solution:

1. **UNDERSTAND.** Read and reread the problem. Although the key word is "total," since this is repeated addition of the same number we multiply.
2. **TRANSLATE.**

In words:	golfer's total score	=	number of days	·	score each day
	↓	↓	↓	↓	↓
Translate:	golfer's total	=	3	·	(-7)

3. **SOLVE:** $3 \cdot (-7) = -21$
4. **INTERPRET.** Check and state your conclusion: The golfer's total score is -21, or 21 strokes under par.

▣ **Work Practice Problem 17**

PRACTICE PROBLEM 17

A card player had a score of -13 for each of four games. Find her total score.

Answers

13. undefined **14.** 0 **15.** -40
16. 4 **17.** -52

Vocabulary and Readiness Check

Use the choices below to fill in each blank. Each choice may be used more than once.

negative   0

positive   undefined

1. The product of a negative number and a positive number is a ———————— number.
2. The product of two negative numbers is a ———————— number.
3. The quotient of two negative numbers is a ———————— number.
4. The quotient of a negative number and a positive number is a ———————— number.
5. The product of a negative number and zero is ————————.
6. The quotient of 0 and a negative number is ————————.
7. The quotient of a negative number and 0 is ————————.

 2.4 EXERCISE SET

FOR EXTRA HELP

Student Solutions Manual   PH Math/Tutor Center   CD/Video for Review   MathXL®   MyMathLab

Objective A *Multiply. See Examples 1 through 4.*

1. $-6(-2)$

2. $5(-3)$

3. $-4(9)$

4. $-7(-2)$

5. $9(-9)$

6. $-9(9)$

7. $0(-11)$

8. $-6(0)$

Multiply. See Examples 5 through 7.

9. $6(-2)(-4)$

10. $-2(3)(-7)$

11. $-1(-3)(-4)$

12. $-8(-3)(-3)$

13. $-4(4)(-5)$

14. $2(-5)(-4)$

15. $10(-5)(0)(-7)$

16. $3(0)(-4)(-8)$

17. $-5(3)(-1)(-1)$

18. $-2(-1)(3)(-2)$ -12

Evaluate. See Examples 8 and 9.

19. -3^2

20. -2^4

21. $(-3)^3$

22. $(-1)^4$

23. -6^2

24. -4^3

25. $(-4)^3$

26. $(-3)^2$

Objective B *Find each quotient. See Examples 10 through 14.*

27. $-24 \div 3$

28. $90 \div (-9)$

29. $\dfrac{-30}{6}$

30. $\dfrac{56}{-8}$

31. $\dfrac{-77}{-11}$

32. $\dfrac{-32}{4}$

33. $\dfrac{0}{-21}$

34. $\dfrac{-13}{0}$

35. $\dfrac{-10}{0}$

36. $\dfrac{0}{-15}$

37. $\dfrac{56}{-4}$

38. $\dfrac{-24}{-12}$

Objectives **A** **B** **Mixed Practice** *Multiply or divide as indicated.*

39. $-14(0)$ **40.** $0(-100)$ **41.** $-5(3)$ **42.** $-6 \cdot 2$

43. $-9 \cdot 7$ **44.** $-12(13)$ **45.** $-7(-6)$ **46.** $-9(-5)$

47. $-3(-4)(-2)$ **48.** $-7(-5)(-3)$ **49.** $(-7)^2$ **50.** $(-5)^2$

51. $-\dfrac{25}{5}$ 5 **52.** $-\dfrac{30}{5}$ 6 **53.** $-\dfrac{72}{8}$ -9 **54.** $-\dfrac{49}{7}$ -7

55. $-18 \div 3$ **56.** $-15 \div 3$ **57.** $4(-10)(-3)$ **58.** $6(-5)(-2)$

59. $-30(6)(-2)(-3)$ **60.** $-20 \cdot 5 \cdot (-5) \cdot (-3)$ **61.** $3 \cdot (-8) \cdot 0$ **62.** $-(4)(0)$

63. $\dfrac{120}{-20}$ **64.** $\dfrac{63}{-9}$ **65.** $280 \div (-40)$ **66.** $480 \div (-8)$

67. $\dfrac{-12}{-4}$ **68.** $\dfrac{-36}{-3}$ **69.** -1^4 **70.** -2^3

71. $(-2)^5$ **72.** $(-11)^2$ **73.** $-2(3)(5)(-6)$ **74.** $-1(2)(7)(-3)$

75. $(-1)^{32}$ **76.** $(-1)^{33}$ **77.** $-2(-3)(-5)$ **78.** $-2(-2)(-3)(-2)$

79. $-48 \cdot 23$ **80.** $-56 \cdot 43$ **81.** $35 \cdot (-82)$ **82.** $70 \cdot (-23)$

Objective **C** *Evaluate ab for the given replacement values. See Example 15.*

83. $a = -8$ and $b = 7$ **84.** $a = 5$ and $b = -1$ **85.** $a = 9$ and $b = -2$

86. $a = -9$ and $b = -6$ **87.** $a = -7$ and $b = -5$ **88.** $a = -8$ and $b = 8$

Evaluate $\dfrac{x}{y}$ *for the given replacement values. See Example 16.*

89. $x = 5$ and $y = -5$ **90.** $x = 9$ and $y = -3$ **91.** $x = -15$ and $y = 0$

92. $x = 0$ and $y = -5$ **93.** $x = -36$ and $y = -6$ **94.** $x = -10$ and $y = -10$

Evaluate xy and also $\dfrac{x}{y}$ *for the given replacement values.*

95. $x = -8$ and $y = -2$ **96.** $x = 20$ and $y = -5$ **97.** $x = 0$ and $y = -8$ **98.** $x = -3$ and $y = 0$

Objective **D** **Translating** *Translate each phrase; then simplify. See Example 17.*

99. Find the quotient of −54 and 9.

100. Find the quotient of −63 and −3.

101. Find the product of −42 and −6.

102. Find the product of −49 and 5.

Translating *Translate each phrase to an expression. Use x to represent a number. See Example 17.*

103. The product of −71 and a number

104. The quotient of −8 and a number

105. Subtract a number from −16.

106. The sum of a number and −12

107. −29 increased by a number

108. The difference of a number and −10

109. Divide a number by −33.

110. Multiply a number by −17.

Solve. See Example 17.

111. A football team lost four yards on each of three consecutive plays. Represent the total loss as a product of signed numbers and find the total loss.

112. Joe Norstrom lost $400 on each of seven consecutive days in the stock market. Represent his total loss as a product of signed numbers and find his total loss.

113. A deep-sea diver must move up or down in the water in short steps in order to keep from getting a physical condition called the "bends." Suppose a diver moves down from the surface in five steps of 20 feet each. Represent his total movement as a product of signed numbers and find the product.

114. A weather forecaster predicts that the temperature will drop five degrees each hour for the next six hours. Represent this drop as a product of signed numbers and find the total drop in temperature.

The graph shows melting points in degrees Celsius of selected elements. Use this graph to answer Exercises 115 through 118.

115. The melting point of nitrogen is 3 times the melting point of radon. Find the melting point of nitrogen.

116. The melting point of rubidium is −1 times the melting point of mercury. Find the melting point of rubidium.

117. The melting point of argon is −3 times the melting point of potassium. Find the melting point of argon.

118. The melting point of strontium is −11 times the melting point of radon. Find the melting point of strontium.

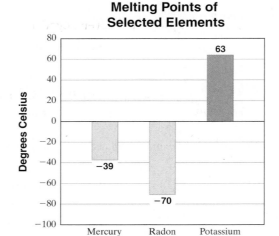

Melting Points of Selected Elements

119. During the first quarter of 2006, Ford Motor Company's North American automotive operations posted a pretax income of −$457 million. If this continues, what will Ford's North American automotive operations' net income be after four quarters? (*Source:* Ford Motor Company)

120. At the end of 2005, United Airlines posted a full year net income of −$21,176 million. If the income rate was consistent over the entire year, how much would you expect United's net income to be for each quarter? (*Source:* United Airlines)

121. In 2001, approximately 626 million prerecorded VHS movie cassettes were shipped to retailers in the United States. In 2005, this number dropped to approximately 50 million cassettes. (*Source: Motion Picture Association: Worldwide Market Research*)

    a. Find the change in the number of VHS cassettes shipped to retailers from 2001 to 2005.

    b. Find the average change per year in the number of VHS cassettes shipped to retailers over this period.

122. In 1987, there were only 27 California Condors in the entire world. Thanks to conservation efforts, in 2006 there were 299 California Condors. (*Source: Arizona Game and Fish*)

    a. Find the change in the number of California Condors from 1987 to 2006.

    b. Find the average change per year in the California Condor population over the period in part a. Round to the nearest whole.

Review

Perform each indicated operation. See Section 1.7.

123. $90 + 12^2 - 5^3$
    124. $3 \cdot (7 - 4) + 2 \cdot 5^2$
    125. $12 \div 4 - 2 + 7$
    126. $12 \div (4 - 2) + 7$

Concept Extensions

Mixed Practice (*Sections 2.2, 2.3, 2.4*) *Perform indicated operations.*

127. $-57 \div 3$
    128. $-9(-11)$
    129. $-8 - 20$

130. $-4 + (-3) + 21$
    131. $-4 - 15 - (-11)$
    132. $-16 - (-2)$

Solve. For Exercises 133 and 134, see the first Concept Check in this section.

133. What is the sign of the product of seven negative numbers?

134. What is the sign of the product of ten negative numbers?

Without actually finding the product, write the list of numbers in Exercises 135 and 136 in order from least to greatest. For help, see a helpful hint box in this section.

135. $(-2)^{12}, (-2)^{17}, (-5)^{12}, (-5)^{17}$

136. $(-1)^{50}, (-1)^{55}, 0^{15}, (-7)^{20}, (-7)^{23}$

137. In your own words, explain how to divide two integers.

138. In your own words, explain how to multiply two integers.

THE BIGGER PICTURE   Operations on Sets of Numbers

Continue your outline from Sections 1.6 and 1.7. Suggestions are once again written to help you complete this part of your outline. Notice that this part of the outline has to do with operations on integers.

I. **Operations on Sets of Numbers**

   A. **Whole Numbers**

     1. **Add or Subtract** (Section 1.3)

     2. **Multiply or Divide** (Sections 1.5, 1.6)

     3. **Exponent** (Section 1.7)

     4. **Order of Operations** (Section 1.7)

   B. **Integers**

     1. **Add:**

$$-5 + (-2) = -7$$ Adding like signs. Add absolute value. Attach the common sign.

$$-5 + 2 = -3$$ Adding unlike signs. Subtract absolute values. Attach the sign of the number with the larger absolute value.

     2. **Subtract:** Add the first number to the opposite of the second number.

$$7 - 10 = 7 + (-10) = -3$$

     3. **Multiply or Divide:** Multiply or divide as usual. If the signs of the two numbers are the same, the answer is positive. If the signs of the two numbers are different, the answer is negative.

$$-5 \cdot 5 = -25, \quad \frac{-32}{-8} = 4$$

Perform the indicated operations.

1. $-9 + 14$

2. $-5(-11)$

3. $5 - 11$

4. $58 - |-70|$

5. $18 + (-30)$

6. $(-9)^2$

7. -9^2

8. $-10 + (-24)$

9. $1 - (-9)$

10. $-15 - 15$

11. $-3(2)(-5)$

12. $\dfrac{|-88|}{-|-8|}$

13. $2 + 4(7 - 9)^3$

14. $-100 - (-20)$

15. $1 + 2(7)$

16. $30 \div 2 \cdot 3$

Integers

Answers

Represent each quantity by an integer.

1. The record low temperature in New Mexico is 50 degrees Fahrenheit below zero. The highest temperature in that state is 122 degrees above zero. Represent each quantity by an integer.

2. Graph the signed numbers on the given number line. $-4, 0, -1, 3$

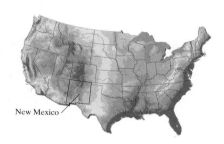

New Mexico

Insert < or > between each pair of numbers to make a true statement.

3. $0 \quad -10$ **4.** $-15 \quad -5$ **5.** $-4 \quad 4$ **6.** $-2 \quad -7$

Simplify.

7. $|-3|$ **8.** $-|-4|$ **9.** $|-9|$ **10.** $-(-5)$

Find the opposite of each number.

11. 11   **12.** -3 **13.** 64   **14.** 0

Perform the indicated operation.

15. $-3 + 15$ **16.** $-9 + (-11)$ **17.** $-8(-6)(-1)$ **18.** $-18 \div 2$

19. $65 + (-55)$ **20.** $1000 - 1002$ **21.** $53 - (-53)$ **22.** $-2 - 1$

Answers

1. _____
2. see number line
3. _____
4. _____
5. _____
6. _____
7. _____
8. _____
9. _____
10. _____
11. _____
12. _____
13. _____
14. _____
15. _____
16. _____
17. _____
18. _____
19. _____
20. _____
21. _____
22. _____

23. $\dfrac{0}{-47}$

24. $\dfrac{-36}{-9}$

25. $-17 - (-59)$

26. $-8 + (-6) + 20$

27. $\dfrac{-95}{-5}$

28. $-9(100)$

29. $-12 - 6 - (-6)$

30. $-4 + (-8) - 16 - (-9)$

31. $\dfrac{-105}{0}$

32. $7(-16)(0)(-3)$

Translating *Translate each phrase; then simplify.*

33. Subtract -8 from -12.

34. Find the sum of -17 and -27.

35. Find the product of -5 and -25.

36. Find the quotient of -100 and -5.

Translating *Translate each phrase to an expression. Use x to represent a number.*

37. Divide a number by -17

38. The sum of -3 and a number

39. A number decreased by -18

40. The product of -7 and a number

Evaluate the expressions below for $x = -3$ and $y = 12$.

41. $x + y$

42. $x - y$

43. $2y - x$

44. $3y + x$

45. $5x$

46. $\dfrac{y}{x}$

Answer blanks: 23.–46.

2.5 ORDER OF OPERATIONS

Objectives

A Simplify Expressions by Using the Order of Operations.

B Evaluate an Algebraic Expression.

C Find the Average of a List of Numbers.

Objective **A** Simplifying Expressions

We first discussed the order of operations in Chapter 1. In this section, you are given an opportunity to practice using the order of operations when expressions contain signed numbers. The rules for the order of operations from Section 1.7 are repeated here.

Order of Operations

1. Perform all operations within parentheses ( ), brackets [ ], or other grouping symbols such as fraction bars, starting with the innermost set.
2. Evaluate any expressions with exponents.
3. Multiply or divide in order from left to right.
4. Add or subtract in order from left to right.

Before simplifying other expressions, make sure you are confident simplifying Examples 1 through 3.

EXAMPLES Find the value of each expression.

1. $(-3)^2 = (-3)(-3) = 9$ The base of the exponent is -3.
2. $-3^2 = -(3)(3) = -9$ The base of the exponent is 3.
3. $2 \cdot 5^2 = 2 \cdot (5 \cdot 5) = 2 \cdot 25 = 50$ The base of the exponent is 5.

◼ **Work Practice Problems 1–3**

PRACTICE PROBLEMS 1–3

Find the value of each expression.
1. $(-2)^4$
2. -2^4
3. $3 \cdot 6^2$

Helpful Hint

When simplifying expressions with exponents, remember that parentheses make an important difference.

$(-3)^2$ and -3^2 **do not** mean the same thing.

$(-3)^2$ means $(-3)(-3) = 9$.

-3^2 means the opposite of $3 \cdot 3$, or -9.

Only with parentheses around it is the -3 squared.

EXAMPLE 4 Simplify: $\dfrac{-6(2)}{-3}$

Solution: First we multiply -6 and 2. Then we divide.

$$\frac{-6(2)}{-3} = \frac{-12}{-3}$$
$$= 4$$

◼ **Work Practice Problem 4**

PRACTICE PROBLEM 4

Simplify: $\dfrac{-25}{5(-1)}$

Answers
1. 16 2. -16 3. 108 4. 5

133

PRACTICE PROBLEM 5

Simplify: $\dfrac{-18 + 6}{-3 - 1}$

EXAMPLE 5 Simplify: $\dfrac{12 - 16}{-1 + 3}$

Solution: We simplify above and below the fraction bar separately. Then we divide.

$$\dfrac{12 - 16}{-1 + 3} = \dfrac{-4}{2}$$
$$= -2$$

■ **Work Practice Problem 5**

PRACTICE PROBLEM 6

Simplify: $30 + 50 + (-4)^3$

EXAMPLE 6 Simplify: $60 + 30 + (-2)^3$

Solution: $60 + 30 + (-2)^3 = 60 + 30 + (-8)$ Write $(-2)^3$ as -8.
$$= 90 + (-8)$$ Add from left to right.
$$= 82$$

■ **Work Practice Problem 6**

PRACTICE PROBLEM 7

Simplify: $-2^3 + (-4)^2 + 1^5$

EXAMPLE 7 Simplify: $-4^2 + (-3)^2 - 1^3$

Solution:

$-4^2 + (-3)^2 - 1^3 = -16 + 9 - 1$ Simplify expressions with exponents.
$$= -7 - 1$$ Add or subtract from left to right.
$$= -8$$

■ **Work Practice Problem 7**

PRACTICE PROBLEM 8

Simplify:
$2(2 - 9) + (-12) - 3$

EXAMPLE 8 Simplify: $3(4 - 7) + (-2) - 5$

Solution:

$3(4 - 7) + (-2) - 5 = 3(-3) + (-2) - 5$ Simplify inside parentheses.
$$= -9 + (-2) - 5$$ Multiply.
$$= -11 - 5$$ Add or subtract from left to right.
$$= -16$$

■ **Work Practice Problem 8**

PRACTICE PROBLEM 9

Simplify:
$(-5) \cdot |-8| + (-3) + 2^3$

EXAMPLE 9 Simplify: $(-3) \cdot |-5| - (-2) + 4^2$

Solution:

$(-3) \cdot |-5| - (-2) + 4^2 = (-3) \cdot 5 - (-2) + 4^2$ Write $|-5|$ as 5.
$$= (-3) \cdot 5 - (-2) + 16$$ Write 4^2 as 16.
$$= -15 - (-2) + 16$$ Multiply.
$$= -13 + 16$$ Add or subtract from left to right.
$$= 3$$

■ **Work Practice Problem 9**

Answers
5. 3 **6.** 16 **7.** 9 **8.** −29 **9.** −35

EXAMPLE 10 Simplify: $-2[-3 + 2(-1 + 6)] - 5$

Solution: Here we begin with the innermost set of parentheses.

$$
\begin{aligned}
-2[-3 + 2(-1 + 6)] - 5 &= -2[-3 + 2(5)] - 5 \quad \text{Write } -1 + 6 \text{ as } 5.\\
&= -2[-3 + 10] - 5 \quad \text{Multiply.}\\
&= -2(7) - 5 \quad \text{Add.}\\
&= -14 - 5 \quad \text{Multiply.}\\
&= -19 \quad \text{Subtract.}
\end{aligned}
$$

▨ **Work Practice Problem 10**

✔ **Concept Check** True or false? Explain your answer. The result of
$-4(3 - 7) - 8(9 - 6)$
is positive because there are four negative signs.

Objective **B** Evaluating Expressions

Now we practice evaluating expressions.

EXAMPLE 11 Evaluate x^2 and $-x^2$ for $x = -11$.

Solution: $x^2 = (-11)^2 = (-11)(-11) = 121$
$-x^2 = -(-11)^2 = -(-11)(-11) = -121$

▨ **Work Practice Problem 11**

EXAMPLE 12 Evaluate $6z^2$ for $z = 2$ and $z = -2$.

Solution: $6z^2 = 6(2)^2 = 6(4) = 24$
$6z^2 = 6(-2)^2 = 6(4) = 24$

▨ **Work Practice Problem 12**

EXAMPLE 13 Evaluate $x + 2y - z$ for $x = 3$ and $y = -5$ and $z = -4$.

Solution: Replace x with 3, y with -5, z with -4 and simplify.

Helpful Hint Remember to rewrite the subtraction sign.

$$
\begin{aligned}
x + 2y - z &= 3 + 2(-5) - (-4) \quad \text{Let } x = 3, y = -5, \text{ and } z = -4.\\
&= 3 + (-10) + 4 \quad \text{Replace } 2(-5) \text{ with its product, } -10.\\
&= -3 \quad \text{Add.}
\end{aligned}
$$

▨ **Work Practice Problem 13**

PRACTICE PROBLEM 10
Simplify:
$-4[-6 + 5(-3 + 5)] - 7$

PRACTICE PROBLEM 11
Evaluate x^2 and $-x^2$ for $x = -15$.

PRACTICE PROBLEM 12
Evaluate $5y^2$ for $y = 4$ and $y = -4$.

PRACTICE PROBLEM 13
Evaluate $x^2 + y$ for $x = -6$ and $y = -3$.

Answers
10. -23 **11.** $225; -225$ **12.** $80; 80$
13. 33

✔ **Concept Check Answer**
false; $-4(3 - 7) - 8(9 - 6) = -8$

PRACTICE PROBLEM 14

Evaluate $4 - x^2$ for $x = -8$.

EXAMPLE 14 Evaluate $7 - x^2$ for $x = -4$.

Solution: Replace x with -4 and simplify carefully!

$$7 - x^2 = 7 - (-4)^2$$
$$= 7 - 16 \qquad (-4)^2 = (-4)(-4) = 16$$
$$= -9 \qquad \text{Subtract.}$$

🖳 **Work Practice Problem 14**

Objective **C** Finding Averages

Recall from Chapter 1 that the average of a list of numbers is

$$\text{average} = \frac{\text{sum of numbers}}{\textit{number} \text{ of numbers}}$$

PRACTICE PROBLEM 15

Find the average of the temperatures for the months October through April.

EXAMPLE 15 The graph shows the monthly normal temperatures for Barrow, Alaska. Use this graph to find the average of the temperatures for months January through May.

Monthly Normal Temperatures for Barrow, Alaska

Solution: By reading the graph, we have

$$\text{average} = \frac{-14 + (-16) + (-14) + (-1) + 20}{5} \qquad \begin{array}{l}\text{There are 5 months}\\\text{from January through}\\\text{May.}\end{array}$$
$$= \frac{-25}{5}$$
$$= -5$$

The average for the temperatures is $-5°F$.

🖳 **Work Practice Problem 15**

Answers

14. -60 **15.** $-6°F$

Simplifying an Expression Containing a Fraction Bar

Recall that even though most calculators follow the order of operations, parentheses must sometimes be inserted. For example, to simplify $\dfrac{-8 + 6}{-2}$ on a calculator, enter parentheses about the expression above the fraction bar so that it is simplified separately.

To simplify $\dfrac{-8 + 6}{-2}$, press the keys

(8 +/- + 6) ÷ 2 +/- = or

((-) 8 + 6) ÷ (-) 2 ENTER

The display will read [1].

Thus, $\dfrac{-8 + 6}{-2} = 1$.

Use a calculator to simplify.

1. $\dfrac{-120 - 360}{-10}$

2. $\dfrac{4750}{-2 + (-17)}$

3. $\dfrac{-316 + (-458)}{28 + (-25)}$

4. $\dfrac{-234 + 86}{-18 + 16}$

Vocabulary and Readiness Check

Use the choices below to fill in each blank. Not all choices will be used.

average	subtraction	division	$-7 - 3(1)$
addition	multiplication	$-7 - 3(-1)$	

1. To simplify $-2 \div 2 \cdot (3)$, which operation should be performed first? _____

2. To simplify $-9 - 3 \cdot 4$, which operation should be performed first? _____

3. The _____ of a list of numbers is $\dfrac{\text{sum of numbers}}{\textit{number} \text{ of numbers}}$.

4. To simplify $5\,[-9 + (-3)] \div 4$, which operation should be performed first? _____

5. To simplify $-2 + 3(10 - 12) \cdot (-8)$, which operation should be performed first? _____

6. To evaluate $x - 3y$ for $x = -7$ and $y = -1$, replace x with -7 and y with -1 and evaluate _____.

2.5 EXERCISE SET

FOR EXTRA HELP

Student Solutions Manual · PH Math/Tutor Center · CD/Video for Review · MathXL® · MyMathLab

Objective Ⓐ *Simplify. See Examples 1 through 10.*

1. $(-5)^3$

2. -2^4

3. -4^3

4. $(-2)^4$

5. $8 \cdot 2^2$

6. $5 \cdot 2^3$

7. $8 - 12 - 4$

8. $10 - 23 - 12$

9. $7 + 3(-6)$

10. $-8 + 4(3)$

11. $5(-9) + 2$

12. $7(-6) + 3$

13. $(-10) + 4 \div 2$

14. $(-12) + 6 \div 3$

15. $6 + 7 \cdot 3 - 10$

16. $5 + 9 \cdot 4 - 20$

17. $\dfrac{16 - 13}{-3}$

18. $\dfrac{20 - 15}{-1}$

19. $\dfrac{24}{10 + (-4)}$

20. $\dfrac{88}{-8 - 3}$

21. $5(-3) - (-12)$

22. $7(-4) - (-6)$

23. $[8 + (-4)]^2$

24. $[9 + (-2)]^3$

25. $8 \cdot 6 - 3 \cdot 5 + (-20)$

26. $7 \cdot 6 - 6 \cdot 5 + (-10)$

27. $4 - (-3)^4$

28. $7 - (-5)^2$

29. $|7 + 3| \cdot 2^3$

30. $|-3 + 7| \cdot 7^2$

31. $7 \cdot 6^2 + 4$

32. $10 \cdot 5^3 + 7$

33. $7^2 - (4 - 2^3)$

34. $8^2 - (5 - 2)^4$

35. $|3 - 15| \div 3$

36. $|12 - 19| \div 7$

37. $-(-2)^6$

38. $-(-2)^3$

39. $(5 - 9)^2 \div (4 - 2)^2$

40. $(2 - 7)^2 \div (4 - 3)^4$

41. $|8 - 24| \cdot (-2) \div (-2)$

42. $|3 - 15| \cdot (-4) \div (-16)$

43. $(-12 - 20) \div 16 - 25$

44. $(-20 - 5) \div 5 - 15$

45. $5(5 - 2) + (-5)^2 - 6$

46. $3 \cdot (8 - 3) + (-4) - 10$

47. $(2 - 7) \cdot (6 - 19)$

48. $(4 - 12) \cdot (8 - 17)$

49. $(-36 \div 6) - (4 \div 4)$

50. $(-4 \div 4) - (8 \div 8)$

51. $(10 - 4^2)^2$

52. $(11 - 3^2)^3$

53. $2(8 - 10)^2 - 5(1 - 6)^2$

54. $-3(4 - 8)^2 + 5(14 - 16)^3$

55. $3(-10) \div [5(-3) - 7(-2)]$

56. $12 - [7 - (3 - 6)] + (2 - 3)^3$

57. $\dfrac{(-7)(-3) - (4)(3)}{3[7 \div (3 - 10)]}$

58. $\dfrac{10(-1) - (-2)(-3)}{2[-8 \div (-2 - 2)]}$

59. $-3[5 + 2(-4 + 9)] + 15$

60. $-2[6 + 4(2 - 8)] - 25$

Objective B *Evaluate each expression for $x = -2$, $y = 4$, and $z = -1$. See Examples 11 through 14.*

61. $x + y + z$

62. $x - y - z$

63. $2x - 3y - 4z$

64. $5x - y + 4z$

65. $x^2 - y$

66. $x^2 + z$

67. $\dfrac{5y}{z}$

68. $\dfrac{4x}{y}$

Evaluate each expression for $x = -3$ and $z = -4$. See Examples 11 through 14.

69. x^2

70. z^2

71. $-z^2$

72. $-x^2$

73. $2z^3$

74. $3x^2$

75. $10 - x^2$

76. $3 - z^2$

77. $2x^3 - z$

78. $3z^2 - x$

Objective C *Find the average of each list of numbers. See Example 15.*

79. $-10, 8, -4, 2, 7, -5, -12$

80. $-18, -8, -1, -1, 0, 4$

81. $-17, -26, -20, -13$

82. $-40, -20, -10, -15, -5$

Scores in golf can be 0 (also called par), a positive integer (also called above par), or a negative integer (also called below par). The bar graph shows scores of selected golfers from a tournament. Use this graph for Exercises 83 through 88.

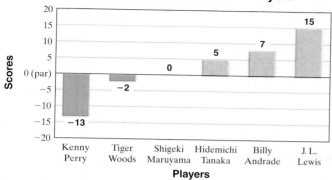

Golf Scores of Selected Players

83. Find the difference between the lowest score shown and the highest score shown.

84. Find the difference between the two lowest scores.

85. Find the average of the scores shown. (*Hint:* Here, the average is the sum of the scores divided by the number of players.)

86. Find the average of the scores for Perry, Woods, Maruyama, and Lewis.

87. Can the average for these scores be greater than the highest score, 15? Explain why or why not.

88. Can the average of all the scores shown be less than the lowest score, −13? Explain why or why not.

Review

Perform each indicated operation. See Sections 1.3, 1.5, and 1.6.

89. $45 \cdot 90$

90. $90 \div 45$

91. $90 - 45$

92. $45 + 90$

Find the perimeter of each figure. See Section 1.3.

93. Square 8 in.

94. Parallelogram 5 cm 3 cm

95. Rectangle 6 ft 9 ft

96. Triangle 17 m 23 m 32 m

Concept Extensions

Insert parentheses where needed so that each expression evaluates to the given number.

97. $2 \cdot 7 - 5 \cdot 3$; evaluates to 12

98. $7 \cdot 3 - 4 \cdot 2$; evaluates to 34

99. $-6 \cdot 10 - 4$; evaluates to −36

100. $2 \cdot 8 \div 4 - 20$; evaluates to −36

101. Are parentheses necessary in the expression $3 + (4 \cdot 5)$? Explain your answer.

102. Are parentheses necessary in the expression $(3 + 4) \cdot 5$? Explain your answer.

103. Discuss the effect parentheses have in an exponential expression. For example, what is the difference between $(-6)^2$ and -6^2?

104. Discuss the effect parentheses have in an exponential expression. For example, what is the difference between $(2 \cdot 4)^2$ and $2 \cdot 4^2$?

Evaluate.

105. $(-12)^4$

106. $(-17)^6$

107. $x^3 - y^2$ for $x = 21$ and $y = -19$

108. $3x^2 + 2x - y$ for $x = -18$ and $y = 2868$

109. $(xy + z)^x$ for $x = 2$, $y = -5$, and $z = 7$

110. $5(ab + 3)^b$ for $a = -2$, $b = 3$

STUDY SKILLS BUILDER

How Are Your Homework Assignments Going?

It is very important in mathematics to keep up with homework. Why? Many concepts build on each other. Often your understanding of a day's concepts depends on an understanding of the previous day's material.

Remember that completing your homework assignment involves a lot more than attempting a few of the problems assigned.

To complete a homework assignment, remember these four things:

- Attempt all of it.
- Check it.
- Correct it.
- If needed, ask questions about it.

Take a moment and review your completed homework assignments. Answer the questions below based on this review.

1. Approximately how much of your homework have you attempted?

2. Approximately how much of your homework have you checked (if possible)?

3. If you are able to check your homework, have you corrected it when errors have been found?

4. When working homework, if you do not understand a concept, what do you do?

2.6 SOLVING EQUATIONS: THE ADDITION AND MULTIPLICATION PROPERTIES

Objectives

A Identify Solutions of Equations.

B Use the Addition Property of Equality to Solve Equations.

C Use the Multiplication Property of Equality to Solve Equation.

In this section, we introduce properties of equations and we use these properties to begin solving equations. Now that we know how to perform operations on integers, this is an excellent way to practice these operations.

First, let's recall the difference between an equation and an expression. From Section 1.8, a combination of operations on variables and numbers is an expression and an equation is of the form "expression = expression."

Equations	Expressions
$3x - 1 = -17$	$3x - 1$
area = length · width	$5(20-3) + 10$
$8 + 16 = 16 + 8$	y^3
$-9a + 11b = 14b + 3$	$-x^2 + y - 2$

Helpful Hint

Simply stated, an equation contains "=" while an expression does not. Also, we *simplify* expressions and *solve* equations.

Objective **A** Identifying Solutions of Equations

Let's practice identifying solutions of equations. Recall from Section 1.8 that a solution of an equation is a number that when substituted for a variable makes the equation a true statement.

For example,

-8 is a solution of

$\dfrac{x}{2} = -4$, because

$\dfrac{-8}{2} = -4$, or

$-4 = -4$ is true.

Also,

-8 is not a solution of

$x + 6 = 2$, because

$-8 + 6 = 2$ is false.

Let's practice determining whether a number is a solution of an equation. In this section, we will be performing operations on integers.

EXAMPLE 1 Determine whether -1 is a solution of the equation $3y + 1 = 3$.

Solution:

$$3y + 1 = 3$$
$$3(-1) + 1 \overset{?}{=} 3$$
$$-3 + 1 \overset{?}{=} 3$$
$$-2 = 3 \quad \text{False.}$$

Since $-2 = 3$ is false, -1 is *not* a solution of the equation.

 Work Practice Problem 1

Now we know how to check whether a number is a solution. But, given an equation, how do we find its **solution?** In other words, how do we find a number that makes the equation true? How do we solve an equation?

PRACTICE PROBLEM 1

Determine whether -2 is a solution of the equation $-4x - 3 = 5$.

Answer

1. yes

141

Objective **B** Using the Addition Property to Solve Equations

To solve an equation, we use properties of equality to write simpler equations, all equivalent to the original equation, until the final equation has the form

$$x = \textbf{number} \quad \text{or} \quad \textbf{number} = x$$

Equivalent equations have the same solution, so the word "number" above represents the solution of the original equation. The first property of equality to help us write simpler, equivalent equations is the **addition property of equality.**

Addition Property of Equality

Let a, b, and c represent numbers. Then

$a = b$ and $a + c = b + c$ are equivalent equations.	Also, $a = b$ and $a - c = b - c$ are equivalent equations.

In other words, the same number may be added to or subtracted from both sides of an equation without changing the solution of the equation. (Recall in Section 2.3 that we defined subtraction as addition of the first number and the opposite of the second number. Because of this, the addition property of equality also allows us to subtract the same number from both sides.)

A good way to visualize a true equation is to picture a balanced scale. Since it is balanced, each side of the scale weighs the same amount. Similarly, in a true equation the expressions on each side have the same value. Picturing our balanced scale, if we add the same weight to each side, the scale remains balanced.

PRACTICE PROBLEM 2

Solve the equation
for y: $y - 6 = -2$

EXAMPLE 2 Solve: $x - 2 = -1$ for x.

Solution: To solve the equation for x, we need to rewrite the equation in the form $x = $ number. In other words, our goal is to get x alone on one side of the equation. To do so, we add 2 to both sides of the equation.

$$x - 2 = -1$$
$$x - 2 + 2 = -1 + 2 \quad \text{Add 2 to both sides of the equation.}$$
$$x + 0 = 1 \quad \text{Replace } -2 + 2 \text{ with 0.}$$
$$x = 1 \quad \text{Simplify by replacing } x + 0 \text{ with } x.$$

Check: To check, we replace x with 1 in the *original* equation.

$$x - 2 = -1 \quad \text{Original equation}$$
$$1 - 2 \stackrel{?}{=} -1 \quad \text{Replace } x \text{ with 1.}$$
$$-1 = -1 \quad \text{True}$$

Since $-1 = -1$ is a true statement, 1 is the solution of the equation.

Work Practice Problem 2

Answer
2. 4

Helpful Hint
Note that it is always a good idea to check the solution in the *original* equation to see that it makes the equation a true statement.

Let's visualize how we used the addition property of equality to solve an equation. Picture the equation, $x - 2 = 1$, as a balanced scale. The left side of the equation has the same value as the right side.

If the same weight is added to each side of a scale, the scale remains balanced. Likewise, if the same number is added to each side of an equation, the left side continues to have the same value as the right side.

EXAMPLE 3 Solve: $-8 = n + 1$

Solution: To get n alone on one side of the equation, we subtract 1 from both sides of the equation.

$$-8 = n + 1$$
$$-8 - 1 = n + 1 - 1 \quad \text{Subtract 1 from both sides.}$$
$$-9 = n + 0 \quad \text{Replace } 1 - 1 \text{ with 0.}$$
$$-9 = n \quad \text{Simplify.}$$

Check:

$$-8 = n + 1$$
$$-8 \stackrel{?}{=} -9 + 1 \quad \text{Replace } n \text{ with } -9.$$
$$-8 = -8 \quad \text{True}$$

The solution is -9.

Work Practice Problem 3

Helpful Hint
Remember that we can get the variable alone on either side of the equation. For example, the equations $-9 = n$ and $n = -9$ both have the solution of -9.

✔ Concept Check What number should be added to or subtracted from both sides of the equation in order to solve the equation $-3 = y + 2$?

PRACTICE PROBLEM 3
Solve: $-2 = z + 8$

Answer
3. -10

✔ Concept Check Answer
Subtract 2 from both sides.

PRACTICE PROBLEM 4

Solve: $10x = -2 + 9x$

EXAMPLE 4   Solve: $7x = 6x + 4$

Solution: Subtract $6x$ from both sides so that variable terms will be on the same side of the equation.

$$7x = 6x + 4$$
$$7x - 6x = 6x + 4 - 6x \qquad \text{Subtract } 6x \text{ from both sides.}$$
$$1x = 4 \quad \text{or} \quad x = 4 \qquad \text{Simplify.}$$

Check to see that 4 is the solution.

🔲 **Work Practice Problem 4**

Objective **C** **Using the Multiplication Property to Solve Equations**

Although the addition property of equality is a powerful tool for helping us solve equations, it cannot help us solve all types of equations. For example, it cannot help us solve an equation such as $2x = 6$. To solve this equation, we use a second property of equality called the **multiplication property of equality.**

Multiplication Property of Equality

Let a, b, and c represent numbers and let $c \neq 0$. Then

$a = b$	Also, $a = b$
and $a \cdot c = b \cdot c$	and $\dfrac{a}{c} = \dfrac{b}{c}$
are equivalent equations.	are equivalent equations.

In other words, both sides of an equation may be multiplied or divided by the same nonzero number without changing the solution of the equation. (We will see in Chapter 4 how the multiplication property also allows us to divide both sides of an equation by the same nonzero number.)

To solve an equation like $2x = 6$ for x, notice that 2 is *multiplied* by x. To get x alone, we use the multiplication property of equality to *divide* both sides of the equation by 2, and simplify as follows:

$$2x = 6$$
$$\frac{2 \cdot x}{2} = \frac{6}{2} \qquad \text{Divide both sides by 2.}$$

Then it can be shown that an expression such as $\dfrac{2 \cdot x}{2}$ is equivalent to $\dfrac{2}{2} \cdot x$, so

$$\frac{2 \cdot x}{2} = \frac{6}{2} \qquad \text{can be written as} \qquad \frac{2}{2} \cdot x = \frac{6}{2}$$
$$1 \cdot x = 3 \quad \text{or} \quad x = 3$$

Picturing again our balanced scale, if we multiply or divide the weight on each side by the same nonzero number, the scale (or equation) remains balanced.

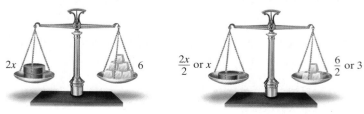

EXAMPLE 5 Solve: $-5x = 15$

Solution: To get x alone, divide both sides by -5.

$$-5x = 15 \qquad \text{Original equation}$$

$$\frac{-5x}{-5} = \frac{15}{-5} \qquad \text{Divide both sides by } -5.$$

$$\frac{-5}{-5} \cdot x = \frac{15}{-5}$$

$$1x = -3 \quad \text{or} \quad x = -3 \qquad \text{Simplify.}$$

Check: To check, replace x with -3 in the original equation.

$$-5x = 15 \qquad \text{Original equation}$$

$$-5(-3) \overset{?}{=} 15 \qquad \text{Let } x = -3.$$

$$15 = 15 \qquad \text{True.}$$

The solution is -3.

▣ **Work Practice Problem 5**

PRACTICE PROBLEM 5
Solve: $3y = -18$

EXAMPLE 6 Solve: $-27 = 3y$

Solution: To get y alone, divide both sides of the equation by 3.

$$-27 = 3y$$

$$\frac{-27}{3} = \frac{3y}{3} \qquad \text{Divide both sides by 3.}$$

$$\frac{-27}{3} = \frac{3}{3} \cdot y$$

$$-9 = 1y \quad \text{or} \quad y = -9$$

Check to see that -9 is the solution.

▣ **Work Practice Problem 6**

PRACTICE PROBLEM 6
Solve: $-32 = 8x$

EXAMPLE 7 Solve: $-12x = -36$

Solution: Divide both sides of the equation by the coefficient of x, which is -12.

$$-12x = -36$$

$$\frac{-12x}{-12} = \frac{-36}{-12}$$

$$\frac{-12}{-12} \cdot x = \frac{-36}{-12}$$

$$x = 3$$

Check: To check, replace x with 3 in the original equation.

$$-12x = -36$$

$$-12(3) \overset{?}{=} -36 \qquad \text{Let } x = 3.$$

$$-36 = -36 \qquad \text{True}$$

Since $-36 = -36$ is a true statement, the solution is 3.

▣ **Work Practice Problem 7**

PRACTICE PROBLEM 7
Solve: $-3y = -27$

Answers
5. -6 **6.** -4 **7.** 9

✔**Concept Check**   Which operation is appropriate for solving each of the following equations, addition or division?

a. $12 = x - 3$

b. $12 = 3x$

The multiplication property also allows us to solve equations like

$$\frac{x}{5} = 2$$

Here, x is *divided* by 5. To get x alone, we use the multiplication property to *multiply* both sides by 5.

$$5 \cdot \frac{x}{5} = 5 \cdot 2 \quad \text{Multiply both sides by 5.}$$

Then it can be shown that

$$5 \cdot \frac{x}{5} = 5 \cdot 2 \text{ can be written as } \frac{5}{5} \cdot x = 5 \cdot 2$$

$$1 \cdot x = 10 \quad \text{or} \quad x = 10$$

PRACTICE PROBLEM 8

Solve: $\dfrac{x}{-4} = 7$

EXAMPLE 8   Solve: $\dfrac{x}{3} = -2$

Solution: To get x alone, multiply both sides by 3.

$$\frac{x}{3} = -2$$

$$3 \cdot \frac{x}{3} = 3 \cdot (-2) \qquad \text{Multiply both sides by 3.}$$

$$\frac{3}{3} \cdot x = 3 \cdot (-2)$$

$$1x = -6 \quad \text{or} \quad x = -6 \quad \text{Simplify.}$$

Check: Replace x with -6 in the original equation.

$$\frac{x}{3} = -2 \qquad \text{Original equation}$$

$$\frac{-6}{3} \stackrel{?}{=} -2 \qquad \text{Let } x = -6.$$

$$-2 = -2 \qquad \text{True}$$

The solution is -6.

▣ **Work Practice Problem 8**

Answer

8. -28

✔ **Concept Check answers**

a. Addition   **b.** Division

Vocabulary and Readiness Check

Use the choices below to fill in each blank. Some choices may be used more than once.

equation   multiplication   addition

expression   solution   equivalent

1. A combination of operations on variables and numbers is called an _____.
2. A statement of the form "expression = expression" is called an _____.
3. An _____ contains an equals sign (=) while an _____ does not.
4. An _____ may be simplified and evaluated while an _____ may be solved.
5. A _____ of an equation is a number that when substituted for a variable makes the equation a true statement.
6. _____ equations have the same solution.
7. By the _____ property of equality, the same number may be added to or subtracted from both sides of an equation without changing the solution of the equation.
8. By the _____ property of equality, the same nonzero number may be multiplied or divided by both sides of an equation without changing the solution of the equation.

2.6 EXERCISE SET

FOR EXTRA HELP

Student Solutions Manual   PH Math/Tutor Center   CD/Video for Review   Math XL MathXL®   MyMathLab MyMathLab

Objective A *Decide whether the given number is a solution of the given equation. See Example 1.*

1. Is 6 a solution of $x - 8 = -2$?

2. Is 9 a solution of $y - 16 = -7$?

3. Is -5 a solution of $x + 12 = 17$?

4. Is -7 a solution of $a + 23 = -16$?

5. Is -8 a solution of $-9f = 64 - f$?

6. Is -6 a solution of $-3k = 12 - k$?

7. Is 3 a solution of $5(c - 5) = -10$?

8. Is 1 a solution of $2(b - 3) = 10$?

Objective B *Solve. Check each solution. See Examples 2 through 4.*

9. $a + 5 = 23$

10. $f + 4 = -6$

11. $d - 9 = -21$

12. $s - 7 = -15$

13. $7 = y - 2$

14. $1 = y + 7$

15. $11x = 10x - 17$

16. $14z = 13z - 15$

Objective C *Solve. Check each solution. See Examples 5 through 8.*

17. $5x = 20$

18. $6y = 48$

19. $-3z = 12$

20. $-2x = 26$

21. $\dfrac{n}{7} = -2$

22. $\dfrac{n}{11} = -5$

23. $2z = -34$

24. $7y = -21$

25. $-4y = 0$

26. $-9x = 0$

27. $-10x = -10$

28. $-31x = -31$

Objectives B C **Mixed Practice** *Solve. See Examples 2 through 8.*

29. $5x = -35$

30. $3y = -27$

31. $n - 5 = -55$

32. $n - 4 = -48$

33. $-15 = y + 10$

34. $-36 = y + 12$

35. $\dfrac{x}{-6} = -6$

36. $\dfrac{x}{-9} = -9$

37. $11n = 10n + 21$

38. $17z = 16z + 8$

39. $-12y = -144$

40. $-11x = -121$

41. $\dfrac{n}{4} = -20$

42. $\dfrac{n}{5} = -20$

43. $-64 = 32y$

44. $-81 = 27x$

Mixed Review–Translating *Translate each phrase to an algebraic expression. Use x to represent "a number."
See Section 1.8.*

45. A number decreased by -2

46. A number increased by -5

47. The product of -6 and a number

48. The quotient of a number and -20.

49. The sum of -15 and a number

50. -32 multiplied by a number

51. -8 divided by a number

52. Subtract a number from -18.

Concept Extensions

Solve.

53. $n - 42,860 = -1,286$

54. $\dfrac{y}{-18} = 1098$

55. $-38x = 15,542$

56. $n + 961 = 120$

57. Explain the differences between an equation and an expression.

58. Explain the differences between the addition property of equality and the multiplication property of equality.

59. Write an equation that can be solved using the addition property of equality.

60. Write an equation that can be solved using the multiplication property of equality.

CHAPTER 2   Group Activity

Magic Squares
Sections 2.1–2.3

A magic square is a set of numbers arranged in a square table so that the sum of the numbers in each column, row, and diagonal is the same. For instance, in the magic square below, the sum of each column, row, and diagonal is 15. Notice that no number is used more than once in the magic square.

2	9	4
7	5	3
6	1	8

The properties of magic squares have been known for a very long time and once were thought to be good luck charms. The ancient Egyptians and Greeks understood their patterns. A magic square even made it into a famous work of art. The engraving titled *Melencolia I,* created by German artist Albrecht Dürer in 1514, features the following four-by-four magic square on the building behind the central figure.

16	3	2	13
5	10	11	8
9	6	7	12
4	15	14	1

Exercises

1. Verify that what is shown in the Dürer engraving is, in fact, a magic square. What is the common sum of the columns, rows, and diagonals?
2. Negative numbers can also be used in magic squares. Complete the following magic square:

	−1	
0		−4

3. Use the numbers −16, −12, −8, −4, 0, 4, 8, 12, and 16 to form a magic square:

Chapter 2 Vocabulary Check

Fill in each blank with one of the words or phrases listed below.

inequality symbols   addition   solution   is less than   integers

expression   average   negative   absolute value   equation

positive   opposites   is greater than   multiplication

1. Two numbers that are the same distance from 0 on the number line but are on opposite sides of 0 are called _____.

2. The _____ of a number is that number's distance from 0 on a number line.

3. The _____ are ..., $-3, -2, -1, 0, 1, 2, 3, \ldots$.

4. The _____ numbers are numbers less than zero.

5. The _____ numbers are numbers greater than zero.

6. The symbols " $<$ " and " $>$ " are called _____.

7. A _____ of an equation is a number that when substituted for a variable makes the equation a true statement.

8. The _____ of a list of numbers is $\dfrac{\text{sum of numbers}}{\text{number of numbers}}$.

9. A combination of operations on variables and numbers is called an _____.

10. A statement of the form "expression = expression" is called an _____.

11. The sign "$<$" means _____ and "$>$" means _____.

12. By the _____ property of equality, the same number may be added to or subtracted from both sides of an equation without changing the solution of the equation.

13. By the _____ property of equality, the same nonzero number may be multiplied or divided by both sides of an equation without changing the solution of the equation.

> **Helpful Hint**
>
> Are you preparing for your test? Don't forget to take the Chapter 2 Test on page 157. Then check your answers at the back of the text and use the Chapter Test Prep Video CD to see the fully worked-out solutions to any of the exercises you want to review.

2 Chapter Highlights

DEFINITIONS AND CONCEPTS	EXAMPLES
Section 2.1 Introduction to Integers	
Together, positive numbers, negative numbers, and 0 are called **signed numbers**.	$-432, -10, 0, 15$
The **integers** are ..., $-3, -2, -1, 0, 1, 2, 3, \ldots$.	
The **absolute value** of a number is that number's distance from 0 on a number line. The symbol for absolute value is $\lvert\ \rvert$.	$\lvert -2 \rvert = 2$ $\lvert 2 \rvert = 2$

DEFINITIONS AND CONCEPTS	EXAMPLES

Section 2.1   Introduction to Integers (*continued*)

Two numbers that are the same distance from 0 on the number line but are on opposite sides of 0 are called **opposites.**

If a is a number, then $-(-a) = a$.

5 and -5 are opposites.

$-(-11) = 11$. Do not confuse with $-|-3| = {}^-3$

Section 2.2   Adding Integers

ADDING TWO NUMBERS WITH THE SAME SIGN

Step 1.   Add their absolute values.

Step 2.   Use their common sign as the sign of the sum.

ADDING TWO NUMBERS WITH DIFFERENT SIGNS

Step 1.   Find the larger absolute value minus the smaller absolute value.

Step 2.   Use the sign of the number with the larger absolute value as the sign of the sum.

Add:

$$-3 + (-2) = -5$$
$$-7 + (-15) = -22$$

$$-6 + 4 = -2$$
$$17 + (-12) = 5$$
$$-32 + (-2) + 14 = -34 + 14$$
$$= -20$$

Section 2.3   Subtracting Integers

SUBTRACTING TWO NUMBERS

If a and b are numbers, then $a - b = a + (-b)$.

Subtract:

$$-35 - 4 = -35 + (-4) = -39$$
$$3 - 8 = 3 + (-8) = -5$$
$$-10 - (-12) = -10 + 12 = 2$$
$$7 - 20 - 18 - (-3) = 7 + (-20) + (-18) + (+3)$$
$$= -13 + (-18) + 3$$
$$= -31 + 3$$
$$= -28$$

Section 2.4   Multiplying and Dividing Integers

MULTIPLYING NUMBERS

The product of two numbers having the same sign is a positive number.

The product of two numbers having unlike signs is a negative number.

DIVIDING NUMBERS

The quotient of two numbers having the same sign is a positive number.

The quotient of two numbers having unlike signs is a negative number.

Multiply:

$$(-7)(-6) = 42$$
$$9(-4) = -36$$

Evaluate:

$$(-3)^2 = (-3)(-3) = 9$$

Divide:

$$-100 \div (-10) = 10$$

$$\frac{14}{-2} = -7, \quad \frac{0}{-3} = 0, \quad \frac{22}{0} \text{ is undefined.}$$

DEFINITIONS AND CONCEPTS	EXAMPLES

Section 2.5 Order of Operations

ORDER OF OPERATIONS

1. Perform all operations within parentheses ( ), brackets [ ], or other grouping symbols such as fraction bars, starting with the innermost set.
2. Evaluate any expressions with exponents.
3. Multiply or divide in order from left to right.
4. Add or subtract in order from left to right.

Simplify:

$$3 + 2 \cdot (-5) = 3 + (-10)$$
$$= -7$$

$$\frac{-2(5 - 7)}{-7 + |-3|} = \frac{-2(-2)}{-7 + 3}$$
$$= \frac{4}{-4}$$
$$= -1$$

Section 2.6 Solving Equations: The Addition and Multiplication Properties

ADDITION PROPERTY OF EQUALITY

Let a, b, and c represent numbers.

If $a = b$, then

$$a + c = b + c \quad \text{and} \quad a - c = b - c$$

In other words, the same number may be added to or subtracted from both sides of an equation without changing the solution of the equation.

Solve: $\quad x + 8 = 1$

$x + 8 - 8 = 1 - 8$ Subtract 8 from both sides.

$\qquad\qquad x = -7$ Simplify.

The solution is -7.

MULTIPLICATION PROPERTY OF EQUALITY

Let a, b, and c represent numbers and let $c \neq 0$.

If $a = b$, then

$$a \cdot c = b \cdot c \quad \text{and} \quad \frac{a}{c} = \frac{b}{c}$$

In other words, both sides of an equation may be multiplied or divided by the same nonzero number without changing the solution of the equation.

Solve: $\quad -6y = 30$

$\dfrac{-6y}{-6} = \dfrac{30}{-6}$ Divide both sides by -6.

$\qquad y = -5$ Simplify.

The solution is -5.

 STUDY SKILLS BUILDER

Are You Prepared for a Test on Chapter 2?

Below I have listed some *common trouble areas* for students in Chapter 2. After studying for your test—but before taking your test—read these.

- Don't forget the difference between $-(-5)$ and $-|-5|$.

  $-(-5) = 5$ The opposite of -5 is 5.

  $-|-5| = -5$ The opposite of the absolute value of -5 is the opposite of 5, which is -5.

- Remember how to simplify $(-7)^2$ and -7^2.

  $(-7)^2 = (-7)(-7) = 49$

  $-7^2 = -(7)(7) = -49$

- Don't forget order of operations.

  $1 + 3(4 - 6) = 1 + 3(-2)$ Simplify inside parentheses.

  $\qquad\qquad\quad = 1 + (-6)$ Multiply.

  $\qquad\qquad\quad = -5$ Add.

Remember: This is simply a checklist of common trouble spots. For a review of Chapter 2, see the Highlights and Chapter Review at the end of this chapter.

2 CHAPTER REVIEW

(2.1) *Represent each quantity by an integer.*

1. A gold miner is working 1572 feet down in a mine.

2. Mount Hood, in Oregon, has an elevation of 11,239 feet.

Graph each integer in the list on the same number line.

3. $-3, -5, 0, 7$

4. $-6, -1, 0, 5$

Simplify.

5. $|-11|$

6. $|0|$

7. $-|8|$

8. $-(-9)$

9. $-|-16|$

10. $-(-2)$

Insert < or > between each pair of integers to make a true statement.

11. $-18 \quad -20$

12. $-5 \quad 5$

13. $|-123| \quad -|-198|$

14. $8 - |-12| \quad -|-16|$

Find the opposite of each integer.

15. -18

16. 42

Answer true or false for each statement.

17. If $a < b$, then a must be a negative number.

18. The absolute value of an integer is always 0 or a positive number.

19. A negative number is always less than a positive number.

20. If a is a negative number, then $-a$ is a positive number.

Evaluate.

21. $|y|$ if $y = -2$

22. $|-x|$ if $x = -3$

23. $-|-z|$ if $z = -5$

24. $-|-n|$ if $n = -10$

Elevator Shaft Heights and Depths

Elevator shafts in some buildings extend not only above ground, but in many cases below ground to accommodate basements, underground parking, etc. The bar graph shows four such elevators and their shaft distance above and below ground. Use the bar graph to answer Exercises 25 and 26.

25. Which elevator shaft extends the farthest below ground?

26. Which elevator shaft extends the highest above ground?

(2.2) *Add.*

27. $5 + (-3)$

28. $18 + (-4)$

29. $-12 + 16$

30. $-23 + 40$

31. $-8 + (-15)$

32. $-5 + (-17)$

33. $-24 + 3$

34. $-89 + 19$

35. $15 + (-15)$

36. $-24 + 24$

37. $-43 + (-108)$

38. $-100 + (-506)$

39. The temperature at 5 a.m. on a day in January was $-15°$ Celsius. By 6 a.m. the temperature had fallen 5 degrees. Use a signed number to represent the temperature at 6 a.m.

40. A diver starts out at 127 feet below the surface and then swims downward another 23 feet. Use a signed number to represent the diver's current depth.

41. During the 2006 PGA Masters tournament, the winner, Phil Mickelson, has scores of $-2, 0, -2$, and -3 over four rounds of golf. What was his total score for the tournament? (*Source*: Professional Golfer's Association)

42. During the 2006 Women's World Cup of Golf in Sun City, South Africa, the winners, the Swedish team, had a score of -7. The fourth-place finisher, the U.S. team, had a score that was 8 more than the winning score. What was the U.S. team's score in Women's World Cup of Golf? (*Source:* Ladies' Professional Golf Association)

(2.3) *Subtract.*

43. $12 - 4$

44. $-12 - 4$

45. $-7 - 17$

46. $7 - 17$

47. $7 - (-13)$

48. $-6 - (-14)$

49. $16 - 16$

50. $-16 - 16$

51. $-12 - (-12)$

52. $-5 - (-12)$

53. $-(-5) - 12 + (-3)$

54. $-8 + (-12) - 10 - (-3)$

Solve.

55. If the elevation of Lake Superior is 600 feet above sea level and the elevation of the Caspian Sea is 92 feet below sea level, find the difference of the elevations.

56. Josh Weidner has $142 in his checking account. He writes a check for $125, makes a deposit for $43, and then writes another check for $85. Represent the balance in his account by an integer.

57. Some roller coasters travel above and below ground. One such roller coaster is Tremors, located in Silverwood Theme Park, Athol, Idaho. If this coaster rises to a height of 85 feet above ground, then drops 99 feet, how many feet below ground are you at the end of the drop? (*Source:* ultimaterollercoaster.com)

58. Go to the bar graph for Review Exercises 25 and 26 and find the total length of the elevator shaft for Elevator C.

Answer true or false for each statement.

59. $|-5| - |-6| = 5 - 6$

60. $|-5 - (-6)| = 5 + 6$

(2.4) *Multiply.*

61. $-3(-7)$

62. $-6(3)$

63. $-4(16)$

64. $-5(-12)$

65. $(-5)^2$

66. $(-1)^5$

67. $12(-3)(0)$

68. $-1(6)(2)(-2)$

Divide.

69. $-15 \div 3$

70. $\dfrac{-24}{-8}$

71. $\dfrac{0}{-3}$

72. $\dfrac{-46}{0}$

73. $\dfrac{100}{-5}$

74. $\dfrac{-72}{8}$

75. $\dfrac{-38}{-1}$

76. $\dfrac{45}{-9}$

77. A football team lost 5 yards on each of two consecutive plays. Represent the total loss by a product of integers, and find the product.

78. A race horse bettor lost \$50 on each of four consecutive races. Represent the total loss by a product of integers, and find the product.

79. A person has a debt of \$1024 and is ordered to pay it back in four equal payments. Represent the amount of each payment by a quotient of integers, and find the quotient.

80. Overnight, the temperature dropped 45 degrees Fahrenheit. If this took place over a time period of nine hours, represent the average temperature drop each hour by a quotient of integers. Then find the quotient.

(2.5) *Simplify.*

81. $(-7)^2$

82. -7^2

83. $5 - 8 + 3$

84. $-3 + 12 + (-7) - 10$

85. $-10 + 3 \cdot (-2)$

86. $5 - 10 \cdot (-3)$

87. $16 \div (-2) \cdot 4$

88. $-20 \div 5 \cdot 2$

89. $16 + (-3) \cdot 12 \div 4$

90. $(-12) + 10 \div (-5)$

91. $4^3 - (8 - 3)^2$

92. $(-3)^3 - 90$

93. $\dfrac{(-4)(-3) - (-2)(-1)}{-10 + 5}$

94. $\dfrac{4(12 - 18)}{-10 \div (-2 - 3)}$

Find the average of each list of numbers.

95. $-18, 25, -30, 7, 0, -2$

96. $-45, -40, -30, -25$

Evaluate each expression for $x = -2$ and $y = 1$.

97. $2x - y$ **98.** $y^2 + x^2$ **99.** $\dfrac{3x}{6}$ **100.** $\dfrac{5y - x}{-y}$

(2.6) *For Exercises 101 and 102, answer "yes" or "no."*

101. Is -5 a solution of $2n - 6 = 16$?   **102.** Is -2 a solution of $2(c - 8) = -20$?

Solve.

103. $n - 7 = -20$ **104.** $-5 = n + 15$ **105.** $10x = -30$ **106.** $-8x = 72$

107. $9y = 8y - 13$ **108.** $6x - 31 = 7x$ **109.** $\dfrac{n}{-4} = -11$ **110.** $\dfrac{x}{-2} = 13$

111. $n + 12 = -7$ **112.** $n - 40 = -2$ **113.** $-36 = -6x$ **114.** $-40 = 8y$

Mixed Review

Perform the indicated operations.

115. $-6 + (-9)$ **116.** $-16 - 3$ **117.** $-4(-12)$

118. $\dfrac{84}{-4}$ **119.** $-76 - (-97)$ **120.** $-9 + 4$

121. Joe owed his mother \$32. He gave her \$23. Write his financial situation as a signed number.

122. The temperature at noon on a Monday in December was $-11°C$. By noon on Tuesday, it had warmed by $17°C$. What was the temperature at noon on Tuesday?

123. The top of the mountain has an altitude of 12,923 feet. The bottom of the valley is 195 feet below sea level. Find the difference between these two elevations.

124. Wednesday's lowest temperature was $-18°C$. The cold weather continued and by Friday it had dropped another $9°C$. What was the temperature on Friday?

Simplify.

125. $(3 - 7)^2 \div (6 - 4)^3$ **126.** $3(4 + 2) + (-6) - 3^2$

127. $2 - 4 \cdot 3 + 5$ **128.** $4 - 6 \cdot 5 + 1$

129. $\dfrac{-|-14| - 6}{7 + 2(-3)}$ **130.** $5(7 - 6)^3 - 4(2 - 3)^2 + 2^4$

Solve.

131. $n - 9 = -30$ **132.** $n + 18 = 1$

133. $-4x = -48$ **134.** $9x = -81$

135. $\dfrac{n}{-2} = 100$ **136.** $\dfrac{y}{-1} = -3$

2 CHAPTER TEST

Remember to use the Chapter Test Prep Video CD to see the fully worked-out solutions to any of the exercises you want to review.

Simplify each expression.

1. $-5 + 8$

2. $18 - 24$

3. $5 \cdot (-20)$

4. $(-16) \div (-4)$

5. $(-18) + (-12)$

6. $-7 - (-19)$

7. $(-5) \cdot (-13)$

8. $\dfrac{-25}{-5}$

9. $|-25| + (-13)$

10. $14 - |-20|$

11. $|5| \cdot |-10|$

12. $\dfrac{|-10|}{-|-5|}$

13. $(-8) + 9 \div (-3)$

14. $-7 + (-32) - 12 + 5$

15. $(-5)^3 - 24 \div (-3)$

16. $(5 - 9)^2 \cdot (8 - 2)^3$

17. $-(-7)^2 \div 7 \cdot (-4)$

18. $3 - (8 - 2)^3$

1. _____

2. _____

3. _____

4. _____

5. _____

6. _____

7. _____

8. _____

9. _____

10. _____

11. _____

12. _____

13. _____

14. _____

15. _____

16. _____

17. _____

18. _____

19. _____

20. _____

21. _____

22. _____

23. _____

24. _____

25. _____

26. _____

27. _____

28. _____

29. _____

30. _____

31. a. _____

  b. _____

32. _____

33. _____

34. _____

35. _____

19. $\dfrac{4}{2} - \dfrac{8^2}{16}$

20. $\dfrac{-3(-2) + 12}{-1(-4 - 5)}$

21. $\dfrac{|25 - 30|^2}{2(-6) + 7}$

22. $5(-8) - [6 - (2 - 4)] + (12 - 16)^2$

Evaluate each expression for $x = 0$, $y = -3$, *and* $z = 2$.

23. $7x + 3y - 4z$

24. $10 - y^2$

25. $\dfrac{3z}{2y}$

26. Mary Dunstan, a diver, starts at sea level and then makes 4 successive descents of 22 feet. After the descents, what is her elevation?

27. Aaron Hawn has $129 in his checking account. He writes a check for $79, withdraws $40 from an ATM, and then deposits $35. Represent the new balance in his account by an integer.

28. Mt. Washington in New Hampshire has an elevation of 6288 feet above sea level. The Romanche Gap in the Atlantic Ocean has an elevation of 25,354 feet below sea level. Represent the difference in elevation between these two points by an integer.
(*Source:* National Geographic Society and Defense Mapping Agency)

29. Lake Baykal in Siberian Russia is the deepest lake in the world with a maximum depth of 5315 feet. The elevation of the lake's surface is 1495 feet above sea level. What is the elevation (with respect to sea level) of the deepest point in the lake?
(*Source:* U.S. Geological Survey)

30. Find the average of $-12, -13, 0, 9$.

31. Translate the following phrases into mathematical expressions. Use x to represent "a number."
a. The product of a number and 17
b. Twice a number subtracted from 20

Solve.

32. $-9n = -45$

33. $\dfrac{n}{-7} = 4$

34. $x - 16 = -36$

35. $9x = 8x - 4$

Find the place value of the digit 3 in each whole number.

1. 396,418

2. 4308

3. 93,192

4. 693,298

5. 534,275,866

6. 267,301,818

7. Insert $<$ or $>$ to make a true statement.
   a. -7 7
   b. 0 -4
   c. -9 -11

8. Insert $<$ or $>$ to make a true statement
   a. 12 4
   b. 13 31
   c. 82 79

9. Add:
   $13 + 2 + 7 + 8 + 9$

10. Add:
   $11 + 3 + 9 + 16$

11. Subtract: $7826 - 505$
   Check by adding.

12. Subtract: $3285 - 272$
   Check by adding.

13. The radius of Jupiter is 43,441 miles. The radius of Saturn is 7257 miles less than the radius of Jupiter. Find the radius of Saturn.
   (*Source:* National Space Science Data Center)

14. C. J. Dufour wants to buy a digital camera. She has $762 in her savings account. If the camera costs $237, how much money will she have in her account after buying the camera?

15. Round 568 to the nearest ten.

16. Round 568 to the nearest hundred.

17. Round each number to the nearest hundred to find an estimated difference.
   4725
   -2879

18. Round each number to the nearest thousand to find an estimated difference.
   8394
   -2913

19. Rewrite each using the distributive property.
   a. $5(6 + 5)$
   b. $20(4 + 7)$
   c. $2(7 + 9)$

20. Rewrite each using the distributive property.
   a. $5(2 + 12)$
   b. $9(3 + 6)$
   c. $4(8 + 1)$

21. Multiply: 631×125

22. Multiply: 299×104

23. Find each quotient. Check by multiplying.
   a. $42 \div 7$
   b. $\dfrac{64}{8}$
   c. $3\overline{)21}$

24. Find each quotient. Check by multiplying.
   a. $\dfrac{35}{5}$
   b. $64 \div 8$
   c. $4\overline{)48}$

Answers

1. _____
2. _____
3. _____
4. _____
5. _____
6. _____
7. a. _____
   b. _____
   c. _____
8. a. _____
   b. _____
   c. _____
9. _____
10. _____
11. _____
12. _____
13. _____
14. _____
15. _____
16. _____
17. _____
18. _____
19. a. _____
    b. _____
    c. _____
20. a. _____
    b. _____
    c. _____
21. _____
22. _____
23. a. _____
    b. _____
    c. _____
24. a. _____
    b. _____
    c. _____

25. _____

26. _____

27. _____

28. _____

29. _____

30. _____

31. _____

32. _____

33. _____

34. _____

35. _____

36. _____

37. _____

38. _____

39. a. _____

b. _____

c. _____

40. a. _____

b. _____

41. _____

42. _____

43. _____

44. _____

45. _____

46. _____

47. _____

48. _____

49. _____

50. _____

25. Divide: $3705 \div 5$
Check by multiplying.

26. Divide: $3648 \div 8$
Check by multiplying.

27. As part of a promotion, Becky Foster receives 238 cards, each good for one free song download from MSN Music. If Becky wants to share them evenly with 19 friends, how many download cards will each friend receive? How many will be left over?

28. Mrs. Mallory's first grade class is going to the zoo. She pays a total of \$324 for 36 admission tickets. How much did each ticket cost?

Evaluate.

29. 9^2

30. 5^3

31. 6^1

32. 4^1

33. $5 \cdot 6^2$

34. $2^3 \cdot 7$

35. Simplify: $\dfrac{7 - 2 \cdot 3 + 3^2}{5(2 - 1)}$

36. Simplify: $\dfrac{6^2 + 4 \cdot 4 + 2^3}{37 - 5^2}$

37. Evaluate $x + 6$ if x is 8.

38. Evaluate $5 + x$ if x is 9.

39. Simplify:
  a. $|-9|$
  b. $|3|$
  c. $|0|$

40. Simplify:
  a. $|4|$
  b. $|-7|$

41. Add: $-2 + 5$

42. Add: $8 + (-3)$

43. Evaluate $2a - b$ for $a = 8$ and $b = -6$.

44. Evaluate $x - y$ for $x = -2$ and $y = -7$.

45. Multiply: $-7 \cdot 3$

46. Multiply: $5(-2)$

47. Multiply: $0 \cdot (-4)$

48. Multiply: $-6 \cdot 9$

49. Simplify: $3(4 - 7) + (-2) - 5$

50. Simplify: $4 - 8(7 - 3) - (-1)$

3

Solving Equations and Problem Solving

In this chapter, we continue making the transition from arithmetic to algebra. Recall that in algebra, letters (called variables) represent unknown quantities. Using variables is a very powerful method for solving problems that cannot be solved with arithmetic alone. This chapter introduces operations on algebraic expressions, and we continue solving variable equations.

Roads and highways can lead travelers on an exciting journey through every period of U.S. history. The *Santa Fe Trail Scenic and Historic Byway* in Colorado retraces the famous road taken by wagon trains of the Old West. The *Seward Highway* in Alaska is named for the U.S. Secretary of State who arranged the purchase of Alaska from Russia in 1867. The *Selma to Montgomery March Byway* in Alabama traces the historic route taken by Dr. Martin Luther King, Jr. in 1965. The horizontal bar graph below shows the states with the greatest number of roadway miles. Section 3.4, Example 3, Exercises 33, 34, and the Mixed Review Exercises 93, 94, page 207, all have to do with roadway miles or miles between cities.

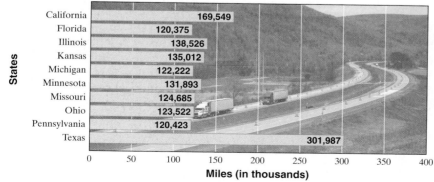

Total Roadway Miles of the Top Ten States

States	Miles (in thousands)
California	169,549
Florida	120,375
Illinois	138,526
Kansas	135,012
Michigan	122,222
Minnesota	131,893
Missouri	124,685
Ohio	123,522
Pennsylvania	120,423
Texas	301,987

0 50 100 150 200 250 300 350 400

Source: U.S. Federal Highway Administration

Objectives

Ⓐ Use Properties of Numbers to Combine Like Terms.

Ⓑ Use Properties of Numbers to Multiply Expressions.

Ⓒ Simplify Expressions by Multiplying and then Combining Like Terms.

Ⓓ Find the Perimeter and Area of Figures.

3.1 SIMPLIFYING ALGEBRAIC EXPRESSIONS

Recall from Section 1.8 that a combination of numbers, letters (variables), and operation symbols is called an **algebraic expression** or simply an **expression.** For example,

Algebraic Expressions

$$4 \cdot x, \quad n + 7, \quad \text{and} \quad 3y - 5 - x$$

are expressions.

Recall that if two variables or a number and a variable are next to each other, with no operation sign between them, the indicated operation is multiplication. For example,

$$3y \quad \text{means} \quad 3 \cdot y$$

and

$$xy \text{ or } x(y) \quad \text{means} \quad x \cdot y$$

Also, the meaning of an exponent remains the same when the base is a variable. For example,

$$y^2 = \underbrace{y \cdot y}_{2 \text{ factors of } y} \qquad \text{and} \qquad x^4 = \underbrace{x \cdot x \cdot x \cdot x}_{4 \text{ factors of } x}$$

Just as we can add, subtract, multiply, and divide numbers, we can add, subtract, multiply, and divide algebraic expressions. In previous sections we evaluated algebraic expressions like $x + 3$, $4x$, and $x + 2y$ for particular values of the variables. In this section, we explore working with variable expressions without evaluating them. We begin with a definition of a term.

Objective Ⓐ Combining Like Terms

The addends of an algebraic expression are called the **terms** of the expression.

$$\underbrace{x + 3}_{2 \text{ terms}}$$

$$\underbrace{3y^2 + (-6y) + 4}_{3 \text{ terms}}$$

A term that is only a number has a special name. It is called a **constant term,** or simply a **constant.** A term that contains a variable is called a **variable term.**

The number factor of a variable term is called the **numerical coefficient.** A numerical coefficient of 1 is usually not written.

Helpful Hint

Recall that $1 \cdot$ any number = that number. This means that

$$1 \cdot x = x \quad \text{or that} \quad 1x = x$$

Thus x can always be replaced by $1x$ or $1 \cdot x$.

162

Terms with the same variable factors, except that they may have different numerical coefficients, are called **like terms.**

Like Terms	Unlike Terms
$3x, -4x$	$5x, x^2$
$-6y, 2y, y$	$7x, 7y$

✔**Concept Check**   True or false? The terms $-7xz^2$ and $3z^2x$ are like terms. Explain.

A sum or difference of like terms can be simplified using the **distributive property.** Recall from Section 1.5 that the distributive property says that multiplication distributes over addition (and subtraction). Using variables, we can write the distributive property as follows:

$$(\overparen{a + b})c = ac + bc.$$

If we write the right side of the equation first, then the left side, we have the following:

Distributive Property

If a, b, and c are numbers, then

$$ac + bc = (a + b)c$$

Also,

$$ac - bc = (a - b)c$$

The distributive property guarantees that, no matter what number x is, $7x + 2x$ (for example) has the same value as $(7 + 2)x$, or $9x$. We then have that

$$7x + 2x = (7 + 2)x = 9x$$

This is an example of **combining like terms.** An algebraic expression is **simplified** when all like terms have been combined.

EXAMPLE 1 Simplify each expression by combining like terms.

a. $4x + 6x$ **b.** $y - 5y$ **c.** $3x^2 + 5x^2 - 2$

Solution: Add or subtract like terms.

a. $4x + 6x = (4 + 6)x$
$\qquad\qquad = 10x$

b. $y - 5y = \overset{\text{Understood 1}}{1y} - 5y$
$\qquad\quad = (1 - 5)y$
$\qquad\quad = -4y$

c. $3x^2 + 5x^2 - 2 = (3 + 5)x^2 - 2$
$\qquad\qquad\qquad = 8x^2 - 2$

🔲 **Work Practice Problem 1**

PRACTICE PROBLEM 1

Simplify each expression by combining like terms.
a. $8m - 14m$
b. $6a + a$
c. $-y^2 + 3y^2 + 7$

Answers
1. a. $-6m$ **b.** $7a$ **c.** $2y^2 + 7$

✔ **Concept Check Answer**
True

Helpful Hint

In this section, we are simplifying expressions. Try not to confuse the two processes below.

Expression: $5y - 8y$	Equation: $8n = -40$
Simplify the expression:	Solve the equation:
$5y - 8y = (5 - 8)y$	$8n = -40$
$\qquad = -3y$	$\dfrac{8n}{8} = \dfrac{-40}{8}$ Divide both sides by 8.
	$n = -5$ The solution is -5.

The commutative and associative properties of addition and multiplication can also help us simplify expressions. We presented these properties in Sections 1.3 and 1.5 and state them again using variables.

Properties of Addition and Multiplication

If a, b, and c are numbers, then

$a + b = b + a$ Commutative property of addition
$a \cdot b = b \cdot a$ Commutative property of multiplication

That is, the **order** of adding or multiplying two numbers can be changed without changing their sum or product.

$(a + b) + c = a + (b + c)$ Associative property of addition
$(a \cdot b) \cdot c = a \cdot (b \cdot c)$ Associative property of multiplication

That is, the **grouping** of numbers in addition or multiplication can be changed without changing their sum or product.

Helpful Hint

• Examples of these properties are

$2 + 3 = 3 + 2$ Commutative property of addition
$7 \cdot 9 = 9 \cdot 7$ Commutative property of multiplication
$(1 + 8) + 10 = 1 + (8 + 10)$ Associative property of addition
$(4 \cdot 2) \cdot 3 = 4 \cdot (2 \cdot 3)$ Associative property of addition

• These properties are not true for subtraction or division.

PRACTICE PROBLEM 2

Simplify: $6z + 5 + z - 4$

EXAMPLE 2   Simplify: $2y - 6 + 4y + 8$

Solution: We begin by writing subtraction as the opposite of addition.

$$
\begin{aligned}
2y - 6 + 4y + 8 &= 2y + (-6) + 4y + 8 && \text{Apply the commutative property} \\
&= 2y + 4y + (-6) + 8 && \text{of addition.} \\
&= (2 + 4)y + (-6) + 8 && \text{Apply the distributive property.} \\
&= 6y + 2 && \text{Simplify.}
\end{aligned}
$$

◼ **Work Practice Problem 2**

Answer
2. $7z + 1$

EXAMPLES Simplify each expression by combining like terms.

3. $6x + 2x - 5 = 8x - 5$

4. $4x + 2 - 5x + 3 = 4x + 2 + (-5x) + 3$
$$= 4x + (-5x) + 2 + 3$$
$$= -1x + 5 \quad \text{or} \quad -x + 5$$

5. $2x - 5 + 3y + 4x - 10y + 11$
$$= 2x + (-5) + 3y + 4x + (-10y) + 11$$
$$= 2x + 4x + 3y + (-10y) + (-5) + 11$$
$$= 6x - 7y + 6$$

Work Practice Problems 3–5

As we practice combining like terms, keep in mind that some of the steps may be performed mentally.

Objective **B** Multiplying Expressions

We can also use properties of numbers to multiply expressions such as $3(2x)$. By the associative property of multiplication, we can write the product $3(2x)$ as $(3 \cdot 2)x$, which simplifies to $6x$.

EXAMPLES Multiply.

6. $5(3y) = (5 \cdot 3)y$ Apply the associative property of multiplication.
$$= 15y$$ Multiply.

7. $-2(4x) = (-2 \cdot 4)x$ Apply the associative property of multiplication.
$$= -8x$$ Multiply.

Work Practice Problems 6–7

We can use the distributive property to combine like terms, which we have done, and also to multiply expressions such as $2(3 + x)$. By the distributive property, we have

$$2(3 + x) = 2 \cdot 3 + 2 \cdot x \quad \text{Apply the distributive property.}$$
$$= 6 + 2x \quad \text{Multiply.}$$

EXAMPLE 8 Use the distributive property to multiply: $6(x + 4)$

Solution: By the distributive property,

$$5(x + 4) = 5 \cdot x + 5 \cdot 4 \quad \text{Apply the distributive property.}$$
$$= 5x + 20 \quad \text{Multiply.}$$

Work Practice Problem 8

✔ **Concept Check**   What's wrong with the following?

$8(a - b) = 8a - b$

PRACTICE PROBLEMS 3–5

Simplify each expression by combining like terms.

3. $6y + 12y - 6$

4. $7y - 5 + y + 8$

5. $-7y + 2 - 2y - 9x + 12 - x$

PRACTICE PROBLEMS 6–7

Multiply.

6. $6(4a)$

7. $-8(9x)$

PRACTICE PROBLEM 8

Use the distributive property to multiply: $8(y + 2)$

Answers

3. $18y - 6$ **4.** $8y + 3$
5. $-9y - 10x + 14$
6. $24a$ **7.** $-72x$ **8.** $8y + 16$

✔ **Concept Check Answer**

Did not distribute the 8;
$8(a - b) = 8a - 8b$

PRACTICE PROBLEM 9

Multiply: $3(7a - 5)$

EXAMPLE 9 Multiply: $-3(5a + 2)$

Solution: By the distributive property,

$$-3(5a + 2) = -3(5a) + (-3)(2) \qquad \text{Apply the distributive property.}$$
$$= (-3 \cdot 5)a + (-6) \qquad \text{Use the associative property and multiply.}$$
$$= -15a - 6 \qquad \text{Multiply.}$$

◼ **Work Practice Problem 9**

PRACTICE PROBLEM 10

Multiply: $6(5 - y)$

EXAMPLE 10 Multiply: $8(x - 4)$

Solution:

$$8(x - 4) = 8 \cdot x - 8 \cdot 4$$
$$= 8x - 32$$

◼ **Work Practice Problem 10**

Objective C Simplifying Expressions

Next we will **simplify** expressions by first using the distributive property to multiply and then **combining** any like terms.

PRACTICE PROBLEM 11

Simplify: $5(2y - 3) - 8$

Helpful Hint
2 is *not* distributed to the -15 since it is not within the parentheses.

EXAMPLE 11 Simplify: $2(3 + 7x) - 15$

Solution: First we use the distributive property to remove parentheses.

$$2(3 + 7x) - 15 = 2(3) + 2(7x) - 15 \qquad \text{Apply the distributive property.}$$
$$= 6 + 14x - 15 \qquad \text{Multiply.}$$
$$= 14x + (-9) \quad \text{or} \quad 14x - 9 \qquad \text{Combine like terms.}$$

◼ **Work Practice Problem 11**

PRACTICE PROBLEM 12

Simplify:
$-7(x - 1) + 5(2x + 3)$

EXAMPLE 12 Simplify: $-2(x - 5) + 4(2x + 2)$

Solution: First we use the distributive property to remove parentheses.

$$-2(x - 5) + 4(2x + 2) = -2(x) - (-2)(5) + 4(2x) + 4(2) \qquad \text{Apply the distributive property.}$$
$$= -2x + 10 + 8x + 8 \qquad \text{Multiply.}$$
$$= 6x + 18 \qquad \text{Combine like terms.}$$

◼ **Work Practice Problem 12**

PRACTICE PROBLEM 13

Simplify:
$-(y + 1) + 3y - 12$

EXAMPLE 13 Simplify: $-(x + 4) + 5x + 16$

Solution: The expression $-(x + 4)$ means $-1(x + 4)$.

$$-(x + 4) + 5x + 16 = -1(x + 4) + 5x + 16 \qquad \text{Apply the distributive property.}$$
$$= -1 \cdot x + (-1)(4) + 5x + 16 \qquad \text{Multiply.}$$
$$= -x + (-4) + 5x + 16 \qquad \text{Combine like terms.}$$
$$= 4x + 12$$

◼ **Work Practice Problem 13**

Answers

9. $21a - 15$ **10.** $30 - 6y$
11. $10y - 23$ **12.** $3x + 22$
13. $2y - 13$

Objective D Finding Perimeter and Area

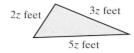 **EXAMPLE 14** Find the perimeter of the triangle.

2z feet   3z feet   5z feet

Solution: Recall that the perimeter of a figure is the distance around the figure. To find the perimeter, then, we find the sum of the lengths of the sides. We use the letter P to represent perimeter.

$$P = 2z + 3z + 5z$$
$$= 10z$$

> **Helpful Hint**
> Don't forget to insert proper units.

The perimeter is $10z$ feet.

■ **Work Practice Problem 14**

EXAMPLE 15 **Finding the Area of a Deck**

Find the area of the rectangular deck.

(2x − 7) meters

5 meters

Solution: Recall how to find the area of a rectangle. **A**rea = **L**ength · **W**idth, or if A represents area, l represents length, and w represents width, we have $A = l \cdot w$.

$$A = l \cdot w$$

$$= 5(2x - 7) \quad \text{Let length} = 5 \text{ and width} = (2x - 7).$$
$$= 10x - 35 \quad \text{Multiply.}$$

The area is $(10x - 35)$ *square* meters.

■ **Work Practice Problem 15**

> **Helpful Hint**
> Don't forget . . .
>
> | Area: | Perimeter: |
> |---|---|
> | • surface enclosed | • distance around |
> | • measured in square units | • measured in units |

Vocabulary and Readiness Check

Use the choices below to fill in each blank. Some choices may be used more than once.

numerical coefficient	combine like terms	like	term	variable	associative
constant	expression	unlike	distributive	commutative	

1. $14y^2 + 2x - 23$ is called a(n) _____ while $14y^2$, $2x$, and -23 are each called a(n) _____.

2. To multiply $3(-7x + 1)$, we use the _____ property.

3. To simplify an expression like $y + 7y$, we _____.

4. By the _____ properties, the *order* of adding or multiplying two numbers can be changed without changing their sum or product.

5. The term $5x$ is called a(n) _____ term while the term 7 is called a(n) _____ term.

6. The term z has an understood _____ of 1.

7. By the _____ properties, the grouping of adding or multiplying numbers can be changed without changing their sum or product.

8. The terms $-x$ and $5x$ are _____ terms and the terms $5x$ and $5y$ are _____ terms.

9. For the term $-3x^2y$, -3 is called the _____.

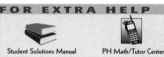

3.1 EXERCISE SET

FOR EXTRA HELP

Student Solutions Manual   PH Math/Tutor Center   CD/Video for Review   Math XL MathXL®   MyMathLab MyMathLab

Objective A *Simplify each expression by combining like terms. See Examples 1 through 5.*

1. $3x + 5x$

2. $8y + 3y$

3. $2n - 3n$

4. $7z - 10z$

5. $4c + c - 7c$

6. $5b - 8b - b$

7. $4x - 6x + x - 5x$

8. $8y + y - 2y - 8y$

9. $3a + 2a + 7a - 5$

10. $5b - 4b + b - 15$

Objective B *Multiply. See Examples 6 and 7.*

11. $6(7x)$

12. $4(4x)$

13. $-3(11y)$

14. $-3(21z)$

15. $12(6a)$

16. $13(5b)$

Multiply. See Examples 8 through 10.

17. $2(y + 3)$

18. $3(x + 1)$

19. $3(a - 6)$

20. $4(y - 6)$

21. $-4(3x + 7)$

22. $-8(8y + 10)$

Objective C *Simplify each expression. First use the distributive property to multiply and remove parentheses. See Examples 11 through 13.*

23. $2(x + 4) - 7$

24. $5(6 - y) - 2$

25. $8 + 5(3c - 1)$

26. $10 + 4(6d - 2)$

27. $-4(6n - 5) + 3n$

28. $-3(5 - 2b) - 4b$

29. $3 + 6(w + 2) + w$

30. $8z + 5(6 + z) + 20$

31. $2(3x + 1) + 5(x - 2)$

32. $3(5x - 2) + 2(3x + 1)$

33. $-(2y - 6) + 10$

34. $-(5x - 1) - 10$

Objectives A B C **Mixed Practice** *Simplify each expression. See Examples 1 through 13.*

35. $18y - 20y$

36. $x + 12x$

37. $z - 8z$

38. $-12x + 8x$

39. $9d - 3c - d$

40. $8r + s - 7s$

41. $2y - 6 + 4y - 8$

42. $a + 4 - 7a - 5$

43. $5q + p - 6q - p$

44. $m - 8n + m + 8n$

45. $2(x + 1) + 20$

46. $5(x - 1) + 18$

47. $5(x - 7) - 8x$

48. $3(x + 2) - 11x$

49. $-5(z + 3) + 2z$

50. $-8(1 + v) + 6v$

51. $8 - x + 4x - 2 - 9x$

52. $5y - 4 + 9y - y + 15$

53. $-7(x + 5) + 5(2x + 1)$

54. $-2(x + 4) + 8(3x - 1)$

55. $3r - 5r + 8 + r$

56. $6x - 4 + 2x - x + 3$

57. $-3(n - 1) - 4n$

58. $5(c + 2) + 7c$

59. $4(z - 3) + 5z - 2$

60. $8(m + 3) - 20 + m$

61. $6(2x - 1) - 12x$

62. $5(2a + 3) - 10a$

63. $-(4x - 5) + 5$

64. $-(7y - 2) + 6$

65. $-(4xy - 10) + 2(3xy + 5)$

66. $-(12ab - 10) + 5(3ab - 2)$

67. $3a + 4(a + 3)$

68. $b + 2(b - 5)$

69. $5y - 2(y - 1) + 3$

70. $3x - 4(x + 2) + 1$

Objective **D** *Find the perimeter of each figure. See Example 14.*

△ **71.**   2y meters   6 meters
        4y meters
    3y meters   5y meters
           16 meters

△ **72.**   3x feet   x feet
                     7 feet
    5x feet   4x feet
           12 feet

△ **73.**   2a feet 2a feet
        2a feet   2a feet
    6 feet   6 feet
           5a feet

△ **74.**   3z meters
    1 meter   1 meter
           5z meters

△ **75.**   Each side:
             $(-5x + 11)$
             inches

76.   Each side:
             $(9y + 1)$
             kilometers

Find the area of each rectangle. See Example 15.

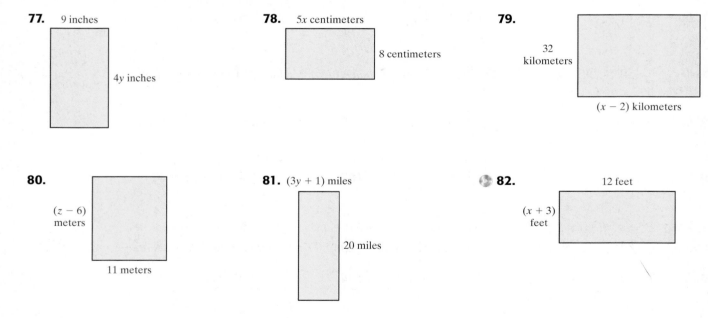

77. 9 inches
         4y inches

78. 5x centimeters
         8 centimeters

79. 32 kilometers
         $(x - 2)$ kilometers

80. $(z - 6)$ meters
         11 meters

81. $(3y + 1)$ miles
         20 miles

82. 12 feet
         $(x + 3)$ feet

Objectives **A** **B** **C** **D** **Mixed Practice** *Solve. See Examples 1 through 15.*

83. Find the area of a rectangular movie screen that is 50 feet long and 40 feet high.

84. Find the area of a 60-meter by 25-meter rectangular swimming pool.

85. A decorator wishes to put a wallpaper border around a rectangular room that measures 14 feet by 18 feet. Find the room's perimeter.

86. How much fencing will a rancher need for a rectangular cattle lot that measures 80 feet by 120 feet?

87. Find the perimeter of a triangular garden that measures 5 feet by x feet by $(2x + 1)$ feet.

88. Find the perimeter of a triangular picture frame that measures x inches by x inches by $(x - 14)$ inches.

Review

Perform each indicated operation. See Sections 2.2 and 2.3.

89. $-13 + 10$

90. $-7 - (-4)$

91. $-4 - (-12)$

92. $-15 + 23$

93. $-4 + 4$

94. $8 + (-8)$

Concept Extensions

If the expression on the left side of the equal sign is equivalent to the right, write "correct." If not, write "incorrect" and then write an expression that is equivalent to the left side. See the Concept Check in Objective B.

95. $5(3x - 2) \stackrel{?}{=} 15x - 2$

96. $2(xy) \stackrel{?}{=} 2x \cdot 2y$

97. $7x - (x + 2) \stackrel{?}{=} 7x - x - 2$

98. $4(y - 3) + 11 \stackrel{?}{=} 4y - 6 + 11$

Review commutative, associative, and distributive properties. Then identify which property allows us to write the equivalent expression on the right side of the equal sign.

99. $6(2x - 3) + 5 = 12x - 18 + 5$

100. $9 + 7x + (-2) = 7x + 9 + (-2)$

101. $-7 + (4 + y) = (-7 + 4) + y$

102. $(x + y) + 11 = 11 + (x + y)$

Write the expression that represents the area of each composite figure. Then simplify to find the total area.

△ **103.**

△ **104.**

Simplify.

▦ **105.** $9684q - 686 - 4860q + 12{,}960$

▦ **106.** $76(268x + 592) - 2960$

107. If x is a whole number, which expression is the greatest: $2x$ or $5x$? Explain your answer.

108. If x is a whole number, which expression is the greatest: $-2x$ or $-5x$? Explain your answer.

109. Explain what makes two terms "like terms."

110. Explain how to combine like terms.

 STUDY SKILLS BUILDER

Are You Familiar with Your Textbook Supplements?

There are many student supplements available for additional study. Below, I have listed some of these. See the Preface of this text or your instructor for further information.

Chapter Test Prep Video CD. This material is found in your textbook and is fully explained there. The CD contains video clip solutions to the Chapter Test exercises in this text and are excellent help when studying for chapter tests.

Lecture Video CDs. These video segments are keyed to each sections of the text. The material is presented by me, Elayn Martin-Gay, and I have placed a video icon by the exercises in the text that I have worked on the video.

The Student Solutions Manual. This contains worked out solutions to odd-numbered exercises as well as every exercise in the Integrated Reviews, Chapter Reviews, Chapter Tests, and Cumulative Reviews.

Prentice Hall Tutor Center. Mathematics questions may be phoned, faxed, or emailed to this center.

MyMathLab, MathXL, and Interact Math. These are computer and Internet tutorials. This supplement may already be available to you somewhere on campus, for example at your local learning resource lab. Take a moment and find the name and location of any such lab on campus.

As usual, your instructor is your best source of information.

Let's see how you are doing with textbook supplements:

1. Name one way the Chapter Test Prep Video can help you prepare for a chapter test.

2. List any textbook supplements that you have found useful.

3. Have you located and visited a learning resource lab located on your campus?

4. List the textbook supplements that are currently housed in your campus's learning resource lab.

3.2 SOLVING EQUATIONS: REVIEW OF THE ADDITION AND MULTIPLICATION PROPERTIES

Objectives

A Use the Addition Property or the Multiplication Property to Solve Equations.

B Use Both Properties to Solve Equations.

C Translate Word Phrases to Mathematical Expressions.

Objective A Using the Addition Property or the Multiplication Property

In this section, we continue solving equations using the properties first introduced in Section 2.6.

First, let's recall the difference between an **expression** and an **equation**. Remember—an equation contains an equals sign and an expression does not.

	Equations	Expressions	
equals signs	$7x = 6x + 4$ $3(3y - 5) = 10y$	$7x - 6x + 4$ $y - 1 + 11y - 21$	no equals signs

Thus far in this text, we have

Solved some equations (Sections 1.8 and 2.6)

and

Simplified some expressions (Section 3.1)

As we will see in this section, the ability to simplify expressions will help us as we solve more equations.

The addition and multiplication properties are reviewed below.

Addition Property of Equality

Let a, b, and c represent numbers. Then

$$a = b$$
and $a + c = b + c$
are equivalent equations.

Also, $a = b$
and $a - c = b - c$
are equivalent equations.

Multiplication Property of Equality

Let a, b, and c represent numbers and let $c \neq 0$. Then

$$a = b$$
and $a \cdot c = b \cdot c$
are equivalent equations.

Also, $a = b$
and $\dfrac{a}{c} = \dfrac{b}{c}$
are equivalent equations.

In other words, the same number may be added to or subtracted from *both* sides of an equation without changing the solution of the equation. Also, *both* sides of an equation may be multiplied or divided by the same nonzero number without changing the solution of the equation.

Many equations in this section will contain expressions that can be simplified. If one or both sides of an equation can be simplified, do that first.

PRACTICE PROBLEM 1

Solve: $x + 6 = 1 - 3$

EXAMPLE 1 Solve: $y - 5 = -2 - 6$

Solution: First we simplify the right side of the equation.

$y - 5 = -2 - 6$
$y - 5 = -8$ Combine like terms.

Next we get y alone by using the addition property of equality. We add 5 to both sides of the equation.

$y - 5 + 5 = -8 + 5$ Add 5 to both sides.
$y = -3$ Simplify.

Check: To see that -3 is the solution, replace y with -3 in the original equation.

$y - 5 = -2 - 6$
$-3 - 5 \stackrel{?}{=} -2 - 6$ Replace y with -3.
$-8 = -8$ True

Since $-8 = -8$ is true, the solution is -3.

■ **Work Practice Problem 1**

PRACTICE PROBLEM 2

Solve: $10 = 2m - 4m$

EXAMPLE 2 Solve: $3y - 7y = 12$

Solution: First, simplify the left side of the equation by combining like terms.

$3y - 7y = 12$
$-4y = 12$ Combine like terms.

Next, we get y alone by using the multiplication property of equality and dividing both sides by -4.

$\dfrac{-4y}{-4} = \dfrac{12}{-4}$ Divide both sides by -4.
$y = -3$ Simplify.

Check: Replace y with -3 in the original equation.

$3y - 7y = 12$
$3(-3) - 7(-3) \stackrel{?}{=} 12$
$-9 + 21 \stackrel{?}{=} 12$
$12 = 12$ True

The solution is -3.

■ **Work Practice Problem 2**

PRACTICE PROBLEM 3

Solve: $-8 + 6 = \dfrac{a}{3}$

EXAMPLE 3 Solve: $\dfrac{z}{-4} = 11 - 5$

Solution: Simplify the right side of the equation first.

$\dfrac{z}{-4} = 11 - 5$
$\dfrac{z}{-4} = 6$

Copyright 2008 Pearson Education Inc

Next, to get z alone, multiply both sides by -4.

$$-4 \cdot \frac{z}{-4} = -4 \cdot 6 \qquad \text{Multiply both sides by } -4.$$

$$\frac{-4}{-4} \cdot z = -4 \cdot 6$$

$$1z = -24 \quad \text{or} \quad z = -24$$

Check to see that -24 is the solution.

Work Practice Problem 3

EXAMPLE 4 Solve: $5x + 2 - 4x = 7 - 19$

Solution: First we simplify each side of the equation separately.

$$5x + 2 - 4x = 7 - 19$$
$$\underbrace{5x - 4x} + 2 = \underbrace{7 - 19}$$
$$1x + 2 = -12 \qquad \text{Combine like terms.}$$

To get x alone on the left side, we subtract 2 from both sides.

$$1x + 2 - 2 = -12 - 2 \qquad \text{Subtract 2 from both sides.}$$
$$1x = -14 \quad \text{or} \quad x = -14 \qquad \text{Simplify.}$$

Check to see that -14 is the solution.

Work Practice Problem 4

EXAMPLE 5 Solve: $8x - 9x = 12 - 17$

Solution: First combine like terms on each side of the equation.

$$8x - 9x = 12 - 17$$
$$-x = -5$$

Recall that $-x$ means $-1x$ and divide both sides by -1.

$$\frac{-1x}{-1} = \frac{-5}{-1} \qquad \text{Divide both sides by } -1.$$
$$x = 5 \qquad \text{Simplify.}$$

Check to see that the solution is 5.

Work Practice Problem 5

EXAMPLE 6 Solve: $3(3x - 5) = 10x$

Solution: First we multiply on the left side to remove the parentheses.

$$3(3x - 5) = 10x$$
$$3 \cdot 3x - 3 \cdot 5 = 10x \qquad \text{Use the distributive property.}$$
$$9x - 15 = 10x$$

Now we subtract $9x$ from both sides.

$$9x - 15 - 9x = 10x - 9x \qquad \text{Subtract } 9x \text{ from both sides.}$$
$$-15 = 1x \quad \text{or} \quad x = -15 \qquad \text{Simplify.}$$

Work Practice Problem 6

PRACTICE PROBLEM 4
Solve:
$-6y - 1 + 7y = 17 + 2$

PRACTICE PROBLEM 5
Solve: $-4 - 10 = 4y - 5y$

PRACTICE PROBLEM 6
Solve: $13x = 4(3x - 1)$

Answers
4. 20 **5.** 14 **6.** -4

Objective B Using Both Properties to Solve Equations

We now solve equations in one variable using more than one property of equality. To solve an equation such as $2x - 6 = 18$, we first get the variable term $2x$ alone on one side of the equation.

PRACTICE PROBLEM 7

Solve: $5y + 2 = 17$

EXAMPLE 7 Solve: $2x - 6 = 18$

Solution: We start by adding 6 to both sides to get the variable term $2x$ alone.

$$2x - 6 = 18$$
$$2x - 6 + 6 = 18 + 6 \quad \text{Add 6 to both sides.}$$
$$2x = 24 \quad \text{Simplify.}$$

To finish solving, we divide both sides by 2.

$$\frac{2x}{2} = \frac{24}{2} \quad \text{Divide both sides by 2.}$$
$$x = 12 \quad \text{Simplify.}$$

Helpful Hint Don't forget to check the proposed solution in the *original* equation.

Check:
$$2x - 6 = 18$$
$$2(12) - 6 \stackrel{?}{=} 18 \quad \text{Replace } x \text{ with 12 and simplify.}$$
$$24 - 6 \stackrel{?}{=} 18$$
$$18 = 18 \quad \text{True}$$

The solution is 12.

◻ **Work Practice Problem 7**

Don't forget, if one or both sides of an equation can be simplified, do that first.

PRACTICE PROBLEM 8

Solve:
$-4(x + 2) - 60 = 2 - 10$

EXAMPLE 8 Solve: $2 - 6 = -5(x + 4) - 39$

Solution: First, simplify each side of the equation.

$$2 - 6 = -5(x + 4) - 39$$
$$2 - 6 = -5x - 20 - 39 \quad \text{Use the distributive property.}$$
$$-4 = -5x - 59 \quad \text{Combine like terms on each side.}$$
$$-4 + 59 = -5x - 59 + 59 \quad \text{Add 59 to both sides to get the variable term alone.}$$
$$55 = -5x \quad \text{Simplify.}$$
$$\frac{55}{-5} = \frac{-5x}{-5} \quad \text{Divide both sides by } -5.$$
$$-11 = x \quad \text{or} \quad x = -11 \quad \text{Simplify.}$$

Check to see that -11 is the solution.

◻ **Work Practice Problem 8**

Answers
7. 3   8. -15

Objective ⓒ Translating Word Phrases into Expressions

Section 3.4 in this chapter contains a formal introduction to problem solving. To prepare for this section, let's once again review writing phrases as algebraic expressions using the following key words and phrases as a guide:

Addition	Subtraction	Multiplication	Division
sum	difference	product	quotient
plus	minus	times	divided
added to	subtracted from	multiply	shared equally among
more than	less than	twice	per
increased by	decreased by	of	divided by
total	less	twice/double/triple	divided into

EXAMPLE 9 Write each phrase as an algebraic expression. Use x to represent "a number."

a. a number increased by -5
b. the product of -7 and a number
c. a number less 20
d. the quotient of -18 and a number
e. a number subtracted from -2

Solution:

a. In words: a number   increased by -5

   Translate: x $+$ (-5) or $x - 5$

b. In words: the product of

   -7 and   a number

   Translate: -7 $\cdot$ x or $-7x$

c. In words: a number   less   20

   Translate: x $-$ 20

d. In words: the quotient of

   -18 and   a number

   Translate: -18 $\div$ x or $\dfrac{-18}{x}$ or $-\dfrac{18}{x}$

e. In words: a number   subtracted from -2

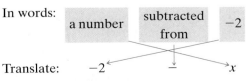

   Translate: -2 $-$ x

▣ **Work Practice Problem 9**

PRACTICE PROBLEM 9

Write each phrase as an algebraic expression. Use x to represent "a number."

a. the sum of -3 and a number
b. -5 decreased by a number
c. three times a number
d. a number subtracted from 83
e. the quotient of a number and -4

Answers
9. a. $-3 + x$ b. $-5 - x$ c. $3x$
d. $83 - x$ e. $\dfrac{x}{-4}$ or $-\dfrac{x}{4}$

Helpful Hint

As we reviewed in Chapter 1, don't forget that order is important when subtracting. Notice the translation order of numbers and variables below.

Phrase	Translation
a number less 9	$x - 9$
a number subtracted from 9	$9 - x$

PRACTICE PROBLEM 10

Translate each phrase into an algebraic expression. Let x be the unknown number.

a. The product of 5 and a number, decreased by 25

b. Twice the sum of a number and 3

c. The quotient of 39 and twice a number

EXAMPLE 10 Write each phrase as an algebraic expression. Let x be the unknown number.

a. Twice a number, increased by -9

b. Three times the difference of a number and 11

c. The quotient of 5 times a number and 17

Solution:

a. In words:

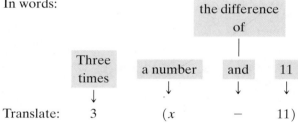

Twice a number	increased by	-9
↓	↓	↓

Translate: $2x$ $+$ (-9) or $2x - 9$

b. In words:

	the difference of	
Three times	a number and 11	
↓	↓ ↓ ↓	

Translate: 3 $(x$ $-$ $11)$

c. In words:

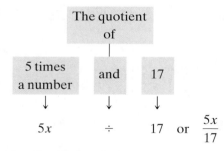

The quotient of		
5 times a number	and	17
↓	↓	↓

      $5x$ $\div$ 17   or $\dfrac{5x}{17}$

▨ **Work Practice Problem 10**

Answers

10. a. $5x - 25$ **b.** $2(x + 3)$

c. $39 \div 2x$ or $\dfrac{39}{2x}$

Vocabulary and Readiness Check

Use the choices below to fill in each blank.

equation	multiplication	equivalent	expression
solving	addition	simplifying	

1. The equations $-3x = 51$ and $\dfrac{-3x}{-3} = \dfrac{51}{-3}$ are called _____ equations.

2. The difference between an equation and an expression is that a(n) _____ contains an equal sign, while an _____ does not.

3. The process of writing $-3x + 10x$ as $7x$ is called _____ the expression.

4. For the equation $-5x - 1 = -21$, the process of finding that 4 is the solution is called _____ the equation.

5. By the _____ property of equality, $x = -2$ and $x + 7 = -2 + 7$ are equivalent equations.

6. By the _____ property of equality, $y = 8$ and $3 \cdot y = 3 \cdot 8$ are equivalent equations.

3.2 EXERCISE SET

FOR EXTRA HELP

Student Solutions Manual   PH Math/Tutor Center   CD/Video for Review   MathXL   MyMathLab

Objective A *Solve each equation. First combine any like terms on each side of the equation. See Examples 1 through 5.*

1. $x - 3 = -1 + 4$

2. $x + 7 = 2 + 3$

3. $-7 + 10 = m - 5$

4. $1 - 8 = n + 2$

5. $2w - 12w = 40$

6. $10y - y = 45$

7. $24 = t + 3t$

8. $100 = 15y + 5y$

9. $2z = 12 - 14$

10. $-3x = 11 - 2$

11. $4 - 10 = \dfrac{z}{-3}$

12. $20 - 22 = \dfrac{z}{-4}$

13. $-3x - 3x = 50 - 2$

14. $5y - 9y = -14 + (-14)$

15. $\dfrac{x}{5} = -26 + 16$

16. $\dfrac{y}{3} = 32 - 52$

17. $7x + 7 - 6x = 10$

18. $-3 + 5x - 4x = 13$

19. $-8 - 9 = 3x + 5 - 2x$

20. $-7 + 10 = 4x - 6 - 3x$

Solve. First multiply to remove parentheses. See Example 6.

21. $2(5x - 3) = 11x$

22. $6(3x + 1) = 19x$

23. $3y = 2(y + 12)$

24. $17x = 4(4x - 6)$

25. $21y = 5(4y - 6)$

26. $28z = 9(3z - 2)$

27. $-3(-4 - 2z) = 3z$

28. $-2(-1 - 3y) = 7y$

Objective B *Solve each equation. See Examples 7 and 8.*

29. $2x - 8 = 0$

30. $3y - 12 = 0$

31. $7y + 3 = 24$

32. $5m + 1 = 46$

33. $-7 = 2x - 1$

34. $-11 = 3t - 2$

35. $6(6y - 4) = 12y$

36. $4(3y - 5) = 14y$

37. $11(x - 6) = -4 - 7$

38. $5(x - 6) = -2 - 8$

39. $-3(x + 1) - 10 = 12 + 8$

40. $-2(x + 5) - 2 = -8 - 4$

41. $y - 20 = 6y$

42. $x - 63 = 10x$

43. $22 - 42 = 4(x - 1) - 4$

44. $35 - (-3) = 3(x - 2) + 17$

Objectives Ⓐ Ⓑ **Mixed Practice** *Solve each equation. See Examples 1 through 8.*

45. $-2 - 3 = -4 + x$

46. $7 - (-10) = x - 5$

47. $y + 1 = -3 + 4$

48. $y - 8 = -5 - 1$

49. $3w - 12w = -27$

50. $y - 6y = 20$

51. $-4x = 20 - (-4)$

52. $6x = 5 - 35$

53. $18 - 11 = \dfrac{x}{-5}$

54. $9 - 14 = \dfrac{x}{-12}$

55. $9x - 12 = 78$

56. $8x - 8 = 32$

57. $10 = 7t - 12t$

58. $-30 = t + 9t$

59. $5 - 5 = 3x + 2x$

60. $-42 + 20 = -2x + 13x$

61. $50y = 7(7y + 4)$

62. $65y = 8(8y - 9)$

63. $8x = 2(6x + 10)$

64. $10x = 6(2x - 3)$

65. $7x + 14 - 6x = -4 - 10$

66. $-10x + 11x + 5 = 9 - 5$

67. $\dfrac{x}{-4} = -1 - (-8)$

68. $\dfrac{y}{-6} = 6 - (-1)$

69. $23x + 8 - 25x = 7 - 9$

70. $8x - 4 - 6x = 12 - 22$

71. $-3(x + 9) - 41 = 4 - 60$

72. $-4(x + 7) - 30 = 3 - 37$

Objective Ⓒ **Translating** *Write each phrase as a variable expression. Use x to represent "a number."*
See Examples 9 and 10.

73. The sum of -7 and a number

74. Negative eight plus a number

75. Eleven subtracted from a number

76. A number subtracted from twelve

77. The product of -13 and a number

78. Twice a number

79. A number divided by -12

80. The quotient of negative six and a number

81. The product of -11 and a number, increased by 5

82. The quotient of -20 and a number, decreased by three

83. Negative ten decreased by 7 times a number

84. Twice a number decreased by thirty

85. Seven added to the product of 4 and a number

86. The product of 7 and a number, added to 100

87. Twice a number, decreased by 17

88. A number decreased by 14 and increased by 5 times the number

89. The product of -6 and the sum of a number and 15

90. Twice the sum of a number and -5

91. The quotient of 45 and the product of a number and -5

92. The quotient of ten times a number, and -4

93. The quotient of seventeen and a number, increased by -15

94. The difference of -9 times a number, and 1

Review

The trumpeter swan is the largest waterfowl in the United States. Although it was thought to be nearly extinct at the beginning of the twentieth century, recent conservation efforts have been succeeding. Use the bar graph to answer Exercises 95 through 98. See Section 1.3.

95. During what year shown is the number of trumpeter swans the greatest?

96. During what year shown is the number of trumpeter swans the least?

97. Use the length of the bar to estimate the number of trumpeter swans in 2005.

98. Describe any trends shown in this graph.

Trumpeter Swan Population in North America

Source: U.S. Fish and Wildlife Service

Concept Extensions

99. In your own words, explain the addition property of equality.

100. Write an equation that can be solved using the addition property of equality.

101. Are the equations below equivalent? Why or why not?

$$x + 7 = 4 + (-9)$$
$$x + 7 = 5$$

102. Write 2 equivalent equations.

103. Why does the multiplication property of equality not allow us to divide both sides of an equation by zero?

104. Is the equation $-x = 6$ solved for the variable? Explain why or why not.

Solve.

105. $\dfrac{y}{72} = -86 - (-1029)$

106. $\dfrac{x}{-13} = 4^6 - 5^7$

107. $\dfrac{x}{-2} = 5^2 - |-10| - (-9)$

108. $\dfrac{y}{10} = (-8)^2 - |20| + (-2)^2$

109. $|-13| + 3^2 = 100y - |-20| - 99y$

110. $4(x - 11) + |90| - |-86| + 2^5 = 5x$

Expressions and Equations

Answers

7. _____

8. _____

9. _____

10. _____

11. _____

12. _____

13. _____

14. _____

15. _____

16. _____

17. _____

18. _____

19. _____

20. _____

For the table below, identify each as an expression or an equation.

	Expression or Equation	
1.	$7x - 5y + 14$	
2.	$7x = 35 + 14$	
3.	$3(x - 2) = 5(x + 1) - 17$	
4.	$-9(2x + 1) - 4(x - 2) + 14$	

Fill in each blank with "simplify" or "solve."

5. To _____ an expression, we combine any like terms.

6. To _____ an equation, we use the properties of equality to find any value of the variable that makes the equation a true statement.

Simplify each expression by combining like terms.

7. $7x + x$

8. $6y - 10y$

9. $2a + 5a - 9a - 2$

10. $6a - 12 - a - 14$

Multiply and simplify if possible.

11. $-2(4x + 7)$

12. $-3(2x - 10)$

13. $5(y + 2) - 20$

14. $12x + 3(x - 6) - 13$

15. Find the area of the rectangle.

Rectangle 3 meters

$(4x - 2)$ meters

16. Find the perimeter of the triangle.

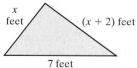

x feet $(x + 2)$ feet

7 feet

Solve and check.

17. $12 = 11x - 14x$

18. $8y + 7y = -45$

19. $x - 12 = -45 + 23$

20. $6 - (-5) = x + 5$

Solve and check.

21. $\frac{x}{3} = -14 + 9$

22. $\frac{z}{4} = -23 - 7$

23. $-6 + 2 = 4x + 1 - 3x$

24. $5 - 8 = 5x + 10 - 4x$

25. $6(3x - 4) = 19x$

26. $25x = 6(4x - 9)$

27. $-36x - 10 + 37x = -12 - (-14)$

28. $-8 + (-14) = -80y + 20 + 81y$

29. $3x - 16 = -10$

30. $4x - 21 = -13$

31. $-8z - 2z = 26 - (-4)$

32. $-12 + (-13) = 5x - 10x$

33. $-4(x + 8) - 11 = 3 - 26$

34. $-6(x - 2) + 10 = -4 - 10$

Translating *Write each phrase as an algebraic expression. Use x to represent "a number."*

35. The difference of a number and 10

36. The sum of −20 and a number

37. The product of 10 and a number

38. The quotient of 10 and a number

39. Five added to the product of −2 and a number

40. The product of −4 and the difference of a number and 1

21. _____

22. _____

23. _____

24. _____

25. _____

26. _____

27. _____

28. _____

29. _____

30. _____

31. _____

32. _____

33. _____

34. _____

35. _____

36. _____

37. _____

38. _____

39. _____

40. _____

Objectives

A Solve Linear Equations Using the Addition and Multiplication Properties.

B Solve Linear Equations Containing Parentheses.

C Write Numerical Sentences as Equations.

PRACTICE PROBLEM 1

Solve: $7x + 12 = 3x - 4$

3.3 SOLVING LINEAR EQUATIONS IN ONE VARIABLE

In this chapter, the equations we are solving are called **linear equations in one variable** or **first-degree equations in one variable**. For example, an equation such as $5x - 2 = 6x$ is a linear equation in one variable. It is called linear or first degree because the exponent on each x is 1 and there is no variable below a fraction bar. It is an equation in one variable because it contains one variable, x.

Let's continue solving linear equations in one variable.

Objective **A** Solving Equations Using the Addition and Multiplication Properties

If an equation contains variable terms on both sides, we use the addition property of equality to get all the variable terms on one side and all the constants or numbers on the other side.

EXAMPLE 1 Solve: $3a - 6 = a + 4$

Solution: Although it makes no difference which side you choose, let's move variable terms to the left side and constants to the right side.

$$3a - 6 = a + 4$$
$$3a - 6 + 6 = a + 4 + 6 \quad \text{Add 6 to both sides.}$$
$$3a = a + 10 \quad \text{Simplify.}$$
$$3a - a = a + 10 - a \quad \text{Subtract } a \text{ from both sides.}$$
$$2a = 10 \quad \text{Simplify.}$$
$$\frac{2a}{2} = \frac{10}{2} \quad \text{Divide both sides by 2.}$$
$$a = 5 \quad \text{Simplify.}$$

Check:
$$3a - 6 = a + 4 \quad \text{original equation}$$
$$3 \cdot 5 - 6 \overset{?}{=} 5 + 4 \quad \text{Replace } a \text{ with 5.}$$
$$15 - 6 \overset{?}{=} 9 \quad \text{Simplify.}$$
$$9 = 9 \quad \text{True}$$

The solution is 5.

Work Practice Problem 1

Helpful Hint Make sure you understand which property to use to solve an equation.

Addition
$$x + 2 = 10$$
To undo addition of 2, we subtract 2 from both sides.
$$x + 2 - 2 = 10 - 2 \quad \text{Use addition property of equality.}$$
$$x = 8$$
Check: $x + 2 = 10$
$$8 + 2 = 10$$
$$10 = 10$$

Understood multiplication
$$2x = 10$$
To undo multiplication of 2, we divide both sides by 2.
$$\frac{2x}{2} = \frac{10}{2} \quad \text{Use multiplication property of equality.}$$
$$x = 5$$
Check: $2x = 10$
$$2 \cdot 5 = 10$$
$$10 = 10$$

Answer
1. -4

184

EXAMPLE 2 Solve: $17 - 7x + 3 = -3x + 21 - 3x$

Solution: First, simplify both sides of the equation.

$17 - 7x + 3 = -3x + 21 - 3x$

$20 - 7x = -6x + 21$ Simplify.

Next, move variable terms on one side of the equation and constants, or numbers, to the other side. To begin, let's add $6x$ to both sides.

$20 - 7x + 6x = -6x + 21 + 6x$ Add $6x$ to each side.

$20 - x = 21$ Simplify.

$20 - x - 20 = 21 - 20$ Subtract 20 from both sides.

$-1x = 1$ Simplify. Recall that $-x$ means $-1x$.

$\dfrac{-1x}{-1} = \dfrac{1}{-1}$ Divide both sides by -1.

$x = -1$ Simplify.

Check: $17 - 7x + 3 = -3x + 21 - 3x$

$17 - 7(-1) + 3 \overset{?}{=} -3(-1) + 21 - 3(-1)$

$17 + 7 + 3 \overset{?}{=} 3 + 21 + 3$

$27 = 27$ True

The solution is -1.

🔲 **Work Practice Problem 2**

Objective B Solving Equations Containing Parentheses

Recall from the previous section that if an equation contains parentheses, we will first use the distributive property to remove them.

EXAMPLE 3 Solve: $7(x - 2) = 9x - 6$

Solution: First we apply the distributive property.

$7(x - 2) = 9x - 6$

$7x - 14 = 9x - 6$ Apply the distributive property.

Next we move variable terms to one side of the equation and constants to the other side.

$7x - 14 - 9x = 9x - 6 - 9x$ Subtract $9x$ from both sides.

$-2x - 14 = -6$ Simplify.

$-2x - 14 + 14 = -6 + 14$ Add 14 to both sides.

$-2x = 8$ Simplify.

$\dfrac{-2x}{-2} = \dfrac{8}{-2}$ Divide both sides by -2.

$x = -4$ Simplify.

Check to see that -4 is the solution.

🔲 **Work Practice Problem 3**

✔ **Concept Check**   In Example 3, the solution is -4. To check this solution, what equation should you use?

PRACTICE PROBLEM 2

Solve:

$40 - 5y + 5 = -2y - 10 - 4y$

PRACTICE PROBLEM 3

Solve: $6(a - 5) = 4a + 4$

Answers

2. -55 **3.** 17

✔ **Concept Check Answer**

$7(x - 2) = 9x - 6$

You may want to use the following steps to solve equations.

Steps for Solving an Equation

Step 1: If parentheses are present, use the distributive property.

Step 2: Combine any like terms on each side of the equation.

Step 3: Use the addition property of equality to rewrite the equation so that variable terms are on one side of the equation and constant terms are on the other side.

Step 4: Use the multiplication property of equality to divide both sides by the numerical coefficient of the variable to solve.

Step 5: Check the solution in the *original equation*.

PRACTICE PROBLEM 4

Solve: $4(x + 3) + 1 = 13$

EXAMPLE 4 Solve: $3(2x - 6) + 6 = 0$

Solution: $3(2x - 6) + 6 = 0$

Step 1: $6x - 18 + 6 = 0$ — Apply the distributive property.

Step 2: $6x - 12 = 0$ — Combine like terms on the left side of the equation.

Step 3: $6x - 12 + 12 = 0 + 12$ — Add 12 to both sides.

$6x = 12$ — Simplify.

Step 4: $\dfrac{6x}{6} = \dfrac{12}{6}$ — Divide both sides by 6.

$x = 2$ — Simplify.

Check:

Step 5: $3(2x - 6) + 6 = 0$

$3(2 \cdot 2 - 6) + 6 \stackrel{?}{=} 0$

$3(4 - 6) + 6 \stackrel{?}{=} 0$

$3(-2) + 6 \stackrel{?}{=} 0$

$-6 + 6 \stackrel{?}{=} 0$

$0 = 0$ True

The solution is 2.

Work Practice Problem 4

Objective C Writing Numerical Sentences as Equations

Next, we practice translating sentences into equations. Below are key words and phrases that translate to an equals sign: (*Note:* For a review of key words and phrases that translate to addition, subtraction, multiplication, and division, see Sections 1.8 and 3.2.)

Key Words or Phrases	Examples	Symbols
equals	3 equals 2 plus 1	$3 = 2 + 1$
gives	the quotient of 10 and -5 gives -2	$\dfrac{10}{-5} = -2$
is/was	17 minus 12 is 5	$17 - 12 = 5$
yields	11 plus 2 yields 13	$11 + 2 = 13$
amounts to	twice -15 amounts to -30	$2(-15) = -30$
is equal to	-24 is equal to 2 times -12	$-24 = 2(-12)$

Answer

4. 0

Translate each sentence into an equation.

a. The product of 7 and 6 is 42.

b. Twice the sum of 3 and 5 is equal to 16.

c. The quotient of −45 and 5 yields −9.

Solution:

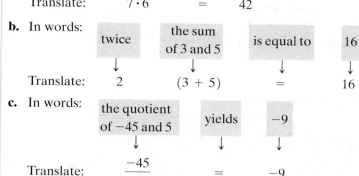

a. In words:

the product of 7 and 6	is	42
↓	↓	↓

Translate: $7 \cdot 6$ $=$ 42

b. In words:

twice	the sum of 3 and 5	is equal to	16
↓	↓	↓	↓

Translate: 2 $(3 + 5)$ $=$ 16

c. In words:

the quotient of −45 and 5	yields	−9
↓	↓	↓

Translate: $\dfrac{-45}{5}$ $=$ -9

◼ **Work Practice Problem 5**

PRACTICE PROBLEM 5

Translate each sentence into an equation.

a. The difference of 110 and 80 is 30.

b. The product of 3 and the sum of −9 and 11 amounts to 6.

c. The quotient of 24 and −6 yields −4.

Answers

5. a. $110 - 80 = 30$

b. $3(-9 + 11) = 6$ **c.** $\dfrac{24}{-6} = -4$

CALCULATOR EXPLORATIONS   Checking Equations

A calculator can be used to check possible solutions of equations. To do this, replace the variable by the possible solution and evaluate each side of the equation separately. For example, to see whether 7 is a solution of the equation $52x = 15x + 259$, replace x with 7 and use your calculator to evaluate each side separately.

Equation: $52x = 15x + 259$

$52 \cdot 7 \overset{?}{=} 15 \cdot 7 + 259$ *Replace x with 7.*

Evaluate left side: $\boxed{52}$ $\boxed{\times}$ $\boxed{7}$ $\boxed{=}$ or $\boxed{\text{ENTER}}$.
Display: $\boxed{364}$.

Evaluate right side: $\boxed{15}$ $\boxed{\times}$ $\boxed{7}$ $\boxed{+}$ $\boxed{259}$ $\boxed{=}$
or $\boxed{\text{ENTER}}$. Display: $\boxed{364}$.

Since the left side equals the right side, 7 is a solution of the equation $52x = 15x + 259$.

Use a calculator to determine whether the numbers given are solutions of each equation.

1. $76(x - 25) = -988$;   12

2. $-47x + 862 = -783$;   35

3. $x + 562 = 3x + 900$;   −170

4. $55(x + 10) = 75x + 910$;   −18

5. $29x - 1034 = 61x - 362$;   −21

6. $-38x + 205 = 25x + 120$;   25

Vocabulary and Readiness Check

Use the choices below to fill in each blank. Some choices may be used more than once.

addition	multiplication	combine like terms
$5(2x + 6) - 1 = 39$	$3x - 9 + x - 16$	distributive

1. An example of an expression is _____ while an example of an equation is _____.

2. To solve $\dfrac{x}{-7} = -10$, we use the _____ property of equality.

3. To solve $x - 7 = -10$, we use the _____ property of equality.

Use the order of the Steps for Solving an Equation in this section to answer Exercises 4 through 6.

4. To solve $9x - 6x = 10 + 6$, first _____.

5. To solve $5(x - 1) = 25$, first use the _____ property.

6. To solve $4x + 3 = 19$, first use the _____ property of equality.

3.3 EXERCISE SET

FOR EXTRA HELP

Student Solutions Manual · PH Math/Tutor Center · CD/Video for Review · MathXL® · MyMathLab

Objective A *Solve each equation. See Examples 1 and 2.*

1. $3x - 7 = 4x + 5$

2. $7x - 1 = 8x + 4$

 3. $10x + 15 = 6x + 3$

4. $5x - 3 = 2x - 18$

5. $19 - 3x = 14 + 2x$

6. $4 - 7m = -3m + 4$

7. $-14x - 20 = -12x + 70$

8. $57y + 140 = 54y - 100$

9. $x + 20 + 2x = -10 - 2x - 15$

10. $2x + 10 + 3x = -12 - x - 20$

11. $40 + 4y - 16 = 13y - 12 - 3y$

12. $19x - 2 - 7x = 31 + 6x - 15$

Objective B *Solve each equation. See Examples 3 and 4.*

13. $35 - 17 = 3(x - 2)$

14. $22 - 42 = 4(x - 1)$

15. $3(x - 1) - 12 = 0$

16. $2(x + 5) + 8 = 0$

17. $2(y - 3) = y - 6$

18. $3(z + 2) = 5z + 6$

19. $-2(y + 4) = 2$

20. $-1(y + 3) = 10$

21. $2t - 1 = 3(t + 7)$

22. $-4 + 3c = 4(c + 2)$

23. $3(5c + 1) - 12 = 13c + 3$

24. $4(3t + 4) - 20 = 3 + 5t$

Mixed Practice (*Sections 1.8, 3.2, 3.3*) *Solve each equation.*

25. $-4x = 44$

26. $-3x = 51$

27. $x + 9 = 2$

28. $y - 6 = -11$

29. $8 - b = 13$

30. $7 - z = 15$

31. $-20 - (-50) = \dfrac{x}{9}$

32. $-2 - 10 = \dfrac{z}{10}$

33. $3r + 4 = 19$

34. $7y + 3 = 38$

35. $-7c + 1 = -20$

36. $-2b + 5 = -7$

37. $8y - 13y = -20 - 25$

38. $4x - 11x = -14 - 14$

39. $6(7x - 1) = 43x$

40. $5(3y - 2) = 16y$

41. $-4 + 12 = 16x - 3 - 15x$

42. $-9 + 20 = 19x - 4 - 18x$

43. $-10(x + 3) + 28 = -16 - 16$

44. $-9(x + 2) + 25 = -19 - 19$

45. $4x + 3 = 2x + 11$

46. $6y - 8 = 3y + 7$

47. $-2y - 10 = 5y + 18$

48. $7n + 5 = 12n - 10$

49. $-8n + 1 = -6n - 5$

50. $10w + 8 = w - 10$

51. $9 - 3x = 14 + 2x$

52. $4 - 7m = -3m$

53. $9a + 29 + 7 = 0$

54. $10 + 4v + 6 = 0$

55. $7(y - 2) = 4y - 29$

56. $2(z - 2) = 5z + 17$

57. $12 + 5t = 6(t + 2)$

58. $4 + 3c = 2(c + 2)$

59. $3(5c - 1) - 2 = 13c + 3$

60. $4(2t + 5) - 21 = 7t - 6$

61. $10 + 5(z - 2) = 4z + 1$

62. $14 + 4(w - 5) = 6 - 2w$

63. $7(6 + w) = 6(2 + w)$

64. $6(5 + c) = 5(c - 4)$

Objective **C** **Translating** *Write each sentence as an equation. See Example 5.*

65. The sum of -42 and 16 is -26.

66. The difference of -30 and 10 equals -40.

67. The product of -5 and -29 gives 145.

68. The quotient of -16 and 2 yields -8.

69. Three times the difference of -14 and 2 amounts to -48.

70. Negative 2 times the sum of 3 and 12 is -30.

71. The quotient of 100 and twice 50 is equal to 1.

72. Seventeen subtracted from -12 equals -29.

Review

The following bar graph shows the number of U.S. federal individual income tax returns that are filed electronically during the years shown (some years are projected). Electronically filed returns include Telefile and online returns. Use this graph to answer Exercises 73 through 76. See Section 2.1.

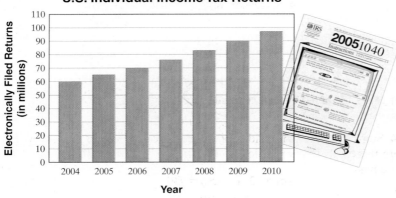

**Total Electronically Filed
U.S. Individual Income Tax Returns**

Source: IRS Compliance Research Division

73. Approximate the number of electronically filed returns projected for 2010.

74. Approximate the number of electronically filed returns projected for 2008.

75. By how much is the number of electronically filed returns expected to increase from 2006 to 2009?

76. Describe any trends shown in this graph.

Evaluate each expression for $x = 3$, $y = -1$, and $z = 0$. See Section 2.5.

77. $x^3 - 2xy$ **78.** $y^3 + 3xyz$ **79.** $y^5 - 4x^2$ **80.** $(-y)^3 + 3xyz$

Concept Extensions

Using the steps for solving an equation, choose the next operation for solving the given equation.

81. $2x - 5 = -7$

   a. Add 7 to both sides.
   b. Add 5 to both sides.
   c. Divide both sides by 2.

82. $3x + 2x = -x - 4$

   a. Add 4 to both sides.
   b. Subtract $2x$ from both sides.
   c. Add $3x$ and $2x$.

83. $-3x = -12$

   a. Divide both sides by -3.
   b. Add 12 to both sides.
   c. Add $3x$ to both sides.

84. $9 - 5x = 15$

   a. Divide both sides by -5.
   b. Subtract 15 from both sides.
   c. Subtract 9 from both sides.

85. A classmate shows you his steps for solving the given equation. His solution does not check, but he is unable to find the error. Check this solution, find the error, and correct it.

$$2(3x - 5) = 5x - 7$$
$$6x - 5 = 5x - 7$$
$$6x - 5 + 5 = 5x - 7 + 5$$
$$6x = 5x - 2$$
$$6x - 5x = 5x - 2 - 5x$$
$$x = -2$$

Solve.

86. $(-8)^2 + 3x = 5x + 4^3$ **87.** $3^2 \cdot x = (-9)^3$ **88.** $2^3(x + 4) = 3^2(x + 4)$ ▦ **89.** $x + 45^2 = 54^2$

 90. A classmate tries to solve $3x = 39$ by subtracting 3 from both sides of the equation. Will this step solve the equation for x? Why or why not?

91. A classmate tries to solve $2 + x = 20$ by dividing both sides by 2. Will this step solve the equation for x? Why or why not?

THE BIGGER PICTURE   Operations on Sets of Numbers and Solving Equations

Continue your outline from Sections 1.6, 1.7, and 2.4. Suggestions are once again written to help you complete this part of your outline. Notice that this part of the outline has to do with solving equations.

I. Operations on Sets of Numbers

  A. Whole Numbers

   1. Add or Subtract (Section 1.3)

   2. Multiply or Divide (Sections 1.5, 1.6)

   3. Exponent (Section 1.7)

   4. Order of Operations (Section 1.7)

  B. Integers

   1. Add (Section 2.2)

   2. Subtract (Section 2.3)

   3. Multiply or Divide (Section 2.4)

II. Solving Equations

  A. Equations in General: Simplify both sides of the equation by removing parentheses and combining any like terms. Then use the addition property to write variable terms on one side, constants (or numbers) on the other side. Then use the multiplication property to solve for the variable by dividing both sides of the equation by the coefficient of the variable.

Solve: $2(x - 5) = 80$

$$2x - 10 = 80 \qquad \text{Use the distributive property.}$$
$$2x - 10 + 10 = 80 + 10 \qquad \text{Add 10 to both sides.}$$
$$2x = 90 \qquad \text{Simplify.}$$
$$\frac{2x}{2} = \frac{90}{2} \qquad \text{Divide both sides by 2.}$$
$$x = 45 \qquad \text{Simplify.}$$

Solve.

1. $-8x = 40$

2. $x - 14 = -3$

3. $2x = -14$

4. $5y + 7 - 4y - 10 = 100$

5. $-3n + n = -100 - 50$

6. $8x + 9 = -79$

7. $4x - 5 = 2x - 17$

8. $30 + 5(2n - 4) = 100$

3.4 LINEAR EQUATIONS IN ONE VARIABLE AND PROBLEM SOLVING

Objective **A** Writing Sentences as Equations

Now that we have practiced solving equations for a variable, we can extend considerably our problem-solving skills. We begin by writing sentences as equations using the following key words and phrases as a guide:

Addition	Subtraction	Multiplication	Division	Equals Sign
sum	difference	product	quotient	equals
plus	minus	times	divide	gives
added to	subtracted from	multiply	shared equally among	is/was
more than	less than	twice	per	yields
increased by	decreased by	of	divided by	amounts to
total	less	double	divided into	is equal to

Notice that these sentences contain unknown numbers, which we will represent with x.

EXAMPLE 1 Write each sentence as an equation. Use x to represent "a number."

a. Twenty increased by a number is 5.
b. Twice a number equals −10.
c. A number minus 11 amounts to 168.
d. Three times the sum of a number and 5 is −30.
e. The quotient of twice a number and 8 is equal to 2.

Solution:

a. In words:

twenty	increased by	a number	is	5
↓	↓	↓	↓	↓

Translate: 20 $+$ x $=$ 5

b. In words:

twice a number	equals	−10
↓	↓	↓

Translate: $2x$ $=$ -10

c. In words:

a number	minus	11	amounts to	168
↓	↓	↓	↓	↓

Translate: x $-$ 11 $=$ 168

d. In words:

three times	the sum of a number and 5	is	−30
↓	↓	↓	↓

Translate: 3 $(x + 5)$ $=$ -30

e. In words:

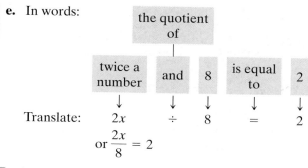

Translate: $2x$ $\div$ 8 $=$ 2

$$\text{or } \frac{2x}{8} = 2$$

■ **Work Practice Problem 1**

Objective B Using Problem-Solving Steps to Solve Problems

Our main purpose for studying arithmetic and algebra is to solve problems. In previous sections, we have prepared for problem solving by writing phrases as algebraic expressions and sentences as equations. We now draw upon this experience as we solve problems. The following problem-solving steps will be used throughout this text.

Problem-Solving Steps

1. **UNDERSTAND** the problem. During this step, become comfortable with the problem. Some ways of doing this are as follows:
   - Read and reread the problem.
   - Construct a drawing.
   - Propose a solution and check. Pay careful attention to how you check your proposed solution. This will help when writing an equation to model the problem.
   - Choose a variable to represent the unknown. Use this variable to represent any other unknowns.
2. **TRANSLATE** the problem into an equation.
3. **SOLVE** the equation.
4. **INTERPRET** the results: *Check* the proposed solution in the stated problem and *state* your conclusion.

The first problem that we solve consists of finding an unknown number.

EXAMPLE 2 **Finding an Unknown Number**

Twice a number plus 3 is the same as the number minus 6. Find the unknown number.

Solution:

1. **UNDERSTAND** the problem. To do so, we read and reread the problem.
   Let's propose a solution to help us understand. Suppose the unknown number is 5. Twice this number plus 3 is $2 \cdot 5 + 3$ or 13. Is this the same as the number minus 6, or $5 - 6$, or -1? Since 13 is not the same as -1, we know that 5 is not the solution. However, remember that the purpose of proposing a solution is not to guess correctly, but to better understand the problem.
   Now let's choose a variable to represent the unknown. Let's let

   $x =$ unknown number

Continued on next page

PRACTICE PROBLEM 2

Translate "the sum of a number and 2 equals 6 added to three times the number" into an equation and solve.

Answer
2. -2

2. **TRANSLATE** the problem into an equation.

In words:

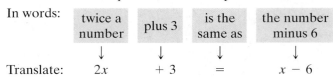

twice a number	plus 3	is the same as	the number minus 6
↓	↓	↓	↓

Translate: $\qquad 2x \qquad + 3 \qquad = \qquad x - 6$

3. **SOLVE** the equation. To solve the equation, we first subtract x from both sides.

$$2x + 3 = x - 6$$
$$2x + 3 - x = x - 6 - x$$
$$x + 3 = -6 \qquad \text{Simplify.}$$
$$x + 3 - 3 = -6 - 3 \qquad \text{Subtract 3 from both sides.}$$
$$x = -9 \qquad \text{Simplify.}$$

4. **INTERPRET** the results. First, *Check* the proposed solution in the stated problem. Twice "-9" is -18 and $-18 + 3$ is -15. This is equal to the number minus 6, or "-9" $- 6$, or -15. Then *state* your conclusion: The unknown number is -9.

▣ **Work Practice Problem 2**

✔ **Concept Check** Suppose you have solved an equation involving perimeter to find the length of a rectangular table. Explain why you would want to recheck your math if you obtain the result of -5.

PRACTICE PROBLEM 3

The distance by road from Cincinnati, Ohio, to Denver, Colorado, is 71 miles *less* than the distance from Denver to San Francisco, California. If the total of these two distances is 2399 miles, find the distance from Denver to San Francisco.

EXAMPLE 3 **Determining Distances**

The distance by road from Chicago, Illinois, to Los Angeles, California, is 1091 miles *more* than the distance from Chicago to Boston, Massachusetts. If the total of these two distances is 3017 miles, find the distance from Chicago to Boston. (*Source: World Almanac* 2006)

1. **UNDERSTAND** the problem. We read and reread the problem.

Let's propose and check a solution to help us better understand the problem. Suppose the distance from Chicago to Boston is 600 miles. Since the distance from Chicago to Los Angeles is 1091 *more* miles, then this distance is $600 + 1091 = 1691$ miles. With these numbers, the total of the distances is $600 + 1691 = 2291$ miles. This is less than the given total of 3017 miles, so we are incorrect. But not only do we have a better understanding of this exercise, we also know that the distance from Boston to Chicago is greater than 600 miles since this proposed solution led to a total too small. Now let's choose a variable to represent an unknown. Then use this variable to represent any other unknown quantities. Let

x = distance from Chicago to Boston

Then

$x + 1091$ = distance from Chicago to Los Angeles

since that distance is 1091 more miles.

2. **TRANSLATE** the problem into an equation.

In words:

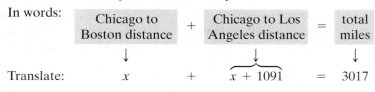

Chicago to Boston distance	+	Chicago to Los Angeles distance	=	total miles
↓		↓		↓

Translate: $\qquad x \qquad + \qquad \underbrace{x + 1091} \qquad = \qquad 3017$

Answer

3. 1235 miles

✔ **Concept Check Answer**

length cannot be negative

3. SOLVE the equation:

$$x + x + 1091 = 3017$$

$$2x + 1091 = 3017 \qquad \text{Combine like terms.}$$

$$2x + 1091 - 1091 = 3017 - 1091 \qquad \text{Subtract 1091 from both sides.}$$

$$2x = 1926 \qquad \text{Simplify.}$$

$$\frac{2x}{2} = \frac{1926}{2} \qquad \text{Divide both sides by 2.}$$

$$x = 963 \qquad \text{Simplify.}$$

4. INTERPRET the results. First *Check* the proposed solution in the stated problem. Since x represents the distance from Chicago to Boston, this is 963 miles. The distance from Chicago to Los Angeles is $x + 1091 = 963 + 1091 = 2054$ miles. To check, notice that the total number of miles is $963 + 2054 = 3017$ miles, the given total of miles. Also, 2054 is 1091 more miles than 963, so the solution checks. Then, *state* your conclusion: The distance from Chicago to Boston is 963 miles.

 Work Practice Problem 3

EXAMPLE 4 **Calculating Separate Costs**

A sales person at an electronics store sold a computer system and software for $2100, receiving four times as much money for the computer system as for the software. Find the price of each.

Solution:

1. UNDERSTAND the problem. We read and reread the problem. Then we choose a variable to represent an unknown. We use this variable to represent any other unknown quantities. We let

$x =$ the software price

$4x =$ the computer system price

2. TRANSLATE the problem into an equation.

In words:

software price	and	computer price	is	2100
↓	↓	↓	↓	↓
x	$+$	$4x$	$=$	2100

Translate:

3. SOLVE the equation:

$$x + 4x = 2100$$

$$5x = 2100 \qquad \text{Combine like terms.}$$

$$\frac{5x}{5} = \frac{2100}{5} \qquad \text{Divide both sides by 5.}$$

$$x = 420 \qquad \text{Simplify.}$$

4. INTERPRET the results. *Check* the proposed solution in the stated problem. The software sold for $420. The computer system sold for $4x = 4(\$420) = \$1680$. Since $\$420 + \$1680 = \$2100$, the total price, and $1680 is four times $420, the solution checks. *State* your conclusion: The software sold for $420, and the computer system sold for $1680.

 Work Practice Problem 4

PRACTICE PROBLEM 4

A woman's $57,000 estate is to be divided so that her husband receives twice as much as her son. How much will each receive?

Answer
4. husband: $38,000; son: $19,000

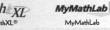
Objective A Translating *Write each sentence as an equation. Use x to represent "a number." See Example 1.*

1. A number added to −5 is −7.

2. Five subtracted from a number equals 10.

3. Three times a number yields 27.

4. The quotient of 8 and a number is −2.

5. A number subtracted from −20 amounts to 104.

6. Two added to twice a number gives −14.

7. Twice a number gives 108.

8. Five times a number is equal to −75.

9. The product of 5 and the sum of −3 and a number is −20.

10. Twice the sum of −17 and a number is −14.

Objective B Translating *Translate each to an equation. Then solve the equation. See Example 2.*

11. Three times a number, added to 9 is 33. Find the number.

12. Twice a number, subtracted from 60 is 20. Find the number.

13. The sum of 3, 4, and a number amounts to 16. Find the number.

14. The sum of 7, 9, and a number is 40. Find the number.

15. The difference of a number and 3 is equal to the quotient of 10 and 5. Find the number.

16. Eight decreased by a number equals the quotient of 15 and 5. Find the number.

17. Thirty less a number is equal to the product of 3 and the sum of the number and 6. Find the number.

18. The product of a number and 3 is twice the sum of that number and 5. Find the number.

19. 40 subtracted from five times a number is 8 more than the number. Find the number.

20. Five times the sum of a number and 2 is 11 less than the number times 8. Find the number.

21. Three times the difference of some number and 5 amounts to the quotient of 108 and 12. Find the number.

22. Seven times the difference of some number and 1 gives the quotient of 70 and 10. Find the number.

23. The product of 4 and a number is the same as 30 less twice that same number. Find the number.

24. Twice a number equals 25 less triple that same number. Find the number.

Solve. See Examples 3 and 4.

25. Based on the 2000 Census, Florida has 28 fewer electoral votes for president than California. If the total number of electoral votes for these two states is 82, find the number for each state. (*Source: The World Almanac* 2006)

26. In the 2004 presidential election, George W. Bush received 34 more electoral votes than John Kerry. If a total of 538 electoral votes were cast for the two candidates, find how many votes each candidate received. (*Source:* Voter News Service)

27. A falcon, when diving, can travel five times as fast as a pheasant's top speed. If the total speeds for these two birds is 222 miles per hour, find the fastest speed of the falcon and the fastest speed of the pheasant. (*Source: Fantastic Book of Comparisons*)

28. Norway has had three times as many rulers as Liechtenstein. If the total rulers for both countries is 56, find the number of rulers for Norway and the number for Liechtenstein.

Norway   Liechtenstein

29. The country with the most universities* is India, followed by the United States. If India has 2649 more universities than the United States and their combined total is 14,165, find the number of universities in India and the number in the United States [*Includes all further education establishments.] (*Source: The Top 10 of Everything*)

30. The average life expectancy for an elephant is 24 years longer than the life expectancy for a chimpanzee. If the total of these life expectancies is 130 years, find the life expectancy of each.

31. An Xbox 360 game system and several games are sold for $560. The cost of the Xbox 360 is 3 times as much as the cost of the games. Find the cost of the Xbox 360 and the cost of the games.

32. In a recent year, the two top-selling PC games were *World of Warcraft* and *The Sims 2: University Expansion Pack*. The average price of *World of Warcraft* is $14 more than the average price of *Sims*. If the total of these two prices is $80, find the price of each game. (*Source:* The NPD Group, Inc.)

33. By air, the distance from New York City to London is 2001 miles *less* than the distance from Los Angeles to Tokyo. If the total of these two distances is 8939 miles, find the distance from Los Angeles to Tokyo.

34. By air, the distance from Melbourne, Australia, to Cairo, Egypt, is 2338 miles *more* than the distance from Madrid, Spain, to Bangkok, Thailand. If the total of these distances is 15,012 miles, find the distance from Madrid to Bangkok.

35. The two NCAA stadiums with the largest capacities are Michigan Stadium (Univ. of Michigan) and Neyland Stadium (Univ. of Tennessee). Michigan Stadium has a capacity of 4647 more than Neyland. If the combined capacity for the two stadiums is 210,355, find the capacity for each stadium. (*Source: National Collegiate Athletic Association*)

36. A National Hot Rod Association (NHRA) top fuel dragster has a top speed of 95 mph faster than an Indy Racing League car. If the combined top speeds for these two cars is 565 mph, find the top speed of each car. (*Source: USA Today*)

37. In 2020, China is projected to be the country with the greatest number of visiting tourists. This number is twice the number of tourists projected for Spain. If the total number of tourists for these two countries is projected to be 210 million, find the number projected for each. (*Source: The State of the World Atlas* by Dan Smith)

38. California contains the largest state population of native Americans. This population is three times the native American population of Washington state. If the total of these two populations is 412 thousand, find the native American population in each of these two states. (*Source: U.S. Census Bureau*)

39. In Germany, about twice as many cars are manufactured per day than in Spain. If the total number of these cars manufactured per day is 19,827, find the number manufactured in Spain and the number manufactured in Germany.

40. A Toyota Camry is traveling twice as fast as a Dodge truck. If their combined speed is 105 miles per hour, find the speed of the car and find the speed of the truck.

41. Anthony Tedesco sold his used mountain bike and accessories for $270. If he received five times as much money for the bike as he did for the accessories, find how much money he received for the bike.

42. A tractor and a plow attachment are worth $1200. The tractor is worth seven times as much money as the plow. Find the value of the tractor and the value of the plow.

43. During the 2006 Women's NCAA Division I basketball championship game, the Maryland Terrapins scored 3 more points than the Duke Blue Devils. Together, both teams scored a total of 153 points. How many points did the 2006 Champion Maryland Terrapins score during this game? (*Source:* National Collegiate Athletic Association)

44. During the 2006 Men's NCAA Division I basketball championship game, the UCLA Bruins scored 16 fewer points than the Florida Gators. Together, both teams scored a total of 130 points. How many points did the 2006 Champion Florida Gators score during this game? (*Source:* National Collegiate Athletic Association)

Review

Round each number to the given place value. See Section 1.4.

45. 586 to the nearest ten

46. 82 to the nearest ten

47. 1026 to the nearest hundred

48. 52,333 to the nearest thousand

49. 2986 to the nearest thousand

50. 101,552 to the nearest hundred

Concept Extensions

 51. Solve Example 3 again, but this time let *x* be the distance from Chicago to Los Angeles. Did you get the same results? Explain why or why not.

In real estate, a house's selling price P is found by adding the real estate agent's commission C to the amount A that the seller of the house receives: $P = A + C$.

52. A house sold for $230,000. The owner's real estate agent received a commission of $13,800. How much did the seller receive? (*Hint:* Substitute the known values into the equation, then solve the equation for the remaining unknown.)

53. A homeowner plans to use a real estate agent to sell his house. He hopes to sell the house for $165,000 and keep $156,750 of that. If everything goes as he has planned, how much will his real estate agent receive as a commission?

In retailing, the retail price P of an item can be computed using the equation $P = C + M$, where C is the wholesale cost of the item and M is the amount of markup.

54. The retail price of a computer system is $999 after a markup of $450. What is the wholesale cost of the computer system? (*Hint:* Substitute the known values into the equation, then solve the equation for the remaining unknown.)

55. Slidell Feed and Seed sells a bag of cat food for $12. If the store paid $7 for the cat food, what is the markup on the cat food?

STUDY SKILLS BUILDER

How Well Do You Know Your Textbook?

The questions below will determine whether you are familiar with your textbook. For help, see Section 1.1 in this text.

1. What does the 🔘 icon mean?

2. What does the ✎ icon mean?

3. What does the △ icon mean?

4. Where can you find a review for each chapter? What answers to this review can be found in the back of your text?

5. Each chapter contains an overview of the chapter along with examples. What is this feature called?

6. Each chapter contains a review of vocabulary. What is this feature called?

7. There is a CD in your text. What content is contained on this CD?

8. What is the location of the section that is entirely devoted to study skills?

9. There are Practice Problems that are contained in the margin of the text. What are they and how can they be used?

CHAPTER 3   Group Activity

Modeling Equation Solving with Addition and Subtraction

Sections 3.1–3.4

We can use positive counters ● and negative counters ● to help us model the equation-solving process. We also need to use an object that represents a variable. We use small slips of paper with the variable name written on them.

Recall that taking a ● and ● together creates a neutral or zero pair. After a neutral pair has been formed, it can be removed from or added to an equation model without changing the overall value. We also need to remember that we can add or remove the same number of positive or negative counters from both sides of an equation without changing the overall value.

We can represent the equation $x + 5 = 2$ as follows:

To get the variable by itself, we must remove 5 black counters from both sides of the model. Because there are only 2 counters on the right side, we must add 5 negative counters to both sides of the model. Then we can remove neutral pairs: 5 from the left side and 2 from the right side.

We are left with the following model, which represents the solution, $x = -3$.

$$\boxed{x} = \bullet\bullet\bullet$$

Similarly, we can represent the equation $x - 4 = -6$ as follows:

$$\boxed{x}\ \substack{\bullet\bullet \\ \bullet\bullet} = \substack{\bullet\bullet\bullet \\ \bullet\bullet\bullet}$$

To get the variable by itself, we must remove 4 red counters from both sides of the model

$$\boxed{x}\ \substack{\bullet\bullet \\ \bullet\bullet} = \substack{\bullet\bullet\bullet \\ \bullet\bullet\bullet}$$

We are left with the following model, which represents the solution, $x = -2$.

$$\boxed{x} = \substack{\bullet \\ \bullet}$$

Use the counter model to solve each equation.

1. $x - 3 = -7$ **2.** $x - 1 = -9$

3. $x + 2 = 8$ **4.** $x + 4 = 5$

5. $x + 8 = 3$ **6.** $x - 5 = -1$

7. $x - 2 = 1$ **8.** $x - 5 = 10$

9. $x + 3 = -7$ **10.** $x + 8 = -2$

Chapter 3 Vocabulary Check

Fill in each blank with one of the words or phrases listed below.

variable	addition	constant	algebraic expression	equation
terms	simplified	multiplication	evaluating the expression	solution
like	combined	numerical coefficient	distributive	

1. An algebraic expression is _____ when all like terms have been _____.
2. Terms that are exactly the same, except that they may have different numerical coefficients, are called _____ terms.
3. A letter used to represent a number is called a _____.
4. A combination of operations on variables and numbers is called an _____.
5. The addends of an algebraic expression are called the _____ of the expression.
6. The number factor of a variable term is called the _____.
7. Replacing a variable in an expression by a number and then finding the value of the expression is called _____ for the variable.
8. A term that is a number only is called a _____.
9. An _____ is of the form expression = expression.
10. A _____ of an equation is a value for the variable that makes an equation a true statement.
11. To multiply $-3(2x + 1)$, we use the _____ property.
12. By the _____ property of equality, we may multiply or divide both sides of an equation by any nonzero number without changing the solution of the equation.
13. By the _____ property of equality, the same number may be added to or subtracted from both sides of an equation without changing the solution of the equation.

Helpful Hint

Are you preparing for your test? Don't forget to take the Chapter 3 Test on page 208. Then check your answers at the back of the text and use the Chapter Test Prep Video CD to see the fully worked-out solutions to any of the exercises you want to review.

3 Chapter Highlights

DEFINITIONS AND CONCEPTS	EXAMPLES
Section 3.1 Simplifying Algebraic Expressions	

The addends of an algebraic expression are called the **terms** of the expression.	$5x^2 + (-4x) + (-2)$ 3 terms
The number factor of a variable term is called the **numerical coefficient.**	**Term** **Numerical Coefficient** $7x$ — 7 $-6y$ — -6 x or $1x$ — 1
Terms that are exactly the same, except that they may have different numerical coefficients, are called **like terms.**	$5x + 11x = (5 + 11)x = 16x$ like terms
An algebraic expression is **simplified** when all like terms have been **combined.**	$y - 6y = (1 - 6)y = -5y$
Use the **distributive property** to multiply an algebraic expression by a term.	Simplify: $-4(x + 2) + 3(5x - 7)$ $= -4(x) + (-4)(2) + 3(5x) + 3(-7)$ $= -4x + (-8) + 15x + (-21)$ $= 11x + (-29)$ or $11x - 29$

201

DEFINITIONS AND CONCEPTS	EXAMPLES

Section 3.2 Solving Equations: Review of the Addition and Multiplication Properties

ADDITION PROPERTY OF EQUALITY

Let a, b, and c represent numbers. Then

$a = b$	Also, $a = b$
and $a + c = b + c$	and $a - c = b - c$
are equivalent equations.	are equivalent equations.

In other words, the same number may be added to or subtracted from both sides of an equation without changing the solution of the equation.

Solve for x:

$$x + 8 = 2 + (-1)$$
$$x + 8 = 1$$
$$x + 8 - 8 = 1 - 8 \quad \text{Subtract 8 from both sides.}$$
$$x = -7 \quad \text{Simplify.}$$

The solution is -7.

MULTIPLICATION PROPERTY OF EQUALITY

Let a, b, and c represent numbers and let $c \neq 0$. Then

$a = b$	Also, $a = b$
and $a \cdot c = b \cdot c$	and $\dfrac{a}{c} = \dfrac{b}{c}$
are equivalent equations.	are equivalent equations.

In other words, both sides of an equation may be multiplied or divided by the same nonzero number without changing the solution of the equation.

Solve:
$$y - 7y = 30$$
$$-6y = 30 \quad \text{Combine like terms.}$$
$$\frac{-6y}{-6} = \frac{30}{-6} \quad \text{Divide both sides by } -6.$$
$$y = -5 \quad \text{Simplify.}$$

The solution is -5.

Section 3.3 Solving Linear Equations in One Variable

STEPS FOR SOLVING AN EQUATION

Step 1. If parentheses are present, use the distributive property.

Step 2. Combine any like terms on each side of the equation.

Step 3. Use the addition property of equality to rewrite the equation so that variable terms are on one side of the equation and constant terms are on the other side.

Step 4. Use the multiplication property of equality to divide both sides by the numerical coefficient of the variable to solve.

Step 5. Check the solution in the *original equation*.

Solve for x: $5(3x - 1) + 15 = -5$

Step 1. $15x - 5 + 15 = -5$ Apply the distributive property.

Step 2. $15x + 10 = -5$ Combine like terms.

Step 3. $15x + 10 - 10 = -5 - 10$ Subtract 10 from both sides.
$$15x = -15$$

Step 4. $\dfrac{15x}{15} = \dfrac{-15}{15}$ Divide both sides by 15.
$$x = -1$$

Step 5. Check to see that -1 is the solution.

Section 3.4 Linear Equations in One Variable and Problem Solving

PROBLEM-SOLVING STEPS

1. UNDERSTAND the problem. Some ways of doing this are

    Read and reread the problem.

    Construct a drawing.

    Choose a variable to represent an unknown in the problem.

The incubation period for a golden eagle is three times the incubation period for a hummingbird. If the total of their incubation periods is 60 days, find the incubation period for each bird. (*Source: Wildlife Fact File*, International Masters Publishers)

1. UNDERSTAND the problem. Then choose a variable to represent an unknown. Let

    x = incubation period of a hummingbird

    $3x$ = incubation period of a golden eagle

DEFINITIONS AND CONCEPTS	**EXAMPLES**
Section 3.4 Linear Equations in One Variable and Problem Solving (*continued*)	

2. TRANSLATE the problem into an equation.	**2.** TRANSLATE.
	$$\underset{x}{\text{incubation of hummingbird}} + \underset{3x}{\text{incubation of golden eagle}} \underset{=}{\text{is}} \underset{60}{60}$$
3. SOLVE the equation.	**3.** SOLVE: $$x + 3x = 60$$ $$4x = 60$$ $$\frac{4x}{4} = \frac{60}{4}$$ $$x = 15$$
4. INTERPRET the results. *Check* the proposed solution in the stated problem and *state* your conclusion.	**4.** INTERPRET the solution in the stated problem. The incubation period for a hummingbird is 15 days. The incubation period for a golden eagle is $3x = 3 \cdot 15 = 45$ days. Since 15 days + 45 days = 60 days and 45 is 3(15), the solution checks. *State* your conclusion: The incubation period for a hummingbird is 15 days. The incubation period for a golden eagle is 45 days.

STUDY SKILLS BUILDER

Are You Prepared for a Test on Chapter 3?

Below I have listed some *common trouble areas* for students in Chapter 3. After studying for your test, but before taking your test, read these.

- Be careful when evaluating expressions. For example, evaluate $3x - y$ when $x = -2$ and $y = -3$.

$$3x - y = 3(-2) - (-3) \quad \text{Let } x = -2 \text{ and } y = -3.$$
$$= -6 - (-3) \quad \text{Multiply.}$$
$$= -6 + 3$$
$$= -3 \quad \text{Add.}$$

- Remember the distributive property.

$$5(4x - 3) + 2 = 5 \cdot 4x - 5 \cdot 3 + 2 \quad \text{Use the distributive property.}$$
$$= 20x - 15 + 2 \quad \text{Simplify.}$$
$$= 20x - 13 \quad \text{Combine like terms.}$$

- Don't forget the steps for solving a linear equation.

$$2(3x - 2) + 16 = 6$$
$$6x - 4 + 16 = 6 \quad \text{Apply the distributive property.}$$
$$6x + 12 = 6 \quad \text{Combine like terms.}$$
$$6x + 12 - 12 = 6 - 12 \quad \text{Subtract 12 from both sides.}$$
$$6x = -6 \quad \text{Simplify.}$$
$$\frac{6x}{6} = \frac{-6}{6} \quad \text{Divide both sides by 6.}$$
$$1 \cdot x = -1 \quad \text{Simplify.}$$
$$x = -1$$

Remember: This is simply a checklist of common trouble areas. For a review of Chapter 3 see the Highlights and Chapter Review at the end of this chapter.

3 CHAPTER REVIEW

(3.1) *Simplify each expression by combining like terms.*

1. $3y + 7y - 15$　　　**2.** $2y - 10 - 8y$　　　**3.** $8a + a - 7 - 15a$　　　**4.** $y + 3 - 9y - 1$

Multiply.

5. $2(x + 5)$

6. $-3(y + 8)$

Simplify.

7. $7x + 3(x - 4) + x$

8. $-(3m + 2) - m - 10$

9. $3(5a - 2) - 20a + 10$

10. $6y + 3 + 2(3y - 6)$

11. $6y - 7 + 11y - y + 2$

12. $10 - x + 5x - 12 - 3x$

Find the perimeter of each figure.

△ **13.**

2x yards
3 yards
Rectangle

△ **14.**

Square 5y meters

Find the area of each figure.

△ **15.**

(2x − 1) yards
3 yards
Rectangle

△ **16.**

(x − 2) centimeters
(5x + 4) centimeters
10 centimeters
Rectangle
Rectangle
7 centimeters

(3.2) *Solve each equation.*

17. $z - 5 = -7$

18. $3x + 10 = 4x$

19. $3y = -21$

20. $-3a = -15$

204

21. $\dfrac{x}{-6} = 2$

22. $\dfrac{y}{-15} = -3$

23. $n + 18 = 10 - (-2)$

24. $c - 5 = -13 + 7$

25. $7x + 5 - 6x = -20$

26. $17x = 2(8x - 4)$

27. $5x + 7 = -3$

28. $-14 = 9y + 4$

29. $\dfrac{z}{4} = -8 - (-6)$

30. $-1 + (-8) = \dfrac{x}{5}$

31. $6y - 7y = 100 - 105$

32. $19x - 16x = 45 - 60$

33. $9(2x - 7) = 19x$

34. $-5(3x + 3) = -14x$

35. $3x - 4 = 11$

36. $6y + 1 = 73$

37. $2(x + 4) - 10 = -2(7)$

38. $-3(x - 6) + 13 = 20 - 1$

Translating *Translate each phrase into an algebraic expression. Let x represent "a number."*

39. The product of -5 and a number

40. Three subtracted from a number

41. The sum of -5 and a number

42. The quotient of -2 and a number

43. Eleven added to twice a number

44. The product of -5 and a number, decreased by 50

45. The quotient of 70 and the sum of a number and 6

46. Twice the difference of a number and 13

(3.3) *Solve each equation.*

47. $2x + 5 = 7x - 100$

48. $-6x - 4 = x + 66$

49. $2x + 7 = 6x - 1$

50. $5x - 18 = -4x$

51. $5(n - 3) = 7 + 3n$

52. $7(2 + x) = 4x - 1$

53. $6x + 3 - (-x) = -20 + 5x - 7$

54. $x - 25 + 2x = -5 + 2x - 10$

55. $3(x - 4) = 5x - 8$

56. $4(x - 3) = -2x - 48$

57. $6(2n - 1) + 18 = 0$

58. $7(3y - 2) - 7 = 0$

59. $95x - 14 = 20x - 10 + 10x - 4$

60. $32z + 11 - 28z = 50 + 2z - (-1)$

Translating *Write each sentence as an equation.*

61. The difference of 20 and -8 is 28.

62. Five times the sum of 2 and -6 yields -20.

63. The quotient of -75 and the sum of 5 and 20 is equal to -3.

64. Nineteen subtracted from -2 amounts to -21.

(3.4) Translating *Write each sentence as an equation using x as the variable.*

65. Twice a number minus 8 is 40.

66. The product of some number and 6 is equal to the sum of the number and 2.

67. Twelve subtracted from the quotient of a number and 2 is 10.

68. The difference of a number and 3 is the quotient of the number and 4.

Solve.

69. Five times a number subtracted from 40 is the same as three times the number. Find the number.

70. The product of a number and 3 is twice the difference of that number and 8. Find the number.

71. In an election the incumbent received 14,000 votes of the 18,500 votes cast. Of the remaining votes, the Democratic candidate received 272 more than the Independent candidate. Find how many votes the Democratic candidate received.

72. Rajiv Puri has twice as many movies on DVDs as he has video tapes. Find the number of DVDs if he has a total of 126 movie recordings.

Mixed Review

Simplify.

73. $9x - 20x$

74. $-5(7x)$

75. $12x + 5(2x - 3) - 4$

76. $-7(x + 6) - 2(x - 5)$

Solve.

77. $c - 5 = -13 + 7$

78. $7x + 5 - 6x = -20$

79. $-7x + 3x = -50 - 2$

80. $-x + 8x = -38 - 4$

81. $9x + 12 - 8x = -6 + (-4)$

82. $-17x + 14 + 20x - 2x = 5 - (-3)$

83. $5(2x - 3) = 11x$

84. $\dfrac{y}{-3} = -1 - 5$

85. $12y - 10 = -70$

86. $-6(x - 3) = x + 4$

87. $4n - 8 = 2n + 14$

88. $9(3x - 4) + 63 = 0$

89. $-5z + 3z - 7 = 8z - 1 - 6$

90. $4x - 3 + 6x = 5x - 3 - 30$

91. Three times a number added to twelve is 27. Find the number.

92. Twice the sum of a number and four is ten. Find the number.

93. Out of the 50 states, Hawaii has the least number of roadway miles followed by Delaware. If Delaware has 1585 more roadway miles than Hawaii and the total number of roadway miles for both states is 10,203, find the number of roadway miles for each state.

94. North and South Dakota both have over 80,000 roadway miles. North Dakota has 3094 more miles and the total number of roadway miles for both states is 170,470. Find the number of roadway miles for North Dakota and for South Dakota.

3 CHAPTER TEST

Remember to check your answers and use the Chapter Test Prep Video to view solutions.

1. Simplify $7x - 5 - 12x + 10$ by combining like terms.

2. Multiply: $-2(3y + 7)$

3. Simplify: $-(3z + 2) - 5z - 18$

△ **4.** Write an expression that represents the perimeter of the equilateral triangle. Simplify the expression. (A triangle with three sides of equal length.)

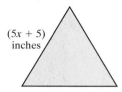

$(5x + 5)$ inches

5. Write an expression that represents the area of the rectangle. Simplify the expression.

4 meters

Rectangle $(3x - 1)$ meters

Solve each equation.

6. $12 = y - 3y$

7. $\dfrac{x}{2} = -5 - (-2)$

8. $5x + 12 - 4x - 14 = 22$

9. $-4x + 7 = 15$

10. $2(x - 6) = 0$

11. $-4(x - 11) - 34 = 10 - 12$

12. $5x - 2 = x - 10$

13. $4(5x + 3) = 2(7x + 6)$

14. $6 + 2(3n - 1) = 28$

Translate the following phrases into mathematical expressions. If needed, use x to represent "a number."

15. The sum of -23 and a number.

16. Three times a number, subtracted from -2.

Answers

1. _____

2. _____

3. _____

4. _____

5. _____

6. _____

7. _____

8. _____

9. _____

10. _____

11. _____

12. _____

13. _____

14. _____

15. _____

16. _____

Translate each sentence into an equation. If needed, use x to represent "a number."

17. The sum of twice 5 and −15 is −5.

18. Six added to three times a number equals −30.

Solve.

19. The difference of three times a number and five times the same number is 4. Find the number.

20. In a championship basketball game, Paula Zimmerman made twice as many free throws as Maria Kaminsky. If the total number of free throws made by both women was 12, find how many free throws Paula made.

21. In a 10-kilometer race, there are 112 more men entered than women. Find the number of female runners if the total number of runners in the race is 600.

17. _____

18. _____

19. _____

20. _____

21. _____

Answers

1. Write 308,063,557 in words.

2. Write 276,004 in words.

△ 3. Find the perimeter of the polygon shown.

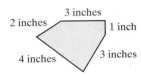

4. Find the perimeter of the rectangle shown.

1. _____

5. Subtract: $900 - 174$. Check by adding.

6. Subtract: $17,801 - 8216$. Check by adding.

2. _____

3. _____

4. _____

7. Round 248,982 to the nearest hundred.

8. Round 844,497 to the nearest thousand.

5. _____

6. _____

9. Multiply: 25×8

10. Multiply: 395×74

7. _____

11. Divide and check: $1872 \div 9$

12. Divide and check: $3956 \div 46$

8. _____

9. _____

13. Simplify: $2 \cdot 4 - 3 \div 3$

14. Simplify: $8 \cdot 4 + 9 \div 3$

10. _____

11. _____

15. Evaluate $x^2 + z - 3$ for $x = 5$ and $z = 4$.

16. Evaluate $2a^2 + 5 - c$ for $a = 2$ and $c = 3$.

12. _____

13. _____

14. _____

17. Determine which numbers in the set $\{26, 40, 20\}$ are solutions of the equation $2n - 30 = 10$.

18. Insert $<$ or $>$ to make a true statement.
    a. -14 0
    b. $-(-7)$ -8

15. _____

16. _____

19. Add using a number line: $5 + (-2)$

20. Add using a number line: $-3 + (-4)$

17. _____

18. a. _____

    b. _____

Add.

21. $-15 + (-10)$

22. $3 + (-7)$

19. _____

20. _____

21. _____

22. _____

210

23. $-2 + 5$

24. $21 + 15 + (-19)$

Subtract.

25. $-4 - 10$

26. $-2 - 3$

27. $6 - (-5)$

28. $19 - (-10)$

29. $-11 - (-7)$

30. $-16 - (-13)$

Divide.

31. $\dfrac{-12}{6}$

32. $\dfrac{-30}{-5}$

33. $-20 \div (-4)$

34. $26 \div (-2)$

35. $\dfrac{48}{-3}$

36. $\dfrac{-120}{12}$

Find the value of each expression.

37. $(-3)^2$

38. -2^5

39. -3^2

40. $(-5)^2$

41. Simplify: $2y - 6 + 4y + 8$

42. Simplify: $6x + 2 - 3x + 7$

43. Determine whether -1 is a solution of the equation $3y + 1 = 3$.

44. Determine whether 2 is a solution of $5x - 3 = 7$.

45. Solve: $-12x = -36$

46. Solve: $-3y = 15$

47. Solve: $2x - 6 = 18$

48. Solve: $3a + 5 = -1$

49. A salesperson at an electronics store sold a computer system and software for $2100, receiving four times as much money for the computer system as for the software. Find the price of each.

50. Rose Daunis is thinking of a number. Two times the number plus four is the same amount as three times the number minus seven. Find Rose's number.

23. _____

24. _____

25. _____

26. _____

27. _____

28. _____

29. _____

30. _____

31. _____

32. _____

33. _____

34. _____

35. _____

36. _____

37. _____

38. _____

39. _____

40. _____

41. _____

42. _____

43. _____

44. _____

45. _____

46. _____

47. _____

48. _____

49. _____

50. _____

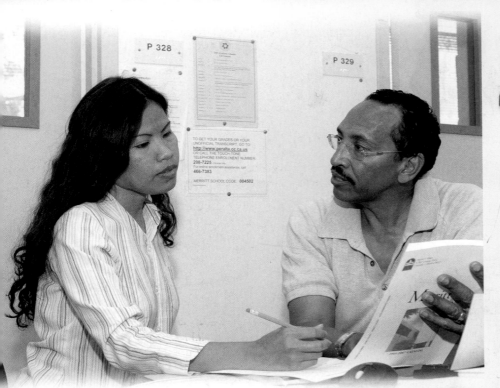

4

Fractions and Mixed Numbers

Fractions are numbers and, like whole numbers and integers, they can be added, subtracted, multiplied, and divided. Fractions are very useful and appear frequently in everyday language, in common phrases such as "half an hour," "quarter of a pound," and "third of a cup." This chapter reviews the concept of fractions and mixed numbers and demonstrates how to add, subtract, multiply, and divide these numbers.

The following graph is called a circle graph or pie chart. Each sector (shaped like a piece of pie) shows the fraction of entering college freshmen in the United States who expect to major in each discipline shown. The whole circle represents the entire class of college freshmen. In Section 4.2, Exercises 92–95, Section 4.3, Exercise 85, and Section 4.5, Exercises 95–98, we study this circle graph further and use the information to project the number of college freshmen in a certain major.

College Freshmen Majors

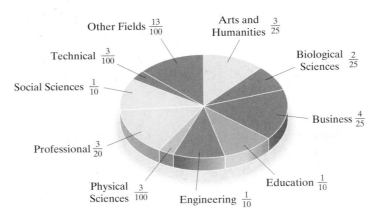

Other Fields $\frac{13}{100}$

Arts and Humanities $\frac{3}{25}$

Technical $\frac{3}{100}$

Biological Sciences $\frac{2}{25}$

Social Sciences $\frac{1}{10}$

Business $\frac{4}{25}$

Professional $\frac{3}{20}$

Physical Sciences $\frac{3}{100}$

Engineering $\frac{1}{10}$

Education $\frac{1}{10}$

Source: The Higher Education Research Institute

4.1 INTRODUCTION TO FRACTIONS AND MIXED NUMBERS

Objectives

A Identify the Numerator and the Denominator of a Fraction.

B Write a Fraction to Represent Parts of Figures or Real-Life Data.

C Graph Fractions on a Number Line.

D Review Division Properties for 0 and 1.

E Write Mixed Numbers as Improper Fractions.

F Write Improper Fractions as Mixed Numbers or Whole Numbers.

Objective **A** Identifying Numerators and Denominators

Whole numbers are used to count whole things or units, such as cars, horses, dollars, and people. To refer to a part of a whole, **fractions** can be used. Here are some examples of **fractions.** Study these examples for a moment.

$\frac{1}{2}$ of a cup $\frac{2}{3}$ of a foot $\frac{5}{6}$ of a pizza

In a fraction, the top number is called the **numerator** and the bottom number is called the **denominator.** The bar between the numbers is called the **fraction bar.**

Names	Fraction	Meaning
numerator $\longrightarrow$ denominator $\longrightarrow$	$\dfrac{5}{6}$	$\longleftarrow$ number of parts being considered $\longleftarrow$ number of equal parts in the whole

EXAMPLES Identify the numerator and the denominator of each fraction.

1. $\dfrac{3}{7}$ ← numerator
   ← denominator

2. $\dfrac{13}{5x}$ ← numerator
   ← denominator

⬛ **Work Practice Problems 1–2**

Helpful Hint

$\dfrac{3}{7}$ ← Remember that the bar in a fraction means division. Since division by 0 is undefined, a fraction with a denominator of 0 is undefined. For example, $\dfrac{3}{0}$ is undefined.

Objective **B** Writing Fractions to Represent Parts of Figures or Real-Life Data

One way to become familiar with the concept of fractions is to visualize fractions with shaded figures. We can then write a fraction to represent the shaded area of the figure (or diagram).

PRACTICE PROBLEMS 1–2

Identify the numerator and the denominator of each fraction.

1. $\dfrac{11}{2}$

2. $\dfrac{10y}{17}$

Answers

1. numerator = 11, denominator = 2
2. numerator = 10y, denominator = 17

PRACTICE PROBLEMS 3–4

Write a fraction to represent the shaded part of each figure.

3.

4.

EXAMPLES Write a fraction to represent the shaded part of each figure.

3. In this figure, 2 of the 5 equal parts are shaded. Thus, the fraction is $\frac{2}{5}$.

$\dfrac{2}{5}$ ← number of parts shaded
      ← number of equal parts

4. In this figure, 3 of the 10 rectangles are shaded. Thus, the fraction is $\frac{3}{10}$.

$\dfrac{3}{10}$ ← number of parts shaded
      ← number of equal parts

🔲 **Work Practice Problems 3–4**

PRACTICE PROBLEMS 5–6

Write a fraction to represent the part of the whole shown.

5. Just consider this part of the syringe

6. |← Whole part →|

EXAMPLES Write a fraction to represent the shaded part of the diagram.

5.

The fraction is $\frac{3}{10}$.

10 equal parts

3 shaded parts

6.

The fraction is $\frac{1}{3}$.

3 equal parts

1 part shaded

🔲 **Work Practice Problems 5–6**

PRACTICE PROBLEM 7

Draw and shade a part of a figure to represent the fraction.

7. $\frac{2}{3}$ of a figure

EXAMPLES Draw a figure and then shade a part of it to represent each fraction.

7. $\frac{5}{6}$ of a figure

We will use a geometric figure such as a rectangle. Since the denominator is 6, we divide it into 6 equal parts. Then we shade 5 of the equal parts.

5 parts shaded

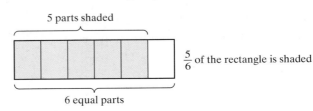

$\frac{5}{6}$ of the rectangle is shaded

6 equal parts

Answers

3. $\frac{3}{8}$ **4.** $\frac{1}{6}$ **5.** $\frac{7}{10}$ **6.** $\frac{9}{16}$

7. answers may vary; for example,

8. $\frac{3}{8}$ of a figure

If you'd like, our figure can consist of 8 triangles of the same size. We will shade 3 of the triangles.

3 triangles shaded

8 triangles

$\frac{3}{8}$ of the figure or diagram is shaded

▣ **Work Practice Problems 7–8**

✔**Concept Check** If ⬜⬜⬜⬜⬜ represents $\frac{6}{7}$ of a whole diagram, sketch the whole diagram.

EXAMPLE 9 **Writing Fractions from Real-Life Data**

Of the eight planets in our solar system, (Pluto is now a dwarf planet), three are closer to the Sun than Mars. What fraction of the planets are closer to the Sun than Mars?

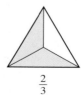

Solution: The fraction of planets closer to the Sun than Mars is:

$\frac{3}{8}$ ← number of planets closer
← number of planets in our solar system

Thus, $\frac{3}{8}$ of the planets in our solar system are closer to the Sun than Mars.

▣ **Work Practice Problem 9**

A **proper fraction** is a fraction whose numerator is less than its denominator. Proper fractions are less than 1. For example, the shaded portion of the triangle is represented by $\frac{2}{3}$.

An **improper fraction** is a fraction whose numerator is greater than or equal to its denominator. Improper fractions are greater than or equal to 1.

$\frac{2}{3}$

PRACTICE PROBLEM 8

Draw and shade a part of a figure to represent the fraction.

8. $\frac{7}{11}$ of a figure

PRACTICE PROBLEM 9

Of the eight planets in our solar system, five are farther from the Sun than Earth is. What fraction of the planets are farther from the Sun than Earth is?

Answers

8. answers may vary; for example,

9. $\frac{5}{8}$

✔ **Concept Check Answer**

⬜⬜⬜⬜⬜⬜⬜

The shaded part of the group of circles below is $\frac{9}{4}$. The shaded part of the rectangle is $\frac{6}{6}$. Recall that $\frac{6}{6}$ simplifies to 1 and notice that the entire rectangle (1 whole figure) is shaded below.

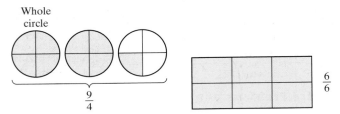

A **mixed number** contains a whole number and a fraction. Mixed numbers are greater than 1. Earlier, we wrote the shaded part of the group of circles below as the improper fraction $\frac{9}{4}$. Now let's write the shaded part as a mixed number. The shaded part of the group of circles' area is $2\frac{1}{4}$. Read this as "two and one-fourth."

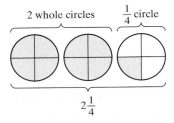

> Note: The mixed number $2\frac{1}{4}$, diagramed above, represents $2 + \frac{1}{4}$.
>
> The mixed number $-3\frac{1}{5}$ represents $-\left(3 + \frac{1}{5}\right)$ or $-3 - \frac{1}{5}$. We review this later in this chapter.

PRACTICE PROBLEMS 10–11

Represent the shaded part of each figure group as both an improper fraction and a mixed number.

10.

11.

EXAMPLES Represent the shaded part of each figure group as both an improper fraction and a mixed number.

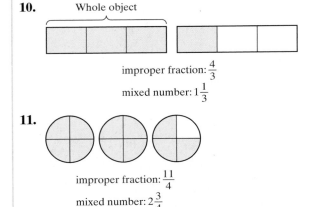

10. Whole object

improper fraction: $\frac{4}{3}$

mixed number: $1\frac{1}{3}$

11.

improper fraction: $\frac{11}{4}$

mixed number: $2\frac{3}{4}$

■ **Work Practice Problems 10–11**

Answers

10. $\frac{8}{3}, 2\frac{2}{3}$ **11.** $\frac{5}{4}, 1\frac{1}{4}$

✔ **Concept Check Answer**

3, answers may vary

✔**Concept Check**   If you were to round $2\frac{3}{4}$, shown in Example 11 above, to the nearest a whole number, would you choose 2 or 3? Why?

Objective C Graphing Fractions on a Number Line

Another way to visualize fractions is to graph them on a number line. To do this, think of 1 unit on the number line as a whole. To graph $\frac{2}{5}$, for example, divide the distance from 0 to 1 into 5 equal parts. Then start at 0 and count 2 parts to the right.

Notice that the graph of $\frac{2}{5}$ lies between 0 and 1. This means

$$0 < \frac{2}{5} \left(\text{or } \frac{2}{5} > 0 \right) \text{ and also } \frac{2}{5} < 1$$

EXAMPLE 12 Graph each proper fraction on a number line.

a. $\frac{3}{4}$ **b.** $\frac{1}{2}$ **c.** $\frac{3}{6}$

Solution:

a. To graph $\frac{3}{4}$, divide the distance from 0 to 1 into 4 parts. Then start at 0 and count over 3 parts.

b.

c.

Work Practice Problem 12

The fractions in Example 12 are all proper fractions. Notice that the value of each is less than 1. This is always true for proper fractions since the numerator of a proper fraction is less than the denominator.

On the next page, we graph improper fractions. Notice that improper fractions are greater than or equal to 1. This is always true since the numerator of an improper fraction is greater than or equal to the denominator.

PRACTICE PROBLEM 12

Graph each proper fraction on a number line.

a. $\frac{5}{7}$ **b.** $\frac{2}{3}$ **c.** $\frac{4}{6}$

Answers

12. **a.**

b.

c.

PRACTICE PROBLEM 13

Graph each improper fraction on a number line.

a. $\dfrac{8}{3}$ b. $\dfrac{5}{4}$ c. $\dfrac{7}{7}$

EXAMPLE 13 Graph each improper fraction on a number line.

a. $\dfrac{7}{6}$ b. $\dfrac{9}{5}$ c. $\dfrac{6}{6}$ d. $\dfrac{3}{1}$

Solution:

a.

b.

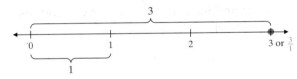

c.

d. Each 1-unit distance has 1 equal part. Count over 3 parts.

■ **Work Practice Problem 13**

Note: We will graph mixed numbers at the end of this chapter.

Objective **D** **Reviewing Division Properties for 0 and 1**

Before we continue further, don't forget from Section 1.6 that the fraction bar indicates division. Let's review some division properties for 1 and 0.

$$\dfrac{9}{9} = 1 \text{ because } 1 \cdot 9 = 9 \qquad \dfrac{-11}{1} = -11 \text{ because } -11 \cdot 1 = -11$$

$$\dfrac{0}{6} = 0 \text{ because } 0 \cdot 6 = 0 \qquad \dfrac{6}{0} \text{ is undefined because there is no number that when multiplied by 0 gives 6.}$$

In general, we can say the following.

Let n be any integer except 0.

$$\dfrac{n}{n} = 1 \qquad \dfrac{0}{n} = 0$$

$$\dfrac{n}{1} = n \qquad \dfrac{n}{0} \text{ is undefined.}$$

Answers

13. a.

b.

c.

EXAMPLES Simplify.

14. $\dfrac{5}{5} = 1$ **15.** $\dfrac{-2}{-2} = 1$ **16.** $\dfrac{0}{-5} = 0$

17. $\dfrac{-5}{1} = -5$ **18.** $\dfrac{41}{1} = 41$ **19.** $\dfrac{19}{0}$ is undefined

■ **Work Practice Problems 14–19**

Notice from Example 18 that we can have negative fractions. In fact,

$$\dfrac{-5}{1} = -5, \qquad \dfrac{5}{-1} = -5, \qquad \text{and} \quad -\dfrac{5}{1} = -5$$

Because all of the fractions equal -5, we have

$$\dfrac{-5}{1} = \dfrac{5}{-1} = -\dfrac{5}{1}$$

This means that the negative sign in a fraction can be written in the numerator, the denominator, or in front of the fraction. Remember this as we work with negative fractions.

Objective **E** Writing Mixed Numbers as Improper Fractions

Earlier in this section, mixed numbers and improper fractions were both used to represent the shaded part of the figure groups. For example,

$1\dfrac{2}{3}$ or $\dfrac{5}{3}$ Thus $1\dfrac{2}{3} = \dfrac{5}{3}$.

The following steps may be used to write a mixed number as an improper fraction:

Writing a Mixed Number as an Improper Fraction

To write a mixed number as an improper fraction:

Step 1: Multiply the denominator of the fraction by the whole number.

Step 2: Add the numerator of the fraction to the product from Step 1.

Step 3: Write the sum from Step 2 as the numerator of the improper fraction over the original denominator.

For example,

$$1\dfrac{2}{3} = \dfrac{\overset{\text{Step 1}}{3 \cdot 1} + \overset{\text{Step 2}}{2}}{3} = \dfrac{3 + 2}{3} = \dfrac{5}{3} \quad \text{or} \quad 1\dfrac{2}{3} = \dfrac{5}{3}, \text{ as stated above.}$$

(Step 3)

EXAMPLE 20 Write each as an improper fraction.

a. $4\dfrac{2}{9} = \dfrac{9 \cdot 4 + 2}{9} = \dfrac{36 + 2}{9} = \dfrac{38}{9}$ **b.** $1\dfrac{8}{11} = \dfrac{11 \cdot 1 + 8}{11} = \dfrac{11 + 8}{11} = \dfrac{19}{11}$

■ **Work Practice Problem 20**

PRACTICE PROBLEMS 14–19

Simplify.

14. $\dfrac{9}{9}$ **15.** $\dfrac{-6}{-6}$

16. $\dfrac{0}{-1}$ **17.** $\dfrac{4}{1}$

18. $\dfrac{-13}{0}$ **19.** $\dfrac{-13}{1}$

Helpful Hint
Remember, for example, that

$$-\dfrac{2}{3} = \dfrac{-2}{3} = \dfrac{2}{-3}$$

PRACTICE PROBLEM 20

Write each as an improper fraction.

a. $5\dfrac{2}{7}$ **b.** $6\dfrac{2}{3}$ **c.** $10\dfrac{9}{10}$ **d.** $4\dfrac{1}{5}$

Answers

14. 1 **15.** 1 **16.** 0 **17.** 4
18. undefined **19.** -13

20. a. $\dfrac{37}{7}$ **b.** $\dfrac{20}{3}$ **c.** $\dfrac{109}{10}$ **d.** $\dfrac{21}{5}$

Objective F Writing Improper Fractions as Mixed Numbers or Whole Numbers

Just as there are times when an improper fraction is preferred, sometimes a mixed or a whole number better suits a situation. To write improper fractions as mixed or whole numbers, we use division. Recall once again from Section 1.6 that the fraction bar means division. This means that the fraction

$$\frac{5}{3} \begin{array}{l}\text{numerator}\\ \text{denominator}\end{array} \quad \text{means } 3\overline{)5} \begin{array}{l}\uparrow\quad\uparrow\\ \text{numerator}\\ \text{denominator}\end{array}$$

Writing an Improper Fraction as a Mixed Number or a Whole Number

To write an improper fraction as a mixed number or a whole number:

Step 1: Divide the denominator into the numerator.

Step 2: The whole number part of the mixed number is the quotient. The fraction part of the mixed number is the remainder over the original denominator.

$$\text{quotient}\frac{\text{remainder}}{\text{original denominator}}$$

For example,

Step 1
$$\frac{5}{3} : 3\overline{)5} \atop \underline{3} \atop 2 \atop 1$$

Step 2
$$\frac{5}{3} = 1\frac{2}{3} \begin{array}{l}\leftarrow \text{remainder}\\ \leftarrow \text{original denominator}\end{array}$$
$$\uparrow \atop \text{quotient}$$

PRACTICE PROBLEM 21

Write each as a mixed number or a whole number.

a. $\frac{9}{5}$ **b.** $\frac{23}{9}$ **c.** $\frac{48}{4}$

d. $\frac{62}{13}$ **e.** $\frac{51}{7}$ **f.** $\frac{21}{20}$

EXAMPLE 21 Write each as a mixed number or a whole number.

a. $\frac{30}{7}$ **b.** $\frac{16}{15}$ **c.** $\frac{84}{6}$

Solution:

a. $\frac{30}{7} : 7\overline{)30} \atop \underline{28} \atop 2$ $\frac{30}{7} = 4\frac{2}{7}$

b. $\frac{16}{15} : 15\overline{)16} \atop \underline{15} \atop 1$ $\frac{16}{15} = 1\frac{1}{15}$

c. $\frac{84}{6} : 6\overline{)84} \atop \underline{6} \atop 24 \atop \underline{24} \atop 0$ $\frac{84}{6} = 14$ Since the remainder is 0, the result is the whole number 14.

> **Helpful Hint**
>
> When the remainder is 0, the improper fraction is a whole number. For example, $\frac{92}{4} = 23$.
>
> $$4\overline{)92} \atop \underline{8} \atop 12 \atop \underline{12} \atop 0 \atop 23$$

Answers

21. **a.** $1\frac{4}{5}$ **b.** $2\frac{5}{9}$ **c.** 12 **d.** $4\frac{10}{13}$

e. $7\frac{2}{7}$ **f.** $1\frac{1}{20}$

▢ **Work Practice Problem 21**

Vocabulary and Readiness Check

Use the choices below to fill in each blank.

improper   fraction   proper

is undefined   mixed number $= 0$

≥ 1 denominator $= 1$

< 1 numerator

1. The number $\frac{17}{31}$ is called a _____. The number 31 is called its _____ and 17 is called its _____.

2. If we simplify each fraction, $\frac{-9}{-9}$ _____, $\frac{0}{-4}$ _____, and we say $\frac{-4}{0}$ _____.

3. The fraction $\frac{8}{3}$ is called a(n) _____ fraction, the fraction $\frac{3}{8}$ is called a(n) _____ fraction, and $10\frac{3}{8}$ is called a(n) _____.

4. The value of an improper fraction is always _____, and the value of a proper fraction is always _____.

4.1 EXERCISE SET

FOR EXTRA HELP

Student Solutions Manual   PH Math/Tutor Center   CD/Video for Review   Math XL MathXL®   MyMathLab MyMathLab

Objective A *Identify the numerator and the denominator of each fraction and identify each fraction as proper or improper. See Examples 1 and 2.*

 1. $\frac{1}{2}$

2. $\frac{1}{4}$

3. $\frac{10}{3}$

4. $\frac{53}{21}$

5. $\frac{15}{15}$

6. $\frac{26}{26}$

Objective B *Write a proper or improper fraction to represent the shaded part of each. If an improper fraction is appropriate, write the shaded figure as (a) an improper fraction and (b) a mixed number. See Examples 3 through 6 and 10 and 11.*

7.

8.

9.

10.

11.

12.

13.

14.

15.

16.

17.

18.

19.

20.

21.

22.

23.

24.

25.

1 whole mile

26.

1 whole inch

Objective **B** *Draw and shade a part of a diagram to represent each fraction. See Examples 7 and 8.*

27. $\frac{1}{5}$ of a diagram

28. $\frac{1}{16}$ of a diagram

29. $\frac{6}{7}$ of a diagram

30. $\frac{7}{9}$ of a diagram

31. $\frac{4}{4}$ of a diagram

32. $\frac{6}{6}$ of a diagram

Write each fraction. See Example 9.

33. Of the 131 students at a small private school, 42 are freshmen. What fraction of the students are freshmen?

34. Of the 63 employees at a new biomedical engineering firm, 22 are men. What fraction of the executives are men?

35. Use Exercise 33 to answer a and b.

   a. How many students are *not* freshmen?

   b. What fraction of the students are *not* freshmen?

36. Use Exercise 34 to answer a and b.

   a. How many of the employees are women?

   b. What fraction of the employees are women?

37. As of 2006, the United States has had 43 different presidents. A total of eight U.S. presidents were born in the state of Virginia, more than any other state. What fraction of U.S. presidents were born in Virginia? (*Source: 2006 World Almanac and Book of Facts*)

38. Of the eight planets in our solar system, four have days that are longer than the 24-hour Earth day. What fraction of the planets have longer days than Earth has? (*Source:* National Space Science Data Center)

39. The Atlantic hurricane season of 2005 rewrote the record books. There were 28 tropical storms, 15 of which turned into hurricanes. What fraction of the 2005 Atlantic tropical storms escalated to hurricanes?

40. There are 12 inches in a foot. What fractional part of a foot does 5 inches represent?

41. There are 31 days in the month of March. What fraction of the month does 11 days represent?

Mon.	Tue.	Wed.	Thu.	Fri.	Sat.	Sun.
					1	2
3	4	5	6	7	8	9
10	11	12	13	14	15	16
17	18	19	20	21	22	23
24	25	26	27	28	29	30
31						

42. There are 60 minutes in an hour. What fraction of an hour does 37 minutes represent?

43. In a prealgebra class containing 31 students, there are 18 freshmen, 10 sophomores, and 3 juniors. What fraction of the class is sophomores?

44. In a sports team with 20 children, there are 9 boys and 11 girls. What fraction of the team is boys?

45. Thirty-three out of the fifty total states in the United States contain federal Indian reservations.

   a. What fraction of the states contain federal Indian reservations?

   b. How many states do not contain federal Indian reservations?

   c. What fraction of the states do not contain federal Indian reservations? (*Source:* Tiller Research, Inc., Albuquerque, NM)

46. Consumer fireworks are legal in 40 out of the 50 total states in the United States.

   a. In what fraction of the states are consumer fireworks legal?

   b. In how many states are consumer fireworks illegal?

   c. In what fraction of the states are consumer fireworks illegal? (*Source:* United States Fireworks Safety Council)

47. A bag contains 50 red or blue marbles. If 21 marbles are blue,

    a. What fraction of the marbles are blue?

    b. How many marbles are red?

    c. What fraction of the marbles are red?

48. An art dealer is taking inventory. His shop contains a total of 37 pieces, which are all sculptures, watercolor paintings, or oil paintings. If there are 15 watercolor paintings and 17 oil paintings, answer each question.

    a. What fraction of the inventory is watercolor paintings?

    b. What fraction of the inventory is oil paintings?

    c. How many sculptures are there?

    d. What fraction of the inventory is sculptures?

Objective C *Graph each fraction on a number line. See Examples 12 and 13.*

49. $\dfrac{1}{4}$

50. $\dfrac{1}{3}$

51. $\dfrac{4}{7}$

52. $\dfrac{5}{6}$

53. $\dfrac{8}{5}$

54. $\dfrac{9}{8}$

55. $\dfrac{7}{3}$

56. $\dfrac{13}{7}$

Objective D *Simplify by dividing. See Examples 14 through 19.*

57. $\dfrac{12}{12}$

58. $\dfrac{-3}{-3}$

59. $\dfrac{-5}{1}$

60. $\dfrac{-20}{1}$

61. $\dfrac{0}{-2}$

62. $\dfrac{0}{-8}$

63. $\dfrac{-8}{-8}$

64. $\dfrac{-14}{-14}$

65. $\dfrac{-9}{0}$

66. $\dfrac{-7}{0}$

67. $\dfrac{3}{1}$

68. $\dfrac{5}{5}$

Objective E *Write each mixed number as an improper fraction. See Example 20.*

69. $2\dfrac{1}{3}$

70. $1\dfrac{13}{17}$

71. $3\dfrac{3}{5}$

72. $2\dfrac{5}{9}$

73. $6\dfrac{5}{8}$

74. $7\dfrac{3}{8}$

75. $11\dfrac{6}{7}$

76. $12\dfrac{2}{5}$

77. $9\dfrac{7}{20}$

78. $10\dfrac{14}{27}$

79. $166\dfrac{2}{3}$

80. $114\dfrac{2}{7}$

Objective **F** *Write each improper fraction as a mixed number or a whole number. See Example 21.*

81. $\dfrac{17}{5}$

82. $\dfrac{13}{7}$

83. $\dfrac{37}{8}$

84. $\dfrac{64}{9}$

85. $\dfrac{47}{15}$

86. $\dfrac{65}{12}$

87. $\dfrac{225}{15}$

88. $\dfrac{196}{14}$

89. $\dfrac{182}{175}$

90. $\dfrac{149}{143}$

91. $\dfrac{737}{112}$

92. $\dfrac{901}{123}$

Review

Simplify. See Section 1.7.

93. 3^2

94. 4^3

95. 5^3

96. 3^4

Concept Extensions

Write each fraction in two other equivalent ways by inserting the negative sign in different places.

97. $-\dfrac{11}{2} = \quad = $

98. $-\dfrac{21}{4} = \quad = $

99. $\dfrac{-13}{15} = \quad = $

100. $\dfrac{45}{-57} = \quad = $

101. In your own words, explain why $\dfrac{0}{10} = 0$ and $\dfrac{10}{0}$ are undefined.

Solve. See the Concept Checks in this section.

102. If ◯◯◯◯ represents $\dfrac{4}{9}$ of a whole diagram, sketch the whole diagram.

103. If △ △ represents $\dfrac{1}{3}$ of a whole diagram, sketch the whole diagram.

104. Round the mixed number $7\dfrac{1}{8}$ to the nearest whole number.

105. Round the mixed number $5\dfrac{11}{12}$ to the nearest whole number.

106. The Gap Corporation owns stores with four different brand names, as shown on the bar graph. What fraction of the stores owned by The Gap Corporation are named "Old Navy"?

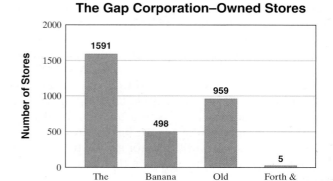

The Gap Corporation–Owned Stores

Number of Stores

The Gap: 1591
Banana Republic: 498
Old Navy: 959
Forth & Towne: 5

Store Names

107. The Public Broadcasting Service (PBS) provides programming to the noncommercial public TV stations of the United States. The bar graph shows a breakdown of the public television licensees by type. Each licensee operates one or more PBS member TV stations. What fraction of the public television licensees are universities or colleges? (*Source:* The Public Broadcast Service)

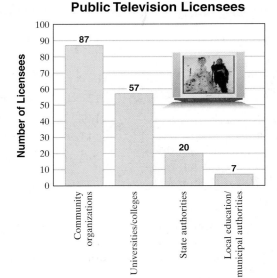

Public Television Licensees

108. Habitat for Humanity is a nonprofit organization that helps provide affordable housing to families in need. Habitat for Humanity does its work of building and renovating houses through 1651 local affiliates in the United States and 634 international affiliates. What fraction of the total Habitat for Humanity affiliates are located in the United States? (*Hint:* First find the total number of affiliates.) (*Source:* Habitat for Humanity International)

109. The United States Marine Corps (USMC) has five principal training centers in California, three in North Carolina, two in South Carolina, one in Arizona, one in Hawaii, and one in Virginia. What fraction of the total USMC principal training centers are located in California? (*Source:* U.S. Department of Defense)

STUDY SKILLS BUILDER

Are You Organized?

Have you ever had trouble finding a completed assignment? When it's time to study for a test, are your notes neat and organized? Have you ever had trouble reading your own mathematics handwriting? (Be honest—I have.)

When any of these things happen, it's time to get organized. Here are a few suggestions:

Write your notes and complete your homework assignment in a notebook with pockets (spiral or ring binder.) Take class notes in this notebook, and then follow the notes with your completed homework assignment. When you receive graded papers or handouts, place them in the notebook pocket so that you will not lose them.

Remember to mark (possibly with an exclamation point) any note(s) that seem extra important to you. Also remember to mark (possibly with a question mark) any notes or homework that you are having trouble with. Don't forget to see your instructor or a math tutor to help you with the concepts or exercises that you are having trouble understanding.

Also, if you are having trouble reading your own handwriting, *slow down* and write your mathematics work clearly!

Exercises

1. Have you been completing your assignments on time?

2. Have you been correcting any exercises you may be having difficulty with?

3. If you are having trouble with a mathematical concept or correcting any homework exercises, have you visited your instructor, a tutor, or your campus math lab?

4. Are you taking lecture notes in your mathematics course? (By the way, these notes should include worked-out examples solved by your instructor.)

5. Is your mathematics course material (handouts, graded papers, lecture notes) organized?

6. If your answer to Exercise 5 is no, take a moment and review your course material. List at least two ways that you might better organize it. Then read the Study Skills Builder on organizing a notebook in Section 2.3.

4.2 FACTORS AND SIMPLEST FORM

Objectives

A) Write a Number as a Product of Prime Numbers.

B) Write a Fraction in Simplest Form.

C) Determine Whether Two Fractions Are Equivalent.

D) Solve Problems by Writing Fractions in Simplest Form.

Objective A) Writing a Number as a Product of Prime Numbers

Recall from Section 1.5 that since $12 = 2 \cdot 2 \cdot 3$, the numbers 2 and 3 are called *factors* of 12. A **factor** is any number that divides a number evenly (with a remainder of 0).

To perform operations on fractions, it is necessary to be able to factor a number. Remember that factoring a number means writing a number as a product. We first practice writing a number as a product of prime numbers.

A **prime number** is a natural number greater than 1 whose only factors are 1 and itself. The first few prime numbers are 2, 3, 5, 7, 11, 13, 17, 19, 23, 29, ...
A **composite number** is a natural number greater than 1 that is not prime.

Helpful Hint

The natural number 1 is neither prime nor composite.

When a composite number is written as a product of prime numbers, this product is called the **prime factorization** of the number. For example, the prime factorization of 12 is $2 \cdot 2 \cdot 3$ because

$12 = \underbrace{2 \cdot 2 \cdot 3}$

This product is 12 and each number is a prime number.

Because multiplication is commutative, the order of the factors is not important. We can write the factorization $2 \cdot 2 \cdot 3$ as $2 \cdot 3 \cdot 2$ or $3 \cdot 2 \cdot 2$. Any of these is called the prime factorization of 12.

Every whole number greater than 1 has exactly one prime factorization.

One method for finding the prime factorization of a number is by using a factor tree, as shown in the next example.

EXAMPLE 1 Write the prime factorization of 45.

Solution: We can begin by writing 45 as the product of two numbers, say 5 and 9.

The number 5 is prime but 9 is not, so we write 9 as $3 \cdot 3$.

A factor tree

Each factor is now a prime number, so the prime factorization of 45 is $3 \cdot 3 \cdot 5$ or $3^2 \cdot 5$.

■ **Work Practice Problem 1**

PRACTICE PROBLEM 1

Use a factor tree to find the prime factorization of each number.

a. 30   **b.** 56   **c.** 72

Answers

1. a. $2 \cdot 3 \cdot 5$ **b.** $2^3 \cdot 7$ **c.** $2^3 \cdot 3^2$

✔ **Concept Check**   True or false? Two different numbers can have exactly the same prime factorization. Explain your answer.

PRACTICE PROBLEM 2

Write the prime factorization of 60.

EXAMPLE 2   Write the prime factorization of 80.

Solution: Write 80 as a product of two numbers. Continue this process until all factors are prime.

All factors are now prime, so the prime factorization of 80 is

$$2 \cdot 2 \cdot 2 \cdot 2 \cdot 5 \quad \text{or} \quad 2^4 \cdot 5.$$

▨ **Work Practice Problem 2**

Helpful Hint

It makes no difference which factors you start with. The prime factorization of a number will be the same.

Same factors as in Example 2

There are a few quick **divisibility tests** to determine whether a number is divisible by the primes 2, 3, or 5. (A number is divisible by 2, for example, if 2 divides it evenly so that the remainder is 0.)

Divisibility Tests

A whole number is divisible by:

- **2** if the last digit is 0, 2, 4, 6, or 8.

    132 is divisible by 2 since the last digit is a 2.

- **3** if the sum of the digits is divisible by 3.

    144 is divisible by 3 since $1 + 4 + 4 = 9$ is divisible by 3.

- **5** if the last digit is 0 or 5.

    1115 is divisible by 5 since the last digit is a 5.

Answer

2. $2^2 \cdot 3 \cdot 5$

✔ **Concept Check Answer**

false; answers may vary

Helpful
Hint

Here are a few other divisibility tests you may want to use. A whole number is divisible by:

- **4** if its last two digits are divisible by 4.

    1712 is divisible by 4.

- **6** if it's divisible by 2 and 3.

    9858 is divisible by 6.

- **9** if the sum of its digits is divisible by 9.

    5238 is divisible by 9 since $5 + 2 + 3 + 8 = 18$ is divisible by 9.

When finding the prime factorization of larger numbers, you may want to use the procedure shown in Example 3.

EXAMPLE 3 Write the prime factorization of 252.

Solution: For this method, we divide prime numbers into the given number. Since the ones digit of 252 is 2, we know that 252 is divisible by 2.

$$\begin{array}{r} 126 \\ 2\overline{)252} \end{array}$$

126 is divisible by 2 also.

$$\begin{array}{r} 63 \\ 2\overline{)126} \\ 2\overline{)252} \end{array}$$

63 is not divisible by 2 but is divisible by 3. Divide 63 by 3 and continue in this same manner until the quotient is a prime number.

$$\begin{array}{r} 7 \\ 3\overline{)\;21} \\ 3\overline{)\;63} \\ 2\overline{)126} \\ 2\overline{)252} \end{array}$$

Helpful
Hint

The order of choosing prime numbers does not matter. For consistency, we use the order 2, 3, 5, 7, ...

The prime factorization of 252 is $2 \cdot 2 \cdot 3 \cdot 3 \cdot 7$ or $2^2 \cdot 3^2 \cdot 7$.

Work Practice Problem 3

In this text, we will write the factorization of a number from the smallest factor to the largest factor.

✔**Concept Check** True or false? The prime factorization of 117 is $9 \cdot 13$. Explain your reasoning.

PRACTICE PROBLEM 3

Write the prime factorization of 297.

Answer

3. $3^3 \cdot 11$

✔ **Concept Check Answer**

No; $3 \cdot 3 \cdot 13$

Objective B Writing Fractions in Simplest Form

Fractions that represent the same portion of a whole or the same point on a number line are called **equivalent fractions.** Study the table below to see two ways to visualize equivalent fractions.

Equivalent Fractions	
Figures	**Number Line**
When we shade $\frac{1}{3}$ and $\frac{2}{6}$ on the same-sized figures,	When we graph $\frac{1}{3}$ and $\frac{2}{6}$ on a number line,
notice that both $\frac{1}{3}$ and $\frac{2}{6}$ represent the same portion of a whole. These fractions are called **equivalent fractions** and we write $\frac{1}{3} = \frac{2}{6}$.	notice that both $\frac{1}{3}$ and $\frac{2}{6}$ correspond to the same point. These fractions are called **equivalent fractions,** and we write $\frac{1}{3} = \frac{2}{6}$.
Thus, $\frac{1}{3} = \frac{2}{6}$ and $\frac{1}{3}$ and $\frac{2}{6}$ are equivalent.	

$$\frac{2}{3} \qquad \frac{4}{6} \qquad \frac{8}{12}$$

For example, $\frac{2}{3}, \frac{4}{6}$, and $\frac{8}{12}$ all represent the same shaded portion of the rectangle's area, so they are equivalent fractions. To show that these fractions are equivalent, we place an equals sign between them. In other words,

$$\frac{2}{3} = \frac{4}{6} = \frac{8}{12}$$

There are many equivalent forms of a fraction. A special equivalent form of a fraction is called **simplest form.**

Simplest Form of a Fraction

A fraction is written in **simplest form** or **lowest terms** when the numerator and the denominator have no common factors other than 1.

For example, the fraction $\frac{2}{3}$ is in simplest form because 2 and 3 have no common factor other than 1. The fraction $\frac{4}{6}$ is *not* in simplest form because 4 and 6 both have a factor of 2. That is, 2 is a common factor of 4 and 6. The process of writing a fraction in simplest form is called **simplifying** the fraction.

To simplify $\frac{4}{6}$ and write it as $\frac{2}{3}$, let's first study a few properties. Recall from Section 4.1 that any nonzero whole number n divided by itself is 1.

Any nonzero number n divided by itself is 1.

$$\frac{5}{5} = 1, \quad \frac{17}{17} = 1, \quad \frac{24}{24} = 1, \text{ or, in general, } \frac{n}{n} = 1$$

Also, in general, if $\frac{a}{b}$ and $\frac{c}{d}$ are fractions (with b and d not 0), the following is true.

$$\frac{a \cdot c}{b \cdot d} = \frac{a}{b} \cdot \frac{c}{d}*$$

These properties allow us to do the following:

$$\frac{4}{6} = \frac{2 \cdot 2}{2 \cdot 3} = \frac{2}{2} \cdot \frac{2}{3} = 1 \cdot \frac{2}{3} = \frac{2}{3}$$ When 1 is multiplied by a number, the result is the same number.

$\llcorner$ This is 1

EXAMPLE 4   Write in simplest form: $\frac{12}{20}$

Solution: Notice that 12 and 20 have a common factor of 4.

$$\frac{12}{20} = \frac{4 \cdot 3}{4 \cdot 5} = \frac{4}{4} \cdot \frac{3}{5} = 1 \cdot \frac{3}{5} = \frac{3}{5}$$

Since 3 and 5 have no common factors (other than 1), $\frac{3}{5}$ is in simplest form.

Work Practice Problem 4

PRACTICE PROBLEM 4

Write in simplest form: $\frac{30}{45}$

If you have trouble finding common factors, write the prime factorization of the numerator and the denominator.

EXAMPLE 5   Write in simplest form: $\frac{42x}{66}$

Solution: Let's write the prime factorizations of 42 and 66. Remember that $42x$ means $42 \cdot x$.

$$\frac{42x}{66} = \frac{2 \cdot 3 \cdot 7 \cdot x}{2 \cdot 3 \cdot 11} = \frac{2}{2} \cdot \frac{3}{3} \cdot \frac{7x}{11} = 1 \cdot 1 \cdot \frac{7x}{11} = \frac{7x}{11}$$

Work Practice Problem 5

PRACTICE PROBLEM 5

Write in simplest form: $\frac{39x}{51}$

In the example above, you may have saved time by noticing that 42 and 66 have a common factor of 6.

$$\frac{42x}{66} = \frac{6 \cdot 7x}{6 \cdot 11} = \frac{6}{6} \cdot \frac{7x}{11} = 1 \cdot \frac{7x}{11} = \frac{7x}{11}$$

Helpful Hint

Writing the prime factorizations of the numerator and the denominator is helpful in finding any common factors.

The method for simplifying negative fractions is the same as for positive fractions.

EXAMPLE 6   Write in simplest form: $-\frac{10}{27}$

Solution:

$$-\frac{10}{27} = -\frac{2 \cdot 5}{3 \cdot 3 \cdot 3}$$ Prime factorizations of 10 and 27.

Since 10 and 27 have no common factors, $-\frac{10}{27}$ is already in simplest form.

Work Practice Problem 6

*Note: We will study this concept further in the next section.

PRACTICE PROBLEM 6

Write in simplest form: $-\frac{9}{50}$

Answers

4. $\frac{2}{3}$ **5.** $\frac{13x}{17}$ **6.** $-\frac{9}{50}$

PRACTICE PROBLEM 7

Write in simplest form: $\dfrac{49}{112}$

EXAMPLE 7 Write in simplest form: $\dfrac{30}{108}$

Solution:

$$\frac{30}{108} = \frac{2 \cdot 3 \cdot 5}{2 \cdot 2 \cdot 3 \cdot 3 \cdot 3} = \frac{2}{2} \cdot \frac{3}{3} \cdot \frac{5}{2 \cdot 3 \cdot 3} = 1 \cdot 1 \cdot \frac{5}{18} = \frac{5}{18}$$

Work Practice Problem 7

We can use a shortcut procedure with common factors when simplifying.

$$\frac{4}{6} = \frac{\overset{1}{\cancel{2}} \cdot 2}{\underset{1}{\cancel{2}} \cdot 3} = \frac{1 \cdot 2}{1 \cdot 3} = \frac{2}{3}$$ Divide out the common factor of 2 in the numerator and denominator.

This procedure is possible because dividing out a common factor in the numerator and denominator is the same as removing a factor of 1 in the product.

Writing a Fraction in Simplest Form

To write a fraction in simplest form, write the prime factorization of the numerator and the denominator and then divide both by all common factors.

PRACTICE PROBLEM 8

Write in simplest form: $-\dfrac{64}{20}$

EXAMPLE 8 Write in simplest form: $-\dfrac{72}{26}$

Solution:

$$-\frac{72}{26} = -\frac{\overset{1}{\cancel{2}} \cdot 2 \cdot 2 \cdot 3 \cdot 3}{\underset{1}{\cancel{2}} \cdot 13} = -\frac{1 \cdot 2 \cdot 2 \cdot 3 \cdot 3}{1 \cdot 13} = -\frac{36}{13}$$

Work Practice Problem 8

✔**Concept Check** Which is the correct way to simplify the fraction $\dfrac{15}{25}$? Or are both correct? Explain.

a. $\dfrac{15}{25} = \dfrac{3 \cdot \overset{1}{\cancel{5}}}{5 \cdot \underset{1}{\cancel{5}}} = \dfrac{3}{5}$ **b.** $\dfrac{1\overset{1}{\cancel{5}}}{2\underset{1}{\cancel{5}}} = \dfrac{11}{21}$

In this chapter, we will simplify and perform operations on fractions containing variables. When the denominator of a fraction contains a variable, such as $\dfrac{6x^2}{60x^3}$, we will assume that the variable does not represent 0. Recall that the denominator of a fraction cannot be 0.

PRACTICE PROBLEM 9

Write in simplest form: $\dfrac{7a^3}{56a^2}$

EXAMPLE 9 Write in simplest form: $\dfrac{6x^2}{60x^3}$

Solution: Notice that 6 and 60 have a common factor of 6. Let's also use the definition of an exponent to factor x^2 and x^3.

$$\frac{6x^2}{60x^3} = \frac{\overset{1}{\cancel{6}} \cdot \overset{1}{\cancel{x}} \cdot \overset{1}{\cancel{x}}}{\underset{1}{\cancel{6}} \cdot 10 \cdot \underset{1}{\cancel{x}} \cdot \underset{1}{\cancel{x}} \cdot x} = \frac{1 \cdot 1 \cdot 1}{1 \cdot 10 \cdot 1 \cdot 1 \cdot x} = \frac{1}{10x}$$

Work Practice Problem 9

Answers

7. $\dfrac{7}{16}$ 8. $-\dfrac{16}{5}$ 9. $\dfrac{a}{8}$

✔ **Concept Check Answers**

a. correct **b.** incorrect

Helpful Hint

Be careful when all factors of the numerator or denominator are divided out. In Example 9, the numerator was $1 \cdot 1 \cdot 1 = 1$, so the final result was $\dfrac{1}{10x}$.

Objective C Determining Whether Two Fractions Are Equivalent

Recall from Objective B that two fractions are equivalent if they represent the same part of a whole. One way to determine whether two fractions are equivalent is to see whether they simplify to the same fraction.

EXAMPLE 10 Determine whether $\dfrac{16}{40}$ and $\dfrac{10}{25}$ are equivalent.

Solution: Simplify each fraction.

$$\frac{16}{40} = \frac{\overset{1}{\cancel{8}} \cdot 2}{\underset{1}{\cancel{8}} \cdot 5} = \frac{1 \cdot 2}{1 \cdot 5} = \frac{2}{5}$$

$$\frac{10}{25} = \frac{2 \cdot \overset{1}{\cancel{5}}}{5 \cdot \underset{1}{\cancel{5}}} = \frac{2 \cdot 1}{5 \cdot 1} = \frac{2}{5}$$

Since these fractions are the same, $\dfrac{16}{40} = \dfrac{10}{25}$.

Work Practice Problem 10

There is a shortcut method you may use to check or test whether two fractions are equivalent. In the example above, we learned that the fractions are equivalent, or

$$\frac{16}{40} = \frac{10}{25}$$

In this example above, we call $25 \cdot 16$ and $40 \cdot 10$ **cross products** because they are the products one obtains by multiplying diagonally across the equal sign, as shown below.

Cross Products

$25 \cdot 16 \qquad\qquad 40 \cdot 10$

$$\frac{16}{40} = \frac{10}{25}$$

Notice that these cross products are equal

$$25 \cdot 16 = 400, \quad 40 \cdot 10 = 400$$

In general, this is true for equivalent fractions.

Equality of Fractions

$8 \cdot 6 \qquad\qquad 24 \cdot 2$

$$\frac{6}{24} \overset{?}{=} \frac{2}{8}$$

Since the cross products ($8 \cdot 6 = 48$ and $24 \cdot 2 = 48$) are equal, the fractions are equal.

Note: If the cross products are not equal, the fractions are not equal.

PRACTICE PROBLEM 10

Determine whether $\dfrac{7}{9}$ and $\dfrac{21}{27}$ are equivalent.

Answer
10. equivalent

PRACTICE PROBLEM 11

Determine whether $\frac{4}{13}$ and $\frac{5}{18}$ are equivalent.

EXAMPLE 11 Determine whether $\frac{8}{11}$ and $\frac{19}{26}$ are equivalent.

Solution: Let's check cross products.

$$26 \cdot 8 = 208 \qquad \frac{8}{11} \overset{?}{=} \frac{19}{26} \qquad 11 \cdot 19 = 209$$

Since $208 \neq 209$, then $\frac{8}{11} \neq \frac{19}{26}$.

Helpful Hint
"Not equal to" symbol

◻ **Work Practice Problem 11**

Objective ⒟ Solving Problems by Writing Fractions in Simplest Form

Many real-life problems can be solved by writing fractions. To make the answers clearer, these fractions should be written in simplest form.

PRACTICE PROBLEM 12

There are six national parks (including historic) in Washington state. See Example 12 and determine what fraction of the United States' national parks are located in Washington. Write the fraction in simplest form.

EXAMPLE 12 Calculating Fraction of Parks in Wyoming State

There are currently 58 national parks in the United States. Two of these parks are located in the state of Wyoming. What fraction of the United States' national parks can be found in Wyoming? Write the fraction in simplest form. (*Source:* National Park Service)

Solution: First we determine the fraction of parks found in Wyoming state.

$$\frac{2}{58} \quad \begin{array}{l} \leftarrow \text{ national parks in Wyoming} \\ \leftarrow \text{ total national parks} \end{array}$$

Next we simplify the fraction.

$$\frac{2}{58} = \frac{\overset{1}{\cancel{2}}}{\underset{1}{\cancel{2}} \cdot 29} = \frac{1}{1 \cdot 29} = \frac{1}{29}$$

Thus, $\frac{1}{29}$ of the United States' national parks are in Wyoming state.

◻ **Work Practice Problem 12**

Answers
11. not equivalent   **12.** $\frac{3}{29}$

CALCULATOR EXPLORATIONS Simplifying Fractions

Scientific Calculator

Many calculators have a fraction key, such as $\boxed{a\;b/c}$, that allows you to simplify a fraction on the calculator. For example, to simplify $\dfrac{324}{612}$, enter

$$\boxed{3}\;\boxed{2}\;\boxed{4}\;\boxed{a\;b/c}\;\boxed{6}\;\boxed{1}\;\boxed{2}\;\boxed{=}$$

The display will read

$$\boxed{\qquad\qquad 9\;|\;17}$$

which represents $\dfrac{9}{17}$, the original fraction simplified.

Graphing Calculator

Graphing calculators also allow you to simplify fractions. The fraction option on a graphing calculator may be found under the $\boxed{\text{MATH}}$ menu.

To simplify $\dfrac{324}{612}$, enter

$$\boxed{3}\;\boxed{2}\;\boxed{4}\;\boxed{\div}\;\boxed{6}\;\boxed{1}\;\boxed{2}\;\boxed{\text{MATH}}\;\boxed{\text{ENTER}}\;\boxed{\text{ENTER}}$$

The display will read

$$\boxed{324/612 \blacktriangleright \text{Frac } 9/17}$$

Helpful Hint

The Calculator Explorations boxes in this chapter provide only an introduction to fraction keys on calculators. Any time you use a calculator, there are both advantages and limitations to its use. Never rely solely on your calculator. It is very important that you understand how to perform all operations on fractions by hand in order to progress through later topics. For further information, talk to your instructor.

Use your calculator to simplify each fraction.

1. $\dfrac{128}{224}$

2. $\dfrac{231}{396}$

3. $\dfrac{340}{459}$

4. $\dfrac{999}{1350}$

5. $\dfrac{432}{810}$

6. $\dfrac{225}{315}$

7. $\dfrac{54}{243}$

8. $\dfrac{455}{689}$

Vocabulary and Readiness Check

Use the choices below to fill in each blank.

    cross products   equivalent   composite

    simplest form   prime factorization   prime

1. The number 40 equals $2 \cdot 2 \cdot 2 \cdot 5$. Since each factor is prime, we call $2 \cdot 2 \cdot 2 \cdot 5$ the _____ of 40.

2. A natural number, other than 1, that is not prime is called a _____ number.

3. A natural number that has exactly two different factors, 1 and itself, is called a _____ number.

4. In $\frac{11}{48}$, since 11 and 48 have no common factors other than 1, $\frac{11}{48}$ is in _____.

5. Fractions that represent the same portion of a whole are called _____ fractions.

6. In the statement $\frac{5}{12} = \frac{15}{36}$, $5 \cdot 36$ and $12 \cdot 15$ are called _____.

4.2 EXERCISE SET

FOR EXTRA HELP

Student Solutions Manual   PH Math/Tutor Center   CD/Video for Review   MathXL® MathXL®   MyMathLab MyMathLab

Objective A *Write the prime factorization of each number. See Examples 1 through 3.*

1. 20

2. 12

3. 48

4. 75

5. 81

6. 64

7. 162

8. 128

9. 110

10. 130

11. 85

12. 93

13. 240

14. 836

15. 828

16. 504

Objective B *Write each fraction in simplest form. See Examples 4 through 9.*

17. $\frac{3}{12}$

18. $\frac{5}{30}$

19. $\frac{4x}{42}$

20. $\frac{9y}{48}$

21. $\frac{14}{16}$

22. $\frac{22}{34}$

23. $\frac{20}{30}$

24. $\frac{70}{80}$

25. $\frac{35a}{50a}$

26. $\frac{25z}{55z}$

27. $-\frac{63}{81}$

28. $-\frac{21}{49}$

29. $\frac{30x^2}{36x}$

30. $\frac{45b}{80b^2}$

31. $\frac{27}{64}$

32. $\frac{32}{63}$

33. $\frac{25xy}{40y}$

34. $\frac{36y}{42yz}$

35. $-\frac{40}{64}$

36. $-\frac{28}{60}$

37. $\frac{36x^3y^2}{24xy}$

38. $\frac{60a^2b}{36ab^3}$

39. $\frac{90}{120}$

40. $\frac{60}{150}$

41. $\frac{40xy}{64xyz}$

42. $\dfrac{28abc}{60ac}$ **43.** $\dfrac{66}{308}$ **44.** $\dfrac{65}{234}$ **45.** $-\dfrac{55}{85y}$ **46.** $-\dfrac{78}{90x}$

47. $\dfrac{189z}{216z}$ **48.** $\dfrac{144y}{162y}$ **49.** $\dfrac{224a^3b^4c^2}{16ab^4c^2}$ **50.** $\dfrac{270x^4y^3z^3}{15x^3y^3z^3}$

Objective Ⓒ *Determine whether each pair of fractions is equivalent. See Examples 10 and 11.*

51. $\dfrac{2}{6}$ and $\dfrac{4}{12}$ **52.** $\dfrac{3}{6}$ and $\dfrac{5}{10}$ 🔘 **53.** $\dfrac{7}{11}$ and $\dfrac{5}{8}$

54. $\dfrac{2}{5}$ and $\dfrac{4}{11}$ **55.** $\dfrac{10}{15}$ and $\dfrac{6}{9}$ **56.** $\dfrac{4}{10}$ and $\dfrac{6}{15}$

🔘 **57.** $\dfrac{3}{9}$ and $\dfrac{6}{18}$ **58.** $\dfrac{2}{8}$ and $\dfrac{7}{28}$ **59.** $\dfrac{10}{13}$ and $\dfrac{13}{15}$

60. $\dfrac{16}{20}$ and $\dfrac{9}{16}$ **61.** $\dfrac{8}{18}$ and $\dfrac{12}{24}$ **62.** $\dfrac{6}{21}$ and $\dfrac{14}{35}$

Objective Ⓓ *Solve. Write each fraction in simplest form. See Example 12.*

63. A work shift for an employee at Starbucks consists of 8 hours. What fraction of the employee's work shift is represented by 2 hours?

🔘 **64.** Two thousand baseball caps were sold one year at the U.S. Open Golf Tournament. What fractional part of this total does 200 caps represent?

65. There are 5280 feet in a mile. What fraction of a mile is represented by 2640 feet?

66. There are 100 centimeters in 1 meter. What fraction of a meter is 20 centimeters?

67. Fifteen out of the total fifty states in the United States have Ritz-Carlton hotels. (*Source:* Ritz-Carlton Hotel Company, LLC)

   a. What fraction of states can claim at least one Ritz-Carlton hotel?

   b. How many states do not have a Ritz-Carlton hotel?

   c. Write the fraction of states without a Ritz-Carlton hotel.

68. There are 78 national monuments in the United States. Ten of these monuments are located in New Mexico. (*Source:* National Park Service)

   a. What fraction of the national monuments in the United States can be found in New Mexico?

   b. How many of the national monuments in the United States are found outside New Mexico?

   c. Write the fraction of national monuments found in states other than New Mexico.

69. The outer wall of the Pentagon is 24 inches wide. Ten inches is concrete, 8 inches is brick, and 6 inches is limestone. What fraction of the wall is concrete?

🔘 **70.** There are 35 students in a biology class. If 10 students made an A on the first test, what fraction of the students made an A?

Limestone (6 in.)

Brick (8 in.)

Concrete (10 in.)

71. As Internet usage grows in the United States, more and more state governments are placing services online. Twenty-two out of the total fifty states have Web sites that allow residents to file their state income tax online directly to the state.

   a. How many states do not have this type of Web site?

   b. What fraction of states do not have this type of Web site? (*Source:* Federation of Tax Administrators)

72. Chris Callac just bought a brand new 2006 Toyota Camry for $22,000. His old car was traded in for $10,000.

   a. How much of his purchase price was not covered by his trade-in?

   b. What fraction of the purchase price was not covered by the trade-in?

73. As of March 2006, a total of 449 individuals from around the world had flown in space. Of these, 291 were United States astronauts. What fraction of individuals who have flown into space were Americans? (*Source:* NASA)

74. Worldwide, Hallmark employs 18,000 full-time employees. About 4500 employees work at the Hallmark headquarters in Kansas City, Missouri. What fraction of Hallmark employees work in Kansas City? (*Source:* Hallmark Cards, Inc.)

Review

Evaluate each expression using the given replacement numbers. See Section 2.5.

75. $\dfrac{x^3}{9}$ when $x = -3$

76. $\dfrac{y^3}{5}$ when $y = -5$

77. $2y$ when $y = -7$

78. $-5a$ when $a = -4$

Concept Extensions

79. In your own words, define equivalent fractions.

80. Given a fraction, say $\dfrac{15}{40}$, how many fractions are there that are equivalent to it, but in simplest form or lowest terms? Explain your answer.

Write each fraction in simplest form.

81. $\dfrac{3975}{6625}$

82. $\dfrac{9506}{12,222}$

There are generally considered to be eight basic blood types. The table shows the number of people with the various blood types in a typical group of 100 blood donors. Use the table to answer Exercises 83 through 86. Write each answer in simplest form.

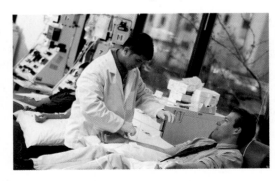

Distribution of Blood Types in Blood Donors	
Blood Type	**Number of People**
O Rh-positive	37
O Rh-negative	7
A Rh-positive	36
A Rh-negative	6
B Rh-positive	9
B Rh-negative	1
AB Rh-positive	3
AB Rh-negative	1
(*Source:* American Red Cross Biomedical Services)	

83. What fraction of blood donors have blood type A Rh-positive?

84. What fraction of blood donors have an O blood type?

85. What fraction of blood donors have an AB blood type?

86. What fraction of blood donors have a B blood type?

Find the prime factorization of each number.

87. 34,020

88. 131,625

89. In your own words, define a prime number.

90. The number 2 is a prime number. All other even natural numbers are composite numbers. Explain why.

91. Two students have different prime factorizations for the same number. Is this possible? Explain.

The following graph is called a circle graph or pie chart. Each sector (shaped like a piece of pie) shows the fraction of entering college freshmen who expect to major in each discipline shown. The whole circle represents the entire class of college freshmen. Use this graph to answer Exercises 92 through 95.

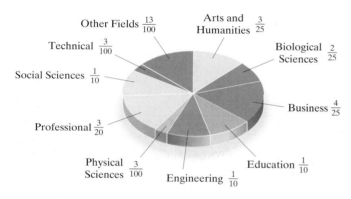

Other Fields $\frac{13}{100}$ Arts and Humanities $\frac{3}{25}$
Technical $\frac{3}{100}$
Biological Sciences $\frac{2}{25}$
Social Sciences $\frac{1}{10}$
Business $\frac{4}{25}$
Professional $\frac{3}{20}$
Physical Sciences $\frac{3}{100}$
Engineering $\frac{1}{10}$
Education $\frac{1}{10}$

Source: The Higher Education Research Institute

92. What fraction of entering college freshmen plan to major in education?

93. What fraction of entering college freshmen plan to major in biological sciences?

94. Why is the Social Sciences sector the same size as the Engineering sector?

95. Why is the Physical Sciences sector smaller than the Business sector?

Use the following numbers for Exercises 96 through 99.

8691   786   1235   2235   85   105   22   222   900   1470

96. List the numbers divisible by both 2 and 3.

97. List the numbers that are divisible by both 3 and 5.

98. The answers to Exercise 96 are also divisible by what number? Tell why.

99. The answers to Exercise 97 are also divisible by what number? Tell why.

4.3 MULTIPLYING AND DIVIDING FRACTIONS

Objective **A** Multiplying Fractions

Let's use a diagram to discover how fractions are multiplied. For example, to multiply $\frac{1}{2}$ and $\frac{3}{4}$, we find $\frac{1}{2}$ of $\frac{3}{4}$. To do this, we begin with a diagram showing $\frac{3}{4}$ of a rectangle's area shaded.

$\frac{3}{4}$ of the rectangle's area is shaded.

To find $\frac{1}{2}$ of $\frac{3}{4}$, we heavily shade $\frac{1}{2}$ of the part that is already shaded.

By counting smaller rectangles, we see that $\frac{3}{8}$ of the larger rectangle is now heavily shaded, so that

$$\frac{1}{2} \text{ of } \frac{3}{4} \text{ is } \frac{3}{8}, \text{ or } \frac{1}{2} \cdot \frac{3}{4} = \frac{3}{8} \quad \text{Notice that } \frac{1}{2} \cdot \frac{3}{4} = \frac{1 \cdot 3}{2 \cdot 4} = \frac{3}{8}.$$

Multiplying Fractions

To multiply two fractions, multiply the numerators and multiply the denominators.

If a, b, c, and d represent numbers, and b and d are not 0, we have

$$\frac{a}{b} \cdot \frac{c}{d} = \frac{a \cdot c}{b \cdot d}$$

PRACTICE PROBLEMS 1–2

Multiply.

1. $\frac{3}{7} \cdot \frac{5}{11}$ **2.** $\frac{1}{3} \cdot \frac{1}{9}$

EXAMPLES Multiply.

1. $\frac{2}{3} \cdot \frac{5}{11} = \frac{2 \cdot 5}{3 \cdot 11} = \frac{10}{33}$ Multiply numerators. Multiply denominators.

This fraction is in simplest form since 10 and 33 have no common factors other than 1.

2. $\frac{1}{4} \cdot \frac{1}{2} = \frac{1 \cdot 1}{4 \cdot 2} = \frac{1}{8}$ This fraction is in simplest form.

☐ **Work Practice Problems 1–2**

Answers

1. $\frac{15}{77}$ **2.** $\frac{1}{27}$

240

EXAMPLE 3 Multiply and simplify: $\dfrac{6}{7} \cdot \dfrac{14}{27}$

Solution:

$$\dfrac{6}{7} \cdot \dfrac{14}{27} = \dfrac{6 \cdot 14}{7 \cdot 27}$$

We can simplify by finding the prime factorizations and using our shortcut procedure of dividing out common factors in the numerator and denominator.

$$\dfrac{6 \cdot 14}{7 \cdot 27} = \dfrac{2 \cdot \overset{1}{\cancel{3}} \cdot 2 \cdot \overset{1}{\cancel{7}}}{\underset{1}{\cancel{7}} \cdot \underset{1}{\cancel{3}} \cdot 3 \cdot 3} = \dfrac{2 \cdot 2}{3 \cdot 3} = \dfrac{4}{9}$$

▢ **Work Practice Problem 3**

Helpful Hint

Remember that the shortcut procedure above is the same as removing factors of 1 in the product.

$$\dfrac{6 \cdot 14}{7 \cdot 27} = \dfrac{2 \cdot 3 \cdot 2 \cdot 7}{7 \cdot 3 \cdot 3 \cdot 3} = \dfrac{7}{7} \cdot \dfrac{3}{3} \cdot \dfrac{2 \cdot 2}{3 \cdot 3} = 1 \cdot 1 \cdot \dfrac{4}{9} = \dfrac{4}{9}$$

EXAMPLE 4 Multiply and simplify: $\dfrac{23}{32} \cdot \dfrac{4}{7}$

Solution: Notice that 4 and 32 have a common factor of 4.

$$\dfrac{23}{32} \cdot \dfrac{4}{7} = \dfrac{23 \cdot 4}{32 \cdot 7} = \dfrac{23 \cdot \overset{1}{\cancel{4}}}{\underset{1}{\cancel{4}} \cdot 8 \cdot 7} = \dfrac{23}{8 \cdot 7} = \dfrac{23}{56}$$

▢ **Work Practice Problem 4**

After multiplying two fractions, always check to see whether the product can be simplified.

EXAMPLE 5 Multiply: $-\dfrac{1}{4} \cdot \dfrac{1}{2}$

Solution: Recall that the product of a negative number and a positive number is a negative number.

$$-\dfrac{1}{4} \cdot \dfrac{1}{2} = -\dfrac{1 \cdot 1}{4 \cdot 2} = -\dfrac{1}{8}$$

▢ **Work Practice Problem 5**

EXAMPLES Multiply.

6. $\left(-\dfrac{6}{13}\right)\left(-\dfrac{26}{30}\right) = \dfrac{6 \cdot 26}{13 \cdot 30} = \dfrac{\overset{1}{\cancel{6}} \cdot \overset{1}{\cancel{13}} \cdot 2}{\underset{1}{\cancel{13}} \cdot \underset{1}{\cancel{6}} \cdot 5} = \dfrac{2}{5}$ The product of two negative numbers is a positive number.

7. $\dfrac{1}{3} \cdot \dfrac{2}{5} \cdot \dfrac{9}{16} = \dfrac{1 \cdot 2 \cdot 9}{3 \cdot 5 \cdot 16} = \dfrac{1 \cdot \overset{1}{\cancel{2}} \cdot \overset{1}{\cancel{3}} \cdot 3}{\underset{1}{\cancel{3}} \cdot 5 \cdot \underset{1}{\cancel{2}} \cdot 8} = \dfrac{3}{40}$

▢ **Work Practice Problems 6–7**

PRACTICE PROBLEM 3

Multiply and simplify: $\dfrac{6}{77} \cdot \dfrac{7}{8}$

PRACTICE PROBLEM 4

Multiply and simplify: $\dfrac{4}{27} \cdot \dfrac{3}{8}$

Helpful Hint

Don't forget that we may identify common factors that are not prime numbers.

PRACTICE PROBLEM 5

Multiply.

5. $\dfrac{1}{2} \cdot \left(-\dfrac{11}{28}\right)$

PRACTICE PROBLEMS 6–7

Multiply.

6. $\left(-\dfrac{4}{11}\right)\left(-\dfrac{33}{16}\right)$

7. $\dfrac{1}{6} \cdot \dfrac{3}{10} \cdot \dfrac{25}{16}$

Answers

3. $\dfrac{3}{44}$ **4.** $\dfrac{1}{18}$ **5.** $-\dfrac{11}{56}$ **6.** $\dfrac{3}{4}$ **7.** $\dfrac{5}{64}$

We multiply fractions in the same way if variables are involved.

PRACTICE PROBLEM 8

Multiply: $\dfrac{2}{3} \cdot \dfrac{3y}{2}$

EXAMPLE 8 Multiply: $\dfrac{3x}{4} \cdot \dfrac{8}{5x}$

Solution: Notice that 8 and 4 have a common factor of 4.

$$\frac{3x}{4} \cdot \frac{8}{5x} = \frac{3 \cdot \overset{1}{\cancel{x}} \cdot \overset{2}{\cancel{4}} \cdot 2}{\underset{1}{\cancel{4}} \cdot 5 \cdot \underset{1}{\cancel{x}}} = \frac{3 \cdot 1 \cdot 1 \cdot 2}{1 \cdot 5 \cdot 1} = \frac{6}{5}$$

■ **Work Practice Problem 8**

Helpful Hint   Recall that when the denominator of a fraction contains a variable, such as $\dfrac{8}{5x}$, we assume that the variable does not represent 0.

PRACTICE PROBLEM 9

Multiply: $\dfrac{a^3}{b^2} \cdot \dfrac{b}{a^2}$

EXAMPLE 9 Multiply: $\dfrac{x^2}{y} \cdot \dfrac{y^3}{x}$

Solution:

$$\frac{x^2}{y} \cdot \frac{y^3}{x} = \frac{x^2 \cdot y^3}{y \cdot x} = \frac{\overset{1}{\cancel{x}} \cdot x \cdot \overset{1}{\cancel{y}} \cdot y \cdot y}{\underset{1}{\cancel{y}} \cdot \underset{1}{\cancel{x}}} = \frac{x \cdot y \cdot y}{1} = xy^2$$

■ **Work Practice Problem 9**

Objective **B** **Evaluating Expressions with Fractional Bases**

The base of an exponential expression can also be a fraction.

$$\left(\frac{1}{3}\right)^4 = \underbrace{\frac{1}{3} \cdot \frac{1}{3} \cdot \frac{1}{3} \cdot \frac{1}{3}}_{\frac{1}{3} \text{ is a factor 4 times.}} = \frac{1 \cdot 1 \cdot 1 \cdot 1}{3 \cdot 3 \cdot 3 \cdot 3} = \frac{1}{81}$$

PRACTICE PROBLEM 10

Evaluate.

a. $\left(\dfrac{3}{4}\right)^3$ **b.** $\left(-\dfrac{4}{5}\right)^2$

EXAMPLE 10 Evaluate.

a. $\left(\dfrac{2}{5}\right)^4 = \dfrac{2}{5} \cdot \dfrac{2}{5} \cdot \dfrac{2}{5} \cdot \dfrac{2}{5} = \dfrac{2 \cdot 2 \cdot 2 \cdot 2}{5 \cdot 5 \cdot 5 \cdot 5} = \dfrac{16}{625}$

b. $\left(-\dfrac{1}{4}\right)^2 = \left(-\dfrac{1}{4}\right) \cdot \left(-\dfrac{1}{4}\right) = \dfrac{1 \cdot 1}{4 \cdot 4} = \dfrac{1}{16}$ The product of two negative numbers is a positive number.

■ **Work Practice Problem 10**

Objective **C** **Dividing Fractions**

Before we can divide fractions, we need to know how to find the **reciprocal** of a fraction.

Reciprocal of a Fraction

Two numbers are **reciprocals** of each other if their product is 1. The reciprocal of the fraction $\dfrac{a}{b}$ is $\dfrac{b}{a}$ because $\dfrac{a}{b} \cdot \dfrac{b}{a} = \dfrac{a \cdot b}{b \cdot a} = 1.$

Answers

8. y **9.** $\dfrac{a}{b}$ **10. a.** $\dfrac{27}{64}$ **b.** $\dfrac{16}{25}$

Every number has a reciprocal except 0. The number 0 has no reciprocal because there is no number such that $0 \cdot a = 1$.

For example,

The reciprocal of $\dfrac{2}{5}$ is $\dfrac{5}{2}$ because $\dfrac{2}{5} \cdot \dfrac{5}{2} = \dfrac{10}{10} = 1$.

The reciprocal of 5 is $\dfrac{1}{5}$ because $5 \cdot \dfrac{1}{5} = \dfrac{5}{1} \cdot \dfrac{1}{5} = \dfrac{5}{5} = 1$.

The reciprocal of $-\dfrac{7}{11}$ is $-\dfrac{11}{7}$ because $-\dfrac{7}{11} \cdot -\dfrac{11}{7} = \dfrac{77}{77} = 1$.

Division of fractions has the same meaning as division of whole numbers. For example,

$10 \div 5$ means: How many 5s are there in 10?

There are two 5s in 10, so
$10 \div 5 = 2$.

$\dfrac{3}{4} \div \dfrac{1}{8}$ means: How many $\dfrac{1}{8}$s are there in $\dfrac{3}{4}$?

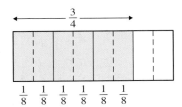

There are six $\dfrac{1}{8}$s in $\dfrac{3}{4}$, so $\dfrac{3}{4} \div \dfrac{1}{8} = 6$.

We use reciprocals to divide fractions.

Dividing Fractions

To divide two fractions, multiply the first fraction by the reciprocal of the second fraction.

If a, b, c, and d represent numbers, and b, c, and d are not 0, then

$$\frac{a}{b} \div \frac{c}{d} = \frac{a}{b} \cdot \frac{d}{c} = \frac{a \cdot d}{b \cdot c}$$

$\llcorner$ reciprocal $\lrcorner$

For example,

multiply by reciprocal

$$\frac{3}{4} \div \frac{1}{8} = \frac{3}{4} \cdot \frac{8}{1} = \frac{3 \cdot 8}{4 \cdot 1} = \frac{3 \cdot 2 \cdot \overset{1}{\cancel{4}}}{\underset{1}{\cancel{4}} \cdot 1} = \frac{6}{1} \text{ or } 6$$

After dividing fractions, *always* check to see whether the result can be simplified.

PRACTICE PROBLEMS 11–12

Divide and simplify.

11. $\dfrac{8}{7} \div \dfrac{2}{9}$

12. $\dfrac{4}{9} \div \dfrac{1}{2}$

EXAMPLES Divide and simplify.

11. $\dfrac{5}{16} \div \dfrac{3}{4} = \dfrac{5}{16} \cdot \dfrac{4}{3} = \dfrac{5 \cdot 4}{16 \cdot 3} = \dfrac{5 \cdot \overset{1}{\cancel{4}}}{\cancel{4} \cdot 4 \cdot 3} = \dfrac{5}{12}$

12. $\dfrac{2}{5} \div \dfrac{1}{2} = \dfrac{2}{5} \cdot \dfrac{2}{1} = \dfrac{2 \cdot 2}{5 \cdot 1} = \dfrac{4}{5}$

Work Practice Problems 11–12

Helpful Hint

When dividing by a fraction, do not look for common factors to divide out until you rewrite the division as multiplication.

Do not try to divide out these two 2s.

$\dfrac{1}{2} \div \dfrac{\mathbf{2}}{3} = \dfrac{1}{2} \cdot \dfrac{3}{2} = \dfrac{3}{4}$

PRACTICE PROBLEM 13

Divide: $-\dfrac{10}{4} \div \dfrac{2}{9}$

EXAMPLE 13 Divide: $-\dfrac{5}{16} \div -\dfrac{3}{4}$

Solution: Recall that the quotient (or product) of two negative numbers is a positive number.

$-\dfrac{5}{16} \div -\dfrac{3}{4} = -\dfrac{5}{16} \cdot -\dfrac{4}{3} = \dfrac{5 \cdot \overset{1}{\cancel{4}}}{\underset{1}{\cancel{4}} \cdot 4 \cdot 3} = \dfrac{5}{12}$

Work Practice Problem 13

PRACTICE PROBLEM 14

Divide: $\dfrac{3y}{4} \div 5y^3$

EXAMPLE 14 Divide: $\dfrac{2x}{3} \div 3x^2$

Solution:

$\dfrac{2x}{3} \div 3x^2 = \dfrac{2x}{3} \div \dfrac{3x^2}{1} = \dfrac{2x}{3} \cdot \dfrac{1}{3x^2} = \dfrac{2 \cdot \overset{1}{\cancel{x}} \cdot 1}{3 \cdot 3 \cdot \underset{1}{\cancel{x}} \cdot x} = \dfrac{2}{9x}$

Work Practice Problem 14

PRACTICE PROBLEM 15

Simplify: $\left(-\dfrac{2}{3} \cdot \dfrac{9}{14}\right) \div \dfrac{7}{15}$

EXAMPLE 15 Simplify: $\left(\dfrac{4}{7} \cdot \dfrac{3}{8}\right) \div -\dfrac{3}{4}$

Solution: Remember to perform the operations inside the ( ) first.

$\left(\dfrac{4}{7} \cdot \dfrac{3}{8}\right) \div -\dfrac{3}{4} = \left(\dfrac{\overset{1}{\cancel{4}} \cdot 3}{7 \cdot 2 \cdot \underset{1}{\cancel{4}}}\right) \div -\dfrac{3}{4} = \dfrac{3}{14} \div -\dfrac{3}{4}$

Now divide.

$\dfrac{3}{14} \div -\dfrac{3}{4} = \dfrac{3}{14} \cdot -\dfrac{4}{3} = -\dfrac{\overset{1}{\cancel{3}} \cdot \overset{1}{\cancel{2}} \cdot 2}{\underset{1}{\cancel{2}} \cdot 7 \cdot \underset{1}{\cancel{3}}} = -\dfrac{2}{7}$

Work Practice Problem 15

Answers

11. $\dfrac{36}{7}$ **12.** $\dfrac{8}{9}$ **13.** $-\dfrac{45}{4}$

14. $\dfrac{3}{20y^2}$ **15.** $-\dfrac{45}{49}$

✔ **Concept Check** Which is the correct way to divide $\frac{3}{5}$ by $\frac{5}{12}$? Explain.

a. $\frac{3}{5} \div \frac{5}{12} = \frac{5}{3} \cdot \frac{5}{12}$ **b.** $\frac{3}{5} \div \frac{5}{12} = \frac{3}{5} \cdot \frac{12}{5}$

Objective D Multiplying and Dividing with Fractional Replacement Values

Recall the difference between an expression and an equation. For example, xy and $x \div y$ are expressions. They contain no equals signs. In Example 16, we practice *simplifying* expressions given fractional replacement values.

EXAMPLE 16 If $x = \frac{7}{8}$ and $y = -\frac{1}{3}$, evaluate **(a)** xy and **(b)** $x \div y$.

Solution: Replace x with $\frac{7}{8}$ and y with $-\frac{1}{3}$.

a. $xy = \frac{7}{8} \cdot -\frac{1}{3}$

$= -\frac{7 \cdot 1}{8 \cdot 3}$

$= -\frac{7}{24}$

b. $x \div y = \frac{7}{8} \div -\frac{1}{3}$

$= \frac{7}{8} \cdot -\frac{3}{1}$

$= -\frac{7 \cdot 3}{8 \cdot 1}$

$= -\frac{21}{8}$

PRACTICE PROBLEM 16

If $x = -\frac{3}{4}$ and $y = \frac{9}{2}$, evaluate **(a)** xy, and **(b)** $x \div y$.

Helpful Hint
"of" usually translates to multiplication.

■ **Work Practice Problem 16**

EXAMPLE 17 Is $-\frac{2}{3}$ a solution of the equation $-\frac{1}{2}x = \frac{1}{3}$?

Solution: To check whether a number is a solution of an equation, recall that we replace the variable with the given number and see if a true statement results.

$-\frac{1}{2} \cdot x = \frac{1}{3}$ Recall that $-\frac{1}{2}x$ means $-\frac{1}{2} \cdot x$.

$-\frac{1}{2} \cdot -\frac{2}{3} \stackrel{?}{=} \frac{2}{3}$ Replace x with $-\frac{2}{3}$.

$\frac{1 \cdot \overset{1}{\cancel{2}}}{\underset{1}{\cancel{2}} \cdot 3} \stackrel{?}{=} \frac{1}{3}$ The product of two negative numbers is a positive number.

$\frac{1}{3} = \frac{1}{3}$ True

Since we have a true statement, $-\frac{2}{3}$ is a solution.

PRACTICE PROBLEMS 17

Is $-\frac{9}{8}$ a solution of the equation $2x = -\frac{9}{4}$?

■ **Work Practice Problem 17**

Objective E Solving Problems by Multiplying Fractions

To solve real-life problems that involve multiplying fractions, we use our four problem-solving steps from Chapter 3. In Example 18, a new key word that implies multiplication is used. That key word is "**of.**"

Answers

16. a. $-\frac{27}{8}$ b. $-\frac{1}{6}$ 17. yes

✔ **Concept Check Answers**
a. incorrect b. correct

PRACTICE PROBLEM 18

Hershey Park is an amusement park in Hershey, Pennsylvania. Of its 60 rides, $\frac{1}{6}$ of them are roller coasters. How many roller coasters are in HersheyPark?

EXAMPLE 18 Finding the Number of Roller Coasters in an Amusement Park

Cedar Point is an amusement park located in Sandusky, Ohio. Its collection of 68 rides is the largest in the world. Of the rides, $\frac{4}{17}$ are roller coasters. How many roller coasters are in Cedar Point's collection of rides? (*Source:* Wikipedia)

Solution:

1. UNDERSTAND the problem. To do so, read and reread the problem. We are told that $\frac{4}{17}$ of Cedar Point's rides are roller coasters. The word "of" here means multiplication.

2. TRANSLATE.

In words:	Number of roller coasters	is	$\frac{4}{17}$	of	total rides at Cedar Point
	↓	↓	↓	↓	↓
Translate:	Number of roller coasters	=	$\frac{4}{17}$	·	68

3. SOLVE: Before we solve, let's estimate a reasonable answer. The fraction $\frac{4}{17}$ is less than $\frac{1}{2}$ (draw a diagram, if needed), and $\frac{1}{2}$ of 68 rides is 34 rides, so the number of roller coasters should be less than 34.

$$\frac{4}{17} \cdot 68 = \frac{4}{17} \cdot \frac{68}{1} = \frac{4 \cdot 68}{17 \cdot 1} = \frac{4 \cdot \overset{1}{\cancel{17}} \cdot 4}{\underset{1}{\cancel{17}} \cdot 1} = \frac{16}{1} \quad \text{or} \quad 16$$

4. INTERPRET. *Check* your work. From our estimate, our answer is reasonable. *State* your conclusion: The number of roller coasters at Cedar Point is 16.

🔲 **Work Practice Problem 18**

Helpful Hint

To help visualize a fractional part of a whole number, look at the diagram below.

$\frac{1}{5}$ of 60 = ?

$\frac{1}{5}$ of 60 is 12.

Answer

18. 10 roller coasters

Vocabulary and Readiness Check

Use the choices below to fill in each blank. Not all choices will be used.

multiplication $\dfrac{a \cdot d}{b \cdot c}$ $\dfrac{a \cdot c}{b \cdot d}$ $\dfrac{2 \cdot 2 \cdot 2}{7}$ $\dfrac{2}{7} \cdot \dfrac{2}{7} \cdot \dfrac{2}{7}$

division 0 reciprocals

1. To multiply two fractions, we write $\dfrac{a}{b} \cdot \dfrac{c}{d} = $ _____ .

2. Two numbers are _____ of each other if their product is 1.

3. The expression $\dfrac{2^3}{7} = $ _____ while $\left(\dfrac{2}{7}\right)^3 = $ _____ .

4. Every number has a reciprocal except _____ .

5. To divide two fractions, we write $\dfrac{a}{b} \div \dfrac{c}{d} = $ _____ .

6. The word "of" indicates _____ .

4.3 EXERCISE SET

Objective A *Multiply. Write the product in simplest form. See Examples 1 through 9.*

1. $\dfrac{6}{11} \cdot \dfrac{3}{7}$ **2.** $\dfrac{5}{9} \cdot \dfrac{7}{4}$ **3.** $-\dfrac{2}{7} \cdot \dfrac{5}{8}$ **4.** $\dfrac{4}{15} \cdot -\dfrac{1}{20}$

 5. $-\dfrac{1}{2} \cdot -\dfrac{2}{15}$ **6.** $-\dfrac{3}{11} \cdot -\dfrac{11}{12}$ **7.** $\dfrac{18x}{20} \cdot \dfrac{36}{99}$ **8.** $\dfrac{5}{32} \cdot \dfrac{64y}{100}$

9. $3a^2 \cdot \dfrac{1}{4}$ **10.** $-\dfrac{2}{3} \cdot 6y^3$ **11.** $\dfrac{x^3}{y^3} \cdot \dfrac{y^2}{x}$ **12.** $\dfrac{a}{b^3} \cdot \dfrac{b}{a^3}$

13. $0 \cdot \dfrac{8}{9}$ **14.** $\dfrac{11}{12} \cdot 0$ **15.** $-\dfrac{17y}{20} \cdot \dfrac{4}{5y}$ **16.** $-\dfrac{13x}{20} \cdot \dfrac{5}{6x}$

17. $\dfrac{11}{20} \cdot \dfrac{1}{7} \cdot \dfrac{5}{22}$ **18.** $\dfrac{27}{32} \cdot \dfrac{10}{13} \cdot \dfrac{16}{30}$

Objective B *Evaluate. See Example 10.*

19. $\left(\dfrac{1}{5}\right)^3$ **20.** $\left(\dfrac{8}{9}\right)^2$ **21.** $\left(-\dfrac{2}{3}\right)^2$

22. $\left(-\dfrac{1}{2}\right)^4$ **23.** $\left(-\dfrac{2}{3}\right)^3 \cdot \dfrac{1}{2}$ **24.** $\left(-\dfrac{3}{4}\right)^3 \cdot \dfrac{1}{3}$

Objective C *Divide. Write all quotients in simplest form. See Examples 11 through 14.*

25. $\dfrac{2}{3} \div \dfrac{5}{6}$ **26.** $\dfrac{5}{8} \div \dfrac{3}{4}$ **27.** $-\dfrac{6}{15} \div \dfrac{12}{5}$ **28.** $-\dfrac{4}{15} \div -\dfrac{8}{3}$

29. $-\dfrac{8}{9} \div \dfrac{x}{2}$ **30.** $\dfrac{10}{11} \div -\dfrac{4}{5x}$ **31.** $\dfrac{11y}{20} \div \dfrac{3}{11}$ **32.** $\dfrac{9z}{20} \div \dfrac{2}{9}$

33. $-\dfrac{2}{3} \div 4$ **34.** $-\dfrac{5}{6} \div 10$ **35.** $\dfrac{1}{5x} \div \dfrac{5}{x^2}$ **36.** $\dfrac{3}{y^2} \div \dfrac{9}{y^3}$

Objectives **A** **B** **C** **Mixed Practice** *Perform each indicated operation. See Examples 1 through 15.*

37. $\dfrac{2}{3} \cdot \dfrac{5}{9}$ **38.** $\dfrac{8}{15} \cdot \dfrac{5}{32}$ **39.** $\dfrac{3x}{7} \div \dfrac{5}{6x}$ **40.** $\dfrac{2}{5y} \div \dfrac{5y}{11}$

41. $\dfrac{16}{27y} \div \dfrac{8}{15y}$ **42.** $\dfrac{12y}{21} \div \dfrac{4y}{7}$ **43.** $-\dfrac{5}{28} \cdot \dfrac{35}{25}$ **44.** $\dfrac{24}{45} \cdot -\dfrac{5}{8}$

45. $\left(-\dfrac{3}{4}\right)^2$ **46.** $\left(-\dfrac{1}{2}\right)^5$ **47.** $\dfrac{x^2}{y} \cdot \dfrac{y^3}{x}$ **48.** $\dfrac{b}{a^2} \cdot \dfrac{a^3}{b^3}$

49. $7 \div \dfrac{2}{11}$ **50.** $-100 \div \dfrac{1}{2}$ **51.** $-3x \div \dfrac{x^2}{12}$ **52.** $-7x^2 \div \dfrac{14x}{3}$

53. $\left(\dfrac{2}{7} \div \dfrac{7}{2}\right) \cdot \dfrac{3}{4}$ **54.** $\dfrac{1}{2} \cdot \left(\dfrac{5}{6} \div \dfrac{1}{12}\right)$ **55.** $-\dfrac{19}{63y} \cdot 9y^2$ **56.** $16a^2 \cdot -\dfrac{31}{24a}$

57. $-\dfrac{2}{3} \cdot -\dfrac{6}{11}$ **58.** $-\dfrac{1}{5} \cdot -\dfrac{6}{7}$ **59.** $\dfrac{4}{8} \div \dfrac{3}{16}$ **60.** $\dfrac{9}{2} \div \dfrac{16}{15}$

61. $\dfrac{21x^2}{10y} \div \dfrac{14x}{25y}$ **62.** $\dfrac{17y^2}{24x} \div \dfrac{13y}{18x}$ **63.** $\left(1 \div \dfrac{3}{4}\right) \cdot \dfrac{2}{3}$ **64.** $\left(33 \div \dfrac{2}{11}\right) \cdot \dfrac{5}{9}$

65. $\dfrac{a^3}{2} \div 30a^3$ **66.** $15c^3 \div \dfrac{3c^2}{5}$ **67.** $\dfrac{ab^2}{c} \cdot \dfrac{c}{ab}$ **68.** $\dfrac{ac}{b} \cdot \dfrac{b^3}{a^2c}$

69. $\left(\dfrac{1}{2} \cdot \dfrac{2}{3}\right) \div \dfrac{5}{6}$ **70.** $\left(\dfrac{3}{4} \cdot \dfrac{8}{9}\right) \div \dfrac{2}{5}$ **71.** $-\dfrac{4}{7} \div \left(\dfrac{4}{5} \cdot \dfrac{3}{7}\right)$ **72.** $\dfrac{5}{8} \div \left(\dfrac{4}{7} \cdot -\dfrac{5}{16}\right)$

Objective **D** *Given the following replacement values, evaluate (**a**) xy and (**b**) x ÷ y. See Example 16.*

73. $x = \dfrac{2}{5}$ and $y = \dfrac{5}{6}$ **74.** $x = \dfrac{8}{9}$ and $y = \dfrac{1}{4}$ **75.** $x = -\dfrac{4}{5}$ and $y = \dfrac{9}{11}$ **76.** $x = \dfrac{7}{6}$ and $y = -\dfrac{1}{2}$

    a.   **b.**   **a.**   **b.**   **a.**   **b.**   **a.**   **b.**

Determine whether the given replacement values are solutions of the given equations. See Example 17.

77. Is $-\dfrac{5}{18}$ a solution to $3x = -\dfrac{5}{6}$?   **78.** Is $\dfrac{9}{11}$ a solution to $\dfrac{2}{3}y = \dfrac{6}{11}$?

79. Is $\dfrac{2}{5}$ a solution to $-\dfrac{1}{2}z = \dfrac{1}{10}$?   **80.** Is $\dfrac{3}{5}$ a solution to $5x = \dfrac{1}{3}$?

Objective **E** **Translating** *Solve. Write each answer in simplest form. For Exercises 81 through 84, recall that "of" translates to multiplication. See Example 18.*

81. Find $\dfrac{1}{4}$ of 200.   **82.** Find $\dfrac{1}{5}$ of 200.   **83.** Find $\dfrac{5}{6}$ of 24.   **84.** Find $\dfrac{5}{8}$ of 24.

Solve. See Example 18.

85. In the United States, $\frac{4}{25}$ of college freshmen major in business. A community college in Pennsylvania has a freshman enrollment of approximately 800 students. How many of these freshmen might we project are majoring in business?

86. In a recent year, movie theater owners received a total of $7660 million in movie admission tickets. About $\frac{7}{10}$ of this amount was for R-rated movies. Find the amount of money received from R-rated movies. (*Source:* Motion Picture Association of America)

87. The Oregon National Historic Trail is 2,170 miles long. It begins in Independence, Missouri, and ends in Oregon City, Oregon. Manfred Coulon has hiked $\frac{2}{5}$ of the trail before. How many miles has he hiked?

(*Source:* National Park Service)

88. Each turn of a screw sinks it $\frac{3}{16}$ of an inch deeper into a piece of wood. Find how deep the screw is after 8 turns.

89. The radius of a circle is one-half of its diameter as shown. If the diameter of a circle is $\frac{3}{8}$ of an inch, what is its radius?

90. A patient was told that no more than $\frac{1}{5}$ of his calories should come from fat. If his diet consists of 3000 calories a day, how many of these calories can come from fat?

91. A special on a cruise to the Bahamas is advertised to be $\frac{2}{3}$ of the regular price. If the regular price is $2757, what is the sale price?

92. The Gonzales family recently sold their house for $102,000, but $\frac{3}{50}$ of this amount goes to the real estate companies that helped them sell their house. How much money do the Gonzales pay to the real estate companies?

93. There have been about 180 different contestants on the reality television show *Survivor* over 12 seasons. Some of these contestants have appeared in two different series and/or seasons. If the number of returning contestants was $\frac{1}{9}$ of the total number of participants in the first 12 seasons, how many contestants participated more than once? (*Source:* Survivor.com)

94. In 2005, about $\frac{23}{50}$ of the people who saw movies in theaters were in the 16–24 age bracket. If the local cinema boasted that its total attendance for 2005 was 854,000 people, how many might the cinema owners believe were between the ages of 16 and 24? (*Source:* Motion Picture Association of America)

Find the area of each rectangle. Recall that area = length · width.

△ **95.**

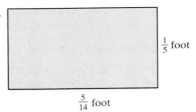

$\frac{1}{5}$ foot

$\frac{5}{14}$ foot

△ **96.**

$\frac{1}{2}$ mile

$\frac{3}{8}$ mile

*Recall from Section 4.2 that the following graph is called a **circle graph** or **pie chart**. Each sector (shaped like a piece of pie) shows the fractional part of a car's total mileage that falls into a particular category. The whole circle represents a car's total mileage.*

Shopping $\frac{3}{25}$

Vacation/other $\frac{3}{50}$

Work $\frac{8}{25}$

Social/recreational $\frac{13}{100}$

Medical $\frac{1}{100}$

Family business $\frac{1}{5}$

Visit friends $\frac{3}{25}$

School/church $\frac{2}{50}$

Source: The American Automobile Manufacturers Association and The National Automobile Dealers Association

In one year, a family drove 12,000 miles in the family car. Use the circle graph to determine how many of these miles might be expected to fall in the categories shown in Exercises 97 through 100.

97. Work

98. Shopping

99. Family business

100. Medical

Review

Perform each indicated operation. See Section 1.3.

101.
   27
   76
 + 98

102.
   811
    42
 + 69

103.
   968
 − 772

104.
   882
 − 773

Concept Extensions

105. In your own words, describe how to divide fractions.

106. In your own words, explain how to multiply fractions.

Simplify.

107. $\frac{42}{25} \cdot \frac{125}{36} \div \frac{7}{6}$

108. $\left(\frac{8}{13} \cdot \frac{39}{16} \cdot \frac{8}{9} \right)^2 \div \frac{1}{2}$

109. Approximately $\frac{3}{25}$ of the U.S. population lives in the state of California. If the U.S. population is approximately 290,810,000, find the approximate population of California. (*Source: World Almanac 2006*)

110. In 2006, there were approximately 10,600 commercial radio stations broadcasting in the United States. Of these stations, $\frac{32}{265}$ were news/talk stations. How many radio stations were news/talk stations in 2006? (*Source:* Corporation for Public Broadcasting)

111. If $\frac{3}{4}$ of 36 students on a first bus are girls and $\frac{2}{3}$ of the 30 students on a second bus are *boys,* how many students on the two buses are girls?

112. The FedEx Express air fleet includes 252 Cessnas. These Cessnas make up $\frac{42}{113}$ of the FedEx fleet. How many aircraft make up the entire FedEx Express air fleet? (*Source:* FedEx Corporation)

 THE BIGGER PICTURE Operations on Sets of Numbers

Continue your outline from Sections 1.6, 1.7, 2.4, and 3.3. Suggestions are once again written to help you complete this part of your outline, Section I.C. Fractions.

I. Operations on Sets of Numbers

   A. Whole Numbers

     1. Add or Subtract (Section 1.3)

     2. Multiply or Divide (Sections 1.5, 1.6)

     3. Exponent (Section 1.7)

     4. Order of Operations (Section 1.7)

   B. Integers

     1. Add (Section 2.2)

     2. Subtract (Section 2.3)

     3. Multiply or Divide (Section 2.4)

   C. Fractions

     1. Simplify: Factor the numerator and denominator. Then divide out factors of 1 by dividing out common factors in the numerator and denominator.

     Simplify: $\dfrac{20}{28} = \dfrac{\cancel{4} \cdot 5}{\cancel{4} \cdot 7} = \dfrac{5}{7}$

     2. Multiply: Numerator times numerator over denominator times denominator. $\dfrac{5}{9} \cdot \dfrac{2}{7} = \dfrac{10}{63}$

     3. Divide: First fraction times the reciprocal of the second fraction.

$$\frac{2}{11} \div \frac{3}{4} = \frac{2}{11} \cdot \frac{4}{3} = \frac{8}{33}$$

II. Solving Equations

   A. Equations in General (Section 3.3)

Perform the indicated operations.

1. $\dfrac{2}{3} \cdot \dfrac{8}{9}$ **2.** $-\dfrac{2}{3} \div \dfrac{8}{9}$

3. $\dfrac{12}{1} \cdot \dfrac{1}{9}$ **4.** $\dfrac{9x}{2} \div \dfrac{15x}{8}$

5. $80 \div 20 \cdot 2$ **6.** $3^2 \cdot 2^3$

7. $\dfrac{11}{20} \cdot \dfrac{5}{8} \cdot \dfrac{4}{33}$ **8.** $20 \div \dfrac{1}{2}$

9. $3 + 4(18 - 16)^3$ **10.** $100 - 76$

Solve.

11. $-12x + 7 = 7$ **12.** $9n - 14 = 6n - 32$

A Add or Subtract Like Fractions.

B Add or Subtract Given Fractional Replacement Values.

C Solve Problems by Adding or Subtracting Like Fractions.

D Find the Least Common Denominator of a List of Fractions.

E Write Equivalent Fractions.

Fractions with the same denominator are called **like fractions.** Fractions that have different denominators are called **unlike fractions.**

Like Fractions	Unlike Fractions
$\frac{2}{5}$ and $\frac{3}{5}$ ↑ ↑ —— same denominator	$\frac{2}{5}$ and $\frac{3}{4}$ ↑ ↑ —— different denominators
$\frac{5}{21}, \frac{16}{21}$, and $\frac{7}{21}$ ↑ ↑ ↑ —— same denominator	$\frac{5}{7}$ and $\frac{5}{9}$ ↑ ↑ —— different denominators

Objective **A** Adding or Subtracting Like Fractions

To see how we add like fractions (fractions with the same denominator), study one or both illustrations below.

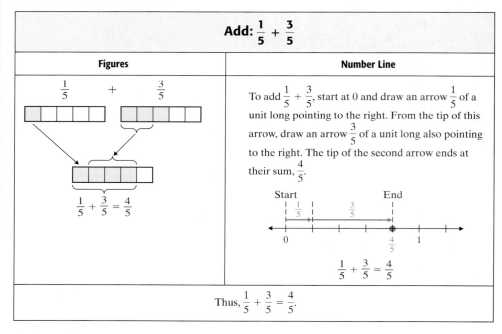

Add: $\frac{1}{5} + \frac{3}{5}$	
Figures	**Number Line**

Notice that the numerator of the sum is the sum of the numerators. Also, the denominator of the sum is the **common denominator.** This is how we add fractions. Similar illustrations can be shown for subtracting fractions.

Adding or Subtracting Like Fractions (Fractions with the Same Denominator)

If a, b, and c, are numbers and b is not 0, then

$$\frac{a}{b} + \frac{c}{b} = \frac{a+c}{b} \qquad \text{and also} \qquad \frac{a}{b} - \frac{c}{b} = \frac{a-c}{b}$$

In other words, to add or subtract fractions with the same denominator, add or subtract their numerators and write the sum or difference over the **common** denominator.

For example,

$$\frac{1}{4} + \frac{2}{4} = \frac{1+2}{4} = \frac{3}{4} \quad \text{Add the numerators.}$$
Keep the denominator.

$$\frac{4}{5} - \frac{2}{5} = \frac{4-2}{5} = \frac{2}{5} \quad \text{Subtract the numerators.}$$
Keep the denominator.

Helpful Hint

As usual, don't forget to write all answers in simplest form.

EXAMPLES Add and simplify.

1. $\frac{2}{7} + \frac{3}{7} = \frac{2+3}{7} = \frac{5}{7}$ ← Add the numerators.
← Keep the common denominator.

2. $\frac{3}{16} + \frac{7}{16} = \frac{3+7}{16} = \frac{10}{16} = \frac{\overset{1}{\cancel{2}} \cdot 5}{\cancel{2} \cdot 8} = \frac{5}{8}$

3. $\frac{7}{8} + \frac{6}{8} + \frac{3}{8} = \frac{7+6+3}{8} = \frac{16}{8} \text{ or } 2$

■ Work Practice Problems 1–3

✔ Concept Check Find and correct the error in the following:

$$\frac{1}{5} + \frac{1}{5} = \frac{2}{10}$$

EXAMPLES Subtract and simplify.

4. $\frac{8}{9} - \frac{1}{9} = \frac{8-1}{9} = \frac{7}{9}$ ← Subtract the numerators.
← Keep the common denominator.

5. $\frac{7}{8} - \frac{5}{8} = \frac{7-5}{8} = \frac{2}{8} = \frac{\overset{1}{\cancel{2}}}{\cancel{2} \cdot 4} = \frac{1}{4}$

■ Work Practice Problems 4–5

From our earlier work, we know that

$$\frac{-12}{6} = \frac{12}{-6} = -\frac{12}{6} \quad \text{since these all simplify to } -2.$$

In general, the following is true:

$$\frac{-a}{b} = \frac{a}{-b} = -\frac{a}{b} \quad \text{as long as } b \text{ is not 0.}$$

EXAMPLE 6 Add: $-\frac{11}{8} + \frac{6}{8}$

Solution: $-\frac{11}{8} + \frac{6}{8} = \frac{-11 + 6}{8}$

$$= \frac{-5}{8} \quad \text{or} \quad -\frac{5}{8}$$

■ Work Practice Problem 6

PRACTICE PROBLEMS 1–3
Add and simplify.

1. $\frac{6}{13} + \frac{2}{13}$

2. $\frac{5}{8} + \frac{1}{8}$

3. $\frac{20}{11} + \frac{6}{11} + \frac{7}{11}$

PRACTICE PROBLEMS 4–5
Subtract and simplify.

4. $\frac{11}{12} - \frac{6}{12}$ **5.** $\frac{7}{15} - \frac{2}{15}$

PRACTICE PROBLEM 6

Add: $-\frac{8}{17} + \frac{4}{17}$

Answers

1. $\frac{8}{13}$ **2.** $\frac{3}{4}$ **3.** 3 **4.** $\frac{5}{12}$ **5.** $\frac{1}{3}$

6. $-\frac{4}{17}$

✔ Concept Check Answer
We don't add denominators together; correct solution: $\frac{1}{5} + \frac{1}{5} = \frac{2}{5}$.

PRACTICE PROBLEM 7

Subtract: $\dfrac{2}{5} - \dfrac{7y}{5}$

EXAMPLE 7 Subtract: $\dfrac{3x}{4} - \dfrac{7}{4}$

Solution: $\dfrac{3x}{4} - \dfrac{7}{4} = \dfrac{3x - 7}{4}$

Recall from Section 3.1 that the terms in the numerator are unlike terms and cannot be combined.

◼ **Work Practice Problem 7**

PRACTICE PROBLEM 8

Subtract: $\dfrac{4}{11} - \dfrac{6}{11} - \dfrac{3}{11}$

EXAMPLE 8 Subtract: $\dfrac{3}{7} - \dfrac{6}{7} - \dfrac{3}{7}$

Solution: $\dfrac{3}{7} - \dfrac{6}{7} - \dfrac{3}{7} = \dfrac{3 - 6 - 3}{7} = \dfrac{-6}{7}$ or $-\dfrac{6}{7}$

◼ **Work Practice Problem 8**

> **Helpful Hint**
>
> Recall that $\dfrac{-6}{7} = -\dfrac{6}{7}$ $\left(\text{Also, } \dfrac{6}{-7} = -\dfrac{6}{7}, \text{ if needed.}\right)$

Objective B Adding and Subtracting Given Fractional Replacement Values

PRACTICE PROBLEM 9

Evaluate $x + y$ if $x = -\dfrac{10}{12}$ and $y = \dfrac{5}{12}$.

EXAMPLE 9 Evaluate $y - x$ if $x = -\dfrac{3}{10}$ and $y = -\dfrac{8}{10}$.

Solution: Be very careful when replacing x and y with replacement values.

$$y - x = -\dfrac{8}{10} - \left(-\dfrac{3}{10}\right) \quad \text{Replace } x \text{ with } -\dfrac{3}{10} \text{ and } y \text{ with } -\dfrac{8}{10}.$$

$$= \dfrac{-8 - (-3)}{10}$$

$$= \dfrac{-5}{10} = \dfrac{-1 \cdot 5}{2 \cdot 5} = \dfrac{-1}{2} \text{ or } -\dfrac{1}{2}$$

◼ **Work Practice Problem 9**

✔ **Concept Check** Fill in each blank with the best choice given.

expression   equation   simplified   solved

An _____ contains an equals sign and may be _____ for the variable. An _____ does not contain an equals sign but may be _____.

Objective C Solving Problems by Adding or Subtracting Like Fractions

Many real-life problems involve finding the perimeters of square or rectangular-shaped figures such as pastures, swimming pools, and so on. We can use our knowledge of adding fractions to find perimeters.

Answers

7. $\dfrac{2 - 7y}{5}$ 8. $-\dfrac{5}{11}$ 9. $-\dfrac{5}{12}$

✔ **Concept Check Answer**

equation solved expression simplified

EXAMPLE 10 Find the perimeter of the rectangle.

$\frac{2}{15}$ inch

$\frac{4}{15}$ inch

Solution: Recall that perimeter means distance around and that opposite sides of a rectangle are the same length.

$\frac{4}{15}$ inch

$\frac{2}{15}$ inch $\frac{2}{15}$ inch

$\frac{4}{15}$ inch

$$\text{Perimeter} = \frac{2}{15} + \frac{4}{15} + \frac{2}{15} + \frac{4}{15} = \frac{2+4+2+4}{15}$$

$$= \frac{12}{15} = \frac{\overset{1}{\cancel{3}} \cdot 4}{\underset{1}{\cancel{3}} \cdot 5} = \frac{4}{5}$$

The perimeter of the rectangle is $\frac{4}{5}$ inch.

▢ **Work Practice Problem 10**

PRACTICE PROBLEM 10

Find the perimeter of the square.

$\frac{3}{20}$ mile

We can combine our skills in adding and subtracting fractions with our four problem-solving steps from Section 3.4 to solve many kinds of real-life problems.

EXAMPLE 11 **Calculating Distance**

The distance from home to the World Gym is $\frac{7}{8}$ of a mile and from home to the post office is $\frac{3}{8}$ of a mile. How much farther is it from home to the World Gym than from home to the post office?

PRACTICE PROBLEM 11

A jogger ran $\frac{13}{4}$ miles on Monday and $\frac{11}{4}$ miles on Wednesday. How much farther did he run on Monday than on Wednesday?

Home

$\frac{7}{8}$ mile   WORLD GYM

$\frac{3}{8}$ mile   Post office

Solution:

1. **UNDERSTAND.** Read and reread the problem. The phrase "How much farther" tells us to subtract distances.

2. **TRANSLATE.**

In words:	distance farther	is	home to World Gym distance	minus	home to post office distance
	↓	↓	↓	↓	↓
Translate:	distance farther	=	$\frac{7}{8}$	−	$\frac{3}{8}$

3. **SOLVE:** $\dfrac{7}{8} - \dfrac{3}{8} = \dfrac{7-3}{8} = \dfrac{4}{8} = \dfrac{\overset{1}{\cancel{4}}}{2 \cdot \underset{1}{\cancel{4}}} = \dfrac{1}{2}$

Continued on next page

Answers

10. $\frac{3}{5}$ mi **11.** $\frac{1}{2}$ mi

4. INTERPRET. *Check* your work. *State* your conclusion: The distance from home to the World Gym is $\frac{1}{2}$ mile farther than from home to the post office.

▣ **Work Practice Problem 11**

Objective Ⓓ **Finding the Least Common Denominator**

In the next section, we will add and subtract fractions that have different, or unlike, denominators. To do so, we first write them as equivalent fractions with a common denominator.

Although any common denominator can be used to add or subtract unlike fractions, we will use the **least common denominator (LCD).** The LCD of a list of fractions is the same as the **least common multiple (LCM)** of the denominators. Why do we use this number as the common denominator? Since the LCD is the *smallest* of all common denominators, operations are usually less tedious with this number.

The **least common denominator (LCD)** of a list of fractions is the smallest positive number divisible by all the denominators in the list. (The least common denominator is also the **least common multiple (LCM) of the denominators.**)

For example, the LCD of $\frac{1}{4}$ and $\frac{3}{10}$ is 20 because 20 is the smallest positive number divisible by both 4 and 10.

Finding the LCD: Method 1

One way to find the LCD is to see whether the larger denominator is divisible by the smaller denominator. If so, the larger number is the LCD. If not, then check consecutive multiples of the larger denominator until the LCD is found.

Method 1: Finding the LCD of a List of Fractions Using Multiples of the Largest Number

Step 1: Write the multiples of the largest denominator (starting with the number itself) until a multiple common to all denominators in the list is found.

Step 2: The multiple found in Step 1 is the LCD.

PRACTICE PROBLEM 12

Find the LCD of $\frac{7}{8}$ and $\frac{11}{16}$.

EXAMPLE 12 Find the LCD of $\frac{3}{7}$ and $\frac{5}{14}$.

Solution: We write the multiples of 14 until we find one that is also a multiple of 7.

$14 \cdot 1 = 14$ A multiple of 7

The LCD is 14.

▣ **Work Practice Problem 12**

Answer
12. 16

EXAMPLE 13 Find the LCD of $\frac{11}{12}$ and $\frac{7}{20}$.

Solution: We write the multiples of 20 until we find one that is also a multiple of 12.

$20 \cdot 1 = 20$ Not a multiple of 12

$20 \cdot 2 = 40$ Not a multiple of 12

$20 \cdot 3 = 60$ A multiple of 12

The LCD is 60.

Work Practice Problem 13

Method 1 for finding multiples works fine for smaller numbers, but may get tedious for larger numbers. For this reason, let's study a second method that uses prime factorization.

Finding the LCD: Method 2

For example, to find the LCD of $\frac{11}{12}$ and $\frac{7}{20}$, such as in Example 13, let's look at the prime factorization of each denominator.

$12 = 2 \cdot 2 \cdot 3$

$20 = 2 \cdot 2 \cdot 5$

Recall that the LCD must be a multiple of both 12 and 20. Thus, to build the LCD, we will circle the greatest number of factors for each different prime number. The LCD is the product of the circled factors.

Prime Number Factors

$12 = (2 \cdot 2) \quad (3)$

$20 = 2 \cdot 2 \quad (5)$ Circle either pair of 2s, but not both.

$LCD = 2 \cdot 2 \cdot 3 \cdot 5 = 60$

The number 60 is the smallest number that both 12 and 20 divide into evenly. This method 2 is summarized below:

Method 2: Finding the LCD of a List of Denominators Using Prime Factorization

Step 1: Write the prime factorization of each denominator.

Step 2: For each different prime factor in Step 1, circle the *greatest* number of times that factor occurs in any one factorization.

Step 3: The LCD is the product of the circled factors.

EXAMPLE 14 Find the LCD of $-\frac{23}{72}$ and $\frac{17}{60}$.

Solution: First we write the prime factorization of each denominator.

$72 = 2 \cdot 2 \cdot 2 \cdot 3 \cdot 3$

$60 = 2 \cdot 2 \cdot 3 \cdot 5$

For the prime factors shown, we circle the greatest number of factors found in either factorization.

$72 = (2 \cdot 2 \cdot 2) \cdot (3 \cdot 3)$

$60 = 2 \cdot 2 \cdot 3 \cdot (5)$

Continued on next page

PRACTICE PROBLEM 13

Find the LCD of $\frac{23}{25}$ and $\frac{1}{30}$.

PRACTICE PROBLEM 14

Find the LCD of $\frac{-3}{40}$ and $\frac{11}{108}$.

Answers

13. 150   **14.** 1080

The LCD is the product of the circled factors.
$$\text{LCD} = 2 \cdot 2 \cdot 2 \cdot 3 \cdot 3 \cdot 5 = 360$$
The LCD is 360.

Work Practice Problem 14

PRACTICE PROBLEM 15

Find the LCD of $\dfrac{7}{20}, \dfrac{1}{24}$, and $\dfrac{13}{45}$.

EXAMPLE 15 Find the LCD of $\dfrac{1}{15}, \dfrac{5}{18}$, and $\dfrac{53}{54}$.

Solution:
$$15 = 3 \cdot \boxed{5}$$
$$18 = \boxed{2} \cdot 3 \cdot 3$$
$$54 = 2 \cdot \boxed{3 \cdot 3 \cdot 3}$$

The LCD is $2 \cdot 3 \cdot 3 \cdot 3 \cdot 5$ or 270.

Work Practice Problem 15

PRACTICE PROBLEM 16

Find the LCD of $\dfrac{7}{y}$ and $\dfrac{6}{11}$.

EXAMPLE 16 Find the LCD of $\dfrac{3}{5}, \dfrac{2}{x}$, and $\dfrac{7}{x^3}$.

Solution:
$$5 = \boxed{5}$$
$$x = x$$
$$x^3 = \boxed{x \cdot x \cdot x}$$

$$\text{LCD} = 5 \cdot x \cdot x \cdot x = 5x^3$$

Work Practice Problem 16

✔ **Concept Check** True or false? The LCD of the fractions $\dfrac{1}{6}$ and $\dfrac{1}{8}$ is 48.

Objective **E** Writing Equivalent Fractions

To add or subtract unlike fractions in the next section, we first write equivalent fractions with the LCD as the denominator.

To write $\dfrac{1}{3}$ as an equivalent fraction with a denominator of 6, we multiply by 1 in the form of $\dfrac{2}{2}$. Why? Because $3 \cdot 2 = 6$, so the new denominator will become 6, as shown below.

$$\frac{1}{3} = \frac{1}{3} \cdot 1 = \frac{1}{3} \cdot \frac{2}{2} = \frac{1 \cdot 2}{3 \cdot 2} = \frac{2}{6}$$
$$\frac{2}{2} = 1$$

So $\dfrac{1}{3} = \dfrac{2}{6}$.

To write an equivalent fraction,

$$\frac{a}{b} = \frac{a}{b} \cdot \frac{c}{c} = \frac{a \cdot c}{b \cdot c}$$

where a, b, and c are nonzero numbers.

✔ **Concept Check** Which of the following is *not* equivalent to $\frac{3}{4}$?

a. $\frac{6}{8}$　　**b.** $\frac{18}{24}$　　**c.** $\frac{9}{14}$　　**d.** $\frac{30}{40}$

EXAMPLE 17 Write $\frac{3}{4}$ as an equivalent fraction with a denominator of 20.

$$\frac{3}{4} = \frac{}{20}$$

Solution: In the denominators, since $4 \cdot 5 = 20$, we will multiply by 1 in the form

of $\frac{5}{5}$.

$$\frac{3}{4} = \frac{3}{4} \cdot \frac{5}{5} = \frac{3 \cdot 5}{4 \cdot 5} = \frac{15}{20}$$

Thus, $\frac{3}{4} = \frac{15}{20}$.

▢ **Work Practice Problem 17**

Helpful Hint

To check Example 17, write $\frac{15}{20}$ in simplest form.

$$\frac{15}{20} = \frac{3 \cdot \overset{1}{\cancel{5}}}{4 \cdot \cancel{5}} = \frac{3}{4}, \text{ the original fraction.}$$
$$\phantom{\frac{15}{20} = \frac{3 \cdot \cancel{5}}{4 \cdot }}{}_1$$

If the original fraction is in lowest terms, we can check our work by writing the new equivalent fraction in simplest form. This form should be the original fraction.

✔ **Concept Check** True or false? When the fraction $\frac{2}{9}$ is rewritten as an equivalent fraction with 27 as the denominator, the result is $\frac{2}{27}$.

EXAMPLE 18 Write an equivalent fraction with the given denominator.

$$\frac{2}{5} = \frac{}{15}$$

Solution: Since $5 \cdot 3 = 15$, we multiply by 1 in the form of $\frac{3}{3}$.

$$\frac{2}{5} = \frac{2}{5} \cdot \frac{3}{3} = \frac{2 \cdot 3}{5 \cdot 3} = \frac{6}{15}$$

Then $\frac{2}{5}$ is equivalent to $\frac{6}{15}$. They both represent the same part of a whole.

▢ **Work Practice Problem 18**

PRACTICE PROBLEM 17

Write $\frac{7}{8}$ as an equivalent fraction with a denominator of 56.

$$\frac{7}{8} = \frac{}{56}$$

PRACTICE PROBLEM 18

Write an equivalent fraction with the given denominator.

$$\frac{1}{4} = \frac{}{20}$$

Answers

17. $\frac{49}{56}$　**18.** $\frac{5}{20}$

✔ **Concept Check Answers**

c

false; the correct result would be $\frac{6}{27}$

PRACTICE PROBLEM 19

Write an equivalent fraction with the given denominator.

$$\frac{3x}{7} = \frac{}{42}$$

EXAMPLE 19 Write an equivalent fraction with the given denominator.

$$\frac{9x}{11} = \frac{}{44}$$

Solution: Since $11 \cdot 4 = 44$, we multiply by 1 in the form of $\frac{4}{4}$.

$$\frac{9x}{11} = \frac{9x}{11} \cdot \frac{4}{4} = \frac{9x \cdot 4}{11 \cdot 4} = \frac{36x}{44}$$

Then $\frac{9x}{11}$ is equivalent to $\frac{36x}{44}$.

☐ **Work Practice Problem 19**

PRACTICE PROBLEM 20

Write an equivalent fraction with the given denominator.

$$4 = \frac{}{6}$$

EXAMPLE 20 Write an equivalent fraction with the given denominator.

$$3 = \frac{}{7}$$

Solution: Recall that $3 = \frac{3}{1}$. Since $1 \cdot 7 = 7$, multiply by 1 in the form $\frac{7}{7}$.

$$\frac{3}{1} = \frac{3}{1} \cdot \frac{7}{7} = \frac{3 \cdot 7}{1 \cdot 7} = \frac{21}{7}$$

☐ **Work Practice Problem 20**

Don't forget that when the denominator of a fraction contains a variable, such as $\frac{8}{3x}$, we will assume that the variable does not represent 0. Recall that the denominator of a fraction cannot be 0.

PRACTICE PROBLEM 21

Write an equivalent fraction with the given denominator.

$$\frac{9}{4x} = \frac{}{36x}$$

EXAMPLE 21 Write an equivalent fraction with the given denominator.

$$\frac{8}{3x} = \frac{}{24x}$$

Solution: Since $3x \cdot 8 = 24x$, multiply by 1 in the form $\frac{8}{8}$.

$$\frac{8}{3x} = \frac{8}{3x} \cdot \frac{8}{8} = \frac{8 \cdot 8}{3x \cdot 8} = \frac{64}{24x}$$

☐ **Work Practice Problem 21**

✔**Concept Check**   What is the first step in writing $\frac{3}{10}$ as an equivalent fraction whose denominator is 100?

Answers

19. $\frac{18x}{42}$ **20.** $\frac{24}{6}$ **21.** $\frac{81}{36x}$

✔ **Concept Check Answer**
answers may vary

Vocabulary and Readiness Check

least common denominator (LCD)   like $-\dfrac{a}{b}$ $\dfrac{a-c}{b}$ $\dfrac{a+c}{b}$ $-\dfrac{a}{-b}$

perimeter   unlike

equivalent

1. The fractions $\dfrac{9}{11}$ and $\dfrac{13}{11}$ are called _____ fractions while $\dfrac{3}{4}$ and $\dfrac{1}{3}$ are called _____ fractions.

2. $\dfrac{a}{b} + \dfrac{c}{b} =$ _____ and $\dfrac{a}{b} - \dfrac{c}{b} =$ _____ .

3. As long as b is not 0, $\dfrac{-a}{b} = \dfrac{a}{-b} =$ _____ .

4. The distance around a figure is called its _____ .

5. The smallest positive number divisible by all the denominators of a list of fractions is called the _____ .

6. Fractions that represent the same portion of a whole are called _____ fractions.

4.4 EXERCISE SET

FOR EXTRA HELP

Student Solutions Manual   PH Math/Tutor Center   CD/Video for Review   MathXL®   MyMathLab

Objective A *Add and simplify. See Examples 1 through 3, and 6.*

1. $\dfrac{5}{11} + \dfrac{2}{11}$

2. $\dfrac{9}{17} + \dfrac{2}{17}$

3. $\dfrac{2}{9} + \dfrac{4}{9}$

4. $\dfrac{3}{10} + \dfrac{2}{10}$

5. $-\dfrac{6}{20} + \dfrac{1}{20}$

6. $-\dfrac{3}{8} + \dfrac{1}{8}$

7. $-\dfrac{3}{14} + \left(-\dfrac{4}{14}\right)$

8. $-\dfrac{5}{24} + \left(-\dfrac{7}{24}\right)$

9. $\dfrac{2}{9x} + \dfrac{4}{9x}$

10. $\dfrac{3}{10y} + \dfrac{2}{10y}$

11. $-\dfrac{7x}{18} + \dfrac{3x}{18} + \dfrac{2x}{18}$

12. $-\dfrac{7z}{15} + \dfrac{3z}{15} + \dfrac{1z}{15}$

Subtract and simplify. See Examples 4, 5, 7, and 8.

13. $\dfrac{10}{11} - \dfrac{4}{11}$

14. $\dfrac{9}{13} - \dfrac{5}{13}$

15. $\dfrac{7}{8} - \dfrac{1}{8}$

16. $\dfrac{5}{6} - \dfrac{1}{6}$

17. $\dfrac{1}{y} - \dfrac{4}{y}$

18. $\dfrac{4}{z} - \dfrac{7}{z}$

19. $-\dfrac{27}{33} - \left(-\dfrac{8}{33}\right)$

20. $-\dfrac{37}{45} - \left(-\dfrac{18}{45}\right)$

21. $\dfrac{20}{21} - \dfrac{10}{21} - \dfrac{17}{21}$

22. $\dfrac{27}{28} - \dfrac{5}{28} - \dfrac{28}{28}$

23. $\dfrac{7a}{4} - \dfrac{3}{4}$

24. $\dfrac{18b}{5} - \dfrac{3}{5}$

Mixed Practice *Perform the indicated operation. See Examples 1 through 8.*

25. $-\dfrac{9}{100} + \dfrac{99}{100}$

26. $-\dfrac{15}{200} + \dfrac{85}{200}$

27. $-\dfrac{13x}{28} - \dfrac{13x}{28}$

28. $-\dfrac{15}{26y} - \dfrac{15}{26y}$

29. $\dfrac{9x}{15} + \dfrac{1}{15}$

30. $\dfrac{2x}{15} + \dfrac{7}{15}$

31. $\dfrac{7x}{16} - \dfrac{15x}{16}$

32. $\dfrac{3}{16z} - \dfrac{15}{16z}$

33. $\dfrac{9}{12} - \dfrac{7}{12} - \dfrac{10}{12}$

34. $\dfrac{1}{8} - \dfrac{15}{8} + \dfrac{2}{8}$

35. $\dfrac{x}{4} + \dfrac{3x}{4} - \dfrac{2x}{4} + \dfrac{x}{4}$

36. $\dfrac{9y}{8} + \dfrac{2y}{8} + \dfrac{5y}{8} - \dfrac{4y}{8}$

Objective **B** *Evaluate each expression for the given replacement values. See Example 9.*

37. $x + y$; $x = \dfrac{3}{4}$, $y = \dfrac{2}{4}$

38. $x - y$; $x = \dfrac{7}{8}$, $y = \dfrac{9}{8}$

39. $x - y$; $x = -\dfrac{1}{5}$, $y = -\dfrac{3}{5}$

40. $x + y$; $x = -\dfrac{1}{6}$, $y = \dfrac{5}{6}$

Objective **C** *Find the perimeter of each figure. (Hint: Recall that perimeter means distance around.) See Example 10.*

△ **41.**

△ **42.**

△ **43.**
$\dfrac{5}{12}$ meter Rectangle
$\dfrac{7}{12}$ meter

△ **44.**
Square $\dfrac{1}{6}$ centimeter

Solve. Write each answer in simplest form. See Example 11.

The map of the world below shows the fraction of the world's surface land area taken up by each continent. In other words, the continent of Africa makes up $\dfrac{20}{100}$ of the land in the world. Use this map for Exercises 45 through 48.

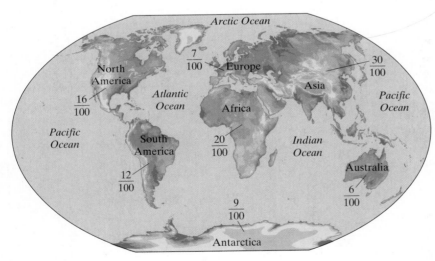

45. Find the fractional part of the world's land area within the continents of North America and South America.

46. Find the fractional part of the world's land area within the continents of Asia and Africa.

47. How much greater is the fractional part of the continent of Antarctica than the fractional part of the continent of Europe?

48. How much greater is the fractional part of the continent of Asia than the continent of Australia?

Solve.

49. A railroad inspector must inspect $\dfrac{19}{20}$ of a mile of railroad track. If she has already inspected $\dfrac{5}{20}$ of a mile, how much more does she need to inspect?

50. Scott Davis has run $\dfrac{11}{8}$ miles already and plans to complete $\dfrac{16}{8}$ miles. To do this, how much farther must he run?

51. As of 2006, the fraction of states in the United States with maximum interstate highway speed limits up to and including 70 mph was $\frac{37}{50}$. The fraction of states with 70 mph speed limits was $\frac{17}{50}$. What fraction of states had speed limits that were less than 70 mph? (*Source:* Insurance Institute for Highway Safety)

52. When people take aspirin, $\frac{31}{50}$ of the time it is used to treat some type of pain. Approximately $\frac{7}{50}$ of all aspirin use is for treating headaches. What fraction of aspirin use is for treating pain other than headaches? (*Source:* Bayer Market Research)

Objective **D** *Find the LCD of each list of fractions. See Examples 12 through 16.*

53. $\frac{2}{9}, \frac{6}{15}$

54. $\frac{7}{12}, \frac{3}{20}$

55. $-\frac{1}{36}, \frac{1}{24}$

56. $-\frac{1}{15}, \frac{1}{90}$

57. $\frac{2}{25}, \frac{3}{15}, \frac{5}{6}$

58. $\frac{3}{4}, \frac{1}{14}, \frac{13}{20}$

59. $-\frac{7}{24}, -\frac{5}{x}$

60. $-\frac{11}{y}, -\frac{13}{70}$

61. $\frac{23}{18}, \frac{1}{21}$

62. $\frac{45}{24}, \frac{2}{45}$

63. $\frac{4}{3}, \frac{8}{21}, \frac{3}{56}$

64. $\frac{12}{11}, \frac{20}{33}, \frac{12}{121}$

Objective **E** *Write each fraction as an equivalent fraction with the given denominator. See Examples 17 through 21.*

65. $\frac{2}{3} = \frac{}{21}$

66. $\frac{5}{6} = \frac{}{24}$

67. $\frac{4}{7} = \frac{}{35}$

68. $\frac{3}{5} = \frac{}{100}$

69. $\frac{1}{2} = \frac{}{50}$

70. $\frac{1}{5} = \frac{}{50}$

71. $\frac{14x}{17} = \frac{}{68}$

72. $\frac{19z}{21} = \frac{}{126}$

73. $\frac{2y}{3} = \frac{}{12}$

74. $\frac{3x}{2} = \frac{}{12}$

75. $\frac{5}{9} = \frac{}{36a}$

76. $\frac{7}{6} = \frac{}{36a}$

The table shows the fraction of the population in selected countries that used cell phones in a recent year. Use this table to answer Exercises 77 through 80.

77. Complete the table by writing each fraction as an equivalent fraction with a denominator of 100.

78. Which of these countries has the largest fraction of cell phone users?

79. Which of these countries shown has the smallest fraction of cell phone users?

80. In which of these countries do over $\frac{3}{4}$ of the population use cell phones? (*Hint:* Write $\frac{3}{4}$ as an equivalent fraction with a denominator of 100.)

Country	Fraction of Population Using Cell Phones	Equivalent Fraction with a Denominator of 100
Denmark	$\frac{89}{100}$	
Finland	$\frac{9}{10}$	
Israel	$\frac{24}{25}$	
Spain	$\frac{23}{25}$	
Japan	$\frac{17}{25}$	
Norway	$\frac{91}{100}$	
Singapore	$\frac{4}{5}$	
South Korea	$\frac{69}{100}$	
India	$\frac{2}{25}$	
United States	$\frac{7}{10}$	

(*Source:* International Telecommunication and World Almanac, 2006)

Review

Simplify. See Section 1.8.

81. 3^2
82. 4^3
83. 5^3
84. 3^4

85. 7^2
86. 5^4
87. $2^3 \cdot 3$
88. $4^2 \cdot 5$

Concept Extensions

Find and correct the error. See the first Concept Check in this section.

89. $\dfrac{2}{7} + \dfrac{9}{7} = \dfrac{11}{14}$

90. $\dfrac{3}{4} - \dfrac{1}{4} = \dfrac{2}{8} = \dfrac{1}{4}$

Solve.

91. In your own words, explain how to add like fractions.

92. In your own words, explain how to subtract like fractions.

93. Use the map of the world for Exercises 45 through 48 and find the sum of all the continents' fractions. Explain your answer.

94. Mike Cannon jogged $\dfrac{3}{8}$ of a mile from home and then rested. Then he continued jogging further from home for another $\dfrac{3}{8}$ of a mile until he discovered his watch had fallen off. He walked back along the same path for $\dfrac{4}{8}$ of a mile until he found his watch. Find how far he was from his home.

Write each fraction as an equivalent fraction with the indicated denominator.

95. $\dfrac{37x}{165} = \dfrac{}{3630}$

96. $\dfrac{108}{215y} = \dfrac{}{4085y}$

97. In your own words, explain how to find the LCD of two fractions.

98. In your own words, explain how to write a fraction as an equivalent fraction with a given denominator.

Solve. See the Concept Checks in this section.

99. Which of the following are equivalent to $\dfrac{2}{3}$?

   a. $\dfrac{10}{15}$ **b.** $\dfrac{40}{60}$

   c. $\dfrac{16}{20}$ **d.** $\dfrac{200}{300}$

100. True or False? When the fraction $\dfrac{7}{12}$ is rewritten with a denominator of 48, the result is $\dfrac{11}{48}$. If false, give the correct fraction.

4.5 ADDING AND SUBTRACTING UNLIKE FRACTIONS

Objective A Adding and Subtracting Unlike Fractions

In this section we add and subtract fractions with unlike denominators. To add or subtract these unlike fractions, we first write the fractions as equivalent fractions with a common denominator and then add or subtract the like fractions. Recall from the previous section that the common denominator we use is called the **least common denominator (LCD).**

To begin, let's add the unlike fractions $\frac{3}{4} + \frac{1}{6}$.

The LCD of these fractions is 12. So we write each fraction as an equivalent fraction with a denominator of 12, then add as usual. This addition process is shown next and also illustrated by figures.

<table>
<tr><td colspan="2" align="center">Add: $\frac{3}{4} + \frac{1}{6}$</td><td align="center">The LCD is 12.</td></tr>
</table>

Figures	Algebra
$\frac{3}{4}$ + $\frac{1}{6}$ $\frac{9}{12}$ + $\frac{2}{12}$ $\frac{9}{12} + \frac{2}{12} = \frac{11}{12}$	$\frac{3}{4} = \frac{3}{4} \cdot \frac{3}{3} = \frac{9}{12}$ and $\frac{1}{6} = \frac{1}{6} \cdot \frac{2}{2} = \frac{2}{12}$ Remember $\frac{3}{3} = 1$ and $\frac{2}{2} = 1$. Now we can add just as we did in Section 4.4. $\frac{3}{4} + \frac{1}{6} = \frac{9}{12} + \frac{2}{12} = \frac{11}{12}$

Thus, the sum is $\frac{11}{12}$.

Adding or Subtracting Unlike Fractions

Step 1: Find the least common denominator (LCD) of the fractions.

Step 2: Write each fraction as an equivalent fraction whose denominator is the LCD.

Step 3: Add or subtract the like fractions.

Step 4: Write the sum or difference in simplest form.

PRACTICE PROBLEM 1

Add: $\dfrac{2}{7} + \dfrac{8}{21}$

EXAMPLE 1 Add: $\dfrac{2}{5} + \dfrac{4}{15}$

Solution:

Step 1: The LCD of the fractions is 15. In later examples, we shall simply say, for example, that the LCD of 5 and 15 is 15.

Step 2: $\dfrac{2}{5} = \dfrac{2}{5} \cdot \dfrac{3}{3} = \dfrac{6}{15}, \quad \dfrac{4}{15} = \dfrac{4}{15}$ ← This fraction already has a denominator of 15.

$\uparrow$
└── Multiply by 1 in the form $\dfrac{3}{3}$

Step 3: $\dfrac{2}{5} + \dfrac{4}{15} = \dfrac{6}{15} + \dfrac{4}{15} = \dfrac{10}{15}$

Step 4: Write in simplest form.

$$\dfrac{10}{15} = \dfrac{2 \cdot \overset{1}{\cancel{5}}}{3 \cdot \underset{1}{\cancel{5}}} = \dfrac{2}{3}$$

◻ **Work Practice Problem 1**

When the fractions contain variables, we add and subtract the same way.

PRACTICE PROBLEM 2

Add: $\dfrac{5y}{6} + \dfrac{2y}{9}$

EXAMPLE 2 Add: $\dfrac{2x}{15} + \dfrac{3x}{10}$

Solution:

Step 1: The LCD of the denominators 15 and 10 is 30.

Step 2: $\dfrac{2x}{15} = \dfrac{2x}{15} \cdot \dfrac{2}{2} = \dfrac{4x}{30} \qquad \dfrac{3x}{10} = \dfrac{3x}{10} \cdot \dfrac{3}{3} = \dfrac{9x}{30}$

Step 3: $\dfrac{2x}{15} + \dfrac{3x}{10} = \dfrac{4x}{30} + \dfrac{9x}{30} = \dfrac{13x}{30}$

Step 4: $\dfrac{13x}{30}$ is in simplest form.

◻ **Work Practice Problem 2**

PRACTICE PROBLEM 3

Add: $-\dfrac{1}{5} + \dfrac{9}{20}$

EXAMPLE 3 Add: $-\dfrac{1}{6} + \dfrac{1}{2}$

Solution: The LCD of the denominators 6 and 2 is 6.

$$-\dfrac{1}{6} + \dfrac{1}{2} = \dfrac{-1}{6} + \dfrac{1 \cdot 3}{2 \cdot 3}$$

$$= \dfrac{-1}{6} + \dfrac{3}{6}$$

$$= \dfrac{2}{6}$$

Helpful Hint

Recall that

$$-\dfrac{1}{6} = \dfrac{-1}{6} = \dfrac{1}{-6}$$

Next, simplify $\dfrac{2}{6}$.

$$\dfrac{2}{6} = \dfrac{\overset{1}{\cancel{2}}}{\underset{1}{\cancel{2}} \cdot 3} = \dfrac{1}{3}$$

◻ **Work Practice Problem 3**

Answers

1. $\dfrac{2}{3}$ **2.** $\dfrac{19y}{18}$ **3.** $\dfrac{1}{4}$

✔ **Concept Check** Find and correct the error in the following:

$$\frac{2}{9} + \frac{4}{11} = \frac{6}{20} = \frac{3}{10}$$

EXAMPLE 4 Subtract: $\frac{2}{3} - \frac{10}{11}$

Solution:

Step 1: The LCD of 3 and 11 is 33.

Step 2: $\frac{2}{3} = \frac{2}{3} \cdot \frac{11}{11} = \frac{22}{33}$ $\quad$ $\frac{10}{11} = \frac{10}{11} \cdot \frac{3}{3} = \frac{30}{33}$

Step 3: $\frac{2}{3} - \frac{10}{11} = \frac{22}{33} - \frac{30}{33} = \frac{-8}{33}$ or $-\frac{8}{33}$

Step 4: $-\frac{8}{33}$ is in simplest form.

◼ **Work Practice Problem 4**

EXAMPLE 5 Find: $-\frac{3}{4} - \frac{1}{14} + \frac{6}{7}$

Solution: The LCD of 4, 14, and 7 is 28.

$$-\frac{3}{4} - \frac{1}{14} + \frac{6}{7} = -\frac{3 \cdot 7}{4 \cdot 7} - \frac{1 \cdot 2}{14 \cdot 2} + \frac{6 \cdot 4}{7 \cdot 4}$$

$$= -\frac{21}{28} - \frac{2}{28} + \frac{24}{28}$$

$$= \frac{1}{28}$$

◼ **Work Practice Problem 5**

✔ **Concept Check** Find and correct the error in the following:

$$\frac{7}{12} - \frac{3}{4} = \frac{4}{8} = \frac{1}{2}$$

EXAMPLE 6 Subtract: $2 - \frac{x}{3}$

Solution: Recall that $2 = \frac{2}{1}$. The LCD of the denominators 1 and 3 is 3.

$$\frac{2}{1} - \frac{x}{3} = \frac{2 \cdot 3}{1 \cdot 3} - \frac{x}{3}$$

$$= \frac{6}{3} - \frac{x}{3}$$

$$= \frac{6 - x}{3}$$

Helpful Hint

The expression $\frac{6 - x}{3}$ from Example 6 *does not simplify* to $2 - x$. The number 3 must be a factor of both terms in the numerator (not just 6) in order to simplify.

The numerator $6 - x$ cannot be simplified further since 6 and $-x$ are unlike terms.

◼ **Work Practice Problem 6**

PRACTICE PROBLEM 4

Subtract: $\frac{5}{7} - \frac{9}{10}$

PRACTICE PROBLEM 5

Find: $\frac{5}{8} - \frac{1}{3} - \frac{1}{12}$

PRACTICE PROBLEM 6

Subtract: $5 - \frac{y}{4}$

Answers

4. $-\frac{13}{70}$ 5. $\frac{5}{24}$ 6. $\frac{20 - y}{4}$

✔ **Concept Check Answer**

When adding unlike fractions, we don't add the denominators. Correct solution:

$$\frac{2}{9} + \frac{4}{11} = \frac{22}{99} + \frac{36}{99} = \frac{58}{99}$$

$$\frac{7}{12} - \frac{3}{4} = \frac{7}{12} - \frac{9}{12} = -\frac{2}{12} = -\frac{1}{6}$$

Objective B Writing Fractions in Order

One important application of the least common denominator is to use the LCD to help order or compare fractions.

PRACTICE PROBLEM 7

Insert < or > to form a true sentence.

$\dfrac{5}{8} \quad \dfrac{11}{20}$

EXAMPLE 7 Insert < or > to form a true sentence.

$\dfrac{3}{4} \quad \dfrac{9}{11}$

Solution: The LCD for these fractions is 44. Let's write each fraction as an equivalent fraction with a denominator of 44.

$$\dfrac{3}{4} = \dfrac{3 \cdot 11}{4 \cdot 11} = \dfrac{33}{44} \qquad \dfrac{9}{11} = \dfrac{9 \cdot 4}{11 \cdot 4} = \dfrac{36}{44}$$

Since $33 < 36$, then

$$\dfrac{33}{44} < \dfrac{36}{44} \text{ or}$$

$$\dfrac{3}{4} < \dfrac{9}{11}$$

■ Work Practice Problem 7

PRACTICE PROBLEM 8

Insert < or > to form a true sentence.

$-\dfrac{17}{20} \quad -\dfrac{4}{5}$

EXAMPLE 8 Insert < or > to form a true sentence.

$-\dfrac{2}{7} \quad -\dfrac{1}{3}$

Solution: The LCD is 21.

$$-\dfrac{2}{7} = -\dfrac{2 \cdot 3}{7 \cdot 3} = -\dfrac{6}{21} \text{ or } \dfrac{-6}{21} \qquad -\dfrac{1}{3} = -\dfrac{1 \cdot 7}{3 \cdot 7} = -\dfrac{7}{21} \text{ or } \dfrac{-7}{21}$$

Since $-6 > -7$, then

$$-\dfrac{6}{21} > -\dfrac{7}{21} \text{ or}$$

$$-\dfrac{2}{7} > -\dfrac{1}{3}$$

■ Work Practice Problem 8

Objective C Evaluating Expressions Given Fractional Replacement Values

PRACTICE PROBLEM 9

Evaluate $x - y$ if $x = \dfrac{5}{11}$ and $y = \dfrac{4}{9}$.

EXAMPLE 9 Evaluate $x - y$ if $x = \dfrac{7}{18}$ and $y = \dfrac{2}{9}$.

Solution: Replace x with $\dfrac{7}{18}$ and y with $\dfrac{2}{9}$ in the expression $x - y$.

$$x - y = \dfrac{7}{18} - \dfrac{2}{9}$$

The LCD of the denominators 18 and 9 is 18. Then

$$\dfrac{7}{18} - \dfrac{2}{9} = \dfrac{7}{18} - \dfrac{2 \cdot 2}{9 \cdot 2}$$
$$= \dfrac{7}{18} - \dfrac{4}{18}$$
$$= \dfrac{3}{18} = \dfrac{1}{6} \quad \text{Simplified}$$

■ Work Practice Problem 9

Answers

7. > 8. < 9. $\dfrac{1}{99}$

Objective **D** **Solving Problems by Adding or Subtracting Unlike Fractions**

Very often, real-world problems involve adding or subtracting unlike fractions.

EXAMPLE 10 **Finding Total Weight**

A freight truck has $\frac{1}{4}$ ton of computers, $\frac{1}{3}$ ton of televisions, and $\frac{3}{8}$ ton of small appliances. Find the total weight of its load.

Solution:

1. UNDERSTAND. Read and reread the problem. The phrase "total weight" tells us to add.
2. TRANSLATE.

In words:	total weight	is	weight of computers	plus	weight of televisions	plus	weight of appliances
	↓	↓	↓	↓	↓	↓	↓
Translate:	total weight	=	$\frac{1}{4}$	+	$\frac{1}{3}$	+	$\frac{3}{8}$

3. SOLVE: The LCD is 24.

$$\frac{1}{4} + \frac{1}{3} + \frac{3}{8} = \frac{1}{4} \cdot \frac{6}{6} + \frac{1}{3} \cdot \frac{8}{8} + \frac{3}{8} \cdot \frac{3}{3}$$
$$= \frac{6}{24} + \frac{8}{24} + \frac{9}{24}$$
$$= \frac{23}{24}$$

4. INTERPRET. *Check* the solution. *State* your conclusion: The total weight of the truck's load is $\frac{23}{24}$ ton.

▪ **Work Practice Problem 10**

PRACTICE PROBLEM 10

To repair her sidewalk, a homeowner must pour small amounts of cement in three different locations. She needs $\frac{3}{5}$ of a cubic yard, $\frac{3}{10}$ of a cubic yard, and $\frac{1}{15}$ of a cubic yard for these locations. Find the total amount of cement the homeowner needs.

Answer

10. $\frac{29}{30}$ cu yd

PRACTICE PROBLEM 11

Find the difference in length of two boards if one board is $\frac{3}{4}$ of a foot long and the other is $\frac{2}{3}$ of a foot long.

EXAMPLE 11 **Calculating Flight Time**

A flight from Tucson to Phoenix, Arizona, requires $\frac{5}{12}$ of an hour. If the plane has been flying $\frac{1}{4}$ of an hour, find how much time remains before landing.

Solution:

1. UNDERSTAND. Read and reread the problem. The phrase "how much time remains" tells us to subtract.

2. TRANSLATE.

In words:	time remaining	is	flight time from Tucson to Phoenix	minus	flight time already passed
	↓	↓	↓	↓	↓
Translate:	time remaining	$=$	$\frac{5}{12}$	$-$	$\frac{1}{4}$

3. SOLVE: The LCD is 12.

$$\frac{5}{12} - \frac{1}{4} = \frac{5}{12} - \frac{1}{4} \cdot \frac{3}{3}$$
$$= \frac{5}{12} - \frac{3}{12}$$
$$= \frac{2}{12} = \frac{\overset{1}{\cancel{2}}}{\underset{1}{\cancel{2}} \cdot 6} = \frac{1}{6}$$

4. INTERPRET. *Check* the solution. *State* your conclusion: The flight time remaining is $\frac{1}{6}$ of an hour.

🔲 **Work Practice Problem 11**

Answer

11. $\frac{1}{12}$ ft

CALCULATOR EXPLORATIONS   Performing Operations on Fractions

Scientific Calculator

Many calculators have a fraction key, such as $\boxed{a_{b/c}}$, that allows you to enter fractions, perform operations on fractions, and will give the result as a fraction. If your calculator has a fraction key, use it to calculate

$$\frac{3}{5} + \frac{4}{7}$$

Enter the keystrokes

$\boxed{3}\ \boxed{a_{b/c}}\ \boxed{5}\ \boxed{+}\ \boxed{4}\ \boxed{a_{b/c}}\ \boxed{7}\ \boxed{=}$

The display should read $\boxed{1\_6 \mid 35}$

which represents the mixed number $1\frac{6}{35}$. Let's write the result as a fraction. To convert from mixed number notation to fractional notation, press

$\boxed{2^{nd}}\ \boxed{d/c}$

The display now reads $\boxed{41 \mid 35}$

which represents $\frac{41}{35}$, the sum in fractional notation.

Graphing Calculator

Graphing calculators also allow you to perform operations on fractions and will give exact fractional results. The fraction option on a graphing calculator may be found under the $\boxed{MATH}$ menu. To perform the addition above, try the keystrokes.

$\boxed{3}\ \boxed{\div}\ \boxed{5}\ \boxed{+}\ \boxed{4}\ \boxed{\div}\ \boxed{7}\ \boxed{MATH}\ \boxed{ENTER}$
$\boxed{ENTER}$

The display should read

$\boxed{3/5 + 4/7 \blacktriangleright Frac\ 41/35}$

Use a calculator to add the following fractions. Give each sum as a fraction.

1. $\frac{1}{16} + \frac{2}{5}$ 2. $\frac{3}{20} + \frac{2}{25}$ 3. $\frac{4}{9} + \frac{7}{8}$

4. $\frac{9}{11} + \frac{5}{12}$ 5. $\frac{10}{17} + \frac{12}{19}$ 6. $\frac{14}{31} + \frac{15}{21}$

Vocabulary and Readiness Check

Use the choices below to fill in each blank. Any numerical answers are not listed.

expression   least common denominator   >

equation   equivalent   <

1. To add or subtract unlike fractions, we first write the fractions as _____ fractions with a common denominator. The common denominator we use is called the _____.

2. The LCD for $\frac{1}{6}$ and $\frac{5}{8}$ is _____.

3. $\frac{1}{6} + \frac{5}{8} = \frac{1}{6} \cdot \frac{4}{4} + \frac{5}{8} \cdot \frac{3}{3} = \frac{\ }{\ } + \frac{\ }{\ } = \underline{\ }.$

4. $\frac{1}{6} - \frac{5}{8} = \frac{1}{6} \cdot \frac{4}{4} - \frac{5}{8} \cdot \frac{3}{3} = \underline{\ } - \underline{\ } = \underline{\ }.$

5. $x - y$ is an _____ while $3x = \frac{1}{5}$ is an _____.

6. Since $-10 < -1$, we know that $-\frac{10}{13}$ _____ $-\frac{1}{13}$.

Objective A *Add or subtract as indicated. See Examples 1 through 6.*

1. $\dfrac{2}{3} + \dfrac{1}{6}$

2. $\dfrac{5}{6} + \dfrac{1}{12}$

3. $\dfrac{1}{2} - \dfrac{1}{3}$

4. $\dfrac{2}{3} - \dfrac{1}{4}$

5. $-\dfrac{2}{11} + \dfrac{2}{33}$

6. $-\dfrac{5}{9} + \dfrac{1}{3}$

7. $\dfrac{3}{14} - \dfrac{3}{7}$

8. $\dfrac{2}{15} - \dfrac{2}{5}$

9. $\dfrac{11x}{35} + \dfrac{2x}{7}$

10. $\dfrac{2y}{5} + \dfrac{3y}{25}$

11. $2 - \dfrac{y}{12}$

12. $5 - \dfrac{y}{20}$

13. $\dfrac{5}{12} - \dfrac{1}{9}$

14. $\dfrac{7}{12} - \dfrac{5}{18}$

15. $-7 + \dfrac{5}{7}$

16. $-10 + \dfrac{7}{10}$

17. $\dfrac{5a}{11} + \dfrac{4a}{9}$

18. $\dfrac{7x}{18} + \dfrac{2x}{9}$

 19. $\dfrac{2y}{3} - \dfrac{1}{6}$

20. $\dfrac{5z}{6} - \dfrac{1}{12}$

21. $\dfrac{1}{2} + \dfrac{3}{x}$

22. $\dfrac{2}{5} + \dfrac{3}{x}$

23. $-\dfrac{2}{11} - \dfrac{2}{33}$

24. $-\dfrac{5}{9} - \dfrac{1}{3}$

25. $\dfrac{9}{14} - \dfrac{3}{7}$

26. $\dfrac{4}{5} - \dfrac{2}{15}$

27. $\dfrac{11y}{35} - \dfrac{2}{7}$

28. $\dfrac{2b}{5} - \dfrac{3}{25}$

29. $\dfrac{1}{9} - \dfrac{5}{12}$

30. $\dfrac{5}{18} - \dfrac{7}{12}$

 31. $\dfrac{7}{15} - \dfrac{5}{12}$

32. $\dfrac{5}{8} - \dfrac{3}{20}$

33. $\dfrac{5}{7} - \dfrac{1}{8}$

34. $\dfrac{10}{13} - \dfrac{7}{10}$

 35. $\dfrac{7}{8} + \dfrac{3}{16}$

36. $\dfrac{7}{18} + \dfrac{2}{9}$

37. $\dfrac{3}{9} - \dfrac{5}{9}$

38. $\dfrac{1}{13} - \dfrac{4}{13}$

39. $-\dfrac{2}{5} + \dfrac{1}{3} - \dfrac{3}{10}$

40. $-\dfrac{1}{3} - \dfrac{1}{4} + \dfrac{2}{5}$

41. $\dfrac{5}{11} + \dfrac{y}{3}$

42. $\dfrac{5z}{13} + \dfrac{3}{26}$

43. $-\dfrac{5}{6} - \dfrac{3}{7}$

44. $-\dfrac{1}{2} - \dfrac{3}{29}$

45. $\dfrac{x}{2} + \dfrac{x}{4} + \dfrac{2x}{16}$

46. $\dfrac{z}{4} + \dfrac{z}{8} + \dfrac{2z}{16}$

47. $\dfrac{7}{9} - \dfrac{1}{6}$

48. $\dfrac{9}{16} - \dfrac{3}{8}$

49. $\dfrac{2a}{3} + \dfrac{6a}{13}$

50. $\dfrac{3y}{4} + \dfrac{y}{7}$

51. $\dfrac{7}{30} - \dfrac{5}{12}$

52. $\dfrac{7}{30} - \dfrac{3}{20}$

 53. $\dfrac{5}{9} + \dfrac{1}{y}$

54. $\dfrac{1}{12} - \dfrac{5}{x}$

55. $\dfrac{6}{5} - \dfrac{3}{4} + \dfrac{1}{2}$

56. $\dfrac{6}{5} + \dfrac{3}{4} - \dfrac{1}{2}$

57. $\dfrac{4}{5} + \dfrac{4}{9}$

58. $\dfrac{11}{12} - \dfrac{7}{24}$

59. $\dfrac{5}{9x} + \dfrac{1}{8}$

60. $\dfrac{3}{8} + \dfrac{5}{12x}$

61. $-\dfrac{9}{12} + \dfrac{17}{24} - \dfrac{1}{6}$

62. $-\dfrac{5}{14} + \dfrac{3}{7} - \dfrac{1}{2}$

63. $\dfrac{3x}{8} + \dfrac{2x}{7} - \dfrac{5}{14}$

64. $\dfrac{9x}{10} - \dfrac{1}{2} + \dfrac{x}{5}$

Objective B *Insert < or > to form a true sentence. See Examples 7 and 8.*

65. $\dfrac{2}{7} \quad \dfrac{3}{10}$

66. $\dfrac{5}{9} \quad \dfrac{6}{11}$

67. $\dfrac{5}{6} \quad -\dfrac{13}{15}$

68. $-\dfrac{7}{8} \quad -\dfrac{5}{6}$

69. $-\dfrac{3}{4} \quad -\dfrac{11}{14}$

70. $-\dfrac{2}{9} \quad -\dfrac{3}{13}$

Objective C *Evaluate each expression if* $x = \dfrac{1}{3}$ *and* $y = \dfrac{3}{4}$. *See Example 9.*

71. $x + y$

72. $x - y$

73. xy

74. $x \div y$

75. $2y + x$

76. $2x + y$

Objective D *Find the perimeter of each geometric figure. (Hint: Recall that perimeter means distance around.)*

△ **77.**

△ **78.**

△ **79.**

△ **80.**

Translating *Translate each phrase to an algebraic expression. Use "x" to represent "a number." See Examples 10 and 11.*

81. The sum of a number and $\dfrac{1}{2}$

82. A number increased by $-\dfrac{2}{5}$

83. A number subtracted from $-\dfrac{3}{8}$

84. The difference of a number and $\dfrac{7}{20}$

Solve. See Examples 10 and 11.

85. Find the inner diameter of the washer.

$\frac{3}{16}$ inch $\frac{3}{16}$ inch

1 inch

86. Find the inner diameter of the tubing.

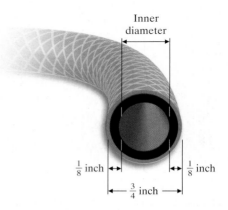

$\frac{1}{8}$ inch $\frac{1}{8}$ inch

$\frac{3}{4}$ inch

87. The slowest mammal is the three-toed sloth from South America. The sloth has an average ground speed of $\frac{1}{10}$ mph. In the trees, it can accelerate to $\frac{17}{100}$ mph. How much faster can a sloth travel in the trees? (*Source: The Guiness Book of World Records*)

88. Killer bees have been known to chase people for up to $\frac{1}{4}$ of a mile, while domestic European honeybees will normally chase a person for no more than 100 feet, or $\frac{5}{264}$ of a mile. How much farther will a killer bee chase a person than a domestic honeybee? (*Source:* Coachella Valley Mosquito & Vector Control District)

89. Given the following diagram, find its total length.

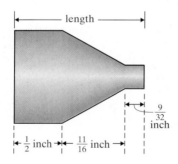

length

$\frac{9}{32}$ inch

$\frac{1}{2}$ inch $\frac{11}{16}$ inch

90. Given the following diagram, find its total width.

width

$\frac{11}{16}$ inch

$\frac{5}{8}$ inch

$\frac{11}{16}$ inch

91. About $\frac{13}{20}$ of American students ages 10 to 17 name math, science, or art as their favorite subject in school. Art is the favorite subject for about $\frac{4}{25}$ of the American students ages 10 to 17. For what fraction of students this age is math or science their favorite subject? (*Source:* Peter D. Hart Research Associates for the National Science Foundation)

92. Together, the United States' and Japan's postal services handle $\frac{49}{100}$ of the world's mail volume. Japan's postal service alone handles $\frac{3}{50}$ of the world's mail. What fraction of the world's mail is handled by the postal service of the United States? (*Source:* United States Postal Service)

The table below shows the fraction of the Earth's water area taken up by each ocean. Use this table for Exercises 93 and 94.

Fraction of Earth's Water Area per Ocean	
Ocean	**Fraction**
Artic	$\frac{1}{25}$
Atlantic	$\frac{13}{50}$
Pacific	$\frac{1}{2}$
Indian	$\frac{1}{5}$

93. What fraction of the world's water surface area is accounted for by the Pacific and Atlantic Oceans?

94. What fraction of the world's water surface area is accounted for by the Arctic and Indian Oceans?

Each sector (shaped like a piece of pie) in the circle graph below shows the fraction of entering college freshmen who expect to major in each discipline shown. The whole circle represents the entire class of college freshmen. Use this graph to answer Exercises 95 through 98.

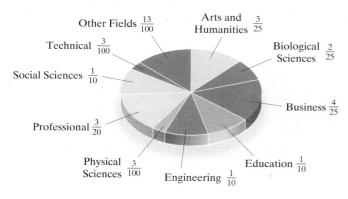

Other Fields $\frac{13}{100}$

Arts and Humanities $\frac{3}{25}$

Technical $\frac{3}{100}$

Biological Sciences $\frac{2}{25}$

Social Sciences $\frac{1}{10}$

Business $\frac{4}{25}$

Professional $\frac{3}{20}$

Physical Sciences $\frac{3}{100}$

Engineering $\frac{1}{10}$

Education $\frac{1}{10}$

Source: The Higher Education Research Institute

95. What fraction of beginning freshmen plan to major in arts and humanities or education?

96. What fraction of beginning freshmen plan to major in a technical or professional field?

97. What fraction of beginning freshmen do *not* plan to major in business?

98. What fraction of beginning freshmen do *not* plan to major in engineering?

Review

Use order of operations to simplify.

99. $-50 \div 5 \cdot 2$ **100.** $8 - 6 \cdot 4 - 7$ **101.** $(8 - 6) \cdot (4 - 7)$ **102.** $50 \div (5 \cdot 2)$

Concept Extensions

For Exercises 103 and 104 below, do the following:

a. *Draw three rectangles of the same size and represent each fraction in the sum or difference, one fraction per rectangle, by shading.*

b. *Using these rectangles as estimates, determine whether there is an error in the sum.*

c. *If there is an error, correctly calculate the sum.*

See the Concept Check in this section.

103. $\frac{3}{5} + \frac{4}{5} \stackrel{?}{=} \frac{7}{10}$ **104.** $\frac{5}{8} - \frac{3}{4} \stackrel{?}{=} \frac{2}{4}$

Subtract from left to right.

105. $\dfrac{2}{3} - \dfrac{1}{4} - \dfrac{2}{540}$

106. $\dfrac{9}{10} - \dfrac{7}{200} - \dfrac{1}{3}$

Perform each indicated operation.

107. $\dfrac{30}{55} + \dfrac{1000}{1760}$

108. $\dfrac{19}{26} - \dfrac{968}{1352}$

109. In your own words, describe how to add or subtract two fractions with different denominators.

110. Find the sum of the fractions in the circle graph on page 275. Did the sum surprise you? Why or why not?

111. In a recent year, about $\dfrac{24}{53}$ of the total number of pieces of mail delivered by the United States Postal Service was first class mail. That same year, about $\dfrac{51}{106}$ of the total number of pieces of mail delivered by the United States Postal Service was standard mail. Which of these two categories account for a greater portion of the mail handled by volume? (*Source*: United States Postal Service)

STUDY SKILLS BUILDER

Are You Satisfied with Your Performance on a Particular Quiz or Exam?

If not, don't forget to analyze your quiz or exam and look for common errors. Were most of your errors a result of:

- *Carelessness?* Did you turn in your quiz or exam before the allotted time expired? If so, resolve next time to use the entire time allotted. Any extra time can be spent checking your work.

- *Running out of time?* If so, make a point to better manage your time on your next quiz or exam. Try completing any questions that you are unsure of last and delay checking your work until all questions have been answered.

- *Not understanding a concept?* If so, review that concept and correct your work. Try to understand how this happened so that you make sure it doesn't happen before the next quiz or exam.

- *Test conditions?* When studying for a quiz or exam, make sure you place yourself in conditions similar to test conditions. For example, before your next quiz or exam, use a few sheets of blank paper and take a sample test without the aid of your notes or text. (See

your instructor or use the Chapter Test at the end of each chapter.)

Exercises

1. Have you corrected all your previous quizzes and exams?

2. List any errors you have found common to two or more of your graded papers.

3. Is one of your common errors not understanding a concept? If so, are you making sure you understand all the concepts for the next quiz or exam?

4. Is one of your common errors making careless mistakes? If so, are you now taking all the time allotted to check over your work so that you can minimize the number of careless mistakes?

5. Are you satisfied with your grades thus far on quizzes and tests?

6. If your answer to Exercise 5 is no, are there any more suggestions you can make to your instructor or yourself to help? If so, list them here and share these with your instructor.

Sections 4.1–4.5

Summary on Fractions and Operations on Fractions

Use a fraction to represent the shaded area of each figure. If the fraction is improper, also write the fraction as a mixed number.

1.

2.

Solve.

3. In a survey, 73 people out of 85 get fewer than 8 hours of sleep each night. What fraction of people in the survey get fewer than 8 hours of sleep?

4. Sketch a diagram to represent $\frac{9}{13}$.

Simplify.

5. $\frac{11}{-11}$ **6.** $\frac{17}{1}$ **7.** $\frac{0}{-3}$ **8.** $\frac{7}{0}$

Write the prime factorization of each composite number. Write any repeated factors using exponents.

9. 65   **10.** 70   **11.** 315   **12.** 441

Write each fraction in simplest form.

13. $\frac{2}{14}$ **14.** $\frac{24}{20}$ **15.** $-\frac{56}{60}$ **16.** $-\frac{72}{80}$

17. $\frac{54x}{135}$ **18.** $\frac{90}{240y}$ **19.** $\frac{165z^3}{210z}$ **20.** $\frac{245ab}{385a^2b^3}$

Determine whether each pair of fractions is equivalent.

21. $\frac{7}{8}$ and $\frac{9}{10}$ **22.** $\frac{10}{12}$ and $\frac{15}{18}$

23. Of the 50 states, 2 states are not adjacent to any other states.
   a. What fraction of the states are not adjacent to other states?
   b. How many states are adjacent to other states?
   c. What fraction of the states are adjacent to other states?

24. In a recent year, 460 new films were released. Of these, 92 were rated R. (*Source:* Motion Picture Association of America)
   a. What fraction were rated R?
   b. How many films were rated other than R?
   c. What fraction of films were rated other than R?

1. _____
2. _____
3. _____
4. _____
5. _____
6. _____
7. _____
8. _____
9. _____
10. _____
11. _____
12. _____
13. _____
14. _____
15. _____
16. _____
17. _____
18. _____
19. _____
20. _____
21. _____
22. _____
23. a.   b.   c.
24. a.   b.   c.

25. _____ **26.** _____

27. _____ **28.** _____

29. _____ **30.** _____

31. _____ **32.** _____

33. _____ **34.** _____

35. _____ **36.** _____

37. _____

38. _____

39. _____

40. _____

41. _____

42. _____

43. _____

44. _____

45. _____

46. _____

47. _____

48. _____

49. _____

50. _____

51. _____

52. _____

53. _____

Find the LCM of each list of numbers.

25. 5, 6   **26.** 2, 14   **27.** 6, 18, 30

Write each fraction as an equivalent fraction with the indicated denominator.

28. $\dfrac{7}{9} = \dfrac{}{36}$ **29.** $\dfrac{11}{15} = \dfrac{}{75}$ **30.** $\dfrac{5}{6} = \dfrac{}{48}$

The following summary will help you with the following review of operations on fractions.

Operations on Fractions

Let a, b, c, and d be integers.

Addition: $\dfrac{a}{b} + \dfrac{c}{b} = \dfrac{a+c}{b}$ Subtraction: $\dfrac{a}{b} - \dfrac{c}{b} = \dfrac{a-c}{b}$

$(b \neq 0)$ ↑ ↑ $(b \neq 0)$ ↑ ↑
     common denominator   **common denominator**

Multiplication: $\dfrac{a}{b} \cdot \dfrac{c}{d} = \dfrac{a \cdot c}{b \cdot d}$ Division: $\dfrac{a}{b} \div \dfrac{c}{d} = \dfrac{a}{b} \cdot \dfrac{d}{c} = \dfrac{a \cdot d}{b \cdot c}$

$(b \neq 0, d \neq 0)$ $(b \neq 0, d \neq 0, c \neq 0)$

Perform each indicated operation.

31. $\dfrac{1}{5} + \dfrac{3}{5}$ **32.** $\dfrac{1}{5} - \dfrac{3}{5}$ **33.** $\dfrac{1}{5} \cdot \dfrac{3}{5}$ **34.** $\dfrac{1}{5} \div \dfrac{3}{5}$

35. $\dfrac{2}{3} \div \dfrac{5}{6}$ **36.** $\dfrac{2a}{3} \cdot \dfrac{5}{6a}$ **37.** $\dfrac{2}{3y} - \dfrac{5}{6y}$ **38.** $\dfrac{2x}{3} + \dfrac{5x}{6}$

39. $-\dfrac{1}{7} \cdot -\dfrac{7}{18}$ **40.** $-\dfrac{4}{9} \cdot -\dfrac{3}{7}$ **41.** $-\dfrac{7z}{8} \div 6z^2$ **42.** $-\dfrac{9}{10} \div 5$

43. $\dfrac{7}{8} + \dfrac{1}{20}$ **44.** $\dfrac{5}{12} - \dfrac{1}{9}$ **45.** $\dfrac{2}{9} + \dfrac{1}{18} + \dfrac{1}{3}$ **46.** $\dfrac{3y}{10} + \dfrac{y}{5} + \dfrac{6}{25}$

Translating *Translate each to an expression. Use x to represent a number.*

47. $\dfrac{2}{3}$ of a number   **48.** The quotient of a number and $-\dfrac{1}{5}$

49. A number subtracted from $-\dfrac{8}{9}$ **50.** $\dfrac{6}{11}$ increased by a number

Solve.

51. Find $\dfrac{2}{3}$ of 1530.   **52.** A contractor is using 18 acres of his land to sell $\dfrac{3}{4}$-acre lots. How many lots can he sell?

53. Suppose that the cross-section of a piece of pipe looks like the diagram shown. What is the inner diameter?

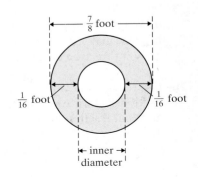

4.6 COMPLEX FRACTIONS AND REVIEW OF ORDER OF OPERATIONS

Objectives

A Simplify Complex Fractions.

B Review the Order of Operations.

C Evaluate Expressions Given Replacement Values.

Objective **A** Simplifying Complex Fractions

Thus far, we have studied operations on fractions. We now practice simplifying fractions whose numerators or denominators themselves contain fractions. These fractions are called **complex fractions.**

Complex Fraction

A fraction whose numerator or denominator or both numerator and denominator contain fractions is called a **complex fraction.**

Examples of complex fractions are

$$\dfrac{\dfrac{x}{4}}{\dfrac{3}{2}} \qquad \dfrac{\dfrac{1}{2}+\dfrac{3}{8}}{\dfrac{3}{4}-\dfrac{1}{6}} \qquad \dfrac{\dfrac{y}{5}-2}{\dfrac{3}{10}} \qquad \dfrac{-4z}{\dfrac{3}{5}}$$

Method 1 for Simplifying Complex Fractions

Two methods are presented to simplify complex fractions. The first method makes use of the fact that a fraction bar means division.

EXAMPLE 1 Simplify: $\dfrac{\dfrac{x}{4}}{\dfrac{3}{2}}$

Solution: Since a fraction bar means division, the complex fraction $\dfrac{\dfrac{x}{4}}{\dfrac{3}{2}}$ can be written as $\dfrac{x}{4} \div \dfrac{3}{2}$. Then divide as usual to simplify.

$$\dfrac{x}{4} \div \dfrac{3}{2} = \dfrac{x}{4} \cdot \dfrac{2}{3} \qquad \text{Multiply by the reciprocal.}$$

$$= \dfrac{x \cdot \overset{1}{\cancel{2}}}{\underset{1}{\cancel{2}} \cdot 2 \cdot 3}$$

$$= \dfrac{x}{6}$$

PRACTICE PROBLEM 1

Simplify: $\dfrac{\dfrac{7y}{10}}{\dfrac{1}{5}}$

☐ **Work Practice Problem 1**

EXAMPLE 2 Simplify: $\dfrac{\dfrac{1}{2}+\dfrac{3}{8}}{\dfrac{3}{4}-\dfrac{1}{6}}$

Solution: Recall the order of operations. Since the fraction bar is considered a grouping symbol, we simplify the numerator and the denominator of the complex fraction separately. Then we divide.

$$\dfrac{\dfrac{1}{2}+\dfrac{3}{8}}{\dfrac{3}{4}-\dfrac{1}{6}} = \dfrac{\dfrac{1\cdot4}{2\cdot4}+\dfrac{3}{8}}{\dfrac{3\cdot3}{4\cdot3}-\dfrac{1\cdot2}{6\cdot2}} = \dfrac{\dfrac{4}{8}+\dfrac{3}{8}}{\dfrac{9}{12}-\dfrac{2}{12}} = \dfrac{\dfrac{7}{8}}{\dfrac{7}{12}}$$

PRACTICE PROBLEM 2

Simplify: $\dfrac{\dfrac{1}{2}+\dfrac{1}{6}}{\dfrac{3}{4}-\dfrac{2}{3}}$

Answers

1. $\dfrac{7y}{2}$ 2. $\dfrac{8}{1}$ or 8

Continued on next page

279

Thus,

$$\frac{\frac{1}{2}+\frac{3}{8}}{\frac{3}{4}-\frac{1}{6}} = \frac{\frac{7}{8}}{\frac{7}{12}}$$

$$= \frac{7}{8} \div \frac{7}{12} \quad \text{Rewrite the quotient using the } \div \text{ sign.}$$

$$= \frac{7}{8} \cdot \frac{12}{7} \quad \text{Multiply by the reciprocal.}$$

$$= \frac{\overset{1}{\cancel{7}} \cdot 3 \cdot \overset{1}{\cancel{4}}}{2 \cdot \underset{1}{\cancel{4}} \cdot \underset{1}{\cancel{7}}} \quad \text{Multiply.}$$

$$= \frac{3}{2} \quad \text{Simplify.}$$

◼ **Work Practice Problem 2**

Method 2 for Simplifying Complex Fractions The second method for simplifying complex fractions is to multiply the numerator and the denominator of the complex fraction by the LCD of all the fractions in its numerator and its denominator. This has the effect of leaving sums and differences of integers in the numerator and the denominator as we shall see in the example below.

Let's use this second method to simplify the complex fraction in Example 2 again.

PRACTICE PROBLEM 3

Use Method 2 to simplify:
$$\frac{\frac{1}{2}+\frac{1}{6}}{\frac{3}{4}-\frac{2}{3}}$$

EXAMPLE 3 Simplify: $\dfrac{\frac{1}{2}+\frac{3}{8}}{\frac{3}{4}-\frac{1}{6}}$

Solution: The complex fraction contains fractions with denominators 2, 8, 4, and 6. The LCD is 24. By the fundamental property of fractions, we can multiply the numerator and the denominator of the complex fraction by 24. Notice below that by the distributive property, this means that we multiply each term in the numerator and denominator by 24.

$$\frac{\frac{1}{2}+\frac{3}{8}}{\frac{3}{4}-\frac{1}{6}} = \frac{24\left(\frac{1}{2}+\frac{3}{8}\right)}{24\left(\frac{3}{4}-\frac{1}{6}\right)}$$

$$= \frac{\left(\overset{12}{\cancel{24}} \cdot \frac{1}{\cancel{2}}\right)+\left(\overset{3}{\cancel{24}} \cdot \frac{3}{\cancel{8}}\right)}{\left(\overset{6}{\cancel{24}} \cdot \frac{3}{\cancel{4}}\right)-\left(\overset{4}{\cancel{24}} \cdot \frac{1}{\cancel{6}}\right)} \quad \text{Apply the distributive property. Then divide out common factors to aid in multiplying.}$$

$$= \frac{12+9}{18-4} \quad \text{Multiply.}$$

$$= \frac{21}{14}$$

$$= \frac{\overset{1}{\cancel{7}} \cdot 3}{\cancel{7} \cdot 2} = \frac{3}{2} \quad \text{Simplify.}$$

◼ **Work Practice Problem 3**

The simplified result is the same, of course, no matter which method is used.

Answer

3. $\frac{8}{1}$ or 8

EXAMPLE 4 Simplify: $\dfrac{\dfrac{y}{5} - 2}{\dfrac{3}{10}}$

Solution: Use the second method and multiply the numerator and the denominator of the complex fraction by the LCD of all fractions. Recall that $2 = \dfrac{2}{1}$. The LCD of the denominators 5, 1, and 10 is 10.

$$\frac{\dfrac{y}{5} - \dfrac{2}{1}}{\dfrac{3}{10}} = \frac{10\left(\dfrac{y}{5} - \dfrac{2}{1}\right)}{10\left(\dfrac{3}{10}\right)} \qquad \text{Multiply the numerator and denominator by 10.}$$

$$= \frac{\left(\overset{2}{\cancel{10}} \cdot \dfrac{y}{\cancel{5}_1}\right) - \left(10 \cdot \dfrac{2}{1}\right)}{\underset{1}{\cancel{10}} \cdot \dfrac{3}{\cancel{10}_1}} \qquad \begin{array}{l}\text{Apply the distributive property. Then divide out}\\ \text{common factors to aid in multiplying.}\end{array}$$

$$= \frac{2y - 20}{3} \qquad \text{Multiply.}$$

PRACTICE PROBLEM 4

Simplify: $\dfrac{\dfrac{3}{4}}{\dfrac{x}{5} - 1}$

Helpful Hint
Don't forget to multiply the numerator and the denominator of the complex fraction by the same number—the LCD.

◻ **Work Practice Problem 4**

Objective **B** Reviewing the Order of Operations

At this time, it is probably a good idea to review the order of operations on expressions containing fractions. Before we do so, let's review how we perform operations on fractions.

	Review of Operations on Fractions	
Operation	**Procedure**	**Example**
Multiply	Multiply the numerators and multiply the denominators.	$\dfrac{5}{9} \cdot \dfrac{1}{2} = \dfrac{5 \cdot 1}{9 \cdot 2} = \dfrac{5}{18}$
Divide	Multiply the first fraction by the reciprocal of the second fraction.	$\dfrac{2}{3} \div \dfrac{11}{13} = \dfrac{2}{3} \cdot \dfrac{13}{11} = \dfrac{2 \cdot 13}{3 \cdot 11} = \dfrac{26}{33}$
Add or Subtract	1. Write each fraction as an equivalent fraction whose denominator is the LCD. 2. Add or subtract numerators and write the result over the common denominator.	$\dfrac{3}{4} + \dfrac{1}{8} = \dfrac{3}{4} \cdot \dfrac{2}{2} + \dfrac{1}{8} = \dfrac{6}{8} + \dfrac{1}{8} = \dfrac{7}{8}$

Now let's review order of operations.

Order of Operations

1. Perform all operations within parentheses ( ), brackets [ ], or other grouping symbols such as fraction bars, starting with the innermost set.
2. Evaluate any expressions with exponents.
3. Multiply or divide in order from left to right.
4. Add or subtract in order from left to right.

Answer

4. $\dfrac{15}{4x - 20}$

PRACTICE PROBLEM 5

Simplify: $\left(\dfrac{2}{3}\right)^3 - 2$

EXAMPLE 5 Simplify: $\left(\dfrac{4}{5}\right)^2 - 1$

Solution: According to the order of operations, first evaluate $\left(\dfrac{4}{5}\right)^2$.

$$\left(\dfrac{4}{5}\right)^2 - 1 = \dfrac{16}{25} - 1 \qquad \text{Write } \left(\dfrac{4}{5}\right)^2 \text{ as } \dfrac{16}{25}.$$

Next, combine the fractions. The LCD of 25 and 1 is 25.

$$\dfrac{16}{25} - 1 = \dfrac{16}{25} - \dfrac{25}{25} \qquad \text{Write 1 as } \dfrac{25}{25}.$$

$$= \dfrac{-9}{25} \text{ or } -\dfrac{9}{25} \qquad \text{Subtract.}$$

◻ **Work Practice Problem 5**

PRACTICE PROBLEM 6

Simplify: $\left(-\dfrac{1}{2} + \dfrac{1}{5}\right)\left(\dfrac{7}{8} + \dfrac{1}{8}\right)$

EXAMPLE 6 Simplify: $\left(\dfrac{1}{4} + \dfrac{2}{3}\right)\left(\dfrac{11}{12} + \dfrac{1}{4}\right)$

Solution: First perform operations inside parentheses. Then multiply.

$$\left(\dfrac{1}{4} + \dfrac{2}{3}\right)\left(\dfrac{11}{12} + \dfrac{1}{4}\right) = \left(\dfrac{1 \cdot 3}{4 \cdot 3} + \dfrac{2 \cdot 4}{3 \cdot 4}\right)\left(\dfrac{11}{12} + \dfrac{1 \cdot 3}{4 \cdot 3}\right) \qquad \text{Each LCD is 12.}$$

$$= \left(\dfrac{3}{12} + \dfrac{8}{12}\right)\left(\dfrac{11}{12} + \dfrac{3}{12}\right)$$

$$= \left(\dfrac{11}{12}\right)\left(\dfrac{14}{12}\right) \qquad \text{Add.}$$

$$= \dfrac{11 \cdot \overset{1}{\cancel{2}} \cdot 7}{\underset{1}{\cancel{2}} \cdot 6 \cdot 12} \qquad \text{Multiply.}$$

$$= \dfrac{77}{72} \qquad \text{Simplify.}$$

◻ **Work Practice Problem 6**

✔ **Concept Check** What should be done first to simplify the expression $\dfrac{1}{5} \cdot \dfrac{5}{2} - \left(\dfrac{2}{3} + \dfrac{4}{5}\right)^2$?

Objective **C** **Evaluating Algebraic Expressions**

EXAMPLE 7 Evaluate $2x + y^2$ if $x = -\dfrac{1}{2}$ and $y = \dfrac{1}{3}$.

PRACTICE PROBLEM 7

Evaluate $-\dfrac{3}{5} - xy$ if $x = \dfrac{3}{10}$ and $y = \dfrac{2}{3}$.

Solution: Replace x and y with the given values and simplify.

$$2x + y^2 = 2\left(-\dfrac{1}{2}\right) + \left(\dfrac{1}{3}\right)^2 \qquad \text{Replace } x \text{ with } -\dfrac{1}{2} \text{ and } y \text{ with } \dfrac{1}{3}.$$

$$= 2\left(-\dfrac{1}{2}\right) + \dfrac{1}{9} \qquad \text{Write } \left(\dfrac{1}{3}\right)^2 \text{ as } \dfrac{1}{9}.$$

$$= -1 + \dfrac{1}{9} \qquad \text{Multiply.}$$

$$= -\dfrac{9}{9} + \dfrac{1}{9} \qquad \text{The LCD is 9.}$$

$$= -\dfrac{8}{9} \qquad \text{Add.}$$

Answers

5. $-\dfrac{46}{27}$ 6. $-\dfrac{3}{10}$ 7. $-\dfrac{4}{5}$

✔ **Concept Check Answer**

Add inside parentheses.

◻ **Work Practice Problem 7**

If you find it difficult replacing a variable with a number, try the following. First, replace the variable with a set of parentheses, then place the replacement number between the parentheses.

If $x = \dfrac{4}{5}$, find $2x + x^2$.

$$2x + x^2 = 2(\ ) + (\ )^2$$
$$= 2\left(\dfrac{4}{5}\right) + \left(\dfrac{4}{5}\right)^2 \ldots$$

then continue simplifying.

Vocabulary and Readiness Check

Use the choices below to fill in each blank.

addition   multiplication   evaluate the exponential expression

subtraction   division   complex

1. A fraction whose numerator or denominator or both numerator and denominator contain fractions is called a _____ fraction.

2. To simplify $-\dfrac{1}{2} + \dfrac{2}{3} \cdot \dfrac{7}{8}$, which operation do we perform first? _____

3. To simplify $-\dfrac{1}{2} \div \dfrac{2}{3} \cdot \dfrac{7}{8}$, which operation do we perform first? _____

4. To simplify $\dfrac{7}{8} \cdot \left(\dfrac{1}{2} - \dfrac{2}{3}\right)$, which operation do we perform first? _____

5. To simplify $\dfrac{1}{3} \div \dfrac{1}{4} \cdot \left(\dfrac{9}{11} + \dfrac{3}{8}\right)^3$, which operation do we perform first? _____

6. To simplify $9 - \left(-\dfrac{3}{4}\right)^2$, which operation do we perform first? _____

4.6 EXERCISE SET

FOR EXTRA HELP

Student Solutions Manual   PH Math/Tutor Center   CD/Video for Review   Math XL   MathXL®   MyMathLab   MyMathLab

Objective A *Simplify each complex fraction. See Examples 1 through 4.*

1. $\dfrac{\frac{1}{8}}{\frac{3}{4}}$

2. $\dfrac{\frac{5}{12}}{\frac{15}{12}}$

3. $\dfrac{\frac{2}{3}}{\frac{2}{7}}$

4. $\dfrac{\frac{9}{25}}{\frac{6}{25}}$

5. $\dfrac{\frac{2x}{27}}{\frac{4}{9}}$

6. $\dfrac{\frac{3y}{11}}{\frac{1}{2}}$

7. $\dfrac{\frac{3}{4} + \frac{2}{5}}{\frac{1}{2} + \frac{3}{5}}$

8. $\dfrac{\frac{7}{6} + \frac{2}{3}}{\frac{3}{2} - \frac{8}{9}}$

 9. $\dfrac{\frac{3x}{4}}{5 - \frac{1}{8}}$

10. $\dfrac{\frac{3}{10} + 2}{\frac{2}{5y}}$

Objective **B** *Use the order of operations to simplify each expression. See Examples 5 and 6.*

11. $\dfrac{1}{5} + \dfrac{1}{3} \cdot \dfrac{1}{4}$

12. $\dfrac{1}{2} + \dfrac{1}{6} \cdot \dfrac{1}{3}$

13. $\dfrac{5}{6} \div \dfrac{1}{3} \cdot \dfrac{1}{4}$

14. $\dfrac{7}{8} \div \dfrac{1}{4} \cdot \dfrac{1}{7}$

15. $2^2 - \left(\dfrac{1}{3}\right)^2$

16. $3^2 - \left(\dfrac{1}{2}\right)^2$

17. $\left(\dfrac{2}{9} + \dfrac{4}{9}\right)\left(\dfrac{1}{3} - \dfrac{9}{10}\right)$

18. $\left(\dfrac{1}{5} - \dfrac{1}{10}\right)\left(\dfrac{1}{5} + \dfrac{1}{10}\right)$

19. $\left(\dfrac{7}{8} - \dfrac{1}{2}\right) \div \dfrac{3}{11}$

20. $\left(-\dfrac{2}{3} - \dfrac{7}{3}\right) \div \dfrac{4}{9}$

21. $2 \cdot \left(\dfrac{1}{4} + \dfrac{1}{5}\right) + 2$

22. $\dfrac{2}{5} \cdot \left(5 - \dfrac{1}{2}\right) - 1$

23. $\left(\dfrac{3}{4}\right)^2 \div \left(\dfrac{3}{4} - \dfrac{1}{12}\right)$

24. $\left(\dfrac{8}{9}\right)^2 \div \left(2 - \dfrac{2}{3}\right)$

25. $\left(\dfrac{2}{5} - \dfrac{3}{10}\right)^2$

26. $\left(\dfrac{3}{2} - \dfrac{4}{3}\right)^3$

27. $\left(\dfrac{3}{4} + \dfrac{1}{8}\right)^2 - \left(\dfrac{1}{2} + \dfrac{1}{8}\right)$

28. $\left(\dfrac{1}{6} + \dfrac{1}{3}\right)^3 + \left(\dfrac{2}{5} \cdot \dfrac{3}{4}\right)^2$

Objective **C** *Evaluate each expression if $x = -\dfrac{1}{3}$, $y = \dfrac{2}{5}$, and $z = \dfrac{5}{6}$. See Example 7.*

29. $5y - z$

30. $2z - x$

31. $\dfrac{x}{z}$

32. $\dfrac{y + x}{z}$

33. $x^2 - yz$

34. $(1 + x)(1 + z)$

Objectives **A** **B** **Mixed Practice** *Simplify the following. See Examples 1 through 6.*

35. $\dfrac{\frac{5a}{24}}{\frac{1}{12}}$

36. $\dfrac{\frac{7}{10}}{\frac{14z}{25}}$

37. $\left(\dfrac{3}{2}\right)^3 + \left(\dfrac{1}{2}\right)^3$

38. $\left(\dfrac{5}{21} \div \dfrac{1}{2}\right) + \left(\dfrac{1}{7} \cdot \dfrac{1}{3}\right)$

39. $\left(-\dfrac{1}{2}\right)^2 + \dfrac{1}{5}$

40. $\left(-\dfrac{3}{4}\right)^2 + \dfrac{3}{8}$

41. $\dfrac{2 + \frac{1}{6}}{1 - \frac{4}{3}}$

42. $\dfrac{3 - \frac{1}{2}}{4 + \frac{1}{5}}$

43. $\left(1 - \dfrac{2}{5}\right)^2$

44. $\left(-\dfrac{1}{2}\right)^2 - \left(\dfrac{3}{4}\right)^2$

45. $\left(\dfrac{3}{4} - 1\right)\left(\dfrac{1}{8} + \dfrac{1}{2}\right)$

46. $\left(\dfrac{1}{10} + \dfrac{3}{20}\right)\left(\dfrac{1}{5} - 1\right)$

47. $\left(-\dfrac{2}{9} - \dfrac{7}{9}\right)^4$

48. $\left(\dfrac{5}{9} - \dfrac{2}{3}\right)^2$

49. $\dfrac{\dfrac{1}{3} - \dfrac{5}{6}}{\dfrac{3}{4} + \dfrac{1}{2}}$

50. $\dfrac{\dfrac{7}{10} + \dfrac{1}{2}}{\dfrac{4}{5} + \dfrac{3}{4}}$

51. $\left(\dfrac{3}{4} \div \dfrac{6}{5}\right) - \left(\dfrac{3}{4} \cdot \dfrac{6}{5}\right)$

52. $\left(\dfrac{1}{2} \cdot \dfrac{2}{7}\right) - \left(\dfrac{1}{2} \div \dfrac{2}{7}\right)$

53. $\dfrac{\dfrac{x}{3} + 2}{5 + \dfrac{1}{3}}$

54. $\dfrac{1 - \dfrac{x}{4}}{2 + \dfrac{3}{8}}$

Review

Perform each indicated operation. If the result is an improper fraction, also write the improper fraction as a mixed number. See Sections 4.1 and 4.5.

55. $3 + \dfrac{1}{2}$

56. $2 + \dfrac{2}{3}$

57. $9 - \dfrac{5}{6}$

58. $4 - \dfrac{1}{5}$

Concept Extensions

59. Calculate $\dfrac{2^3}{3}$ and $\left(\dfrac{2}{3}\right)^3$. Do both of these expressions simplify to the same number? Explain why or why not.

60. Calculate $\left(\dfrac{1}{2}\right)^2 \cdot \left(\dfrac{3}{4}\right)^2$ and $\left(\dfrac{1}{2} \cdot \dfrac{3}{4}\right)^2$. Do both of these expressions simplify to the same number? Explain why or why not.

Recall that to find the average of two numbers, find their sum and divide by 2. For example, the average of $\dfrac{1}{2}$ and $\dfrac{3}{4}$ is $\dfrac{\dfrac{1}{2} + \dfrac{3}{4}}{2}$. Find the average of each pair of numbers.

61. $\dfrac{1}{2}, \dfrac{3}{4}$

62. $\dfrac{3}{5}, \dfrac{9}{10}$

63. $\dfrac{1}{4}, \dfrac{2}{14}$

64. $\dfrac{5}{6}, \dfrac{7}{9}$

65. Two positive numbers, a and b, are graphed below. Where should the graph of their average lie?

Answer true or false for each statement.

66. It is possible for the average of two numbers to be greater than both numbers.

67. It is possible for the average of two numbers to be less than both numbers.

68. The sum of two negative fractions is always a negative number.

69. The sum of a negative fraction and a positive fraction is always a positive number.

70. It is possible for the sum of two fractions to be a whole number.

71. It is possible for the difference of two fractions to be a whole number.

72. What operation should be performed first to simplify

$$\frac{1}{5} \cdot \frac{5}{2} - \left(\frac{2}{3} + \frac{4}{5}\right)^2$$

Explain your answer.

73. A student is to evaluate $x - y$ when $x = \frac{1}{5}$ and $y = -\frac{1}{7}$. This student is asking you if he should evaluate $\frac{1}{5} - \frac{1}{7}$. What do you tell this student and why?

Each expression contains one addition, one subtraction, one multiplication, and one division. Write the operations in the order that they should be performed. Do not actually simplify. See the Concept Check in this section.

74. $[9 + 3(4 - 2)] \div \dfrac{10}{21}$

75. $[30 - 4(3 + 2)] \div \dfrac{5}{2}$

76. $\dfrac{1}{3} \div \left(\dfrac{2}{3}\right)\left(\dfrac{4}{5}\right) - \dfrac{1}{4} + \dfrac{1}{2}$

77. $\left(\dfrac{5}{6} - \dfrac{1}{3}\right) \cdot \dfrac{1}{3} + \dfrac{1}{2} \div \dfrac{9}{8}$

Evaluate each expression if $x = \dfrac{3}{4}$ and $y = -\dfrac{4}{7}$.

78. $\dfrac{2 + x}{y}$

79. $4x + y$

80. $x^2 + 7y$

81. $\dfrac{\dfrac{9}{14}}{x + y}$

4.7 OPERATIONS ON MIXED NUMBERS

Objectives

A Graph Positive and Negative Fractions and Mixed Numbers.

B Multiply or Divide Mixed or Whole Numbers.

C Add or Subtract Mixed Numbers.

D Solve Problems Containing Mixed Numbers.

E Perform Operations on Negative Mixed Numbers.

Objective A Graphing Fractions and Mixed Numbers

Let's review graphing fractions and practice graphing mixed numbers on a number line. This will help us visualize rounding and estimating operations with mixed numbers.

Recall that $5\frac{2}{3}$ means $5 + \frac{2}{3}$ and

$-4\frac{1}{6}$ means $-4 - \frac{1}{6}$ or $-4 + \left(-\frac{1}{6}\right)$

EXAMPLE 1 Graph the numbers on a number line:

$\frac{1}{2}, -\frac{3}{4}, 2\frac{2}{3}, -3, -3\frac{1}{8}$

Solution: Remember that $2\frac{2}{3}$ means $2 + \frac{2}{3}$.

Also, $-3\frac{1}{8}$ means $-3 - \frac{1}{8}$, so $-3\frac{1}{8}$ lies to the left of -3.

Work Practice Problem 1

✔ **Concept Check** Which of the following are equivalent to 9?

a. $7\frac{6}{3}$ **b.** $8\frac{4}{4}$ **c.** $8\frac{9}{9}$ **d.** $\frac{18}{2}$ **e.** all of these

Objective B Multiplying or Dividing with Mixed Numbers or Whole Numbers

When multiplying or dividing a fraction and a mixed or a whole number, remember that mixed and whole numbers can be written as improper fractions.

Multiplying or Dividing Fractions and Mixed Numbers or Whole Numbers

To multiply or divide with mixed numbers or whole numbers, first write any mixed or whole numbers as improper fractions and then multiply or divide as usual.

(*Note:* If an exercise contains a mixed number, we will write the answer as a mixed number, if possible.)

PRACTICE PROBLEM 1

Graph the numbers on a number line.

$-5, -4\frac{1}{2}, 2\frac{3}{4}, \frac{1}{8}, -\frac{1}{2}$

Answer

1.

✔ **Concept Check Answer**

e

287

PRACTICE PROBLEM 2

Multiply and simplify: $1\frac{2}{3}\cdot\frac{11}{15}$

EXAMPLE 2 Multiply: $3\frac{1}{3}\cdot\frac{7}{8}$

Solution: Recall from Section 4.1 that the mixed number $3\frac{1}{3}$ can be written as the fraction $\frac{10}{3}$. Then

$$3\frac{1}{3}\cdot\frac{7}{8}=\frac{10}{3}\cdot\frac{7}{8}=\frac{\overset{1}{\cancel{2}}\cdot5\cdot7}{3\cdot\underset{1}{\cancel{2}}\cdot4}=\frac{35}{12}\quad\text{or}\quad2\frac{11}{12}$$

◼ **Work Practice Problem 2**

Don't forget that a whole number can be written as a fraction by writing the whole number over 1. For example,

$$20=\frac{20}{1}\quad\text{and}\quad7=\frac{7}{1}$$

PRACTICE PROBLEM 3

Multiply: $\frac{5}{6}\cdot18$

EXAMPLE 3 Multiply: $\frac{3}{4}\cdot20$

$$\frac{3}{4}\cdot20=\frac{3}{4}\cdot\frac{20}{1}=\frac{3\cdot20}{4\cdot1}=\frac{3\cdot\overset{1}{\cancel{4}}\cdot5}{\underset{1}{\cancel{4}}\cdot1}=\frac{15}{1}\quad\text{or}\quad15$$

◼ **Work Practice Problem 3**

When both numbers to be multiplied are mixed or whole numbers, it is a good idea to estimate the product to see if your answer is reasonable. To do this, we first practice rounding mixed numbers to the nearest whole. If the fraction part of the mixed number is $\frac{1}{2}$ or greater, we round the whole number part up. If the fraction part of the mixed number is less than $\frac{1}{2}$, then we do not round the whole number part up. Study the table below for examples.

Mixed Number	Rounding
$5\frac{1}{4}$ $\frac{1}{4}$ is less than $\frac{1}{2}$	Thus, $5\frac{1}{4}$ rounds to 5.
$3\frac{9}{16}$ ← 9 is greater than 8 → Half of 16 is 8.	Thus, $3\frac{9}{16}$ rounds to 4.
$1\frac{3}{7}$ ← 3 is less than $3\frac{1}{2}$. → Half of 7 is $3\frac{1}{2}$.	Thus, $1\frac{3}{7}$ rounds to 1.

PRACTICE PROBLEM 4

Multiply. Check by estimating.
$3\frac{1}{5}\cdot2\frac{3}{4}$

EXAMPLE 4 Multiply. Check by estimating.

$$1\frac{2}{3}\cdot2\frac{1}{4}=\frac{5}{3}\cdot\frac{9}{4}=\frac{5\cdot9}{3\cdot4}=\frac{5\cdot\overset{1}{\cancel{3}}\cdot3}{\underset{1}{\cancel{3}}\cdot4}=\frac{15}{4}\text{ or }3\frac{3}{4}\quad\text{Exact}$$

Let's check by estimating.

$1\frac{2}{3}$ rounds to 2, $2\frac{1}{4}$ rounds to 2, and $2\cdot2=4$ Estimate

The estimate is close to the exact value, so our answer is reasonable.

◼ **Work Practice Problem 4**

Answers
2. $\frac{11}{9}$ or $1\frac{2}{9}$ **3.** 15 **4.** $\frac{44}{5}$ or $8\frac{4}{5}$

EXAMPLE 5 Multiply: $7 \cdot 2\frac{11}{14}$. Check by estimating.

$$7 \cdot 2\frac{11}{14} = \frac{7}{1} \cdot \frac{39}{14} = \frac{7 \cdot 39}{1 \cdot 14} = \frac{\overset{1}{\cancel{7}} \cdot 39}{1 \cdot 2 \cdot \cancel{7}} = \frac{39}{2} \text{ or } 19\frac{1}{2}$$ Exact

To estimate,

$2\frac{11}{14}$ rounds to 3 and $7 \cdot 3 = 21$. Estimate

The estimate is close to the exact value, so our answer is reasonable.

◻ **Work Practice Problem 5**

✔ **Concept Check** Find the error.

$$\cancel{2\frac{1}{4} \cdot \frac{1}{2}} = 2\frac{1 \cdot 1}{4 \cdot 2} = 2\frac{1}{8}$$

EXAMPLES Divide.

6. $\dfrac{3}{4} \div 5 = \dfrac{3}{4} \div \dfrac{5}{1} = \dfrac{3}{4} \cdot \dfrac{1}{5} = \dfrac{3 \cdot 1}{4 \cdot 5} = \dfrac{3}{20}$

7. $\dfrac{11}{18} \div 2\dfrac{5}{6} = \dfrac{11}{18} \div \dfrac{17}{6} = \dfrac{11}{18} \cdot \dfrac{6}{17} = \dfrac{11 \cdot 6}{18 \cdot 17} = \dfrac{11 \cdot \overset{1}{\cancel{6}}}{\cancel{6} \cdot 3 \cdot 17} = \dfrac{11}{51}$

8. $5\dfrac{2}{3} \div 2\dfrac{5}{9} = \dfrac{17}{3} \div \dfrac{23}{9} = \dfrac{17}{3} \cdot \dfrac{9}{23} = \dfrac{17 \cdot 9}{3 \cdot 23} = \dfrac{17 \cdot \overset{1}{\cancel{3}} \cdot 3}{\cancel{3} \cdot 23} = \dfrac{51}{23} \text{ or } 2\dfrac{5}{23}$

◻ **Work Practice Problems 6–8**

Objective **C** Adding or Subtracting Mixed Numbers

We can add or subtract mixed numbers, too, by first writing each mixed number as an improper fraction. But it is often easier to add or subtract the whole-number parts and add or subtract the proper-fraction parts vertically.

Adding or Subtracting Mixed Numbers

To add or subtract mixed numbers, add or subtract the fraction parts and then add or subtract the whole number parts.

EXAMPLE 9 Add: $2\dfrac{1}{3} + 5\dfrac{3}{8}$. Check by estimating.

Solution: The LCD of 3 and 8 is 24.

$$\begin{array}{r} 2\dfrac{1 \cdot 8}{3 \cdot 8} = 2\dfrac{8}{24} \\ + 5\dfrac{3 \cdot 3}{8 \cdot 3} = 5\dfrac{9}{24} \\ \hline 7\dfrac{17}{24} \end{array}$$

← Add the fractions
← Add the whole numbers

To check by estimating, we round as usual. The fraction $2\dfrac{1}{3}$ rounds to 2, $5\dfrac{3}{8}$ rounds to 5, and $2 + 5 = 7$, our estimate.

Our exact answer is close to 7, so our answer is reasonable.

◻ **Work Practice Problem 9**

Helpful Hint

When adding or subtracting mixed numbers and whole numbers, it is a good idea to estimate to see if your answer is reasonable.

For the rest of this section, we leave most of the checking by estimating to you.

PRACTICE PROBLEM 10

Add: $3\frac{5}{14} + 2\frac{6}{7}$

EXAMPLE 10 Add: $3\frac{4}{5} + 1\frac{4}{15}$

Solution: The LCD of 5 and 15 is 15.

$$3\frac{4}{5} = 3\frac{12}{15}$$
$$+1\frac{4}{15} = 1\frac{4}{15} \quad \text{Add the fractions; then add the whole numbers.}$$
$$4\frac{16}{15} \quad \text{Notice that the fraction part is improper.}$$

Since $\frac{16}{15}$ is $1\frac{1}{15}$ we can write the sum as

$$4\frac{16}{15} = 4 + 1\frac{1}{15} = 5\frac{1}{15}$$

Work Practice Problem 10

✔ **Concept Check** Explain how you could estimate the following sum: $5\frac{1}{9} + 14\frac{10}{11}$.

PRACTICE PROBLEM 11

Add: $12 + 3\frac{6}{7} + 2\frac{1}{5}$

EXAMPLE 11 Add: $2\frac{4}{5} + 5 + 1\frac{1}{2}$

Solution: The LCD of 5 and 2 is 10.

$$2\frac{4}{5} = 2\frac{8}{10}$$
$$5 = 5$$
$$+1\frac{1}{2} = 1\frac{5}{10}$$
$$8\frac{13}{10} = 8 + 1\frac{3}{10} = 9\frac{3}{10}$$

Work Practice Problem 11

PRACTICE PROBLEM 12

Subtract: $32\frac{7}{9} - 16\frac{5}{18}$

Answers

10. $6\frac{3}{14}$ **11.** $18\frac{2}{35}$ **12.** $16\frac{1}{2}$

EXAMPLE 12 Subtract: $8\frac{3}{7} - 5\frac{2}{21}$. Check by estimating.

Solution: The LCD of 7 and 21 is 21.

$$8\frac{3}{7} = 8\frac{9}{21} \quad \leftarrow \text{The LCD of 7 and 21 is 21.}$$
$$-5\frac{2}{21} = -5\frac{2}{21}$$
$$3\frac{7}{21} \quad \leftarrow \text{Subtract the fractions.}$$
$$\uparrow$$
Subtract the whole numbers.

Then $3\frac{7}{21}$ simplifies to $3\frac{1}{3}$. The difference is $3\frac{1}{3}$.

To check, $8\frac{3}{7}$ rounds to 8, $5\frac{2}{21}$ rounds to 5, and $8 - 5 = 3$, our estimate.

Our exact answer is close to 3, so our answer is reasonable.

🔲 **Work Practice Problem 12**

When subtracting mixed numbers, borrowing may be needed, as shown in the next example.

EXAMPLE 13 Subtract: $7\frac{3}{14} - 3\frac{6}{7}$

Solution: The LCD of 7 and 14 is 14.

$$7\frac{3}{14} = \quad 7\frac{3}{14} \quad \text{Notice that we cannot subtract } \frac{12}{14} \text{ from } \frac{3}{14}, \text{ so we borrow}$$
$$-3\frac{6}{7} = -3\frac{12}{14} \quad \text{from the whole number 7.}$$

borrow 1 from 7

$$7\frac{3}{14} = 6 + 1\frac{3}{14} = 6 + \frac{17}{14} \text{ or } 6\frac{17}{14}$$

Now subtract.

$$7\frac{3}{14} = \quad 7\frac{3}{14} = \quad 6\frac{17}{14}$$
$$-3\frac{6}{7} = -3\frac{12}{14} = -3\frac{12}{14}$$
$$\overline{3\frac{5}{14}} \leftarrow \text{Subtract the fractions.}$$
$$\uparrow$$
$$\text{Subtract the whole numbers.}$$

🔲 **Work Practice Problem 13**

✔ **Concept Check** In the subtraction problem $5\frac{1}{4} - 3\frac{3}{4}$, $5\frac{1}{4}$ must be rewritten because $\frac{3}{4}$ cannot be subtracted from $\frac{1}{4}$. Why is it incorrect to rewrite $5\frac{1}{4}$ as $5\frac{5}{4}$?

EXAMPLE 14 Subtract: $14 - 8\frac{3}{7}$

Solution: $\quad 14 \quad = \quad 13\frac{7}{7} \quad$ Borrow 1 from 14 and write it as $\frac{7}{7}$.

$$-8\frac{3}{7} = -8\frac{3}{7}$$
$$\overline{5\frac{4}{7}} \leftarrow \text{Subtract the fractions.}$$
$$\uparrow$$
$$\text{Subtract the whole numbers.}$$

🔲 **Work Practice Problem 14**

PRACTICE PROBLEM 13

Subtract: $9\frac{7}{15} - 4\frac{3}{5}$

PRACTICE PROBLEM 14

Subtract: $25 - 10\frac{2}{9}$

Answers

13. $4\frac{13}{15}$ **14.** $14\frac{7}{9}$

✔ **Concept Check Answer**

Rewrite $5\frac{1}{4}$ as $4\frac{5}{4}$ by borrowing from the 5.

Objective D Solving Problems Containing Mixed Numbers

Now that we know how to perform operations on mixed numbers, we can solve real-life problems.

EXAMPLE 15 Finding Legal Lobster Size

Lobster fishermen must measure the upper body shells of the lobsters they catch. Lobsters that are too small are thrown back into the ocean. Each state has its own size standard for lobsters to help control the breeding stock. In 1988, Massachusetts increased its legal lobster size from $3\frac{3}{16}$ inches to $3\frac{7}{32}$ inches. How much of an increase was this?

(*Source:* Peabody Essex Museum, Salem, Massachusetts)

Solution:

1. UNDERSTAND. Read and reread the problem carefully. The word "increase" found in the problem might make you think that we add to solve the problem. But the phrase "how much of an increase" tells us to subtract to find the increase.

2. TRANSLATE.

In words:	increase	is	new lobster size	minus	old lobster size
	↓	↓	↓	↓	↓
Translate:	increase	=	$3\frac{7}{32}$	−	$3\frac{3}{16}$

3. SOLVE: Before we solve, let's estimate. The fraction $3\frac{7}{32}$ rounds to 3, $3\frac{3}{16}$ rounds to 3, and $3 - 3 = 0$. The increase is not 0, but will be very small.

$$3\frac{7}{32} = 3\frac{7}{32}$$
$$-3\frac{3}{16} = 3\frac{6}{32}$$
$$\frac{1}{32}$$

4. INTERPRET. *Check* your work. Our estimate tells us that the exact increase of $\frac{1}{32}$ inch is reasonable. *State* your conclusion: The increase in lobster size is $\frac{1}{32}$ of an inch.

◻ **Work Practice Problem 15**

EXAMPLE 16 Calculating Manufacturing Materials Needed

In a manufacturing process, a metal-cutting machine cuts strips $1\frac{3}{5}$ inches long from a piece of metal stock. How many such strips can be cut from a 48-inch piece of stock?

PRACTICE PROBLEM 15

The measurement around the trunk of a tree just below shoulder height is called its girth. The largest known American beech tree in the United States has a girth of $23\frac{1}{4}$ feet. The largest known sugar maple tree in the United States has a girth of $19\frac{5}{12}$ feet. How much larger is the girth of the largest known American beech tree than the girth of the largest known sugar maple tree? (*Source: American Forests*)

Girth

PRACTICE PROBLEM 16

A designer of women's clothing designs a woman's dress that requires $3\frac{1}{7}$ yards of material. How many dresses can be made from a 44-yard bolt of material?

Answers

15. $3\frac{5}{6}$ ft   **16.** 14 dresses

Solution:

1. UNDERSTAND the problem. To do so, read and reread the problem. Then draw a diagram:

   We want to know how many $1\frac{3}{5}$s there are in 48.

   ⟵ 48 inches ⟶

   $1\frac{3}{5}$ inches

2. TRANSLATE.

   In words: | Number of strips | is | 48 | divided by | $1\frac{3}{5}$ |

   Translate: Number of strips = 48 ÷ $1\frac{3}{5}$

3. SOLVE: Let's estimate a reasonable answer. The mixed number $1\frac{3}{5}$ rounds to 2 and $48 \div 2 = 24$.

$$48 \div 1\frac{3}{5} = 48 \div \frac{8}{5} = \frac{48}{1} \cdot \frac{5}{8} = \frac{48 \cdot 5}{1 \cdot 8} = \frac{\overset{1}{\cancel{8}} \cdot 6 \cdot 5}{1 \cdot \underset{1}{\cancel{8}}} = \frac{30}{1} \text{ or } 30$$

4. INTERPRET. *Check* your work. Since the exact answer of 30 is close to our estimate of 24, our answer is reasonable. *State* your conclusion: Thirty strips can be cut from the 48-inch piece of stock.

▦ **Work Practice Problem 16**

Objective ⓔ Operating on Negative Mixed Numbers

To perform operations on negative mixed numbers, let's first practice writing these numbers as negative fractions and negative fractions as negative mixed numbers.

   To understand negative mixed numbers, we simply need to know that, for example,

$$-3\frac{2}{5} \text{ means } -\left(3\frac{2}{5}\right)$$

Thus, to write a negative mixed number as a fraction, we do the following.

$$-3\frac{2}{5} = -\left(3\frac{2}{5}\right) = -\left(\frac{5 \cdot 3 + 2}{5}\right) = -\left(\frac{17}{5}\right) \quad \text{or} \quad -\frac{17}{5}$$

EXAMPLES   Write each as a fraction.

17. $-1\frac{7}{8} = -\frac{8 \cdot 1 + 7}{8} = -\frac{15}{8}$ Write $1\frac{7}{8}$ as an improper fraction and keep the negative sign.

18. $-23\frac{1}{2} = -\frac{2 \cdot 23 + 1}{2} = -\frac{47}{2}$ Write $23\frac{1}{2}$ as an improper fraction and keep the negative sign.

▦ **Work Practice Problems 17–18**

PRACTICE PROBLEMS 17–18

Write each as a fraction.

17. $-9\frac{3}{7}$ 18. $-5\frac{10}{11}$

Answers

17. $-\frac{66}{7}$ 18. $-\frac{65}{11}$

To write a negative fraction as a negative mixed number, we use a similar procedure. We simply disregard the negative sign, convert the improper fraction to a mixed number, then reinsert the negative sign.

PRACTICE PROBLEMS 19–20

Write each as a mixed number.

19. $-\dfrac{37}{8}$ **20.** $-\dfrac{46}{5}$

EXAMPLES   Write each as a mixed number.

19. $-\dfrac{22}{5} = -4\dfrac{2}{5}$

$$5\overline{)22} \qquad \dfrac{22}{5} = 4\dfrac{2}{5}$$
$$\underline{-20}$$
$$2$$

20. $-\dfrac{9}{4} = -2\dfrac{1}{4}$

$$4\overline{)9} \qquad \dfrac{9}{4} = 2\dfrac{1}{4}$$
$$\underline{-8}$$
$$1$$

◼ **Work Practice Problems 19–20**

We multiply or divide with negative mixed numbers the same way that we multiply or divide with positive mixed numbers. We first write each mixed number as a fraction.

PRACTICE PROBLEMS 21–22

21. $2\dfrac{3}{4} \cdot \left(-3\dfrac{3}{5}\right)$

22. $-4\dfrac{2}{7} \div 1\dfrac{1}{4}$

EXAMPLES   Perform the indicated operations.

21. $-4\dfrac{2}{5} \cdot 1\dfrac{3}{11} = -\dfrac{22}{5} \cdot \dfrac{14}{11} = -\dfrac{22 \cdot 14}{5 \cdot 11} = -\dfrac{2 \cdot \overset{1}{\cancel{11}} \cdot 14}{5 \cdot \cancel{11}} = -\dfrac{28}{5}$ or $-5\dfrac{3}{5}$

22. $-2\dfrac{1}{3} \div \left(-2\dfrac{1}{2}\right) = -\dfrac{7}{3} \div \left(-\dfrac{5}{2}\right) = -\dfrac{7}{3} \cdot \left(-\dfrac{2}{5}\right) = \dfrac{7 \cdot 2}{3 \cdot 5} = \dfrac{14}{15}$

◼ **Work Practice Problems 21–22**

Helpful Hint
Recall that $(-) \cdot (-) = +$

To add or subtract with negative mixed numbers, we must be very careful! Problems arise because recall that

$$-3\dfrac{2}{5} \text{ means } -\left(3\dfrac{2}{5}\right)$$

This means that

$$-3\dfrac{2}{5} = -\left(3\dfrac{2}{5}\right) = -\left(3 + \dfrac{2}{5}\right) = -3 - \dfrac{2}{5} \quad \text{\small This can sometimes be easily overlooked.}$$

To avoid problems, we will add or subtract negative mixed numbers by rewriting as addition and recalling how to add signed numbers.

Answers

19. $-4\dfrac{5}{8}$ **20.** $-9\dfrac{1}{5}$

21. $-9\dfrac{9}{10}$ **22.** $-3\dfrac{3}{7}$

EXAMPLE 23 Add: $6\frac{3}{5} + \left(-9\frac{7}{10}\right)$

Solution: Here we are adding two numbers with different signs. Recall that we then subtract the absolute values and keep the sign of the larger absolute value.

Since $-9\frac{7}{10}$ has the larger absolute value, the answer is negative.

First, subtract absolute values:

$$9\frac{7}{10} = 9\frac{7}{10}$$
$$-6\frac{3\cdot 2}{5\cdot 2} = -6\frac{6}{10}$$
$$\overline{\phantom{-6\frac{3\cdot 2}{5\cdot 2}}}$$
$$3\frac{1}{10}$$

Thus,

$$6\frac{3}{5} + \left(-9\frac{7}{10}\right) = -3\frac{1}{10}$$ The result is negative since $-9\frac{7}{10}$ has the larger absolute value.

🔲 **Work Practice Problem 23**

Add: $6\frac{2}{3} + \left(-12\frac{3}{4}\right)$

EXAMPLE 24 Subtract: $-11\frac{5}{6} - 20\frac{4}{9}$

Solution: Let's write as an equivalent addition: $-11\frac{5}{6} + \left(-20\frac{4}{9}\right)$. Here, we are adding two numbers with like signs. Recall that we add their absolute values and keep the common negative sign.

First, add absolute values:

$$11\frac{5\cdot 3}{6\cdot 3} = 11\frac{15}{18}$$
$$+20\frac{4\cdot 2}{9\cdot 2} = +20\frac{8}{18}$$
$$\overline{\phantom{+20\frac{4\cdot 2}{9\cdot 2}}}$$
$$31\frac{23}{18} \text{ or } 32\frac{5}{18}$$

Since $\frac{23}{18} = 1\frac{5}{18}$

Thus,

$$-11\frac{5}{6} - 20\frac{4}{9} = -32\frac{5}{18}$$

Keep the common sign.

🔲 **Work Practice Problem 24**

Subtract: $-9\frac{2}{7} - 30\frac{11}{14}$

Answers

23. $-6\frac{1}{12}$ **24.** $-40\frac{1}{14}$

CALCULATOR EXPLORATIONS   Converting Between Mixed-Number and Fraction Notation

If your calculator has a fraction key, such as $\boxed{a\ b/c}$, you can use it to convert between mixed-number notation and fraction notation.

To write $13\frac{7}{16}$ as an improper fraction, press

$\boxed{1}\ \boxed{3}\ \boxed{a\ b/c}\ \boxed{7}\ \boxed{a\ b/c}\ \boxed{1}\ \boxed{6}\ \boxed{2nd}\ \boxed{d/c}$

The display will read

$\boxed{215\ |\ 16}$

which represents $\frac{215}{16}$. Thus $13\frac{7}{16} = \frac{215}{16}$.

To convert $\frac{190}{13}$ to a mixed number, press

$\boxed{1}\ \boxed{9}\ \boxed{0}\ \boxed{a\ b/c}\ \boxed{1}\ \boxed{3}\ \boxed{=}$

The display will read

$\boxed{14\_8/13}$

which represents $14\frac{8}{13}$. Thus $\frac{190}{13} = 14\frac{8}{13}$.

Write each mixed number as a fraction and each fraction as a mixed number.

1. $25\frac{5}{11}$ 2. $67\frac{14}{15}$ 3. $107\frac{31}{35}$

4. $186\frac{17}{21}$ 5. $\frac{365}{14}$ 6. $\frac{290}{13}$

7. $\frac{2769}{30}$ 8. $\frac{3941}{17}$

Vocabulary and Readiness Check

Use the choices below to fill in each blank.

round   fraction   whole number

improper   mixed number

1. The number $5\frac{3}{4}$ is called a(n) _____.

2. For $5\frac{3}{4}$, the 5 is called the _____ part and $\frac{3}{4}$ is called the _____ part.

3. To estimate operations on mixed numbers, we _____ mixed numbers to the nearest whole number.

4. The mixed number $2\frac{5}{8}$ written as an _____ fraction is $\frac{21}{8}$.

4.7 EXERCISE SET

Objective A *Graph each list of numbers on the given number line. See Example 1.*

1. $-2, -2\frac{2}{3}, 0, \frac{7}{8}, -\frac{1}{3}$

2. $-1, -1\frac{1}{4}, -\frac{1}{4}, 3\frac{1}{4}, 3$

3. $4, \frac{1}{3}, -3, -3\frac{4}{5}, 1\frac{1}{3}$

4. $3, \frac{3}{8}, -4, -4\frac{1}{3}, -\frac{9}{10}$

Objective B *Choose the best estimate for each product or quotient. See Examples 4 through 8.*

5. $2\frac{11}{12} \cdot 1\frac{1}{4}$
   a. 2   **b.** 3   **c.** 1   **d.** 12

6. $5\frac{1}{6} \cdot 3\frac{5}{7}$
   a. 9   **b.** 15   **c.** 8   **d.** 20

7. $12\frac{2}{11} \div 3\frac{9}{10}$
   a. 3   **b.** 4   **c.** 36   **d.** 9

8. $20\frac{3}{14} \div 4\frac{8}{11}$
   a. 5   **b.** 80   **c.** 4   **d.** 16

Multiply or divide. For Exercises 13 through 16, find an exact answer and an estimated answer. See Examples 2 through 8.

9. $2\frac{2}{3} \cdot \frac{1}{7}$

10. $\frac{5}{9} \cdot 4\frac{1}{5}$

11. $7 \div 1\frac{3}{5}$

12. $9 \div 1\frac{2}{3}$

13. $2\frac{1}{5} \cdot 3\frac{1}{2}$

Exact:

Estimate:

14. $2\frac{1}{4} \cdot 7\frac{1}{8}$

Exact:

Estimate:

15. $3\frac{4}{5} \cdot 6\frac{2}{7}$

Exact:

Estimate:

16. $5\frac{5}{6} \cdot 7\frac{3}{5}$

Exact:

Estimate:

17. $5 \cdot 2\frac{1}{2}$

18. $6 \cdot 3\frac{1}{3}$

19. $3\frac{2}{3} \cdot 1\frac{1}{2}$

20. $2\frac{4}{5} \cdot 2\frac{5}{8}$

21. $2\frac{2}{3} \div \frac{1}{7}$

22. $\frac{5}{9} \div 4\frac{1}{5}$

297

Objective **C** *Choose the best estimate for each sum or difference. See Examples 9 through 13.*

23. $3\frac{7}{8} + 2\frac{1}{5}$
  a. 6 **b.** 5 **c.** 1 **d.** 2

24. $3\frac{7}{8} - 2\frac{1}{5}$
  a. 6 **b.** 5 **c.** 1 **d.** 2

25. $8\frac{1}{3} + 1\frac{1}{2}$
  a. 4 **b.** 10 **c.** 6 **d.** 16

26. $8\frac{1}{3} - 1\frac{1}{2}$
  a. 4 **b.** 10 **c.** 6 **d.** 16

Add. For Exercises 27 through 30, find an exact sum and an estimated sum. See Examples 9 through 11.

27. $4\frac{7}{12}$ $+2\frac{1}{12}$ Exact: Estimate:

28. $7\frac{4}{11}$ $+3\frac{2}{11}$ Exact: Estimate:

29. $10\frac{3}{14}$ $+3\frac{4}{7}$ Exact: Estimate:

30. $12\frac{5}{12}$ $+4\frac{1}{6}$ Exact: Estimate:

31. $9\frac{1}{5}$ $+8\frac{2}{25}$

32. $6\frac{2}{13}$ $+8\frac{7}{26}$

33. $12\frac{3}{14}$ 10 $+25\frac{5}{12}$

34. $8\frac{2}{9}$ 32 $+9\frac{10}{21}$

35. $15\frac{4}{7}$ $+9\frac{11}{14}$

36. $23\frac{3}{5}$ $+8\frac{8}{15}$

37. $3\frac{5}{8}$ $2\frac{1}{6}$ $+7\frac{3}{4}$

38. $4\frac{1}{3}$ $9\frac{2}{5}$ $+3\frac{1}{6}$

Subtract. For Exercises 39 through 42, find an exact difference and an estimated difference. See Examples 12 through 14.

39. $4\frac{7}{10}$ $-2\frac{1}{10}$ Exact: Estimate:

40. $7\frac{4}{9}$ $-3\frac{2}{9}$ Exact: Estimate:

41. $10\frac{13}{14}$ $-3\frac{4}{7}$ Exact: Estimate:

42. $12\frac{5}{12}$ $-4\frac{1}{6}$ Exact: Estimate:

43. $9\frac{1}{5}$ $-8\frac{6}{25}$

44. $5\frac{2}{13}$ $-4\frac{7}{26}$

45. 6 $-2\frac{4}{9}$

46. 8 $-1\frac{7}{10}$

47. $63\frac{1}{6}$ $-47\frac{5}{12}$

48. $86\frac{2}{15}$ $-27\frac{3}{10}$

Objectives **B** **C** **Mixed Practice** *Perform each indicated operation. See Examples 2 through 14.*

49. $2\frac{3}{4}$
$+1\frac{1}{4}$

50. $5\frac{5}{8}$
$+2\frac{3}{8}$

51. $15\frac{4}{7}$
$-9\frac{11}{14}$

52. $23\frac{3}{5}$
$-8\frac{8}{15}$

53. $3\frac{1}{9} \cdot 2$

54. $4\frac{1}{2} \cdot 3$

55. $1\frac{2}{3} \div 2\frac{1}{5}$

56. $5\frac{1}{5} \div 3\frac{1}{4}$

57. $22\frac{4}{9} + 13\frac{5}{18}$

58. $15\frac{3}{25} - 5\frac{2}{5}$

59. $5\frac{2}{3} - 3\frac{1}{6}$

60. $5\frac{3}{8} - 2\frac{13}{16}$

61. $15\frac{1}{5}$
$20\frac{3}{10}$
$+37\frac{2}{15}$

62. $7\frac{3}{7}$
15
$+20\frac{1}{2}$

63. $6\frac{4}{7} - 5\frac{11}{14}$

64. $47\frac{5}{12} - 23\frac{19}{24}$

65. $4\frac{2}{7} \cdot 1\frac{3}{10}$

66. $6\frac{2}{3} \cdot 2\frac{3}{4}$

67. $6\frac{2}{11}$
3
$+4\frac{10}{33}$

68. $3\frac{7}{16}$
$6\frac{1}{2}$
$+9\frac{3}{8}$

Objective **D** **Translating** *Translate each phrase to an algebraic expression. Use x to represent a number. See Examples 15 and 16.*

69. $-5\frac{2}{7}$ decreased by a number

70. The sum of $8\frac{3}{4}$ and a number

71. Multiply $1\frac{9}{10}$ by a number.

72. Divide a number by $-6\frac{1}{11}$.

Solve. Write each answer in simplest form. See Examples 15 and 16.

73. The Gauge Act of 1846 set the standard gauge for U.S. railroads at $56\frac{1}{2}$ inches. (See figure.) If the standard gauge in Spain is $65\frac{9}{10}$ inches, how much wider is Spain's standard gauge than the U.S. standard gauge? (*Source:* San Diego Railroad Museum)

74. The standard railroad track gauge (see figure) in Spain is $65\frac{9}{10}$ inches, while in neighboring Portugal it is $65\frac{11}{20}$ inches. Which gauge is wider and by how much? (*Source:* San Diego Railroad Museum)

Track gauge (U.S. $56\frac{1}{2}$ inches)

$\frac{5}{8}$ inch

Point of measurement of gauge

75. If Tucson's average rainfall is $11\frac{1}{4}$ inches and Yuma's is $3\frac{3}{5}$ inches, how much more rain, on the average, does Tucson get than Yuma?

76. A pair of crutches needs adjustment. One crutch is 43 inches and the other is $41\frac{5}{8}$ inches. Find how much the short crutch should be lengthened to make both crutches the same length.

For Exercises 77 and 78, find the area of each figure.

△**77.**

$1\frac{3}{4}$ yards

2 yards

△**78.**

5 inches

$3\frac{1}{2}$ inches

△**79.** A model for a proposed computer chip measures $\frac{3}{4}$ inch by $1\frac{1}{4}$ inches. Find its area.

$1\frac{1}{4}$ inches

$\frac{3}{4}$ inch

△**80.** The Saltalamachios are planning to build a deck that measures $4\frac{1}{2}$ yards by $6\frac{1}{3}$ yards. Find the area of their proposed deck.

$4\frac{1}{2}$ yards

$6\frac{1}{3}$ yards

For Exercises 81 and 82, find the perimeter of each figure.

△**81.**

$5\frac{1}{3}$ meters

3 meters

5 meters

$7\frac{7}{8}$ meters

△**82.**

$3\frac{1}{4}$ yards $3\frac{1}{4}$ yards

$3\frac{1}{4}$ yards $3\frac{1}{4}$ yards

$3\frac{1}{4}$ yards

83. A homeowner has $15\frac{2}{3}$ feet of plastic pipe. She cuts off a $2\frac{1}{2}$-foot length and then a $3\frac{1}{4}$-foot length. If she now needs a 10-foot piece of pipe, will the remaining piece do? If not, by how much will the piece be short?

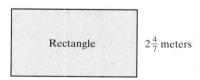

$3\frac{1}{4}$ feet   ? $2\frac{1}{2}$ feet

370 PSI ● 73°F ASTM D-2665 ● MADE IN USA   ASTM D-1785 ● 370 PSI ● 73°F ASTM D-2665 ● MADE IN USA

$15\frac{2}{3}$ feet

84. A trim carpenter cuts a board $3\frac{3}{8}$ feet long from one 6 feet long. How long is the remaining piece?

$3\frac{3}{8}$ feet   ?

6 feet

85. A heart attack patient in rehabilitation walked on a treadmill $12\frac{3}{4}$ miles over 4 days. How many miles is this per day?

86. A local restaurant is selling hamburgers from a booth on Memorial Day. A total of $27\frac{3}{4}$ pounds of hamburger have been ordered. How many quarter-pound hamburgers can this make?

△**87.** The area of the rectangle below is 12 square meters. If its width is $2\frac{4}{7}$ meters, find its length.

| Rectangle | $2\frac{4}{7}$ meters |

△**88.** The perimeter of the square below is $23\frac{1}{2}$ feet. Find the length of each side.

Square

The following table lists some upcoming total eclipses of the Sun that will be visible in North America. The duration of each eclipse is listed in the table. Use the table to answer Exercises 89 through 92.

Total Solar Eclipses Visible from North America	
Date of Eclipse	Duration (in Minutes)
August 1, 2008	$2\frac{9}{20}$
August 21, 2017	$2\frac{2}{3}$
April 8, 2024	$4\frac{7}{15}$
(*Source:* NASA/Goddard Space Flight Center)	

89. What is the total duration for the three eclipses?

90. What is the total duration for the two eclipses occurring in even-numbered years?

91. How much longer will the April 8, 2024, eclipse be than the August 21, 2017, eclipse?

92. How much longer will the August 21, 2017, eclipse be than the August 1, 2008, eclipse?

Objective **E** *Perform the indicated operations. See Examples 17 through 24.*

93. $-4\frac{2}{5} \cdot 2\frac{3}{10}$

94. $-3\frac{5}{6} \div \left(-3\frac{2}{3}\right)$

95. $-5\frac{1}{8} - 19\frac{3}{4}$

96. $17\frac{5}{9} + \left(-14\frac{2}{3}\right)$

97. $-31\frac{2}{15} + 17\frac{3}{20}$

98. $-1\frac{5}{7} \cdot \left(-2\frac{1}{2}\right)$

99. $1\frac{3}{4} \div \left(-3\frac{1}{2}\right)$

100. $-31\frac{7}{8} - \left(-26\frac{5}{12}\right)$

101. $11\frac{7}{8} - 13\frac{5}{6}$

102. $-20\frac{2}{5} + \left(-30\frac{3}{10}\right)$

103. $-7\frac{3}{10} \div (-100)$

104. $-4\frac{1}{4} \div 2\frac{3}{8}$

Review

Multiply. See Section 4.3.

105. $\frac{1}{3}(3x)$

106. $\frac{1}{5}(5y)$

107. $\frac{2}{3}\left(\frac{3}{2}a\right)$

108. $-\frac{9}{10}\left(-\frac{10}{9}m\right)$

Concept Extensions

Solve. See the first Concept Check in this section.

109. Which of the following are equivalent to 10?

    a. $9\frac{5}{5}$ **b.** $9\frac{100}{100}$ **c.** $6\frac{44}{11}$ **d.** $8\frac{13}{13}$

110. Which of the following are equivalent to $7\frac{3}{4}$?

    a. $6\frac{7}{4}$ **b.** $5\frac{11}{4}$ **c.** $7\frac{12}{16}$ **d.** all of them

Solve. See the second Concept Check in this section.

111. A student asked you to check her work below. Is it correct? If not, where is the error?

$$20\frac{2}{3} \div 10\frac{1}{2} \overset{?}{=} 2\frac{1}{3}.$$

112. In your own words, describe how to divide mixed numbers.

113. In your own words, explain how to multiply
  a. fractions
  b. mixed numbers

114. In your own words, explain how to round a mixed number to the nearest whole number.

Solve. See the second Concept Check in this section.

115. A student asked you to check his work below. Is it correct? If not, where is the error?

$$3\frac{2}{3} \cdot 1\frac{1}{7} \stackrel{?}{=} 3\frac{2}{21}$$

Solve.

116. Explain in your own words why $9\frac{13}{9}$ is equal to $10\frac{4}{9}$.

117. In your own words, explain
  a. when to borrow when subtracting mixed numbers, and
  b. how to borrow when subtracting mixed numbers.

THE BIGGER PICTURE   **Operations on Sets of Numbers**

Continue your outline from Sections 1.6, 1.7, 2.4, 3.3, and 4.3. Suggestions are once again written to help you complete this part of your outline.

I. Operations on Sets of Numbers
  A. Whole Numbers
    1. Add or Subtract (Section 1.3)
    2. Multiply or Divide (Sections 1.5, 1.6)
    3. Exponent (Section 1.7)
    4. Order of Operations (Section 1.7)
  B. Integers
    1. Add (Section 2.2)
    2. Subtract (Section 2.3)
    3. Multiply or Divide (Section 2.4)
  C. Fractions
    1. Simplify (Section 4.2)
    2. Multiply (Section 4.3)
    3. Divide (Section 4.3)
    4. Add or Subtract: Must have same denominators. If not, find the LCD, and write each fraction as an equivalent fraction with the LCD as denominator.

$$\frac{2}{5} + \frac{1}{15} = \frac{2}{5} \cdot \frac{3}{3} + \frac{1}{15} = \frac{6}{15} + \frac{1}{15} = \frac{7}{15}$$

II. Solving Equations
  A. Equations in General (Section 3.3)

Perform indicated operations.

1. $\dfrac{3}{17} + \dfrac{2}{17}$

2. $\dfrac{9}{10} - \dfrac{1}{10}$

3. $\dfrac{2}{3} + \dfrac{3}{10}$

4. $\dfrac{23}{24} - \dfrac{11}{12}$

5. $\dfrac{7}{8} + \dfrac{19}{20}$

6. $\dfrac{3^3}{4^3}$

7. $\begin{array}{r} 16 \\ -\ 3\frac{4}{7} \\ \hline \end{array}$

8. $\begin{array}{r} 2\frac{5}{8} \\ 1\frac{1}{6} \\ +5\frac{3}{4} \\ \hline \end{array}$

9. $\dfrac{6}{11} \cdot \dfrac{8}{9}$

10. $2\dfrac{4}{15} \div 1\dfrac{4}{5}$

4.8 SOLVING EQUATIONS CONTAINING FRACTIONS

Objective **A** Solving Equations Containing Fractions

In Chapter 3, we solved linear equations in one variable. In this section, we practice this skill by solving linear equations containing fractions. To help us solve these equations, let's review the properties of equality.

Addition Property of Equality

Let a, b, and c represent numbers. Then

$a = b$	Also, $\quad a = b$
and $a + c = b + c$	and $\quad a - c = b - c$
are equivalent equations.	are equivalent equations.

Multiplication Property of Equality

Let a, b, and c represent numbers and let $c \neq 0$. Then

$a = b$	Also, $a = b$
and $a \cdot c = b \cdot c$	and $\dfrac{a}{c} = \dfrac{b}{c}$
are equivalent equations.	are equivalent equations.

In other words, the same number may be added to or subtracted from both sides of an equation without changing the solution of the equation. Also, the same nonzero number may be multiplied to or divided by both sides of an equation without changing the solution.

Also, don't forget that to solve an equation in x, our goal is to use properties of equality to write simpler equations, all equivalent to the original equation, until the final equation has the form

$$x = \textbf{number} \quad \text{or} \quad \textbf{number} = x$$

PRACTICE PROBLEM 1

Solve: $y - \dfrac{2}{3} = \dfrac{5}{12}$

EXAMPLE 1 Solve: $x - \dfrac{3}{4} = \dfrac{1}{20}$

Solution: To get x by itself, add $\dfrac{3}{4}$ to both sides.

$$x - \frac{3}{4} = \frac{1}{20}$$

$$x - \frac{3}{4} + \frac{3}{4} = \frac{1}{20} + \frac{3}{4} \qquad \text{Add } \frac{3}{4} \text{ to both sides.}$$

$$x = \frac{1}{20} + \frac{3 \cdot 5}{4 \cdot 5} \qquad \text{The LCD of 20 and 4 is 20.}$$

$$x = \frac{1}{20} + \frac{15}{20}$$

$$x = \frac{16}{20}$$

$$x = \frac{\overset{1}{\cancel{4}} \cdot 4}{\underset{1}{\cancel{4}} \cdot 5} = \frac{4}{5} \qquad \text{Write } \frac{16}{20} \text{ in simplest form}$$

Continued on next page

Check: To check, replace x with $\frac{4}{5}$ in the original equation.

$$x - \frac{3}{4} = \frac{1}{20}$$

$$\frac{4}{5} - \frac{3}{4} \overset{?}{=} \frac{1}{20} \quad \text{Replace } x \text{ with } \frac{4}{5}.$$

$$\frac{4 \cdot 4}{5 \cdot 4} - \frac{3 \cdot 5}{4 \cdot 5} \overset{?}{=} \frac{1}{20} \quad \text{The LCD of 5 and 4 is 20.}$$

$$\frac{16}{20} - \frac{15}{20} \overset{?}{=} \frac{1}{20}$$

$$\frac{1}{20} = \frac{1}{20} \quad \text{True}$$

Thus $\frac{4}{5}$ is the solution of $x - \frac{3}{4} = \frac{1}{20}$.

🔲 **Work Practice Problem 1**

EXAMPLE 2   Solve: $\frac{1}{3}x = 7$

Solution: Recall that isolating x means that we want the coefficient of x to be 1. To do so, we use the multiplication property of equality and multiply both sides of the equation by the reciprocal of $\frac{1}{3}$, or 3. Since $\frac{1}{3} \cdot 3 = 1$, we will have isolated x.

$$\frac{1}{3}x = 7$$

$$3 \cdot \frac{1}{3}x = 3 \cdot 7 \quad \text{Multiply both sides by 3.}$$

$$1 \cdot x = 21 \text{ or } x = 21 \quad \text{Simplify.}$$

To check, replace x with 21 in the original equation.

$$\frac{1}{3}x = 7 \quad \text{Original equation}$$

$$\frac{1}{3} \cdot 21 \overset{?}{=} 7 \quad \text{Replace } x \text{ with 21.}$$

$$7 = 7 \quad \text{True}$$

Since $7 = 7$ is a true statement, 21 is the solution of $\frac{1}{3}x = 7$.

🔲 **Work Practice Problem 2**

EXAMPLE 3   Solve: $\frac{3}{5}a = 9$

Solution: Multiply both sides by $\frac{5}{3}$, the reciprocal of $\frac{3}{5}$, so that the coefficient of a is 1.

$$\frac{3}{5}a = 9$$

$$\frac{5}{3} \cdot \frac{3}{5}a = \frac{5}{3} \cdot 9 \quad \text{Multiply both sides by } \frac{5}{3}.$$

$$1a = \frac{5 \cdot 9}{3} \quad \text{Multiply.}$$

$$a = 15 \quad \text{Simplify.}$$

PRACTICE PROBLEM 2

Solve: $\frac{1}{5}y = 2$

PRACTICE PROBLEM 3

Solve: $\frac{5}{7}b = 25$

Answers

2. 10   **3.** 35

Continued on next page

To check, replace a with 15 in the original equation.

$$\frac{3}{5}a = 9$$

$$\frac{3}{5} \cdot 15 \overset{?}{=} 9 \quad \text{Replace } a \text{ with 15.}$$

$$\frac{3 \cdot 15}{5} \overset{?}{=} 9 \quad \text{Multiply.}$$

$$9 = 9 \quad \text{True}$$

Since $9 = 9$ is true, 15 is the solution of $\frac{3}{5}a = 9$.

■ **Work Practice Problem 3**

PRACTICE PROBLEM 4

Solve: $-\frac{7}{10}x = \frac{2}{5}$

EXAMPLE 4 Solve: $\frac{3}{4}x = -\frac{1}{8}$

Solution: Multiply both sides of the equation by $\frac{4}{3}$, the reciprocal of $\frac{3}{4}$.

$$\frac{3}{4}x = -\frac{1}{8}$$

$$\frac{4}{3} \cdot \frac{3}{4}x = \frac{4}{3} \cdot -\frac{1}{8} \quad \text{Multiply both sides by } \frac{4}{3}.$$

$$1x = -\frac{4 \cdot 1}{3 \cdot 8} \quad \text{Multiply.}$$

$$x = -\frac{1}{6} \quad \text{Simplify.}$$

To check, replace x with $-\frac{1}{6}$ in the original equation.

$$\frac{3}{4}x = -\frac{1}{8} \quad \text{Original equation}$$

$$\frac{3}{4} \cdot -\frac{1}{6} \overset{?}{=} -\frac{1}{8} \quad \text{Replace } x \text{ with } -\frac{1}{6}.$$

$$-\frac{1}{8} = -\frac{1}{8} \quad \text{True}$$

Since we arrived at a true statement, $-\frac{1}{6}$ is the solution of $\frac{3}{4}x = -\frac{1}{8}$.

■ **Work Practice Problem 4**

PRACTICE PROBLEM 5

Solve: $5x = -\frac{3}{4}$

EXAMPLE 5 Solve: $3y = -\frac{2}{11}$

Solution: We can either divide both sides by 3 or multiply both sides by the reciprocal of 3, which is $\frac{1}{3}$.

$$3y = -\frac{2}{11}$$

$$\frac{1}{3} \cdot 3y = \frac{1}{3} \cdot -\frac{2}{11} \quad \text{Multiply both sides by } \frac{1}{3}.$$

$$1y = -\frac{1 \cdot 2}{3 \cdot 11} \quad \text{Multiply.}$$

$$y = -\frac{2}{33} \quad \text{Simplify.}$$

Check to see that the solution is $-\frac{2}{33}$.

■ **Work Practice Problem 5**

Answers

4. $-\frac{4}{7}$ **5.** $-\frac{3}{20}$

Objective B Solving Equations by Multiplying by the LCD

Solving equations with fractions can be tedious. If an equation contains fractions, it is often helpful to first multiply both sides of the equation by the LCD of the fractions. This has the effect of eliminating the fractions in the equation, as shown in the next example.

Let's solve the equation in Example 4 again. This time, we will multiply both sides by the LCD.

EXAMPLE 6 Solve: $\dfrac{3}{4}x = -\dfrac{1}{8}$

Solution: First, multiply both sides of the equation by the LCD of the fractions $\dfrac{3}{4}$ and $-\dfrac{1}{8}$. The LCD of the denominators is 8.

$$\frac{3}{4}x = -\frac{1}{8}$$

$$8 \cdot \frac{3}{4}x = 8 \cdot -\frac{1}{8} \qquad \text{Multiply both sides by 8.}$$

$$\frac{8 \cdot 3}{1 \cdot 4}x = -\frac{\overset{1}{8} \cdot 1}{1 \cdot \underset{1}{8}} \qquad \text{Multiply the fractions.}$$

$$\frac{2 \cdot \overset{1}{4} \cdot 3}{1 \cdot \underset{1}{4}}x = -\frac{1 \cdot 1}{1 \cdot 1} \qquad \text{Simplify.}$$

$$6x = -1$$

$$\frac{6x}{6} = \frac{-1}{6} \qquad \text{Divide both sides by 6.}$$

$$x = \frac{-1}{6} \text{ or } -\frac{1}{6} \qquad \text{Simplify.}$$

As seen in Example 4, the solution is $-\dfrac{1}{6}$.

■ **Work Practice Problem 6**

EXAMPLE 7 Solve: $\dfrac{x}{6} + 1 = \dfrac{4}{3}$

Solution: First multiply both sides of the equation by the LCD of the fractions. The LCD of the denominators 6 and 3 is 6.

$$\frac{x}{6} + 1 = \frac{4}{3}$$

$$6\left(\frac{x}{6} + 1\right) = 6\left(\frac{4}{3}\right) \qquad \text{Multiply both sides by 6.}$$

$$\overset{1}{6}\left(\frac{x}{\underset{1}{6}}\right) + 6(1) = \overset{2}{6}\left(\frac{4}{\underset{1}{3}}\right) \qquad \text{Apply the distributive property.}$$

$$x + 6 = 8 \qquad \text{Simplify.}$$

$$x + 6 + (-6) = 8 + (-6) \qquad \text{Add } -6 \text{ to both sides.}$$

$$x = 2 \qquad \text{Simplify.}$$

Continued on next page

PRACTICE PROBLEM 6

Solve: $\dfrac{11}{15}x = -\dfrac{3}{5}$

PRACTICE PROBLEM 7

Solve: $\dfrac{y}{8} + \dfrac{3}{4} = 2$

Answers

6. $-\dfrac{9}{11}$ **7.** 10

To check, replace x with 2 in the original equation.

$$\frac{x}{6} + 1 = \frac{4}{3} \quad \text{Original equation}$$

$$\frac{2}{6} + 1 \stackrel{?}{=} \frac{4}{3} \quad \text{Replace } x \text{ with 2.}$$

$$\frac{1}{3} + \frac{3}{3} \stackrel{?}{=} \frac{4}{3} \quad \text{Simplify } \frac{2}{6}. \text{ The LCD of 3 and 1 is 3.}$$

$$\frac{4}{3} = \frac{4}{3} \quad \text{True}$$

Since we arrived at a true statement, 2 is the solution of $\frac{x}{6} + 1 = \frac{4}{3}$.

◼ **Work Practice Problem 7**

Let's review the steps for solving equations in x. An extra step is now included to handle equations containing fractions.

Solving an Equation in x

Step 1: If fractions are present, multiply both sides of the equation by the LCD of the fractions.

Step 2: If parentheses are present, use the distributive property.

Step 3: Combine any like terms on each side of the equation.

Step 4: Use the addition property of equality to rewrite the equation so that variable terms are on one side of the equation and constant terms are on the other side.

Step 5: Divide both sides of the equation by the numerical coefficient of x to solve.

Step 6: Check the answer in the **original equation.**

PRACTICE PROBLEM 8

Solve: $\frac{x}{5} - x = \frac{1}{5}$

Helpful Hint Don't forget to multiply *both* sides of the equation by the LCD.

EXAMPLE 8 Solve: $\frac{z}{5} - \frac{z}{3} = 6$

Solution: $\frac{z}{5} - \frac{z}{3} = 6$

$$15\left(\frac{z}{5} - \frac{z}{3}\right) = 15(6) \quad \text{Multiply both sides by the LCD, 15.}$$

$$\overset{3}{\cancel{15}}\left(\frac{z}{\cancel{5}}\right) - \overset{5}{\cancel{15}}\left(\frac{z}{\cancel{3}}\right) = 15(6) \quad \text{Apply the distributive property.}$$

$$3z - 5z = 90 \quad \text{Simplify.}$$
$$-2z = 90 \quad \text{Combine like terms.}$$
$$\frac{-2z}{-2} = \frac{90}{-2} \quad \text{Divide both sides by } -2, \text{ the coefficient of } z.$$
$$z = -45 \quad \text{Simplify.}$$

To check, replace z with -45 in the **original equation** to see that a true statement results.

◼ **Work Practice Problem 8**

Answer

8. $-\frac{1}{4}$

EXAMPLE 9 Solve: $\dfrac{x}{2} = \dfrac{x}{3} + \dfrac{1}{2}$

Solution: First multiply both sides by the LCD, 6.

$$\frac{x}{2} = \frac{x}{3} + \frac{1}{2}$$

$$6\left(\frac{x}{2}\right) = 6\left(\frac{x}{3} + \frac{1}{2}\right) \qquad \text{Multiply both sides by the LCD, 6.}$$

$$\overset{3}{6}\left(\frac{x}{2}\right)_{1} = \overset{2}{6}\left(\frac{x}{3}\right)_{1} + \overset{3}{6}\left(\frac{1}{2}\right)_{1} \qquad \text{Apply the distributive property.}$$

$$3x = 2x + 3 \qquad \text{Simplify.}$$

$$3x + (-2x) = 2x + 3 + (-2x) \qquad \text{Add } (-2x) \text{ to both sides.}$$

$$x = 3 \qquad \text{Simplify.}$$

To check, replace x with 3 in the original equation to see that a true statement results.

◼ **Work Practice Problem 9**

Objective C Review of Adding and Subtracting Fractions

Make sure you understand the difference between **solving an equation** containing fractions and **adding or subtracting two fractions.** To solve an equation containing fractions, we use the multiplication property of equality and multiply both sides by the LCD of the fractions, thus eliminating the fractions. This method does not apply to adding or subtracting fractions. The multiplication property of equality applies only to equations. To add or subtract unlike fractions, we write each fraction as an equivalent fraction using the LCD of the fractions as the denominator. See the next example for a review.

EXAMPLE 10 Add: $\dfrac{x}{3} + \dfrac{2}{5}$

Solution: This expression is not an equation. Here, we are adding two unlike fractions. To add unlike fractions, we need to find the LCD. The LCD of the denominators 3 and 5 is 15. Write each fraction as an equivalent fraction with a denominator of 15.

$$\frac{x}{3} + \frac{2}{5} = \frac{x}{3} \cdot \frac{5}{5} + \frac{2}{5} \cdot \frac{3}{3} = \frac{x \cdot 5}{3 \cdot 5} + \frac{2 \cdot 3}{5 \cdot 3}$$

$$= \frac{5x}{15} + \frac{6}{15}$$

$$= \frac{5x + 6}{15}$$

◼ **Work Practice Problem 10**

✔ **Concept Check** Which of the following are equations and which are expressions?

a. $\dfrac{1}{2} + 3x = 5$ **b.** $\dfrac{2}{3}x - \dfrac{x}{5}$

c. $\dfrac{x}{12} + \dfrac{5x}{24}$ **d.** $\dfrac{x}{5} = \dfrac{1}{10}$

PRACTICE PROBLEM 9

Solve: $\dfrac{y}{2} = \dfrac{y}{5} + \dfrac{3}{2}$

PRACTICE PROBLEM 10

Subtract: $\dfrac{9}{10} - \dfrac{y}{3}$

Answers

9. 5 **10.** $\dfrac{27 - 10y}{30}$

✔ **Concept Check Answers**

equations: a, d; expressions: b, c

4.8 EXERCISE SET

Objective A *Solve each equation. Check your proposed solution. See Example 1.*

1. $x + \dfrac{1}{3} = -\dfrac{1}{3}$

2. $x + \dfrac{1}{9} = -\dfrac{7}{9}$

3. $y - \dfrac{3}{13} = -\dfrac{2}{13}$

4. $z - \dfrac{5}{14} = \dfrac{4}{14}$

5. $3x - \dfrac{1}{5} - 2x = \dfrac{1}{5} + \dfrac{2}{5}$

6. $5x + \dfrac{1}{11} - 4x = \dfrac{2}{11} - \dfrac{5}{11}$

7. $x - \dfrac{1}{12} = \dfrac{5}{6}$

8. $y - \dfrac{8}{9} = \dfrac{1}{3}$

9. $\dfrac{2}{5} + y = -\dfrac{3}{10}$

10. $\dfrac{1}{2} + a = -\dfrac{3}{8}$

11. $7z + \dfrac{1}{16} - 6z = \dfrac{3}{4}$

12. $9x - \dfrac{2}{7} - 8x = \dfrac{11}{14}$

13. $-\dfrac{2}{9} = x - \dfrac{5}{6}$

14. $-\dfrac{1}{4} = y - \dfrac{7}{10}$

Solve each equation. See Examples 2 through 5.

15. $7x = 2$

16. $-5x = 4$

17. $\dfrac{1}{4}x = 3$

18. $\dfrac{1}{3}x = 6$

19. $\dfrac{2}{9}y = -6$

20. $\dfrac{4}{7}x = -8$

21. $-\dfrac{4}{9}z = -\dfrac{3}{2}$

22. $-\dfrac{11}{10}x = -\dfrac{2}{7}$

23. $7a = \dfrac{1}{3}$

24. $2z = -\dfrac{5}{12}$

25. $-3x = -\dfrac{6}{11}$

26. $-4z = -\dfrac{12}{25}$

Objective B *Solve each equation. See Examples 6 through 9.*

27. $\dfrac{5}{9}x = -\dfrac{3}{18}$

28. $\dfrac{3}{5}y = -\dfrac{7}{20}$

29. $\dfrac{x}{3} + 2 = \dfrac{7}{3}$

30. $\dfrac{x}{5} - 1 = \dfrac{7}{5}$

31. $\dfrac{x}{5} - \dfrac{5}{10} = 1$

32. $\dfrac{x}{3} - x = -6$

33. $\dfrac{1}{2} - \dfrac{3}{5} = \dfrac{x}{10}$

34. $\dfrac{2}{3} - \dfrac{1}{4} = \dfrac{x}{12}$

310

35. $\dfrac{x}{3} = \dfrac{x}{5} - 2$

36. $\dfrac{a}{2} = \dfrac{a}{7} + \dfrac{5}{2}$

Objective C *Add or subtract as indicated. See Example 10.*

37. $\dfrac{x}{7} - \dfrac{4}{3}$

38. $-\dfrac{5}{9} + \dfrac{y}{8}$

39. $\dfrac{y}{2} + 5$

40. $2 + \dfrac{7x}{3}$

41. $\dfrac{3x}{10} + \dfrac{x}{6}$

42. $\dfrac{9x}{8} - \dfrac{5x}{6}$

Objectives A B C Mixed Practice *Solve. If no equation is given, perform the indicated operation.*

43. $\dfrac{3}{8}x = \dfrac{1}{2}$

44. $\dfrac{2}{5}y = \dfrac{3}{10}$

45. $\dfrac{2}{3} - \dfrac{x}{5} = \dfrac{4}{15}$

46. $\dfrac{4}{5} + \dfrac{x}{4} = \dfrac{21}{20}$

47. $\dfrac{9}{14}z = \dfrac{27}{20}$

48. $\dfrac{5}{16}a = \dfrac{5}{6}$

49. $-3m - 5m = \dfrac{4}{7}$

50. $30n - 34n = \dfrac{3}{20}$

51. $\dfrac{x}{4} + 1 = \dfrac{1}{4}$

52. $\dfrac{y}{7} - 2 = \dfrac{1}{7}$

53. $\dfrac{5}{9} - \dfrac{2}{3}$

54. $\dfrac{8}{11} - \dfrac{1}{2}$

55. $\dfrac{1}{5}y = 10$

56. $\dfrac{1}{4}x = -2$

57. $\dfrac{5}{7}y = -\dfrac{15}{49}$

58. $-\dfrac{3}{4}x = \dfrac{9}{2}$

59. $\dfrac{x}{2} - x = -2$

60. $\dfrac{y}{3} = -4 + y$

61. $-\dfrac{5}{8}y = \dfrac{3}{16} - \dfrac{9}{16}$

62. $-\dfrac{7}{9}x = -\dfrac{5}{18} - \dfrac{4}{18}$

63. $17x - 25x = \dfrac{1}{3}$

64. $27x - 30x = \dfrac{4}{9}$

65. $\dfrac{7}{6}x = \dfrac{1}{4} - \dfrac{2}{3}$

66. $\dfrac{5}{4}y = \dfrac{1}{2} - \dfrac{7}{10}$

67. $\dfrac{b}{4} = \dfrac{b}{12} + \dfrac{2}{3}$

68. $\dfrac{a}{6} = \dfrac{a}{3} + \dfrac{1}{2}$

69. $\dfrac{x}{3} + 2 = \dfrac{x}{2} + 8$

70. $\dfrac{y}{5} - 2 = \dfrac{y}{3} - 4$

Review

Round each number to the given place value. See Section 1.4.

71. 57,236 to the nearest hundred

72. 576 to the nearest hundred

73. 327 to the nearest ten

74. 2333 to the nearest ten

Concept Extensions

75. Explain why the method for eliminating fractions in an **equation** does not apply to simplifying **expressions** containing fractions.

Solve.

76. $\dfrac{14}{11} + \dfrac{3x}{8} = \dfrac{x}{2}$

77. $\dfrac{19}{53} = \dfrac{353x}{1431} + \dfrac{23}{27}$

78. The area of the rectangle is $\dfrac{5}{12}$ square inches. Find its length, x.

$\frac{3}{4}$ inch

x

CHAPTER 4   Group Activity

Lobster Classification

Sections 4.1, 4.7, 4.8

This activity may be completed by working in groups or individually.

Lobsters are normally classified by weight. Use the weight classification table to answer the questions in this activity.

Classification of Lobsters	
Class	**Weight (in Pounds)**
Chicken	1 to $1\frac{1}{8}$
Quarter	$1\frac{1}{4}$
Half	$1\frac{1}{2}$ to $1\frac{3}{4}$
Select	$1\frac{3}{4}$ to $2\frac{1}{2}$
Large select	$2\frac{1}{2}$ to $3\frac{1}{2}$
Jumbo	Over $3\frac{1}{2}$

(*Source:* The Maine Lobster Promotion Council)

1. A lobster fisher has kept four lobsters from a lobster trap. Classify each lobster if they have the following weights:

a. $1\frac{7}{8}$ pounds

b. $1\frac{9}{16}$ pounds

c. $2\frac{3}{4}$ pounds

d. $2\frac{3}{8}$ pounds

2. A recipe requires 5 pounds of lobster. Using the minimum weight for each class, decide whether a chicken, half, and select lobster will be enough for the recipe, and explain your reasoning. If not, suggest a better choice of lobsters to meet the recipe requirements.

3. A lobster market customer has selected two chickens, a select, and a large select. What is the most that these four lobsters could weigh? What is the least that these four lobsters could weigh?

4. A lobster market customer wishes to buy three quarters. If lobsters sell for $7 per pound, how much will the customer owe for her purchase?

5. Why do you think there is no classification for lobsters weighing under 1 pound?

STUDY SKILLS BUILDER

Tips for Studying for an Exam

To prepare for an exam, try the following study techniques:

- Start the study process days before your exam.
- Make sure that you are up-to-date on your assignments.
- If there is a topic that you are unsure of, use one of the many resources that are available to you. For example,

    See your instructor.

    Visit a learning resource center on campus.

    Read the textbook material and examples on the topic.

    View a video on the topic.

- Reread your notes and carefully review the Chapter Highlights at the end of any chapter.
- Work the review exercises at the end of the chapter. Check your answers and correct any mistakes. If you have trouble, use a resource listed above.
- Find a quiet place to take the Chapter Test found at the end of the chapter. Do not use any resources when taking this sample test. This way, you will have a clear indication of how prepared you are for your exam.

Check your answers and make sure that you correct any missed exercises.

- Get lots of rest the night before the exam. It's hard to show how well you know the material if your brain is foggy from lack of sleep.

Good luck, and keep a positive attitude.

Let's see how you did on your last exam.

1. How many days before your last exam did you start studying for that exam?
2. Were you up-to-date on your assignments at that time or did you need to catch up on assignments?
3. List the most helpful text supplement (if you used one).
4. List the most helpful campus supplement (if you used one).
5. List your process for preparing for a mathematics test.
6. Was this process helpful? In other words, were you satisfied with your performance on your exam?
7. If not, what changes can you make in your process that will make it more helpful to you?

Chapter 4 Vocabulary Check

Fill in each blank with one of the words or phrases listed below.

mixed number	complex fraction	like	numerator	prime factorization
composite number	equivalent	cross products	least common denominator	denominator
prime number	improper fraction	simplest form	undefined	0
reciprocals	proper fraction			

1. Two numbers are _____ of each other if their product is 1.
2. A _____ is a natural number greater than 1 that is not prime.
3. Fractions that represent the same portion of a whole are called _____ fractions.
4. An _____ is a fraction whose numerator is greater than or equal to its denominator.
5. A _____ is a natural number greater than 1 whose only factors are 1 and itself.
6. A fraction is in _____ when the numerator and the denominator have no factors in common other than 1.
7. A _____ is one whose numerator is less than its denominator.
8. A _____ contains a whole number part and a fraction part.
9. In the fraction $\frac{7}{9}$, the 7 is called the _____ and the 9 is called the _____.
10. The _____ of a number is the factorization in which all the factors are prime numbers.
11. The fraction $\frac{3}{0}$ is _____.
12. The fraction $\frac{0}{5}$ = _____.
13. Fractions that have the same denominator are called _____ fractions.
14. The LCM of the denominators in a list of fractions is called the _____.
15. A fraction whose numerator or denominator or both numerator and denominator contain fractions is called a _____.
16. In $\frac{a}{b} = \frac{c}{d}$, $a \cdot d$ and $b \cdot c$ are called _____.

> **Helpful Hint** Are you preparing for your test? Don't forget to take the Chapter 4 Test on page 325. Then check your answers at the back of the text and use the Chapter Test Prep Video CD to see the fully worked-out solutions to any of the exercises you want to review.

4 Chapter Highlights

DEFINITIONS AND CONCEPTS	EXAMPLES
Section 4.1 Introduction to Fractions and Mixed Numbers	

A **fraction** is of the form $\dfrac{\text{numerator}}{\text{denominator}}$ ← number of parts being considered ← number of equal parts in the whole	Write a fraction to represent the shaded part of the figure. $\dfrac{3}{8}$ ← number of parts shaded ← number of equal parts
A fraction is called a **proper fraction** if its numerator is less than its denominator.	Proper Fractions: $\dfrac{1}{3}, \dfrac{2}{5}, \dfrac{7}{8}, \dfrac{100}{101}$
A fraction is called an **improper fraction** if its numerator is greater than or equal to its denominator.	Improper Fractions: $\dfrac{5}{4}, \dfrac{2}{2}, \dfrac{9}{7}, \dfrac{101}{100}$
A **mixed number** contains a whole number and a fraction.	Mixed Numbers: $1\dfrac{1}{2}, 5\dfrac{7}{8}, 25\dfrac{9}{10}$

DEFINITIONS AND CONCEPTS	**EXAMPLES**

Section 4.1   Introduction to Fractions and Mixed Numbers (*continued*)

TO WRITE A MIXED NUMBER AS AN IMPROPER FRACTION	$$5\frac{2}{7} = \frac{5 \cdot 7 + 2}{7} = \frac{35 + 2}{7} = \frac{37}{7}$$
1. Multiply the denominator of the fraction by the whole number.	
2. Add the numerator of the fraction to the product from Step 1.	
3. Write this sum from Step 2 as the numerator of the improper fraction over the original denominator.	
TO WRITE AN IMPROPER FRACTION AS A MIXED NUMBER OR A WHOLE NUMBER	$$\frac{17}{3} = 5\frac{2}{3}$$
1. Divide the denominator into the numerator.	$$\begin{array}{r} 5 \\ 3\overline{)17} \\ \underline{15} \\ 2 \end{array}$$
2. The whole number part of the mixed number is the quotient. The fraction is the remainder over the original denominator.	
$$\text{quotient} \frac{\text{remainder}}{\text{original denominator}}$$	

Section 4.2   Factors and Simplest Form

A **prime number** is a natural number that has exactly two different factors, 1 and itself.	$2, 3, 5, 7, 11, 13, 17, \ldots$
A **composite number** is any natural number other than 1 that is not prime.	$4, 6, 8, 9, 10, 12, 14, 15, 16, \ldots$
The **prime factorization** of a number is the factorization in which all the factors are prime numbers.	Write the prime factorization of 60. $$60 = 6 \cdot 10$$ $$= 2 \cdot 3 \cdot 2 \cdot 5 \quad \text{or} \quad 2^2 \cdot 3 \cdot 5$$
Fractions that represent the same portion of a whole are called **equivalent fractions.**	$$\frac{3}{4} \quad = \quad \frac{12}{16}$$
A fraction is in **simplest form** or **lowest terms** when the numerator and the denominator have no common factors other than 1.	The fraction $\frac{2}{3}$ is in simplest form.
To write a fraction in simplest form, write the prime factorizations of the numerator and the denominator and then divide both by all common factors.	Write in simplest form: $\frac{30}{36}$ $$\frac{30}{36} = \frac{2 \cdot 3 \cdot 5}{2 \cdot 2 \cdot 3 \cdot 3} = \frac{2}{2} \cdot \frac{3}{3} \cdot \frac{5}{2 \cdot 3} = 1 \cdot 1 \cdot \frac{5}{6} = \frac{5}{6}$$ $$\text{or} \quad \frac{30}{36} = \frac{\cancel{2} \cdot \cancel{3} \cdot 5}{\cancel{2} \cdot 2 \cdot \cancel{3} \cdot 3} = \frac{5}{6}$$

DEFINITIONS AND CONCEPTS	EXAMPLES

Section 4.2   Factors and Simplest Form (*continued*)

Two fractions are equivalent if	Determine whether $\frac{7}{8}$ and $\frac{21}{24}$ are equivalent.
Method 1. They simplify to the same fraction.	$\frac{7}{8}$ is in simplest form
Method 2. Their cross products are equal. $$\begin{matrix} 24 \cdot 7 \\ = 168 \end{matrix} \quad \frac{7}{8} = \frac{21}{24} \quad \begin{matrix} 8 \cdot 21 \\ = 168 \end{matrix}$$ Since $168 = 168$, $\frac{7}{8} = \frac{21}{24}$.	$\frac{21}{24} = \frac{\overset{1}{\cancel{3}} \cdot 7}{\underset{1}{\cancel{3}} \cdot 8} = \frac{1 \cdot 7}{1 \cdot 8} = \frac{7}{8}$ Since both simplify to $\frac{7}{8}$, then $\frac{7}{8} = \frac{21}{24}$.

Section 4.3   Multiplying and Dividing Fractions

To multiply two fractions, multiply the numerators and multiply the denominators.	Multiply. $$\frac{2x}{3} \cdot \frac{5}{7} = \frac{2x \cdot 5}{3 \cdot 7} = \frac{10x}{21}$$ $$\frac{3}{4} \cdot \frac{1}{6} = \frac{3 \cdot 1}{4 \cdot 6} = \frac{\overset{1}{\cancel{3}} \cdot 1}{4 \cdot \underset{1}{\cancel{3}} \cdot 2} = \frac{1}{8}$$
To find the **reciprocal** of a fraction, interchange its numerator and denominator.	The reciprocal of $\frac{3}{5}$ is $\frac{5}{3}$.
To divide two fractions, multiply the first fraction by the reciprocal of the second fraction.	Divide. $$-\frac{3}{10} \div \frac{7}{9} = -\frac{3}{10} \cdot \frac{9}{7} = -\frac{3 \cdot 9}{10 \cdot 7} = -\frac{27}{70}$$

Section 4.4   Adding and Subtracting Like Fractions, Least Common Denominator, and Equivalent Fractions

Fractions that have the same denominator are called **like fractions.**	$-\frac{1}{3}$ and $\frac{2}{3}$; $\frac{5x}{7}$ and $\frac{6}{7}$
To add or subtract like fractions, combine the numerators and place the sum or difference over the common denominator.	$\frac{2}{7} + \frac{3}{7} = \frac{5}{7}$ ← Add the numerators. ← Keep the common denominator. $\frac{7}{8} - \frac{4}{8} = \frac{3}{8}$ ← Subtract the numerators. ← Keep the common denominator.
The least common denominator (LCD) of a list of fractions is the smallest positive number divisible by all the denominators in the list.	The LCD of $\frac{1}{2}$ and $\frac{5}{6}$ is 6 because 6 is the smallest positive number that is divisible by both 2 and 6.

DEFINITIONS AND CONCEPTS	EXAMPLES

Section 4.4   Adding and Subtracting Like Fractions, Least Common Denominator, and Equivalent Fractions

METHOD 1 FOR FINDING THE LCD OF A LIST OF FRACTIONS USING MULTIPLES

Step 1. Write the multiples of the largest denominator (starting with the number itself) until a multiple common to all denominators in the list is found.

Step 2. The multiple found in Step 1 is the LCD.

METHOD 2 FOR FINDING THE LCD OF A LIST OF FRACTIONS USING PRIME FACTORIZATION

Step 1. Write the prime factorization of each denominator.

Step 2. For each different prime factor in Step 1, circle the greatest number of times that factor occurs in any one factorization.

Step 3. The LCD is the product of the circled factors.

Equivalent fractions represent the same portion of a whole.

Find the LCD of $\frac{1}{4}$ and $\frac{5}{6}$ using Method 1.

$6 \cdot 1 = 6$ Not a multiple of 4
$6 \cdot 2 = 12$ A multiple of 4

The LCD is 12.

Find the LCD of $\frac{5}{6}$ and $\frac{11}{20}$ using Method 2.

$6 = 2 \cdot ③$
$20 = ②\cdot②\cdot⑤$

The LCD is

$2 \cdot 2 \cdot 3 \cdot 5 = 60$

Write an equivalent fraction with the indicated denominator.

$$\frac{2}{8} = \frac{}{16}$$

$$\frac{2 \cdot 2}{8 \cdot 2} = \frac{4}{16}$$

Section 4.5   Adding and Subtracting Unlike Fractions

TO ADD OR SUBTRACT FRACTIONS WITH UNLIKE DENOMINATORS

Step 1. Find the LCD.

Step 2. Write each fraction as an equivalent fraction whose denominator is the LCD.

Step 3. Add or subtract the like fractions.

Step 4. Write the sum or difference in simplest form.

Add: $\frac{3}{20} + \frac{2}{5}$

Step 1. The LCD of 20 and 5 is 20.

Step 2. $\frac{3}{20} = \frac{3}{20}; \frac{2}{5} = \frac{2}{5} \cdot \frac{4}{4} = \frac{8}{20}$

Step 3. $\frac{3}{20} + \frac{2}{5} = \frac{3}{20} + \frac{8}{20} = \frac{11}{20}$

Step 4. $\frac{11}{20}$ is in simplest form.

Section 4.6   Complex Fractions and Review of Order of Operations

A fraction whose numerator or denominator or both contain fractions is called a **complex fraction.**

Complex Fractions:

$$\frac{\frac{11}{4}}{\frac{7}{10}}, \quad \frac{\frac{y}{6} - 11}{\frac{4}{3}}$$

DEFINITIONS AND CONCEPTS	**EXAMPLES**

Section 4.6   Complex Fractions and Review of Order of Operations (*continued*)

One method for simplifying complex fractions is to multiply the numerator and the denominator of the complex fraction by the LCD of all fractions in its numerator and its denominator.

$$\frac{\dfrac{y}{6} - 11}{\dfrac{4}{3}} = \frac{6\left(\dfrac{y}{6} - 11\right)}{6\left(\dfrac{4}{3}\right)} = \frac{6\left(\dfrac{y}{6}\right) - 6(11)}{6\left(\dfrac{4}{3}\right)}$$

$$= \frac{y - 66}{8}$$

Section 4.7   Operations on Mixed Numbers

To multiply with mixed numbers or whole numbers, first write any mixed or whole numbers as improper fractions and then multiply as usual.

$$2\frac{1}{3} \cdot \frac{1}{9} = \frac{7}{3} \cdot \frac{1}{9} = \frac{7 \cdot 1}{3 \cdot 9} = \frac{7}{27}$$

To divide with mixed numbers or whole numbers, first write any mixed or whole numbers as fractions and then divide as usual.

$$2\frac{5}{8} \div 3\frac{7}{16} = \frac{21}{8} \div \frac{55}{16} = \frac{21}{8} \cdot \frac{16}{55} = \frac{21 \cdot 16}{8 \cdot 55}$$

$$= \frac{21 \cdot 2 \cdot \overset{1}{\cancel{8}}}{\underset{1}{\cancel{8}} \cdot 55} = \frac{42}{55}$$

To add or subtract with mixed numbers, add or subtract the fractions and then add or subtract the whole numbers.

Add: $2\frac{1}{2} + 5\frac{7}{8}$

$$2\frac{1}{2} = 2\frac{4}{8}$$

$$\underline{+5\frac{7}{8} = 5\frac{7}{8}}$$

$$7\frac{11}{8} = 7 + 1\frac{3}{8} = 8\frac{3}{8}$$

Section 4.8   Solving Equations Containing Fractions

TO SOLVE AN EQUATION IN x

Step 1. If fractions are present, multiply both sides of the equation by the LCD of the fractions.

Step 2. If parentheses are present, use the distributive property.

Step 3. Combine any like terms on each side of the equation.

Step 4. Use the addition property of equality to rewrite the equation so that variable terms are on one side of the equation and constant terms are on the other side.

Step 5. Divide both sides by the numerical coefficient of x to solve.

Step 6. Check the answer in the *original equation*.

Solve: $\dfrac{x}{15} + 2 = \dfrac{7}{3}$

$$15\left(\frac{x}{15} + 2\right) = 15\left(\frac{7}{3}\right) \quad \text{Multiply by the LCD 15.}$$

$$15\left(\frac{x}{15}\right) + 15 \cdot 2 = 15\left(\frac{7}{3}\right)$$

$$x + 30 = 35$$

$$x + 30 + (-30) = 35 + (-30)$$

$$x = 5$$

Check to see that 5 is the solution.

Are you Prepared for a Test on Chapter 4?

Below I have listed some *common trouble areas* for students in Chapter 4. After studying for your test—but before taking your test—read these.

Make sure you remember how to perform different operations on fractions!!! Try to add, subtract, multiply, then divide $\frac{3}{5}$ and $\frac{7}{15}$. Check your results below.

$$\frac{3}{5} + \frac{7}{15} = \frac{3}{5} \cdot \frac{3}{3} + \frac{7}{15} = \frac{9}{15} + \frac{7}{15} = \frac{16}{15} \text{ or } 1\frac{1}{15}$$

To add or subtract, the fractions must have a common denominator.

$$\frac{3}{5} - \frac{7}{15} = \frac{3}{5} \cdot \frac{3}{3} - \frac{7}{15} = \frac{9}{15} - \frac{7}{15} = \frac{2}{15}$$

$$\frac{3}{5} \cdot \frac{7}{15} = \frac{3 \cdot 7}{5 \cdot 15} = \frac{\overset{1}{\cancel{3}} \cdot 7}{5 \cdot \underset{1}{\cancel{3}} \cdot 5} = \frac{7}{25}$$

To multiply, multiply numerators and multiply denominators.

$$\frac{3}{5} \div \frac{7}{15} = \frac{3}{5} \cdot \frac{15}{7} = \frac{3 \cdot 15}{5 \cdot 7} = \frac{3 \cdot 3 \cdot \overset{1}{\cancel{5}}}{\underset{1}{\cancel{5}} \cdot 7} = \frac{9}{7} \text{ or } 1\frac{2}{7}$$

To divide, multiply by the reciprocal.

4 CHAPTER REVIEW

(4.1) *Write a fraction to represent the shaded area. If the fraction is improper, write the shaded area as a mixed number, also.*

1.

2.

3.

4.

5. A basketball player made 11 free throws out of 12 during a game. What fraction of free throws did the player make?

6. A new car lot contained 23 blue cars out of a total of 131 cars.
   a. How many cars on the lot are not blue?
   b. What fraction of cars on the lot are not blue?

Simplify by dividing.

7. $\dfrac{3}{-3}$

8. $\dfrac{-20}{-20}$

9. $\dfrac{0}{-1}$

10. $\dfrac{4}{0}$

Graph each fraction on a number line.

11. $\dfrac{7}{9}$

12. $\dfrac{4}{7}$

13. $\dfrac{5}{4}$

14. $\dfrac{7}{5}$

Write each improper fraction as a mixed number or a whole number.

15. $\dfrac{15}{4}$

16. $\dfrac{39}{13}$

Write each mixed number as an improper fraction.

17. $2\dfrac{1}{5}$

18. $3\dfrac{8}{9}$

(4.2) *Write each fraction in simplest form.*

19. $\dfrac{12}{28}$

20. $\dfrac{15}{27}$

21. $-\dfrac{25x}{75x^2}$

22. $-\dfrac{36y^3}{72y}$

23. $\dfrac{29ab}{32abc}$

24. $\dfrac{18xyz}{23xy}$

25. $\dfrac{45x^2y}{27xy^3}$

26. $\dfrac{42ab^2c}{30abc^3}$

27. There are 12 inches in a foot. What fractional part of a foot does 8 inches represent?

28. Six out of 15 cars are white. What fraction of cars are *not* white?

Determine whether each two fractions are equivalent.

29. $\dfrac{10}{34}$ and $\dfrac{4}{14}$

30. $\dfrac{30}{50}$ and $\dfrac{9}{15}$

(4.3) *Multiply.*

31. $\dfrac{3}{5} \cdot \dfrac{1}{2}$

32. $-\dfrac{6}{7} \cdot \dfrac{5}{12}$

33. $-\dfrac{24x}{5} \cdot -\dfrac{15}{8x^3}$

34. $\dfrac{27y^3}{21} \cdot \dfrac{7}{18y^2}$

35. $\left(-\dfrac{1}{3}\right)^3$

36. $\left(-\dfrac{5}{12}\right)^2$

Divide.

37. $-\dfrac{3}{4} \div \dfrac{3}{8}$

38. $\dfrac{21a}{4} \div \dfrac{7a}{5}$

39. $-\dfrac{9}{2} \div -\dfrac{1}{3}$

40. $-\dfrac{5}{3} \div 2y$

41. Evaluate $x \div y$ if $x = \dfrac{9}{7}$ and $y = \dfrac{3}{4}$.

42. Evaluate ab if $a = -7$ and $b = \dfrac{9}{10}$.

Find the area of each figure.

△ **43.**

Rectangle $\frac{7}{8}$ feet

$\frac{11}{6}$ feet

△ **44.**

Square $\frac{2}{3}$ meter

(4.4) *Add or subtract as indicated.*

45. $\dfrac{7}{11} + \dfrac{3}{11}$

46. $\dfrac{4}{9} + \dfrac{2}{9}$

47. $\dfrac{1}{12} - \dfrac{5}{12}$

48. $\dfrac{11x}{15} + \dfrac{x}{15}$

49. $\dfrac{4y}{21} - \dfrac{3}{21}$

50. $\dfrac{4}{15} - \dfrac{3}{15} - \dfrac{2}{15}$

Find the LCD of each list of fractions.

51. $\dfrac{2}{3}, \dfrac{5}{x}$

52. $\dfrac{3}{4}, \dfrac{3}{8}, \dfrac{7}{12}$

Write each fraction as an equivalent fraction with the given denominator.

53. $\dfrac{2}{3} = \dfrac{?}{30}$

54. $\dfrac{5}{8} = \dfrac{?}{56}$

55. $\dfrac{7a}{6} = \dfrac{?}{42}$

56. $\dfrac{9b}{4} = \dfrac{?}{20}$

57. $\dfrac{4}{5x} = \dfrac{?}{50x}$

58. $\dfrac{5}{9y} = \dfrac{?}{18y}$

Solve.

59. One evening Mark Alorenzo did $\dfrac{3}{8}$ of his homework before supper, another $\dfrac{2}{8}$ of it while his children did their homework, and $\dfrac{1}{8}$ after his children went to bed. What part of his homework did he do that evening?

△ **60.** The Simpsons will be fencing in their land, which is in the shape of a rectangle. In order to do this, they need to find its perimeter. Find the perimeter of their land.

$\frac{3}{16}$ mile

$\frac{9}{16}$ mile

(4.5) *Add or subtract as indicated.*

61. $\dfrac{7}{18} + \dfrac{2}{9}$

62. $\dfrac{4}{13} - \dfrac{1}{26}$

63. $-\dfrac{1}{3} + \dfrac{1}{4}$

64. $-\dfrac{2}{3} + \dfrac{1}{4}$

65. $\dfrac{5x}{11} + \dfrac{2}{55}$

66. $\dfrac{4}{15} + \dfrac{b}{5}$

67. $\dfrac{5y}{12} - \dfrac{2y}{9}$

68. $\dfrac{7x}{18} + \dfrac{2x}{9}$

69. $\dfrac{4}{9} + \dfrac{5}{y}$

70. $-\dfrac{9}{14} - \dfrac{3}{7}$

71. $\dfrac{4}{25} + \dfrac{23}{75} + \dfrac{7}{50}$

72. $\dfrac{2}{3} - \dfrac{2}{9} - \dfrac{1}{6}$

Find the perimeter of each figure.

△ **73.**

$\dfrac{2}{9}$ meter   Rectangle

$\dfrac{5}{6}$ meter

△ **74.**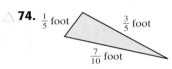

$\dfrac{1}{5}$ foot $\dfrac{3}{5}$ foot

$\dfrac{7}{10}$ foot

75. In a group of 100 blood donors, typically $\dfrac{9}{25}$ have type A Rh-positive blood and $\dfrac{3}{50}$ have type A Rh-negative blood. What fraction have type A blood?

76. Find the difference in length of two scarves if one scarf is $\dfrac{5}{12}$ of a yard long and the other is $\dfrac{2}{3}$ of a yard long.

$\dfrac{2}{3}$ of a yard

$\dfrac{5}{12}$ of a yard

(4.6) *Simplify each complex fraction.*

77. $\dfrac{\dfrac{2x}{5}}{\dfrac{7}{10}}$

78. $\dfrac{\dfrac{3y}{7}}{\dfrac{11}{7}}$

79. $\dfrac{\dfrac{2}{5} - \dfrac{1}{2}}{\dfrac{3}{4} - \dfrac{7}{10}}$

80. $\dfrac{\dfrac{5}{6} - \dfrac{1}{4}}{\dfrac{-1}{12y}}$

Evaluate each expression if $x = \dfrac{1}{2}$, $y = -\dfrac{2}{3}$, and $z = \dfrac{4}{5}$.

81. $\dfrac{x}{y + z}$

82. $\dfrac{x + y}{z}$

Evaluate each expression. Use the order of operations to simplify.

83. $\dfrac{5}{13} \div \dfrac{1}{2} \cdot \dfrac{4}{5}$

84. $\dfrac{2}{27} - \left(\dfrac{1}{3}\right)^2$

85. $\dfrac{9}{10} \cdot \dfrac{1}{3} - \dfrac{2}{5} \cdot \dfrac{1}{11}$

86. $-\dfrac{2}{7} \cdot \left(\dfrac{1}{5} + \dfrac{3}{10}\right)$

(4.7) *Perform operations as indicated. Simplify your answers. Estimate where noted.*

87.
$$7\frac{3}{8}$$
$$9\frac{5}{6}$$
$$+3\frac{1}{12}$$
_____

88.
$$8\frac{1}{5}$$
$$-5\frac{3}{11}$$
_____

Exact:

Estimate:

89. $1\frac{5}{8} \cdot 3\frac{1}{5}$

Exact:

Estimate:

90. $6\frac{3}{4} \div 1\frac{2}{7}$

91. A truck traveled 341 miles on $15\frac{1}{2}$ gallons of gas. How many miles might we expect the truck to travel on 1 gallon of gas?

92. There are $7\frac{1}{3}$ grams of fat in each ounce of hamburger. How many grams of fat are in a 5-ounce hamburger patty?

Find the unknown measurements.

△ **93.**

△ **94.**

Perform the indicated operations.

95. $-12\frac{1}{7} + \left(-15\frac{3}{14}\right)$

96. $-3\frac{1}{5} \div \left(-2\frac{7}{10}\right)$

97. $-2\frac{1}{4} \cdot 1\frac{3}{4}$

98. $23\frac{7}{8} - 24\frac{7}{10}$

(4.8) *Solve each equation.*

99. $a - \frac{2}{3} = \frac{1}{6}$

100. $9x + \frac{1}{5} - 8x = -\frac{7}{10}$

101. $-\frac{3}{5}x = 6$

102. $\frac{2}{9}y = -\frac{4}{3}$

103. $\frac{x}{7} - 3 = -\frac{6}{7}$

104. $\frac{y}{5} + 2 = \frac{11}{5}$

105. $\frac{1}{6} + \frac{x}{4} = \frac{17}{12}$

106. $\frac{x}{5} - \frac{5}{4} = \frac{x}{2} - \frac{1}{20}$

Mixed Review

Perform indicated operations. Write each answer in simplest form. Estimate where noted.

107. $\dfrac{6}{15} \cdot \dfrac{5}{8}$

108. $\dfrac{5x^2}{y} \div \dfrac{10x^3}{y^3}$

109. $\dfrac{3}{10} - \dfrac{1}{10}$

110. $\dfrac{7}{8x} \cdot -\dfrac{2}{3}$

111. $\dfrac{2x}{3} + \dfrac{x}{4}$

112. $-\dfrac{5}{11} + \dfrac{2}{55}$

113. $-1\dfrac{3}{5} \div \dfrac{1}{4}$

114. $\begin{array}{r} 2\dfrac{7}{8} \\ +9\dfrac{1}{2} \\ \hline \end{array}$ Exact:
   Estimate:

115. $\begin{array}{r} 12\dfrac{1}{7} \\ -9\dfrac{3}{5} \\ \hline \end{array}$ Exact:
   Estimate:

116. Simplify: $\dfrac{2 + \dfrac{3}{4}}{1 - \dfrac{1}{8}}$

Solve.

117. Evaluate: $-\dfrac{3}{8} \cdot \left(\dfrac{2}{3} - \dfrac{4}{9} \right)$

118. $11x - \dfrac{2}{7} - 10x = -\dfrac{13}{14}$

119. $-\dfrac{3}{5}x = \dfrac{4}{15}$

120. $\dfrac{x}{12} + \dfrac{5}{6} = -\dfrac{3}{4}$

Solve.

121. A ribbon $5\dfrac{1}{2}$ yards long is cut from a reel of ribbon with 50 yards on it. Find the length of the piece remaining on the reel.

△ **122.** A slab of natural granite is purchased and a rectangle with length $5\dfrac{1}{2}$ feet and width $7\dfrac{4}{11}$ feet is cut from it. Find the area of the rectangle.

$7\dfrac{4}{11}$ feet

$5\dfrac{1}{2}$ feet

4 CHAPTER TEST

 Remember to use the Chapter Test Prep Video CD to see the fully worked-out solutions to any of the exercises you want to review.

Write a fraction to represent the shaded area.

1.

Write the mixed number as an improper fraction.

2. $7\frac{2}{3}$

Write the improper fraction as a mixed number.

3. $\frac{75}{4}$

Write each fraction in simplest form.

4. $\frac{24}{210}$

5. $-\frac{42x}{70}$

Determine whether these fractions are equivalent.

6. $\frac{5}{7}$ and $\frac{8}{11}$

7. $\frac{6}{27}$ and $\frac{14}{63}$

Find the prime factorization of each number.

8. 84

9. 495

Perform each indicated operation and write the answers in simplest form.

10. $\frac{4}{4} \div \frac{3}{4}$

11. $-\frac{4}{3} \cdot \frac{4}{4}$

12. $\frac{7x}{9} + \frac{x}{9}$

13. $\frac{1}{7} - \frac{3}{x}$

14. $\frac{xy^3}{z} \cdot \frac{z}{xy}$

15. $-\frac{2}{3} \cdot -\frac{8}{15}$

16. $\frac{9a}{10} + \frac{2}{5}$

17. $-\frac{8}{15y} - \frac{2}{15y}$

18. $\frac{3a}{8} \cdot \frac{16}{6a^3}$

19. $\frac{11}{12} - \frac{3}{8} + \frac{5}{24}$

20. $3\frac{7}{8}$
$7\frac{2}{5}$
$+2\frac{3}{4}$

21. 19
$-2\frac{3}{11}$

Answers

1. _____

2. _____

3. _____

4. _____

5. _____

6. _____

7. _____

8. _____

9. _____

10. _____

11. _____

12. _____

13. _____

14. _____

15. _____

16. _____

17. _____

18. _____

19. _____

20. _____

21. _____

22. _____

23. _____

24. _____

25. _____

26. _____

27. _____

28. _____

29. _____

30. _____

31. _____

32. _____

33. _____

34. _____

35. _____

36. _____

37. _____

38. _____

39. _____

40. _____

22. $-\dfrac{16}{3} \div -\dfrac{3}{12}$

23. $3\dfrac{1}{3} \cdot 6\dfrac{3}{4}$

24. $-\dfrac{2}{7} \cdot \left(6 - \dfrac{1}{6}\right)$

25. $\dfrac{1}{2} \div \dfrac{2}{3} \cdot \dfrac{3}{4}$

26. $\left(-\dfrac{3}{4}\right)^2 \div \left(\dfrac{2}{3} + \dfrac{5}{6}\right)$

27. Find the average of $\dfrac{5}{6}, \dfrac{4}{3},$ and $\dfrac{7}{12}.$

Simplify each complex fraction.

28. $\dfrac{\dfrac{5x}{7}}{\dfrac{20x^2}{21}}$

29. $\dfrac{5 + \dfrac{3}{7}}{2 - \dfrac{1}{2}}$

Solve.

30. $-\dfrac{3}{8}x = \dfrac{3}{4}$

31. $\dfrac{x}{5} + x = -\dfrac{24}{5}$

32. $\dfrac{2}{3} + \dfrac{x}{4} = \dfrac{5}{12} + \dfrac{x}{2}$

Evaluate each expression for the given replacement values.

33. $-5x; \ x = -\dfrac{1}{2}$

34. $x \div y; \ x = \dfrac{1}{2}, \ y = 3\dfrac{7}{8}$

Solve.

35. A carpenter cuts a piece $2\dfrac{3}{4}$ feet long from a cedar plank that is $6\dfrac{1}{2}$ feet long. How long is the remaining piece?

The circle graph below shows us how the average consumer spends money. For example, $\dfrac{7}{50}$ of your spending goes for food. Use this information for Exercises 36 through 38.

Consumer Spending

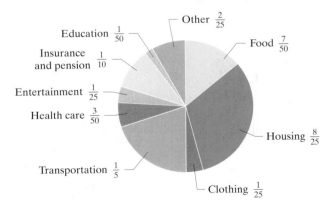

Other $\frac{2}{25}$

Education $\frac{1}{50}$

Insurance and pension $\frac{1}{10}$

Entertainment $\frac{1}{25}$

Health care $\frac{3}{50}$

Transportation $\frac{1}{5}$

Clothing $\frac{1}{25}$

Food $\frac{7}{50}$

Housing $\frac{8}{25}$

Source: U.S. Bureau of Labor Statistics; based on survey

36. What fraction of spending goes for housing and food combined?

37. What fraction of spending goes for education, transportation, and clothing?

38. Suppose your family spent $47,000 on the items in the graph. How much might we expect was spent on health care?

Find the perimeter and area of the figure.

△ **39.**

Rectangle　$\frac{2}{3}$ foot

1 foot

40. During a 258-mile trip, a car used $10\dfrac{3}{4}$ gallons of gas. How many miles would we expect the car to travel on 1 gallon of gas?

Write each number in words.

1. 546

2. 115

3. 27,034

4. 6573

5. Add: 46 + 713

6. Add: 587 + 44

7. Subtract: 543 − 29. Check by adding.

8. Subtract: 995 − 62. Check by adding.

9. Round 278,362 to the nearest thousand.

10. Round 1436 to the nearest ten.

11. A digital video disc (DVD) can hold about 4800 megabytes (MB) of information. How many megabytes can 12 DVDs hold?

12. On a trip across country, Daniel Daunis travels 435 miles per day. How many total miles does he travel in 3 days?

13. Divide and check: 56,717 ÷ 8

14. Divide and check: 4558 ÷ 12

Write using exponential notation.

15. $7 \cdot 7 \cdot 7$

16. $7 \cdot 7$

17. $3 \cdot 3 \cdot 3 \cdot 3 \cdot 9 \cdot 9 \cdot 9$

18. $9 \cdot 9 \cdot 9 \cdot 9 \cdot 5 \cdot 5$

19. Evaluate $2(x - y)$ if $x = 6$ and $y = 3$.

20. Evaluate $8a + 3(b - 5)$ if $a = 5$ and $b = 9$.

21. The world's deepest cave is Krubera (or Voronja) in the country of Georgia, located by the Black Sea in Asia. It has been explored to a depth of 6824 feet below the surface of the Earth. Represent this position using an integer.
(*Source: Guinness World Records* 2006)

22. The temperature on a cold day in Minneapolis, MN, is 21°F below zero. Represent this temperature using an integer.

Answers

1. _____

2. _____

3. _____

4. _____

5. _____

6. _____

7. _____

8. _____

9. _____

10. _____

11. _____

12. _____

13. _____

14. _____

15. _____

16. _____

17. _____

18. _____

19. _____

20. _____

21. _____

22. _____

23. _____

24. _____

25. _____

26. _____

27. _____

28. _____

29. _____

30. _____

31. _____

32. _____

33. _____

34. _____

35. _____

36. _____

37. _____

38. _____

39. a.   b. _____

40. _____

41. _____

42. _____

43. _____

44. _____

45. _____

46. _____

23. Add using a number line: $-7 + 3$

24. Add using a number line: $-3 + 8$

25. Simplify: $7 - 8 - (-5) - 1$

26. Simplify: $6 + (-8) - (-9) + 3$

27. Evaluate: $(-5)^2$

28. Evaluate: -2^4

29. Simplify: $3(4 - 7) + (-2) - 5$

30. Simplify: $(20 - 5^2)^2$

31. Simplify: $2y - 6 + 4y + 8$

32. Simplify: $5x - 1 + x + 10$

Solve.

33. $5x + 2 - 4x = 7 - 19$

34. $9y + 1 - 8y = 3 - 20$

35. $17 - 7x + 3 = -3x + 21 - 3x$

36. $9x - 2 = 7x - 24$

37. Write a fraction to represent the shaded part of the figure.

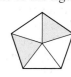

38. Write the prime factorization of 156.

39. Write each as an improper fraction.
   a. $4\dfrac{2}{9}$ **b.** $1\dfrac{8}{11}$

40. Write $7\dfrac{4}{5}$ as an improper fraction.

41. Write in simplest form: $\dfrac{42x}{66}$

42. Write in simplest form: $\dfrac{70}{105y}$

43. Multiply: $3\dfrac{1}{3} \cdot \dfrac{7}{8}$

44. Multiply: $\dfrac{2}{3} \cdot 4$

45. Divide and simplify: $\dfrac{5}{16} \div \dfrac{3}{4}$

46. Divide: $1\dfrac{1}{10} \div 5\dfrac{3}{5}$

5

Decimals

Decimal numbers represent parts of a whole, just like fractions. For example, one penny is 0.01 or $\frac{1}{100}$ of a dollar. In this chapter, we learn to perform arithmetic operations on decimals and to analyze the relationship between fractions and decimals. We also learn how decimals are used in the real world.

A twinkling star that we see from Earth is actually steady starlight that has been bent by traveling through Earth's atmosphere. Telescopes here on Earth, which also require visual focusing through our atmosphere, reveal these same distortions. That's why the Hubble Space Telescope (HST) was conceived and built, to detect light from objects in space before it is absorbed or distorted. The HST was launched in 1990 from the space shuttle *Discovery*, and it orbits at a height of 380 miles above Earth's surface. (*Source:* NASA)

In Section 5.3, Exercise 41, and Section 5.5, Exercises 37 and 38, we examine some facts about the Hubble Space Telescope.

Door   Light shield

Antennas (2)

Telescope truss

Secondary mirror assembly

STIS

Primary mirror

Advanced Camera for Surveys (ACS)

Rigid solar arrays (2)

Fixed-head star trackers (3)

Source: National Aeronautics and Space Administration

Length

13.3 meters or 43.5 feet   (It is nearly the size of a large school bus.)

Diameter:

4.2 meters or 14 feet

Orbit:

28.5 degrees from the equator

Primary Mirror Diameter:

2.4 meters or 94.5 inches

Secondary Mirror Diameter:

0.3 meters or 12 inches

Cost at Launch:

$1.5 billion

A Know the Meaning of Place Value for a Decimal Number and Write Decimals in Words.

B Write Decimals in Standard Form.

C Write Decimals as Fractions.

D Compare Decimals.

E Round Decimals to a Given Place Value.

5.1 INTRODUCTION TO DECIMALS

Objective **A** Decimal Notation and Writing Decimals in Words

Like fractional notation, decimal notation is used to denote a part of a whole. Numbers written in decimal notation are called **decimal numbers,** or simply **decimals.** The decimal 17.758 has three parts.

In Section 1.2, we introduced place value for whole numbers. Place names and place values for the whole number part of a decimal number are exactly the same. Place names and place values for the decimal part are shown below.

> **Helpful Hint**
> Notice that place values to the left of the decimal point end in "s." Place values to the right of the decimal point end in "ths."

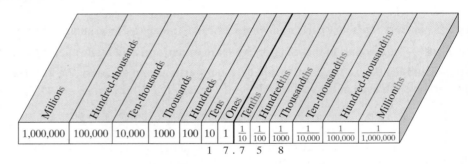

Notice that the value of each place is $\frac{1}{10}$ of the value of the place to its left. For example,

$$1 \cdot \frac{1}{10} = \frac{1}{10} \quad \text{and} \quad \frac{1}{10} \cdot \frac{1}{10} = \frac{1}{100}$$

ones   tenths   tenths   hundredths

The decimal number 17.758 means

1 ten	+	7 ones	+	7 tenths	+	5 hundredths	+	8 thousandths
or 1·10	+	7·1	+	$7 \cdot \frac{1}{10}$	+	$5 \cdot \frac{1}{100}$	+	$8 \cdot \frac{1}{1000}$
or 10	+	7	+	$\frac{7}{10}$	+	$\frac{5}{100}$	+	$\frac{8}{1000}$

Writing (or Reading) a Decimal in Words

Step 1: Write the whole number part in words.

Step 2: Write "and" for the decimal point.

Step 3: Write the decimal part in words as though it were a whole number, followed by the place value of the last digit.

EXAMPLE 1 Write each decimal in words.

a. 0.7   **b.** −50.82   **c.** 21.093

Solution:

a. seven tenths

b. negative fifty and eighty-two hundredths

c. twenty-one and ninety-three thousandths

■ **Work Practice Problem 1**

EXAMPLE 2

Write the decimal in the following sentence in words: The Golden Jubilee Diamond is a 545.67-carat cut diamond. (*Source: The Guinness Book of Records*)

Solution: five hundred forty-five and sixty-seven hundredths

■ **Work Practice Problem 2**

EXAMPLE 3

Write the decimal in the following sentence in words: The oldest known fragments of the Earth's crust are Zircon crystals; they were discovered in Australia and are thought to be 4.276 billion years old. (*Source: The Guinness Book of Records*)

Solution: four and two hundred seventy-six thousandths

■ **Work Practice Problem 3**

Suppose that you are paying for a purchase of $368.42 at Circuit City by writing a check. Checks are usually written using the following format.

PRACTICE PROBLEM 1

Write each decimal in words.

a. 0.06

b. −200.073

c. 0.0829

PRACTICE PROBLEM 2

Write the decimal 87.31 in words.

PRACTICE PROBLEM 3

Write the decimal 52.1085 in words.

Answers

1. a. six hundredths   **b.** negative two hundred and seventy-three thousandths   **c.** eight hundred twenty-nine ten-thousandths
2. eighty-seven and thirty-one hundredths   **3.** fifty-two and one thousand eighty-five ten-thousandths

PRACTICE PROBLEM 4

Fill in the check to CLECO (Central Louisiana Electric Company) to pay for your monthly electric bill of $207.40.

EXAMPLE 4 Fill in the check to Camelot Music to pay for your purchase of $92.98.

Solution:

Work Practice Problem 4

Objective B Writing Decimals in Standard Form

A decimal written in words can be written in standard form by reversing the procedure in Objective A.

PRACTICE PROBLEMS 5–6

Write each decimal in standard form.

5. Five hundred and ninety-six hundredths

6. Thirty-nine and forty-two thousandths

EXAMPLES Write each decimal in standard form.

5. Forty-eight and twenty-six hundredths is

48.26

hundredths place

6. Six and ninety-five thousandths is

6.095

thousandths place

Work Practice Problems 5–6

Helpful Hint

When converting a decimal from words to decimal notation, make sure the last digit is in the correct place by inserting 0s if necessary. For example,

Two and thirty-eight thousandths is 2.038

thousandths place

Objective C Writing Decimals as Fractions

Once you master reading and writing decimals, writing a decimal as a fraction follows naturally.

Decimal	In Words	Fraction
0.7	seven tenths	$\frac{7}{10}$
0.51	fifty-one hundredths	$\frac{51}{100}$
0.009	nine thousandths	$\frac{9}{1000}$
0.05	five hundredths	$\frac{5}{100} = \frac{1}{20}$

Answers

4.

5. 500.96 **6.** 39.042

Notice that the number of decimal places in a decimal number is the same as the number of zeros in the denominator of the equivalent fraction. We can use this fact to write decimals as fractions.

$$0.31 = \frac{31}{100} \qquad 0.007 = \frac{7}{1000}$$

2 decimal places ↑ 2 zeros ↑   3 decimal places ↑ 3 zeros ↑

EXAMPLE 7 Write 0.47 as a fraction.

Solution: $0.47 = \dfrac{47}{100}$

2 decimal places ↑ 2 zeros ↑

◻ **Work Practice Problem 7**

EXAMPLE 8 Write 5.9 as a mixed number.

Solution: $5.9 = 5\dfrac{9}{10}$

1 decimal place ↑ 1 zero ↑

◻ **Work Practice Problem 8**

EXAMPLES Write each decimal as a fraction or a mixed number. Write your answer in simplest form.

9. $0.125 = \dfrac{125}{1000} = \dfrac{1}{8}$

10. $43.5 = 43\dfrac{5}{10} = 43\dfrac{\overset{1}{\cancel{5}}}{2\cdot\underset{1}{\cancel{5}}} = 43\dfrac{1}{2\cdot 1} = 43\dfrac{1}{2}$

11. $-105.083 = -105\dfrac{83}{1000}$

◻ **Work Practice Problems 9–11**

Later in the chapter, we write fractions as decimals. If you study Examples 7–11, you already know how to write fractions with denominators of 10, 100, 1000, and so on, as decimals.

Objective D Comparing Decimals

One way to compare positive decimals is by comparing digits in corresponding places. To see why this works, let's compare 0.5 or $\dfrac{5}{10}$ and 0.8 or $\dfrac{8}{10}$. We know

$$\frac{5}{10} < \frac{8}{10} \text{ since } 5 < 8, \text{ so}$$

$$0.5 < 0.8 \text{ since } 5 < 8$$

This leads to the following.

PRACTICE PROBLEM 7
Write 0.051 as a fraction.

PRACTICE PROBLEM 8
Write 29.97 as a mixed number.

PRACTICE PROBLEMS 9–11
Write each decimal as a fraction or mixed number. Write your answer in simplest form.
9. 0.12
10. 64.8
11. −209.986

Answers
7. $\dfrac{51}{1000}$ **8.** $29\dfrac{97}{100}$ **9.** $\dfrac{3}{25}$
10. $64\dfrac{4}{5}$ **11.** $-209\dfrac{493}{500}$

Comparing Two Positive Decimals

Compare digits in the same places from left to right. When two digits are not equal, the number with the larger digit is the larger decimal. If necessary, insert 0s after the last digit to the right of the decimal point to continue comparing.

Compare hundredths place digits

$$28.253 \qquad\qquad 28.263$$
$$\uparrow \qquad\qquad\qquad \uparrow$$
$$5 \quad < \quad 6$$
$$\text{so } 28.253 \quad < \quad 28.263$$

Helpful Hint

For any decimal, writing 0s after the last digit to the right of the decimal point does not change the value of the number.

$$7.6 = 7.60 = 7.600, \text{ and so on}$$

When a whole number is written as a decimal, the decimal point is placed to the right of the ones digit.

$$25 = 25.0 = 25.00, \text{ and so on}$$

PRACTICE PROBLEM 12

Insert $<$, $>$, or $=$ to form a true statement.

26.208   26.28

EXAMPLE 12 Insert $<$, $>$, or $=$ to form a true statement.

0.378   0.368

Solution: 0. 3 78   0. 3 68   The tenths places are the same.

0.3 7 8   0.3 6 8   The hundredths places are different.

Since $7 > 6$, then $0.378 > 0.368$.

☐ **Work Practice Problem 12**

PRACTICE PROBLEM 13

Insert $<$, $>$, or $=$ to form a true statement.

0.12   0.026

EXAMPLE 13 Insert $<$, $>$, or $=$ to form a true statement.

0.052   0.236

Solution: 0. 0 52 $<$ 0. 2 36   0 is smaller than 2 in the tenths place.
          $\uparrow$ $\uparrow$

☐ **Work Practice Problem 13**

We can also use a number line to compare decimals. This is especially helpful when comparing negative decimals. Remember, the number whose graph is to the left is smaller, and the number whose graph is to the right is larger.

$$-1.7 < -1.2 \qquad\qquad 0.5 < 0.8$$

Helpful Hint

If you have trouble comparing two negative decimals, try the following: Compare their absolute values. Then to correctly compare the negative decimals, reverse the direction of the inequality symbol.

$$0.568 < 0.586 \qquad so \qquad -0.568 > -0.586$$

EXAMPLE 14 Insert $<$, $>$, or $=$ to form a true statement.

$-0.0101 \qquad -0.00109$

Solution: Since $0.0101 > 0.00109$, then $-0.0101 \; < \; -0.00109$

Work Practice Problem 14

PRACTICE PROBLEM 14

Insert $<$, $>$, or $=$ to form a true statement.

$-0.039 \qquad -0.0309$

Objective E Rounding Decimals

We **round the decimal part** of a decimal number in nearly the same way as we round whole numbers. The only difference is that we drop digits to the right of the rounding place, instead of replacing these digits with 0s. For example,

36.954 rounded to the nearest hundredth is 36.95.

Rounding Decimals to a Place Value to the Right of the Decimal Point

Step 1: Locate the digit to the right of the given place value.

Step 2: If this digit is 5 or greater, add 1 to the digit in the given place value and drop all digits to its right. If this digit is less than 5, drop all digits to the right of the given place.

EXAMPLE 15 Round 736.2359 to the nearest tenth.

Solution:

Step 1: We locate the digit to the right of the tenths place.

$$736.2\,3\,59$$

tenths place

digit to the right

Step 2: Since this digit to the right is less than 5, we drop it and all digits to its right.

Thus, 736.2359 rounded to the nearest tenth is 736.2.

Work Practice Problem 15

The same steps for rounding can be used when the decimal is negative.

PRACTICE PROBLEM 15

Round 482.7817 to the nearest thousandth.

Answers

14. $-0.039 < -0.0309$ **15.** 482.782

PRACTICE PROBLEM 16

Round −0.032 to the nearest hundredth.

EXAMPLE 16 Round −0.027 to the nearest hundredth.

Solution:

Step 1: Locate the digit to the right of the hundredths place.

hundredths place

−0.02 7

digit to the right

Step 2: Since this digit to the right is 5 or greater, we add 1 to the hundredths digit and drop all digits to its right.

Thus, −0.027 is −0.03 rounded to the nearest hundredth.

▣ Work Practice Problem 16

The following number line illustrates the rounding of negative decimals.

−0.03　　(−0.027)　−0.025　　　　−0.02

In Section 5.3, we will introduce a formula for the distance around a circle. This distance around a circle is given a special name called **circumference.**

The symbol π is the Greek letter pi, pronounced "pie." We use π to denote the following constant:

$$\pi = \frac{\text{circumference of a circle}}{\text{diameter of a circle}}$$

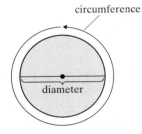

circumference

diameter

The value π is an **irrational number.** This means if we try to write it as a decimal, it neither ends nor repeats in a pattern.

PRACTICE PROBLEM 17

$\pi \approx 3.14159265$. Round π to the nearest ten thousandth.

EXAMPLE 17 $\pi \approx 3.14159265$. Round π to the nearest hundredth.

hundredths place ⎯⎯⎯⎯ 1 is less than 5.

3.14159265

⎯⎯ Delete these digits.

Thus, 3.14159265 rounded to the nearest hundredth is 3.14. In other words, $\pi \approx 3.14$.

▣ Work Practice Problem 17

Answers

16. −0.03　**17.** $\pi \approx 3.1416$

Rounding often occurs with money amounts. Since there are 100 cents in a dollar, each cent is $\frac{1}{100}$ of a dollar. This means that if we want to round to the nearest cent, we round to the nearest hundredth of a dollar.

✔**Concept Check**   1756.0894 rounded to the nearest *ten* is

a. 1756.1   **b.** 1760.0894   **c.** 1760   **d.** 1750

EXAMPLE 18 **Determining State Taxable Income**

A high school teacher's taxable income is $41,567.72. The tax tables in the teacher's state use amounts rounded to the nearest dollar. Round the teacher's income to the nearest whole dollar.

Solution: Rounding to the nearest whole dollar means rounding to the ones place.

ones place ─────┐ ┌───── 7 is greater than 5.

$41,567.72

│ └──── Delete these digits.
Add 1.

Thus, the teacher's income rounded to the nearest dollar is $41,568.

🔲 **Work Practice Problem 18**

PRACTICE PROBLEM 18

Water bills in Mexia are always rounded to the nearest dollar. Round a water bill of $24.62 to the nearest dollar.

Answer

18. $25

✔ **Concept Check Answer**

c

Vocabulary and Readiness Check

Use the choices below to fill in each blank.

words   decimals   tenths   after

tens   circumference   and   standard form

1. The number "twenty and eight hundredths" is written in _____ and "20.08" is written in _____ .
2. Another name for the distance around a circle is its _____ .
3. Like fractions, _____ are used to denote part of a whole.
4. When writing a decimal number in words, the decimal point is written as _____ .
5. The place value _____ is to the right of the decimal point while _____ is to the left of the decimal point.
6. The decimal point in a whole number is _____ the last digit.

5.1 EXERCISE SET

FOR EXTRA HELP

 Student Solutions Manual   PH Math/Tutor Center   CD/Video for Review   Math XL MathXL®   MyMathLab MyMathLab

Objective A *Write each decimal number in words. See Examples 1 through 3.*

1. 5.62
2. 9.57
3. 16.23
4. 47.65
5. −0.205
6. −0.495
7. 167.009
8. 233.056
9. 3000.04
10. 5000.02
11. 105.6
12. 410.3

13. The Akashi Kaikyo Bridge, between Kobe and Awaji-Shima, Japan, is approximately 2.43 miles long.

14. The English Channel Tunnel is 31.04 miles long. (*Source: Railway Directory & Year Book*)

Fill in each check for the described purchase. See Example 4.

15. Your monthly car loan of $321.42 to R. W. Financial.

16. Your part of the monthly apartment rent, which is $213.70. You pay this to Amanda Dupre.

17. Your cell phone bill of $59.68 to Bell South.

Your Preprinted Name Your Preprinted Address	60-8124/7233 1000613331	1409
	DATE	
PAY TO THE ORDER OF		$
		DOLLARS
FIRST STATE BANK OF FARTHINGTON FARTHINGTON, IL 64422		
MEMO		
⑈621497260⑈ 1000613331⑈ 1409		

18. Your grocery bill of $87.49 at Albertsons.

Your Preprinted Name Your Preprinted Address	60-8124/7233 1000613331	1410
	DATE	
PAY TO THE ORDER OF		$
		DOLLARS
FIRST STATE BANK OF FARTHINGTON FARTHINGTON, IL 64422		
MEMO		
⑈621497260⑈ 1000613331⑈ 1410		

Objective B *Write each decimal number in standard form. See Examples 5 and 6.*

19. Two and eight tenths

20. Five and one tenth

21. Nine and eight hundredths

22. Twelve and six hundredths

23. Negative seven hundred five and six hundred twenty-five thousandths

24. Negative eight hundred four and three hundred ninety-nine thousandths

25. Forty-six ten-thousandths

26. Eighty-three ten-thousandths

Objective C *Write each decimal as a fraction or a mixed number. Write your answer in simplest form. See Examples 7 through 11.*

27. 0.7

28. 0.9

29. 0.27

30. 0.39

31. 0.4

32. 0.8

33. 5.4

34. 6.8

35. -0.058

36. -0.024

37. 7.008

38. 9.005

39. 15.802

40. 11.406

41. 0.3005

42. 0.2006

Objectives A B C Mixed Practice *Fill in the chart. The first row is completed for you.*

	Decimal Number in Standard Form	In Words	Fraction
	0.37	thirty-seven hundredths	$\frac{37}{100}$
43.		eight tenths	
44.		five tenths	
45.	0.077		
46.	0.019		

Objective D *Insert* <, >, *or* = *between each pair of numbers to form a true statement. See Examples 12 through 14.*

47. 0.15   0.16

48. 0.12   0.15

49. −0.57   −0.54

50. −0.59   −0.52

51. 0.098   0.1

52. 0.0756   0.2

53. 0.54900   0.549

54. 0.98400   0.984

55. 167.908   167.980

56. 519.3405   519.3054

57. −1.062   −1.07

58. −18.1   −18.01

59. −7.052   7.0052

60. 0.01   −0.1

61. −0.023   −0.024

62. −0.562   −0.652

Objective E *Round each decimal to the given place value. See Examples 15 through 18.*

63. 0.57, nearest tenth

64. 0.64, nearest tenth

65. 98,207.23, nearest ten

66. 68,934.543, nearest ten

67. −0.234, nearest hundredth

68. −0.892, nearest hundredth

69. 0.5942, nearest thousandth

70. 63.4523, nearest thousandth

Recall that the number π, written as a decimal, neither ends nor repeats in a pattern. Given that $\pi \approx 3.14159265$, round π to the given place values below. (We study π further in Section 5.3.) See Example 17.

71. tenth

72. ones

73. thousandth

74. hundred thousandth

Round each monetary amount to the nearest cent or dollar as indicated. See Example 18.

75. $26.95, to the nearest dollar

76. $14,769.52, to the nearest dollar

77. $0.1992, to the nearest cent

78. $0.7633, to the nearest cent

Round each number to the given place value. See Example 18.

79. At this writing, the smallest personal computer, developed by OQO in the United States, measures 10.414 centimeters across. Round this number to the nearest tenth. (*Source: Guinness World Records 2006*)

80. A large tropical cockroach of the family Dictyoptera is the fastest-moving insect. This insect was clocked at a speed of 3.36 miles per hour. Round this number to the nearest tenth. (*Source:* University of California, Berkeley)

81. During the 2006 Boston Marathon, Edith Hunkeler of Switzerland was the winner in the women's wheelchair division, with a finishing time of 1.7283 hours. Round this time to the nearest hundredth of an hour. (*Source:* Boston Athletic Association)

82. The population density of the state of Louisiana is 102.5794 people per square mile. Round this population density to the nearest tenth. (*Source:* U.S. Census Bureau)

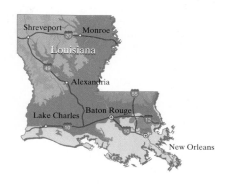

83. A used biology textbook is priced at $47.89. Round this price to the nearest dollar.

84. A used office desk is advertised at $19.95 by Drawley's Office Furniture. Round this price to the nearest dollar.

85. The length of a day on Mars, a full rotation about its axis, is 24.6229 hours. Round this figure to the nearest thousandth. (*Source:* National Space Science Data Center)

86. Venus makes a complete orbit around the Sun every 224.695 days. Round this figure to the nearest whole day. (*Source:* National Space Science Data Center)

Review

Perform each indicated operation. See Section 1.3.

87. $3452 + 2314$

88. $8945 + 4536$

89. $82 - 47$

90. $4002 - 3897$

Concept Extensions

Solve. See the Concept Check in this section.

91. 2849.1738 rounded to the nearest hundred is
  a. 2849.17   **c.** 2850
  b. 2800   **d.** 2849.174

92. 146.059 rounded to the nearest ten is
  a. 146.0   **c.** 140
  b. 146.1   **d.** 150

93. 2849.1738 rounded to the nearest hundredth is
  a. 2849.17   **c.** 2850
  b. 2800   **d.** 2849.18

94. 146.059 rounded to the nearest tenth is
  a. 146.0   **c.** 140
  b. 146.1   **d.** 150

95. In your own words, describe how to write a decimal as a fraction or a mixed number.

96. Explain how to identify the value of the 9 in the decimal 486.3297.

97. Write $7\dfrac{12}{100}$ as a decimal.

98. Write $17\dfrac{268}{1000}$ as a decimal.

99. Write 0.00026849576 as a fraction.

100. Write 0.00026849576 in words.

101. Write a 5-digit number that rounds to 1.7.

102. Write a 4-digit number that rounds to 26.3.

103. Write a decimal number that is greater than 48.1, but less than 48.2.

104. Which number(s) rounds to 0.26?
0.26559   0.26499   0.25786   0.25186

105. Which number(s) rounds to 0.06?
0.0612   0.066   0.0586   0.0506

Write these numbers from smallest to largest.

106. 0.9
0.1038
0.10299
0.1037

107. 0.01
0.0839
0.09
0.1

108. The all-time top six movies (those that earned the most money in the United States) along with the approximate amount of money they have earned are listed in the table. Estimate the total amount of money that these movies have earned by first rounding each earning to the nearest hundred million. (*Source:* Movie Web)

Top All-Time American Movies	
Movie	**Gross Domestic Earnings**
Titanic (1997)	$600.8 million
Star Wars: A New Hope (1977)	$461.0 million
Shrek 2 (2004)	$437.2 million
E.T. (1982)	$433.0 million
Star Wars: The Phantom Menace (1999)	$431.1 million
Pirates of the Caribbean: Dead Man's Chest (2006)	$421.4 million

109. In 2005, American manufacturers shipped approximately 33.8 million music videos to retailers. The value of these shipments was approximately $602.2 million. Estimate the value of an individual music video by rounding 602.2 and 33.8 to the nearest ten, then dividing. (*Source:* Recording Industry Association of America)

 STUDY SKILLS BUILDER

Are You Getting All the Mathematics Help That You Need?

Remember that, in addition to your instructor, there are many places to get help with your mathematics course. For example,

- This text has an accompanying video lesson for every section and worked out solutions to every Chapter Test exercise on video.
- The back of the book contains answers to odd-numbered exercises and selected solutions.
- A *Student Solutions Manual* is available that contains worked-out solutions to odd-numbered exercises as well as solutions to every exercise in the Integrated Reviews, Chapter Reviews, Chapter Tests, and Cumulative Reviews.

- Don't forget to check with your instructor for other local resources available to you, such as a tutor center.

Exercises

1. List items you find helpful in the text and all student supplements to this text.
2. List all the campus help that is available to you for this course.
3. List any help (besides the textbook) from Exercises 1 and 2 above that you are using.
4. List any help (besides the textbook) that you feel you should try.
5. Write a goal for yourself that includes trying anything you listed in Exercise 4 during the next week.

5.2 ADDING AND SUBTRACTING DECIMALS

Objective A Adding or Subtracting Decimals

Adding or subtracting decimals is similar to adding or subtracting whole numbers. We add or subtract digits in corresponding place values from right to left, carrying or borrowing if necessary. To make sure that digits in corresponding place values are added or subtracted, we line up the decimal points vertically.

Adding or Subtracting Decimals

Step 1: Write the decimals so that the decimal points line up vertically.

Step 2: Add or subtract as with whole numbers.

Step 3: Place the decimal point in the sum or difference so that it lines up vertically with the decimal points in the problem.

In this section, we will insert zeros in decimals numbers so that place value digits line up neatly. For instance, see Example 1.

EXAMPLE 1 Add: $23.85 + 1.604$

Solution: First we line up the decimal points vertically.

$$
\begin{array}{r}
23.85\textcolor{gray}{0} \quad \text{\small Insert one 0 so that digits line up neatly.}\\
+\ \ 1.604 \\
\uparrow \\
\text{\small line up decimal points}
\end{array}
$$

Then we add the digits from right to left as for whole numbers.

$$
\begin{array}{r}
\overset{1}{}23.850 \\
+\ \ 1.604 \\
\hline
25.454
\end{array}
$$

— Place the decimal point in the sum so that all decimal points line up.

■ **Work Practice Problem 1**

> **Helpful Hint**
>
> Recall that 0s may be placed after the last digit to the right of the decimal point without changing the value of the decimal. This may be used to help line up place values when adding decimals.
>
> | | | |
> |---|---|---|
> | 3.2 | becomes | $3.2\textcolor{gray}{00}$ Insert two 0s. |
> | 15.567 | | 15.567 |
> | $+\ 0.11$ | | $+\ 0.11\textcolor{gray}{0}$ Insert one 0. |
> | | | 18.877 Add. |

Objectives

A Add or Subtract Decimals.

B Estimate when Adding or Subtracting Decimals.

C Evaluate Expressions with Decimal Replacement Values.

D Simplify Expressions Containing Decimals.

E Solve Problems That Involve Adding or Subtracting Decimals.

PRACTICE PROBLEM 1

Add.

a. $19.52 + 5.371$

b. $40.08 + 17.612$

c. $0.125 + 422.8$

Answers

1. a. 24.891 **b.** 57.692 **c.** 422.925

PRACTICE PROBLEM 2

Add.

a. $34.567 + 129.43 + 2.8903$

b. $11.21 + 46.013 + 362.526$

EXAMPLE 2 Add: $763.7651 + 22.001 + 43.89$

Solution: First we line up the decimal points.

$$
\begin{array}{r}
\overset{1\ \ 1\ \ 1}{763.7651} \\
22.0010 \quad \text{Insert one 0.} \\
+\ \ 43.8900 \quad \text{Insert two 0s.} \\
\hline
829.6561 \quad \text{Add.}
\end{array}
$$

■ **Work Practice Problem 2**

Helpful Hint

Don't forget that the decimal point in a whole number is positioned after the last digit.

EXAMPLE 3 Add: $45 + 2.06$

PRACTICE PROBLEM 3

Add: $19 + 26.072$

Solution:

$$
\begin{array}{r}
45.00 \quad \text{Insert a decimal point and two 0s.} \\
+\ \ 2.06 \quad \text{Line up decimal points.} \\
\hline
47.06 \quad \text{Add.}
\end{array}
$$

■ **Work Practice Problem 3**

✔ **Concept Check** What is wrong with the following calculation of the sum of 7.03, 2.008, 19.16, and 3.1415?

$$
\begin{array}{r}
7.03 \\
2.008 \\
19.16 \\
+3.1415 \\
\hline
3.6042
\end{array}
$$

PRACTICE PROBLEM 4

Add: $7.12 + (-9.92)$

EXAMPLE 4 Add: $3.62 + (-4.78)$

Solution: Recall from Chapter 2 that to add two numbers with different signs we find the difference of the larger absolute value and the smaller absolute value. The sign of the answer is the same as the sign of the number with the larger absolute value.

$$
\begin{array}{r}
4.78 \\
-3.62 \\
\hline
1.16 \quad \text{Subtract the absolute values.}
\end{array}
$$

Thus, $3.62 + (-4.78) = -1.16$

The sign of the number with the larger absolute value; -4.78 has the larger absolute value.

■ **Work Practice Problem 4**

Subtracting decimals is similar to subtracting whole numbers. We line up digits and subtract from right to left, borrowing when needed.

Answers

2. a. 166.8873 **b.** 419.749

3. 45.072 **4.** -2.8

✔ **Concept Check Answer**

The decimal places are not lined up properly.

EXAMPLE 5 Subtract: $3.5 - 0.068$. Check your answer.

Solution:
$$\begin{array}{r} \overset{9}{\overset{4\ \cancel{10}10}{3.\cancel{5}\cancel{0}\cancel{0}}} \\ -\,0.0\,6\,8 \\ \hline 3.4\,3\,2 \end{array}$$ Insert two 0s.
Line up decimal points.
Subtract.

Check: Recall that we can check a subtraction problem by adding.

$$\begin{array}{r} 3.432 \\ +\,0.068 \\ \hline 3.500 \end{array}$$ Difference
Subtrahend
Minuend

■ **Work Practice Problem 5**

EXAMPLE 6 Subtract: $85 - 17.31$. Check your answer.

Solution:
$$\begin{array}{r} \overset{7\ 14\ \overset{9}{\cancel{10}}10}{\cancel{8}\cancel{5}.\cancel{0}\cancel{0}} \\ -\,1\,7.3\,1 \\ \hline 6\,7.6\,9 \end{array}$$

Check:
$$\begin{array}{r} 67.69 \\ +\,17.31 \\ \hline 85.00 \end{array}$$ Difference
Subtrahend
Minuend

■ **Work Practice Problem 6**

EXAMPLE 7 Subtract 3 from 6.98.

Solution:
$$\begin{array}{r} 6.98 \\ -\,3.00 \\ \hline 3.98 \end{array}$$ Insert two 0s.

Check:
$$\begin{array}{r} 3.98 \\ +\,3.00 \\ \hline 6.98 \end{array}$$ Difference
Subtrahend
Minuend

■ **Work Practice Problem 7**

EXAMPLE 8 Subtract: $-5.8 - 1.7$

Solution: Recall from Chapter 2 that to subtract 1.7 we add the opposite of 1.7, or -1.7. Thus

$$-5.8 - 1.7 = -5.8 + (-1.7)$$ To subtract, add the opposite of 1.7 which is -1.7.

Add the absolute values.
$$= -7.5.$$
Use the common negative sign.

■ **Work Practice Problem 8**

EXAMPLE 9 Subtract: $-2.56 - (-4.01)$

Solution: $-2.56 - (-4.01) = -2.56 + 4.01$ To subtract, add the opposite of -4.01, which is 4.01.

Add the absolute values.
$$= 1.45$$
The answer is positive since 4.01 has the larger absolute value.

■ **Work Practice Problem 9**

PRACTICE PROBLEM 5

Subtract. Check your answers.
a. $6.7 - 3.92$
b. $9.72 - 4.068$

PRACTICE PROBLEM 6

Subtract. Check your answers.
a. $73 - 29.31$
b. $210 - 68.22$

PRACTICE PROBLEM 7

Subtract: 19 from 25.91

PRACTICE PROBLEM 8

Subtract: $-5.4 - 9.6$

PRACTICE PROBLEM 9

Subtract: $-1.05 - (-7.23)$

Answers
5. a. 2.78 **b.** 5.652
6. a. 43.69 **b.** 141.78
7. 6.91 **8.** -15 **9.** 6.18

Objective B Estimating when Adding or Subtracting Decimals

To help avoid errors, we can also estimate to see if our answer is reasonable when adding or subtracting decimals. Although only one estimate is needed per operation, we show two for variety.

EXAMPLE 10 Add or subtract as indicated. Then estimate to see if the answer is reasonable by rounding the given numbers and adding or subtracting the rounded numbers.

a. $27.6 + 519.25$

Exact		Estimate 1		Estimate 2
$\overset{1}{27.60}$	rounds to	30		30
$+519.25$	rounds to	$+500$	or	$+520$
546.85		530		550

Since the exact answer is close to either estimate, it is reasonable. (In the first estimate, each number is rounded to the place value of the leftmost digit. In the second estimate, each number is rounded to the nearest ten.)

b. $11.01 - 0.862$

Exact		Estimate 1		Estimate 2
$1\!\!\!/1.\!\!\!/0\!\!\!/1\!\!\!/0$	rounds to	10		11
-0.862	rounds to	-1	or	-1
10.148		9		10

In the first estimate, we rounded the first number to the nearest ten and the second number to the nearest one. In the second estimate, we rounded both numbers to the nearest one. Both estimates show us that our answer is reasonable.

■ **Work Practice Problem 10**

Helpful Hint Remember that estimates are used for our convenience to quickly check the reasonableness of an answer.

✔ **Concept Check** Why shouldn't the sum $21.98 + 42.36$ be estimated as $30 + 50 = 80$?

Objective C Using Decimals as Replacement Values

Let's review evaluating expressions with given replacement values. This time the replacement values are decimals.

EXAMPLE 11 Evaluate $x - y$ for $x = 2.8$ and $y = 0.92$.

Solution: Replace x with 2.8 and y with 0.92 and simplify.

$$x - y = 2.8 - 0.92$$
$$= 1.88$$

2.80
-0.92
1.88

■ **Work Practice Problem 11**

EXAMPLE 12 Is 2.3 a solution of the equation $6.3 = x + 4$?

Solution: Replace x with 2.3 in the equation $6.3 = x + 4$ to see if the result is a true statement.

$6.3 = x + 4$
$6.3 \overset{?}{=} 2.3 + 4$ Replace x with 2.3.
$6.3 = 6.3$ True

Since $6.3 = 6.3$ is a true statement, 2.3 is a solution of $6.3 = x + 4$.

▢ **Work Practice Problem 12**

PRACTICE PROBLEM 12

Is 12.1 a solution of the equation $y - 4.3 = 7.8$?

Objective D Simplify Expressions Containing Decimals

EXAMPLE 13 Simplify by combining like terms:

$11.1x - 6.3 + 8.9x - 4.6$

Solution: $11.1x - 6.3 + 8.9x - 4.6 = 11.1x + 8.9x + (-6.3) + (-4.6)$
$= 20x + (-10.9)$
$= 20x - 10.9$

▢ **Work Practice Problem 13**

PRACTICE PROBLEM 13

Simplify by combining like terms:
$-4.3y + 7.8 - 20.1y + 14.6$

Objective E Solving Problems by Adding or Subtracting Decimals

Decimals are very common in real-life problems.

EXAMPLE 14 **Calculating the Cost of Owning an Automobile**

Find the total monthly cost of owning and operating a certain automobile given the expenses shown.

Monthly car payment:	$256.63
Monthly insurance cost:	$47.52
Average gasoline bill per month:	$95.33

Solution:

1. **UNDERSTAND.** Read and reread the problem. The phrase "total monthly cost" tells us to add.

2. **TRANSLATE.**

In words:	total monthly cost	is	car payment	plus	insurance cost	plus	gasoline bill
	↓	↓	↓	↓	↓	↓	↓
Translate:	total monthly cost	=	$256.63	+	$47.52	+	$95.33

3. **SOLVE:** Let's also estimate by rounding each number to the nearest ten.

$\overset{1\ 1\ 1}{256.63}$	rounds to	260	
47.52	rounds to	50	
+ 95.33	rounds to	100	
$399.48	Exact	410	Estimate

PRACTICE PROBLEM 14

Find the total monthly cost of owning and operating a certain automobile given the expenses shown.

Monthly car payment:	$563.52
Monthly insurance cost:	$52.68
Average gasoline bill per month:	$127.50

Continued on next page

Answers
12. yes   **13.** $-24.4y + 22.4$ **14.** $743.70

4. INTERPRET. *Check* your work. Since our estimate is close to our exact answer, our answer is reasonable. *State* your conclusion: The total monthly cost is $399.48.

🔲 **Work Practice Problem 14**

The next bar graph has horizontal bars. To visualize the value represented by a bar, see how far it extends to the right. The value of each bar is labeled, and we will study bar graphs further in a later chapter.

PRACTICE PROBLEM 15

Use the bar graph in Example 15. How much greater is the average height in the Netherlands than the average height in Czechoslovakia?

EXAMPLE 15 Comparing Average Heights

The bar graph shows the current average heights for adults in various countries. How much greater is the average height in Denmark than the average height in the United States?

Average Adult Height

Netherlands 72.6 inches
Denmark 72.2 inches
Norway 71.9 inches
Sweden 71.8 inches
Germany 71.6 inches
USA 70.8 inches
Czechoslovakia[1] 70.8 inches

[1]Average for Czech Republic, Slovakia
Source: USA Today

* The 〰 means that some numbers are purposefully missing on the axis.

Solution:

1. UNDERSTAND. Read and reread the problem. Since we want to know "how much greater," we subtract.

2. TRANSLATE.

In words:	How much greater	is	Denmark's average height	minus	U.S. average height
	↓	↓	↓	↓	↓
Translate:	How much greater	=	72.2	−	70.8

3. SOLVE: We estimate by rounding each number to the nearest whole.

$$
\begin{array}{r}
\overset{1\ \ 12}{7\cancel{2}.\cancel{2}} \\
-\ 70.8 \\
\hline
1.4
\end{array}
$$

72.2 rounds to 72
−70.8 rounds to −71
1.4 Exact   1 Estimate

4. INTERPRET. *Check* your work. Since our estimate is close to our exact answer, 1.4 inches is reasonable. *State* your conclusion: The average height in Denmark is 1.4 inches greater than the average U.S. height.

🔲 **Work Practice Problem 15**

Answer
15. 1.8 in.

▦ **CALCULATOR EXPLORATIONS** Decimals

Entering Decimal Numbers

To enter a decimal number, find the key marked ⟨ · ⟩.
To enter the number 2.56, for example, press the keys
⟨ 2 ⟩⟨ · ⟩⟨ 5 ⟩⟨ 6 ⟩.
The display will read ⟨   2.56 ⟩.

Operations on Decimal Numbers

Operations on decimal numbers are performed in the same way as operations on whole or signed numbers. For example, to find $8.625 - 4.29$, press the keys ⟨ 8.625 ⟩⟨ − ⟩⟨ 4.29 ⟩⟨ = ⟩ or ⟨ ENTER ⟩.
The display will read ⟨   4.335 ⟩. (Although entering 8.625, for example, requires pressing more than one key, we group numbers together here for easier reading.)

Use a calculator to perform each indicated operation.

1. $315.782 + 12.96$

2. $29.68 + 85.902$

3. $6.249 - 1.0076$

4. $5.238 - 0.682$

5.
$$\begin{array}{r} 12.555 \\ 224.987 \\ 5.2 \\ +\,622.65 \end{array}$$

6.
$$\begin{array}{r} 47.006 \\ 0.17 \\ 313.259 \\ +\,139.088 \end{array}$$

Vocabulary and Readiness Check

Use the choices below to fill in each blank. Not all choices will be used.

minuend	vertically	like	true
difference	subtrahend	last	false

1. The decimal point in a whole number is positioned after the _____ digit.

2. In $89.2 - 14.9 = 74.3$, the number 74.3 is called the _____ , 89.2 is the _____ , and 14.9 is the _____ .

3. To simplify an expression, we combine any _____ terms.

4. True or false: If we replace x with 11.2 and y with -8.6 in the expression $x - y$, we have $11.2 - 8.6$. _____

5. To add or subtract decimals, we line up the decimal points _____ .

Objectives A B **Mixed Practice** *Add or subtract. See Examples 1 through 4, and 10. For those exercises marked, also estimate to see if the answer is reasonable.*

1. 5.6 + 2.1

2. 3.6 + 4.1

3. 8.2 + 2.15

4. 5.17 + 3.7

5. 24.6 + 2.39 + 0.0678

6. 32.4 + 1.58 + 0.0934

7. −2.6 + (−5.97)

8. −18.2 + (−10.8)

9. 18.56 + (−8.23)

10. 4.38 + (−6.05)

11.
$$\begin{array}{r} 234.89 \\ + \ 230.67 \end{array}$$
Exact: Estimate:

12.
$$\begin{array}{r} 734.89 \\ + \ 640.56 \end{array}$$
Exact: Estimate:

13.
$$\begin{array}{r} 100.009 \\ 6.08 \\ + \quad 9.034 \end{array}$$
Exact: Estimate:

14.
$$\begin{array}{r} 200.89 \\ 7.49 \\ + \quad 62.83 \end{array}$$
Exact: Estimate:

15. Find the sum of 39, 3.006, and 8.403

16. Find the sum of 65, 5.0903, and 6.9003

Subtract and check. See Examples 5 through 10. For those exercises marked, also estimate to see if the answer is reasonable.

17. 12.6 − 8.2

18. 8.9 − 3.1

19. 18 − 2.7

20. 28 − 3.3

21.
$$\begin{array}{r} 654.9 \\ - \quad 56.67 \end{array}$$

22.
$$\begin{array}{r} 863.23 \\ - \quad 39.453 \end{array}$$

23. 5.9 − 4.07
Exact:
Estimate:

24. 6.4 − 3.04
Exact:
Estimate:

25.
$$\begin{array}{r} 1000 \\ - \quad 123.4 \end{array}$$
Exact:

26.
$$\begin{array}{r} 2000 \\ - \quad 327.47 \end{array}$$
Exact:

27. 200 − 5.6

28. 800 − 8.9

Estimate:

Estimate:

29. −1.12 − 5.2

30. −8.63 − 5.6

31. 5.21 − 11.36

32. 8.53 − 17.84

33. −2.6 − (−5.7)

34. −9.4 − (−10.4)

35. 3 − 0.0012

36. 7 − 0.097

37. Subtract 6.7 from 23.

38. Subtract 9.2 from 45.

Objective A *Perform the indicated operation. See Examples 1 through 10.*

39. 0.9 + 2.2

40. 0.7 + 3.4

41. −6.06 + 0.44

42. −5.05 + 0.88

43. $500.21 - 136.85$ **44.** $600.47 - 254.68$ **45.** $50.2 - 600$ **46.** $40.3 - 700$

 47. Subtract 61.9 from 923.5   **48.** Subtract 45.8 from 845.9   **49.** Add 100.009 and 6.08 and 9.034

50. Add 200.89 and 7.49 and 62.83   **51.** $-0.003 + 0.091$ **52.** $-0.004 + 0.085$

53. $-102.4 - 78.04$ **54.** $-36.2 - 10.02$ **55.** $-2.9 - (-1.8)$ **56.** $-6.5 - (-3.3)$

Objective **C** *Evaluate each expression for $x = 3.6$, $y = 5$, and $z = 0.21$. See Example 11.*

57. $x + z$ **58.** $y + x$ **59.** $x - z$

60. $y - z$ **61.** $y - x + z$ **62.** $x + y + z$

Determine whether the given values are solutions to the given equations. See Example 12.

63. Is 7 a solution to $x + 2.7 = 9.3$?   **64.** Is 3.7 a solution to $x + 5.9 = 8.6$?

65. Is -11.4 a solution to $27.4 + y = 16$?   **66.** Is -22.9 a solution to $45.9 + z = 23$?

67. Is 1 a solution to $2.3 + x = 5.3 - x$?   **68.** Is 0.9 a solution to $1.9 - x = x + 0.1$?

Objective **D** *Simplify by combining like terms. See Example 13.*

69. $30.7x + 17.6 - 23.8x - 10.7$ **70.** $14.2z + 11.9 - 9.6z - 15.2$

71. $-8.61 + 4.23y - 2.36 - 0.76y$ **72.** $-8.96x - 2.31 - 4.08x + 9.68$

Objective **E** *Solve. See Examples 14 and 15.*

73. Microsoft stock opened the day at $21.90 per share, and the closing price the same day was $22.07. By how much did the price of each share change?

74. A pair of eyeglasses costs a total of $347.89. The frames of the glasses are $97.23. How much do the lenses of the eyeglasses cost?

75. Find the perimeter.

Square | 7.14 meters

76. Find the perimeter.

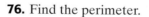

4.2 in.   5.78 in.

7.8 in.

The iPod mini is a miniature version of Apple Computer's popular iPod portable audio player. This mini was introduced in January 2004 with a storage capacity of 4 gigabytes. (This is about 1000 3-minute or 3-megabyte songs.)

77. The top face of the iPod mini shown measures 3.6 inches by 2.0 inches. Find the perimeter of the rectangular face.

78. The face of the larger Apple iPod measures 4.1 inches by 2.4 inches. Find the perimeter of this rectangular face.

Solve.

79. Ann-Margaret Tober bought a book for $32.48. If she paid with two $20 bills, what was her change?

80. Phillip Guillot bought a car part for $18.26. If he paid with two $10 bills, what was his change?

81. The average wind speed at the weather station on Mt. Washington in New Hampshire is 35.2 miles per hour. The highest speed ever recorded at the station is 321.0 miles per hour. How much faster is the highest speed than the average wind speed? (*Source:* National Climatic Data Center)

82. The average annual rainfall in Omaha, Nebraska, is 30.22 inches. The average annual rainfall in New Orleans, Louisiana, is 61.88 inches. On average, how much more rain does New Orleans receive annually than Omaha? (*Source:* National Climatic Data Center)

This bar graph shows the predicted decrease in home video rental revenue for chains such as Blockbuster and Hollywood Video. Use this graph for Exercise 83.

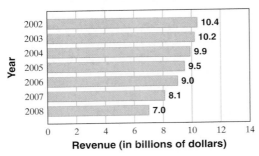

Video Rental Revenues

Year

	Revenue (in billions of dollars)
2002	10.4
2003	10.2
2004	9.9
2005	9.5
2006	9.0
2007	8.1
2008	7.0

Source: Forrester Research; *Note:* Many of these years are projections.

83. Find the decrease in video rental revenue from the year 2002 to 2008.

84. It is predicted that home video *sales* (not shown on the bar graph) will increase from $15.3 billion in 2002 to $24.6 billion in 2008. Find the amount of increase.

85. The top three U.S. movies that made the most money through movie ticket sales are *Titanic* (1997) $600.78 million, *Star Wars IV: A New Hope* (1997) $460.99 million, and *Shrek 2* (2004) $436.72 million. What was the total amount of ticket sales for these three movies? (*Source:* MovieWeb)

86. In 1995, the average credit-card late fee was $12.53. By 2005, the average late fee had increased to $29.45. By how much did the average credit-card late fee increase from 1995 to 2005? (*Source: Money magazine*)

87. The snowiest city in the United States is Blue Canyon, California, which receives an average of 111.6 more inches of snow than the second snowiest city. The second snowiest city in the United States is Marquette, Michigan. Marquette receives an average of 129.2 inches of snow annually. How much snow does Blue Canyon receive on average each year? (*Source:* National Climatic Data Center)

88. The driest city in the world is Aswan, Egypt, which receives an average of only 0.02 inch of rain per year. Yuma, Arizona, is the driest city in the United States. Yuma receives an average of 2.63 more inches of rain each year than Aswan. What is the average annual rainfall in Yuma? (*Source:* National Climatic Data Center)

89. A landscape architect is planning a border for a flower garden shaped like a triangle. The sides of the garden measure 12.4 feet, 29.34 feet, and 25.7 feet. Find the amount of border material needed.

90. A contractor purchased enough railing to completely enclose the newly built deck shown below. Find the amount of railing purchased.

29.34 feet

12.4 feet   25.7 feet

15.7 feet

10.6 feet

The table shows the average speeds for the Daytona 500 Winners for the years shown. Use this table to answer Exercises 91 and 92.

Daytona 500 Winners		
Year	Winner	Average Speed
1966	Richard Petty	160.927
1976	David Pearson	152.181
1986	Geoff Bodine	148.124
1996	Dale Jarrett	154.308
2006	Jimmie Johnson	142.734
(*Source:* Daytona International Speedway)		

91. How much slower was the average Daytona 500 winning speed in 2006 than in 1966?

92. How much faster was the average Daytona 500 winning speed in 1996 than in 1986?

The bar graph shows the top five chocolate-consuming nations in the world. Use this table to answer Exercises 93 through 97.

The World's Top Chocolate-Consuming Countries

Source: Hershey Foods Corporation

93. Which country in the table has the greatest chocolate consumption per person?

94. Which country in the table has the least chocolate consumption per person?

95. How much more is the greatest chocolate consumption than the least chocolate consumption shown in the table?

96. How much more chocolate does the average German consume than the average citizen of the United Kingdom?

97. Make a new chart listing the countries and their corresponding chocolate consumptions in order from greatest to least.

Review

Multiply. See Sections 1.5 and 4.3.

98. $23 \cdot 2$

99. $46 \cdot 3$

100. $\left(\dfrac{2}{3}\right)^2$

101. $\left(\dfrac{1}{5}\right)^3$

Concept Extensions

Solve. See the first Concept Check in this section.

102. A friend asks you to check his calculation to the right. Is it correct? If not, explain your friends' error and correct the calculation.

$$\begin{array}{r} \overset{1}{9}.2 \\ \overset{1}{8}.63 \\ +\,4.005 \\ \hline 4.960 \end{array}$$

Find the unknown length in each figure.

△ **103.**

2.3 inches   ?   2.3 inches

├────── 10.68 inches ──────┤

△ **104.**

├─ 5.26 ─┤├─ 7.82 ─┤├─ ? ─┤
meters ┊ meters ┊ meters

├────── 17.67 meters ──────┤

Let's review the values of these common U.S. coins in order to answer the following exercises.

Penny   Nickel   Dime   Quarter

$0.01 $0.05 $0.10 $0.25

For Exercises 105 and 106, write the value of each group of coins. To do so, it is usually easiest to start with the coin(s) of greatest value and end with the coin(s) of least value.

105.

106.

107. Name the different ways that coins can have a value of $0.17 given that you may use no more than 10 coins.

108. Name the different ways that coin(s) can have a value of $0.25 given that there are no pennies.

109. Why shouldn't the sum

$$82.95 + 51.26$$

be estimated as $90 + 60 = 150$?
See the second Concept Check in this section.

110. Explain how adding or subtracting decimals is similar to adding or subtracting whole numbers.

111. Laser beams can be used to measure the distance to the moon. One measurement showed the distance to the moon to be 256,435.235 miles. A later measurement showed that the distance is 256,436.012 miles. Find how much farther away the moon is in the second measurement as compared to the first.

Combine like terms and simplify.

112. $-8.689 + 4.286x - 14.295 - 12.966x + 30.861x$

113. $14.271 - 8.968x + 1.333 - 201.815x + 101.239x$

114. Can the sum of two negative decimals ever be a positive decimal? Why or why not?

5.3 MULTIPLYING DECIMALS AND CIRCUMFERENCE OF A CIRCLE

Objective A Multiplying Decimals

Multiplying decimals is similar to multiplying whole numbers. The only difference is that we place a decimal point in the product. To discover where a decimal point is placed in the product, let's multiply 0.6×0.03. We first write each decimal as an equivalent fraction and then multiply.

$$0.6 \quad \times \quad 0.03 \quad = \frac{6}{10} \times \frac{3}{100} = \frac{18}{1000} = 0.018$$

1 decimal place   2 decimal places   3 decimal places

A Multiply Decimals.

B Estimate when Multiplying Decimals.

C Multiply Decimals by Powers of 10.

D Evaluate Expressions with Decimal Replacement Values

E Find the Circumference of Circle.

F Solve Problems by Multiplying Decimals.

Notice that $1 + 2 = 3$, the number of decimal places in the product. Now let's multiply 0.03×0.002.

$$0.03 \quad \times \quad 0.002 \quad = \frac{3}{100} \times \frac{2}{1000} = \frac{6}{100,000} = 0.00006$$

2 decimal places   3 decimal places   5 decimal places

Again, we see that $2 + 3 = 5$, the number of decimal places in the product.

Instead of writing decimals as fractions each time we want to multiply, we notice a pattern from these examples and state a rule that we can use:

Multiplying Decimals

Step 1: Multiply the decimals as though they are whole numbers.

Step 2: The decimal point in the product is placed so that the number of decimal places in the product is equal to the *sum* of the number of decimal places in the factors.

EXAMPLE 1 Multiply: 23.6×0.78

Solution:
$$
\begin{array}{r}
23.6 \\
\times\, 0.78 \\
\hline
1888 \\
16520 \\
\hline
18.408
\end{array}
$$
23.6   1 decimal place
$\times 0.78$ 2 decimal places

Since $1 + 2 = 3$, insert the decimal point in the product so that there are 3 decimal places.

Work Practice Problem 1

EXAMPLE 2 Multiply: 0.0531×16

Solution:
$$
\begin{array}{r}
0.0531 \\
\times\quad 16 \\
\hline
3186 \\
5310 \\
\hline
0.8496
\end{array}
$$
0.0531   4 decimal places
$\times\ 16$ 0 decimal places

4 decimal places $(4 + 0 = 4)$

Work Practice Problem 2

PRACTICE PROBLEM 1
Multiply: 34.8×0.62

PRACTICE PROBLEM 2
Multiply: 0.0641×27

Answers
1. 21.576   **2.** 1.7307

✔ **Concept Check** True or false? The number of decimal places in the product of 0.261 and 0.78 is 6. Explain.

EXAMPLE 3 Multiply: $(-2.6)(0.8)$

Solution: Recall that the product of a negative number and a positive number is a negative number.

$(-2.6)(0.8) = -2.08$

▣ **Work Practice Problem 3**

Objective B Estimating when Multiplying Decimals

Just as for addition and subtraction, we can estimate when multiplying decimals to check the reasonableness of our answer.

EXAMPLE 4 Multiply: 28.06×1.95. Then estimate to see whether the answer is reasonable by rounding each factor, then multiplying the rounded numbers.

Solution:

Exact:	Estimate 1	Estimate 2
28.06	28 Rounded to ones or	30 Rounded to tens
$\times\ 1.95$	$\times\ 2$	$\times\ 2$
14030	56	60
252540		
280600		
54.7170		

The answer 54.7170 is reasonable.

▣ **Work Practice Problem 4**

As shown in Example 4, estimated results will vary depending on what estimates are used. Notice that estimating results is a good way to see whether the decimal point has been correctly placed.

Objective C Multiplying Decimals by Powers of 10

There are some patterns that occur when we multiply a number by a power of 10 such as 10, 100, 1000, 10,000, and so on.

$23.6951 \times 10 = 236.951$ Move the decimal point *1 place* to the *right*.
1 zero

$23.6951 \times 100 = 2369.51$ Move the decimal point *2 places* to the *right*.
2 zeros

$23.6951 \times 100,000 = 2,369,510.$ Move the decimal point *5 places* to the *right* (insert a 0).
5 zeros

Notice that we move the decimal point the same number of places as there are zeros in the power of 10.

Multiplying Decimals by Powers of 10 such as 10, 100, 1000, 10,000 . . .

Move the decimal point to the *right* the same number of places as there are *zeros* in the power of 10.

EXAMPLES Multiply.

5. $7.68 \times 10 = 76.8$ 7.68

6. $23.702 \times 100 = 2370.2$ 23.702

7. $(-76.3)(1000) = -76,300$ 76.300

Work Practice Problems 5–7

PRACTICE PROBLEMS 5–7

Multiply.

5. 46.8×10

6. 203.004×100

7. $(-2.33)(1000)$

There are also powers of 10 that are less than 1. The decimals 0.1, 0.01, 0.001, 0.0001, and so on are examples of powers of 10 less than 1. Notice the pattern when we multiply by these powers of 10:

$569.2 \times 0.1 = 56.92$ Move the decimal point *1 place* to the *left*.
    1 decimal place

$569.2 \times 0.01 = 5.692$ Move the decimal point *2 places* to the *left*.
    2 decimal places

$569.2 \times 0.0001 = 0.05692$ Move the decimal point *4 places* to the *left* (insert one 0).
    4 decimal places

Multiplying Decimals by Powers of 10 such as 0.1, 0.01, 0.001, 0.0001 . . .

Move the decimal point to the *left* the same number of places as there are *decimal places* in the power of 10.

EXAMPLES Multiply.

8. $42.1 \times 0.1 = 4.21$ 42.1

9. $76,805 \times 0.01 = 768.05$ 76,805.

10. $(-9.2)(-0.001) = 0.0092$ 0009.2

Work Practice Problems 8–10

PRACTICE PROBLEMS 8–10

Multiply.

8. 6.94×0.1

9. 3.9×0.01

10. $(-7682)(-0.001)$

Many times we see large numbers written, for example, in the form 297.9 million rather than in the longer standard notation. The next example shows us how to interpret these numbers.

Answers

5. 468   **6.** 20,300.4   **7.** −2330
8. 0.694   **9.** 0.039   **10.** 7.682

PRACTICE PROBLEM 11

In 2004, there were 76.2 million families in the United States. Write this number in standard notation. (*Source:* U.S. Census Bureau)

EXAMPLE 11

In 2010, the population of the United States is projected to be 308.9 million. Write this number in standard notation. (*Source:* U.S. Census Bureau)

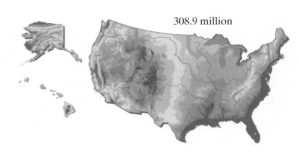

308.9 million

Solution: 308.9 million = 308.9 × 1 million

= 308.9 × 1,000,000 = 308,900,000

⬛ **Work Practice Problem 11**

Objective D Using Decimals as Replacement Values

Now let's practice working with variables.

PRACTICE PROBLEM 12

Evaluate $7y$ for $y = -0.028$.

EXAMPLE 12 Evaluate xy for $x = 2.3$ and $y = 0.44$.

Solution: Recall that xy means $x \cdot y$.

$$xy = (2.3)(0.44)$$

$$\begin{array}{r} 2.3 \\ \times\ 0.44 \\ \hline 92 \\ 920 \\ \hline \end{array}$$

$$= 1.012 \longleftarrow \quad 1.012$$

⬛ **Work Practice Problem 12**

PRACTICE PROBLEM 13

Is -5.5 a solution of the equation $-6x = 33$?

EXAMPLE 13 Is -9 a solution of the equation $3.7y = -3.33$?

Solution: Replace y with -9 in the equation $3.7y = -3.33$ to see if a true equation results.

$$3.7y = -3.33$$

$$3.7(-9) \stackrel{?}{=} -3.33 \quad \text{Replace } y \text{ with } -9.$$

$$-33.3 = -3.33 \quad \text{False}$$

Since $-33.3 = -3.33$ is a false statement, -9 is **not** a solution of $3.7y = -3.33$.

⬛ **Work Practice Problem 13**

Objective E Finding the Circumference of a Circle

Recall from Section 1.3 that the distance around a polygon is called its perimeter. The distance around a circle is given a special name called the **circumference,** and this distance depends on the radius or the diameter of the circle.

Circumference of a Circle

Circumference = $2 \cdot \pi \cdot$ **radius** or Circumference = $\pi \cdot$ **diameter**

$$C = 2\pi r \qquad \text{or} \qquad C = \pi d$$

Answers
11. 76,200,000　**12.** −0.196　**13.** yes

In Section 5.1, we learned about the symbol π as the Greek letter pi, pronounced "pie." It is a constant between 3 and 4. A decimal approximation for π is 3.14. Also, a fraction approximation for π is $\frac{22}{7}$.

EXAMPLE 14 Circumference of a Circle

Find the circumference of a circle whose radius is 5 inches. Then use the approximation 3.14 for π to approximate the circumference.

Solution: Let $r = 5$ in the formula $C = 2\pi r$.

$$C = 2\pi r$$
$$= 2\pi \cdot 5$$
$$= 10\pi$$

5 inches

Next, replace π with the approximation 3.14.

$$C = 10\pi$$
(is approximately) $\longrightarrow \approx 10(3.14)$
$$= 31.4$$

The **exact** circumference or distance around the circle is 10π inches, which is **approximately** 31.4 inches.

📖 **Work Practice Problem 14**

Objective F Solving Problems by Multiplying Decimals

The solutions to many real-life problems are found by multiplying decimals. We continue using our four problem-solving steps to solve such problems.

EXAMPLE 15 Finding the Total Cost of Materials for a Job

A college student is hired to paint a billboard with paint costing $2.49 per quart. If the job requires 3 quarts of paint, what is the total cost of the paint?

Solution:

1. UNDERSTAND. Read and reread the problem. The phrase "total cost" might make us think addition, but since this problem requires repeated addition, let's multiply.

2. TRANSLATE.

In words:	Total cost	is	cost per quart of paint	times	number of quarts
	↓	↓	↓	↓	↓
Translate:	Total cost	=	2.49	×	3

3. SOLVE. We can estimate to check our calculations. The number 2.49 rounds to 2 and $2 \times 3 = 6$.

$$\begin{array}{r} {\scriptstyle 1\ 2} \\ 2.49 \\ \times \quad 3 \\ \hline 7.47 \end{array}$$

4. INTERPRET. *Check* your work. Since 7.47 is close to our estimate of 6, our answer is reasonable. *State* your conclusion: The total cost of the paint is $7.47.

📖 **Work Practice Problem 15**

Vocabulary and Readiness Check

Use the choices below to fill in each blank.

circumference   left   sum   zeros

decimal places   right   product   factor

1. When multiplying decimals, the number of decimal places in the product is equal to the _____ of the number of decimal places in the factors.
2. In $8.6 \times 5 = 43$, the number 43 is called the _____, while 8.6 and 5 are each called a _____.
3. When multiplying a decimal number by powers of 10, such as 10, 100, 1000, and so on, we move the decimal point in the number to the _____ the same number of places as there are _____ in the power of 10.
4. When multiplying a decimal number by powers of 10, such as 0.1, 0.01, and so on, we move the decimal point in the number to the _____ the same number of places as there are _____ in the power of 10.
5. The distance around a circle is called its _____.

5.3 EXERCISE SET

FOR EXTRA HELP

Student Solutions Manual   PH Math/Tutor Center   CD/Video for Review   MathXL®   MyMathLab

Objectives Ⓐ Ⓑ Mixed Practice *Multiply. See Examples 1 through 4. For those exercises marked, also estimate to see if the answer is reasonable.*

1. 0.17×8

2. 0.23×9

3.
$$\begin{array}{r} 1.2 \\ \times\,0.5 \\ \hline \end{array}$$

4.
$$\begin{array}{r} 6.8 \\ \times\,0.3 \\ \hline \end{array}$$

5. $(-2.3)(7.65)$

6. $(4.7)(-9.02)$

7. $(-5.73)(-9.6)$

8. $(-7.84)(-3.5)$

9. 6.8×4.2
Exact:
Estimate:

10. 8.3×2.7
Exact:
Estimate:

11.
$$\begin{array}{r} 0.347 \\ \times\,\,\,\,\,0.3 \\ \hline \end{array}$$

12.
$$\begin{array}{r} 0.864 \\ \times\,\,\,\,\,0.4 \\ \hline \end{array}$$

13.
$$\begin{array}{r} 1.0047 \\ \times\,\,\,\,\,\,8.2 \\ \hline \end{array}$$
Exact: Estimate:

14.
$$\begin{array}{r} 2.0005 \\ \times\,\,\,\,\,\,5.5 \\ \hline \end{array}$$
Exact: Estimate:

15.
$$\begin{array}{r} 490.2 \\ \times\,0.023 \\ \hline \end{array}$$

16.
$$\begin{array}{r} 300.9 \\ \times\,0.032 \\ \hline \end{array}$$

Objective Ⓒ *Multiply. See Examples 5 through 10.*

17. 6.5×10

18. 7.2×100

19. 8.3×0.1

20. 23.4×0.1

21. $(-7.093)(1000)$

22. $(-1.123)(1000)$

23. 7.093×100

24. 0.5×100

25. $(-9.83)(-0.01)$

26. $(-4.72)(-0.01)$

27. 25.23×0.001

28. 36.41×0.001

Objectives Ⓐ Ⓑ Ⓒ Mixed Practice *Multiply. See Examples 1 through 10.*

29. 0.123×0.4

30. 0.216×0.3

31. $(147.9)(100)$

32. $(345.2)(100)$

33. 8.6×0.15

34. 0.42×5.7

35. $(937.62)(-0.01)$

36. $(-0.001)(562.01)$

37. 562.3×0.001 **38.** 993.5×0.001

39. $\begin{array}{r} 6.32 \\ \times\ 5.7 \\ \hline \end{array}$

40. $\begin{array}{r} 9.21 \\ \times\ 3.8 \\ \hline \end{array}$

Write each number in standard notation. See Example 11.

41. The cost of the Hubble Space Telescope at launch was $1.5 billion. (*Source:* NASA)

42. About 40.4 million American households own at least one dog. (*Source:* American Pet Products Manufacturers Association)

43. The Blue Streak is the oldest roller coaster at Cedar Point, an amusement park in Sandusky, Ohio. Since 1964, it has given more than 49.8 million rides. (*Source:* Cedar Fair, L.P.)

44. There are 166.2 thousand full service restaurants in the United States. (*Source:* National Restaurant Association)

Objective **D** *Evaluate each expression for $x = 3$, $y = -0.2$, and $z = 5.7$. See Example 12.*

45. xy **46.** yz **47.** $xz - y$ **48.** $-5y + z$

Determine whether the given value is a solution of each given equation. See Example 13.

49. Is 14.2 a solution of $0.6x = 4.92$?

50. Is 1414 a solution of $100z = 14.14$?

51. Is -4 a solution of $3.5y = -14$?

52. Is -3.6 a solution of $0.7x = -2.52$?

Objective **E** *Find the circumference of each circle. Then use the approximation 3.14 for π and approximate each circumference. See Example 14.*

53.

10 centimeters

54.

22 inches

55.

9.1 yards

56.

5.9 kilometers

Objectives **E** **F** **Mixed Practice** *Solve. See Examples 14 and 15. For circumference applications find the exact circumference and then use 3.14 for π to approximate the circumference.*

57. A 1-ounce serving of cream cheese contains 6.2 grams of saturated fat. How much saturated fat is in 4 ounces of cream cheese? (*Source: Home and Garden Bulletin No. 72;* U.S. Department of Agriculture)

58. A 3.5-ounce serving of lobster meat contains 0.1 gram of saturated fat. How much saturated fat do 3 servings of lobster meat contain? (*Source:* The National Institute of Health)

59. Recall that the top face of the Apple iPod mini (see Section 5.2) measures 3.6 inches by 2.0 inches. Find the area of the face of the iPod mini.

60. Recall from Section 5.2 that the face of the regular Apple iPod measures 4.1 inches by 2.4 inches. Find the area of the face of this iPod.

61. In 1893, the first ride called a Ferris wheel was constructed by Washington Gale Ferris. Its diameter was 250 feet. Find its circumference. Give an exact answer and an approximation using 3.14 for π. (*Source: The Handy Science Answer Book*, Visible Ink Press, 1994)

62. The radius of Earth is approximately 3950 miles. Find the distance around Earth at the equator. Give an exact answer and an approximation using 3.14 for π. (*Hint:* Find the circumference of a circle with radius 3950 miles.)

63. The London Eye, built for the Millennium celebration in London, resembles a gigantic ferris wheel with a diameter of 135 meters. If Adam Hawn rides the Eye for one revolution, find how far he travels. Give an exact answer and an approximation using 3.14 for π. (*Source:* Londoneye.com)

64. The world's longest suspension bridge is the Akashi Kaikyo Bridge in Japan. This bridge has two circular caissons, which are underwater foundations. If the diameter of a caisson is 80 meters, find its circumference. Give an exact answer and an approximation using 3.14 for π. (*Source: Scientific American; How Things Work Today*)

65. A meter is a unit of length in the metric system that is approximately equal to 39.37 inches. Sophia Wagner is 1.65 meters tall. Find her approximate height in inches.

66. The doorway to a room is 2.15 meters tall. Approximate this height in inches. (*Hint:* See Exercise 65.)

67. An electrician for Central Power and Light worked 40 hours last week. Calculate his pay before taxes for last week if his hourly wage is $17.88.

68. An assembly line worker worked 20 hours last week. Her hourly rate is $19.52 per hour. Calculate her pay before taxes.

69. **a.** Approximate the circumference of each circle.

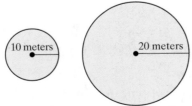

   10 meters   20 meters

  b. If the radius of a circle is doubled, is its corresponding circumference doubled?

70. **a.** Approximate the circumference of each circle.

   16 inches   32 inches

  b. If the diameter of a circle is doubled, is its corresponding circumference doubled?

71. In 2005, American farmers received an average of $3.40 per bushel of wheat. How much did a farmer receive for selling 100 bushels of wheat? (*Source:* National Agricultural Statistics Service)

72. In 2005, American farmers received an average of $5.50 per bushel of soybeans. How much did a farmer receive for selling 10,000 bushels of soybeans? (*Source:* National Agricultural Statistics Service)

This table shows currency exchange rates for various countries on June 19, 2006. To find the amount of foreign currency equivalent to an amount of U.S. dollars, multiply the U.S. dollar amount by the exchange rate listed in the table. Use this table to answer Exercises 73 through 76.

Foreign Currency Exchange Rates	
Country	**Exchange Rate**
Australian Dollar	1.3559
Canadian Dollar	1.1232
Sri Lanka Rupee	103.550
Japanese Yen	115.26
European Union Euro	0.7951
Chinese Yuan	8.0035

73. How many Japanese yen are equivalent to 675 U.S. dollars?

74. Suppose you wish to exchange 500 U.S. dollars into Chinese yuan. How much money, in Chinese yuan, would you receive?

75. The Robinson family is taking a vacation to Europe. How much European Union euros can they "buy" with 900 U.S. dollars?

76. An Australian tourist spent $85 for souvenir tee shirts in San Francisco. What did he pay for the tee shirts in Australian dollars?

Review

Divide. See Sections 1.6 and 4.3.

77. $2920 \div 365$

78. $2916 \div 6$

79. $-\dfrac{24}{7} \div \dfrac{8}{21}$

80. $\dfrac{162}{25} \div -\dfrac{9}{75}$

Concept Extensions

Mixed Practice (*Sections 5.2, 5.3*) *Perform the indicated operations.*

81. $3.6 + 0.04$

82. -3.6×0.04

83. $3.6 - 0.04$

84. $100 - 48.6$

85. -0.221×0.5

86. $7.2 + 0.14 + 98.6$

87. Find how far radio waves travel in 20.6 seconds. (Radio waves travel at a speed of $1.86 \times 100{,}000$ miles per second.)

88. If it takes radio waves approximately 8.3 minutes to travel from the Sun to the Earth, find approximately how far it is from the Sun to the Earth. (*Hint:* See Exercise 87.)

89. In your own words, explain how to find the number of decimal places in a product of decimal numbers.

90. In your own words, explain how to multiply by a power of 10.

91. Write down two decimal numbers whose product will contain 5 decimal places. Without multiplying, explain how you know your answer is correct.

STUDY SKILLS BUILDER

How Are Your Homework Assignments Going?

Remember that it is important to keep up with homework. Why? Many concepts in mathematics build on each other. Often, your understanding of a day's lecture depends on an understanding of the previous day's material.

To complete a homework assignment, remember these four things:

- Attempt all of it.
- Check it.
- Correct it.
- If needed, ask questions about it.

Take a moment and review your completed homework assignments. Answer the exercises below based on this review.

1. Approximate the fraction of your homework you have attempted.

2. Approximate the fraction of your homework you have checked (if possible).

3. If you are able to check your homework, have you corrected it when errors have been found?

4. When working homework, if you do not understand a concept, what do you personally do?

PRACTICE PROBLEM 1

Divide: $370.4 \div 8$. Check your answer.

PRACTICE PROBLEM 2

Divide: $48\overline{)34.08}$. Check your answer.

5.4 DIVIDING DECIMALS

Objective **A** Dividing Decimals

Dividing decimal numbers is similar to dividing whole numbers. The only difference is that we place a decimal point in the quotient. If the divisor is a whole number, we place the decimal point in the quotient directly above the decimal point in the dividend, and then divide as with whole numbers. Recall that division can be checked by multiplication.

EXAMPLE 1 Divide: $270.2 \div 7$. Check your answer.

Solution: We divide as usual. The decimal point in the quotient is directly above the decimal point in the dividend.

```
                    ┌─ Write the decimal point.
               38.6 ← quotient
 divisor → 7)270.2 ← dividend
            −21↓
              60
            −56 ↓
               4 2
             −4 2
                 0
```

Check:
$$\begin{array}{r} {}^{6\,4} \\ 38.6 \\ \times\ 7 \\ \hline 270.2 \end{array}$$

The quotient is 38.6.

⬛ **Work Practice Problem 1**

EXAMPLE 2 Divide: $32\overline{)8.32}$

Solution: We divide as usual. The decimal point in the quotient is directly above the decimal point in the dividend.

```
                0.26 ← quotient
 divisor → 32)8.32 ← dividend
             −64
              192
            −192
                0
```

Check:
$$\begin{array}{r} 0.26 \quad \text{quotient} \\ \times\ 32 \quad \text{divisor} \\ \hline 52 \\ 7\ 80 \\ \hline 8.32 \quad \text{dividend} \end{array}$$

⬛ **Work Practice Problem 2**

Sometimes to continue dividing we need to insert zeros after the last digit in the dividend.

EXAMPLE 3 Divide: $-5.98 \div 115$

Solution: Recall that a negative number divided by a positive number gives a negative quotient.

$$
\begin{array}{r}
0.052 \\
115\overline{)5.980} \quad \leftarrow \text{Insert one } 0. \\
\underline{-5\,75} \\
230 \\
\underline{-230} \\
0
\end{array}
$$

Thus $-5.98 \div 115 = -0.052$.

■ **Work Practice Problem 3**

PRACTICE PROBLEM 3
Divide and check.
a. $-15.89 \div 14$
b. $-2.808 \div (-104)$

If the divisor is not a whole number, before we divide we need to move the decimal point to the right until the divisor is a whole number.

$$1.5\overline{)64.85}$$

divisor ↗   ↖ dividend

To understand how this works, let's rewrite

$1.5\overline{)64.85}$ as $\dfrac{64.85}{1.5}$

and then multiply by 1 in the form of $\dfrac{10}{10}$. We use the form $\dfrac{10}{10}$ so that the denominator (divisor) becomes a whole number.

$$\frac{64.85}{1.5} = \frac{64.85}{1.5} \cdot 1 = \frac{64.85}{1.5} \cdot \frac{10}{10} = \frac{64.85 \cdot 10}{1.5 \cdot 10} = \frac{648.5}{15},$$

which can be written as $15.\overline{)648.5}$. Notice that

$1.5\overline{)64.85}$ is equivalent to $15.\overline{)648.5}$

The decimal points in the dividend and the divisor were both moved one place to the right, and the divisor is now a whole number. This procedure is summarized next:

Dividing by a Decimal

Step 1: Move the decimal point in the divisor to the right until the divisor is a whole number.

Step 2: Move the decimal point in the dividend to the right the *same number of places* as the decimal point was moved in Step 1.

Step 3: Divide. Place the decimal point in the quotient directly over the moved decimal point in the dividend.

Answers
3. a. -1.135 b. 0.027

PRACTICE PROBLEM 4

Divide: $166.88 \div 5.6$

EXAMPLE 4 Divide: $10.764 \div 2.3$

Solution: We move the decimal points in the divisor and the dividend one place to the right so that the divisor is a whole number.

$$2.3\overline{)10.764} \qquad \text{becomes} \qquad \begin{array}{r} 4.68 \\ 23.\overline{)107.64} \\ -92 \downarrow \\ \hline 15\,6 \\ -13\,8 \downarrow \\ \hline 1\,84 \\ -1\,84 \\ \hline 0 \end{array}$$

◻ **Work Practice Problem 4**

PRACTICE PROBLEM 5

Divide: $1.976 \div 0.16$

EXAMPLE 5 Divide: $5.264 \div 0.32$

Solution:

$$0.32\overline{)5.264} \qquad \text{becomes} \qquad \begin{array}{r} 16.45 \\ 32\overline{)526.40} \quad \text{Insert one 0.} \\ -32 \downarrow \\ \hline 206 \\ -192 \downarrow \\ \hline 14\,4 \\ -12\,8 \downarrow \\ \hline 1\,60 \\ -1\,60 \\ \hline 0 \end{array}$$

◻ **Work Practice Problem 5**

✔ **Concept Check** Is it always true that the number of decimal places in a quotient equals the sum of the decimal places in the dividend and divisor?

PRACTICE PROBLEM 6

Divide $23.4 \div 0.57$. Round the quotient to the nearest hundredth.

EXAMPLE 6 Divide: $17.5 \div 0.48$. Round the quotient to the nearest hundredth.

Solution: First we move the decimal points in the divisor and the dividend two places. Then we divide and round the quotient to the nearest hundredth.

$$\begin{array}{r} \text{hundredths place} \\ 36.458 \approx 36.46 \\ 48.\overline{)1750.000} \quad \text{"is approximately"} \\ -144 \downarrow \\ \hline 310 \\ -288 \downarrow \\ \hline 22\,0 \\ -19\,2 \downarrow \\ \hline 2\,80 \\ -2\,40 \downarrow \\ \hline 400 \\ -384 \\ \hline 16 \end{array}$$

When rounding to the nearest hundredth, carry the division process out to one more decimal place, the thousandths place.

◻ **Work Practice Problem 6**

Answers

4. 29.8   **5.** 12.35   **6.** 41.05

✔ **Concept Check Answer**

no

✔**Concept Check**   If a quotient is to be rounded to the nearest thousandth, to what place should the division be carried out? (Assume that the division carries out to your answer.)

Objective B Estimating when Dividing Decimals

Just as for addition, subtraction, and multiplication of decimals, we can estimate when dividing decimals to check the reasonableness of our answer.

EXAMPLE 7   Divide: $272.356 \div 28.4$. Then estimate to see whether the proposed result is reasonable.

Solution:

Exact:	Estimate 1		Estimate 2

$$
\begin{array}{r}
9.59 \\
284.\overline{)2723.56} \\
-\,2556 \\
\hline
1675 \\
-\,1420 \\
\hline
2556 \\
-\,2556 \\
\hline
0
\end{array}
$$

$$
\begin{array}{r}
9 \\
30\overline{)270}
\end{array}
\quad \text{or} \quad
\begin{array}{r}
10 \\
30\overline{)300}
\end{array}
$$

The estimate is 9 or 10, so 9.59 is reasonable.

▦ **Work Practice Problem 7**

Objective C Dividing Decimals by Powers of 10

As with multiplication, there are patterns that occur when we divide decimals by powers of 10 such as 10, 100, 1000, and so on.

$$\frac{569.2}{10} = 56.92$$ Move the decimal point *1 place* to the *left.*

— 1 zero

$$\frac{569.2}{10,000} = 0.05692$$ Move the decimal point *4 places* to the *left.*

— 4 zeros

This pattern suggests the following rule:

Dividing Decimals by Powers of 10 such as 10, 100, or 1000

Move the decimal point of the dividend to the *left* the same number of places as there are *zeros* in the power of 10.

EXAMPLES   Divide.

8. $\frac{786.1}{1000} = 0.7861$ Move the decimal point *3 places* to the *left.*

— 3 zeros

9. $-\frac{0.12}{10} = -0.012$ Move the decimal point *1 place* to the *left.*

— 1 zero

▦ **Work Practice Problems 8–9**

PRACTICE PROBLEM 7

Divide: $713.7 \div 91.5$. Then estimate to see whether the proposed answer is reasonable.

PRACTICE PROBLEMS 8–9

Divide.

8. $\frac{362.1}{1000}$ **9.** $-\frac{0.49}{10}$

Answers

7. 7.8   **8.** 0.3621   **9.** −0.049

✔ **Concept Check Answer**

ten-thousandths place

Objective D Using Decimals as Replacement Values

Evaluate $x \div y$ for $x = 0.035$ and $y = 0.02$.

EXAMPLE 10 Evaluate $x \div y$ for $x = 2.5$ and $y = 0.05$.

Solution: Replace x with 2.5 and y with 0.05.

$$x \div y = 2.5 \div 0.05 \qquad 0.05\overline{)2.5} \text{ becomes } 5\overline{)250} \;^{50}$$
$$= 50$$

◻ **Work Practice Problem 10**

PRACTICE PROBLEM 11

Is 39 a solution of the equation
$$\frac{x}{100} = 3.9?$$

EXAMPLE 11 Is 720 a solution of the equation $\frac{y}{100} = 7.2$?

Solution: Replace y with 720 to see if a true statement results.

$$\frac{y}{100} = 7.2 \qquad \text{Original equation}$$
$$\frac{720}{100} \overset{?}{=} 7.2 \qquad \text{Replace } y \text{ with 720.}$$
$$7.2 = 7.2 \qquad \text{True}$$

Since $7.2 = 7.2$ is a true statement, 720 is a solution of the equation.

◻ **Work Practice Problem 11**

Objective E Solving Problems by Dividing Decimals

Many real-life problems involve dividing decimals.

PRACTICE PROBLEM 12

A bag of fertilizer covers 1250 square feet of lawn. Tim Parker's lawn measures 14,800 square feet. How many bags of fertilizer does he need? If he can buy only whole bags of fertilizer, how many whole bags does he need?

EXAMPLE 12 Calculating Materials Needed for a Job

A gallon of paint covers a 250-square-foot area. If Betty Adkins wishes to paint a wall that measures 1450 square feet, how many gallons of paint does she need? If she can buy only gallon containers of paint, how many gallon containers does she need?

Solution:

1. UNDERSTAND. Read and reread the problem. We need to know how many 250s are in 1450, so we divide.

2. TRANSLATE.

In words:	number of gallons	is	square feet	divided by	square feet per gallon
	↓	↓	↓	↓	↓
Translate:	number of gallons	=	1450	÷	250

3. SOLVE. Let's see if our answer is reasonable by estimating. The dividend 1450 rounds to 1500 and the divisor 250 rounds to 300. Then $1500 \div 300 = 5$.

$$\begin{array}{r} 5.8 \\ 250\overline{)1450.0} \\ -1250 \\ \hline 200\,0 \\ -200\,0 \\ \hline 0 \end{array}$$

4. INTERPRET. *Check* your work. Since our estimate is close to our answer of 5, our answer is reasonable. *State* your conclusion: Betty needs 5.8 gallons of paint. If she can buy only gallon containers of paint, she needs 6 gallon containers to complete the job.

◻ **Work Practice Problem 12**

CALCULATOR EXPLORATIONS   Estimation

Calculator errors can easily be made by pressing an incorrect key or by not pressing a correct key hard enough. Estimation is a valuable tool that can be used to check calculator results.

EXAMPLE   Use estimation to determine whether the calculator result is reasonable or not. (For example, a result that is not reasonable can occur if proper keys are not pressed.)

Simplify: $82.064 \div 23$
Calculator display: | 35.68 |

Solution: Round each number to the nearest 10. Since $80 \div 20 = 4$, the calculator display 35.68 is not reasonable.

Use estimation to determine whether each result is reasonable or not.

1. 102.62×41.8 Result: 428.9516

2. $174.835 \div 47.9$ Result: 3.65

3. $1025.68 - 125.42$ Result: 900.26

4. $562.781 + 2.96$ Result: 858.781

Vocabulary and Readiness Check

Use the choices below to fill in each blank. Some choices may be used more than once, and some not used at all.

dividend	divisor	quotient	true
zeros	left	right	false

1. In $6.5 \div 5 = 1.3$, the number 1.3 is called the _____, 5 is the _____, and 6.5 is the _____.

2. To check a division exercise, we can perform the following multiplication: quotient · _____ = _____.

3. To divide a decimal number by a power of 10, such as 10, 100, 1000, and so on, we move the decimal point in the number to the _____ the same number of places as there are _____ in the power of 10.

4. True or false: If we replace x with -12.6 and y with 0.3 in the expression $y \div x$, we have $0.3 \div (-12.6)$. _____

5.4 EXERCISE SET

FOR EXTRA HELP

Student Solutions Manual

PH Math/Tutor Center

CD/Video for Review

Math XL
MathXL®

MyMathLab
MyMathLab

Objectives Ⓐ Ⓑ **Mixed Practice** *Divide. See Examples 1 through 5 and 7. For those exercises marked, also estimate to see if the answer is reasonable.*

1. $6\overline{)27.6}$

2. $4\overline{)23.6}$

3. $5\overline{)0.47}$

4. $6\overline{)0.51}$

5. $0.06\overline{)18}$

6. $0.04\overline{)20}$

7. $0.82\overline{)4.756}$

8. $0.92\overline{)3.312}$

9. $5.5\overline{)36.3}$
Exact:
Estimate:

10. $2.2\overline{)21.78}$
Exact:
Estimate:

11. $7.434 \div 18$

12. $8.304 \div 16$

13. $36 \div (-0.06)$

14. $36 \div (-0.04)$

15. Divide -4.2 by -0.6.

16. Divide -3.6 by -0.9.

17. $0.27\overline{)1.296}$

18. $0.34\overline{)2.176}$

19. $0.02\overline{)42}$

20. $0.03\overline{)24}$

21. $4.756 \div 0.82$

22. $3.312 \div 0.92$

23. $-36.3 \div -6.6$

24. $-21.78 \div -9.9$

25. $7.2\overline{)70.56}$
Exact:
Estimate:

26. $6.3\overline{)54.18}$
Exact:
Estimate:

27. $5.4\overline{)51.84}$

28. $7.7\overline{)33.88}$

29. $\dfrac{1.215}{0.027}$

30. $\dfrac{3.213}{0.051}$

31. $0.25\overline{)13.648}$

32. $0.75\overline{)49.866}$

33. $3.78\overline{)0.02079}$

34. $2.96\overline{)0.01332}$

Divide. Round the quotients as indicated. See Example 6.

35. Divide: $0.549 \div 0.023$. Round the quotient to the nearest hundredth.

36. Divide: $0.0453 \div 0.98$. Round the quotient to the nearest thousandth.

37. Divide: $68.39 \div 0.6$. Round the quotient to the nearest tenth.

38. Divide: $98.83 \div 3.5$. Round the quotient to the nearest tenth.

Objective Ⓒ *Divide. See Examples 8 and 9.*

39. $\dfrac{83.397}{100}$

40. $\dfrac{64.423}{100}$

41. $\dfrac{26.87}{10}$

42. $\dfrac{13.49}{10}$

43. $12.9 \div (-1000)$

44. $13.49 \div (-10,000)$

Objectives Ⓐ Ⓒ **Mixed Practice** *Divide. See Examples 1 through 5, 8, and 9.*

45. $7\overline{)88.2}$

46. $9\overline{)130.5}$

47. $\dfrac{13.1}{10}$

48. $\dfrac{17.7}{10}$

49. $\dfrac{456.25}{10,000}$

50. $\dfrac{986.11}{10,000}$

51. $1.239 \div 3$

52. $0.54 \div 12$

53. Divide 4.2 by -0.6.

54. Divide 3.6 by -0.9.

55. $-1.296 \div 0.27$

56. $-2.176 \div 0.34$

57. Divide 42 by 0.02.

58. Divide 24 by 0.03.

59. Divide -18 by -0.6.

60. Divide 20 by 0.4.

61. Divide 87 by -0.0015.

62. Divide 35 by -0.0007.

63. $-1.104 \div 1.6$

64. $-2.156 \div 0.98$

65. $-2.4 \div (-100)$

66. $-86.79 \div (-1000)$

67. $\dfrac{4.615}{0.071}$

68. $\dfrac{23.8}{0.035}$

Objective Ⓓ *Evaluate each expression for $x = 5.65$, $y = -0.8$, and $z = 4.52$. See Example 10.*

69. $z \div y$

70. $z \div x$

71. $x \div y$

72. $y \div 2$

Determine whether the given values are solutions of the given equations. See Example 11.

73. $\dfrac{x}{4} = 3.04$; $x = 12.16$

74. $\dfrac{y}{8} = 0.89$; $y = 7.12$

75. $\dfrac{z}{100} = 0.8$; $z = 8$

76. $\dfrac{x}{10} = 0.23$; $x = 23$

Objective Ⓔ *Solve. See Example 12.*

77. Josef Jones is painting the walls of a room. The walls have a total area of 546 square feet. A quart of paint covers 52 square feet. If he must buy paint in whole quarts, how many quarts does he need?

78. A page of a book contains about 1.5 kilobytes of information. If a computer disk can hold 740 kilobytes of information, how many pages of a book can be stored on one computer disk? Round to the nearest tenth of a page.

79. There are approximately 39.37 inches in 1 meter. How many meters, to the nearest tenth of a meter, are there in 200 inches?

80. There are 2.54 centimeters in 1 inch. How many inches are there in 50 centimeters? Round to the nearest tenth.

81. In the United States, an average child will wear down 730 crayons by his or her tenth birthday. Find the number of boxes of 64 crayons this is equivalent to. Round to the nearest tenth. (*Source:* Binney & Smith Inc.)

82. In 2005, American farmers received an average of $41.96 per hundred pounds of turkey. What was the average price per pound for turkeys? (*Source:* National Agricultural Statistics Service)

A child is to receive a dose of 0.5 teaspoon of cough medicine every 4 hours. If the bottle contains 4 fluid ounces, answer Exercises 83 through 86.

83. A fluid ounce equals 6 teaspoons. How many teaspoons are in 4 fluid ounces?

84. The bottle of medicine contains how many doses for the child?

85. If the child takes a dose every four hours, how many days will the medicine last?

86. If the child takes a dose every six hours, how many days will the medicine last?

87. Americans aged 18–22 drive, on average, 12,900 miles per year. About how many miles each week is that? Round to the nearest tenth. (*Note:* There are 52 weeks in a year.) (*Source:* U.S. Department of Energy)

88. Javier Xeron was interested in the gas mileage on his "new" used car. He filled the tank, drove 423.8 miles, and filled the tank again. When he refilled the tank, it took 19.35 gallons of gas. Calculate the miles per gallon for Javier's car. Round to the nearest tenth

89. The leading money winner in men's professional golf in 2005 was Tiger Woods. He earned approximately $10,628,000. Suppose he had earned this working 40 hours each week for a year. Determine his hourly wage to the nearest cent. (*Note:* There are 52 weeks in a year.) (*Source:* Professional Golf Association)

90. The book *Harry Potter and the Half-Blood Prince* was released to the public on July 16, 2005. Booksellers in the United State sold approximately 6900 thousand copies in the first 24 hours after release. If the same number of books were sold each hour, calculate the number of books sold each hour in the United States for that first day.

Review

Perform the indicated operation. See Sections 4.3 and 4.5.

91. $\dfrac{3}{5} \cdot \dfrac{7}{10}$

92. $\dfrac{3}{5} \div \dfrac{7}{10}$

93. $\dfrac{3}{5} - \dfrac{7}{10}$

94. $-\dfrac{3}{4} - \dfrac{1}{14}$

Concept Extensions

Mixed Practice (*Sections 5.2, 5.3, 5.4*) *Perform the indicated operation.*

95. $1.278 \div 0.3$

96. 1.278×0.3

97. $1.278 + 0.3$

98. $1.278 - 0.3$

99. $(-8.6)(3.1)$

100. $7.2 + 0.05 + 49.1$

101. $\begin{array}{r} 1000 \\ -\ \ \ 95.71 \\ \hline \end{array}$

102. $\dfrac{87.2}{-10,000}$

Choose the best estimate.

103. 8.62×41.7
  a. 36
  b. 32
  c. 360
  d. 3.6

104. $1.437 + 20.69$
  a. 34
  b. 22
  c. 3.4
  d. 2.2

105. $78.6 \div 97$
  a. 7.86
  b. 0.786
  c. 786
  d. 7860

106. $302.729 - 28.697$
  a. 270
  b. 20
  c. 27
  d. 300

Recall from Section 1.7 that the average of a list of numbers is their total divided by how many numbers there are in the list. Use this procedure to find the average of the test scores listed in Exercises 107 and 108. If necessary, round to the nearest tenth.

107. 86, 78, 91, 87

108. 56, 75, 80

△ **109.** The area of a rectangle is 38.7 square feet. If its width is 4.5 feet, find its length.

△ **110.** The perimeter of a square is 180.8 centimeters. Find the length of a side.

111. When dividing decimals, describe the process you use to place the decimal point in the quotient.

112. In your own words, describe how to quickly divide a number by a power of 10 such as 10, 100, 1000, etc.

To convert wind speeds in miles per hour to knots, divide by 1.15. Use this information and the Saffir-Simpson Hurricane Intensity chart below to answer Exercises 113 and 114. Round to the nearest tenth.

Saffir-Simpson Hurricane Intensity Scale				
Category	**Wind Speed**	**Barometric Pressure [inches of mercury (Hg)]**	**Storm Surge**	**Damage Potential**
1 (Weak)	75–95 mph	≥ 28.94 in.	4–5 ft	Minimal damage to vegetation
2 (Moderate)	96–110 mph	28.50–28.93 in.	6–8 ft	Moderate damage to houses
3 (Strong)	111–130 mph	27.91–28.49 in.	9–12 ft	Extensive damage to small buildings
4 (Very Strong)	131–155 mph	27.17–27.90 in.	13–18 ft	Extreme structural damage
5 (Devastating)	>155 mph	<27.17 in.	>18 ft	Catastrophic building failures possible

113. The chart gives wind speeds in miles per hour. What is the range of wind speeds for a Category 1 hurricane in knots?

114. What is the range of wind speeds for a Category 4 hurricane in knots?

115. Don Larson is building a horse corral that's shaped like a rectangle with dimensions of 24.28 meters by 15.675 meters. He plans to make a four-wire fence; that is, he will string four wires around the corral. How much wire will he need?

116. Takeesha Bethel signed up for a new credit card that guarantees her no interest charges on transferred balances for a year. She transferred over a $2523.86 balance from her old credit card. Her minimum payment is $185.35 per month. If she only pays the minimum, will she pay off her balance before interest charges start again?

THE BIGGER PICTURE   Operations on Sets of Numbers

Continue your outline from Sections 1.6, 1.7, 2.4, 3.3, 4.3, and 4.7. Suggestions are once again written to help you complete this part of your outline.

I. Operations on Sets of Numbers

  A. Whole Numbers

    1. Add or Subtract (Section 1.3)

    2. Multiply or Divide (Sections 1.5, 1.6)

    3. Exponent (Section 1.7)

    4. Order of Operations (Section 1.7)

  B. Integers

    1. Add (Section 2.2)

    2. Subtract (Section 2.3)

    3. Multiply or Divide (Section 2.4)

  C. Fractions

    1. Simplify (Section 4.2)

    2. Multiply (Section 4.3)

    3. Divide (Section 4.3)

    4. Add or Subtract (Sections 4.4, 4.5)

  D. Decimals

    1. Add or Subtract: Line up decimal points.

$$\begin{array}{r} 1.27 \\ +0.6 \\ \hline 1.87 \end{array}$$

    2. Multiply:

$$\begin{array}{rl} 2.56 & \text{2 decimal places} \\ \times\ 3.2 & \text{1 decimal place} \\ \hline 512 & 2 + 1 = 3 \\ 7680 & \\ \hline 8.192 & \text{3 decimal places} \end{array}$$

    3. Divide:

$$\begin{array}{r} 0.7 \\ 8\overline{)5.6} \end{array} \qquad \begin{array}{r} 1.31 \\ 0.6\,\overline{)0.786} \end{array}$$

II. Solving Equations

  A. Equations in General (Section 3.3)

Perform the indicated operations.

1. $3.6 + 8.092 + 10.48$

2. $7 - 3.049$

3. 91.332×100

4. $-\dfrac{68}{10}$

5. $\begin{array}{r} 5.2 \\ \times\ 0.27 \end{array}$

6. $9\overline{)77.94}$

7. $0.35\overline{)0.01785}$

8. $2.3 - (0.4)^2$

9. $\dfrac{8}{15} - \dfrac{2}{5}$

10. $-\dfrac{8}{15} \cdot \dfrac{2}{5}$

Sections 5.1–5.4

Operations on Decimals

Perform the indicated operation.

1. $1.6 + 0.97$

2. $3.2 + 0.85$

3. $9.8 - 0.9$

4. $10.2 - 6.7$

5. $\begin{array}{r} 0.8 \\ \times\,0.2 \\ \hline \end{array}$

6. $\begin{array}{r} 0.6 \\ \times\,0.4 \\ \hline \end{array}$

7. $8\overline{)2.16}$

8. $6\overline{)3.12}$

9. $(9.6)(-0.5)$

10. $(-8.7)(-0.7)$

11. $\begin{array}{r} 123.6 \\ -\,\,\,48.04 \\ \hline \end{array}$

12. $\begin{array}{r} 325.2 \\ -\,\,\,36.08 \\ \hline \end{array}$

13. $-25 + 0.026$

14. $0.125 + (-44)$

15. $29.24 \div (-3.4)$

16. $-10.26 \div (-1.9)$

17. -2.8×100

18. 1.6×1000

19. $\begin{array}{r} 96.21 \\ 7.028 \\ +121.7 \\ \hline \end{array}$

20. $\begin{array}{r} 0.268 \\ 1.93 \\ +142.881 \\ \hline \end{array}$

21. $-25.76 \div -46$

22. $-27.09 \div 43$

23. $\begin{array}{r} 12.004 \\ \times\,\,\,\,\,2.3 \\ \hline \end{array}$

24. $\begin{array}{r} 28.006 \\ \times\,\,\,\,\,5.2 \\ \hline \end{array}$

1. _____
2. _____
3. _____
4. _____
5. _____
6. _____
7. _____
8. _____
9. _____
10. _____
11. _____
12. _____
13. _____
14. _____
15. _____
16. _____
17. _____
18. _____
19. _____
20. _____
21. _____
22. _____
23. _____
24. _____

25. _____

26. _____

27. _____

28. _____

29. _____

30. _____

31. _____

32. _____

33. _____

34. _____

35. _____

36. _____

37. _____

38. _____

39. _____

25. Subtract 4.6 from 10.

26. Subtract 18 from 0.26.

27. $-268.19 - 146.25$

28. $-860.18 - 434.85$

29. $\dfrac{2.958}{-0.087}$

30. $\dfrac{-1.708}{0.061}$

31. $160 - 43.19$

32. $120 - 101.21$

33. 15.62×10

34. $15.62 \div 10$

35. $15.62 + 10$

36. $15.62 - 10$

37. Estimate the distance in miles between Garden City, Kansas, and Wichita, Kansas, by rounding each given distance to the nearest ten.

38. It costs $3.30 to send a 2-pound package locally via parcel post at a U.S. Post Office. To send the same package as Priority Mail, it costs $4.20. How much more does it cost to send a package as Priority Mail? (_Source:_ United States Postal Service)

39. In 2005, Americans spent a total of $23.324 billion on DVD home entertainment and $1.556 on VHS home entertainment. Find the total spent on DVD or VHS home entertainment. Write the total in billions of dollars and also in standard notation.

5.5 FRACTIONS, DECIMALS, AND ORDER OF OPERATIONS

Objective A Writing Fractions as Decimals

To write a fraction as a decimal, we interpret the fraction bar to mean division and find the quotient.

> **Writing Fractions as Decimals**
>
> To write a fraction as a decimal, divide the numerator by the denominator.

EXAMPLE 1   Write $\frac{1}{4}$ as a decimal.

Solution: $\frac{1}{4} = 1 \div 4$

$$
\begin{array}{r}
0.25 \\
4{\overline{\smash{)}\,1.00}} \\
\underline{-\ 8} \\
20 \\
\underline{-\ 20} \\
0
\end{array}
$$

Thus, $\frac{1}{4}$ written as a decimal is 0.25.

🖳 **Work Practice Problem 1**

PRACTICE PROBLEM 1

a. Write $\frac{2}{5}$ as a decimal.

b. Write $\frac{9}{40}$ as a decimal.

EXAMPLE 2   Write $-\frac{5}{8}$ as a decimal.

Solution: $-\frac{5}{8} = -(5 \div 8) = -0.625$

$$
\begin{array}{r}
0.625 \\
8{\overline{\smash{)}\,5.000}} \\
\underline{-4\ 8} \\
20 \\
\underline{-16} \\
40 \\
\underline{-40} \\
0
\end{array}
$$

🖳 **Work Practice Problem 2**

PRACTICE PROBLEM 2

Write $-\frac{3}{8}$ as a decimal.

EXAMPLE 3   Write $\frac{2}{3}$ as a decimal.

Solution:

$$
\begin{array}{r}
0.666\ldots \\
3{\overline{\smash{)}\,2.000}} \\
\underline{-1\ 8} \\
20 \\
\underline{-18} \\
20 \\
\underline{-18} \\
2
\end{array}
$$

This pattern will continue because $\frac{2}{3} = 0.6666\ldots$

Remainder is 2, then 0 is brought down.

Remainder is 2, then 0 is brought down.

Remainder is 2.

PRACTICE PROBLEM 3

a. Write $\frac{5}{6}$ as a decimal.

b. Write $\frac{2}{9}$ as a decimal.

Answers

1. a. 0.4 **b.** 0.225   **2.** −0.375
3. a. $0.8\overline{3}$ **b.** $0.\overline{2}$

Continued on next page

Notice that the digit 2 keeps occurring as the remainder. This will continue so that the digit 6 will keep repeating in the quotient. We place a bar over the digit 6 to indicate that it repeats.

$$\frac{2}{3} = 0.666\ldots = 0.\overline{6}$$

We can also write a decimal approximation for $\frac{2}{3}$. For example, $\frac{2}{3}$ rounded to the nearest hundredth is 0.67. This can be written as $\frac{2}{3} \approx 0.67$.

◼ **Work Practice Problem 3**

PRACTICE PROBLEM 4

Write $\frac{28}{13}$ as a decimal. Round to the nearest thousandth.

EXAMPLE 4 Write $\frac{22}{7}$ as a decimal. (Recall that the fraction $\frac{22}{7}$ is an approximation for π.) Round to the nearest hundredth.

Solution: $3.142 \approx 3.14$ Carry the division out to the thousandths place.

$$
\begin{array}{r}
3.142 \\
7)\overline{22.000} \\
-21 \\
\hline
1\,0 \\
-\ 7 \\
\hline
30 \\
-28 \\
\hline
20 \\
-14 \\
\hline
6
\end{array}
$$

The fraction $\frac{22}{7}$ in decimal form is approximately 3.14.

◼ **Work Practice Problem 4**

PRACTICE PROBLEM 5

Write $3\frac{5}{16}$ as a decimal.

EXAMPLE 5 Write $2\frac{3}{16}$ as a decimal.

Solution:

Option 1. Write the fractional part only as a decimal.

$$\frac{3}{16} \longrightarrow
\begin{array}{r}
0.1875 \\
16)\overline{3.0000} \\
-1\,6 \\
\hline
1\,40 \\
-1\,28 \\
\hline
120 \\
-112 \\
\hline
80 \\
-80 \\
\hline
0
\end{array}
$$

Thus $2\frac{3}{16} = 2.1875$

Option 2. Write $2\frac{3}{16}$ as an improper fraction, and divide.

$$2\frac{3}{16} = \frac{35}{16} \longrightarrow
\begin{array}{r}
2.1875 \\
16)\overline{35.0000} \\
-32 \\
\hline
3\,0 \\
-1\,6 \\
\hline
1\,40 \\
-1\,28 \\
\hline
120 \\
-112 \\
\hline
80 \\
-80 \\
\hline
0
\end{array}
$$

Thus $2\frac{3}{16} = 2.1875$

◼ **Work Practice Problem 5**

Some fractions may be written as decimals using our knowledge of decimals. From Section 5.1, we know that if the denominator of a fraction is 10, 100, 1000, or so on, we can immediately write the fraction as a decimal. For example,

$$\frac{4}{10} = 0.4, \qquad \frac{12}{100} = 0.12, \text{ and so on}$$

EXAMPLE 6 Write $\frac{4}{5}$ as a decimal.

Solution: Let's write $\frac{4}{5}$ as an equivalent fraction with a denominator of 10.

$$\frac{4}{5} = \frac{4}{5} \cdot \frac{2}{2} = \frac{8}{10} = 0.8$$

▯ **Work Practice Problem 6**

EXAMPLE 7 Write $\frac{1}{25}$ as a decimal.

Solution: $\frac{1}{25} = \frac{1}{25} \cdot \frac{4}{4} = \frac{4}{100} = 0.04$

▯ **Work Practice Problem 7**

✔ **Concept Check**   Suppose you are writing the fraction $\frac{9}{16}$ as a decimal. How do you know you have made a mistake if your answer is 1.735?

Objective B Comparing Decimals and Fractions

Now we can compare decimals and fractions by writing fractions as equivalent decimals.

EXAMPLE 8 Insert $<$, $>$, or $=$ to form a true statement.

$$\frac{1}{8} \qquad 0.12$$

Solution: First we write $\frac{1}{8}$ as an equivalent decimal. Then we compare decimal places.

$$\begin{array}{r} 0.125 \\ 8\overline{)1.000} \\ -8 \\ \hline 20 \\ -16 \\ \hline 40 \\ -40 \\ \hline 0 \end{array}$$

Original numbers	$\frac{1}{8}$	0.12
Decimals	0.125	0.120
Compare	0.125 > 0.12	

Thus, $\frac{1}{8} > 0.12$

▯ **Work Practice Problem 8**

PRACTICE PROBLEM 9

Insert $<$, $>$, or $=$ to form a true statement.

a. $\dfrac{1}{2}$ 0.54   **b.** $0.\overline{4}$ $\dfrac{4}{9}$

c. $\dfrac{5}{7}$ 0.72

EXAMPLE 9 Insert $<$, $>$, or $=$ to form a true statement.

$$0.\overline{7} \qquad \frac{7}{9}$$

Solution: We write $\dfrac{7}{9}$ as a decimal and then compare.

$$
\begin{array}{r}
0.77\ldots = 0.\overline{7} \\
9\overline{)7.00} \\
-\,6\,3 \\
\hline
70 \\
-\,63 \\
\hline
7
\end{array}
$$

Original numbers	$0.\overline{7}$	$\dfrac{7}{9}$
Decimals	$0.\overline{7}$	$0.\overline{7}$
Compare	$0.\overline{7} = 0.\overline{7}$	

Thus, $0.\overline{7} = \dfrac{7}{9}$

▢ **Work Practice Problem 9**

PRACTICE PROBLEM 10

Write the numbers in order from smallest to largest.

a. $\dfrac{1}{3}, 0.302, \dfrac{3}{8}$ **b.** $1.26, 1\dfrac{1}{4}, 1\dfrac{2}{5}$

c. $0.4, 0.41, \dfrac{5}{7}$

EXAMPLE 10 Write the numbers in order from smallest to largest.

$$\frac{9}{20}, \frac{4}{9}, 0.456$$

Solution:

Original numbers	$\dfrac{9}{20}$	$\dfrac{4}{9}$	0.456
Decimals	0.450	$0.444\ldots$	0.456
Compare in order	2nd	1st	3rd

Written in order, we have

1st   2nd   3rd
↓   ↓   ↓

$$\frac{4}{9}, \frac{9}{20}, 0.456$$

▢ **Work Practice Problem 10**

Objective C Simplifying Expressions with Decimals and Fractions

In the remaining examples, we will review the order of operations by simplifying expressions that contain decimals.

Order of Operations

1. Perform all operations within parentheses ( ), brackets [ ], or other grouping symbols such as fraction bars.

2. Evaluate any expressions with exponents.

3. Multiply or divide in order from left to right.

4. Add or subtract in order from left to right.

Answers

9. a. $<$ **b.** $=$ **c.** $<$

10. a. $0.302, \dfrac{1}{3}, \dfrac{3}{8}$ **b.** $1\dfrac{1}{4}, 1.26, 1\dfrac{2}{5}$

c. $0.4, 0.41, \dfrac{5}{7}$

EXAMPLE 11 Simplify: $723.6 \div 1000 \times 10$

Solution: Multiply or divide in order from left to right.

$$
\begin{aligned}
723.6 \div 1000 \times 10 &= 0.7236 \times 10 \quad \text{Divide.} \\
&= 7.236 \quad \text{Multiply.}
\end{aligned}
$$

◻ **Work Practice Problem 11**

PRACTICE PROBLEM 11
Simplify: $897.8 \div 100 \times 10$

EXAMPLE 12 Simplify: $-0.5(8.6 - 1.2)$

Solution: According to the order of operations, we simplify inside the parentheses first.

$$
\begin{aligned}
-0.5(8.6 - 1.2) &= -0.5(7.4) \quad \text{Subtract.} \\
&= -3.7 \quad \text{Multiply.}
\end{aligned}
$$

◻ **Work Practice Problem 12**

PRACTICE PROBLEM 12
Simplify: $-8.69(3.2 - 1.8)$

EXAMPLE 13 Simplify: $(-1.3)^2 + 2.4$

Solution: Recall the meaning of an exponent.

$$
\begin{aligned}
(-1.3)^2 &= (-1.3)(-1.3) + 2.4 \quad \text{Use the definition of an exponent.} \\
&= 1.69 + 2.4 \quad \text{Multiply. The product of two negative numbers is a positive number} \\
&= 4.09 \quad \text{Add.}
\end{aligned}
$$

◻ **Work Practice Problem 13**

PRACTICE PROBLEM 13
Simplify: $(-0.7)^2 + 2.1$

EXAMPLE 14 Simplify: $\dfrac{5.68 + (0.9)^2 \div 100}{0.2}$

Solution: First we simplify the numerator of the fraction. Then we divide.

$$
\begin{aligned}
\frac{5.68 + (0.9)^2 \div 100}{0.2} &= \frac{5.68 + 0.81 \div 100}{0.2} \quad \text{Simplify } (0.9)^2. \\
&= \frac{5.68 + 0.0081}{0.2} \quad \text{Divide.} \\
&= \frac{5.6881}{0.2} \quad \text{Add.} \\
&= 28.4405 \quad \text{Divide.}
\end{aligned}
$$

◻ **Work Practice Problem 14**

PRACTICE PROBLEM 14
Simplify: $\dfrac{20.06 - (1.2)^2 \div 10}{0.02}$

Objective D Solving Area Problems Containing Fractions and Decimals

Sometimes real-life problems contain both fractions and decimals. In the next example, we review the area of a triangle, and when values are substituted, the result may be an expression containing both fractions and decimals.

Answers
11. 89.78 **12.** -12.166
13. 2.59 **14.** 995.8

PRACTICE PROBLEM 15

Find the area of the triangle.

EXAMPLE 15 The area of a triangle is Area $= \frac{1}{2} \cdot$ base $\cdot$ height. Find the area of the triangle shown.

Solution:

$$\text{Area} = \frac{1}{2} \cdot \text{base} \cdot \text{height}$$

$$= \frac{1}{2} \cdot 5.6 \cdot 3$$

$$= 0.5 \cdot 5.6 \cdot 3 \qquad \text{Write } \frac{1}{2} \text{ as the decimal } 0.5.$$

$$= 8.4$$

The area of the triangle is 8.4 square feet.

▣ **Work Practice Problem 15**

Objective Ⓔ Using Decimals as Replacement Values

PRACTICE PROBLEM 16

Evaluate $1.7y - 2$ for $y = 2.3$.

EXAMPLE 16 Evaluate $-2x + 5$ for $x = 3.8$.

Solution: Replace x with 3.8 in the expression $-2x + 5$ and simplify.

$$-2x + 5 = -2(3.8) + 5 \qquad \text{Replace } x \text{ with 3.8.}$$

$$= -7.6 + 5 \qquad \text{Multiply.}$$

$$= -2.6 \qquad \text{Add.}$$

▣ **Work Practice Problem 16**

Answers

15. 7.35 sq m   **16.** 1.91

Vocabulary and Readiness Check

Answer each exercise "true" or "false."

1. The number $0.\overline{5}$ means 0.555.

2. To write $\frac{9}{19}$ as a decimal, perform the division $19\overline{)9}$.

3. $(-1.2)^2$ means $(-1.2)(-1.2)$ or -1.44.

4. To simplify $8.6(4.8 - 9.6)$, we first subtract.

5.5 EXERCISE SET

FOR EXTRA HELP

Student Solutions Manual   PH Math/Tutor Center   CD/Video for Review   MathXL MathXL®   MyMathLab MyMathLab

Objective Ⓐ *Write each number as a decimal. See Examples 1 through 7.*

1. $\frac{1}{5}$

2. $\frac{1}{20}$

3. $\frac{17}{25}$

4. $\frac{13}{25}$

 5. $\frac{3}{4}$

6. $\frac{3}{8}$

7. $-\frac{2}{25}$

8. $-\frac{3}{25}$

9. $\frac{9}{4}$

10. $\frac{8}{5}$

 11. $\frac{11}{12}$

12. $\frac{5}{12}$

13. $\frac{17}{40}$

14. $\frac{19}{25}$

15. $\frac{9}{20}$

16. $\frac{31}{40}$

17. $-\frac{1}{3}$

18. $-\frac{7}{9}$

19. $\frac{7}{16}$

20. $\frac{9}{16}$

21. $\frac{7}{11}$

22. $\frac{9}{11}$

23. $5\frac{17}{20}$

24. $4\frac{7}{8}$

25. $\frac{78}{125}$

26. $\frac{159}{375}$

Round each number as indicated. See Example 4.

27. Round your decimal answer to Exercise 17 to the nearest hundredth.

28. Round your decimal answer to Exercise 18 to the nearest hundredth.

29. Round your decimal answer to Exercise 19 to the nearest hundredth.

30. Round your decimal answer to Exercise 20 to the nearest hundredth.

31. Round your decimal answer to Exercise 21 to the nearest tenth.

32. Round your decimal answer to Exercise 22 to the nearest tenth.

Write each fraction as a decimal. If necessary, round to the nearest hundredth. See Examples 1 through 7.

33. Of the U.S. mountains that are over 14,000 feet in elevation, $\frac{56}{91}$ are located in Colorado. (*Source:* U.S. Geological Survey)

34. About $\frac{21}{50}$ of all blood donors have type A blood. (*Source:* American Red Cross Biomedical Services)

35. The United States contains the greatest fraction of people who use the Internet, with about $\frac{67}{94}$ people using it. (*Source:* UCLA Center for Communication Policy)

36. Hungary has the lowest fraction of people using the Internet, with only $\frac{7}{40}$ people using it. (*Source:* UCLA Center for Communication Policy)

37. When first launched, the Hubble Space Telescope's primary mirror was out of shape on the edges by $\frac{1}{50}$ of a human hair. This very small defect made it difficult to focus faint objects being viewed. Because the HST was in low Earth orbit, it was serviced by a shuttle and the defect was corrected.

38. The two mirrors currently in use in the Hubble Space Telescope were ground so that they do not deviate from a perfect curve by more than $\frac{1}{800,000}$ of an inch. Do not round this off.

Objective B *Insert $<$, $>$, or $=$ to form a true statement. See Examples 8 and 9.*

39. 0.562   0.569

40. 0.983   0.988

41. 0.215 $\frac{43}{200}$

42. $\frac{29}{40}$ 0.725

43. -0.0932 -0.0923

44. -0.00563 -0.00536

45. $0.\overline{6}$ $\frac{5}{6}$

46. $0.\overline{1}$ $\frac{2}{17}$

47. $\frac{51}{91}$ $0.56\overline{4}$

48. $0.58\overline{3}$ $\frac{6}{11}$

49. $\frac{4}{7}$ 0.14

50. $\frac{5}{9}$ 0.557

51. 1.38 $\frac{18}{13}$

52. 0.372 $\frac{22}{59}$

53. 7.123 $\frac{456}{64}$

54. 12.713 $\frac{89}{7}$

Write the numbers in order from smallest to largest. See Example 10.

55. 0.34, 0.35, 0.32

56. 0.47, 0.42, 0.40

57. 0.49, 0.491, 0.498

58. 0.72, 0.727, 0.728

59. 5.23, $\frac{42}{8}$, 5.34

60. 7.56, $\frac{67}{9}$, 7.562

61. $\frac{5}{8}$, 0.612, 0.649

62. $\frac{5}{6}$, 0.821, 0.849

Objective C *Simplify each expression. See Examples 11 through 14.*

63. $(0.3)^2 + 0.5$

64. $(-2.5)(3) - 4.7$

65. $\dfrac{1 + 0.8}{-0.6}$

66. $(-0.05)^2 + 3.13$

67. $(-2.3)^2(0.3 + 0.7)$

68. $(8.2)(100) - (8.2)(10)$

69. $(5.6 - 2.3)(2.4 + 0.4)$

70. $\dfrac{0.222 - 2.13}{12}$

71. $\dfrac{(4.5)^2}{100}$

72. $0.9(5.6 - 6.5)$

73. $\dfrac{7 + 0.74}{-6}$

74. $(1.5)^2 + 0.5$

Find the value of each expression. Give the result as a decimal. See Examples 11 through 14.

75. $\dfrac{1}{5} - 2(7.8)$

76. $\dfrac{3}{4} - (9.6)(5)$

77. $\dfrac{1}{4}(-9.6 - 5.2)$

78. $\dfrac{3}{8}(4.7 - 5.9)$

Objective *Find the area of each triangle or rectangle. See Example 15.*

 79.

9 inches

5.7 inches

△ **80.**

4.4 feet

17 feet

△ **81.**

0.62 yard

$\frac{2}{5}$ yard

△ **82.**

1.2 miles

$\frac{7}{8}$ mile

Objective 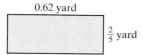 *Evaluate each expression for $x = 6$, $y = 0.3$, and $z = -2.4$. See Example 16.*

83. z^2 **84.** y^2 **85.** $x - y$ **86.** $x - z$ **87.** $4y - z$ **88.** $\frac{x}{y} + 2z$

Review

Simplify. See Sections 1.7 and 4.6.

89. $\frac{9}{10} + \frac{16}{25}$ **90.** $\frac{4}{11} - \frac{19}{22}$ **91.** $\left(\frac{2}{5}\right)\left(\frac{5}{2}\right)^2$ **92.** $\left(\frac{2}{3}\right)^2\left(\frac{3}{2}\right)^3$

Concept Extensions

Without calculating, describe each number as < 1, $= 1$, or > 1. See the Concept Check in this section.

93. 1.0   **94.** 1.0000   **95.** 1.00001   **96.** $\frac{101}{99}$ **97.** $\frac{99}{100}$ **98.** $\frac{99}{99}$

In 2005, there were 10,661 commercial radio stations in the United States. The most popular formats are listed in the table along with their counts. Use this graph to answer Exercises 99 through 102.

99. Write the fraction of radio stations with a country music format as a decimal. Round to the nearest thousandth.

100. Write the fraction of radio stations with a news/talk format as a decimal. Round to the nearest hundredth.

101. Estimate, by rounding each number in the table to the nearest hundred, the total number of stations with the top six formats in 2005.

102. Use your estimate from Exercise 101 to write the fraction of radio stations accounted for by the top six formats as a decimal. Round to the nearest hundredth.

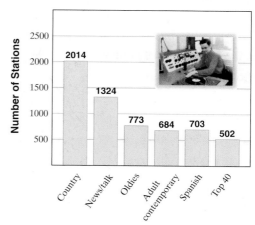

Top Commercial Radio Station Formats in 2004

Number of Stations

2014, 1324, 773, 684, 703, 502

Country, News/talk, Oldies, Adult contemporary, Spanish, Top 40

Format (Total stations: 10,661)

103. Describe two ways to write fractions as decimals.

104. Describe two ways to write mixed numbers as decimals.

5.6 EQUATIONS CONTAINING DECIMALS

Objective **A** Solving Equations Containing Decimals

In this section, we continue our work with decimals and algebra by solving equations containing decimals. First, we review the steps given earlier for solving an equation.

Steps for Solving an Equation in x

Step 1: If fractions are present, multiply both sides of the equation by the LCD of the fractions.

Step 2: If parentheses are present, use the distributive property.

Step 3: Combine any like terms on each side of the equation.

Step 4: Use the addition property of equality to rewrite the equation so that variable terms are on one side of the equation and constant terms are on the other side.

Step 5: Divide both sides by the numerical coefficient of x to solve.

Step 6: Check the answer in the **original equation.**

PRACTICE PROBLEM 1

Solve: $z + 0.9 = 1.3$

EXAMPLE 1 Solve: $x - 1.5 = 8$

Solution: Steps 1 through 3 are not needed for this equation, so we begin with Step 4. To get x alone on one side of the equation, add 1.5 to both sides.

$$x - 1.5 = 8 \quad \text{Original equation}$$
$$x - 1.5 + 1.5 = 8 + 1.5 \quad \text{Add 1.5 to both sides.}$$
$$x = 9.5 \quad \text{Simplify.}$$

Check: To check, replace x with 9.5 in the *original equation*.

$$x - 1.5 = 8 \quad \text{Original equation}$$
$$9.5 - 1.5 \overset{?}{=} 8 \quad \text{Replace } x \text{ with 9.5.}$$
$$8 = 8 \quad \text{True}$$

Since $8 = 8$ is a true statement, 9.5 is a solution of the equation.

■ **Work Practice Problem 1**

PRACTICE PROBLEM 2

Solve: $0.17x = -0.34$

EXAMPLE 2 Solve: $-2y = 6.7$

Solution: Steps 1 through 4 are not needed for this equation, so we begin with Step 5. To solve for y, divide both sides by the coefficient of y, which is -2.

$$-2y = 6.7 \quad \text{Original equation}$$
$$\frac{-2y}{-2} = \frac{6.7}{-2} \quad \text{Divide both sides by } -2.$$
$$y = -3.35 \quad \text{Simplify.}$$

Check: To check, replace y with -3.35 in the original equation.

$$-2y = 6.7 \quad \text{Original equation}$$
$$-2(-3.35) \overset{?}{=} 6.7 \quad \text{Replace } y \text{ with } -3.35.$$
$$6.7 = 6.7 \quad \text{True}$$

Thus -3.35 is a solution of the equation $-2y = 6.7$.

■ **Work Practice Problem 2**

Answers
1. 0.4   2. −2

EXAMPLE 3 Solve: $1.2x + 5.8 = 8.2$

Solution: We begin with Step 4 and get the variable term alone by subtracting 5.8 from both sides.

$$1.2x + 5.8 = 8.2$$
$$1.2x + 5.8 - 5.8 = 8.2 - 5.8 \quad \text{Subtract 5.8 from both sides.}$$
$$1.2x = 2.4 \quad \text{Simplify.}$$
$$\frac{1.2x}{1.2} = \frac{2.4}{1.2} \quad \text{Divide both sides by 1.2.}$$
$$x = 2 \quad \text{Simplify.}$$

To check, replace x with 2 in the original equation. The solution is 2.

Work Practice Problem 3

PRACTICE PROBLEM 3
Solve: $2.9 = 1.7 + 0.3x$

EXAMPLE 4 Solve: $7x + 3.2 = 4x - 1.6$

Solution: We start with Step 4 to get variable terms on one side and numerical terms on the other.

$$7x + 3.2 = 4x - 1.6$$
$$7x + 3.2 - 3.2 = 4x - 1.6 - 3.2 \quad \text{Subtract 3.2 from both sides.}$$
$$7x = 4x - 4.8 \quad \text{Simplify.}$$
$$7x - 4x = 4x - 4.8 - 4x \quad \text{Subtract } 4x \text{ from both sides.}$$
$$3x = -4.8 \quad \text{Simplify.}$$
$$\frac{\overset{1}{\cancel{3}}x}{\underset{1}{\cancel{3}}} = -\frac{4.8}{3} \quad \text{Divide both sides by 3.}$$
$$x = -1.6 \quad \text{Simplify.}$$

Check to see that -1.6 is the solution.

Work Practice Problem 4

PRACTICE PROBLEM 4
Solve: $8x + 4.2 = 10x + 11.6$

EXAMPLE 5 Solve: $5(x - 0.36) = -x + 2.4$

Solution: First use the distributive property to distribute the factor 5.

$$5(x - 0.36) = -x + 2.4 \quad \text{Original equation}$$
$$5x - 1.8 = -x + 2.4 \quad \text{Apply the distributive property.}$$

Next, get x alone on the left side of the equation by adding 1.8 to both sides of the equation and then adding x to both sides of the equation.

$$5x - 1.8 + 1.8 = -x + 2.4 + 1.8 \quad \text{Add 1.8 to both sides.}$$
$$5x = -x + 4.2 \quad \text{Simplify.}$$
$$5x + x = -x + 4.2 + x \quad \text{Add } x \text{ to both sides.}$$
$$6x = 4.2 \quad \text{Simplify.}$$
$$\frac{6x}{6} = \frac{4.2}{6} \quad \text{Divide both sides by 6.}$$
$$x = 0.7 \quad \text{Simplify.}$$

To verify that 0.7 is the solution, replace x with 0.7 in the original equation.

Work Practice Problem 5

PRACTICE PROBLEM 5
Solve: $6.3 - 5x = 3(x + 2.9)$

Instead of solving equations with decimals, sometimes it may be easier to first rewrite the equation so that it contains integers only. Recall that multiplying a decimal by a power of 10, such as 10, 100, or 1000, has the effect of moving the decimal point to the right. We can use the multiplication property of equality to multiply both sides of the equation through by an appropriate power of 10. The resulting equivalent equation will contain integers only.

Answers
3. 4   **4.** −3.7   **5.** −0.3

PRACTICE PROBLEM 6

Solve: $0.2y + 2.6 = 4$

EXAMPLE 6 Solve: $0.5y + 2.3 = 1.65$

Solution: Multiply the equation through by 100. This will move the decimal point in each term two places to the right.

$$0.5y + 2.3 = 1.65 \qquad \text{Original equation}$$
$$100(0.5y + 2.3) = 100(1.65) \qquad \text{Multiply both sides by 100.}$$
$$100(0.5y) + 100(2.3) = 100(1.65) \qquad \text{Apply the distributive property.}$$
$$50y + 230 = 165 \qquad \text{Simplify.}$$

Now the equation contains integers only. Finish solving by subtracting 230 from both sides.

$$50y + 230 = 165$$
$$50y + 230 - 230 = 165 - 230 \qquad \text{Subtract 230 from both sides.}$$
$$50y = -65 \qquad \text{Simplify.}$$
$$\frac{50y}{50} = \frac{-65}{50} \qquad \text{Divide both sides by 50.}$$
$$y = -1.3 \qquad \text{Simplify.}$$

Check to see that -1.3 is the solution by replacing y with -1.3 in the original equation.

▢ **Work Practice Problem 6**

Answer

6. 7

✔ **Concept Check Answer**

Multiply by 1000.

✔ **Concept Check** By what number would you multiply both sides of the following equation to make calculations easier? Explain your choice.

$$1.7x + 3.655 = -14.2$$

5.6 EXERCISE SET

Objective Ⓐ *Solve each equation. See Examples 1 and 2.*

1. $x + 1.2 = 7.1$

2. $y - 0.5 = 9$

3. $-5y = 2.15$

4. $-0.4x = 50$

5. $6.2 = y - 4$

6. $9.7 = x + 11.6$

7. $3.1x = -13.95$

8. $3y = -25.8$

Solve each equation. See Examples 3 through 5.

9. $-3.5x + 2.8 = -11.2$

10. $7.1 - 0.2x = 6.1$

11. $6x + 8.65 = 3x + 10$

12. $7x - 9.64 = 5x + 2.32$

13. $2(x - 1.3) = 5.8$

14. $5(x + 2.3) = 19.5$

Solve each equation by first multiplying both sides through by an appropriate power of 10 so that the equation contains integers only. See Example 6.

15. $0.4x + 0.7 = -0.9$

16. $0.7x + 0.1 = 1.5$

17. $7x - 10.8 = x$

18. $3y = 7y + 24.4$

19. $2.1x + 5 - 1.6x = 10$

20. $1.5x + 2 - 1.2x = 12.2$

Solve. See Examples 1 through 6.

21. $y - 3.6 = 4$

22. $x + 5.7 = 8.4$

23. $-0.02x = -1.2$

24. $-9y = -0.162$

25. $6.5 = 10x + 7.2$

26. $2x - 4.2 = 8.6$

27. $2.7x - 25 = 1.2x + 5$

28. $9y - 6.9 = 6y - 11.1$

29. $200x - 0.67 = 100x + 0.81$

30. $2.3 + 500x = 600x - 0.2$

31. $3(x + 2.71) = 2x$

32. $7(x + 8.6) = 6x$

33. $8x - 5 = 10x - 8$

34. $24y - 10 = 20y - 17$

35. $1.2 + 0.3x = 0.9$

36. $1.5 = 0.4x + 0.5$

37. $-0.9x + 2.65 = -0.5x + 5.45$

38. $-50x + 0.81 = -40x - 0.48$

39. $4x + 7.6 = 2(3x - 3.2)$

40. $4(2x - 1.6) = 5x - 6.4$

41. $0.7x + 13.8 = x - 2.16$

42. $y - 5 = 0.3y + 4.1$

Review

Simplify each expression by combining like terms. If parentheses are present, first use the distributive property. See Section 3.1.

43. $2x - 7 + x - 9$

44. $x + 14 - 5x - 17$

Perform the indicated operation. See Sections 4.3 and 4.5.

45. $\dfrac{6x}{5} \cdot \dfrac{1}{2x^2}$

46. $\dfrac{x}{3} + \dfrac{2x}{7}$

47. $5\dfrac{1}{3} \div 9\dfrac{1}{6}$

48. $50 - 14\dfrac{9}{13}$

Concept Extensions

Mixed Practice (*Sections 5.2 and 5.6*) *This section of exercises contains equations and expressions. If the exercise contains an equation, solve it for the variable. If the exercise contains an expression, simplify it by combining any like terms.*

49. $b + 4.6 = 8.3$

50. $y - 15.9 = -3.8$

51. $2x - 0.6 + 4x - 0.01$

52. $-x - 4.1 - x - 4.02$

53. $5y - 1.2 - 7y + 8$

54. $9a - 5.6 - 3a + 6$

55. $2.8 = z - 6.3$

56. $9.7 = x + 4.3$

57. $4.7x + 8.3 = -5.8$

58. $2.8x + 3.4 = -13.4$

59. $7.76 + 8z - 12z + 8.91$

60. $9.21 + x - 4x + 11.33$

61. $5(x - 3.14) = 4x$

62. $6(x + 1.43) = 5x$

63. $2.6y + 8.3 = 4.6y - 3.4$

64. $8.4z - 2.6 = 5.4z + 10.3$

65. $9.6z - 3.2 - 11.7z - 6.9$

66. $-3.2x + 12.6 - 8.9x - 15.2$

67. Explain in your own words the property of equality that allows us to multiply an equation through by a power of 10.

68. Construct an equation whose solution is 1.4.

Solve.

69. $-5.25x = -40.33575$

70. $7.68y = -114.98496$

71. $1.95y + 6.834 = 7.65y - 19.8591$

72. $6.11x + 4.683 = 7.51x + 18.235$

5.7 DECIMAL APPLICATIONS: MEAN, MEDIAN, AND MODE

Objectives

A Find the Mean of a List of Numbers.

B Find the Median of a List of Numbers.

C Find the Mode of a List of Numbers.

Objective **A** Finding the Mean

Sometimes we want to summarize data by displaying them in a graph, but sometimes it is also desirable to be able to describe a set of data, or a set of numbers, by a single "middle" number. Three such **measures of central tendency** are the **mean, the median,** and the **mode.**

The most common measure of central tendency is the mean (sometimes called the "arithmetic mean" or the "average"). Recall that we first introduced finding the average of a list of numbers in Section 1.6.

The **mean (average)** of a set of number items is the sum of the items divided by the number of items.

$$\text{mean} = \frac{\text{sum of items}}{\text{number of items}}$$

EXAMPLE 1 Finding the Mean Time in an Experiment

Seven students in a psychology class conducted an experiment on mazes. Each student was given a pencil and asked to successfully complete the same maze. The timed results are below:

Student	Ann	Thanh	Carlos	Jesse	Melinda	Ramzi	Dayni
Time (Seconds)	13.2	11.8	10.7	16.2	15.9	13.8	18.5

a. Who completed the maze in the shortest time? Who completed the maze in the longest time?

b. Find the mean time.

c. How many students took longer than the mean time? How many students took shorter than the mean time?

Solution:

a. Carlos completed the maze in 10.7 seconds, the shortest time. Dayni completed the maze in 18.5 seconds, the longest time.

b. To find the mean (or average), we find the sum of the items and divide by 7, the number of items.

$$\text{mean} = \frac{13.2 + 11.8 + 10.7 + 16.2 + 15.9 + 13.8 + 18.5}{7}$$

$$= \frac{100.1}{7} = 14.3$$

c. Three students, Jesse, Melinda, and Dayni, had times longer than the mean time. Four students, Ann, Thanh, Carlos, and Ramzi, had times shorter than the mean time.

Work Practice Problem 1

✔ **Concept Check** Estimate the mean of the following set of data:

5, 10, 10, 10, 10, 15

Often in college, the calculation of a **grade point average** (GPA) is a **weighted mean** and is calculated as shown in Example 2.

PRACTICE PROBLEM 1

Find the mean of the following test scores: 87, 75, 96, 91, and 78.

Answer
1. 85.4

✔ **Concept Check Answer**
10

PRACTICE PROBLEM 2

Find the grade point average if the following grades were earned in one semester.

Grade	Credit Hours
A	2
B	4
C	5
D	2
A	2

EXAMPLE 2 **Calculating Grade Point Average (GPA)**

The following grades were earned by a student during one semester. Find the student's grade point average.

Course	Grade	Credit Hours
College mathematics	A	3
Biology	B	3
English	A	3
PE	C	1
Social studies	D	2

Solution: To calculate the grade point average, we need to know the point values for the different possible grades. The point values of grades commonly used in colleges and universities are given below:

A: 4,   B: 3,   C: 2,   D: 1,   F: 0

Now, to find the grade point average, we multiply the number of credit hours for each course by the point value of each grade. The grade point average is the sum of these products divided by the sum of the credit hours.

Course	Grade	Point Value of Grade	Credit Hours	Point Value Credit Hours
College mathematics	A	4	3	12
Biology	B	3	3	9
English	A	4	3	12
PE	C	2	1	2
Social studies	D	1	2	2
		Totals:	12	37

$$\text{grade point average} = \frac{37}{12} \approx 3.08 \text{ rounded to two decimal places}$$

The student earned a grade point average of 3.08.

▨ **Work Practice Problem 2**

Objective **B** **Finding the Median**

You may have noticed that a very low number or a very high number can affect the mean of a list of numbers. Because of this, you may sometimes want to use another measure of central tendency. A second measure of central tendency is called the **median.** The median of a list of numbers is not affected by a low or high number in the list.

> The **median** of a set of numbers in numerical order is the middle number. If the number of items is odd, the median is the middle number. If the number of items is even, the median is the mean of the two middle numbers.

PRACTICE PROBLEM 3

Find the median of the list of numbers: 5, 11, 14, 23, 24, 35, 38, 41, 43

EXAMPLE 3 Find the median of the following list of numbers:

25, 54, 56, 57, 60, 71, 98

Solution: Because this list is in numerical order, the median is the middle number, 57.

▨ **Work Practice Problem 3**

Answers
2. 2.67   **3.** 24

EXAMPLE 4 Find the median of the following list of scores: 67, 91, 75, 86, 55, 91

Solution: First we list the scores in numerical order and then find the middle number.

55, 67, 75, 86, 91, 91

Since there is an even number of scores, there are two middle numbers, 75 and 86. The median is the mean of the two middle numbers.

$$\text{median} = \frac{75 + 86}{2} = 80.5$$

The median is 80.5.

◻ **Work Practice Problem 4**

Helpful Hint
Don't forget to write the numbers in order from smallest to largest before finding the median.

Objective C Finding the Mode

The last common measure of central tendency is called the **mode**.

The **mode** of a set of numbers is the number that occurs most often. (It is possible for a set of numbers to have more than one mode or to have no mode.)

EXAMPLE 5 Find the mode of the list of numbers:

11, 14, 14, 16, 31, 56, 65, 77, 77, 78, 79

Solution: There are two numbers that occur the most often. They are 14 and 77. This list of numbers has two modes, 14 and 77.

◻ **Work Practice Problem 5**

EXAMPLE 6 Find the median and the mode of the following set of numbers. These numbers were high temperatures for 14 consecutive days in a city in Montana.

76, 80, 85, 86, 89, 87, 82, 77, 76, 79, 82, 89, 89, 92

Solution: First we write the numbers in numerical order.

76, 76, 77, 79, 80, 82, 82, 85, 86, 87, 89, 89, 89, 92

Since there is an even number of items, the median is the mean of the two middle numbers, 82 and 85.

$$\text{median} = \frac{82 + 85}{2} = 83.5$$

The mode is 89, since 89 occurs most often.

◻ **Work Practice Problem 6**

✔**Concept Check** True or false? Every set of numbers *must* have a mean, median, and mode. Explain your answer.

Helpful Hint
Don't forget that it is possible for a list of numbers to have no mode. For example, the list

2, 4, 5, 6, 8, 9

has no mode. There is no number or numbers that occur more often than the others.

PRACTICE PROBLEM 4
Find the median of the list of scores:
36, 91, 78, 65, 95, 95, 88, 71

PRACTICE PROBLEM 5
Find the mode of the list of numbers:
14, 10, 10, 13, 15, 15, 15, 17, 18, 18, 20

PRACTICE PROBLEM 6
Find the median and the mode of the list of numbers:
26, 31, 15, 15, 26, 30, 16, 18, 15, 35

Answers
4. 83 **5.** 15 **6.** median: 22; mode: 15

✔ **Concept Check Answer**
false; a set of numbers may have no mode.

Vocabulary and Readiness Check

Use the choices below to fill in each blank. Some choices may be used more than once.

mean   mode   grade point average

median   average

1. Another word for "mean" is _____.

2. The number that occurs most often in a set of numbers is called the _____.

3. The _____ of a set of number items is $\dfrac{\text{sum of items}}{\text{number of items}}$.

4. The _____ of a set of numbers is the middle number. If the number of numbers is even, it is the _____ of the two middle numbers.

5. An example of weighted mean is a calculation of _____.

5.7 EXERCISE SET

FOR EXTRA HELP

 Student Solutions Manual   PH Math/Tutor Center   CD/Video for Review   MathXL®   MyMathLab

Objectives A B C Mixed Practice *For each set of numbers, find the mean, median, and mode. If necessary, round the mean to one decimal place. See Examples 1 and 3 through 6.*

1. 15, 23, 24, 18, 25

2. 45, 36, 28, 46, 52

3. 7.6, 8.2, 8.2, 9.6, 5.7, 9.1

4. 4.9, 7.1, 6.8, 6.8, 5.3, 4.9

5. 0.5, 0.2, 0.2, 0.6, 0.3, 1.3, 0.8, 0.1, 0.5

6. 0.6, 0.6, 0.8, 0.4, 0.5, 0.3, 0.7, 0.8, 0.1

7. 231, 543, 601, 293, 588, 109, 334, 268

8. 451, 356, 478, 776, 892, 500, 467, 780

The ten tallest buildings in the world are listed in the following table. Use this table to answer Exercises 9 through 14. If necessary, round results to one decimal place. See Examples 1 and 3 through 6.

9. Find the mean height of the five tallest buildings.

10. Find the median height of the five tallest buildings.

11. Find the median height of the eight tallest buildings.

12. Find the mean height of the eight tallest buildings.

Building	Height (in Feet)
Taipei 101	1679
Petronas Tower 1, Kuala Lumpur	1483
Petronas Tower 2, Kuala Lumpur	1483
Sears Tower, Chicago	1450
Jin Mao Building, Shanghai	1380
Two International Finance Centre	1362
Citic Plaza, Guangzhou	1283
Shun Hing Square, Shenzhen	1260
Empire State Building, New York	1250
Central Plaza, Hong Kong	1227
(*Source:* Council on Tall Buildings and Urban Habitat)	

13. Given the building heights, explain how you know, without calculating, that the answer to Exercise 10 is more than the answer to Exercise 11.

14. Given the building heights, explain how you know, without calculating, that the answer to Exercise 12 is less than the answer to Exercise 9.

For Exercises 15 through 18, the grades are given for a student for a particular semester. Find the grade point average. If necessary, round the grade point average to the nearest hundredth. See Example 2.

15.

Grade	Credit Hours
B	3
C	3
A	4
C	4

16.

Grade	Credit Hours
D	1
F	1
C	4
B	5

17.

Grade	Credit Hours
A	3
A	3
A	4
B	3
C	1

18.

Grade	Credit Hours
B	2
B	2
C	3
A	3
B	3

During an experiment, the following times (in seconds) were recorded:

7.8, 6.9, 7.5, 4.7, 6.9, 7.0

19. Find the mean. Round to the nearest tenth.

20. Find the median.

21. Find the mode.

In a mathematics class, the following test scores were recorded for a student: 93, 85, 89, 79, 88, 91.

22. Find the mean. Round to the nearest hundredth.

23. Find the median.

24. Find the mode.

The following pulse rates were recorded for a group of 15 students:

78, 80, 66, 68, 71, 64, 82, 71, 70, 65, 70, 75, 77, 86, 72.

25. Find the mean.

26. Find the median.

27. Find the mode.

28. How many pulse rates were higher than the mean?

29. How many pulse rates were lower than the mean?

Review

Write each fraction in simplest form. See Section 4.2.

30. $\dfrac{12}{20}$

31. $\dfrac{6}{18}$

32. $\dfrac{4x}{36}$

33. $\dfrac{18}{30y}$

34. $\dfrac{35a^3}{100a^2}$

35. $\dfrac{55y^2}{75y^2}$

Concept Extensions

Find the missing numbers in each set of numbers.

36. 16, 18, _____, _____, _____. The mode is 21. The median is 20.

37. _____, _____, _____, 40, _____. The mode is 35. The median is 37. The mean is 38.

 38. Write a list of numbers for which you feel the median would be a better measure of central tendency than the mean.

 39. Without making any computations, decide whether the median of the following list of numbers will be a whole number. Explain your reasoning.

36, 77, 29, 58, 43

STUDY SKILLS BUILDER

Tips for Studying for an Exam

To prepare for an exam, try the following study techniques.

- Start the study process days before your exam.
- Make sure that you are up-to-date on your assignments.
- If there is a topic that you are unsure of, use one of the many resources that are available to you. For example,
    See your instructor.
    Visit a learning resource center on campus.
    Read the textbook material and examples on the topic.
    View a video on the topic.
- Reread your notes and carefully review the Chapter Highlights at the end of any chapter.
- Work the review exercises at the end of the chapter. Check your answers and correct any mistakes. If you have trouble, use a resource listed above.
- Find a quiet place to take the Chapter Test found at the end of the chapter. Do not use any resources when taking this sample test. This way, you will have a clear indication of how prepared you are for your exam.

Check your answers and make sure that you correct any missed exercises.

- Get lots of rest the night before the exam. It's hard to show how well you know the material if your brain is foggy from lack of sleep.

Good luck and keep a positive attitude.

Let's see how you did on your last exam.

1. How many days before your last exam did you start studying?

2. Were you up-to-date on your assignments at that time or did you need to catch up on assignments?

3. List the most helpful text supplement (if you used one).

4. List the most helpful campus supplement (if you used one).

5. List your process for preparing for a mathematics test.

6. Was this process helpful? In other words, were you satisfied with your performance on your exam?

7. If not, what changes can you make in your process that will make it more helpful to you?

CHAPTER 5   Group Activity

Maintaining a Checking Account
(Sections 5.1, 5.2, 5.3, 5.4)

This activity may be completed by working in groups or individually.

A checking account is a convenient way of handling money and paying bills. To open a checking account, the bank or savings and loan association requires a customer to make a deposit. Then the customer receives a checkbook that contains checks, deposit slips, and a register for recording checks written and deposits made. It is important to record all payments and deposits that affect the account. It is also important to keep the checkbook balance current by subtracting checks written and adding deposits made.

About once a month checking customers receive a statement from the bank listing all activity that the account has had in the last month. The statement lists a beginning balance, all checks and deposits, any service charges made against the account, and an ending balance. Because it may take several days for checks that a customer has written to clear the banking system, the check register may list checks that do not appear on the monthly bank statement. These checks are called **outstanding checks.** Deposits that are recorded in the check register but do not appear on the statement are called **deposits in transit.** Because of these differences, it is important to balance, or reconcile, the checkbook against the monthly statement. The steps for doing so are listed below.

Balancing or Reconciling a Checkbook

Step 1: Place a check mark in the checkbook register next to each check and deposit listed on the monthly bank statement. Any entries in the register without a check mark are outstanding checks or deposits in transit.

Step 2: Find the ending checkbook register balance and add to it any outstanding checks and any interest paid on the account.

Step 3: From the total in Step 2, subtract any deposits in transit and any service charges.

Step 4: Compare the amount found in Step 3 with the ending balance listed on the bank statement. If they are the same, the checkbook balances with the bank statement. Be sure to update the check register with service charges and interest.

Step 5: If the checkbook does not balance, recheck the balancing process. Next, make sure that the running checkbook register balance was calculated correctly. Finally, compare the checkbook register with the statement to make sure that each check was recorded for the correct amount.

For the checkbook register and monthly bank statement given:

a. *update the checkbook register*   **b.** *list the outstanding checks and deposits in transit*
c. *balance the checkbook—be sure to update the register with any interest or service fees*

Checkbook Register

#	Date	Description	Payment	✔	Deposit	Balance
						425.86
114	4/1	Market Basket	30.27			
115	4/3	May's Texaco	8.50			
	4/4	Cash at ATM	50.00			
116	4/6	UNO Bookstore	121.38			
	4/7	Deposit			100.00	
117	4/9	MasterCard	84.16			
118	4/10	Blockbuster	6.12			
119	4/12	Kroger	18.72			
120	4/14	Parking sticker	18.50			
	4/15	Direct deposit			294.36	
121	4/20	Rent	395.00			
122	4/25	Student fees	20.00			
	4/28	Deposit			75.00	

First National Bank Monthly Statement 4/30

Date	Number	Amount
BEGINNING BALANCE:		425.86
CHECKS AND ATM WITHDRAWALS		
4/3	114	30.27
4/4	ATM	50.00
4/11	117	84.16
4/13	115	8.50
4/15	119	18.72
4/22	121	395.00
DEPOSITS		
4/7		100.00
4/15	Direct deposit	294.36
SERVICE CHARGES		
Low balance fee		7.50
INTEREST		
Credited 4/30		1.15
ENDING BALANCE:		227.22

Chapter 5 Vocabulary Check

Fill in each blank with one of the choices listed below. Some choices may be used more than once.

vertically	decimal	and	right triangle
standard form	mean	median	circumference
sum	denominator	numerator	mode

1. Like fractional notation, _____ notation is used to denote a part of a whole.
2. To write fractions as decimals, divide the _____ by the _____.
3. To add or subtract decimals, write the decimals so that the decimal points line up _____.
4. When writing decimals in words, write "_____" for the decimal point.
5. When multiplying decimals, the decimal point in the product is placed so that the number of decimal places in the product is equal to the _____ of the number of decimal places in the factors.
6. The _____ of a set of numbers is the number that occurs most often.
7. The distance around a circle is called the _____.
8. The _____ of a set of numbers in numerical order is the middle number. If there are an even number of numbers, the median is the _____ of the two middle numbers.
9. The _____ of a list of numbers of items is $\dfrac{\text{sum of items}}{\text{number of items}}$.
10. When 2 million is written as 2,000,000, we say it is written in _____.

Helpful Hint

Are you preparing for your test? Don't forget to take the Chapter 5 Test on page 408. Then check your answers at the back of the text and use the Chapter Test Prep Video CD to see the fully worked-out solutions to any of the exercises you want to review.

5 Chapter Highlights

DEFINITIONS AND CONCEPTS	EXAMPLES

Section 5.1   Introduction to Decimals

PLACE-VALUE CHART

hundreds	tens	ones	decimal point	tenths	hundredths	thousandths	ten-thousandths	hundred-thousandths
		4	↑	2	6	5		
100	10	1		$\frac{1}{10}$	$\frac{1}{100}$	$\frac{1}{1000}$	$\frac{1}{10,000}$	$\frac{1}{100,000}$

4.265 means

$$4 \cdot 1 + 2 \cdot \frac{1}{10} + 6 \cdot \frac{1}{100} + 5 \cdot \frac{1}{1000}$$

or

$$4 + \frac{2}{10} + \frac{6}{100} + \frac{5}{1000}$$

DEFINITIONS AND CONCEPTS	EXAMPLES

Section 5.1   Introduction to Decimals (*continued*)

WRITING (OR READING) A DECIMAL IN WORDS

Step 1.   Write the whole number part in words.

Step 2.   Write "and" for the decimal point.

Step 3.   Write the decimal part in words as though it were a whole number, followed by the place value of the last digit.

A decimal written in words can be written in standard form by reversing the above procedure.

Write 3.08 in words.
Three and eight hundredths

Write "negative four and twenty-one thousandths" in standard form.

$$-4.021$$

TO ROUND DECIMALS TO A PLACE VALUE TO THE RIGHT OF THE DECIMAL POINT

Step 1.   Locate the digit to the right of the given place value.

Step 2.   If this digit is 5 or greater, add 1 to the digit in the given place value and delete all digits to its right. If this digit is less than 5, delete all digits to the right of the given place value.

Round 86.1256 to the nearest hundredth.

hundredths place

Step 1.   86.12 5 6

digit to the right

Step 2.   Since the digit to the right is 5 or greater, we add 1 to the digit in the hundredths place and delete all digits to its right.

86.1256 rounded to the nearest hundredth is 86.13.

Section 5.2   Adding and Subtracting Decimals

TO ADD OR SUBTRACT DECIMALS

Step 1.   Write the decimals so that the decimal points line up vertically.

Step 2.   Add or subtract as with whole numbers.

Step 3.   Place the decimal point in the sum or difference so that it lines up vertically with the decimal points in the problem.

Add: $4.6 + 0.28$ Subtract: $2.8 - 1.04$

```
    4.60                  2.8 0
  + 0.28                - 1.04
  ------                -------
    4.88                  1.76
```

Section 5.3   Multiplying Decimals and Circumference of a Circle

TO MULTIPLY DECIMALS

Step 1.   Multiply the decimals as though they are whole numbers.

Step 2.   The decimal point in the product is placed so that the number of decimal places in the product is equal to the *sum* of the number of decimal places in the factors.

Multiply: 1.48×5.9

```
    1.4 8   ← 2 decimal places
  × 5.9     ← 1 decimal place
  -------
    1 3 3 2
    7 4 0 0
  ---------
    8.7 3 2  ← 3 decimal places
```

DEFINITIONS AND CONCEPTS	EXAMPLES

Section 5.3   Multiplying Decimals and Circumference of a Circle (*continued*)

The **circumference** of a circle is the distance around the circle.

$C = \pi d$ or $C = 2\pi r$

where $\pi \approx 3.14$ or $\pi \approx \dfrac{22}{7}$.

or

Find the exact circumference of a circle with radius 5 miles and an approximation by using 3.14 for π.

$$C = 2\pi r$$
$$= 2\pi(5)$$
$$= 10\pi$$
$$\approx 10(3.14)$$
$$= 31.4$$

The circumference is exactly 10π miles and approximately 31.4 miles.

Section 5.4   Dividing Decimals

TO DIVIDE DECIMALS

Step 1. If the divisor is not a whole number, move the decimal point in the divisor to the right until the divisor is a whole number.

Step 2. Move the decimal point in the dividend to the right the *same number of places* as the decimal point was moved in Step 1.

Step 3. Divide. The decimal point in the quotient is directly over the moved decimal point in the dividend.

Divide: $1.118 \div 2.6$

$$
\begin{array}{r}
0.43 \\
2.6\overline{)1.118} \\
-1\,04 \\
\hline
78 \\
-78 \\
\hline
0
\end{array}
$$

Section 5.5   Fractions, Decimals, and Order of Operations

To **write fractions as decimals,** divide the numerator by the denominator.

Write $\dfrac{3}{8}$ as a decimal.

$$
\begin{array}{r}
0.375 \\
8\overline{)3.000} \\
-24 \\
\hline
60 \\
-56 \\
\hline
40 \\
-40 \\
\hline
0
\end{array}
$$

ORDER OF OPERATIONS

1. Perform all operations within parentheses ( ), brackets [ ], or grouping symbols such as fraction bars.

2. Evaluate any expressions with exponents.

3. Multiply or divide in order from left to right.

4. Add or subtract in order from left to right.

Simplify.

$$-1.9(12.8 - 4.1) = -1.9(8.7) \quad \text{Subtract.}$$
$$= -16.53 \quad \text{Multiply.}$$

DEFINITIONS AND CONCEPTS	**EXAMPLES**

Section 5.6   Equations Containing Decimals

STEPS FOR SOLVING AN EQUATION IN x

Step 1. If fractions are present, multiply both sides of the equation by the LCD of the fractions.

Step 2. If parentheses are present, use the distributive property.

Step 3. Combine any like terms on each side of the equation.

Step 4. Use the addition property of equality to rewrite the equation so that variable terms are on one side of the equation and constant terms are on the other side.

Step 5. Divide both sides by the numerical coefficient of x to solve.

Step 6. Check your answer in the *original equation.*

Solve:

$$3(x + 2.6) = 10.92$$
$$3x + 7.8 = 10.92 \qquad \text{Apply the distributive property.}$$
$$3x + 7.8 - 7.8 = 10.92 - 7.8 \qquad \text{Subtract 7.8 from both sides.}$$
$$3x = 3.12 \qquad \text{Simplify.}$$
$$\frac{3x}{3} = \frac{3.12}{3} \qquad \text{Divide both sides by 3.}$$
$$x = 1.04 \qquad \text{Simplify.}$$

Check 1.04 in the original equation.

Section 5.7   Decimal Applications: Mean, Median, and Mode

The **mean** (or **average**) of a set of number items is

$$\text{mean} = \frac{\text{sum of items}}{\text{number of items}}$$

The **median** of a set of numbers in numerical order is the middle number. If the number of items is even, the median is the mean of the two middle numbers.

The **mode** of a set of numbers is the number that occurs most often. (A set of numbers may have no mode or more than one mode.)

Find the mean, median, and mode of the following set of numbers: 33, 35, 35, 43, 68, 68

$$\text{mean} = \frac{33 + 35 + 35 + 43 + 68 + 68}{6} = 47$$

The median is the mean of the two middle numbers, 35 and 43

$$\text{median} = \frac{35 + 43}{2} = 39$$

There are two modes because there are two numbers that occur twice:

35 and 68

STUDY SKILLS BUILDER

Are You Prepared for a Test on Chapter 5?

Below I have listed some *common trouble areas* for students in Chapter 5. After studying for your test—but before taking your test—read these.

- Don't forget the order of operations. To simplify $-0.7 + 1.3(5 - 0.1)$, should you add, subtract, or multiply first? First, perform the subtraction within parentheses, then multiply, and finally add.

$$-0.7 + 1.3(5 - 0.1) = -0.7 + 1.3(4.9) \quad \text{Subtract.}$$
$$= -0.7 + 6.37 \quad \text{Multiply.}$$
$$= 5.67 \quad \text{Add.}$$

- If you are having trouble with ordering or operations on decimals, don't forget that you can insert 0s after the last digit to the right of the decimal point as needed.

Addition	Addition with zeros inserted	Subtraction	Subtraction with zeros inserted
8.1	8.100	7	$\overset{9}{\overset{6\ \cancel{10}\ 10}{\cancel{7.00}}}$
0.6	0.600	$-\ 0.28$	$-\ 0.28$
$+\ 23.003$	$+\ 23.003$		6.72
	31.703		

- Place in order from smallest to largest: 0.108, 0.18, 0.0092

  If we insert zeros, we have: 0.1080, 0.1800, 0.0092

  The decimals in order are: 0.0092, 0.1080, 0.1800 or 0.0092, 0.108, 0.18.

- Do you remember that a set of numbers can have no mode, 1 mode, or even more than 1 mode?

2, 5, 8, 9	no mode
2, 2, 8, 9	mode: 2
2, 2, 3, 3, 5, 7, 7	mode: 2, 3, 7

- Do you remember how to find the median of an even-numbered set of numbers?

$$2, \overbrace{5, 8}, 9 \qquad \frac{5 + 8}{2} = 6.5 \qquad \text{The median is the average of the two ``middle'' numbers.}$$

5 CHAPTER REVIEW

(5.1) *Determine the place value of the number 4 in each decimal.*

1. 23.45

2. 0.000345

Write each decimal in words.

3. -23.45

4. 0.00345

5. 109.23

6. 200.000032

Write each decimal in standard form.

7. Eight and six hundredths

8. Negative five hundred three and one hundred two thousandths

9. Sixteen thousand twenty-five and fourteen ten-thousandths

10. Fourteen and eleven thousandths

Write each decimal as a fraction or a mixed number.

11. 0.16

12. −12.023

Write each fraction or mixed number as a decimal.

13. $\dfrac{231}{100,000}$

14. $25\dfrac{1}{4}$

Insert <, >, *or* = *between each pair of numbers to make a true statement.*

15. 0.49   0.43

16. 0.973   0.9730

17. −38.0027   −38.00056

18. −0.230505   −0.23505

Round each decimal to the given place value.

19. 0.623, nearest tenth

20. 0.9384, nearest hundredth

21. −42.895, nearest hundredth

22. 16.34925, nearest thousandth

Write each number in standard notation.

23. Saturn is a distance of about 887 million miles from the Sun.

24. The tail of a comet can be over 600 thousand miles long.

(5.2) *Add.*

25. 8.6 + 9.5

26. 3.9 + 1.2

27. −6.4 + (−0.88)

28. −19.02 + 6.98

29. 200.49 + 16.82 + 103.002

30. 0.00236 + 100.45 + 48.29

Subtract.

31. 4.9 − 3.2

32. 5.23 − 2.74

33. −892.1 − 432.4

34. 0.064 − 10.2

35. 100 − 34.98

36. 200 − 0.00198

Solve.

37. Find the total distance between Grove City and Jerome.

38. Evaluate $x - y$ for $x = 1.2$ and $y = 6.9$.

39. Find the perimeter.

6.2 inches

Rectangle   4.9 inches

40. Find the perimeter.

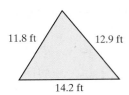

11.8 ft   12.9 ft

14.2 ft

(5.3) *Multiply.*

41. 7.2×10

42. 9.345×1000

43. -34.02×2.3

44. $-839.02 \times (-87.3)$

Find the exact circumference of each circle. Then use the approximation 3.14 for π and approximate the circumference.

45.

7 meters

46.

20 inches

(5.4) *Divide. Round the quotient to the nearest thousandth if necessary.*

47. $3\overline{)0.2631}$

48. $20\overline{)316.5}$

49. $-21 \div (-0.3)$

50. $-0.0063 \div 0.03$
-0.21

51. $0.34\overline{)2.74}$

52. $19.8\overline{)601.92}$

53. $\dfrac{23.65}{1000}$

54. $\dfrac{93}{-10}$

55. There are approximately 3.28 feet in 1 meter. Find how many meters are in 24 feet to the nearest tenth of a meter.

←——1 meter——→

←—— ≈3.28 feet ——→

56. George Strait pays \$69.71 per month to pay back a loan of \$3136.95. In how many months will the loan be paid off?

(5.5) *Write each fraction as a decimal. Round to the nearest thousandth if necessary.*

57. $\dfrac{4}{5}$

58. $-\dfrac{12}{13}$

59. $2\dfrac{1}{3}$

60. $\dfrac{13}{60}$

Insert $<$, $>$, *or* $=$ *to make a true statement.*

61. 0.392   0.392

62. $\dfrac{4}{7}$ 0.625

63. 0.293 $\dfrac{5}{17}$

64. -0.0231 -0.0221

Write the numbers in order from smallest to largest.

65. 0.837, 0.839, 0.832

66. 0.685, 0.626, $\dfrac{5}{8}$

67. $\dfrac{3}{7}$, 0.42, 0.43

68. $\dfrac{18}{11}$, 1.63, $\dfrac{19}{12}$

Simplify each expression.

69. $-7.6 \times 1.9 + 2.5$

70. $(-2.3)^2 - 1.4$

71. $\dfrac{7 + 0.74}{-0.06}$

72. $0.9(6.5 - 5.6)$

73. $\dfrac{(1.5)^2 + 0.5}{0.05}$

74. $0.0726 \div 10 \times 1000$

Find each area.

75.

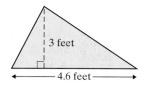

3 feet

4.6 feet

76.

2.1 inches

5.2 inches

(5.6) *Solve.*

77. $x + 3.9 = 4.2$

78. $70 = y - 22.81$

79. $2x = 17.2$

80. $-1.1y = 88$

81. $3x - 0.78 = 1.2 + 2x$

82. $-x + 0.6 - 2x = -4x - 0.9$

83. $-1.3x - 9.4 = -0.4x + 8.6$

84. $3(x - 1.1) = 5x - 5.3$

(5.7) *Find the mean, median, and any mode(s) for each list of numbers. If necessary, round to the nearest tenth.*

85. 13, 23, 33, 14, 6

86. 45, 86, 21, 60, 86, 64, 45

87. 14,000, 20,000, 12,000, 20,000, 36,000, 45,000

88. 560, 620, 123, 400, 410, 300, 400, 780, 430, 450

For Exercises 89 and 90, the grades are given for a student for a particular semester. Find each grade point average. If necessary, round the grade point average to the nearest hundredth.

89.

Grade	Credit Hours
A	3
A	3
C	2
B	3
C	1

90.

Grade	Credit Hours
B	3
B	4
C	2
D	2
B	3

Mixed Review

91. Write 200.0032 in words.

92. Write negative sixteen and nine hundredths in standard form.

93. Write 0.0847 as a fraction or a mixed number.

94. Write the numbers $\frac{6}{7}, \frac{8}{9}, 0.75$ in order from smallest to largest.

Write each fraction as a decimal. Round to the nearest thousandth, if necessary.

95. $-\dfrac{7}{100}$

96. $\dfrac{9}{80}$ (Do not round.)

97. $\dfrac{8935}{175}$

Insert <, >, or = to make a true statement.

98. -402.000032 $\quad$ -402.00032

99. $\dfrac{6}{11}$ $\quad$ 0.55

Round each decimal to the given place value.

100. 86.905, nearest hundredth

101. 3.11526, nearest thousandth

Round each money amount to the nearest dollar.

102. $123.46, nearest dollar

103. $3645.52, nearest dollar

Add or subtract as indicated.

104. $3.2 - 4.9$

105. $9.12 - 3.86$

106. $-102.06 + 89.3$

107. $-4.021 + (-10.83) + (-0.056)$

Multiply or divide as indicated. Round to the nearest thousandth, if necessary.

108.   2.54
         $\times\ 3.2$

109. $(-3.45)(2.1)$

110. $0.005\overline{)24.5}$

111. $2.3\overline{)54.98}$

Solve.

△ **112.** Tomaso is going to fertilize his lawn, a rectangle that measures 77.3 feet by 115.9 feet. Approximate the area of the lawn by rounding each measurement to the nearest ten feet.

113. Estimate the cost of the items to see whether the groceries can be purchased with a $5 bill.

Simplify each expression.

114. $\dfrac{(3.2)^2}{100}$

115. $(2.6 + 1.4)(4.5 - 3.6)$

Find the mean, median, and any mode(s) for each list of numbers. If needed round answers to two decimal places.

116. 73, 82, 95, 68, 54

117. 952, 327, 566, 814, 327, 729

Answers

 Remember to use the Chapter Test Prep Video CD to see the fully worked-out solutions to any of the exercises you want to review.

Write each decimal as indicated.

1. 45.092, in words

2. Three thousand and fifty-nine thousandths, in standard form

Perform each indicated operation. Round the result to the nearest thousandth if necessary.

3. 2.893 + 4.21 + 10.492

4. −47.92 − 3.28

5. 9.83 − 30.25

6. 10.2 × 4.01

7. (−0.00843) ÷ (−0.23)

Round each decimal to the indicated place value.

8. 34.8923, nearest tenth

9. 0.8623, nearest thousandth

Insert <, >, or = between each pair of numbers to form a true statement.

10. 25.0909   25.9090

11. $\dfrac{4}{9}$ 0.445

Write each decimal as a fraction or a mixed number.

12. 0.345

13. −24.73

Write each fraction as a decimal. If necessary, round to the nearest thousandth.

14. $-\dfrac{13}{26}$

15. $\dfrac{16}{17}$

Simplify.

16. $(-0.6)^2 + 1.57$

17. $\dfrac{0.23 + 1.63}{-0.3}$

18. $2.4x − 3.6 − 1.9x − 9.8$

1. _____

2. _____

3. _____

4. _____

5. _____

6. _____

7. _____

8. _____

9. _____

10. _____

11. _____

12. _____

13. _____

14. _____

15. _____

16. _____

17. _____

18. _____

Solve.

19. $0.2x + 1.3 = 0.7$ **20.** $2(x + 5.7) = 6x - 3.4$

Find the mean, median, and mode of each list of numbers.

21. 26, 32, 42, 43, 49   **22.** 8, 10, 16, 16, 14, 12, 12, 13

Find the grade point average. If necessary, round to the nearest hundredth.

23.

Grade	Credit Hours
A	3
B	3
C	3
B	4
A	1

Solve.

24. At its farthest, Pluto is 4,583 million miles from the Sun. Write this number using standard notation.

△ **25.** Find the area.

1.1 miles

4.2 miles

△ **26.** Find the exact circumference of the circle. Then use the approximation 3.14 for π and approximate the circumference.

9 miles

27. Vivian Thomas is going to put insecticide on her lawn to control grubworms. The lawn is a rectangle that measures 123.8 feet by 80 feet. The amount of insecticide required is 0.02 ounces per square foot.

    a. Find the area of her lawn.
    b. Find how much insecticide Vivian needs to purchase.

28. Find the total distance from Bayette to Center City.

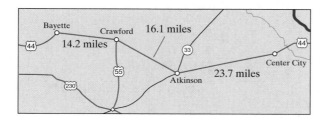

Bayette
Crawford   16.1 miles
44   14.2 miles
33   44
55   Center City
230   Atkinson   23.7 miles

19. _____

20. _____

21. _____

22. _____

23. _____

24. _____

25. _____

26. _____

27. a. _____

    b. _____

28. _____

Write each number in words.

Answers

1. 72

2. 107

3. 546

4. 5026

5. Add: 46 + 713

6. Find the perimeter.

7. Subtract: 543 − 29. Then check by adding.

8. Divide: 3268 ÷ 27

9. Round 278,362 to the nearest thousand.

10. Find all the factors of 30.

11. Multiply: 236 × 86

12. Multiply: 236 × 86 × 0

13. Find each quotient and then check the answer by multiplying.

   a. $1\overline{)7}$

   b. $12 \div 1$

   c. $\dfrac{6}{6}$

   d. $9 \div 9$

   e. $\dfrac{20}{1}$

   f. $18\overline{)18}$

14. Find the average of 25, 17, 19, and 39.

15. Simplify: $2 \cdot 4 - 3 \div 3$

16. Simplify: $77 \div 11 \cdot 7$

Evaluate.

17. 9^2

18. 5^3

Answers

1. _____

2. _____

3. _____

4. _____

5. _____

6. _____

7. _____

8. _____

9. _____

10. _____

11. _____

12. _____

13. a. _____

  b. _____

  c. _____

  d. _____

  e. _____

  f. _____

14. _____

15. _____

16. _____

17. _____

18. _____

19. 3^4

20. 10^3

21. Evaluate $\dfrac{x - 5y}{y}$ for $x = 35$ and $y = 5$.

22. Evaluate $\dfrac{2a + 4}{c}$ for $a = 7$ and $c = 3$.

23. Find the opposite of each number.
    a. 13   **b.** -2 **c.** 0

24. Find the opposite of each number.
    a. -7 **b.** 4   **c.** -1

25. Add: $-2 + (-21)$

26. Add: $-7 + (-15)$

Find the value of each expression.

27. $5 \cdot 6^2$

28. $4 \cdot 2^3$

29. -7^2

30. $(-2)^5$

31. $(-5)^2$

32. -3^2

Represent the shaded part as an improper fraction and a mixed number.

33.

34.

35.

36.

37. Find the prime factorization of 252.

38. Find the difference of 87 and 25.

19. _____

20. _____

21. _____

22. _____

23. a. _____

   b. _____

   c. _____

24. a. _____

   b. _____

   c. _____

25. _____

26. _____

27. _____

28. _____

29. _____

30. _____

31. _____

32. _____

33. _____

34. _____

35. _____

36. _____

37. _____

38. _____

39. _____

40. _____

41. _____

42. _____

43. _____

44. _____

45. _____

46. _____

47. _____

48. _____

49. _____

50. _____

51. _____

52. _____

39. Write $-\dfrac{72}{26}$ in simplest form.

40. Write $9\dfrac{7}{8}$ as an improper fraction.

41. Determine whether $\dfrac{16}{40}$ and $\dfrac{10}{25}$ are equivalent.

42. Insert $<$ or $>$ to form a true statement. $\dfrac{4}{7}\quad\dfrac{5}{9}$

Multiply.

43. $\dfrac{2}{3}\cdot\dfrac{5}{11}$

44. $2\dfrac{5}{8}\cdot\dfrac{4}{7}$

45. $\dfrac{1}{4}\cdot\dfrac{1}{2}$

46. $7\cdot 5\dfrac{2}{7}$

Solve.

47. $\dfrac{z}{-4}=11-5$

48. $6x-12-5x=-20$

49. Add: $763.7651+22.001+43.89$

50. Add: $89.27+14.361+127.2318$

51. Multiply: 23.6×0.78

52. Multiply: 43.8×0.645

6

Ratio, Proportion, and Triangle Applications

Having studied fractions in Chapter 4, we are ready to explore the useful notations of ratio and proportion. *Ratio* is another name for *quotient* and is usually written in fraction form. A proportion is an equation with two equal ratios.

Hybrid cars are becoming more widespread in America each year. New models are being offered by an ever-growing number of auto manufacturers. Concern for the environment, rising costs of gasoline, and financial incentives offered by the federal and state governments make hybrid cars a more attractive alternative to regular gasoline engine vehicles. While these vehicles are not the all-electric cars that futurists used to predict, the innovative use of electric and gasoline power makes these hybrids very energy efficient. We explore the efficiency of a hybrid car in Section 6.3, Exercises 11, 12.

Average Miles per Gallon for Selected Hybrid Models

Accord 2WD Automatic — 29.4
Camry 2WD ECT-1 — 37.0
Civic I 2WD CVT — 45.7
Civic I 2WD Manual — 48.3
Escapade 2WD eCVT — 31.9
Highlander 2WD eCVT — 25.9
Insight 2WD CVT — 55.4
Insight 2WD Manual — 66.0
Mariner 4WD eCVT — 28.9
Prius I (THS) 2WD eCVT — 45.6
Prius II (HSD) 2WD eCVT — 47.8
RX 400h 4WD eCVT — 24.8

Hybrid Model (vertical axis)
Miles per Gallon (horizontal axis): 0, 10, 20, 30, 40, 50, 60, 70, 80

Source: http://www.greenhybrid.com

413

6.1 RATIOS AND RATES

Objective **A** Writing Ratios as Fractions

A **ratio** is the quotient of two quantities. A ratio, in fact, is no different from a fraction, except that a ratio is sometimes written using notation other than fractional notation. For example, the ratio of 1 to 2 can be written as

$$1 \text{ to } 2 \quad \text{or} \quad \frac{1}{2} \quad \text{or} \quad 1 : 2$$

fractional notation   colon notation

These ratios are all read as, "the ratio of 1 to 2."

✔ **Concept Check** How should each ratio be read aloud?

a. $\frac{8}{5}$ **b.** $\frac{5}{8}$

In this section, we write ratios using fractional notation. If the fraction happens to be an improper fraction, do not write the fraction as a mixed number. Why? The mixed number form is not a ratio or quotient of two quantities.

Writing a Ratio as a Fraction

The order of the quantities is important when writing ratios. To write a ratio as a fraction, write the *first number* of the ratio as the *numerator* of the fraction and the *second number* as the *denominator*.

Helpful Hint

The ratio of 6 to 11 is $\frac{6}{11}$, *not* $\frac{11}{6}$.

PRACTICE PROBLEM 1

Write the ratio of 19 to 30 using fractional notation.

EXAMPLE 1 Write the ratio of 21 to 29 using fractional notation.

Solution: The ratio is $\frac{21}{29}$.

☐ **Work Practice Problem 1**

To simplify a ratio, we just write the fraction in simplest form. Common factors as well as common units can be divided out.

PRACTICE PROBLEM 2

Write the ratio of $16 to $12 as a fraction in simplest form.

EXAMPLE 2 Write the ratio of $15 to $10 as a fraction in simplest form.

Solution:

$$\frac{\$15}{\$10} = \frac{15}{10} = \frac{3 \cdot \overset{1}{\cancel{5}}}{2 \cdot \underset{1}{\cancel{5}}} = \frac{3}{2}$$

☐ **Work Practice Problem 2**

Answers

1. $\frac{19}{30}$ 2. $\frac{4}{3}$

✔ **Concept Check Answers**

a. "the ratio of eight to five"
b. "the ratio of five to eight"

Helpful
Hint

In the previous example, although $\frac{3}{2} = 1\frac{1}{2}$, a ratio is a quotient of *two* quantities. For that reason, ratios are not written as mixed numbers.

If a ratio contains decimal numbers or mixed numbers, we simplify by writing the ratio as a ratio of whole numbers.

EXAMPLE 3 Write the ratio of 2.5 to 3.15 as a fraction in simplest form.

Solution: The ratio in fraction form is

$$\frac{2.5}{3.15}$$

Now let's clear the ratio of decimals.

$$\frac{2.5}{3.15} = \frac{2.5}{3.15} \cdot 1 = \frac{2.5}{3.5} \cdot \frac{100}{100} = \frac{2.5 \cdot 100}{3.15 \cdot 100} = \frac{250}{315} = \frac{50}{63} \quad \text{Simplest form}$$

Work Practice Problem 3

EXAMPLE 4 Write the ratio of $2\frac{5}{8}$ to $8\frac{3}{4}$ as a fraction in simplest form.

Solution: The ratio in fraction form is $\dfrac{2\frac{5}{8}}{8\frac{3}{4}}$.

To simplify, remember that the fraction bar means division.

$$\frac{2\frac{5}{8}}{8\frac{3}{4}} = 2\frac{5}{8} \div 8\frac{3}{4} = \frac{21}{8} \div \frac{35}{4} = \frac{21}{8} \cdot \frac{4}{35} = \frac{3 \cdot \cancel{7} \cdot \cancel{4}}{2 \cdot \cancel{4} \cdot 5 \cdot \cancel{7}} = \frac{3}{10} \quad \text{Simplest form.}$$

Work Practice Problem 4

EXAMPLE 5 Writing a Ratio from a Circle Graph

The circle graph in the margin shows the part of a car's total mileage that falls into a particular category. Write the ratio of family business miles to total miles as a fraction in simplest form.

Solution:

$$\frac{\text{family business miles}}{\text{total miles}} = \frac{3000}{15,000} = \frac{3000}{15,000} = \frac{\cancel{3000}}{5 \cdot \cancel{3000}} = \frac{1}{5}$$

Work Practice Problem 5

PRACTICE PROBLEM 3

Write the ratio of 1.68 to 4.8 as a fraction in simplest form.

PRACTICE PROBLEM 4

Write the ratio of $2\frac{2}{3}$ to $1\frac{13}{15}$ as a fraction in simplest form.

PRACTICE PROBLEM 5

Use the circle graph below to write the ratio of work miles to total miles as a fraction in simplest form.

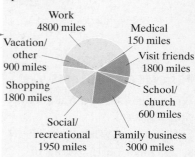

Work
4800 miles
Medical
150 miles
Vacation/
other
900 miles
Visit friends
1800 miles
Shopping
1800 miles
School/
church
600 miles
Social/
recreational
1950 miles
Family business
3000 miles

Total yearly mileage: 15,000

Sources: The American Automobile Manufacturers Association and The National Automobile Dealers Association.

Answers

3. $\frac{7}{20}$ **4.** $\frac{10}{7}$ **5.** $\frac{8}{25}$

△ **PRACTICE PROBLEM 6**

Given the triangle shown:

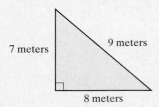

7 meters   9 meters

8 meters

a. Find the ratio of the length of the shortest side to the length of the longest side.

b. Find the ratio of the length of the longest side to the perimeter of the triangle.

△ **EXAMPLE 6** Given the rectangle shown:

a. Find the ratio of its width (shorter side) to its length (longer side).

b. Find the ratio of its length to its perimeter.

7 feet

5 feet

Solution:

a. The ratio of its width to its length is

$$\frac{\text{width}}{\text{length}} = \frac{5 \ \text{feet}}{7 \ \text{feet}} = \frac{5}{7}$$

b. Recall that the perimeter of the rectangle is the distance around the rectangle: $7 + 5 + 7 + 5 = 24$ feet. The ratio of its length to its perimeter is

$$\frac{\text{length}}{\text{perimeter}} = \frac{7 \ \text{feet}}{24 \ \text{feet}} = \frac{7}{24}$$

▢ **Work Practice Problem 6**

Objective **B** **Writing Rates as Fractions**

A special type of ratio is a rate. **Rates** are used to compare *different* kinds of quantities. For example, suppose that a recreational runner can run 3 miles in 33 minutes. If we write this rate as a fraction, we have

$$\frac{3 \ \text{miles}}{33 \ \text{minutes}} = \frac{1 \ \text{mile}}{11 \ \text{minutes}} \quad \text{In simplest form}$$

> **Helpful Hint**
>
> When comparing quantities with different units, write the units as part of the comparison. Units do not divide out unless they are the same.
>
> **Same Units:** $\dfrac{3 \ \text{inches}}{12 \ \text{inches}} = \dfrac{1}{4}$ Units are the same and divide out.
>
> **Different Units:** $\dfrac{2 \ \text{miles}}{20 \ \text{minutes}} = \dfrac{1 \ \text{mile}}{10 \ \text{minutes}}$ Units are still written.

PRACTICE PROBLEMS 7–8

Write each rate as a fraction in simplest form.

7. $1350 for 6 weeks

8. 295 miles on 15 gallons of gasoline

EXAMPLES Write each rate as a fraction in simplest form.

7. $2160 for 12 weeks is $\dfrac{2160 \ \text{dollars}}{12 \ \text{weeks}} = \dfrac{180 \ \text{dollars}}{1 \ \text{week}}$

8. 360 miles on 16 gallons of gasoline is $\dfrac{360 \ \text{miles}}{16 \ \text{gallons}} = \dfrac{45 \ \text{miles}}{2 \ \text{gallons}}$

▢ **Work Practice Problems 7–8**

Answers

6. a. $\dfrac{7}{9}$ **b.** $\dfrac{9}{24}$ **7.** $\dfrac{\$225}{1 \ \text{wk}}$ **8.** $\dfrac{59 \ \text{mi}}{3 \ \text{gal}}$

✔ **Concept Check Answer**

false; a rate compares different kinds of quantities

✔ **Concept Check** True or false? $\dfrac{16 \ \text{gallons}}{4 \ \text{gallons}}$ is a rate. Explain.

Objective C Finding Unit Rates

A **unit rate** is a rate with a denominator of 1. A familiar example of a unit rate is 55 mph, read as "55 **miles per hour.**" This means 55 miles per 1 hour or

$$\frac{55 \text{ miles}}{1 \text{ hour}}$$ Denominator of 1

Helpful Hint

In this context, the word *per* translates to "division."

Writing a Rate as a Unit Rate

To write a rate as a unit rate, divide the numerator of the rate by the denominator.

EXAMPLE 9 Write as a unit rate: $31,500 every 7 months

Solution:

$$\frac{31,500 \text{ dollars}}{7 \text{ months}} \qquad 7\overline{)31,500} = 4,500$$

The unit rate is

$$\frac{4500 \text{ dollars}}{1 \text{ month}} \text{ or } 4500 \text{ dollars/month}$$ Read as, "4500 dollars per month."

◼ **Work Practice Problem 9**

EXAMPLE 10 Write as a unit rate: 337.5 miles every 15 gallons of gas

Solution:

$$\frac{337.5 \text{ miles}}{15 \text{ gallons}} \qquad 15\overline{)337.5} = 22.5$$

The unit rate is

$$\frac{22.5 \text{ miles}}{1 \text{ gallon}} \text{ or } 22.5 \text{ miles/gallon}$$ Read as, "22.5 miles per gallon."

◼ **Work Practice Problem 10**

Objective D Finding Unit Prices

Rates are used extensively in sports, business, medicine, and science. One of the most common uses of rates is in consumer economics. When a unit rate is "money per item," it is also called a **unit price.**

$$\text{unit price} = \frac{\text{price}}{\text{number of units}}$$

PRACTICE PROBLEM 9

Write as a unit rate: 3200 feet every 8 seconds

PRACTICE PROBLEM 10

Write as a unit rate: 78 bushels of fruit from 12 trees

Answers

9. $\frac{400 \text{ ft}}{1 \text{ sec}}$ or 400 ft/sec

10. $\frac{6.5 \text{ bushels}}{1 \text{ tree}}$ or 6.5 bushels/tree

PRACTICE PROBLEM 11

An automobile rental agency charges $170 for 5 days for a certain model car. What is the unit price in dollars per day?

EXAMPLE 11 Finding Unit Price

A store charges $3.36 for a 16-ounce jar of picante sauce. What is the unit price in dollars per ounce?

Solution:

$$\text{unit price} = \frac{\$3.36}{16 \text{ ounces}} = \frac{\$0.21}{1 \text{ ounce}} \quad \text{or} \quad \$0.21 \text{ per ounce}$$

▪ **Work Practice Problem 11**

PRACTICE PROBLEM 12

Approximate each unit price to decide which is the better buy for a bag of nacho chips: 11 ounces for $2.32 or 16 ounces for $3.59.

EXAMPLE 12 Finding the Best Buy

Approximate each unit price to decide which is the better buy: 4 bars of soap for $0.99 or 5 bars of soap for $1.19.

Solution:

$$\frac{\text{unit}}{\text{price}} = \frac{\text{price}}{\text{no. of units}} = \frac{\$0.99}{4 \text{ bars}} \approx \begin{array}{c}\$0.25 \text{ per bar} \\ \text{of soap}\end{array} \qquad \frac{0.247}{4)0.990} \approx 0.25$$

("is approximately")

$$\frac{\text{unit}}{\text{price}} = \frac{\text{price}}{\text{no. of units}} = \frac{\$1.19}{5 \text{ bars}} \approx \begin{array}{c}\$0.24 \text{ per bar} \\ \text{of soap}\end{array} \qquad \frac{0.238}{5)1.190} \approx 0.24$$

Since the 5-bar package has a cheaper price per bar, it is the better buy.

▪ **Work Practice Problem 12**

Answers
11. $34 per day   **12.** 11-oz bag

Vocabulary and Readiness Check

Use the choices below to fill in each blank. Not all choices will be used.

rate	different	denominator
numerator	unit price	true
division	ratio	false

1. A rate with a denominator of 1 is called a _____ rate.

2. When a rate is written as money per item, a unit rate is called a _____.

3. The word *per* translates to "_____."

4. Rates are used to compare _____ types of quantities.

5. To write a rate as a unit rate, divide the _____ of the rate by the _____.

6. The quotient of two quantities is called a _____.

7. Answer true or false: The ratio $\frac{7}{5}$ means the same as the ratio $\frac{5}{7}$. _____

6.1 EXERCISE SET

FOR EXTRA HELP

Student Solutions Manual · PH Math/Tutor Center · CD/Video for Review · MathXL® · MyMathLab

Objective A *Write each ratio as a ratio of whole numbers using fractional notation. Write the fraction in simplest form. See Examples 1 through 6.*

1. 16 to 24

2. 25 to 150

3. 7.7 to 10

4. 8.1 to 10

5. 4.63 to 8.21

6. 9.61 to 7.62

7. 6 ounces to 16 ounces

8. 35 meters to 100 meters

9. \$32 to \$100

10. \$46 to \$102

11. 24 days to 14 days

12. 120 miles to 80 miles

13. $3\frac{1}{2}$ to $12\frac{1}{4}$

14. $3\frac{1}{3}$ to $4\frac{1}{6}$

15. $7\frac{3}{5}$ hours to $1\frac{9}{10}$ hours

16. $25\frac{1}{2}$ days to $2\frac{5}{6}$ days

Find the ratio described in each exercise as a fraction in simplest form. See Examples 5 and 6.

17. More than one billion people per year see Wyland's Whaling Wall murals throughout the United States, Canada, Japan, Australia, New Zealand, and Europe. The mural on the Marketplace Design Center in Philadelphia, "East Coast Humpbacks," measures 125 feet long by 130 feet high. Find the ratio of the height to the perimeter of this mural.

18. Find the ratio of the length to the width of a regulation size basketball court.

50 feet (width)

94 feet (length)

419

19.

American Beverage Consumption—Per person per Year	
Bottled Water	**Soft Drinks**
25 gallons	55 gallons
Source: American Beverage Association	

Use the table to find the ratio of the number of gallons of bottled water to the number of gallons of soft drinks consumed by Americans.

20.

Wal-Mart and Sam's Club Stores in Maine	
Wal-Mart or Wal-Mart Superstores	**Sam's Club Stores**
22	3
Source: Wal-Mart Corporation	

Use the table to find the ratio of Sam's Club Stores to Wal-Mart Stores and Superstores in Maine.

At the Honey Island Parent Teacher Organization meeting one night, there were 125 women and 100 men present.

21. Find the ratio of women to men.

22. Find the ratio of men to the total number of people present.

Blood contains three types of cells: red blood cells, white blood cells, and platelets. For approximately every 600 red blood cells in healthy humans, there are 40 platelets and 1 white blood cell. Use this information for Exercises 23 and 24. (Source: American Red Cross Biomedical Services)

23. Write the ratio of red blood cells to platelet cells.

24. Write the ratio of white blood cells to red blood cells.

25. Find the ratio of the base to the height of the triangular mainsail.

26. Find the ratio of the longest side to the perimeter of the right-triangular-shaped billboard.

27. A large order of McDonald's french fries has 570 calories. Of this total, 270 calories are from fat. Find the ratio of the calories from fat to total calories in a large order of McDonald's french fries. (*Source:* McDonald's Corporation)

28. A McDonald's Quarter Pounder™ with cheese contains 25 grams of fat. A McDonald's Premium Crispy Chicken Ranch BLT™ contains 20 grams of fat. Find the ratio of the amount of fat in a Quarter Pounder with the amount of fat in a Crispy Chicken Ranch BLT sandwich. (*Source:* McDonald's Corporation)

29. Citizens of New Zealand consume the most ice cream per capita of any country in the world. An average New Zealander eats about 56 pints of ice cream per year. In contrast, the average American consumes about 48 pints of ice cream per year. Find the ratio of the average amount of ice cream eaten by New Zealanders to the average amount of ice cream eaten by Americans. (*Source:* Canadian Dairy Information Center)

30. During the 2006 Winter Olympics in Torino, Italy, a total of 256 medals were awarded. Canadian athletes won a total of 24 medals. Find the ratio of medals won by Canada to the total medals awarded. (*Source:* International Olympic Committee)

Objective **B** *Write each rate as a fraction in simplest form. See Examples 7 and 8.*

31. 5 shrubs every 15 feet

32. 14 lab tables for 28 students

33. 15 returns for 100 sales

34. 8 phone lines for 36 employees

35. 6 laser printers for 28 computers

36. 4 inches of rain in 18 hours

37. 18 gallons of pesticide for 4 acres of crops

38. 150 graduate students for 8 advisors

Objective **C** *Write each rate as a unit rate. See Examples 9 and 10.*

39. 330 calories in a 3-ounce serving

40. 275 miles in 11 hours

41. 375 riders in 5 subway cars

42. 18 signs in 6 blocks

43. A hummingbird moves its wings at a rate of 5400 wingbeats a minute. Write this rate in wingbeats per second.

44. A bat moves its wings at a rate of 1200 wingbeats a minute. Write this rate in wingbeats per second.

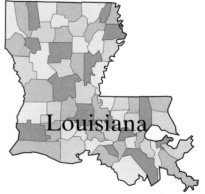

45. $1,000,000 lottery winnings paid over 20 years

46. 400,000 library books for 8000 students

47. The state of Tennessee has 2,738,044 registered voters for two senators. (*Source:* U.S. Bureau of the Census)

48. The state of Louisiana has approximately 4,468,800 residents for 64 parishes. (*Note:* Louisiana is the only U.S. state with parishes instead of counties.) (*Source:* U.S. Bureau of the Census)

49. 12,000 good assembly line products to 40 defective products

50. 5,000,000 lottery tickets for 4 lottery winners

51. Recently, it cost $1,300,000 to build 20 Habitat for Humanity houses in Benton Harbor, Michigan. (*Source:* Habitat World Online)

52. The top-grossing concert tour in North America was the 1994 Rolling Stones tour, which grossed $121,200,000 for 60 shows. (*Source:* Pollstar)

53. Charlie Catlett can assemble 250 computer boards in an 8-hour shift while Suellen Catlett can assemble 402 computer boards in a 12-hour shift.

   a. Find the unit rate of Charlie.
   b. Find the unit rate of Suellen.
   c. Who can assemble computer boards faster, Charlie or Suellen?

54. Jerry Stein laid 713 bricks in 46 minutes while his associate, Bobby Burns, laid 396 bricks in 30 minutes.

   a. Find the unit rate of Jerry.
   b. Find the unit rate of Bobby.
   c. Who is the faster bricklayer?

For Exercises 55 and 56, round the rates to the nearest tenth.

55. One student drove 400 miles in his car on 14.5 gallons of gasoline. His sister drove 270 miles in her truck on 9.25 gallons of gasoline.

    a. Find the unit rate of the car.
    b. Find the unit rate of the truck.
    c. Which vehicle gets better gas mileage?

56. Charlotte Leal is a grocery scanner who can scan an average of 100 items in 3.5 minutes while her cousin Leo can scan 148 items in 5.5 minutes.

    a. Find the unit rate of Charlotte.
    b. Find the unit rate of Leo.
    c. Who is the faster scanner?

Objective **D** *Find each unit price. See Example 11.*

57. $57.50 for 5 compact disks

58. $0.87 for 3 apples

59. $1.19 for 7 bananas

60. $73.50 for 6 lawn chairs

Find each unit price and decide which is the better buy. Round to three decimal places. Assume that we are comparing different sizes of the same brand. See Example 12.

61. Crackers:
    $1.19 for 8 ounces
    $1.59 for 12 ounces

62. Pickles:
    $1.89 for 32 ounces
    $0.89 for 18 ounces

63. Frozen orange juice:
    $1.69 for 16 ounces
    $0.69 for 6 ounces

64. Eggs:
    $0.69 for a dozen
    $2.10 for a flat $\left(2\frac{1}{2} \text{ dozen} \right)$

65. Soy sauce:
    12 ounces for $2.29
    8 ounces for $1.49

66. Shampoo:
    20 ounces for $1.89
    32 ounces for $3.19

67. Napkins:
    100 for $0.59
    180 for $0.93

68. Crackers:
    20 ounces for $2.39
    8 ounces for $0.99

Review

See Section 5.4.

69. $9\overline{)20.7}$

70. $7\overline{)60.2}$

71. $3.7\overline{)0.555}$

72. $4.6\overline{)1.15}$

Concept Extensions

73. Is the ratio $\frac{11}{15}$ the same as the ratio of $\frac{15}{11}$? Explain your answer.

74. Explain why the ratio $\frac{51}{13}$ is incorrect for Exercise 17.

Decide whether each value is a ratio written as a fraction in simplest form. If not, write it as a fraction in simplest form.

75. $\dfrac{6 \text{ inches}}{15 \text{ inches}}$

76. $4\frac{1}{2}$

77. A panty hose manufacturing machine will be repaired if the ratio of defective panty hose to good panty hose is at least 1 to 20. A quality control engineer found 10 defective panty hose in a batch of 200. Determine whether the machine should be repaired.

78. A grocer will refuse a shipment of tomatoes if the ratio of bruised tomatoes to the total batch is at least 1 to 10. A sample is found to contain 3 bruised tomatoes and 33 good tomatoes. Determine whether the shipment should be refused.

79. Fill in the table to calculate miles per gallon.

Beginning Odometer Reading	Ending Odometer Reading	Miles Driven	Gallons of Gas Used	Miles per Gallon (round to the nearest tenth)
79,286	79,543		13.4	
79,543	79,895		15.8	
79,895	80,242		16.1	

Find each unit rate.

80. The longest stairway is the service stairway for the Niesenbahn Cable railway near Spiez, Switzerland. It has 11,674 steps and rises to a height of 7759 feet. Find the unit rate of steps per foot rounded to the nearest tenth of a step. (*Source: Guinness World Records*)

81. In the United States, the total number of students enrolled in public schools is 48,000,000. There are 88,300 public schools. Write a unit rate in students per school. Round to the nearest whole. (*Source: National Center for Education Statistics*)

82. Suppose that the amount of a product decreases, say from an 80-ounce container to a 70-ounce container, but the price of the container remains the same. Does the unit price increase or decrease? Explain why.

 83. In your own words, define the phrase unit rate.

84. In your own words, define the phrase unit price.

85. Should the rate $\dfrac{3 \text{ lights}}{2 \text{ feet}}$ be written as $\dfrac{3}{2}$? Explain why or why not.

86. Find an item in the grocery store and calculate its unit price.

87. In 2006, 20 states had mandatory helmet laws. (*Source:* Bicycle Helmet Safety Institute)

   a. Find the ratio of states with mandatory helmet laws to total U.S. states.

   b. Find the ratio of states with mandatory helmet laws to states without mandatory helmet laws.

   c. Are your ratios for parts **a** and **b** the same? Explain why or why not.

📖 STUDY SKILLS BUILDER

Is Your Notebook Still Organized?

It's never too late to organize your material in a course. Let's see how you are doing.

1. Are all your graded papers in one place in your math notebook or binder?

2. Flip through the pages of your notebook. Are your notes neat and readable?

3. Are your notes complete with no sections missing?

4. Are important notes marked in some way (like an exclamation point) so that you will know to review them before a quiz or test?

5. Are your assignments complete?

6. Do exercises that have given you trouble have a mark (like a question mark) so that you will remember to talk to your instructor or a tutor about them?

7. Describe your attitude toward this course.

8. List ways your attitude can improve and make a commitment to work on at least one of these during the next week.

6.2 PROPORTIONS

Objective **A** Writing Proportions

A **proportion** is a statement that two ratios or rates are equal.

Proportion

A proportion states that two ratios are equal. If $\frac{a}{b}$ and $\frac{c}{d}$ are two ratios, then

$$\frac{a}{b} = \frac{c}{d}$$

is a proportion.

For example,

$$\frac{5}{6} = \frac{10}{12}$$

is a proportion. We can read this as, "5 is to 6 as 10 is to 12."

EXAMPLE 1 Write each sentence as a proportion.

a. 12 diamonds is to 15 rubies as 4 diamonds is to 5 rubies.

b. 5 hits is to 9 at bats as 20 hits is to 36 at bats.

Solution:

a. diamonds $\rightarrow$ $\dfrac{12}{15} = \dfrac{4}{5}$ $\leftarrow$ diamonds
     rubies $\rightarrow$ $\leftarrow$ rubies

b.   hits $\rightarrow$ $\dfrac{5}{9} = \dfrac{20}{36}$ $\leftarrow$ hits
    at bats $\rightarrow$ $\leftarrow$ at bats

◻ **Work Practice Problem 1**

Helpful Hint

Notice in the above examples of proportions that the numerators contain the same units and the denominators contain the same units. In this text, proportions will be written so that this is the case.

Objective **B** Determining Whether Proportions Are True

Like other mathematical statements, a proportion may be either true or false. A proportion is true if its ratios are equal. Since ratios are fractions, one way to determine whether a proportion is true is to write both fractions in simplest form and compare them.

Another way is to compare cross products as we did in Section 4.2.

Using Cross Products to Determine Whether Proportions Are True or False

Cross products

$a \cdot d$ $b \cdot c$

$$\frac{a}{b} = \frac{c}{d}$$

If cross products are *equal*, the proportion is *true*. If cross products are *not equal*, the proportion is *false*.

PRACTICE PROBLEM 1

Write each sentence as a proportion.

a. 24 cups is to 6 cups as 4 cups is to 1 cup.

b. 560 students is to 25 instructors as 112 students is to 5 instructors.

Answers

1. a. $\dfrac{24}{6} = \dfrac{4}{1}$ **b.** $\dfrac{560}{25} = \dfrac{112}{5}$

EXAMPLE 2 Is $\frac{2}{3} = \frac{4}{6}$ a true proportion?

Solution:

Cross products

$2 \cdot 6$ $3 \cdot 4$

$$\frac{2}{3} = \frac{4}{6}$$

$2 \cdot 6 \stackrel{?}{=} 3 \cdot 4$ Are cross products equal?
$12 = 12$ Equal, so proportion is true

Since the cross products are equal, the proportion is true.

▣ **Work Practice Problem 2**

PRACTICE PROBLEM 2

Is $\frac{4}{8} = \frac{10}{20}$ a true proportion?

EXAMPLE 3 Is $\frac{4.1}{7} = \frac{2.9}{5}$ a true proportion?

Solution

Cross products

$4.1 \cdot 5$ $7 \cdot 2.9$

$$\frac{4.1}{7} = \frac{2.9}{5}$$

$4.1 \cdot 5 \stackrel{?}{=} 7 \cdot 2.9$ Are cross products equal?
$20.5 \neq 20.3$ Not equal, so proportion is false

Since the cross products are not equal, $\frac{4.1}{7} \neq \frac{2.9}{5}$. The proportion is false.

▣ **Work Practice Problem 3**

PRACTICE PROBLEM 3

Is $\frac{4.2}{6} = \frac{4.8}{8}$ a true proportion?

EXAMPLE 4 Is $\dfrac{1\frac{1}{6}}{10\frac{1}{2}} = \dfrac{\frac{1}{2}}{4\frac{1}{2}}$ a true proportion?

Solution:

$$\dfrac{1\frac{1}{6}}{10\frac{1}{2}} = \dfrac{\frac{1}{2}}{4\frac{1}{2}}$$

$1\frac{1}{6} \cdot 4\frac{1}{2} \stackrel{?}{=} 10\frac{1}{2} \cdot \frac{1}{2}$ Are cross products equal?

$\frac{7}{6} \cdot \frac{9}{2} \stackrel{?}{=} \frac{21}{2} \cdot \frac{1}{2}$ Write mixed numbers as improper fractions.

$\frac{21}{4} = \frac{21}{4}$ Equal, so proportion is true

Since the cross products are equal, the proportion is true.

▣ **Work Practice Problem 4**

PRACTICE PROBLEM 4

Is $\dfrac{3\frac{3}{10}}{1\frac{5}{6}} = \dfrac{4\frac{1}{5}}{2\frac{1}{3}}$ a true proportion?

✔**Concept Check** Using the numbers in the proportion $\frac{21}{27} = \frac{7}{9}$, write two other true proportions.

Objective C Finding Unknown Numbers in Proportions

When one number of a proportion is unknown, we can use cross products to find the unknown number. For example, to find the unknown number x in the proportion $\frac{2}{3} = \frac{x}{30}$, we use cross products.

Answers
2. yes   **3.** no   **4.** yes

✔ **Concept Check Answer**
possible answers: $\frac{27}{21} = \frac{9}{7}$ and $\frac{9}{27} = \frac{7}{21}$

PRACTICE PROBLEM 5

Solve $\dfrac{2}{5} = \dfrac{x}{25}$ for x.

EXAMPLE 5 Solve $\dfrac{2}{3} = \dfrac{x}{30}$ for x.

Solution: If the cross products are equal, then the proportion is true. We begin, then, by setting cross products equal to each other.

$$\dfrac{2}{3} = \dfrac{x}{30}$$

$2 \cdot 30 = 3 \cdot x$ Set cross products equal.

$60 = 3x$ Multiply.

Recall that to find x, we divide both sides of the equation by 3.

$\dfrac{60}{3} = \dfrac{3x}{3}$ Divide both sides by 3.

$20 = x$ Simplify.

Check: To check, we replace x with 20 in the original proportion to see if the result is a true statement.

$\dfrac{2}{3} = \dfrac{x}{30}$ Original proportion

$\dfrac{2}{3} \overset{?}{=} \dfrac{20}{30}$ Replace x with 20.

$\dfrac{2}{3} = \dfrac{2}{3}$ True

Since $\dfrac{2}{3} = \dfrac{2}{3}$ is a true statement, 20 is the solution.

▢ **Work Practice Problem 5**

PRACTICE PROBLEM 6

Solve $\dfrac{15}{2} = \dfrac{60}{x}$ for x.

EXAMPLE 6 Solve $\dfrac{51}{34} = \dfrac{-3}{x}$ for x.

Solution:

$$\dfrac{51}{34} = \dfrac{-3}{x}$$

$51 \cdot x = 34 \cdot -3$ Set cross products equal.

$51x = -102$ Multiply.

$\dfrac{51x}{51} = \dfrac{-102}{51}$ Divide both sides by 51.

$x = -2$ Simplify.

Check: $\dfrac{51}{34} = \dfrac{-3}{x}$ Original proportion

$\dfrac{51}{34} \overset{?}{=} \dfrac{-3}{-2}$ Replace x with -2.

$\dfrac{51}{34} \overset{?}{=} \dfrac{-3 \cdot -17}{-2 \cdot -17}$

$\dfrac{51}{34} = \dfrac{51}{34}$ True

▢ **Work Practice Problem 6**

Answers

5. 10 **6.** 8

EXAMPLE 7 Solve for y: $\dfrac{\frac{1}{2}}{\frac{4}{5}} = \dfrac{\frac{3}{4}}{y}$

Solution:

$$\dfrac{\frac{1}{2}}{\frac{4}{5}} = \dfrac{\frac{3}{4}}{y}$$

$$\frac{1}{2} \cdot y = \frac{4}{5} \cdot \frac{3}{4} \quad \text{Set cross products equal.}$$

$$\frac{1}{2}y = \frac{3}{5} \quad \text{Multiply.}$$

$$2 \cdot \frac{1}{2}y = 2 \cdot \frac{3}{5} \quad \text{Multiply both sides by 2.}$$

$$y = \frac{6}{5} \quad \text{Simplify.}$$

Verify that $\dfrac{6}{5}$ is the solution.

Check: We will check by cross products.

$$\dfrac{\frac{1}{2}}{\frac{4}{5}} = \dfrac{\frac{3}{4}}{y}$$

$$\dfrac{\frac{1}{2}}{\frac{4}{5}} \stackrel{?}{=} \dfrac{\frac{3}{4}}{\frac{6}{5}} \quad \text{Replace } y \text{ with } \frac{6}{5}.$$

$$\frac{1}{2} \cdot \frac{6}{5} \stackrel{?}{=} \frac{4}{5} \cdot \frac{3}{4} \quad \text{Set cross products equal.}$$

$$\frac{3}{5} = \frac{3}{5} \quad \text{True}$$

True, so the solution is $\dfrac{6}{5}$.

Work Practice Problem 7

EXAMPLE 8 Solve for x: $\dfrac{x}{3} = \dfrac{0.8}{1.5}$

Solution:

$$\dfrac{x}{3} = \dfrac{0.8}{1.5}$$

$$x \cdot 1.5 = 3(0.8) \quad \text{Set cross products equal.}$$

$$1.5x = 2.4 \quad \text{Multiply.}$$

$$\frac{1.5x}{1.5} = \frac{2.4}{1.5} \quad \text{Divide both sides by 1.5.}$$

$$x = 1.6 \quad \text{Simplify.}$$

Check:

$$\dfrac{x}{3} = \dfrac{0.8}{1.5}$$

$$\dfrac{1.6}{3} \stackrel{?}{=} \dfrac{0.8}{1.5} \quad \text{Replace } x \text{ with } 1.6.$$

$$(1.6)(1.5) \stackrel{?}{=} 3(0.8) \quad \text{Set cross products equal.}$$

$$2.4 = 2.4 \quad \text{True}$$

True, so the solution is 1.6.

Work Practice Problem 8

PRACTICE PROBLEM 7

Solve for z:
$$\dfrac{\frac{7}{8}}{z} = \dfrac{\frac{2}{3}}{\frac{4}{7}}$$

PRACTICE PROBLEM 8

Solve for y: $\dfrac{y}{9} = \dfrac{0.6}{1.2}$

Answers

7. $\dfrac{3}{4}$ **8.** 4.5

PRACTICE PROBLEM 9

Solve for z: $\dfrac{17}{z} = \dfrac{8}{10}$

EXAMPLE 9 Solve for y: $\dfrac{14}{y} = \dfrac{12}{16}$

Solution:

$$\dfrac{14}{y} \;\;\asymp\;\; \dfrac{12}{16}$$

$$14 \cdot 16 = y \cdot 12 \qquad \text{Set cross products equal.}$$

$$224 = 12y \qquad \text{Multiply.}$$

$$\dfrac{224}{12} = \dfrac{12y}{12} \qquad \text{Divide both sides by 12.}$$

$$\dfrac{56}{3} = y \qquad \text{Simplify.}$$

Check to see that the solution is $\dfrac{56}{3}$.

☐ **Work Practice Problem 9**

Helpful Hint

In Example 9, the fraction $\dfrac{12}{16}$ may be simplified to $\dfrac{3}{4}$ before solving the equation. The solution will remain the same.

✔ **Concept Check** True or false: the first step in solving the proportion $\dfrac{4}{z} = \dfrac{12}{15}$ yields the equation $4z = 180$. If false, give the correct cross product equation.

PRACTICE PROBLEM 10

Solve for y: $\dfrac{4.5}{1.8} = \dfrac{y}{3}$. Round the solution to the nearest tenth.

EXAMPLE 10 Solve for x: $\dfrac{1.6}{1.1} = \dfrac{x}{0.3}$. Round the solution to the nearest hundredth.

Solution:

$$\dfrac{1.6}{1.1} \;\;\asymp\;\; \dfrac{x}{0.3}$$

$$(1.6)(0.3) = 1.1 \cdot x \qquad \text{Set cross products equal.}$$

$$0.48 = 1.1x \qquad \text{Multiply.}$$

$$\dfrac{0.48}{1.1} = \dfrac{1.1x}{1.1} \qquad \text{Divide both sides by 1.1.}$$

$$0.44 \approx x \qquad \text{Round to the nearest hundredth.}$$

☐ **Work Practice Problem 10**

Answers

9. $\dfrac{85}{4}$ or 21.25 **10.** 7.5

✔ **Concept Check Answer**

false; $60 = 12z$.

Vocabulary and Readiness Check

Use the words and phrases below to fill in each blank.

ratio   cross products   true

false   proportion

1. $\frac{4.2}{8.4} = \frac{1}{2}$ is called a _____ while $\frac{7}{8}$ is called a _____.

2. In $\frac{a}{b} = \frac{c}{d}$, $a \cdot d$ and $b \cdot c$ are called _____.

3. In a proportion, if cross products are equal, the proportion is _____.

4. In a proportion, if cross products are not equal, the proportion is _____.

6.2 EXERCISE SET

FOR EXTRA HELP

Student Solutions Manual   PH Math/Tutor Center   CD/Video for Review   MathXL®   MyMathLab

Objective A **Translating** *Write each sentence as a proportion. See Example 1.*

1. 10 diamonds is to 6 opals as 5 diamonds is to 3 opals.

2. 1 raisin is to 5 cornflakes as 8 raisins is to 40 cornflakes.

3. 20 students is to 5 microscopes as 4 students is to 1 microscope.

4. 4 hit songs is to 16 releases as 1 hit song is to 4 releases.

5. 6 eagles is to 58 sparrows as 3 eagles is to 29 sparrows.

6. 12 errors is to 8 pages as 1.5 errors is to 1 page.

7. $2\frac{1}{4}$ cups of flour is to 24 cookies as $6\frac{3}{4}$ cups of flour is to 72 cookies.

8. $1\frac{1}{2}$ cups milk is to 10 bagels as $\frac{3}{4}$ cup milk is to 5 bagels.

9. 22 vanilla wafers is to 1 cup of cookie crumbs as 55 vanilla wafers is to 2.5 cups of cookie crumbs. (*Source:* Based on data from *Family Circle* magazine)

10. 1 cup of instant rice is to 1.5 cups cooked rice as 1.5 cups of instant rice is to 2.25 cups of cooked rice. (*Source:* Based on data from *Family Circle* magazine)

Objective B *Determine whether each proportion is true or false. See Examples 2 through 4.*

11. $\frac{15}{9} = \frac{5}{3}$

12. $\frac{8}{6} = \frac{20}{15}$

13. $\frac{5}{8} = \frac{4}{7}$

14. $\frac{7}{3} = \frac{9}{5}$

15. $\frac{9}{36} = \frac{2}{8}$

16. $\frac{8}{24} = \frac{3}{9}$

17. $\frac{5}{8} = \frac{625}{1000}$

18. $\frac{30}{50} = \frac{600}{1000}$

19. $\dfrac{0.8}{0.3} = \dfrac{0.2}{0.6}$

20. $\dfrac{0.7}{0.4} = \dfrac{0.3}{0.1}$

21. $\dfrac{4.2}{8.4} = \dfrac{5}{10}$

22. $\dfrac{8}{10} = \dfrac{5.6}{0.7}$

23. $\dfrac{\frac{3}{4}}{\frac{4}{3}} = \dfrac{\frac{1}{2}}{\frac{8}{9}}$

24. $\dfrac{\frac{2}{3}}{\frac{8}{5}} = \dfrac{\frac{1}{7}}{\frac{3}{10}}$

25. $\dfrac{2\frac{2}{5}}{\frac{2}{3}} = \dfrac{\frac{10}{9}}{\frac{1}{4}}$

26. $\dfrac{5\frac{5}{8}}{\frac{5}{3}} = \dfrac{4\frac{1}{2}}{1\frac{1}{5}}$

27. $\dfrac{\frac{4}{5}}{\frac{6}{5}} = \dfrac{\frac{6}{5}}{\frac{5}{9}}$

28. $\dfrac{\frac{6}{7}}{\frac{7}{3}} = \dfrac{\frac{10}{7}}{\frac{7}{5}}$

Objectives A B **Mixed Practice–Translating** *Write each sentence as a proportion. Then determine whether the proportion is a true proportion. See Examples 1 through 4.*

29. ten is to fifteen as four is to six

30. six is to eight as nine is to twelve

31. eleven is to four as five is to two

32. five is to three as seven is to five

33. fifteen hundredths is to three as thirty-five hundredths is to seven

34. one and eight tenths is to two as four and five tenths is to five

35. two thirds is to one fifth as two fifths is to one ninth

36. ten elevenths is to three fourths as one fourth is to one half

Objective C *Solve each proportion for the given variable. Round the solution where indicated. See Examples 5 through 10.*

37. $\dfrac{x}{5} = \dfrac{6}{10}$

38. $\dfrac{x}{3} = \dfrac{12}{9}$

39. $\dfrac{-18}{54} = \dfrac{3}{n}$

40. $\dfrac{25}{100} = \dfrac{-7}{n}$

41. $\dfrac{30}{10} = \dfrac{15}{y}$

42. $\dfrac{16}{20} = \dfrac{z}{35}$

43. $\dfrac{8}{15} = \dfrac{z}{6}$

44. $\dfrac{12}{10} = \dfrac{z}{16}$

45. $\dfrac{24}{x} = \dfrac{60}{96}$

46. $\dfrac{26}{x} = \dfrac{28}{49}$

47. $\dfrac{-3.5}{12.5} = \dfrac{-7}{n}$

48. $\dfrac{-0.2}{0.7} = \dfrac{-8}{n}$

49. $\dfrac{n}{0.6} = \dfrac{0.05}{12}$

50. $\dfrac{7.8}{13} = \dfrac{n}{2.6}$

51. $\dfrac{8}{\frac{1}{3}} = \dfrac{24}{n}$

52. $\dfrac{12}{\frac{3}{4}} = \dfrac{48}{n}$

53. $\dfrac{\frac{1}{3}}{\frac{3}{8}} = \dfrac{\frac{2}{5}}{n}$

54. $\dfrac{\frac{7}{9}}{\frac{8}{27}} = \dfrac{\frac{1}{4}}{n}$

55. $\dfrac{12}{n} = \dfrac{\frac{2}{3}}{\frac{6}{9}}$

56. $\dfrac{24}{n} = \dfrac{\frac{8}{15}}{\frac{5}{9}}$

57. $\dfrac{n}{1\frac{1}{5}} = \dfrac{4\frac{1}{6}}{6\frac{2}{3}}$

58. $\dfrac{n}{3\frac{1}{8}} = \dfrac{7\frac{3}{5}}{2\frac{3}{8}}$

59. $\dfrac{25}{n} = \dfrac{3}{\frac{7}{30}}$

60. $\dfrac{9}{n} = \dfrac{5}{\frac{11}{15}}$

61. $\dfrac{3.2}{0.3} = \dfrac{x}{1.4}$

Round to the nearest tenth.

62. $\dfrac{1.8}{z} = \dfrac{2.5}{8.4}$

Round to the nearest tenth.

63. $\dfrac{z}{5.2} = \dfrac{0.08}{6}$

Round to the nearest hundredth.

64. $\dfrac{4.25}{6.03} = \dfrac{5}{y}$

Round to the nearest hundredth.

65. $\dfrac{7}{18} = \dfrac{x}{5}$

Round to the nearest tenth.

66. $\dfrac{17}{x} = \dfrac{9}{4}$

Round to the nearest thousandth.

67. $\dfrac{43}{17} = \dfrac{8}{z}$

Round to the nearest thousandth.

68. $\dfrac{x}{12} = \dfrac{18}{7}$

Round to the nearest hundredth.

Review

Insert < or > to form a true statement. See Sections 4.7 and 5.1.

69. 8.01   8.1

70. 7.26   7.026

71. $2\frac{1}{2}$ $2\frac{1}{3}$

72. $9\frac{1}{5}$ $9\frac{1}{4}$

Simplify each fraction. See Section 4.2.

73. $\dfrac{75}{125}$

74. $\dfrac{11y}{99y}$

75. $\dfrac{12x}{42}$

76. $\dfrac{28y^2}{42y^3}$

Concept Extensions

Use the numbers in each proportion to write two other true proportions. See the first Concept Check in this section.

77. $\dfrac{9}{15} = \dfrac{3}{5}$

78. $\dfrac{1}{4} = \dfrac{5}{20}$

79. $\dfrac{6}{18} = \dfrac{1}{3}$

80. $\dfrac{2}{7} = \dfrac{4}{14}$

81. If the proportion $\dfrac{a}{b} = \dfrac{c}{d}$ is a true proportion, write two other true proportions using the same letters.

82. Write a true proportion.

83. Explain the difference between a ratio and a proportion.

84. Explain how to find the unknown number in a proportion such as $\frac{n}{18} = \frac{12}{8}$.

For each proportion, solve for the variable.

85. $\dfrac{x}{7} = \dfrac{0}{8}$

86. $\dfrac{0}{2} = \dfrac{y}{3.5}$

87. $\dfrac{z}{1150} = \dfrac{588}{483}$

88. $\dfrac{585}{x} = \dfrac{117}{474}$

89. $\dfrac{222}{1515} = \dfrac{37}{y}$

90. $\dfrac{1425}{1062} = \dfrac{z}{177}$

THE BIGGER PICTURE Operations on Sets of Numbers and Solving Equations

Continue your outline from Sections 1.6, 1.7, 2.4, 3.3, 4.3, 4.7, and 5.4. Suggestions are once again written to help you complete this part of your outline. Notice that this part of the outline has to do with solving a certain type of equation, proportions.

I. Operations on Sets of Numbers

    A. Whole Numbers

        1. Add or Subtract (Section 1.3)

        2. Multiply or Divide (Sections 1.5, 1.6)

        3. Exponent (Section 1.7)

        4. Order of Operations (Section 1.7)

    B. Integers

        1. Add (Section 2.2)

        2. Subtract (Section 2.3)

        3. Multiply or Divide (Section 2.4)

    C. Fractions

        1. Simplify (Section 4.2)

        2. Multiply (Section 4.3)

        3. Divide (Section 4.3)

        4. Add or Subtract (Sections 4.4, 4.5)

    D. Decimals

        1. Add or Subtract (Section 5.2)

        2. Multiply (Section 5.3)

        3. Divide (Section 5.4)

II. Solving Equations

    A. Equations in General (Section 3.3)

    B. Proportions: Set cross products equal to each other. Then solve.

$$\frac{14}{3} = \frac{2}{n}, \text{ or } 14 \cdot n = 3 \cdot 2, \text{ or } 14n = 6, \text{ or } n = \frac{6}{14} = \frac{3}{7}$$

Perform the indicated operations.

1. $\dfrac{7}{20} - \dfrac{1}{10}$

2. $\dfrac{7}{20} \cdot \dfrac{1}{10}$

3. $\dfrac{7}{20} \div \dfrac{1}{10}$

4. $\dfrac{7}{20} + \dfrac{1}{10}$

5. $7.6 + 0.02$

6. $7.6(0.02)$

For each proportion, find the unknown number, n.

7. $\dfrac{4}{n} = \dfrac{50}{100}$

8. $\dfrac{60}{10} = \dfrac{15}{n}$

9. $\dfrac{n}{0.8} = \dfrac{0.06}{12}$

10. $\dfrac{\frac{7}{8}}{\frac{1}{4}} = \dfrac{n}{\frac{5}{6}}$

Sections 6.1–6.2

Ratio, Rate, and Proportion

Write each ratio as a ratio of whole numbers using fractional notation. Write the fraction in simplest form.

1. 27 to 30

2. 18 to 50

3. 9.4 to 10

4. 3.2 to 9.2

5. 8.65 to 6.95

6. 3.6 to 4.2

7. $\frac{7}{2}$ to 13

8. $1\frac{2}{3}$ to $2\frac{3}{4}$

9. 16 inches to 24 inches

10. 5 hours to 40 hours

Find the ratio described in each problem.

11. Find the ratio of the width (shorter side) to the length (longer side) of the sign below.

12 inches

18 inches

12. The circle graph below shows how the top 20 movies of 2004 were rated. Use this graph to answer the questions.

   a. How many top 20 movies were rated PG-13?

   b. Find the ratio of top 20 PG-13 movies to total movies for that year.

Top 20 Movies of 2004

G
1 film

PG
6 films

R
3 films

PG-13
10 films

Source: Internet search and Bryan Gay

Write each rate as a fraction in simplest form.

13. 4 professors for every 20 graduate assistants

14. 6 lights every 20 feet

15. 100 U.S. Senators for 50 states

16. 5 teachers for every 140 students

17. 21 inches every 7 seconds

18. $40 every 5 hours

Answers

1. _____

2. _____

3. _____

4. _____

5. _____

6. _____

7. _____

8. _____

9. _____

10. _____

11. _____

12. a. _____

    b. _____

13. _____

14. _____

15. _____

16. _____

17. _____

18. _____

19. _____

20. _____

21. _____

22. _____

23. _____

24. _____

25. _____

26. _____

27. _____

28. _____

29. _____

30. _____

31. _____

32. _____

33. _____

34. _____

35. _____

36. _____

37. _____

38. _____

19. 76 households with computers for every 100 households

20. 538 electoral votes for 50 states

Write each rate as a unit rate.

21. 560 feet in 4 seconds

22. 195 miles in 3 hours

23. 63 employees per 3 fax lines

24. 85 phone calls for 5 teenagers

25. 156 miles per 6 gallons

26. 112 teachers for 7 computers

27. 8125 books for 1250 college students

28. 2310 pounds for 14 adults

Write each unit price and decide which is the better buy. Round to 3 decimal places.

29. Cat food:
8 pounds for $2.16
18 pounds for $4.99

30. Paper plates:
100 for $1.98
500 for $8.99

31. Microwave popcorn:
3 packs for $2.39
8 packs for $5.99

32. AA batteries:
4 for $3.69
10 for $9.89

Determine whether each proportion is true.

33. $\dfrac{7}{4} = \dfrac{5}{3}$

34. $\dfrac{8.2}{2} = \dfrac{16.4}{4}$

Solve for the variable.

35. $\dfrac{5}{3} = \dfrac{40}{x}$

36. $\dfrac{y}{10} = \dfrac{13}{4}$

37. $\dfrac{6}{11} = \dfrac{z}{5}$

38. $\dfrac{21}{x} = \dfrac{\frac{7}{2}}{3}$

6.3 PROPORTIONS AND PROBLEM SOLVING

Objective **A** Solving Problems by Writing Proportions

Writing proportions is a powerful tool for solving problems in almost every field, including business, chemistry, biology, health sciences, and engineering, as well as in daily life. Given a specified ratio (or rate) of two quantities, a proportion can be used to determine an unknown quantity.

In this section, we use the same problem solving steps that we have used earlier in this text.

EXAMPLE 1 Determining Distances from a Map

On a chamber of commerce map of Abita Springs, 5 miles corresponds to 2 inches. How many miles correspond to 7 inches?

Solution:

1. UNDERSTAND. Read and reread the problem. You may want to draw a diagram.

5 miles	5 miles	5 miles	5 miles	= a little over 15 miles
2 inches	2 inches	2 inches	2 inches	= 7 inches

15 miles — between 15 and 20 miles — 20 miles
6 inches — 7 inches — 8 inches

From the diagram we can see that a reasonable solution should be between 15 and 20 miles.

2. TRANSLATE. We will let x be our unknown number. Since 5 miles corresponds to 2 inches as x miles corresponds to 7 inches, we have the proportion

$$\begin{array}{ccc} \text{miles} & \rightarrow & \dfrac{5}{2} = \dfrac{x}{7} & \leftarrow & \text{miles} \\ \text{inches} & \rightarrow & & \leftarrow & \text{inches} \end{array}$$

3. SOLVE: In earlier sections, we estimated to obtain a reasonable answer. Notice we did this in Step 1 above.

$$\frac{5}{2} \diagdown\!\!\!\diagup \frac{x}{7}$$

$$5 \cdot 7 = 2 \cdot x \qquad \text{Set the cross products equal to each other.}$$

$$35 = 2x \qquad \text{Multiply.}$$

$$\frac{35}{2} = \frac{2x}{2} \qquad \text{Divide both sides by 2.}$$

$$17\frac{1}{2} = x \text{ or } x = 17.5 \quad \text{Simplify.}$$

PRACTICE PROBLEM 1

On an architect's blueprint, 1 inch corresponds to 4 feet. How long is a wall represented by a $4\frac{1}{4}$-inch line on the blueprint?

Answer
1. 17 ft

Continued on next page

435

4. INTERPRET. *Check* your work. This result is reasonable since it is between 15 and 20 miles. *State* your conclusion: 7 inches corresponds to 17.5 miles.

■ **Work Practice Problem 1**

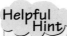

Helpful Hint

We can also solve Example 1 by writing the proportion

$$\frac{2 \text{ inches}}{5 \text{ miles}} = \frac{7 \text{ inches}}{x \text{ miles}}$$

Although other proportions may be used to solve Example 1, we will solve by writing proportions so that the numerators have the same unit measures and the denominators have the same unit measures.

PRACTICE PROBLEM 2

An auto mechanic recommends that 5 ounces of isopropyl alcohol be mixed with a tankful of gas (16 gallons) to increase the octane of the gasoline for better engine performance. At this rate, how many gallons of gas can be treated with a 8-ounce bottle of alcohol?

EXAMPLE 2 Finding Medicine Dosage

The standard dose of an antibiotic is 4 cc (cubic centimeters) for every 25 pounds (lb) of body weight. At this rate, find the standard dose for a 140-lb woman.

Solution:

1. UNDERSTAND. Read and reread the problem. You may want to draw a diagram to estimate a reasonable solution.

From the diagram, we can see that a reasonable solution is a little over 20 cc.

2. TRANSLATE. We will let x be the unknown number. From the problem, we know that 4 cc is to 25 pounds as x cc is to 140 pounds, or

cubic centimeters → $\frac{4}{25} = \frac{x}{140}$ ← cubic centimeters
pounds → $\phantom{\frac{4}{25}}$ ← pounds

3. SOLVE:

$$\frac{4}{25} = \frac{x}{140}$$

$4 \cdot 140 = 25 \cdot x$ Set the cross products equal to each other.

$560 = 25x$ Multiply.

$\frac{560}{25} = \frac{25x}{25}$ Divide both sides by 25.

$22\frac{2}{5} = x$ or $x = 22.4$ Simplify.

4. INTERPRET. *Check* your work. This result is reasonable since it is a little over 20 cc. *State* your conclusion: The standard dose for a 140-lb woman is 22.4 cc.

■ **Work Practice Problem 2**

Answer

2. $25\frac{3}{5}$ or 25.6 gal

⚠ **EXAMPLE 3** **Calculating Supplies Needed to Fertilize a Lawn**

A 50-pound bag of fertilizer covers 2400 square feet of lawn. How many bags of fertilizer are needed to cover a town square containing 15,360 square feet of lawn? Round the answer up to the nearest whole bag.

Solution:

1. UNDERSTAND. Read and reread the problem. Draw a picture.

1 bag covers   How many bags cover ... ?

Since one bag covers 2400 square feet, let's see how many 2400s there are in 15,360. We will estimate. The number 15,360 rounded to the nearest thousand is 15,000 and 2400 rounded to the nearest thousand is 2000. Then

$$15,000 \div 2000 = 7\frac{1}{2} \text{ or } 7.5.$$

2. TRANSLATE. We'll let x be the unknown number. From the problem, we know that 1 bag is to 2400 square feet as x bags is to 15,360 square feet.

    bags $\rightarrow$ $\dfrac{1}{2400} = \dfrac{x}{15,360}$ $\leftarrow$ bags
    square feet $\rightarrow$ $\qquad\qquad\quad$ $\leftarrow$ square feet

3. SOLVE:

    $$\dfrac{1}{2400} = \dfrac{x}{15,360}$$

    $1 \cdot 15,360 = 2400 \cdot x$ Set the cross products equal to each other.

    $15,360 = 2400 \cdot x$ Multiply.

    $\dfrac{15,360}{2400} = \dfrac{2400x}{2400}$ Divide both sides by 2400.

    $6.4 = x$ Simplify.

4. INTERPRET. *Check* that replacing x with 6.4 makes the proportion true. Is the answer reasonable? Yes, since it's close to $7\frac{1}{2}$ or 7.5. Since we must buy whole bags of fertilizer, 7 bags are needed. *State* your conclusion: To cover 15,360 square feet of lawn, 7 bags are needed.

🔲 **Work Practice Problem 3**

✔ **Concept Check** You are told that 12 ounces of ground coffee will brew enough coffee to serve 20 people. How could you estimate how much ground coffee will be needed to serve 95 people?

6.3 EXERCISE SET

FOR EXTRA HELP

Student Solutions Manual

PH Math/Tutor Center

CD/Video for Review

Math XL
MathXL®

MyMathLab

Objective Ⓐ *Solve. See Examples 1 through 3. An NBA basketball player averages 45 baskets for every 100 attempts.*

1. If he attempted 800 field goals, how many baskets did he make?

2. If he completed 225 baskets, how many did he attempt?

It takes a word processor 30 minutes to word process and spell check 4 pages.

3. Find how long it takes her to word process and spell check 22 pages.

4. Find how many pages she can word process and spell check in 4.5 hours.

University Law School accepts 2 out of every 7 applicants.

5. If the school accepted 180 students, find how many applications they received.

6. If the school accepted 150 students, find how many applications were received.

On an architect's blueprint, 1 inch corresponds to 8 feet.

7. Find the length of a wall represented by a line $2\frac{7}{8}$ inches long on the blueprint.

8. Find the length of a wall represented by a line $5\frac{1}{4}$ inches on the blueprint.

A human-factors expert recommends that there be at least 9 square feet of floor space in a college classroom for every student in the class.

9. Find the minimum floor space that 30 students require.

10. Due to a lack of space, a university converts a 21-by-15-foot conference room into a classroom. Find the maximum number of students the room can accommodate.

A Honda Civic Hybrid car averages 627 miles on a 12.3 gallon tank of gas.

11. Manuel Lopez is planning a 1250-mile vacation trip in his Honda Civic Hybrid. Find how many gallons of gas he can expect to burn. Round to the nearest gallon.

12. Ramona Hatch has enough money to put 6.9 gallons of gas in her Honda Civic Hybrid. She is planning on driving home from college for the weekend. If her home is 290 miles away, should she make it home before she runs out of gas?

The scale on an Italian map states that 1 centimeter corresponds to 30 kilometers.

13. Find how far apart Milan and Rome are if their corresponding points on the map are 15 centimeters apart.

14. On the map, a small Italian village is located 0.4 centimeter from the Mediterranean Sea. Find the actual distance.

Milan

Rome

A bag of Scott fertilizer covers 3000 square feet of lawn.

△ **15.** Find how many bags of fertilizer should be purchased to cover a rectangular lawn 260 feet by 180 feet.

△ **16.** Find how many bags of fertilizer should be purchased to cover a square lawn measuring 160 feet on each side.

A Cubs baseball player gets 3 hits in every 8 times at bat.

17. If this Cubs player comes up to bat 40 times in the World Series, find how many hits he would be expected to get.

18. At this rate, if he got 12 hits, find how many times he batted.

A survey reveals that 2 out of 3 people prefer Coke to Pepsi.

19. In a room of 40 people, how many people are likely to prefer Coke? Round the answer to the nearest person.

20. In a college class of 36 students, find how many students are likely to prefer Pepsi.

A self-tanning lotion advertises that a 3-oz bottle will provide four applications.

21. Jen Haddad found a great deal on a 14-oz bottle of the self-tanning lotion she had been using. Based on the advertising claims, how many applications of the self-tanner should Jen expect? Round down to the whole number.

22. The Community College thespians need fake tans for a play they are doing. If the play has a cast of 35, how many ounces of self-tanning lotion should the cast purchase? Round up to the next whole number onces.

The school's computer lab goes through 5 reams of printer paper every 3 weeks.

23. Find out how long a case of printer paper is likely to last (a case of paper holds 8 reams of paper). Round to the nearest week.

24. How many cases of printer paper should be purchased to last the entire semester of 15 weeks? Round to the next case.

A recipe for pancakes calls for 2 cups flour and $1\frac{1}{2}$ cup milk to make a serving for four people.

25. Ming has plenty of flour, but only 4 cups milk. How many servings can he make?

26. The swim team has a weekly breakfast after early practice. How much flour will it take to make pancakes for 18 swimmers?

27. A 16-oz grande Tazo Black Iced Tea at Starbucks has 80 calories. How many calories are there in a 24-oz venti Tazo Black Iced Tea? (*Source:* Starbucks Coffee Company)

28. A 16-oz nonfat Caramel Macchiato at Starbucks has 220 calories. How many calories are there in a 12-oz nonfat Caramel Macchiato? (*Source:* Starbucks Coffee Company)

29. Mosquitos are annoying insects. To eliminate mosquito larvae, a certain granular substance can be applied to standing water in a ratio of 1 tsp per 25 sq ft of standing water.

    a. At this rate, find how many teaspoons of granules must be used for 450 square feet.

    b. If 3 tsp = 1 tbsp, how many tablespoons of granule must be used?

30. Another type of mosquito control is liquid, where 3 oz of pesticide is mixed with 100 oz of water. This mixture is sprayed on roadsides to control mosquito breeding grounds hidden by tall grass.

    a. If one mixture of water with this pesticide can treat 150 feet of roadway, how many ounces of pesticide are needed to treat one mile? (*Hint:* 1 mile = 5280 feet)

    b. If 8 liquid ounces equals one cup, write your answer to part **a** in cups. Round to the nearest cup.

31. The daily supply of oxygen for one person is provided by 625 square feet of lawn. A total of 3750 square feet of lawn would provide the daily supply of oxygen for how many people? (*Source:* Professional Lawn Care Association of America)

32. In the United States, approximately 71 million of the 200 million cars and light trucks in service have driver-side air bags. In a parking lot containing 800 cars and light trucks, how many would be expected to have driver-side air bags? (*Source:* Insurance Institute for Highway Safety)

33. A student would like to estimate the height of the Statue of Liberty in New York City's harbor. The length of the Statue of Liberty's right arm is 42 feet. The student's right arm is 2 feet long and her height is $5\frac{1}{3}$ feet. Use this information to estimate the height of the Statue of Liberty. How close is your estimate to the statue's actual height of 111 feet, 1 inch from heel to top of head? (*Source:* National Park Service)

34. The length of the Statue of Liberty's index finger is 8 feet while the height to the top of the head is about 111 feet. Suppose your measurements are proportionaly the same as this statue and your height is 5 feet.

   a. Use this information to find the proposed length of your index finger. Give an exact measurement and then a decimal rounded to the nearest hundredth.

   b. Measure your index finger and write it as decimal in feet rounded to the nearest hundredth. How close is the length of your index finger to the answer to **a**? Explain why.

42 feet

$5\frac{1}{3}$ feet

2 feet

35. There are 72 milligrams of cholesterol in a 3.5 ounce serving of lobster. How much cholesterol is in 5 ounces of lobster? Round to the nearest tenth of a milligram. (*Source:* The National Institute of Health)

36. There are 76 milligrams of cholesterol in a 3-ounce serving of skinless chicken. How much cholesterol is in 8 ounces of chicken? (*Source:* USDA)

37. Trump World Tower in New York City is 881 feet tall and contains 72 stories. The Empire State Building contains 102 stories. If the Empire State Building has the same number of feet per floor as the Trump World Tower, approximate its height rounded to the nearest foot. (*Source:* skyscrapers.com)

38. Two out of every 5 men blame their poor eating habits on too much fast food. In a room of 40 men, how many would you expect to blame their not eating well on fast food? (*Source:* Healthy Choice Mixed Grills survey)

39. Medication is prescribed in 7 out of every 10 hospital emergency room visits that involve an injury. If a large urban hospital had 620 emergency room visits involving an injury in the past month, how many of these visits would you expect included a prescription for medication? (*Source:* National Center for Health Statistics)

40. Currently in the American population of people aged 65 years old and older, there are 145 women for every 100 men. In a nursing home with 280 male residents over the age of 65, how many female residents over the age of 65 would be expected? (*Source:* U.S. Bureau of the Census)

41. One out of three American adults has worked in the restaurant industry at some point during his or her life. In an office of 84 workers, how many of these people would you expect to have worked in the restaurant industry at some point? (*Source:* National Restaurant Association)

42. One pound of firmly packed brown sugar yields $2\frac{1}{4}$ cups. How many pounds of brown sugar will be required in a recipe that calls for 6 cups of firmly packed brown sugar? (*Source:* Based on data from *Family Circle* magazine)

When making homemade ice cream in a hand-cranked freezer, the tub containing the ice cream mix is surrounded by a brine (water/salt) solution. To freeze the ice cream mix rapidly so that smooth and creamy ice cream results, the brine solution should combine crushed ice and rock salt in a ratio of 5 to 1. Use this for Exercises 43 and 44. (Source: White Mountain Freezers, The Rival Company)

43. A small ice cream freezer requires 12 cups of crushed ice. How much rock salt should be mixed with the ice to create the necessary brine solution?

44. A large ice cream freezer requires $18\frac{3}{4}$ cups of crushed ice. How much rock salt will be needed?

45. The gas/oil ratio for a certain chainsaw is 50 to 1.
   a. How much oil (in gallons) should be mixed with 5 gallons of gasoline?
   b. If 1 gallon equals 128 fluid ounces, write the answer to part **a** in fluid ounces. Round to the nearest whole ounce.

46. The gas/oil ratio for a certain tractor mower is 20 to 1.
   a. How much oil (in gallons) should be mixed with 10 gallons of gas?
   b. If 1 gallon equals 4 quarts, write the answer to part **a** in quarts.

47. The adult daily dosage for a certain medicine is 150 mg (milligrams) of medicine for every 20 pounds of body weight.
   a. At this rate, find the daily dose for a man who weighs 275 pounds.
   b. If the man is to receive 500 mg of this medicine every 8 hours, is he receiving the proper dosage?

48. The adult daily dosage for a certain medicine is 80 mg (milligrams) for every 25 pounds of body weight.
   a. At this rate, find the daily dose for a woman who weighs 190 pounds.
   b. If she is to receive this medicine every 6 hours, find the amount to be given every 6 hours.

Review

Find the prime factorization of each number. See Section 4.2.

49. 200   **50.** 300   **51.** 32   **52.** 81

Concept Extensions

As we have seen earlier, proportions are often used in medicine dosage calculations. The exercises below have to do with liquid drug preparations, where the weight of the drug is contained in a volume of solution. The description of mg and ml below will help. We will study metric units further in Chapter 9.

mg means milligrams (A paper clip weighs about a gram. A milligram is about the weight of $\frac{1}{1000}$ of a paper clip.)

ml means milliliter (A liter is about a quart. A milliliter is about the amount of liquid in $\frac{1}{1000}$ of a quart.)

One way to solve the applications below is to set up the proportion $\frac{mg}{ml} = \frac{mg}{ml}$.

A solution strength of 15 mg of medicine in 1 ml of solution is available.

53. If a patient needs 12 mg of medicine, how many ml do you administer?

54. If a patient needs 33 mg of medicine, how many ml do you administer?

A solution strength of 8 mg of medicine in 1 ml of solution is available.

55. If a patient needs 10 mg of medicine, how many ml do you administer?

56. If a patient needs 6 mg of medicine, how many ml do you administer?

Estimate the following. See the Concept Check in this section.

57. It takes 1.5 cups of milk to make 11 muffins. Estimate the amount of milk needed to make 8 dozen muffins. Explain your calculation.

58. A favorite chocolate chip recipe calls for $2\frac{1}{2}$ cups of flour to make 2 dozen cookies. Estimate the amount of flour needed to make 50 cookies. Explain your calculation.

A board such as the one pictured below will balance if the following proportion is true:

$$\frac{\text{first weight}}{\text{second distance}} = \frac{\text{second weight}}{\text{first distance}}$$

Use this proportion to solve Exercises 59 and 60.

59. Find the distance n that will allow the board to balance.

60. Find the length n needed to lift the weight below.

61. Describe a situation in which writing a proportion might solve a problem related to driving a car.

 STUDY SKILLS BUILDER

How Are You Doing?

If you haven't done so yet, take a few moments and think about how you are doing in this course. Are you working toward your goal of successfully completing this course? Is your performance on homework, quizzes, and tests satisfactory? If not, you might want to see your instructor to see if he/she has any suggestions on how you can improve your performance. Reread Section 1.1 for ideas on places to get help with your mathematics course.

Answer the following.

1. List any textbook supplements you are using to help you through this course.

2. List any campus resources you are using to help you through this course.

3. Write a short paragraph describing how you are doing in your mathematics course.

4. If improvement is needed, list ways that you can work toward improving your situation as described in Exercise 3.

Now that we know how to write ratios and solve proportions, in Section 6.5, we use proportions to help us find unknown sides of similar triangles. In this section, we prepare for work on triangles by studying right triangles and their applications.

First, let's review square roots.

Objective **A** Finding Square Roots

The **square** of a number is the number times itself. For example,

The square of 5 is 25 because 5^2 or $5 \cdot 5 = 25$.
The square of -5 is also 25 because $(-5)^2$ or $(-5)(-5) = 25$.

The reverse process of squaring is finding a **square root.** For example,

A square root of 25 is 5 because $5 \cdot 5$ or $5^2 = 25$.
A square root of 25 is also -5 because $(-5)(-5)$ or $(-5)^2 = 25$.

Every positive number has two square roots. We see above that the square roots of 25 are 5 and -5.

We use the symbol $\sqrt{}$, called a **radical sign,** to indicate the positive square root of a nonnegative number. For example,

$\sqrt{25} = 5$ because $5^2 = 25$ and 5 is positive.
$\sqrt{9} = 3$ because $3^2 = 9$ and 3 is positive.

Square Root of a Number

The square root, $\sqrt{}$, of a positive number a is the positive number b whose square is a. In symbols,

$$\sqrt{a} = b, \quad \text{if } b^2 = a$$

Also, $\sqrt{0} = 0$.

Helpful Hint

Remember that the radical sign $\sqrt{}$ is used to indicate the **positive square root** of a nonnegative number.

EXAMPLES Find each square root.

1. $\sqrt{49} = 7$ because $7^2 = 49$.
2. $\sqrt{36} = 6$ because $6^2 = 36$.
3. $\sqrt{1} = 1$ because $1^2 = 1$.
4. $\sqrt{81} = 9$ because $9^2 = 81$.
5. Find $\sqrt{\dfrac{1}{36}} = \dfrac{1}{6}$ because $\left(\dfrac{1}{6}\right)^2$ or $\dfrac{1}{6} \cdot \dfrac{1}{6} = \dfrac{1}{36}$.
6. Find $\sqrt{\dfrac{4}{25}} = \dfrac{2}{5}$ because $\left(\dfrac{2}{5}\right)^2$ or $\dfrac{2}{5} \cdot \dfrac{2}{5} = \dfrac{4}{25}$.

■ **Work Practice Problems 1–6**

PRACTICE PROBLEMS

Find each square root.

1. $\sqrt{100}$ 2. $\sqrt{64}$
3. $\sqrt{121}$ 4. $\sqrt{0}$
5. $\sqrt{\dfrac{1}{4}}$ 6. $\sqrt{\dfrac{9}{16}}$

Answers

1. 10   2. 8   3. 11   4. 0
5. $\dfrac{1}{2}$ 6. $\dfrac{3}{4}$

Objective **B** Approximating Square Roots

Thus far, we have found square roots of perfect squares. Numbers like $\frac{1}{4}$, 36, $\frac{4}{25}$, and 1 are called **perfect squares** because their square root is a whole number or a fraction. A square root such as $\sqrt{5}$ cannot be written as a whole number or a fraction since 5 is not a perfect square.

Although $\sqrt{5}$ cannot be written as a whole number or a fraction, it can be approximated by estimating, by using a table (as in Appendix A.4), or by using a calculator.

PRACTICE PROBLEM 7

Use Appendix A.4 or a calculator to approximate each square root of 11 to the nearest thousandth.

a. $\sqrt{10}$

b. $\sqrt{62}$

EXAMPLE 7 Use Appendix A.4 or a calculator to approximate each square root fo the nearest thousandth.

a. $\sqrt{43} \approx 6.557$ is approximately

b. $\sqrt{80} \approx 8.944$

◻ **Work Practice Problem 7**

Helpful Hint

$\sqrt{80}$, above, is *approximately* 8.944. This means that if we multiply 8.944 by 8.944, the product is *close* to 80.

$$8.944 \times 8.944 \approx 79.995$$

Objective **C** Using the Pythagorean Theorem

One important application of square roots has to do with right triangles. Recall that a **right triangle** is a triangle in which one of the angles is a right angle, or measures 90° (degrees). The **hypotenuse** of a right triangle is the side opposite the right angle. The **legs** of a right triangle are the other two sides. These are shown in the following figure. The right angle in the triangle is indicated by the small square drawn in that angle.

The following theorem is true for all right triangles.

Pythagorean Theorem

If a and b are the lengths of the legs of a right triangle and c is the length of the hypotenuse, then

$$a^2 + b^2 = c^2$$

In other words, $(\text{leg})^2 + (\text{other leg})^2 = (\text{hypotenuse})^2$.

PRACTICE PROBLEM 8

Find the length of the hypotenuse of the given right triangle.

12 feet

16 feet

EXAMPLE 8 Find the length of the hypotenuse of the given right triangle.

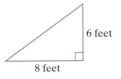

6 feet

8 feet

Solution: Let $a = 6$ and $b = 8$. According to the Pythagorean theorem,

$a^2 + b^2 = c^2$

$6^2 + 8^2 = c^2$ Let $a = 6$ and $b = 8$.

$36 + 64 = c^2$ Evaluate 6^2 and 8^2.

$100 = c^2$ Add.

Answers

7. a. 3.162   **b.** 7.874   **8.** 20 feet

In the equation $c^2 = 100$, the solutions of c are the square roots of 100. Since $10 \cdot 10 = 100$ and $(-10)(-10) = 100$, both 10 and -10 are square roots of 100. Since c represents a length, we are only interested in the positive square root of c^2.

$$c = \sqrt{100}$$
$$= 10$$

The hypotenuse is 10 feet long.

■ **Work Practice Problem 8**

⚠ **EXAMPLE 9** Approximate the length of the hypotenuse of the given right triangle. Round the length to the nearest whole unit.

Solution: Let $a = 17$ and $b = 10$.

$$a^2 + b^2 = c^2$$
$$17^2 + 10^2 = c^2$$
$$289 + 100 = c^2$$
$$389 = c^2$$
$$\sqrt{389} = c \text{ or } c \approx 20 \quad \text{From Appendix A.4 or a calculator}$$

The hypotenuse is exactly $\sqrt{389}$ meters, which is approximately 20 meters.

17 meters, 10 meters

■ **Work Practice Problem 9**

⚠ **EXAMPLE 10** Find the length of the leg in the given right triangle. Give the exact length and a two-decimal-place approximation.

5 inches, 7 inches

Solution: Notice that the hypotenuse measures 7 inches and that the length of one leg measures 5 inches. Thus, let $c = 7$ and a or b be 5. We will let $a = 5$.

$$a^2 + b^2 = c^2$$
$$5^2 + b^2 = 7^2 \qquad \text{Let } a = 5 \text{ and } c = 7.$$
$$25 + b^2 = 49 \qquad \text{Evaluate } 5^2 \text{ and } 7^2.$$
$$b^2 = 24 \qquad \text{Subtract 25 from both sides.}$$
$$b = \sqrt{24} \approx 4.90$$

The length of the leg is exactly $\sqrt{24}$ inches and approximately 4.90 inches.

■ **Work Practice Problem 10**

✔**Concept Check** The following lists are the lengths of the sides of two triangles. Which set forms a right triangle?

a. 8, 15, 17
b. 24, 30, 40

PRACTICE PROBLEM 9

Approximate the length of the hypotenuse of the given right triangle. Round to the nearest whole unit.

7 kilometers, 9 kilometers

PRACTICE PROBLEM 10

Find the length of the leg in the given right triangle. Give the exact length and a two-decimal-place approximation.

13 feet, 7 feet

Answers

9. 11 kilometers
10. $\sqrt{120}$ feet ≈ 10.95 feet

✔ **Concept Check Answer**

a

PRACTICE PROBLEM 11

A football field is a rectangle measuring 100 yards by 53 yards. Draw a diagram and find the length of the diagonal of a football field to the nearest yard.

△ **EXAMPLE 11** Finding the Dimensions of a Park

An inner-city park is in the shape of a square that measures 300 feet on a side. A sidewalk is to be constructed along the diagonal of the park. Find the length of the sidewalk rounded to the nearest whole foot.

Solution: The diagonal is the hypotenuse of a right triangle, which we label c.

$$a^2 + b^2 = c^2$$
$$300^2 + 300^2 = c^2 \qquad \text{Let } a = 300 \text{ and } b = 300.$$
$$90{,}000 + 90{,}000 = c^2 \qquad \text{Evaluate } (300)^2.$$
$$180{,}000 = c^2 \qquad \text{Add.}$$
$$\sqrt{180{,}000} = c \text{ or } c \approx 424$$

The length of the sidewalk is approximately 424 feet.

▣ **Work Practice Problem 11**

Answer

11. ≈113 yards

▦ **CALCULATOR EXPLORATIONS** Finding and Approximating Square Roots

To simplify or approximate square roots using a calculator, locate the key marked ⎡√⎤.

To simplify $\sqrt{64}$, for example, press the keys

⎡64⎤ ⎡√⎤ or ⎡√⎤ ⎡64⎤

The display should read ⎡ 8 ⎤. Then

$$\sqrt{64} = 8$$

To *approximate* $\sqrt{10}$, press the keys

⎡10⎤ ⎡√⎤ or ⎡√⎤ ⎡10⎤

The display should read ⎡3.16227766⎤. This is an *approximation* for $\sqrt{10}$. A three-decimal-place approximation is

$$\sqrt{10} \approx 3.162$$

Is this answer reasonable? Since 10 is between the perfect squares 9 and 16, $\sqrt{10}$ is between $\sqrt{9} = 3$ and $\sqrt{16} = 4$. Our answer is reasonable since 3.162 is between 3 and 4.

Simplify.

1. $\sqrt{1024}$ **2.** $\sqrt{676}$

Approximate each square root. Round each answer to the nearest thousandth.

3. $\sqrt{15}$ **4.** $\sqrt{19}$

5. $\sqrt{97}$ **6.** $\sqrt{56}$

Vocabulary and Readiness Check

Use the choices below to fill in each blank. Not all choices will be used, and some choices will be used more than once.

squaring	b^2	radical	-10	leg
hypotenuse	perfect squares	10	c^2	

1. A square root of 100 is _____ and _____ because $10 \cdot 10 = 100$ and $(-10)(-10) = 100$.
2. $\sqrt{100} =$ _____ because $10 \cdot 10 = 100$ and 10 is positive.
3. The _____ sign is used to denote the positive square root of a nonnegative number.
4. The reverse process of _____ a number is finding a square root of a number.
5. The numbers 9, 1, and $\frac{1}{25}$ are called _____.
6. Label the parts of the right triangle.

7. In the given triangle,

$a^2 +$ _____ $=$ _____.

Objective A *Find each square root. See Examples 1 through 6.*

1. $\sqrt{4}$

2. $\sqrt{9}$

3. $\sqrt{121}$

4. $\sqrt{144}$ 12

5. $\sqrt{\dfrac{1}{81}}$

6. $\sqrt{\dfrac{1}{64}}$

7. $\sqrt{\dfrac{16}{64}}$

8. $\sqrt{\dfrac{36}{81}}$

Objective B *Use Appendix A.4 or a calculator to approximate each square root. Round the square root to the nearest thousandth. See Example 7.*

9. $\sqrt{3}$ 1732

10. $\sqrt{5}$

11. $\sqrt{15}$

12. $\sqrt{17}$

13. $\sqrt{31}$

14. $\sqrt{85}$

15. $\sqrt{26}$

16. $\sqrt{35}$

Objectives A B Mixed Practice *Find each square root. If necessary, round the square root to the nearest thousandth. See Examples 1 through 7.*

17. $\sqrt{256}$

18. $\sqrt{625}$

19. $\sqrt{92}$

20. $\sqrt{18}$

21. $\sqrt{\dfrac{49}{144}}$

22. $\sqrt{\dfrac{121}{169}}$ $\dfrac{11}{13}$

23. $\sqrt{71}$

24. $\sqrt{62}$

Objective C *Find the unknown length in each right triangle. If necessary, approximate the length to the nearest thousandth. See Examples 8 through 11.*

25.

5 in.
?
12 in.

26.

?
15 ft
36 ft

27.

10 cm
12 cm
?

28.

3 yd
9 yd
?

29.

22 m
48 m
x

30.

34 mi
70 mi
?

31.

?
108 mm
45 mm

32.

?
36 km
27 km

Sketch each right triangle and find the length of the side not given. If necessary, approximate the length to the nearest thousandth. (Each length is in units.) See Examples 8 through 11.

33. leg = 3, leg = 4

34. leg = 9, leg = 12

35. leg = 5, hypotenuse = 13

36. leg = 6, hypotenuse = 10

37. leg = 10, leg = 14

38. leg = 2, leg = 16

39. leg = 35, leg = 28

40. leg = 30, leg = 15

41. leg = 30, leg = 30

42. leg = 21, leg = 21

43. hypotenuse = 2, leg = 1

44. hypotenuse = 9, leg = 8

45. leg = 7.5, leg = 4

46. leg = 12, leg = 22.5

Solve. See Example 12.

47. A standard city block is a square with each side measuring 100 yards. Find the length of the diagonal of a city block to the nearest hundredth yard.

48. A section of land is a square with each side measuring 1 mile. Find the length of the diagonal of the section of land to the nearest thousandth mile.

49. Find the height of the tree. Round the height to one decimal place.

50. Find the height of the antenna. Round the height to one decimal place.

51. The playing field for football is a rectangle that is 300 feet long by 160 feet wide. Find, to the nearest foot, the length of a straight-line run that started at one corner and went diagonally to end at the opposite corner.

52. A soccer field is in the shape of a rectangle and its dimensions depend on the age of the players. The dimensions of the soccer field below are the minimum dimensions for international play. Find the length of the diagonal of this rectangle. Round answer to the nearest tenth of a yard.

Review

Write each fraction in simplest form. See Section 4.2.

53. $\dfrac{10}{12}$

54. $\dfrac{10}{15}$

55. $\dfrac{2x}{60}$

56. $\dfrac{35}{75y}$

Perform indicated operations. See Sections 4.3 through 4.5.

57. $\dfrac{3x}{9} - \dfrac{5}{9}$

58. $\dfrac{9}{13y} - \dfrac{12}{13y}$

59. $\dfrac{9}{8} \cdot \dfrac{x}{8}$

60. $\dfrac{7x}{11} \div \dfrac{8x}{11}$

Concept Extensions

Determine what two whole numbers each square root is between without using a calculator or table. Then use a calculator or appendix to check.

61. $\sqrt{38}$

62. $\sqrt{27}$

63. $\sqrt{101}$

64. $\sqrt{85}$

65. Without using a calculator, explain how you know that $\sqrt{105}$ is *not* approximately 9.875.

Does the set form the lengths of the sides of a right triangle? See the Concept Check in this section.

66. 25, 60, 65

67. 20, 45, 50

△ **68.** Find the exact length of *x*. Then give a two-decimal-place approximation.

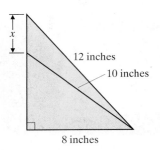

12 inches

10 inches

8 inches

THE BIGGER PICTURE **Operations on Sets of Numbers and Solving Equations**

Continue your outline from Sections 1.6, 1.7, 2.4, 3.3, 4.3, 4.7, 5.4, and 6.2. Suggestions are once again written to help you complete this part of your outline.

I. Operations on Sets of Numbers
  A. Whole Numbers
    1. Add or Subtract (Section 1.3)
    2. Multiply or Divide (Sections 1.5, 1.6)
    3. Exponent (Section 1.7)
    4. Order of Operations (Section 1.7)
    5. Square Root: $\sqrt{25} = 5$ *because* $5 \cdot 5 = 25$
  B. Integers
    1. Add (Section 2.2)
    2. Subtract (Section 2.3)
    3. Multiply or Divide (Section 2.4)
  C. Fractions
    1. Simplify (Section 4.2)
    2. Multiply (Section 4.3)
    3. Divide (Section 4.3)
    4. Add or Subtract (Sections 4.4, 4.5)

  D. Decimals
    1. Add or Subtract (Section 5.2)
    2. Multiply (Section 5.3)
    3. Divide (Section 5.4)
II. Solving Equations
  A. Equations in General (Section 3.3)
  B. Proportions (Section 6.2)

Perform the indicated operations.

1. $\sqrt{81}$

2. $\sqrt{\dfrac{4}{49}}$

3. $\sqrt{9} \cdot \sqrt{25}$

4. $\sqrt{16} \cdot \sqrt{4}$

5. $6 \cdot \sqrt{9} + 3 \cdot \sqrt{4}$

6. $3 \cdot \sqrt{25} + 2 \cdot \sqrt{81}$

7. $4 \cdot \sqrt{49} - 0 \div \sqrt{100}$

8. $7 \cdot \sqrt{36} - 0 \div \sqrt{64}$

9. $\dfrac{\sqrt{4}}{3} + \dfrac{\sqrt{25}}{6}$

10. $\dfrac{\sqrt{9}}{16} + \dfrac{\sqrt{49}}{8}$

6.5 CONGRUENT AND SIMILAR TRIANGLES

Objectives

A Decide Whether Two Triangles Are Congruent.

B Find the Ratio of Corresponding Sides in Similar Triangles.

C Find Unknown Lengths of Sides in Similar Triangles.

Objective A Deciding Whether Two Triangles Are Congruent

Two triangles are **congruent** when they have the same shape and the same size. In congruent triangles, the measures of corresponding angles are equal and the lengths of corresponding sides are equal. The following triangles are congruent:

Since these triangles are congruent, the measures of corresponding angles are equal.

Angles with equal measure: $\angle A$ and $\angle D$, $\angle B$ and $\angle E$, $\angle C$ and $\angle F$. Also, the lengths of corresponding sides are equal.

Equal corresponding sides: $\overline{AB}$ and $\overline{DE}$, $\overline{BC}$ and $\overline{EF}$, $\overline{CA}$ and $\overline{FD}$

Any one of the following may be used to determine whether two triangles are congruent:

Congruent Triangles

Angle-Side-Angle (ASA)

If the measures of two angles of a triangle equal the measures of two angles of another triangle, and the lengths of the sides between each pair of angles are equal, the triangles are congruent.

For example, these two triangles are congruent by Angle-Side-Angle.

Side-Side-Side (SSS)

If the lengths of the three sides of a triangle equal the lengths of the corresponding sides of another triangle, the triangles are congruent.

For example, these two triangles are congruent by Side-Side-Side.

(continued)

Congruent Triangles (continued)

Side-Angle-Side (SAS)

If the lengths of two sides of a triangle equal the lengths of corresponding sides of another triangle, and the measures of the angles between each pair of sides are equal, the triangles are congruent.

For example, these two triangles are congruent by Side-Angle-Side.

PRACTICE PROBLEM 1

a. Determine whether triangle *MNO* is congruent to triangle *RQS*.

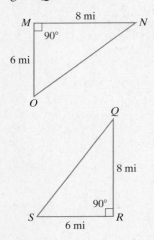

b. Determine whether triangle *GHI* is congruent to triangle *JKL*.

Answers

1. a. congruent **b.** not congruent

EXAMPLE 1 Determine whether triangle *ABC* is congruent to triangle *DEF*.

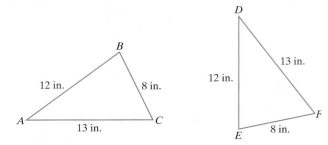

Solution: Since the lengths of all three sides of triangle *ABC* equal the lengths of all three sides of triangle *DEF*, the triangles are congruent.

◼ **Work Practice Problem 1**

In Example 1, notice that as soon as we know that the two triangles are congruent, we know that all three corresponding angles are congruent.

Objective **B** **Finding the Ratios of Corresponding Sides in Similar Triangles**

Two triangles are **similar** when they have the same shape but not necessarily the same size. In similar triangles, the measures of corresponding angles are equal and corresponding sides are in proportion. The following triangles are similar:

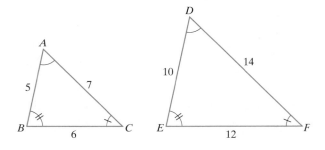

Since these triangles are similar, the measures of corresponding angles are equal.

Angles with equal measure: $\angle A$ and $\angle D$, $\angle B$ and $\angle E$, $\angle C$ and $\angle F$. Also, the lengths of corresponding sides are in proportion.

Sides in proportion: $\dfrac{AB}{DE} = \dfrac{BC}{EF} = \dfrac{CA}{FD}$ or, in this particular case,

$$\frac{AB}{DE} = \frac{5}{10} = \frac{1}{2}, \frac{BC}{EF} = \frac{6}{12} = \frac{1}{2}, \frac{CA}{FD} = \frac{7}{14} = \frac{1}{2}$$

The ratio of corresponding sides is $\dfrac{1}{2}$.

EXAMPLE 2 Find the ratio of corresponding sides for the similar triangles *ABC* and *DEF*.

 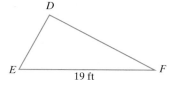

Solution: We are given the lengths of two corresponding sides. Their ratio is

$$\frac{12 \ \text{feet}}{19 \ \text{feet}} = \frac{12}{19}$$

■ **Work Practice Problem 2**

Objective **C** Finding Unknown Lengths of Sides in Similar Triangles

Because the ratios of lengths of corresponding sides are equal, we can use proportions to find unknown lengths in similar triangles.

EXAMPLE 3 Given that the triangles are similar, find the missing length *y*.

Solution: Since the triangles are similar, corresponding sides are in proportion. Thus, the ratio of 2 to 3 is the same as the ratio of 10 to *y*, or

$$\frac{2}{3} = \frac{10}{y}$$

To find the unknown length *y*, we set cross products equal.

$$\frac{2}{3} = \frac{10}{y}$$

$$2 \cdot y = 3 \cdot 10 \qquad \text{Set cross products equal.}$$
$$2y = 30 \qquad \text{Multiply.}$$
$$\frac{2y}{2} = \frac{30}{2} \qquad \text{Divide both sides by 2.}$$
$$y = 15 \qquad \text{Simplify.}$$

The missing length is 15 units.

■ **Work Practice Problem 3**

PRACTICE PROBLEM 2

Find the ratio of corresponding sides for the similar triangles *QRS* and *XYZ*.

PRACTICE PROBLEM 3

Given that the triangles are similar, find the missing length *x*.

Answers

2. $\dfrac{9}{13}$ **3.** $x = \dfrac{10}{3}$ or $3\dfrac{1}{3}$

✔**Concept Check** The following two triangles are similar. Which vertices of the first triangle appear to correspond to which vertices of the second triangle?

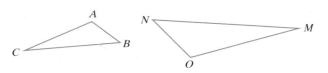

Many applications involve a diagram containing similar triangles. Surveyors, astronomers, and many other professionals continually use similar triangles in their work.

PRACTICE PROBLEM 4

Tammy Shultz, a firefighter, needs to estimate the height of a burning building. She estimates the length of her shadow to be 8 feet long and the length of the building's shadow to be 60 feet long. Find the approximate height of the building if she is 5 feet tall.

EXAMPLE 4 **Finding the Height of a Tree**

Mel Wagstaff is a 6-foot-tall park ranger who needs to know the height of a particular tree. He measures the shadow of the tree to be 69 feet long when his own shadow is 9 feet long. Find the height of the tree.

Solution:

1. UNDERSTAND. Read and reread the problem. Notice that the triangle formed by the Sun's rays, Mel, and his shadow is similar to the triangle formed by the Sun's rays, the tree, and its shadow.

2. TRANSLATE. Write a proportion from the similar triangles formed.

$$\frac{\text{Mel's height}}{\text{height of tree}} \rightarrow \frac{6}{x} = \frac{9}{69} \leftarrow \frac{\text{length of Mel's shadow}}{\text{length of tree's shadow}}$$

$$\text{or } \frac{6}{x} = \frac{3}{23} \quad \text{Simplify } \frac{9}{69}. \text{ (ratio in lowest terms)}$$

3. SOLVE for x:

$$\frac{6}{x} \diagdown \frac{3}{23}$$

$$6 \cdot 23 = x \cdot 3 \quad \text{Set cross products equal.}$$

$$138 = 3x \quad \text{Multiply.}$$

$$\frac{138}{3} = \frac{3x}{3} \quad \text{Divide both sides by 3.}$$

$$46 = x$$

4. INTERPRET. *Check* to see that replacing x with 46 in the proportion makes the proportion true. *State* your conclusion: The height of the tree is 46 feet.

▢ **Work Practice Problem 4**

Answer

4. approximately 37.5 ft

✔ **Concept Check Answer**

A corresponds to O; B corresponds to N; C corresponds to M

Vocabulary and Readiness Check

Answer each question true or false.

1. Two triangles that have the same shape but not necessarily the same sign are congruent.
2. Two triangles are congruent if they have the same shape and size.
3. Congruent triangles are also similar.
4. Similar triangles are also congruent.
5. For the two similar triangles, the ratio of corresponding sides is $\frac{5}{6}$.

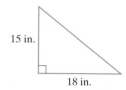

15 in.

18 in.

5 in.

6 in.

6.5 EXERCISE SET

FOR EXTRA HELP

Student Solutions Manual   PH Math/Tutor Center   CD/Video for Review   MathXL®   MyMathLab

Objective **A** *Determine whether each pair of triangles is congruent. See Example 1.*

1.

5 in.   6 in.

7 in.

6 in.   7 in.

5 in.

2.

10 mi   14 mi

14 mi

14 mi   14 mi

10 mi

3.

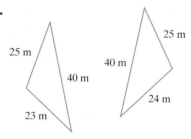

25 m

40 m

23 m

25 m

40 m

24 m

4.

21 cm   17 cm

7 cm

7 cm

21 cm

16 cm

5.

48°   30 m

42°

48°

42°   30 m

6.

15 yd

23°   36°

36°   23°

15 yd

7.

52 m 122° 98 m

52 m 122° 98 m

8.
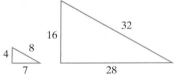
36 km 93° 57 km

57 km 93° 36 km

Objective **B** *Find each ratio of the corresponding sides of the given similar triangles. See Example 2.*

9.

22 11
12 6
14 7

10.

16 32
4 8
7 28

11.

12 8
10.5 7
9 6

12.
6 6 $4\frac{1}{2}$ $4\frac{1}{2}$
8 6

Objective **C** *Given that the pairs of triangles are similar, find the unknown length of the side labeled with a variable. See Example 3.*

13.
3 6 x 9

14.
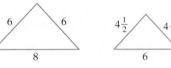
5 60
3 x

15.
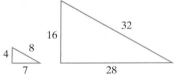
12 18 4 n

16.

4 y
7 14

17.

y 12 3.75 9

18.

9 15 z 22.5

19.

40 30
18 z

20.

14 8
x 9

21.
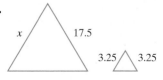
x 17.5
3.25 3.25

22.
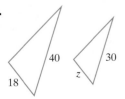
33.2 y
8.3 9.6

23.

$18\frac{1}{3}$ $3\frac{2}{3}$
y 2

24.
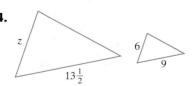
z 6
$13\frac{1}{2}$ 9

25.

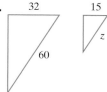

32   15

60   z

26.

26

13   13   x

27.

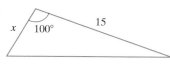

x   100°   15

7   100°   10½

28.

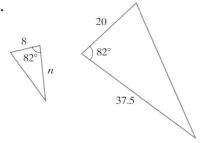

8   82°   n   20   82°   37.5

Solve. See Example 4.

29. Given the following diagram, approximate the height of the Bank One Tower in Oklahoma City, OK. (*Source: The World Almanac,* 2006)

x

25 feet

40 feet   2 feet

30. The tallest tree standing today is a redwood located in the Humboldt Redwoods State Park near Ukiah, California. Given the following diagram, approximate its height. (*Source: Guinness World Records,* 2006)

x

10 feet

73 feet   2 feet

31. Fountain Hills, Arizona, boasts the tallest fountain in the world. The fountain sits in a 28-acre lake and shoots up a column of water every hour. Based on the diagram below, what is the approximate height of the fountain?

28 ft

5 ft

? ft

100 ft

32. Lloyd White, a firefighter, needs to estimate the height of a burning building. He estimates the length of his shadow to be 9 feet long and the length of the building's shadow to be 75 feet long. Find the approximate height of the building if he is 6 feet tall.

33. If a 30-foot tree casts an 18-foot shadow, find the length of the shadow cast by a 24-foot tree.

34. If a 24-foot flagpole casts a 32-foot shadow, find the length of the shadow cast by a 44-foot antenna. Round to the nearest tenth.

35. For the health of his fish, Pete's Sea World uses the standard that a 20-gallon tank should only house 19 neon tetras. Find the number of neon tetras that Pete would place into a 55-gallon tank.

36. A local package express deliveryman is traveling the city expressway at 45 mph when he is forced to slow down due to traffic ahead. His truck slows at the rate of 3 mph every 5 seconds. Find his speed 8 seconds after braking.

Review

Solve. See Section 6.4.

37. Launch Umbilical Tower 1 is the name of the gantry used for the *Apollo* launch that took Neil Armstrong and Buzz Aldrin to the moon. Find the height of the gantry to the nearest whole number.

38. A regulation NCAA basketball court is a 94'-by-50' rectangle. Find, to the nearest foot, the length of a straight-line break that starts at one corner and moves diagonally to end at the opposite corner.

430 feet

200 feet

Perform the indicated operation. See Sections 5.2 through 5.4.

39. $3.6 + 0.41$

40. $0.41 - 3.6$

41. $(0.41)(-3)$

42. $-0.48 \div 3$

Concept Extensions

43. The print area on a particular page measures 7 inches by 9 inches. A printing shop is to copy the page and reduce the print area so that its length is 5 inches. What will its width be? Will the print now fit on a 3-by-5-inch index card?

Given that the pairs of triangles are similar, find the length of the side labeled n. Round your results to 1 decimal place.

 44.

5.2   7.8 n 12.6

 45.

11.6 n 20.8   58.7

46. In your own words, describe any differences in similar triangles and congruent triangles.

47. A triangular park is planned and waiting to be approved by the city zoning commission. A drawing of the park shows sides of length 5 inches, $7\frac{1}{2}$ inches and $10\frac{5}{8}$ inches. If the scale on the drawing is $\frac{1}{4}$ in. = 10 ft, find the actual proposed dimensions of the park.

48. Ben and Joyce Lander draw a triangular deck on their house plans. Joyce measures sides of the deck drawing on the plans to be 3 inches, $4\frac{1}{2}$ inches, and 6 inches. If the scale on the drawing is $\frac{1}{4}$ in. = 1 foot, find the lengths of the sides of the deck they want built.

CHAPTER 6   Group Activity

Investigating Scale Drawings
Sections 6.1, 6.2, and 6.3

Materials:

- ruler
- tape measure
- grid paper (optional)

This activity may be completed by working in groups or individually.

Scale drawings are used by architects, engineers, interior designers, ship builders, and others. In a scale drawing, each unit measurement on the drawing represents a fixed length on the object being drawn. For instance, in an architect's scale drawing, 1 inch on the drawing may represent 10 feet on a building. The scale describes the relationship between the measurements. If the measurements have the same units, the scale can be expressed as a ratio. In this case, the ratio would be 1 : 120, representing 1 inch to 120 inches (or 10 feet).

Use a ruler and the scale drawing of a college building below to answer the following questions.

1. How wide are each of the front doors of the college building?

2. How long is the front of the college building?

3. How tall is the front of the college building?

Now you will draw your own scale floor plan. First choose a room to draw—it can be your math classroom, your living room, your dormitory room, or any room that can be easily measured. Start by using a tape measure to measure the distances around the base of the walls in the room you are drawing.

4. Choose a scale for your floor plan.

5. Convert each measurement in the room you are drawing to the corresponding lengths needed for the scale drawing.

6. Complete your floor plan (you may find it helpful to use grid paper). Mark the locations of doors and windows on your floor plan. Be sure to indicate on the drawing the scale used in your floor plan.

Scale: 1 inch represents 10 feet

Chapter 6 Vocabulary Check

Fill in each blank with one of the words or phrases listed below.

not equal	equal	cross products	rate	right	unit rate	congruent
ratio	unit price	proportion	hypotenuse	Pythagorean	similar	leg

1. A _____ is the quotient of two numbers. It can be written as a fraction, using a colon, or using the word *to*.

2. $\dfrac{x}{2} = \dfrac{7}{16}$ is an example of a _____.

3. A _____ is a rate with a denominator of 1.

4. A _____ is a "money per item" unit rate.

5. A _____ is used to compare different kinds of quantities.

6. In the proportion $\dfrac{x}{2} = \dfrac{7}{16}$, $x \cdot 16$ and $2 \cdot 7$ are called _____.

7. If cross products are _____ the proportion is true.

8. If cross products are _____ the proportion is false.

9. _____ triangles have the same shape and the same size.

10. _____ triangles have exactly the same shape but not necessarily the same size.

11–13. Label the sides of the right triangle. **11.** _____ **13.** _____

12. _____

14. A triangle with one right angle is called a _____ triangle.

15. In the right triangle, $a^2 + b^2 = c^2$ is called the _____ theorem.

Helpful Hint

Are you preparing for your test? Don't forget to take the Chapter 6 Test on page 468. Then check your answers at the back of the text and use the Chapter Test Prep Video CD to see the fully worked-out solutions to any of the exercises you want to review.

6 Chapter Highlights

DEFINITIONS AND CONCEPTS	EXAMPLES
Section 6.1 Ratios and Rates	
A **ratio** is the quotient of two quantities.	The ratio of 3 to 4 can be written as $$\dfrac{3}{4} \quad \text{or} \quad 3\!:\!4$$ $\uparrow$ fraction notation $\qquad$ $\uparrow$ colon notation
Rates are used to compare different kinds of quantities.	Write the rate 12 spikes every 8 inches as a fraction in simplest form. $$\dfrac{12 \text{ spikes}}{8 \text{ inches}} = \dfrac{3 \text{ spikes}}{2 \text{ inches}}$$
A **unit rate** is a rate with a denominator of 1.	Write as a unit rate: 117 miles on 5 gallons of gas $$\dfrac{117 \text{ miles}}{5 \text{ gallons}} = \dfrac{23.4 \text{ miles}}{1 \text{ gallon}} \quad \text{or 23.4 miles per gallon}$$ or 23.4 miles/gallon
A **unit price** is a "money per item" unit rate.	Write as a unit price: \$5.88 for 42 ounces of detergent $$\dfrac{\$5.88}{42 \text{ ounces}} = \dfrac{\$0.14}{1 \text{ ounce}} = \$0.14 \text{ per ounce}$$

DEFINITIONS AND CONCEPTS	**EXAMPLES**

Section 6.2   Proportions

A **proportion** is a statement that two ratios or rates are equal.

$\dfrac{1}{2} = \dfrac{4}{8}$ is a proportion.

Is $\dfrac{6}{10} = \dfrac{9}{15}$ a true proportion?

USING CROSS PRODUCTS TO DETERMINE WHETHER PROPORTIONS ARE TRUE OR FALSE

Cross products

$a \cdot d$ $\qquad$ $b \cdot c$

$$\dfrac{a}{b} = \dfrac{c}{d}$$

Cross products

$6 \cdot 15$ $\qquad$ $10 \cdot 9$

$$\dfrac{6}{10} = \dfrac{9}{15}$$

$6 \cdot 15 \overset{?}{=} 10 \cdot 9$ Are cross products equal?

$90 = 90$

If cross products are equal, the proportion is true.
If $ad = bc$, then the proportion is true.
If cross products are not equal, the proportion is false.
If $ad \neq bc$, then the proportion is false.

Since cross products are equal, the proportion is a true proportion.

To find an unknown value x in a proportion, we set the cross products equal to each other and solve the resulting equation.

Find x: $\dfrac{x}{7} = \dfrac{5}{8}$

$$\dfrac{x}{7} = \dfrac{5}{8}$$

$x \cdot 8 = 7 \cdot 5$ Set the cross products equal to each other.

$8x = 35$ Multiply.

$\dfrac{8x}{8} = \dfrac{35}{8}$ Divide both sides by 8.

$x = 4\dfrac{3}{8}$

Section 6.3   Proportions and Problem Solving

Given a specified ratio (or rate) of two quantities, a proportion can be used to determine an unknown quantity.

On a map, 50 miles corresponds to 3 inches. How many miles correspond to 10 inches?

1. UNDERSTAND. Read and reread the problem.

2. TRANSLATE. We let x represent the unknown number. We are given that 50 miles is to 3 inches as x miles is to 10 inches.

miles $\rightarrow$ $\dfrac{50}{3} = \dfrac{x}{10}$ $\leftarrow$ miles
inches $\rightarrow$ $\phantom{\dfrac{50}{3}}$ $\phantom{\dfrac{x}{10}}$ $\leftarrow$ inches

3. SOLVE:

$$\dfrac{50}{3} = \dfrac{x}{10}$$

$50 \cdot 10 = 3 \cdot x$ Set the cross products equal to each other.

$500 = 3x$ Multiply.

$\dfrac{500}{3} = \dfrac{3x}{3}$ Divide both sides by 3.

$x = 166\dfrac{2}{3}$

4. INTERPRET. *Check* your work. *State* your conclusion:
On the map, $166\dfrac{2}{3}$ miles corresponds to 10 inches.

DEFINITIONS AND CONCEPTS	EXAMPLES

Section 6.4   Square Roots and the Pythagorean Theorem

SQUARE ROOT OF A NUMBER

The square root of a positive number a is the positive number b whose square is a. In symbols,

$$\sqrt{a} = b, \quad \text{if} \quad b^2 = a$$

Also, $\sqrt{0} = 0$.

$$\sqrt{9} = 3, \qquad \sqrt{100} = 10, \qquad \sqrt{1} = 1$$

PYTHAGOREAN THEOREM

If a and b are the lengths of the legs of a right triangle and c is the length of the hypotenuse, then

$$a^2 + b^2 = c^2$$

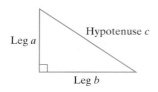

Leg a Hypotenuse c Leg b

Find c. $a = 3$ c $b = 8$

$$a^2 + b^2 = c^2$$
$$3^2 + 8^2 = c^2 \qquad \text{Let } a = 3 \text{ and } b = 8.$$
$$9 + 64 = c^2 \qquad \text{Multiply.}$$
$$73 = c^2 \qquad \text{Simplify.}$$
$$\sqrt{73} = c \text{ or } c \approx 8.5$$

Section 6.5   Congruent and Similar Triangles

Congruent triangles have the same shape and the same size. Corresponding angles are equal, and corresponding sides are equal.

Similar triangles have exactly the same shape but not necessarily the same size. Corresponding angles are equal, and the ratios of the lengths of corresponding sides are equal.

Congruent triangles

$$\frac{AB}{DE} = \frac{3}{9} = \frac{1}{3}, \frac{BC}{EF} = \frac{6}{18} = \frac{1}{3},$$
$$\frac{CA}{FD} = \frac{4}{12} = \frac{1}{3}$$

6 CHAPTER REVIEW

(6.1) *Write each ratio as a fraction in simplest form.*

1. 23 to 37

2. $121 to $143

3. 4.25 yards to 8.75 yards

4. $2\frac{1}{4}$ to $4\frac{3}{8}$

Use the garden shown for Exercises 5 and 6. Write each ratio as a fraction in simplest form.

4.5 meters   2 meters

5. Find the ratio of the garden's length (longer side) to the garden's width (shorter side).

△ **6.** Find the ratio of the width to the perimeter of the rectangular garden.

Write each rate as a fraction in simplest form.

7. 6000 people to 2400 pets

8. 15 pages printed in 6 minutes

Write each rate as a unit rate.

9. 468 miles in 9 hours

10. 180 feet in 12 seconds

11. $6.96 for 4 diskettes

12. 104 bushels of fruit from 8 trees

Find each unit price and decide which is the better buy. Round to 3 decimal places. Assume that we are comparing different sizes of the same brand.

13. Taco sauce: 8 ounces for $0.99 or 12 ounces for $1.69

14. Peanut butter: 18 ounces for $1.49 or 28 ounces for $2.39

(6.2) Translating *Write each sentence as a proportion.*

15. 24 uniforms is to 8 players as 3 uniforms is to 1 player.

16. 12 tires is to 3 cars as 4 tires is to 1 car.

Determine whether each proportion is true.

17. $\dfrac{19}{8} = \dfrac{14}{6}$

18. $\dfrac{3.75}{3} = \dfrac{7.5}{6}$

Find the unknown number x in each proportion.

19. $\dfrac{x}{3} = \dfrac{30}{18}$

20. $\dfrac{x}{9} = \dfrac{7}{3}$

21. $\dfrac{-8}{5} = \dfrac{9}{x}$

22. $\dfrac{-27}{\frac{9}{4}} = \dfrac{x}{-5}$

23. $\dfrac{0.4}{x} = \dfrac{2}{4.7}$

24. $\dfrac{x}{4\frac{1}{2}} = \dfrac{2\frac{1}{10}}{8\frac{2}{5}}$

25. $\dfrac{x}{0.4} = \dfrac{4.7}{3}$ Round to the nearest hundredth.

26. $\dfrac{0.07}{0.3} = \dfrac{7.2}{n}$ Round to the nearest tenth.

(6.3) *Solve.*

The ratio of a quarterback's completed passes to attempted passes is 3 to 7.

27. If he attempts 32 passes, find how many passes he completed. Round to the nearest whole pass.

28. If he completed 15 passes, find how many passes he attempted.

One bag of pesticide covers 4000 square feet of garden.

△ **29.** Find how many bags of pesticide should be purchased to cover a rectangular garden that is 180 feet by 175 feet.

△ **30.** Find how many bags of pesticide should be purchased to cover a square garden that is 250 feet on each side.

On a road map of Texas, 0.75 inch represents 80 miles.

31. Find the distance from Houston to Corpus Christi if the distance on the map is about 2 inches.

32. The distance from El Paso to Dallas is 1025 miles. Find the distance between these cities on the map.

(6.4) *Find each square root. If necessary, round the square root to the nearest thousandth.*

33. $\sqrt{64}$

34. $\sqrt{144}$

35. $\sqrt{12}$

36. $\sqrt{15}$

37. $\sqrt{0}$

38. $\sqrt{1}$

39. $\sqrt{50}$

40. $\sqrt{65}$

41. $\sqrt{\dfrac{4}{25}}$

42. $\sqrt{\dfrac{1}{100}}$

Find the unknown length in each given right triangle. If necessary, round to the nearest tenth.

△ **43.** leg = 12, leg = 5

△ **44.** leg = 20, leg = 21

△ **45.** leg = 9, hypotenuse = 14

△ **46.** leg = 66, leg = 56

△ **47.** Find the length to the nearest hundredth of the diagonal of a square that has a side of length 20 centimeters.

△ **48.** Find the height of the building rounded to the nearest tenth.

126 ft

← 90 ft →

(6.5) *Determine whether each pair of triangles is congruent. If congruent, state the reason why, such as SSS, SAS, or ASA.*

49.

60° 1.7 cm

1.7 cm 60°

85°

85°

50.

12° 14°

154°

12° 14°

154°

Given that the pairs of triangles are similar, find the unknown length x.

51.

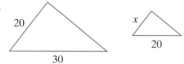

20

30

x

20

52.

x

24

5.8 8

Solve.

53. A housepainter needs to estimate the height of a condominium. He estimates the length of his shadow to be 7 feet long and the length of the building's shadow to be 42 feet long. Find the approximate height of the building if the housepainter is $5\frac{1}{2}$ feet tall.

54. A design company is making a triangular sail for a model sailboat. The model sail is to be the same shape as a life-size sailboat's sail. Use the following diagram to find the unknown lengths x and y.

26 ft 24 ft

10 ft

y 2 in.

x

Mixed Review

Write each ratio as a fraction in simplest form.

55. 15 to 25

56. 3 pints to 81 pints

Write each rate as a fraction in simplest form.

57. 2 teachers for 18 students

58. 6 nurses for 24 patients

Write each rate as a unit rate.

59. 136 miles in 4 hours

60. 12 gallons of milk from 6 cows

For Exercises 61 and 62, write the ratio or rate in simplest form.

61. During the 2006 Winter Olympics in Torino, Italy, a total of 256 medals were awarded. German athletes won a total of 29 medals. Find the ratio of medals won by Germany to the total medals awarded. (*Source:* International Olympic Committee)

62. Paul Crake of Canberra, Australia, holds the record for the Empire State Building Run Up in New York. He ran up 1576 steps in about 9.5 minutes. Find the number of steps per minute. Round to the nearest step. (*Source: Guinness Book of World Records*)

Find each unit price and decide which is the better buy. Round to 3 decimal places. Assume that we are comparing different sizes of the same brand.

63. cold medicine:
$4.94 for 4 oz.
$9.98 for 8 oz.

64. juice:
12 oz for $0.65.
64 oz for $2.98.

Translating *Write each sentence as a proportion.*

65. 2 cups of cookie dough is to 30 cookies as 4 cups of cookie dough is to 60 cookies

66. 5 nickels is to 3 dollars as 20 nickels is to 12 dollars

Find the unknown number x in each proportion.

67. $\dfrac{3}{x} = \dfrac{15}{8}$

68. $\dfrac{5}{4} = \dfrac{x}{20}$

69. $\dfrac{x}{3} = \dfrac{7,5}{6}$

70. $\dfrac{\frac{1}{3}}{25} = \dfrac{x}{30}$

Find each square root. If necessary, approximate and round to the nearest thousandth.

71. $\sqrt{36}$ **72.** $\sqrt{\dfrac{16}{81}}$ **73.** $\sqrt{105}$ **74.** $\sqrt{32}$

Find the unknown length of each given right triangle. If necessary, round to the nearest tenth.

75. leg = 66, leg = 56   **76.** leg = 12, hypotenuse = 24

Given that the pairs of triangles are similar, find the unknown length n.

77.

78.

 STUDY SKILLS BUILDER

Tips for Studying for an Exam

To prepare for an exam, try the following study techniques.

- Start the study process days before your exam.
- Make sure that you are current and up to date on your assignments.
- If there is a topic that you are unsure of, use one of the many resources that are available to you. For example,

   See your instructor.

   Visit a learning resource center on campus where math tutors are available.

   Read the textbook material and examples on the topic.

   View a videotape on the topic.

- Reread your notes and carefully review the Chapter Highlights at the end of the chapter.

- Work the review exercises at the end of the chapter and check your answers. Make sure that you correct any missed exercises. If you have trouble on a topic, use a resource listed above.

- Find a quiet place to take the Chapter Test found at the end of the chapter. Do not use any resources when taking this sample test. This way you will have a clear indication of how prepared you are for your exam. Check your answers and make sure that you correct any missed exercises.

- Get lots of rest the night before the exam. It's hard to show how well you know the material if your brain is foggy from lack of sleep.

Good luck and keep a positive attitude.

6 CHAPTER TEST

 Use the Chapter Test Prep Video CD to see the fully worked-out solutions to any of the exercises you want to review.

Write each ratio or rate as a fraction in simplest form.

1. 4500 trees to 6500 trees

2. 9 inches of rain in 30 days

3. 8.6 to 10

4. $5\frac{7}{8}$ to $9\frac{3}{4}$

5. The world's largest yacht, the Octopus, measures in at 414 feet long. A Boeing 747-400 jumbo jet measures 231 feet long. Find the ratio of the length of the Octopus to the length of a 747-400. (*Source: Power & Motoryacht magazine*)

231 ft

414 ft

Find each unit rate.

6. 650 kilometers in 8 hours

7. 140 students for 5 teachers

8. QRIO (Quest for Curiosity) is the world's first bipedal robot capable of running (moving with both legs off the ground at the same time) at a rate of 108 inches each 12 seconds. (*Source: Guinness World Records*)

QRIO

Find each unit price and decide which is the better buy.

9. Steak sauce:
8 ounces for $1.19
12 ounces for $1.89

10. Jelly:
16 ounces for $1.49
24 ounces for $2.39

Determine whether the proportion is true.

11. $\dfrac{28}{16} = \dfrac{14}{8}$

12. $\dfrac{3.6}{2.2} = \dfrac{1.9}{1.2}$

Answers

1. _____

2. _____

3. _____

4. _____

5. _____

6. _____

7. _____

8. _____

9. _____

10. _____

11. _____

12. _____

Solve each proportion for the given variable.

13. $\dfrac{n}{3} = \dfrac{15}{9}$

14. $\dfrac{8}{x} = \dfrac{11}{6}$

15. $\dfrac{4}{\frac{3}{7}} = \dfrac{y}{\frac{1}{4}}$

16. $\dfrac{1.5}{5} = \dfrac{2.4}{n}$

Solve.

17. On an architect's drawing, 2 inches corresponds to 9 feet. Find the length of a home represented by a line that is 11 inches long.

18. If a car can be driven 80 miles in 3 hours, how long will it take to travel 100 miles?

19. The standard dose of medicine for a dog is 10 grams for every 15 pounds of body weight. What is the standard dose for a dog that weighs 80 pounds?

Find each square root and simplify. Round to the nearest thousandth if necessary.

20. $\sqrt{49}$

21. $\sqrt{157}$

22. $\sqrt{\dfrac{64}{100}}$

Solve.

△ **23.** Approximate to the nearest hundredth of a centimeter the length of the missing side of a right triangle with legs of 4 centimeters each.

△ **24.** Given that the following triangles are similar, find the unknown length *n*.

△ **25.** Tamara Watford, a surveyor, needs to estimate the height of a tower. She estimates the length of her shadow to be 4 feet long and the length of the tower's shadow to be 48 feet long. Find the height of the tower if she is $5\dfrac{3}{4}$ feet tall.

---

13. _____

14. _____

15. _____

16. _____

17. _____

18. _____

19. _____

20. _____

21. _____

22. _____

23. _____

24. _____

25. _____

Chapters 1–6

1. a. _____

   b. _____

   c. _____

   d. _____

2. a. _____

   b. _____

   c. _____

   d. _____

3. _____

4. _____

5. a. _____

   b. _____

6. _____

7. _____

8. _____

9. _____

10. _____

11. _____

12. _____

13. _____

14. _____

15. _____

16. _____

17. _____

18. _____

19. _____

20. _____

1. Subtract. Check each answer by adding.
  a. $12 - 9$
  b. $22 - 7$
  c. $35 - 35$
  d. $70 - 0$

2. Multiply
  a. $20 \cdot 0$
  b. $20 \cdot 1$
  c. $0 \cdot 20$
  d. $1 \cdot 20$

3. Round 248,982 to the nearest hundred.

4. Round 248,982 to the nearest thousand.

Perform indicated operations.

5. **a.** $\begin{array}{r} 25 \\ \times\ 8 \\ \hline \end{array}$ **b.** $\begin{array}{r} 246 \\ \times\ 5 \\ \hline \end{array}$

6. $10{,}468 \div 28$

7. $1 + (-10) + (-8) + 9$

8. $-12(7)$

9. Write the prime factorization of 80.

10. Find $3^2 - 1^2$.

11. Write $\dfrac{12}{20}$ in simplest form.

12. Find $9^2 \cdot 3$.

Multiply.

13. $-\dfrac{6}{13} \cdot -\dfrac{26}{30}$

14. $3\dfrac{3}{8} \cdot 4\dfrac{5}{9}$

Perform the indicated operation and simplify

15. $\dfrac{7}{8} + \dfrac{6}{8} + \dfrac{3}{8}$

16. $\dfrac{7}{10} - \dfrac{3}{10} + \dfrac{4}{10}$

17. Find the LCD of $\dfrac{3}{7}$ and $\dfrac{5}{14}$.

18. Add: $\dfrac{17}{25} + \dfrac{3}{10}$

19. Write an equivalent fraction with the indicated denominator. $\dfrac{3}{4} = \dfrac{}{20}$

20. Determine whether these fractions are equivalent.
$\dfrac{10}{55}, \quad \dfrac{6}{33}$

21. Subtract: $\dfrac{2}{3} - \dfrac{10}{11}$

22. Subtract: $17\dfrac{5}{24} - 9\dfrac{5}{9}$

23. A flight from Tucson to Phoenix, Arizona, requires $\dfrac{5}{12}$ of an hour. If the plane has been flying $\dfrac{1}{4}$ of an hour, find how much time remains before landing.

24. Simplify: $80 \div 8 \cdot 2 + 7$

25. Add: $2\dfrac{1}{3} + 5\dfrac{3}{8}$

26. Find the average of $\dfrac{3}{5}, \dfrac{4}{9}$, and $\dfrac{11}{15}$.

27. Insert $<$ or $>$ to form a true statement.
$$\dfrac{3}{4} \qquad \dfrac{9}{11}$$

Solve.

28. $5y - 8y = 24$

29. $y - 5 = -2 - 6$

30. $3y - 6 = 7y - 6$

31. $3a - 6 = a + 4$

32. $4(y + 1) - 3 = 21$

33. $3(2x - 6) + 6 = 0$

34. Write "seventy-five thousandths" in standard form.

35. Round 736.2359 to the nearest tenth.

36. Round 736.2359 to the nearest thousandth.

37. Add: $23.85 + 1.604$

38. Subtract: $700 - 18.76$

39. Multiply: 0.0531×16

40. Write $\dfrac{3}{8}$ as a decimal.

41. Divide: $-5.98 \div 115$

42. Write 7.9 as an improper fraction.

43. Simplify: $-0.5(8.6 - 1.2)$

44. Find the unknown number n.
$$\dfrac{n}{4} = \dfrac{12}{16}$$

45. Write the numbers in order from smallest to largest.
$$\dfrac{9}{20}, \dfrac{4}{9}, 0.456$$

46. Write the rate as a unit rate.
700 meters in 5 seconds

Write each ratio as a fraction in simplest form.

47. The ratio of \$15 to \$10

48. The ratio of 7 to 21

49. The ratio of 2.5 to 3.15

50. The ratio of 900 to 9000

21. _____

22. _____

23. _____

24. _____

25. _____

26. _____

27. _____

28. _____

29. _____

30. _____

31. _____

32. _____

33. _____

34. _____

35. _____

36. _____

37. _____

38. _____

39. _____

40. _____

41. _____

42. _____

43. _____

44. _____

45. _____

46. _____

47. _____

48. _____

49. _____

50. _____

7

Percent

This chapter is devoted to percent, a concept used virtually every day in ordinary and business life. Understanding percent and using it efficiently depends on understanding ratios, because a percent is a ratio whose denominator is 100. We present techniques to write percents as fractions and as decimals. We then solve problems relating to interest rates, sales tax, discounts, and other real-life situations by writing percent equations.

No one is sure who first concocted ice cream. Historians suspect that ice cream was probably developed independently at various times all over the globe. The current popularity of ice cream is due to two major developments in the nineteenth and early twentieth centuries: the invention of the ice cream freezer and mechanical refrigeration. Ice cream became widely available after 1846, when Nancy Johnson invented the hand-cranked freezer. Turning the freezer handle agitates a container of ice cream mix in a bed of salt and ice until the mix is frozen. By 1926, the first commercially successful continuous process freezer was developed, making the mass production of ice cream feasible. In Section 7.4, Example 1, page 501, we will examine some of the current tastes in ice cream.

Favorite Ice Cream Flavors of Adults

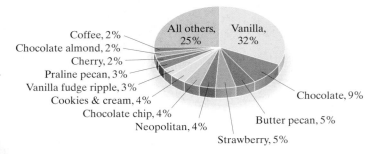

Coffee, 2%
Chocolate almond, 2%
Cherry, 2%
Praline pecan, 3%
Vanilla fudge ripple, 3%
Cookies & cream, 4%
Chocolate chip, 4%
Neopolitan, 4%
Strawberry, 5%
Butter pecan, 5%
Chocolate, 9%
All others, 25%
Vanilla, 32%

Source: International Ice Cream Association

7.1 PERCENTS, DECIMALS, AND FRACTIONS

Objective **A** Understanding Percent

The word **percent** comes from the Latin phrase *per centum*, which means **"per 100."** For example, 53% (53 percent) means 53 per 100. In the square below, 53 of the 100 squares are shaded. Thus, 53% of the figure is shaded.

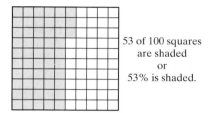

53 of 100 squares
are shaded
or
53% is shaded.

Since 53% means 53 per 100, 53% is the ratio of 53 to 100, or $\frac{53}{100}$.

$$53\% = \frac{53}{100}$$

Also,

$$7\% = \frac{7}{100} \quad \text{7 parts per 100 parts}$$

$$73\% = \frac{73}{100} \quad \text{73 parts per 100 parts}$$

$$109\% = \frac{109}{100} \quad \text{109 parts per 100 parts}$$

Percent

Percent means **per one hundred.** The "%" symbol is used to denote percent.

Percent is used in a variety of everyday situations. For example,

- 58.7% of the U.S. population uses the Internet.
- The store is having a 25%-off sale.
- 78% of us trust our local fire department.
- The enrollment in community colleges is predicted to increase 1.3% each year.

EXAMPLE 1

In a survey of 100 people, 17 people drive blue cars. What percent of people drive blue cars?

Solution: Since 17 people out of 100 drive blue cars, the fraction is $\frac{17}{100}$. Then

$$\frac{17}{100} = 17\%$$

Work Practice Problem 1

PRACTICE PROBLEM 1

Of 100 students in a club, 27 are freshmen. What percent of the students are freshmen?

Answer
1. 27%

PRACTICE PROBLEM 2

31 out of 100 executives are in their forties. What percent of executives are in their forties?

EXAMPLE 2   46 out of every 100 college students live at home. What percent of students live at home? (*Source:* Independent Insurance Agents of America)

Solution:

$$\frac{46}{100} = 46\%$$

⬛ **Work Practice Problem 2**

Objective B Writing Percents as Decimals or Fractions

Since percent means "per hundred," we have that

$$1\% = \frac{1}{100} = 0.01$$

In other words, the percent symbol means "per hundred" or, equivalently, "$\frac{1}{100}$" or "0.01." Thus

Write 87% as a fraction: $87\% = 87 \times \frac{1}{100} = \frac{87}{100}$

or

Write 87% as a decimal: $87\% = 87 \times (0.01) = 0.87$

Results are the same.

Of course, we know that the end results are the same, that is,

$$\frac{87}{100} = 0.87$$

The above gives us two options for converting percents. We can replace the percent symbol, %, by $\frac{1}{100}$ or 0.01 and then multiply.

For consistency, when we

- convert from a percent to a *decimal,* we will drop the % symbol and multiply by 0.01.
- convert from a percent to a *fraction,* we will drop the % symbol and multiply by $\frac{1}{100}$.

Let's practice writing percents as decimals, then writing percents as fractions.

Writing a Percent as a Decimal

Replace the percent symbol with its decimal equivalent, 0.01; then multiply.

$$43\% = 43(0.01) = 0.43$$

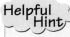 **Helpful Hint**

If it helps, think of writing a percent as a decimal by

Percent → | Remove the % symbol and move decimal point 2 places to the left | → Decimal

Answer

2. 31%

EXAMPLES Write each percent as a decimal.

3. $23\% = 23(0.01) = 0.23$ Replace the percent symbol with 0.01, then multiply.

4. $4.6\% = 4.6(0.01) = 0.046$ Replace the percent symbol with 0.01. Then multiply.

5. $190\% = 190(0.01) = 1.90 \text{ or } 1.9$

6. $0.74\% = 0.74(0.01) = 0.0074$

7. $100\% = 100(0.01) = 1.00 \text{ or } 1$

> **Helpful Hint**
> We just learned that
> $100\% = 1$

■ **Work Practice Problems 3–7**

Write each percent as a decimal.

3. 49%   **4.** 3.1%
5. 175%   **6.** 0.46%
7. 600%

✔**Concept Check** Why is it incorrect to write the percent 0.033% as 3.3 in decimal form?

Now let's write percents as fractions.

Writing a Percent as a Fraction

Replace the percent symbol with its fraction equivalent, $\frac{1}{100}$; then multiply. Don't forget to simplify the fraction if possible.

$$43\% = 43 \cdot \frac{1}{100} = \frac{43}{100}$$

EXAMPLES Write each percent as a fraction or mixed number in simplest form.

8. $40\% = 40 \cdot \frac{1}{100} = \frac{40}{100} = \frac{2 \cdot \overset{1}{\cancel{20}}}{5 \cdot \underset{1}{\cancel{20}}} = \frac{2}{5}$

9. $1.9\% = 1.9 \cdot \frac{1}{100} = \frac{1.9}{100}$. We don't want the numerator of the fraction to contain a decimal, so we multiply by 1 in the form of $\frac{10}{10}$.

$$= \frac{1.9}{100} \cdot \frac{10}{10} = \frac{1.9 \cdot 10}{100 \cdot 10} = \frac{19}{1000}$$

10. $125\% = 125 \cdot \frac{1}{100} = \frac{125}{100} = \frac{5 \cdot \overset{1}{\cancel{25}}}{4 \cdot \underset{1}{\cancel{25}}} = \frac{5}{4} \text{ or } 1\frac{1}{4}$

11. $33\frac{1}{3}\% = 33\frac{1}{3} \cdot \frac{1}{100} = \frac{100}{3} \cdot \frac{1}{100} = \frac{\overset{1}{\cancel{100}} \cdot 1}{3 \cdot \underset{1}{\cancel{100}}} = \frac{1}{3}$

    ⏟ → Write as → an improper fraction.

12. $100\% = 100 \cdot \frac{1}{100} = \frac{100}{100} = 1$

> **Helpful Hint**
> Just as in Example 7, we confirm that $100\% = 1$.

■ **Work Practice Problems 8–12**

Write each percent as a fraction or mixed number in simplest form.

8. 50%
9. 2.3%
10. 150%
11. $66\frac{2}{3}\%$
12. 12%

Answers

3. 0.49 **4.** 0.031 **5.** 1.75 **6.** 0.0046
7. 6 **8.** $\frac{1}{2}$ **9.** $\frac{23}{1000}$ **10.** $\frac{3}{2}$ or $1\frac{1}{2}$
11. $\frac{2}{3}$ **12.** $\frac{3}{25}$

✔ **Concept Check Answer**

To write a percent as a decimal, the decimal point should be moved two places to the left, not to the right. So the correct answer is 0.00033.

Objective C Writing Decimals or Fractions as Percents

To write a decimal or fraction as a percent, we use the result of Examples 7 and 12. In these examples, we found that $1 = 100\%$.

Write 0.38 as a percent: $0.38 = 0.38(1) = 0.38(100\%) = 38.\%$

Write $\frac{1}{4}$ as a percent: $\frac{1}{4} = \frac{1}{4}(1) = \frac{1}{4} \cdot 100\% = \frac{100}{4}\% = 25\%$

First, let's practice writing decimals as percents.

Writing a Decimal as a Percent

Multiply by 1 in the form of 100%.

$0.27 = 0.27(100\%) = 27.\%$

Helpful Hint

If it helps, think of writing a decimal as a percent by reversing the steps in the Helpful Hint on page 474.

Percent ← | Move the decimal point 2 places to the right and attach a % symbol. | ← Decimal

the Helpful Hint on page 474.

PRACTICE PROBLEMS 13–16

Write each decimal as a percent.

13. 0.14   **14.** 1.75

15. 0.057   **16.** 0.5

EXAMPLES Write each decimal as a percent.

13. $0.65 = 0.65(100\%) = 65.\%$ or 65% Multiply by 100%.

14. $1.25 = 1.25(100\%) = 125.\%$ or 125%

15. $0.012 = 0.012(100\%) = 001.2\%$ or 1.2%

16. $0.6 = 0.6(100\%) = 060.\%$ or 60%

Helpful Hint
A zero was inserted as a placeholder.

☐ **Work Practice Problems 13–16**

✔ **Concept Check** Why is it incorrect to write the decimal 0.0345 as 34.5% in percent form?

Now let's write fractions as percents.

Writing a Fraction as a Percent

Multiply by 1 in the form of 100%.

$\frac{1}{8} = \frac{1}{8} \cdot 100\% = \frac{1}{8} \cdot \frac{100}{1}\% = \frac{100}{8}\% = 12\frac{1}{2}\%$ or 12.5%

Helpful Hint

From Examples 7 and 12, we know that

$100\% = 1$

Recall that when we multiply a number by 1, we are not changing the value of that number. This means that when we multiply a number by 100%, we are not changing its value but rather writing the number as an equivalent percent.

Answers

13. 14% **14.** 175% **15.** 5.7%
16. 50%

✔ Concept Check Answer

To change a decimal to a percent, multiply by 100%, or move the decimal point *only* two places to the right. So the correct answer is 3.45%.

EXAMPLES Write each fraction or mixed number as a percent.

17. $\dfrac{7}{20} = \dfrac{7}{20} \cdot 100\% = \dfrac{7}{20} \cdot \dfrac{100}{1}\% = \dfrac{700}{20}\% = 35\%$

18. $\dfrac{2}{3} = \dfrac{2}{3} \cdot 100\% = \dfrac{2}{3} \cdot \dfrac{100}{1}\% = \dfrac{200}{3}\% = 66\dfrac{2}{3}\%$

19. $2\dfrac{1}{4} = \dfrac{9}{4} \cdot 100\% = \dfrac{9}{4} \cdot \dfrac{100}{1}\% = \dfrac{900}{4}\% = 225\%$

Helpful Hint
$\dfrac{200}{3} = 66.\overline{6}$. Thus, another way to write $\dfrac{200}{3}\%$ is $66.\overline{6}\%$.

Work Practice Problems 17–19

PRACTICE PROBLEMS 17–19

Write each fraction or mixed number as a percent.

17. $\dfrac{1}{4}$ **18.** $\dfrac{9}{40}$ **19.** $5\dfrac{1}{2}$

✔**Concept Check** Which digit in the percent 76.4582% represents

a. A tenth percent?

c. A hundredth percent?

b. A thousandth percent?

d. A ten percent?

EXAMPLE 20 Write $\dfrac{1}{12}$ as a percent. Round to the nearest hundredth percent.

Solution:

$$\dfrac{1}{12} = \dfrac{1}{12} \cdot 100\% = \dfrac{1}{12} \cdot \dfrac{100\%}{1} = \dfrac{100}{12}\% \approx 8.33\%$$

"approximately"

$$
\begin{array}{r}
8.333 \approx 8.33 \\
12\overline{)100.000} \\
-96 \\
\hline
4\,0 \\
-3\,6 \\
\hline
40 \\
-36 \\
\hline
40 \\
-36 \\
\hline
4
\end{array}
$$

Thus, $\dfrac{1}{12}$ is approximately 8.33%.

Work Practice Problem 20

PRACTICE PROBLEM 20

Write $\dfrac{3}{17}$ as a percent. Round to the nearest hundredth percent.

Objective **D** Applications with Percents, Decimals, and Fractions

Let's summarize what we have learned so far about percents, decimals, and fractions:

Summary of Converting Percents, Decimals, and Fractions

- *To write a percent as a decimal,* replace the % symbol with its decimal equivalent, 0.01; then multiply.
- *To write a percent as a fraction,* replace the % symbol with its fraction equivalent, $\dfrac{1}{100}$; then multiply.
- *To write a decimal or fraction as a percent,* multiply by 100%.

Answers
17. 25% **18.** $22\dfrac{1}{2}\%$ **19.** 550%
20. 17.65%

✔ **Concept Check Answers**
a. 4 **b.** 8 **c.** 5 **d.** 7

If we let x represent a number, below we summarize using symbols.

Write a percent as a decimal:	Write a percent as a fraction:	Write a number as a percent:
$x\% = x(0.01)$	$x\% = x \cdot \dfrac{1}{100}$	$x = x \cdot 100\%$

PRACTICE PROBLEM 21

A family decides to spend no more than 27.5% of its monthly income on rent. Write 27.5% as a decimal and as a fraction.

EXAMPLE 21 17.8% of automobile thefts in the continental United States occur in the Midwest. Write this percent as a decimal and as a fraction. (*Source:* The American Automobile Manufacturers Association)

Solution:

As a decimal: $17.8\% = 17.8(0.01) = 0.178$.

As a fraction: $17.8\% = 17.8 \cdot \dfrac{1}{100} = \dfrac{17.8}{100} = \dfrac{17.8}{100} \cdot \dfrac{10}{10} = \dfrac{178}{1000} = \dfrac{\overset{1}{\cancel{2}} \cdot 89}{\underset{1}{\cancel{2}} \cdot 500} = \dfrac{89}{500}$.

Thus, 17.8% written as a decimal is 0.178, and written as a fraction is $\dfrac{89}{500}$.

⬛ **Work Practice Problem 21**

PRACTICE PROBLEM 22

Provincetown's budget for waste disposal increased by $1\dfrac{3}{4}$ times over the budget from last year. What percent increase is this?

EXAMPLE 22 An advertisement for a stereo system reads "$\dfrac{1}{4}$ off." What percent off is this?

Solution: Write $\dfrac{1}{4}$ as a percent.

$$\frac{1}{4} = \frac{1}{4} \cdot 100\% = \frac{1}{4} \cdot \frac{100\%}{1} = \frac{100}{4}\% = 25\%$$

Thus, "$\dfrac{1}{4}$ off" is the same as "25% off."

⬛ **Work Practice Problem 22**

Note: It is helpful to know a few basic percent conversions. Appendix A.2 contains a handy reference of percent, decimal, and fraction equivalencies.

Also, Appendix A.3 shows how to find common percents of a number.

Answers

21. $0.275, \dfrac{11}{40}$ **22.** 175%

Vocabulary and Readiness Check

Use the choices below to fill in each blank. Some choices may be used more than once.

$\frac{1}{100}$ 0.01   100%   percent

1. _____ means "per hundred."
2. _____ = 1.
3. The % symbol is read as _____.
4. To write a decimal or a fraction as a percent, multiply by 1 in the form of _____.
5. To write a percent as a *decimal,* drop the % symbol and multiply by _____.
6. To write a percent as a *fraction,* drop the % symbol and multiply by _____.

7.1 EXERCISE SET

Objective A *Solve. See Examples 1 and 2.*

1. In a survey of 100 college students, 96 use the Internet. What percent use the Internet?

 2. A basketball player makes 81 out of 100 attempted free throws. What percent of free throws are made?

One hundred adults were asked to name their favorite sport, and the results are shown in the circle graph.

3. What sport was preferred by most adults? What percent preferred this sport?

4. What sport was preferred by the least number of adults? What percent preferred this sport?

5. What percent of adults preferred football or soccer?

6. What percent of adults preferred basketball or baseball?

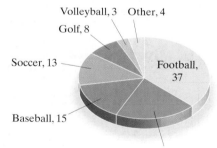

Objective B *Write each percent as a decimal. See Examples 3 through 7.*

7. 41%	**8.** 62%	**9.** 6%	**10.** 3%
11. 100%	**12.** 136%	**13.** 73.6%	**14.** 45.7%
15. 2.8%	**16.** 1.4%	**17.** 0.6%	**18.** 0.9%
19. 300%	**20.** 500%	**21.** 32.58%	**22.** 72.18%

Write each percent as a fraction or mixed number in simplest form. See Examples 8 through 12.

23. 8%	**24.** 22%	**25.** 4%	**26.** 2%

27. 4.5%　　　　**28.** 7.5%　　　　**29.** 175%　　　　**30.** 275%　　　　**31.** 6.25%

32. 8.75%　　　**33.** $10\frac{1}{3}\%$　　　**34.** $7\frac{3}{4}\%$　　　**35.** $22\frac{3}{8}\%$　　　**36.** $21\frac{7}{8}\%$

Objective **C** *Write each decimal as a percent. See Examples 13 through 16.*

37. 0.006　　　**38.** 0.008　　　**39.** 0.22　　　**40.** 0.44　　　**41.** 5.3

42. 2.7　　　　**43.** 0.056　　　**44.** 0.019　　　**45.** 0.2228　　　**46.** 0.1115

47. 3.00　　　**48.** 9.00　　　**49.** 0.7　　　**50.** 0.8

Write each fraction or mixed number as a percent. See Examples 17 through 19.

51. $\frac{7}{10}$　　　**52.** $\frac{9}{10}$　　　**53.** $\frac{4}{5}$　　　**54.** $\frac{2}{5}$　　　**55.** $\frac{34}{50}$

56. $\frac{41}{50}$　　　**57.** $\frac{3}{8}$　　　**58.** $\frac{5}{16}$　　　**59.** $\frac{1}{3}$　　　**60.** $\frac{2}{3}$

61. $4\frac{1}{2}$　　　**62.** $6\frac{1}{5}$　　　**63.** $1\frac{9}{10}$　　　**64.** $2\frac{7}{10}$

Write each fraction as a percent. Round to the nearest hundredth percent. See Example 20.

65. $\frac{9}{11}$　　　**66.** $\frac{11}{12}$　　　**67.** $\frac{4}{15}$　　　**68.** $\frac{10}{11}$

Objectives **A** **B** **C** **Mixed Practice** *Complete each table. See Examples 1 through 20.*

69.

Percent	Decimal	Fraction
60%		
	0.235	
		$\frac{4}{5}$
$33\frac{1}{3}\%$		
		$\frac{7}{8}$
7.5%		

70.

Percent	Decimal	Fraction
	0.525	
		$\frac{3}{4}$
$66\frac{1}{2}\%$		
		$\frac{5}{6}$
100%		
		$\frac{7}{50}$

71.

Percent	Decimal	Fraction
200%		
	2.8	
705%		
		$4\frac{27}{50}$

72.

Percent	Decimal	Fraction
800%		
	3.2	
608%		
		$9\frac{13}{50}$

Objective D *Write each percent as a decimal and a fraction. See Examples 21 and 22.*

73. People take aspirin for a variety of reasons. The most common use of aspirin is to prevent heart disease, accounting for 38% of all aspirin use. (*Source:* Bayer Market Research)

74. Approximately 56% of all U.S. workers use computers on the job. (*Source:* National Center for Education Statistics)

75. In the United States, about 25.2% of the female population age 25 and over graduated from college. (*Source:* U.S. Census Bureau)

76. In the United States, 28.5% of the male population age 25 and over graduated from college. (*Source:* U.S. Census Bureau)

77. Approximately 32.2% of all new cars sold in the United States were manufactured in Japan. (*Source:* American Automobile Manufacturers Association)

78. Approximately 14.8% of new luxury cars are silver, making silver the most popular new vehicle color for that class. (*Source:* Ward's Communications)

In Exercises 79 through 82, write the percent from the circle graph as a decimal and a fraction.

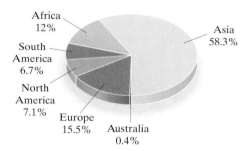

World Population by Continent

Africa 12%
South America 6.7%
North America 7.1%
Europe 15.5%
Australia 0.4%
Asia 58.3%

79. Australia: 0.4%

80. Asia: 58.3%

81. Africa: 12%

82. North America: 7.1%

Solve. See Examples 21 and 22.

83. In a recent year, the U.S. Postal Service handled approximately 0.442 of the world's card and letter mail volume. Write this decimal as a percent. (*Source:* U.S. Postal Service)

84. In the state of California, the Hispanic population is 0.347 of the total population. Write this decimal as a percent. (*Source:* U.S. Census Bureau)

85. The mirrors on the Hubble Space Telescope are able to lock onto a target without deviating more than $\frac{7}{1000}$ of an arc-second. Write this fraction as a percent. (*Source:* NASA)

86. In 2005, $\frac{3}{250}$ of all new cars sold in the United States were hybrids. Write this fraction as a percent. (*Source:* American Automobile Manufacturers Association)

87. In 2005, 3.9% of all households with television sets subscribed to TiVo®. Write this percent as a fraction. (*Source:* Plunkett Research)

88. For the 2006 season, the Tuesday broadcasts of *American Idol* had a 12.9% Nielsen rating. Write this percent as a fraction. (*Source:* Hollywood Reporter)

Review

Perform the indicated operation. See Section 4.6.

89. $\frac{3}{4} - \frac{1}{2} \cdot \frac{8}{9}$

90. $\left(\frac{2}{11} + \frac{5}{11}\right)\left(\frac{2}{11} - \frac{5}{11}\right)$

91. $6\frac{2}{3} - 4\frac{5}{6}$

92. $6\frac{2}{3} \div 4\frac{5}{6}$

Concept Extensions

Solve. See the Concept Checks in this section.

93. Given the percent 52.8647%, round as indicated.

    a. Round to a tenth of a percent.
    b. Round to a hundredth of a percent.

94. Given the percent 0.5269%, round as indicated.

    a. Round to a tenth of a percent.
    b. Round to a hundredth of a percent.

95. Which of the following are correct?

    a. 6.5% = 0.65   **b.** 7.8% = 0.078
    c. 120% = 0.12   **d.** 0.35% = 0.0035

96. Which of the following are correct?

    a. 0.231 = 23.1%   **b.** 5.12 = 0.0512%
    c. 3.2 = 320%   **d.** 0.0175 = 0.175%

Recall that 1 = 100%. This means that 1 whole is 100%. Use this for Exercises 97 and 98. (Source: Some Body, by Dr. Pete Rowen)

97. The four blood types are A, B, O, and AB. (Each blood type can also be further classified as Rh-positive or Rh-negative depending upon whether your blood contains protein or not.) Given the percent blood types for the United States below, calculate the percent of U.S. population with AB blood type.

    45% 40% 11% ?%

98. The top four components of bone are below. Find the missing percent.

    1. Minerals—45%
    2. Living tissue—30%
    3. Water—20%
    4. Other—?

What percent of the figure is shaded?

99.

100.

Fill in the blanks.

101. A fraction written as a percent is greater than 100% when the numerator is _____ than the denominator. (greater/less)

102. A decimal written as a percent is less than 100% when the decimal is _____ than 1. (greater/less)

Write each fraction as a decimal and then write each decimal as a percent. Round the decimal to three decimal places (nearest thousandth) and the percent to the nearest tenth of a percent.

103. $\dfrac{21}{79}$

104. $\dfrac{56}{102}$

The bar graph shows the predicted fastest-growing occupations. Use this graph for Exercises 105 through 108.

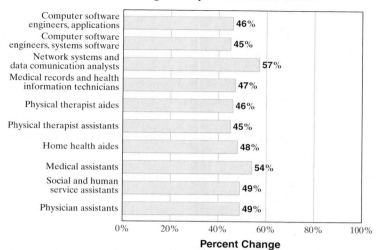

Fastest-Growing Occupations 2002–2012

Source: Bureau of Labor Statistics

105. What occupation is predicted to be the fastest growing?

106. What occupation is predicted to be the second fastest growing?

107. Write the percent change for physician assistants as a decimal.

108. Write the percent change for medical assistants as a decimal.

 109. In your own words, explain how to write a percent as a decimal.

110. In your own words, explain how to write a decimal as a percent.

STUDY SKILLS BUILDER

Are You Satisfied with Your Performance in This Course thus Far?

To see if there is room for improvement, answer these questions:

1. Am I attending all classes and arriving on time?

2. Am I working and checking my homework assignments on time?

3. Am I getting help (from my instructor or a campus learning resource lab) when I need it?

4. In addition to my instructor, am I using the text supplements that might help me?

5. Am I satisfied with my performance on quizzes and exams?

If you answered no to any of these questions, read or reread Section 1.1 for suggestions in these areas. Also, you might want to contact your instructor for additional feedback.

Objectives

A Write Percent Problems as Equations.

B Solve Percent Problems.

7.2 SOLVING PERCENT PROBLEMS WITH EQUATIONS

Sections 7.2 and 7.3 introduce two methods for solving percent problems. It may not be necessary that you study both sections. You may want to check with your instructor for further advice.

To solve percent problems in this section, we will translate the problems into mathematical statements, or equations.

Objective **A** Writing Percent Problems as Equations

Recognizing key words in a percent problem is helpful in writing the problem as an equation. Three key words in the statement of a percent problem and their meanings are as follows:

of means **multiplication** ($\cdot$)
is means **equals** ($=$)
what (or some equivalent) means **the unknown number.**

In our examples, we will let the letter x stand for the unknown number.

PRACTICE PROBLEM 1

Translate: 8 is what percent of 48?

EXAMPLE 1 Translate to an equation:

5 is what percent of 20?

Solution: 5 is what percent of 20?

$$5 = x \cdot 20$$

Work Practice Problem 1

> **Helpful Hint**
> Remember that an equation is simply a mathematical statement that contains an equals sign ($=$).
> $$5 = 20x$$
> ↑ equals sign

PRACTICE PROBLEM 2

Translate: 2.6 is 40% of what number?

EXAMPLE 2 Translate to an equation:

1.2 is 30% of what number?

Solution: 1.2 is 30% of what number?

$$1.2 = 30\% \cdot x$$

Work Practice Problem 2

PRACTICE PROBLEM 3

Translate: What number is 90% of 0.045?

EXAMPLE 3 Translate to an equation:

What number is 25% of 0.008?

Solution: What number is 25% of 0.008?

$$x = 25\% \cdot 0.008$$

Work Practice Problem 3

Answers
1. $8 = x \cdot 48$ **2.** $2.6 = 40\% \cdot x$
3. $x = 90\% \cdot 0.045$

484

EXAMPLES Translate each question to an equation.

4. 38% of 200 is what number?

$$38\% \cdot 200 = x$$

5. 40% of what number is 80?

$$40\% \cdot x = 80$$

6. What percent of 85 is 34?

$$x \cdot 85 = 34$$

■ **Work Practice Problems 4–6**

PRACTICE PROBLEMS 4–6

Translate each question to an equation.

4. 56% of 180 is what number?

5. 12% of what number is 21?

6. What percent of 95 is 76?

✔ **Concept Check** In the equation $2x = 10$, what step is taken to solve the equation?

Objective B Solving Percent Problems

You may have noticed by now that each percent problem has contained three numbers—in our examples, two are known and one is unknown. Each of these numbers is given a special name.

15%	of	60	is	9
15% percent	·	60 base	=	9 amount

We call this equation the **percent equation.**

Percent Equation

$$\text{percent} \cdot \text{base} = \text{amount}$$

Once a percent problem has been written as a percent equation, we can use the equation to find the unknown number, whether it is the percent, the base, or the amount.

Solving Percent Equations for the Amount

EXAMPLE 7 What number is 35% of 60?

Solution:

x	$= 35\% \cdot 60$	Translate to an equation.
x	$= 0.35 \cdot 60$	Write 35% as 0.35.
x	$= 21$	Multiply:

$$\begin{array}{r} 60 \\ \times\,0.35 \\ \hline 300 \\ 1800 \\ \hline 21.00 \end{array}$$

Then 21 is 35% of 40. Is this reasonable? To see, round 35% to 40%. Then 40% or 0.40(60) is 24. Our result is reasonable since 21 is close to 24.

■ **Work Practice Problem 7**

PRACTICE PROBLEM 7

What number is 25% of 90?

Answers

4. $56\% \cdot 180 = x$ **5.** $12\% \cdot x = 21$
6. $x \cdot 95 = 76$ **7.** 22.5

✔ **Concept Check Answer**

Divide both sides of the

CHAPTER 7 | PERCENT

> ### Helpful Hint
> When solving a percent equation, write the percent as a decimal or fraction.

PRACTICE PROBLEM 8
95% of 400 is what number?

EXAMPLE 8 85% of 300 is what number?

Solution:

$85\% \cdot 300 =$	x	Translate to an equation.
$0.85 \cdot 300 =$	x	Write 85% as 0.85.
$255 =$	x	Multiply: $0.85 \cdot 300 = 255$.

Then 85% of 300 is 255. Is this result reasonable? To see, round 85% to 90%. Then 90% of 300 or $0.90(300) = 270$, which is close to 255.

■ **Work Practice Problem 8**

Solving Percent Equations for the Base

PRACTICE PROBLEM 9
15% of what number is 2.4?

EXAMPLE 9 12% of what number is 0.6?

Solution:

$$12\% \cdot x = 0.6 \qquad \text{Translate to an equation.}$$
$$0.12 \cdot x = 0.6 \qquad \text{Write 12\% as 0.12.}$$
$$\frac{0.12 \cdot x}{0.12} = \frac{0.6}{0.12} \qquad \text{Divide both sides by 0.12.}$$
$$x = 5 \qquad 0.12\overline{)0.60}$$
$$\underline{60}$$
$$0$$

Then 12% of 5 is 0.6. Is this reasonable? To see, round 12% to 10%. Then 10% of 0.6 or 0.10 (0.6) is 6, which is close to 5.

■ **Work Practice Problem 9**

PRACTICE PROBLEM 10
18 is $4\frac{1}{2}$% of what number?

EXAMPLE 10 13 is $6\frac{1}{2}$% of what number?

Solution:

$$13 = 6\frac{1}{2}\% \cdot x \qquad \text{Translate to an equation.}$$
$$13 = 0.065 \cdot x \qquad 6\frac{1}{2}\% = 6.5\% = 0.065.$$
$$\frac{13}{0.065} = \frac{0.065 \cdot x}{0.065} \qquad \text{Divide both sides by 0.065.} \quad 0.065\overline{)13.000}$$
$$200 = x \qquad\qquad\qquad\qquad\qquad\qquad \underline{130}$$
$$0$$

Then 13 is $6\frac{1}{2}$% of 200. Check to see if this result is reasonable.

■ **Work Practice Problem 10**

Answers
380 **9.** 16 **10.** 400

Copyright 2008 Pearson Education, Inc.

Solving Percent Equations for the Percent

EXAMPLE 11 What percent of 12 is 9?

Solution:

$$x \cdot 12 = 9 \quad \text{Translate to an equation.}$$

$$\frac{x \cdot 12}{12} = \frac{9}{12} \quad \text{Divide both sides by 12.}$$

$$\text{or} \quad x = 0.75$$

Next, since we are looking for percent, we can write $\frac{9}{12}$ or 0.75 as a percent.

$$x = 75\%$$

Then 75% of 12 is 9. To check, see that $75\% \cdot 12 = 9$.

🔲 **Work Practice Problem 11**

PRACTICE PROBLEM 11
What percent of 90 is 27?

> **Helpful Hint**
> If your unknown in the percent equation is percent, don't forget to convert your answer to a percent.

EXAMPLE 12 78 is what percent of 65?

Solution:

$$78 = x \cdot 65 \quad \text{Translate to an equation.}$$

$$\frac{78}{65} = \frac{x \cdot 65}{65} \quad \text{Divide both sides by 65.}$$

$$1.2 = x$$

$$120\% = x \quad \text{Write 1.2 as a percent.}$$

Then 78 is 120% of 65. Check this result.

🔲 **Work Practice Problem 12**

PRACTICE PROBLEM 12
63 is what percent of 45?

✔ **Concept Check** Consider these problems

1. 75% of 50 =
   a. 50
   b. a number greater than 50
   c. a number less than 50

2. 40% of a number is 10. Is the number
   a. 10
   b. less than 10
   c. greater than 10?

3. 800 is 120% of what number? Is the number
   a. 800
   b. less than 800
   c. greater than 800?

> **Helpful Hint**
> Use the following to see if your answers are reasonable.
>
> (100%) of a number = the number
>
> $\left(\begin{array}{c} \text{a percent} \\ \text{greater than} \\ 100\% \end{array}\right)$ of a number = $\begin{array}{c} \text{a number greater} \\ \text{than the original number} \end{array}$
>
> $\left(\begin{array}{c} \text{a percent} \\ \text{less than } 100\% \end{array}\right)$ of a number = $\begin{array}{c} \text{a number less} \\ \text{than the original number} \end{array}$

Answers
11. 30% **12.** 140%

✔ **Concept Check Answers**
1. c **2.** c **3.** b

Vocabulary and Readiness Check

Use the choices below to fill in each blank.

    percent   amount   of   less

    base   the number   is   greater

1. The word _____ translates to "=" .

2. The word _____ usually translates to "multiplication."

3. In the statement "10% of 90 is 9," the number 9 is called the _____, 90 is called the _____, and 10 is called the _____.

4. 100% of a number = _____.

5. Any "percent greater than 100%" of "a number" = "a number _____ than the original number."

6. Any "percent less than 100%" of "a number" = "a number _____ than the original number."

7.2 EXERCISE SET

FOR EXTRA HELP

 Student Solutions Manual   PH Math/Tutor Center   CD/Video for Review   MathXL®   MyMathLab

Objective A Translating *Translate each to an equation. Do not solve. See Examples 1 through 6.*

1. 18% of 81 is what number?

2. 36% of 72 is what number?

3. 20% of what number is 105?

4. 40% of what number is 6?

5. 0.6 is 40% of what number?

6. 0.7 is 20% of what number?

7. What percent of 80 is 3.8?

8. 9.2 is what percent of 92?

9. What number is 9% of 43?

10. What number is 25% of 55?

11. What percent of 250 is 150?

12. What percent of 375 is 300?

Objective B *Solve. See Examples 7 and 8.*

13. 10% of 35 is what number?

14. 25% of 68 is what number?

15. What number is 14% of 205?

16. What number is 18% of 425?

Solve. See Examples 9 and 10.

17. 1.2 is 12% of what number?

18. 0.22 is 44% of what number?

19. $8\frac{1}{2}\%$ of what number is 51?

20. $4\frac{1}{2}\%$ of what number is 45?

Solve. See Examples 11 and 12.

21. What percent of 80 is 88?

22. What percent of 40 is 60?

23. 17 is what percent of 50?

24. 48 is what percent of 50?

Objectives Ⓐ Ⓑ **Mixed Practice** *Solve. See Examples 1 through 12.*

25. 0.1 is 10% of what number?

26. 0.5 is 5% of what number?

27. 150% of 430 is what number?

28. 300% of 56 is what number?

29. 82.5 is $16\frac{1}{2}$% of what number?

30. 7.2 is $6\frac{1}{4}$% of what number?

31. 2.58 is what percent of 50?

32. 2.64 is what percent of 25?

33. What number is 42% of 60?

34. What number is 36% of 80?

35. What percent of 184 is 64.4?

36. What percent of 120 is 76.8?

37. 120% of what number is 42?

38. 160% of what number is 40?

39. 2.4% of 26 is what number?

40. 4.8% of 32 is what number?

41. What percent of 600 is 3?

42. What percent of 500 is 2?

43. 6.67 is 4.6% of what number?

44. 9.75 is 7.5% of what number?

45. 1575 is what percent of 2500?

46. 2520 is what percent of 3500?

47. 2 is what percent of 50?

48. 2 is what percent of 40?

Review

Find the value of x in each proportion. See Section 6.2.

49. $\dfrac{27}{x} = \dfrac{9}{10}$

50. $\dfrac{35}{x} = \dfrac{7}{5}$

51. $\dfrac{x}{5} = \dfrac{8}{11}$

52. $\dfrac{x}{3} = \dfrac{6}{13}$

Write each sentence as a proportion.

53. 17 is to 12 as x is to 20.

54. 20 is to 25 as x is to 10.

55. 8 is to 9 as 14 is to x.

56. 5 is to 6 as 15 is to x.

Concept Extensions

For each equation, determine the next step taken to find the value of n. See the first Concept Check in this section.

57. $5 \cdot n = 32$

  a. $n = 5 \cdot 32$ **b.** $n = \dfrac{5}{32}$ **c.** $n = \dfrac{32}{5}$ **d.** none of these

58. $n = 0.7 \cdot 12$

  a. $n = 8.4$ **b.** $n = \dfrac{12}{0.7}$ **c.** $n = \dfrac{0.7}{12}$ **d.** none of these

59. $0.06 = n \cdot 7$

  a. $n = 0.06 \cdot 7$ **b.** $n = \dfrac{0.06}{7}$ **c.** $n = \dfrac{7}{0.06}$ **d.** none of these

60. Write a word statement for the equation $20\% \cdot x = 18.6$. Use the phrase "what number" for "x."

61. Write a word statement for the equation $x = 33\frac{1}{3}\% \cdot 24$. Use the phrase "what number" for "x."

For each exercise, determine whether the percent, x, is (a) 100%, (b) greater than 100%, or (c) less than 100%. See the last Concept Check in this section.

62. $x\%$ of 20 is 30

63. $x\%$ of 98 is 98

64. $x\%$ of 120 is 85

For each exercise, determine whether the number, y, is (a) equal to 45, (b) greater than 45, or (c) less than 45.

65. 55% of 45 is y

66. 230% of 45 is y

67. 100% of 45 is y

68. 30% of y is 45

69. 100% of y is 45

70. 180% of y is 45

Solve.

71. In your own words, explain how to solve a percent equation.

72. Write a percent problem that uses the percent 50%.

73. 1.5% of 45,775 is what number?

74. What percent of 75,528 is 27,945.36?

75. 22,113 is 180% of what number?

THE BIGGER PICTURE Operations on Sets of Numbers and Solving Equations

Continue your outline from Sections 1.6, 1.7, 2.4, 3.3, 4.3, 4.7, 5.4, 6.2, and 6.4. Suggestions are once again written to help you complete this part of your outline. Notice that this part of the outline has to do with solving equations.

I. Operations on Sets of Numbers
  A. Whole Numbers
    1. Add or Subtract (Section 1.3)
    2. Multiply or Divide (Sections 1.5, 1.6)
    3. Exponent (Section 1.7)
    4. Order of Operations (Section 1.7)
    5. Square Root (Section 6.4)
  B. Integers
    1. Add (Section 2.2)
    2. Subtract (Section 2.3)
    3. Multiply or Divide (Section 2.4)
  C. Fractions
    1. Simplify (Section 4.2)
    2. Multiply (Section 4.3)
    3. Divide (Section 4.3)
    4. Add or Subtract (Sections 4.4, 4.5)
  D. Decimals
    1. Add or Subtract (Section 5.2)
    2. Multiply (Section 5.3)
    3. Divide (Section 5.4)

II. Solving Equations
  A. Equations in General (Section 3.3)
  B. Proportions (Section 6.2)
  C. Percent Problems
    1. Solved by Equations: Remember that "of" means multiplication and "is" means equals.

    12% of some number is 6 translates to

$$12\% \cdot n = 6 \ or \ 0.12n = 6 \ or \ n = \frac{6}{0.12} \ or \ n = 50$$

Perform the indicated operations.

1. $\dfrac{2}{9} + \dfrac{1}{5}$

2. $42 \div 2 \cdot 3$

3. $-0.03(0.7)$

4. $\sqrt{49} + \sqrt{1}$

Solve.

5. $\dfrac{3}{8} = \dfrac{n}{128}$

6. $\dfrac{7.2}{n} = \dfrac{36}{8}$

7. 215 is what percent of 86?

8. 95% of 48 is what number?

9. 4.2 is what percent of 15?

10. 93.6 is 52% of what number?

7.3 SOLVING PERCENT PROBLEMS WITH PROPORTIONS

Objectives

A Write Percent Problems as Proportions.

B Solve Percent Problems.

There is more than one method that can be used to solve percent problems. (See the note at the beginning of Section 7.2.) In the last section, we used the percent equation. In this section, we will use proportions.

Objective A Writing Percent Problems as Proportions

To understand the proportion method, recall that 70% means the ratio of 70 to 100, or $\frac{70}{100}$.

$\frac{7}{10}$ shaded

70% or $\frac{70}{100}$ shaded

$$70\% = \frac{70}{100} = \frac{7}{10}$$

Since the ratio $\frac{70}{100}$ is equal to the ratio $\frac{7}{10}$, we have the proportion

$$\frac{7}{10} = \frac{70}{100}$$

We call this proportion the **percent proportion.** In general, we can name the parts of this proportion as follows:

Percent Proportion

$$\frac{\text{amount}}{\text{base}} = \frac{\text{percent}}{100} \quad \leftarrow \text{always 100}$$

or

$$\text{amount} \rightarrow \frac{a}{b} = \frac{p}{100} \quad \leftarrow \text{percent}$$
$$\text{base} \rightarrow$$

When we translate percent problems to proportions, the **percent,** p, can be identified by looking for the symbol % or the word *percent*. The **base,** b, usually follows the word *of*. The **amount,** a, is the part compared to the whole.

Helpful Hint

This table may be useful when identifying the parts of a proportion.

Part of Proportion	How It's Identified
Percent	% or percent
Base	Appears after *of*
Amount	Part compared to whole

PRACTICE PROBLEM 1

Translate to a proportion.
27% of what number is 54?

EXAMPLE 1 Translate to a proportion.

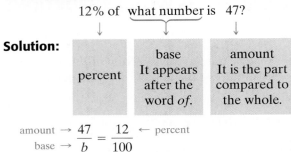

12% of what number is 47?

Solution:

| percent | base
It appears
after the
word *of.* | amount
It is the part
compared to
the whole. |

amount → $\frac{47}{b} = \frac{12}{100}$ ← percent
base →

☐ **Work Practice Problem 1**

PRACTICE PROBLEM 2

Translate to a proportion.
30 is what percent of 90?

EXAMPLE 2 Translate to a proportion.

101 is what percent of 200?

Solution:

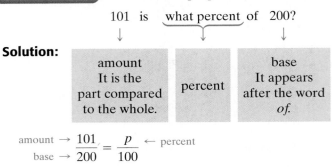

| amount
It is the
part compared
to the whole. | percent | base
It appears
after the word
of. |

amount → $\frac{101}{200} = \frac{p}{100}$ ← percent
base →

☐ **Work Practice Problem 2**

PRACTICE PROBLEM 3

Translate to a proportion.
What number is 25% of 116?

EXAMPLE 3 Translate to a proportion.

What number is 90% of 45?

Solution:

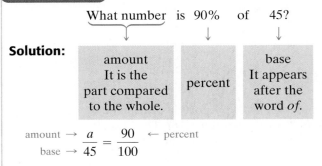

| amount
It is the
part compared
to the whole. | percent | base
It appears
after the
word *of.* |

amount → $\frac{a}{45} = \frac{90}{100}$ ← percent
base →

☐ **Work Practice Problem 3**

PRACTICE PROBLEM 4

Translate to a proportion.
680 is 65% of what number?

EXAMPLE 4 Translate to a proportion.

238 is 40% of what number?

Solution: amount   percent   base

$\frac{238}{b} = \frac{40}{100}$

☐ **Work Practice Problem 4**

Answers

1. $\frac{54}{b} = \frac{27}{100}$ 2. $\frac{30}{90} = \frac{p}{100}$

3. $\frac{a}{116} = \frac{25}{100}$ 4. $\frac{680}{b} = \frac{65}{100}$

EXAMPLE 5 Translate to a proportion.

What percent of 30 is 75?

Solution: percent   base   amount

$$\frac{75}{30} = \frac{p}{100}$$

☐ Work Practice Problem 5

PRACTICE PROBLEM 5

Translate to a proportion.
What percent of 40 is 75?

EXAMPLE 6 Translate to a proportion.

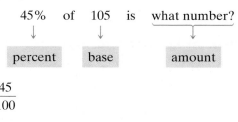

45% of 105 is what number?

Solution: percent   base   amount

$$\frac{a}{105} = \frac{45}{100}$$

☐ Work Practice Problem 6

PRACTICE PROBLEM 6

Translate to a proportion.
46% of 80 is what number?

✔**Concept Check** Consider the statement "78 is what percent of 350?"
Which part of the percent proportion is unknown?

a. the amount
b. the base
c. the percent

Consider another statement: "14 is 10% of some number."
Which part of the percent proportion is unknown?

a. the amount
b. the base
c. the percent

Objective B Solving Percent Problems

The proportions that we have written in this section contain three values that can
change: the percent, the base, and the amount. If any two of these values are known,
we can find the third (the unknown value). To do this, we write a percent proportion
and find the unknown value as we did in Section 6.2.

EXAMPLE 7 **Solving Percent Proportions for the Amount**

What number is 30% of 9?

Solution: amount   percent base

$$\frac{a}{9} = \frac{30}{100}$$

Continued on next page

PRACTICE PROBLEM 7

What number is 8% of 120?

Answers

5. $\frac{75}{40} = \frac{p}{100}$ 6. $\frac{a}{80} = \frac{46}{100}$ 7. 9.6

✔ **Concept Check Answers**

c, b

The proportion in Example 7 contains the ratio $\frac{30}{100}$. A ratio in a proportion may be simplified before solving the proportion. The unknown number in both

$\frac{a}{9} = \frac{30}{100}$ and $\frac{a}{9} = \frac{3}{10}$ is 2.7.

To solve, we set cross products equal to each other.

$$\frac{a}{9} = \frac{30}{100}$$

$a \cdot 100 = 9 \cdot 30$ Set cross products equal.

$100a = 270$ Multiply.

$\dfrac{100a}{100} = \dfrac{270}{100}$ Divide both sides by 100, the coefficient of a.

$a = 2.7$ Simplify.

Thus, 2.7 is 30% of 9.

Work Practice Problem 7

PRACTICE PROBLEM 8

65% of what number is 52?

EXAMPLE 8   **Solving Percent Problems for the Base**

150% of what number is 30?

Solution: percent   base   amount

$\dfrac{30}{b} = \dfrac{150}{100}$ Write the proportion.

$$\frac{30}{b} = \frac{3}{2}$$ Simplify $\frac{150}{100}$ and write as $\frac{3}{2}$.

$30 \cdot 2 = b \cdot 3$ Set cross products equal.

$60 = 3b$ Multiply.

$\dfrac{60}{3} = \dfrac{3b}{3}$ Divide both sides by 3.

$20 = b$ Simplify.

Thus, 150% of 20 is 30.

Work Practice Problem 8

✔ **Concept Check**   When solving a percent problem by using a proportion, describe how you can check the result.

PRACTICE PROBLEM 9

15.4 is 5% of what number?

EXAMPLE 9

20.8 is 40% of what number?

Solution: amount   percent   base

$\dfrac{20.8}{b} = \dfrac{40}{100}$ or $\dfrac{20.8}{b} = \dfrac{2}{5}$ Write the proportion and simplify $\frac{40}{100}$.

$20.8 \cdot 5 = b \cdot 2$ Set cross products equal.

$104 = 2b$ Multiply.

$\dfrac{104}{2} = \dfrac{2b}{2}$ Divide both sides by 2.

$52 = b$ Simplify.

So, 20.8 is 40% of 52.

Work Practice Problem 9

Answers
8. 80   **9.** 308

✔ **Concept Check Answer**
By putting the result into the proportion and checking that the proportion is true

EXAMPLE 10 Solving Percent Problems for the Percent

What percent of 50 is 8?

Solution: percent   base   amount

$$\frac{8}{50} = \frac{p}{100} \quad \text{or} \quad \frac{4}{25} = \frac{p}{100} \qquad \text{Write the proportion and simplify } \frac{8}{50}.$$

$4 \cdot 100 = 25 \cdot p \qquad$ Set cross products equal.

$400 = 25p \qquad$ Multiply.

$$\frac{400}{25} = \frac{25p}{25} \qquad \text{Divide both sides by 25.}$$

$16 = p \qquad$ Simplify.

So, 16% of 50 is 8.

▣ **Work Practice Problem 10**

Helpful Hint
Recall from our percent proportion that this number already is a percent. Just keep the number as is and attach a % symbol.

EXAMPLE 11

504 is what percent of 360?

Solution: amount   percent   base

$$\frac{504}{360} = \frac{p}{100}$$

Let's choose not to simplify the ratio $\frac{504}{360}$.

$504 \cdot 100 = 360 \cdot p \qquad$ Set cross products equal.

$50{,}400 = 360p \qquad$ Multiply.

$$\frac{50{,}400}{360} = \frac{360p}{360} \qquad \text{Divide both sides by 360.}$$

$140 = p \qquad$ Simplify.

Notice that by choosing not to simplify $\frac{504}{360}$, we had larger numbers in our equation. Either way, we find that 504 is 140% of 360.

▣ **Work Practice Problem 11**

Helpful Hint
Use the following to see whether your answers to the above examples and practice problems are reasonable.

100% of a number = the number

$\left(\begin{array}{c} \text{a percent} \\ \text{greater than} \\ 100\% \end{array} \right)$ of a number = $\begin{array}{l} \text{a number larger} \\ \text{than the original number} \end{array}$

$\left(\begin{array}{c} \text{a percent} \\ \text{less than } 100\% \end{array} \right)$ of a number = $\begin{array}{l} \text{a number less} \\ \text{than the original number} \end{array}$

PRACTICE PROBLEM 10
What percent of 40 is 8?

PRACTICE PROBLEM 11
414 is what percent of 180?

Answers
10. 20% **11.** 230%

Vocabulary and Readiness Check

Use the choices below to fill in each blank. These choices will be used more than once.

    amount   base   percent

1. When translating the statement "20% of 15 is 3" to a proportion, the number 3 is called the _____, 15 is the _____, and 20 is the _____.

2. In the question "50% of what number is 28?", which part of the percent proportion is unknown? _____

3. In the question "What number is 25% of 200?", which part of the percent proportion is unknown? _____

4. In the question "38 is what percent of 380?", which part of the percent proportion is unknown? _____

7.3 EXERCISE SET

FOR EXTRA HELP

 Student Solutions Manual   PH Math/Tutor Center   CD/Video for Review   Math XL MathXL®   MyMathLab MyMathLab

Objective A Translating *Translate each to a proportion. Do not solve. See Examples 1 through 6.*

1. 98% of 45 is what number?

2. 92% of 30 is what number?

3. What number is 4% of 150?

4. What number is 7% of 175?

5. 14.3 is 26% of what number?

6. 1.2 is 47% of what number?

7. 35% of what number is 84?

8. 85% of what number is 520?

9. What percent of 400 is 70?

10. What percent of 900 is 216?

11. 8.2 is what percent of 82?

12. 9.6 is what percent of 96?

Objective B *Solve. See Example 7.*

13. 40% of 65 is what number?

14. 25% of 84 is what number?

15. What number is 18% of 105?

16. What number is 60% of 29?

Solve. See Examples 8 and 9.

17. 15% of what number is 90?

18. 55% of what number is 55?

19. 7.8 is 78% of what number?

20. 1.1 is 44% of what number?

Solve. See Examples 10 and 11.

21. 42 is what percent of 35?

22. 147 is what percent of 98?

23. 14 is what percent of 50?

24. 24 is what percent of 50?

Objectives **A** **B** **Mixed Practice** *Solve. See Examples 1 through 11.*

25. 3.7 is 10% of what number?

26. 7.4 is 5% of what number?

27. 2.4% of 70 is what number?

28. 2.5% of 90 is what number?

29. 160 is 16% of what number?

30. 30 is 6% of what number?

31. 394.8 is what percent of 188?

32. 550.4 is what percent of 172?

33. What number is 89% of 62?

34. What number is 53% of 130?

35. What percent of 6 is 2.7?

36. What percent of 5 is 1.6?

37. 140% of what number is 105?

38. 170% of what number is 221?

39. 1.8% of 48 is what number?

40. 7.8% of 24 is what number?

41. What percent of 800 is 4?

42. What percent of 500 is 3?

43. 3.5 is 2.5% of what number?

44. 9.18 is 6.8% of what number?

45. 20% of 48 is what number?

46. 75% of 14 is what number?

47. 2486 is what percent of 2200?

48. 9310 is what percent of 3800?

Review

Add or subtract the fractions. See Sections 4.4, 4.5, and 4.7.

49. $-\dfrac{11}{16} + \left(-\dfrac{3}{16}\right)$

50. $\dfrac{7}{12} - \dfrac{5}{8}$

51. $3\dfrac{1}{2} - \dfrac{11}{30}$

52. $2\dfrac{2}{3} + 4\dfrac{1}{2}$

Add or subtract the decimals. See Section 5.2.

53. $\begin{array}{r} 0.41 \\ + \, 0.29 \\ \hline \end{array}$

54. $\begin{array}{r} 10.78 \\ 4.3 \\ + \;\, 0.21 \\ \hline \end{array}$

55. $\begin{array}{r} 2.38 \\ - \, 0.19 \\ \hline \end{array}$

56. $\begin{array}{r} 16.37 \\ - \;\, 2.61 \\ \hline \end{array}$

Concept Extensions

57. Write a word statement for the proportion $\dfrac{x}{28} = \dfrac{25}{100}$. Use the phrase "what number" for "*x*."

58. Write a percent statement that translates to $\dfrac{16}{80} = \dfrac{20}{100}$

Solve. See the Concept Checks in this section.

Suppose you have finished solving three percent problems using proportions that you set up correctly. Check each answer to see if each makes the proportion a true proportion. If any proportion is not true, solve it to find the correct solution.

59. $\dfrac{a}{64} = \dfrac{25}{100}$

Is the amount equal to 17?

60. $\dfrac{520}{b} = \dfrac{65}{100}$

Is the base equal to 800?

61. $\dfrac{36}{12} = \dfrac{p}{100}$

Is the percent equal to 50 (50%)?

62. In your own words, describe how to identify the percent, the base, and the amount in a percent problem.

63. In your own words, explain how to use a proportion to solve a percent problem.

Solve. Round to the nearest tenth, if necessary.

64. What number is 22.3% of 53,862?

65. What percent of 110,736 is 88,542?

66. 8652 is 119% of what number?

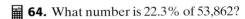

THE BIGGER PICTURE Operations on Sets of Numbers and Solving Equations

Continue your outline from Sections 1.6, 1.7, 2.4, 3.3, 4.3, 4.7, 5.4, 6.2, 6.4, and 7.2. (If you did not cover Section 7.2, pay no attention to the part of the outline numbered II.C.1.) Suggestions are once again written to help you complete this part of your outline. Notice that this part of the outline has to do with solving proportions.

I. Operations on Sets of Numbers

   A. Whole Numbers

      1. Add or Subtract (Section 1.3)

      2. Multiply or Divide (Sections 1.5, 1.6)

      3. Exponent (Section 1.7)

      4. Order of Operations (Section 1.7)

      5. Square Root (Section 6.4)

   B. Integers

      1. Add (Section 2.2)

      2. Subtract (Section 2.3)

      3. Multiply or Divide (Section 2.4)

   C. Fractions

      1. Simplify (Section 4.2)

      2. Multiply (Section 4.3)

      3. Divide (Section 4.3)

      4. Add or Subtract (Sections 4.4, 4.5)

   D. Decimals

      1. Add or Subtract (Section 5.2)

      2. Multiply (Section 5.3)

      3. Divide (Section 5.4)

II. Solving Equations

   A. (Equations in general) (Section 3.3)

   B. Proportions (Section 6.2)

   C. Percent Problems

      1. Solved by Equations (Section 7.2—you may not have covered this section)

      2. Solved by Proportions: Remember that percent, p, is identified by % or percent, base, b, usually appears after "of" and amount, a, is the part compared to the whole. 12% of some number is 6 translates to

$$\frac{6}{b} = \frac{12}{100} \ or \ 6 \cdot 100 = b \cdot 12 \ or \ \frac{600}{12} = b \ or \ 50 = b$$

Perform the indicated operations.

1. $\dfrac{2}{9} + \dfrac{1}{5}$

2. $42 \div 2 \cdot 3$

3. $-0.03 \, (0.7)$

4. $7.9 + 0.1$

Solve.

5. $\dfrac{3}{8} = \dfrac{n}{128}$

6. $\dfrac{7.2}{n} = \dfrac{36}{8}$

7. 215 is what percent of 86?

8. 95% of 48 is what number?

9. 4.2 is what percent of 15?

10. 93.6 is 52% of what number?

Percent and Percent Problems

Write each number as a percent.

1. 0.94   **2.** 0.17   **3.** $\dfrac{3}{8}$ **4.** $\dfrac{7}{2}$

5. 4.7   **6.** 8   **7.** $\dfrac{9}{20}$ **8.** $\dfrac{53}{50}$

9. $6\dfrac{3}{4}$ **10.** $3\dfrac{1}{4}$ **11.** 0.02   **12.** 0.06

Write each percent as a decimal.

13. 71%   **14.** 31%   **15.** 3%   **16.** 4%

17. 224%   **18.** 700%   **19.** 2.9%   **20.** 6.6%

Write each percent as a decimal and as a fraction or mixed number in simplest form. (If necessary when writing as a decimal, round to the nearest thousandth.)

21. 7%   **22.** 5%   **23.** 6.8%   **24.** 11.25%

Answers

1. _____
2. _____
3. _____
4. _____
5. _____
6. _____
7. _____
8. _____
9. _____
10. _____
11. _____
12. _____
13. _____
14. _____
15. _____
16. _____
17. _____
18. _____
19. _____
20. _____
21. _____
22. _____
23. _____
24. _____

25. _____

26. _____

27. _____

28. _____

29. _____

30. _____

31. _____

32. _____

33. _____

34. _____

35. _____

36. _____

37. _____

38. _____

39. _____

40. _____

25. 74%   **26.** 45%   **27.** $16\frac{1}{3}\%$ **28.** $12\frac{2}{3}\%$

Solve each percent problem.

29. 15% of 90 is what number?   **30.** 78 is 78% of what number?

31. 297.5 is 85% of what number?   **32.** 78 is what percent of 65?

33. 23.8 is what percent of 85?   **34.** 38% of 200 is what number?

35. What number is 40% of 85?   **36.** What percent of 99 is 128.7?

37. What percent of 250 is 115?   **38.** What number is 45% of 84?

39. 42% of what number is 63?   **40.** 95% of what number is 58.9?

7.4 APPLICATIONS OF PERCENT

Objective **A** Solving Applications Involving Percent

Percent is used in a variety of everyday situations. The next examples show just a few ways that percent occurs in real-life settings. (Each of these examples shows two ways of solving these problems. If you studied Section 7.2 only, see *Method 1*. If you studied Section 7.3 only, see *Method 2*.)

EXAMPLE 1

Recall that the circle graph from the beginning of this chapter shows the favorite ice cream flavors of adults in the United States. The state of Delaware has an adult population of 637,000. How many of these adults might we predict will choose butter pecan as their favorite ice cream flavor? (*Source:* U.S. Census Bureau)

Solution: *Method 1.* First, we state the problem in words.

In words: What number is 5% of 637,000?

Translate: x = 5% · 637,000

To solve for x, we find 5% · 637,000.

$x = 0.05 \cdot 637,000$ Write 5% as a decimal.

$x = 31,850$

We predict that 31,850 adults in Delaware will choose butter pecan as their favorite ice cream flavor.

Method 2. State the problem in words; then translate.

In words: What number is 5% of 637,000?

amount   percent   base

Translate: amount → $\dfrac{a}{637,000} = \dfrac{5}{100}$ ← percent
base →

Next, we solve for a.

$a \cdot 100 = 637,000 \cdot 5$ Set cross products equal.

$100a = 3,185,000$ Multiply.

$\dfrac{100a}{100} = \dfrac{3,185,000}{100}$ Divide both sides by 100.

$a = 31,850$ Simplify.

We predict that 31,850 adults in Delaware will choose butter pecan as their favorite ice cream flavor.

Work Practice Problem 1

PRACTICE PROBLEM 1

The state of Rhode Island has an adult population of 837,000. How many of these adults might we predict will choose chocolate chip as their favorite ice cream flavor?

Favorite Ice Cream Flavors of Adults

Coffee, 2%
Chocolate almond, 2%
Cherry, 2%
Praline pecan, 3%
Vanilla fudge ripple, 3%
Cookies & cream, 4%
Chocolate chip, 4%
Neopolitan, 4%

All others, 25%
Vanilla, 32%
Chocolate, 9%
Butter pecan, 5%
Strawberry, 5%

Source: International Ice Cream Association

PRACTICE PROBLEM 2

In Florida, about 34,000 new nurses were needed in 2006. If there are currently 130,000 nurses, what percent of new nurses are needed in Florida? Round to the nearest whole percent. (*Source: St. Petersburg Times* and *The Registered Nurse Population*)

EXAMPLE 2 Finding Percent of Nursing Schools with Increases in Enrollment

There is a worldwide shortage of nurses that is projected to be 20% below requirements by 2020. Until 2001, there had also been a continual decline in enrollment in nursing schools. That has recently changed.

Of the total 2593 nursing schools in the United States, 2178 had an increase in applications or enrollment. What percent of nursing schools had an increase? Round to the nearest whole percent. (*Source:* CNN and *Nurse Week*)

Solution: *Method 1.* First, we state the problem in words.

In words: 2178   is   what percent   of   2593?

Translate: 2178   = x ·   2593
         or   2178   = $2593x$

Next, solve for x.

$$\frac{2178}{2593} = \frac{2593x}{2593}$$ Divide both sides by 2593.

$0.84 \approx x$ Divide and round to the nearest hundredth.
$84\% \approx x$ Write as a percent.

About 84% of nursing schools had an increase in applications or enrollment.

Method 2.

In words: 2178   is   what percent   of   2593?

         amount   percent   base

Translate: amount → $\dfrac{2178}{2593}$ = $\dfrac{p}{100}$ ← percent
         base →

Next, solve for p.

$2178 \cdot 100 = 2593 \cdot p$ Set cross products equal.
$217{,}800 = 2593p$ Multiply.

$$\frac{217{,}800}{2593} = \frac{2593p}{2593}$$ Divide both sides by 2593.

$84 \approx p$

About 84% of nursing schools had an increase in applications or enrollment.

◻ **Work Practice Problem 2**

PRACTICE PROBLEM 3

The freshmen class of 864 students is 32% of all students at Euclid University. How many students go to Euclid University?

EXAMPLE 3 Finding the Base Number of Absences

Mr. Percy, the principal at Slidell High School, counted 31 freshmen absent during a particular day. If this is 4% of the total number of freshmen, how many freshmen are there at Slidell High School?

Solution: *Method 1.* First we state the problem in words; then we translate.

In words: 31   is   4%   of   what number?

Translate: 31   =   4%   · x

Answers
2. 26% **3.** 2700

Next, we solve for x.

$$31 = 0.04 \cdot x \qquad \text{Write 4\% as a decimal.}$$

$$\frac{31}{0.04} = \frac{0.04x}{0.04} \qquad \text{Divide both sides by 0.04.}$$

$$775 = x \qquad \text{Simplify.}$$

There are 775 freshmen at Slidell High School.

Method 2.   First we state the problem in words; then we translate.

In words: 31   is   4%   of   what number?

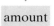

amount   percent   base

Translate: amount $\rightarrow \dfrac{31}{b} = \dfrac{4}{100} \leftarrow$ percent
             base $\rightarrow$

Next, we solve for b.

$$31 \cdot 100 = b \cdot 4 \qquad \text{Set cross products equal.}$$
$$3100 = 4b \qquad \text{Multiply.}$$
$$\frac{3100}{4} = \frac{4b}{4} \qquad \text{Divide both sides by 4.}$$
$$775 = b \qquad \text{Simplify.}$$

There are 775 freshmen at Slidell High School.

🔲 **Work Practice Problem 3**

EXAMPLE 4 **Finding the Base Increase in Licensed Drivers**

From 1994 to 2004, the number of licensed drivers on the road in the United States increased by 14%. In 1994, there were 175 million licensed drivers on the road.

a. Find the increase in licensed drivers from 1994 to 2004.
b. Find the number of licensed drivers on the road in 2004.
   (*Source:* Federal Highway Administration)

Solution: *Method 1.*   First we find the increase in licensed drivers.

In words: What number   is   14%   of   175?

Translate: x =   14%   ·   175

Continued on next page

PRACTICE PROBLEM 4

From 1994 to 2004, the number of registered vehicles on the road in the United States increased by 20%. In 1994, the number of vehicles on the road was 202 million.

a. Find the increase in the number of vehicles on the road in 2004.

b. Find the total number of registered vehicles on the road in 2004.

(*Source:* Federal Highway Administration)

Answers
4. a. 40.4 million   **b.** 242.4 million

Next, we solve for x.

$$x = 0.14 \cdot 175 \qquad \text{Write 14\% as a decimal.}$$
$$x = 24.5 \qquad \text{Multiply.}$$

a. The increase in licensed drivers is 24.5 million.

b. This means that the number of licensed drivers in 2004 was

$$\begin{array}{c} \text{Number of} \\ \text{licensed drivers} \\ \text{in 2004} \end{array} = \begin{array}{c} \text{Number of} \\ \text{licensed drivers} \\ \text{in 1994} \end{array} + \begin{array}{c} \text{Increase} \\ \text{in number of} \\ \text{licensed drivers} \end{array}$$

$$= 175 \text{ million} + 24.5 \text{ million}$$
$$= 199.5 \text{ million}$$

Method 2. First we find the increase in licensed drivers.

In words: <u>What number</u> is 14% of 175?

amount   percent   base

Translate: $\begin{array}{c} \text{amount} \rightarrow \\ \text{base} \rightarrow \end{array} \dfrac{a}{175} = \dfrac{14}{100} \begin{array}{c} \leftarrow \text{percent} \\ \\ \end{array}$

Next, we solve for a.

$$a \cdot 100 = 175 \cdot 14 \qquad \text{Set cross products equal.}$$
$$100a = 2450 \qquad \text{Multiply.}$$
$$\frac{100a}{100} = \frac{2450}{100} \qquad \text{Divide both sides by 100.}$$
$$a = 24.5 \qquad \text{Simplify.}$$

a. The increase in licensed drivers is 24.5 million.

b. This means that the licensed drivers in 2004 was

$$\begin{array}{c} \text{Number of} \\ \text{licensed drivers} \\ \text{in 2004} \end{array} = \begin{array}{c} \text{Number of} \\ \text{licensed drivers} \\ \text{in 1994} \end{array} + \begin{array}{c} \text{Increase} \\ \text{in number of} \\ \text{licensed drivers} \end{array}$$

$$= 175 \text{ million} + 24.5 \text{ million}$$
$$= 199.5 \text{ million}$$

◼ **Work Practice Problem 4**

Objective B Finding Percent Increase and Percent Decrease

We often use percents to show how much an amount has increased or decreased.

Suppose that the population of a town is 10,000 people and then it increases by 2000 people. The **percent of increase** is

$$\begin{array}{c} \text{amount of increase} \rightarrow \\ \text{original amount} \rightarrow \end{array} \dfrac{2000}{10,000} = 0.2 = 20\%$$

In general, we have the following.

Percent of Increase

$$\text{percent of increase} = \frac{\text{amount of increase}}{\text{original amount}}$$

Then write the quotient as a percent.

EXAMPLE 5 Finding Percent Increase

The number of applications for a mathematics scholarship at Yale increased from 34 to 45 in one year. What is the percent increase? Round to the nearest whole percent.

Solution: First we find the amount of increase by subtracting the original number of applicants from the new number of applicants.

$$\text{amount of increase} = 45 - 34 = 11$$

The amount of increase is 11 applicants. To find the percent of increase,

$$\text{percent of increase} = \frac{\text{amount of increase}}{\text{original amount}} = \frac{11}{34} \approx 0.32 = 32\%$$

The number of applications increased by about 32%.

▣ **Work Practice Problem 5**

✔ **Concept Check** A student is calculating the percent increase in enrollment from 180 students one year to 200 students the next year. Explain what is wrong with the following calculations:

~~$$\begin{aligned} \text{Amount of increase} &= 200 - 180 = 20 \\ \text{Percent of increase} &= \frac{20}{200} = 0.1 = 10\% \end{aligned}$$~~

Suppose that your income was $300 a week and then it decreased by $30. The **percent of decrease** is

$$\begin{aligned}\text{amount of decrease} &\rightarrow \\ \text{original amount} &\rightarrow\end{aligned} \frac{\$30}{\$300} = 0.1 = 10\%$$

Percent of Decrease

$$\text{percent of decrease} = \frac{\text{amount of decrease}}{\text{original amount}}$$

Then write the quotient as a percent.

EXAMPLE 6 Finding Percent Decrease

In response to a decrease in sales, a company with 1500 employees reduces the number of employees to 1230. What is the percent decrease?

Solution: First we find the amount of decrease by subtracting 1230 from 1500.

$$\text{amount of decrease} = 1500 - 1230 = 270$$

The amount of decrease is 270. To find the percent of decrease,

$$\text{percent of decrease} = \frac{\text{amount of decrease}}{\text{original amount}} = \frac{270}{1500} = 0.18 = 18\%$$

The number of employees decreased by 18%.

▣ **Work Practice Problem 6**

✔ **Concept Check** An ice cream stand sold 6000 ice cream cones last summer. This year the same stand sold 5400 cones. Was there a 10% increase, a 10% decrease, or neither? Explain.

PRACTICE PROBLEM 5

The number of people attending the local play, *Peter Pan,* increased from 285 on Friday to 333 on Saturday. Find the percent increase in attendance. Round to the nearest tenth of a percent.

Helpful Hint

Make sure that this number is the original number and not the new number.

PRACTICE PROBLEM 6

A town with a population of 20,200 in 1995 decreased to 18,483 in 2005. What was the percent decrease?

Answers
5. 16.8% **6.** 8.5%

✔ **Concept Check Answers**
To find the percent of increase, you have to divide the amount of increase (600) by the original amount (6000); 10% decrease.

Objective A *Solve. See Examples 1 through 4. If necessary, round percents to the nearest tenth and all other answers to the nearest whole.*

1. The Total Gym® provides weight resistance through adjustments of incline. The minimum weight resistance is 4% of the weight of the person using the Total Gym. Find the minimum weight resistance possible for a 220-pound man. (*Source:* Total Gym)

2. The maximum weight resistance for the Total Gym is 60% of the weight of the person using it. Find the maximum weight resistance possible for a 220-pound man. (See Exercise 1 if needed.)

3. An inspector found 24 defective bolts during an inspection. If this is 1.5% of the total number of bolts inspected, how many bolts were inspected?

4. A day care worker found 28 children absent one day during an epidemic of chicken pox. If this was 35% of the total number of children attending the day care center, how many children attend this day care center?

5. A student's cost for last semester at her Community College was $2700. She spent $378 of that on books. What percent of last semester's college costs was spent on books?

6. Pierre Sampeau belongs to his local food cooperative, where he receives a percentage of what he spends each year as a dividend. He spent $3850 last year at the food cooperative store and received a dividend of $154. What percent of his total spending at the food cooperative did he receive as a dividend?

7. Approximately 58% of films are rated R. If 940 films were recently rated, how many were rated R? (*Source*: Motion Picture Association of America)

8. Approximately 11% of films are rated PG-13. If 940 films were recently rated, how many were rated PG-13? (*Source*: Motion Picture Association of America)

9. Of the 535 members of the 109th U.S. Congress, 35 have attended a community college. Determine the percent of the members of the 109th Congress who attended a community college. (*Source:* American Association of Community Colleges)

10. Of the 54,300 veterinarians in private practice in the United States, 23,200 of them are female. Find the percent of female veterinarians in private practice in the United States. (*Source:* American Veterinary Medical Association)

11. A furniture company currently produces 6200 chairs per month. If production decreases by 8%, find the decrease and the new number of chairs produced each month.

12. The enrollment at a local college decreased by 5% over last year's enrollment of 7640. Find the decrease in enrollment and the current enrollment.

13. From 2002 to 2012, the number of people employed as physician assistants in the United States is expected to increase by 49%. The number of people employed as physician assistants in 2002 was 63,000. Find the predicted number of physician assistants in 2012. (*Source:* Bureau of Labor Statistics)

14. The state of North Dakota had the smallest percent increase in population, 0.5%, from the 1990 census to the 2000 census. In 1990, the population of North Dakota was 638,800. What was the population of North Dakota in 2000? (*Source:* U.S. Census Bureau)

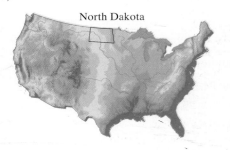

North Dakota

For each food described, find the percent of total calories from fat. If necessary, round to the nearest tenth of a percent.

15. Ranch dressing serving size of 2 tablespoons

	Calories
Total	40
From fat	20

16. Unsweetened cocoa powder serving size of 1 tablespoon

	Calories
Total	20
From fat	5

17.

Nutrition Facts
Serving Size 1 pouch (20g)
Servings Per Container 6

Amount Per Serving

Calories	80
Calories from fat	10

% Daily Value*

Total Fat 1g	**2%**
Sodium 45mg	**2%**
Total Carbohydrate 17g	**6%**
Sugars 9g	
Protein 0g	

Vitamin C	25%

Not a significant source of saturated fat, cholesterol, dietary fiber, vitamin A, calcium and iron.

*Percent Daily Values are based on a 2,000 calorie diet.

Artificial Fruit Snacks

18.

Nutrition Facts
Serving Size $\frac{1}{4}$ cup (33g)
Servings Per Container About 9

Amount Per Serving

Calories 190	Calories from Fat 130

% Daily Value*

Total Fat 16g	**24%**
Saturated Fat 3g	**16%**
Cholesterol 0mg	**0%**
Sodium 135mg	**6%**
Total Carbohydrate 9g	**3%**
Dietary Fiber 1g	**5%**
Sugars 2g	
Protein 5g	

Vitamin A 0% • Vitamin C 0%
Calcium 0% • Iron 8%

Peanut Mixture

19.

Nutrition Facts
Serving Size 18 crackers (29g)
Servings Per Container About 9

Amount Per Serving

Calories 120	Calories from Fat 35

% Daily Value*

Total Fat 4g	**6%**
Saturated Fat 0.5g	**3%**
Polyunsaturated Fat 0g	
Monounsaturated Fat 1.5g	
Cholesterol 0mg	**0%**
Sodium 220mg	**9%**
Total Carbohydrate 21g	**7%**
Dietary Fiber 2g	**7%**
Sugars 3g	
Protein 2g	

Vitamin A 0% • Vitamin C 0%
Calcium 2% • Iron 4%
Phosphorus 10%

Snack Crackers

20.

Nutrition Facts
Serving Size 28 crackers (31g)
Servings Per Container About 6

Amount Per Serving

Calories 130	Calories from Fat 35

% Daily Value*

Total Fat 4g	**6%**
Saturated Fat 2g	**10%**
Polyunsaturated Fat 1g	
Monounsaturated Fat 1g	
Cholesterol 0mg	**0%**
Sodium 470mg	**20%**
Total Carbohydrate 23g	**8%**
Dietary Fiber 1g	**4%**
Sugars 4g	
Protein 2g	

Vitamin A 0% • Vitamin C 0%
Calcium 0% • Iron 2%

Snack Crackers

Solve. Round money amounts to the nearest cent and all other amounts to the nearest tenth. See Example 4.

21. A family paid $26,250 as a down payment for a home. If this represents 15% of the price of the home, find the price of the home.

22. A banker learned that $842.40 is withheld from his monthly check for taxes and insurance. If this represents 18% of his total pay, find the total pay.

23. An owner of a repair service company estimates that for every 40 hours a repairperson is on the job, he can bill for only 78% of the hours. The remaining hours, the repairperson is idle or driving to or from a job. Determine the number of hours per 40-hour week the owner can bill for a repairperson.

24. A manufacturer of electronic components expects 1.04% of its products to be defective. Determine the number of defective components expected in a batch of 28,350 components. Round to the nearest whole component.

25. A car manufacturer announced that next year the price of a certain model of car would increase by 4.5%. This year the price is $19,286. Find the increase in price and the new price.

26. A union contract calls for a 6.5% salary increase for all employees. Determine the increase and the new salary that a worker currently making $58,500 under this contract can expect.

27. The city of Buckeye has a population of approximately 6950 adults. Use the circle graph for Example 1 of this section and predict the number of adults in Buckeye who prefer Vanilla ice cream.

28. The city of Hamlet has a population of approximately 4380 adults. Use the circle graph for Example 1 of this section and predict the number of adults in Hamlet who prefer Strawberry ice cream.

29. The population of Americans aged 65 and older was 35 million in 2000. That population is projected to increase by 80% by 2025. Find the increase and the projected 2025 population. (*Source:* Bureau of the Census)

30. From 2000 to 2010, the number of masters degrees awarded to women is projected to increase by 8.3%. The number of women who received masters degrees in 2000 was 265,000. Find the increase and the predicted number of women to be awarded masters degrees in 2010. (*Source:* U.S. National Center for Education Statistics)

Objective B *Find the amount of increase and the percent increase. See Example 5.*

	Original Amount	New Amount	Amount of Increase	Percent Increase
31.	50	80		
32.	8	12		
33.	65	117		
34.	68	170		

Find the amount of decrease and the percent decrease. See Example 6.

	Original Amount	New Amount	Amount of Decrease	Percent Decrease
35.	8	6		
36.	25	20		
37.	160	40		
38.	200	162		

Solve. Round percents to the nearest tenth, if necessary. See Examples 5 and 6.

39. There are 150 calories in a cup of whole milk and only 84 in a cup of skim milk. In switching to skim milk, find the percent decrease in number of calories per cup.

40. In reaction to a slow economy, the number of employees at a soup company decreased from 530 to 477. What was the percent decrease in the number of employees?

41. The number of cable TV systems recently decreased from 10,845 to 10,700. Find the percent decrease.

42. Before taking a typing course, Geoffry Landers could type 32 words per minute. By the end of the course, he was able to type 76 words per minute. Find the percent increase.

43. In 1940, the average size of a privately owned farm in the United States was 174 acres. By 2005, the average size of a privately owned farm in the United States had increased to 444 acres. What was the percent increase? (*Source:* National Agricultural Statistics Service)

44. In 1995, 722.9 million recorded music CDs were shipped to retailers in the United States. By 2005, this number had decreased to 705.4 million CDs. What was the percent decrease? (*Source:* Recording Industry Association of America)

45. In 2004, approximately 455,000 computer programmers were employed in the United States. By 2014, this number is expected to increase to 464,000 computer programmers. What is the percent increase? (*Source:* Bureau of Labor Statistics)

46. In 1994, there were 784 deaths from boating accidents in the United States. By 2004, the number of deaths from boating accidents had decreased to 676. What was the percent decrease? (*Source:* U.S. Coast Guard)

47. In 2006, there are 3,570 thousand elementary and secondary teachers employed in the United States. This number is expected to increase to 3,769 thousand teachers in 2012. What is the percent increase? (*Source:* National Center for Education Statistics)

48. In 2005, approximately 484,000 correctional officers were employed in the United States. By 2008, this number is expected to increase to 518,000 correctional officers. What is the percent increase? (*Source:* Bureau of Labor Statistics)

49. In 1995, there were 7151 indoor cinema sites in the United States. By 2005, this number had decreased to 5713 sites. What was the percent decrease? (*Source:* National Association of Theater Owners)

50. In 1994, approximately 16,000 occupational therapy assistants and aides were employed in the United States. According to one survey, by 2005, this number is expected to increase to 29,000 assistants and aides. What is the percent increase? (*Source:* Bureau of Labor Statistics)

51. The average soft-drink size has increased from 13.1 oz to 19.9 oz over the past two decades. Find the percent increase. (*Source:* University of North Carolina at Chapel Hill, *Journal for American Medicine*)

52. In 1999, discarded electronics, including obsolete computer equipment, accounted for 75,000 tons of solid waste per year in Massachusetts. By 2006, discarded electronic waste increased to 300,000 tons of waste per year in the state. Find the percent increase. (*Source:* Massachusetts Department of Environmental Protection)

53. The number of Americans who subscribed to cellular phone service was 110,000 thousand in 2000. By 2006, this number had increased to about 212,000 thousand. What was the percent increase? (*Source:* CTIA—The Wireless Association®)

54. The population of Tokyo is expected to decrease from 127,400 thousand in 2005 to 99,900 thousand in 2050. Find the percent decrease. (*Source:* International Programs Center, Bureau of the Census, U.S. Dept. of Commerce)

Review

Perform each indicated operation. See Sections 4.5, 5.2, and 5.3.

55. 0.12×38

56. $29.4 \div 0.7$

57. $9.20 + 1.98$

58. $78 - 19.46$

59. $-\dfrac{3}{8} - \dfrac{5}{12}$

60. $\left(-\dfrac{3}{8}\right)\left(-\dfrac{5}{12}\right)$

61. $2\dfrac{4}{5} \div 3\dfrac{9}{10}$

62. $2\dfrac{4}{5} - 3\dfrac{9}{10}$

Concept Extensions

63. If a number is increased by 100%, how does the increased number compare with the original number? Explain your answer.

64. In your own words, explain what is wrong with the following statement. "Last year we had 80 students attend. This year we have a 50% increase or a total of 160 students attend."

Explain what errors were made by each student when solving percent of increase or decrease problems and then correct the errors. "The population of a certain rural town was 150 in 1980, 180 in 1990, and 150 in 2000."

65. Find the percent of increase in population from 1980 to 1990.

Miranda's solution: Percent of increase $= \dfrac{30}{180} = 0.1\overline{6} \approx 16.7\%$

66. Find the percent of decrease in population from 1990 to 2000.

Jeremy's solution: Percent of decrease $= \dfrac{30}{150} = 0.20 = 20\%$

67. The percent of increase from 1980 to 1990 is the same as the percent of decrease from 1990 to 2000. True or false. Chris's answer: True because they had the same amount of increase as the amount of decrease.

7.5 PERCENT AND PROBLEM SOLVING: SALES TAX, COMMISSION, AND DISCOUNT

Objective **A** Calculating Sales Tax and Total Price

Percents are frequently used in the retail trade. For example, most states charge a tax on certain items when purchased. This tax is called a **sales tax,** and retail stores collect it for the state. Sales tax is almost always stated as a percent of the purchase price.

A 9% sales tax rate on a purchase of a $10 calculator gives a sales tax of

$$\text{sales tax} = 9\% \text{ of } \$10 = 0.09 \cdot \$10.00 = \$0.90$$

The total price to the customer would be

$$\underbrace{\text{purchase price}}_{\$10.00} \quad \text{plus} \quad \underbrace{\text{sales tax}}_{\$0.90} = \$10.90$$

This example suggests the following equations:

Sales Tax and Total Price

$$\text{sales tax} = \text{tax rate} \cdot \text{purchase price}$$
$$\text{total price} = \text{purchase price} + \text{sales tax}$$

In this section we round dollar amounts to the nearest cent.

EXAMPLE 1 Finding Sales Tax and Purchase Price

Find the sales tax and the total price on the purchase of an $85.50 atlas in a city where the sales tax rate is 7.5%.

Solution: The purchase price is $85.50 and the tax rate is 7.5%.

$$\underbrace{\text{sales tax}}_{} = \underbrace{\text{tax rate}}_{} \cdot \underbrace{\text{purchase price}}_{}$$

$$
\begin{aligned}
\text{sales tax} &= 7.5\% \cdot \$85.50 \\
&= 0.075 \cdot \$85.5 \quad \text{\small Write 7.5\% as a decimal.} \\
&\approx \$6.41 \quad \text{\small Rounded to the nearest cent}
\end{aligned}
$$

Thus, the sales tax is $6.41. Next find the total price.

$$\underbrace{\text{total price}}_{} = \underbrace{\text{purchase price}}_{} + \underbrace{\text{sales tax}}_{}$$

$$
\begin{aligned}
\text{total price} &= \$85.50 + \$6.41 \\
&= \$91.91
\end{aligned}
$$

The sales tax on $85.50 is $6.41, and the total price is $91.91.

Work Practice Problem 1

PRACTICE PROBLEM 1

If the sales tax rate is 8.5%, what is the sales tax and the total amount due on a $59.90 Goodgrip tire? (Round the sales tax to the nearest cent.)

Answer
1. tax: $5.09; total: $64.99

511

✔**Concept Check** The purchase price of a textbook is $50 and sales tax is 10%. If you are told by the cashier that the total price is $75, how can you tell that a mistake has been made?

PRACTICE PROBLEM 2

The sales tax on an $18,500 automobile is $1665. Find the sales tax rate.

EXAMPLE 2 **Finding a Sales Tax Rate**

The sales tax on a $406 Sony flat screen digital 27-inch television is $34.51. Find the sales tax rate.

Solution: Let r represent the unknown sales tax rate. Then

$$\text{sales tax} = \text{tax rate} \cdot \text{purchase price}$$

$$\$34.51 = r \cdot \$406$$

$$\frac{34.51}{406} = \frac{r \cdot 406}{406} \qquad \text{Divide both sides by 300.}$$

$$0.085 = r \qquad \text{Simplify.}$$

$$8.5\% = r \qquad \text{Write 0.085 as a percent.}$$

The sales tax rate is 8.5%.

■ **Work Practice Problem 2**

Objective **B** **Calculating Commissions**

A **wage** is payment for performing work. Hourly wage, commissions, and salary are some of the ways wages can be paid. Many people who work in sales are paid a commission. An employee who is paid a **commission** is paid a percent of his or her total sales.

Commission

$$\text{commission} = \text{commission rate} \cdot \text{sales}$$

PRACTICE PROBLEM 3

A sales representative for Office Product Copiers sold $47,632 worth of copy equipment and supplies last month. What is his commission for the month if he is paid a commission of 6.6% of his total sales for the month?

EXAMPLE 3 **Finding the Amount of Commission**

Sherry Souter, a real estate broker for Wealth Investments, sold a house for $214,000 last week. If her commission is 1.5% of the selling price of the home, find the amount of her commission.

Solution:

$$\text{commission} = \text{commission rate} \cdot \text{sales}$$

$$\text{commission} = 1.5\% \cdot \$214,000$$

$$= 0.015 \cdot \$214,000 \qquad \text{Write 1.5\% as 0.015.}$$

$$= \$3210 \qquad \text{Multiply.}$$

Answers

2. 9% **3.** $3143.71

✔ **Concept Check Answer**

Since $10\% = \frac{1}{10}$, the sales tax is $\frac{\$50}{10} = \$5$. The total price should have been $55.

Her commission on the house is $3210.

■ **Work Practice Problem 3**

EXAMPLE 4 Finding a Commission Rate

A salesperson earned $1560 for selling $13,000 worth of electronics equipment. Find the commission rate.

Solution: Let *r* stand for the unknown commission rate. Then

commission	=	commission rate	·	sales
↓		↓		↓
$1560	=	*r*		· $13,000

$$\frac{1560}{13,000} = r \quad \text{Divide 1560 by 13,000, the number multiplied by } r.$$

$$0.12 = r \quad \text{Simplify.}$$

$$12\% = r \quad \text{Write 0.12 as a percent.}$$

The commission rate is 12%.

■ **Work Practice Problem 4**

Objective ◐ **Calculating Discount and Sale Price**

Suppose that an item that normally sells for $40 is on sale for 25% off. This means that the **original price** of $40 is reduced, or **discounted,** by 25% of $40, or $10. The **discount rate** is 25%, the **amount of discount** is $10, and the **sale price** is $40 − $10, or $30. Study the diagram below to visualize these terms.

To calculate discounts and sale prices, we can use the following equations:

Discount and Sale Price

amount of discount = discount rate · original price

sale price = original price − amount of discount

PRACTICE PROBLEM 4
A salesperson earns $645 for selling $4300 worth of appliances. Find the commission rate.

Answer
4. 15%

PRACTICE PROBLEM 5

A discontinued washer and dryer combo is advertised on sale for 35% off the regular price of $700. Find the amount of discount and the sale price.

EXAMPLE 5 **Finding a Discount and a Sale Price**

An electric rice cooker that normally sells for $65 is on sale for 25% off. What is the amount of discount and what is the sale price?

Solution: First we find the amount of discount, or simply the discount.

| amount of discount | = | discount rate | · | original price |

$$\text{amount of discount} = 25\% \cdot \$65$$
$$= 0.25 \cdot \$65 \quad \text{Write 25\% as 0.25.}$$
$$= \$16.25 \quad \text{Multiply.}$$

The discount is $16.25. Next, find the sale price.

| sale price | = | original price | − | discount |

$$\text{sale price} = \$65 - \$16.25$$
$$= \$48.75 \quad \text{Subtract.}$$

The sale price is $48.75.

■ **Work Practice Problem 5**

Answer

5. $245; $455

Vocabulary and Readiness Check

Use the choices below to fill in each blank.

   amount of discount   sale price   sales tax

   commission   total price

1. _____ = tax rate · purchase price.

2. _____ = purchase price + sales tax.

3. _____ = commission rate · sales.

4. _____ = discount rate · original price.

5. _____ = original price − amount of discount.

7.5 EXERCISE SET

Student Solutions Manual   PH Math/Tutor Center   CD/Video for Review   Math XL MathXL®   MyMathLab MyMathLab

Objective A *Solve. See Examples 1 and 2.*

1. What is the sales tax on a jacket priced at $150 if the sales tax rate is 5%?

2. If the sales tax rate is 6%, find the sales tax on a microwave oven priced at $188.

3. The purchase price of a camcorder is $799. What is the total price if the sales tax rate is 7.5%?

4. A stereo system has a purchase price of $426. What is the total price if the sales tax rate is 8%?

5. A new large-screen television has a purchase price of $4790. If the sales tax on this purchase is $335.30, find the sales tax rate.

6. The sales tax on the purchase of a $6800 used car is $374. Find the sales tax rate.

7. The sales tax on a table saw is $10.20.

   a. What is the purchase price of the table saw (before tax) if the sales tax rate is 8.5%? (*Hint:* Use the sales tax equation and insert the replacement values.)

   b. Find the total price of the table saw.

8. The sales tax on a one-half-carat diamond ring is $76.

   a. Find the purchase price of the ring (before tax) if the sales tax rate is 9.5%. (See the hint for Exercise 7a.)

   b. Find the total price of the ring.

9. A gold and diamond bracelet sells for $1800. Find the sales tax and the total price if the sales tax rate is 6.5%.

10. The purchase price of a personal computer is $1890. If the sales tax rate is 8%, what is the sales tax and the total price?

11. The sales tax on the purchase of a futon is $24.25. If the tax rate is 5%, find the purchase price of the futon.

12. The sales tax on the purchase of a TV-DVD combination is $32.85. If the tax rate is 9%, find the purchase price of the TV-DVD.

13. The sales tax is $98.70 on a stereo sound system purchase of $1645. Find the sales tax rate.

14. The sales tax is $103.50 on a necklace purchase of $1150. Find the sales tax rate.

15. A cell phone costs $210, a battery recharger costs $15, and batteries cost $5. What is the sales tax and total price for purchasing these items if the sales tax rate is 7%?

16. Ms. Warner bought a blouse for $35, a skirt for $55, and a blazer for $95. Find the sales tax and the total price she paid, given a sales tax rate of 6.5%.

Objective **B** *Solve. See Examples 3 and 4.*

17. A sales representative for a large furniture warehouse is paid a commission rate of 4%. Find her commission if she sold $1,329,401 worth of furniture last year.

18. Rosie Davis-Smith is a beauty consultant for a home cosmetic business. She is paid a commission rate of 12.8%. Find her commission if she sold $1638 in cosmetics last month.

19. A salesperson earned a commission of $1380.40 for selling $9860 worth of paper products. Find the commission rate.

20. A salesperson earned a commission of $3575 for selling $32,500 worth of books to various bookstores. Find the commission rate.

21. How much commission will Jack Pruet make on the sale of a $325,900 house if he receives 1.5% of the selling price?

22. Frankie Lopez sold $9638 of jewelry this week. Find her commission for the week if she receives a commission rate of 5.6%.

23. A real estate agent earned a commission of $5565 for selling a house. If his rate is 3%, find the selling price of the house. (*Hint:* Use the commission equation and insert the replacement values.)

24. A salesperson earned $1750 for selling fertilizer. If her commission rate is 7%, find the selling price of the fertilizer (*Hint:* See Exercise 23.).

Objective **C** *Find the amount of discount and the sale price. See Example 5.*

	Original Price	Discount Rate	Amount of Discount	Sale Price
25.	$89	10%		
26.	$74	20%		
27.	$196.50	50%		
28.	$110.60	40%		
29.	$410	35%		
30.	$370	25%		
31.	$21,700	15%		
32.	$17,800	12%		

33. A $300 fax machine is on sale for 15% off. Find the amount of discount and the sale price.

34. A $4295 designer dress is on sale for 30% off. Find the amount of discount and the sale price.

Objectives Ⓐ Ⓑ **Mixed Practice** *Complete each table.*

	Purchase Price	Tax Rate	Sales Tax	Total Price
35.	$305	9%		
36.	$243	8%		
37.	$56	5.5%		
38.	$65	8.4%		

	Sale	Commission Rate	Commission
39.	$235,800	3%	
40.	$195,450	5%	
41.	$17,900		$1432
42.	$25,600		$2304

Review

Multiply. See Sections 4.3, 5.3, and 5.5.

43. $2000 \cdot \dfrac{3}{10} \cdot 2$

44. $500 \cdot \dfrac{2}{25} \cdot 3$

45. $400 \cdot \dfrac{3}{100} \cdot 11$

46. $1000 \cdot \dfrac{1}{20} \cdot 5$

47. $600 \cdot 0.04 \cdot \dfrac{2}{3}$

48. $6000 \cdot 0.06 \cdot \dfrac{3}{4}$

Concept Extensions

Solve. See the Concept Check in this section.

49. Your purchase price is $68 and the sales tax rate is 9.5%. Round each amount and use the rounded amounts to estimate the total price. Choose the best estimate.

    a. $105   **b.** $58   **c.** $93   **d.** $77

50. Your purchase price is $200 and the tax rate is 10%. Choose the best estimate of the total price.

    a. $190   **b.** $210   **c.** $220   **d.** $300

Tipping *One very useful application of percent is mentally calculating a tip. Recall that to find 10% of a number, simply move the decimal point one place to the left. To find 20% of a number, just double 10% of the number. To find 15% of a number, find 10% and then add to that number half of the 10% amount. Mentally fill in the chart below. To do so, start by rounding the bill amount to the nearest dollar.*

	Tipping Chart		
Bill Amount	**10%**	**15%**	**20%**
51. $40.21			
52. $15.89			
53. $72.17			
54. $9.33			

55. Suppose that the original price of a shirt is $50. Which is better, a 60% discount or a discount of 30% followed by a discount of 35% of the reduced price? Explain your answer.

56. Which is better, a 30% discount followed by an additional 25% off or a 20% discount followed by an additional 40% off? To see, suppose an item costs $100 and calculate each discounted price. Explain your answer.

57. A diamond necklace sells for $24,966. If the tax rate is 7.5%, find the total price.

58. A house recently sold for $562,560. The commission rate on the sale is 5.5%. If the real estate agent is to receive 60% of the commission, find the amount received by the agent.

STUDY SKILLS BUILDER

Are You Familiar with Your Textbook Supplements?

Below is a review of some of the student supplements available for additional study. Check to see if you are using the ones most helpful to you.

- Chapter Test Prep Videos on CD. This material is found with your textbook and is fully explained there. The CD contains video clip solutions to the Chapter Test exercises in this text and are excellent help when studying for chapter tests.

- Lecture Videos on CD-ROM. These video segments are keyed to each section of the text. The material is presented by me, Elayn Martin-Gay, and I have placed a 💿 by the exercises in the text that I have worked on the video.

- The *Student Solutions Manual*. This contains worked out solutions to odd-numbered exercises as well as every exercise in the Integrated Reviews, Chapter Reviews, Chapter Tests, and Cumulative Reviews.

- Prentice Hall Tutor Center. Mathematic questions may be phoned, faxed, or e-mailed to this center.

- MyMathLab, MathXL, and Interact Math. These are computer and Internet tutorials. This supplement may already be available to you somewhere on campus, for example at your local learning resource lab. Take a moment and find the name and location of any such lab on campus.

  As usual, your instructor is your best source of information.

Let's see how you are doing with textbook supplements.

1. Name one way the Lecture Videos can be helpful to you.

2. Name one way the Chapter Test Prep Video can help you prepare for a chapter test.

3. List any textbook supplements that you have found useful.

4. Have you located and visited a learning resource lab located on your campus?

5. List the textbook supplements that are currently housed in your campus's learning resource lab.

7.6 PERCENT AND PROBLEM SOLVING: INTEREST

Objectives

A Calculate Simple Interest.

B Calculate Compound Interest.

Objective A Calculating Simple Interest

Interest is money charged for using other people's money. When you borrow money, you pay interest. When you loan or invest money, you earn interest. The money borrowed, loaned, or invested is called the **principal amount,** or simply **principal.** Interest is normally stated in terms of a percent of the principal for a given period of time. The **interest rate** is the percent used in computing the interest. Unless stated otherwise, *the rate is understood to be per year.* When the interest is computed on the original principal, it is called **simple interest.** Simple interest is calculated using the following equation:

Simple Interest

$$\text{Simple Interest} = \text{Principal} \cdot \text{Rate} \cdot \text{Time}$$
$$I = P \cdot R \cdot T$$

where the rate is understood to be per year and time is in years.

EXAMPLE 1 Finding Simple Interest

Find the simple interest after 2 years on $500 at an interest rate of 12%.

Solution: In this example, $P = \$500$, $R = 12\%$, and $T = 2$ years. Replace the variables with values in the formula $I = PRT$.

$I = P \cdot R \cdot T$
$I = \$500 \cdot 12\% \cdot 2$ Let $P = \$500$, $R = 12\%$, and $T = 2$.
$ = \$500 \cdot (0.12) \cdot 2$ Write 12% as a decimal.
$ = \$120$ Multiply.

The simple interest is $120.

🔲 **Work Practice Problem 1**

If time is not given in years, we need to convert the given time to years.

EXAMPLE 2 Finding Simple Interest

Ivan Borski borrowed $2400 at 10% simple interest for 8 months to buy a used Toyota Corolla. Find the simple interest he paid.

Solution: Since there are 12 months in a year, we first find what part of a year 8 months is.

$$8 \text{ months} = \frac{8}{12} \text{ year} = \frac{2}{3} \text{ year}$$

Now we find the simple interest.

$I = P \cdot R \cdot T$
$ = \$2400 \cdot (0.10) \cdot \dfrac{2}{3}$ Let $P = \$2400$, $R = 10\%$ or 0.10, and $T = \dfrac{2}{3}$.
$ = \$160$

The interest on Ivan's loan is $160.

🔲 **Work Practice Problem 2**

PRACTICE PROBLEM 1

Find the simple interest after 5 years on $875 at an interest rate of 7%.

PRACTICE PROBLEM 2

A student borrowed $1500 for 9 months on her credit card at a simple interest rate of 20%. How much interest did she pay?

Answers
1. $306.25 2. $225

519

✔**Concept Check** Suppose in Example 2 you had obtained an answer of $16,000. How would you know that you had made a mistake in this problem?

When money is borrowed, the borrower pays the original amount borrowed, or the principal, as well as the interest. When money is invested, the investor receives the original amount invested, or the principal, as well as the interest. In either case, the **total amount** is the sum of the principal and the interest.

Finding the Total Amount of a Loan or Investment

total amount (paid or received) = principal + interest

EXAMPLE 3 Finding the Total Amount of an Investment

An accountant invested $2000 at a simple interest rate of 10% for 2 years. What total amount of money will she have from her investment in 2 years?

Solution: First we find her interest.

$$I = P \cdot R \cdot T$$
$$= \$2000 \cdot (0.10) \cdot 2 \quad \text{Let } P = \$2000, R = 10\% \text{ or } 0.10, \text{ and } T = 2.$$
$$= \$400$$

The interest is $400.

Next, we add the interest to the principal.

total amount	=	principal	+	interest
↓		↓		↓
total amount	=	$2000	+	$400
	=	$2400		

After 2 years, she will have a total amount of $2400.

▢ **Work Practice Problem 3**

✔**Concept Check** Which investment would earn more interest: an amount of money invested at 8% interest for 2 years, or the same amount of money invested at 8% for 3 years? Explain.

Objective B Calculating Compound Interest

Recall that simple interest depends on the original principal only. Another type of interest is compound interest. **Compound interest** is computed on not only the principal, but also on the interest already earned in previous compounding periods. Compound interest is used more often than simple interest.

Let's see how compound interest differs from simple interest. Suppose that $2000 is invested at 7% interest **compounded annually** for 3 years. This means that interest is added to the principal at the end of each year and that next year's interest is computed on this new amount. In this section, we round dollar amounts to the nearest cent.

PRACTICE PROBLEM 3

If $2100 is borrowed at a simple interest rate of 13% for 6 months, find the total amount paid.

Answer
3. $2236.50

✔ Concept Check Answers
$16,000 is too much interest. Answers may vary; 8% for 3 years. Since the interest rate is the same, the longer you keep the money invested, the more interest you earn.

	Amount at Beginning of Year	Principal	·	Rate	·	Time	= Interest	Amount at End of Year
1st year	$2000	$2000	·	0.07	·	1	= $140	$2000 + 140 = $2140
2nd year	$2140	$2140	·	0.07	·	1	= $149.80	$2140 + 149.80 = $2289.80
3rd year	$2289.80	$2289.80	·	0.07	·	1	= $160.29	$2289.80 + 160.29 = $2450.09

The compound interest earned can be found by

total amount	−	original principal	=	compound interest
↓		↓		↓
$2450.09	−	$2000	=	$450.09

The simple interest earned would have been

principal	·	rate	·	time	=	interest
↓		↓		↓		↓
$2000	·	0.07	·	3	=	$420

Since compound interest earns "interest on interest," compound interest earns more than simple interest.

Computing compound interest using the method above can be tedious. We can use a calculator and the compound interest formula below to compute compound interest more quickly.

Compound Interest Formula

The total amount A in an account is given by

$$A = P\left(1 + \frac{r}{n}\right)^{n \cdot t}$$

where P is the principal, r is the interest rate written as a decimal, t is the length of time in years, and n is the number of times compounded per year.

Let's use this formula to check our earlier compound interest calculations.

EXAMPLE 4 $2000 is invested at 7% interest compounded annually. Find the total amount after 3 years.

Solution: "Compounded annually" means 1 time a year, so

$n = 1$. Also, $P =$ $2000, $r = 7\% = 0.07$, and $t = 3$ years.

$$A = P\left(1 + \frac{r}{n}\right)^{n \cdot t}$$

$$= 2000\left(1 + \frac{0.07}{1}\right)^{1 \cdot 3}$$

$$= 2000(1.07)^3 \qquad \longleftarrow$$

$$\approx 2450.09 \qquad \text{Round to 2 decimal places.}$$

The total amount at the end of 3 years is $2450.09.

■ **Work Practice Problem 4**

EXAMPLE 5 **Finding Total Amount Received on an Investment**

$4000 is invested at 5.3% compounded quarterly for 10 years. Find the total amount at the end of 10 years.

Solution: "Compounded quarterly" means 4 times a year, so

$n = 4$. Also, $P =$ $4000, $r = 5.3\% = 0.053$, and $t = 10$ years.

Continued on next page

PRACTICE PROBLEM 4

$3000 is invested at 4% interest compounded annually. Find the total amount after 6 years.

Helpful Hint
Remember order of operations. **First** evaluate $(1.07)^3$, then multiply by 2000.

PRACTICE PROBLEM 5

$5500 is invested at $6\frac{1}{4}\%$ compounded *daily* for 5 years. Find the total amount at the end of 5 years. (Use 1 year = 365 days.)

Answers
4. $3795.96   **5.** $7517.41

$$A = P\left(1 + \frac{r}{n}\right)^{n \cdot t}$$

$$= 4000\left(1 + \frac{0.053}{4}\right)^{4 \cdot 10}$$

$$= 4000(1.01325)^{40}$$

$$\approx 6772.12$$

The total amount after 10 years is \$6772.12.

🔲 **Work Practice Problem 5**

🔢 **CALCULATOR EXPLORATIONS** Compound Interest Formula

For a review of using your calculator to evaluate compound interest, see this box.

Let's review the calculator keys pressed to evaluate the expression in Example 5,

$$4000\left(1 + \frac{0.053}{4}\right)^{40}$$

To evaluate, press the keys

[4000] [×] [( ] [1] [+] [0.053] [÷] [4] [ )] [y^x]
or [∧] [40] [=] or [ENTER]. The display will read
[6772.117549]. Rounded to 2 decimal places, this is 6772.12.

Find the compound interest.

1. \$600, 5 years, 9%, compounded quarterly

2. \$10,000, 15 years, 4%, compounded daily

3. \$1200, 20 years, 11%, compounded annually

4. \$5800, 1 year, 7%, compounded semiannually

5. \$500, 4 years, 6%, compounded quarterly.

6. \$2500, 19 years, 5%, compounded daily.

Vocabulary and Readiness Check

Use the choices below to fill in each blank. Choices may be used more than once.

total amount   simple   compound

1. To calculate _____ interest, use $I = P \cdot R \cdot T$.

2. To calculate _____ interest, use $A = P\left(1 + \dfrac{r}{n}\right)^{n \cdot t}$.

3. _____ interest is computed on not only the original principal, but on interest already earned in previous compounding periods.

4. When interest is computed on the original principal only, it is called _____ interest.

5. _____ (paid or received) = principal + interest.

7.6 EXERCISE SET

FOR EXTRA HELP

Student Solutions Manual   PH Math/Tutor Center   CD/Video for Review   MathXL®   MyMathLab

Objective A *Find the simple interest. See Examples 1 and 2.*

	Principal	Rate	Time
1.	$200	8%	2 years
3.	$160	11.5%	4 years
5.	$5000	10%	$1\frac{1}{2}$ years
7.	$375	18%	6 months
9.	$2500	16%	21 months

	Principal	Rate	Time
2.	$800	9%	3 years
4.	$950	12.5%	5 years
6.	$1500	14%	$2\frac{1}{4}$ years
8.	$775	15%	8 months
10.	$1000	10%	18 months

Solve. See Examples 1 through 3.

11. A company borrows $162,500 for 5 years at a simple interest of 12.5% to buy an airplane. Find the interest paid on the loan.

12. $265,000 is borrowed to buy a house. If the simple interest rate on the 30-year loan is 8.25%, find the interest paid on the loan.

13. A money market fund advertises a simple interest rate of 9%. Find the total amount received on an investment of $5000 for 15 months.

14. The Real Service Company takes out a 270-day (9-month) short-term, simple interest loan of $4500 to finance the purchase of some new equipment. If the interest rate is 14%, find the total amount that the company pays back.

15. Marsha borrows $8500 and agrees to pay it back in 4 years. If the simple interest rate is 17%, find the total amount she pays back.

16. An 18-year-old is given a high school graduation gift of $2000. If this money is invested at 8% simple interest for 5 years, find the total amount.

Objective B *Find the total amount in each compound interest account. See Examples 4 and 5.*

17. $6150 is compounded semiannually at a rate of 14% for 15 years.

18. $2060 is compounded annually at a rate of 15% for 10 years.

19. $1560 is compounded daily at a rate of 8% for 5 years.

20. $1450 is compounded quarterly at a rate of 10% for 15 years.

21. $10,000 is compounded semiannually at a rate of 9% for 20 years.

22. $3500 is compounded daily at a rate of 8% for 10 years.

23. $2675 is compounded annually at a rate of 9% for 1 year.

24. $6375 is compounded semiannually at a rate of 10% for 1 year.

25. $2000 is compounded annually at a rate of 8% for 5 years.

26. $2000 is compounded semiannually at a rate of 8% for 5 years.

27. $2000 is compounded quarterly at a rate of 8% for 5 years.

28. $2000 is compounded daily at a rate of 8% for 5 years.

Review

Find the perimeter of each figure. See Section 1.3.

△ **29.**

Rectangle | 6 yards

10 yards

△ **30.**

18 centimeters

Triangle

16 centimeters   12 centimeters

△ **31.**

Regular pentagon— All sides are same length   7 meters

△ **32.**

Square | 21 miles

Perform the indicated operations. See Sections 4.3 through 4.6.

33. $\dfrac{x}{4} + \dfrac{x}{5}$

34. $-\dfrac{x}{4} \div \left(-\dfrac{x}{5}\right)$

35. $\left(\dfrac{2}{3}\right)\left(-\dfrac{1}{3}\right) - \left(\dfrac{9}{10}\right)\left(\dfrac{2}{5}\right)$

36. $\dfrac{3}{11} \div \dfrac{9}{22} \cdot \dfrac{1}{3}$

Concept Extensions

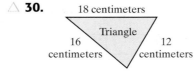 **37.** Review Exercises 25 through 28. As the number of compoundings increases per year, how is the total amount affected?

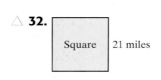 **38.** Explain how to find the amount of interest in a compounded account.

✎ **39.** Compare the following accounts: Account 1: $1000 is invested for 10 years at a simple interest rate of 6%. Account 2: $1000 is compounded semiannually at a rate of 6% for 10 years. Discuss how the interest is computed for each account. Determine which account earns more interest. Why?

CHAPTER 7   Group Activity

Fastest-Growing Occupations

According to U.S. Bureau of Labor Statistics projections, the careers listed below are the top ten fastest-growing jobs, ranked by expected percent increase through the year 2014.

Occupation	Employment in 2004	Percent Increase from 2004 to 2014	Expected Employment In 2014
Home health aides	624,000	56.0%	
Medical Assistants	387,000	52.1%	
Computer software engineers, applications	460,000	48.4%	
Personal and home care aides	701,000	41.0%	
Postsecondary teachers	1,628,000	32.2%	
Registered nurses	2,394,000	29.3%	
Customer service representatives	2,063,000	22.8%	
Accountants and auditors	1,176,000	22.4%	
Nursing aides, orderlies, and attendants	1,455,000	22.3%	
Receptionists and information clerks	1,133,000	21.7%	
(*Source:* Bureau of Labor Statistics)			

What do most of these fast-growing occupations have in common? They require knowledge of math! For some careers, such as registered nurses, accountants, and computer engineers, the ways math is used on the job may be obvious. For other occupations, the use of math may not be quite as apparent. However, tasks common to many jobs—filling in a time sheet, writing up an expense or mileage report, planning a budget, figuring a bill, ordering supplies, and even making a work schedule—all require math.

This activity may be completed by working in groups or individually.

1. List the top five occupations by order of employment figures for 2004.

2. Using the 2004 employment figures and the percent increase from 2004 to 2014, find the expected 2014 employment figures for each occupation listed in the table. Round to the nearest thousand.

3. List the top five occupations by order of employment figures for 2014. Did the order change at all from 2004? Explain.

Chapter 7 Vocabulary Check

Fill in each blank with one of the words or phrases listed below. Some words may be used more than once.

percent   sales tax   is   0.01 $\dfrac{1}{100}$ amount of discount   percent of decrease   total price

base   of   amount   100%   compound interest   percent of increase   sale price   commission

1. In a mathematical statement, _____ usually means "multiplication."
2. In a mathematical statement, _____ means "equals."
3. _____ means "per hundred."
4. _____ is computed not only on the principal, but also on interest already earned in previous compounding periods.
5. In the percent proportion $\dfrac{\rule{3cm}{0.4pt}}{\rule{3cm}{0.4pt}} = \dfrac{\text{percent}}{100}$.
6. To write a decimal or fraction as a percent, multiply by _____.
7. The decimal equivalent of the % symbol is _____.
8. The fraction equivalent of the % symbol is _____.
9. The percent equation is _____ $\cdot$ percent = _____.
10. _____ $= \dfrac{\text{amount of decrease}}{\text{original amount}}$.
11. _____ $= \dfrac{\text{amount of increase}}{\text{original amount}}$.
12. _____ = tax rate $\cdot$ purchase price.
13. _____ = purchase price + sales tax.
14. _____ = commission rate $\cdot$ sales.
15. _____ = discount rate $\cdot$ original price.
16. _____ = original price − amount of discount.

Helpful Hint

Are you preparing for your test? Don't forget to take the Chapter 7 Test on page 533. Then check your answers at the back of the text and use the Chapter Test Prep Video CD to see the fully worked-out solutions to any of the exercises you want to review.

7 Chapter Highlights

DEFINITIONS AND CONCEPTS	EXAMPLES
Section 7.1 Percents, Decimals, and Fractions	
Percent means "per hundred." The % symbol denotes percent.	$51\% = \dfrac{51}{100}$ 51 per 100 $7\% = \dfrac{7}{100}$ 7 per 100
To write a percent as a decimal, replace the % symbol with its decimal equivalent, 0.01, and multiply. **To write a decimal as a percent,** multiply by 100%.	$32\% = 32(0.01) = 0.32$ $0.08 = 0.08(100\%) = 08.\% = 8\%$
To write a percent as a fraction, replace the % symbol with its fraction equivalent, $\dfrac{1}{100}$, and multiply. **To write a fraction as a percent,** multiply by 100%.	$25\% = 25 \cdot \dfrac{1}{100} = \dfrac{25}{100} = \dfrac{\overset{1}{\cancel{25}}}{4 \cdot \cancel{25}} = \dfrac{1}{4}$ $\dfrac{1}{6} = \dfrac{1}{6} \cdot 100\% = \dfrac{1}{6} \cdot \dfrac{100}{1}\% = \dfrac{100}{6}\% = 16\dfrac{2}{3}\%$

DEFINITIONS AND CONCEPTS	**EXAMPLES**

Section 7.2   Solving Percent Problems with Equations

Three key words in the statement of a percent problem are **of,** which means multiplication ($\cdot$) **is,** which means equals ($=$) **what** (or some equivalent word or phrase), which stands for the unknown (x)	Solve: $\begin{array}{ccccc} 6 & \text{is} & 12\% & \text{of} & \text{what number?} \\ \downarrow & \downarrow & \downarrow & \downarrow & \downarrow \end{array}$ $6 = 12\% \cdot x$ $6 = 0.12 \cdot x \quad$ Write 12% as a decimal. $\dfrac{6}{0.12} = \dfrac{0.12 \cdot x}{0.12} \quad$ Divide both sides by 0.12. $50 = x$ Thus, 6 is 12% of 50.

Section 7.3   Solving Percent Problems with Proportions

PERCENT PROPORTION $\dfrac{\text{amount}}{\text{base}} = \dfrac{\text{percent}}{100} \quad \leftarrow \text{always 100}$ or $\begin{array}{l} \text{amount} \rightarrow \\ \text{base} \rightarrow \end{array} \dfrac{a}{b} = \dfrac{p}{100} \leftarrow \text{percent}$	Solve: 20.4 is what percent of 85? $\begin{array}{ccc} \searrow & \downarrow & \searrow \\ \boxed{\text{amount}} & \boxed{\text{percent}} & \boxed{\text{base}} \end{array}$ $\begin{array}{l} \text{amount} \rightarrow \\ \text{base} \rightarrow \end{array} \dfrac{20.4}{85} = \dfrac{p}{100} \leftarrow \text{percent}$ $20.4 \cdot 100 = 85 \cdot p \quad$ Set cross products equal. $2040 = 85 \cdot p \quad$ Multiply. $\dfrac{2040}{85} = \dfrac{85 \cdot p}{85} \quad$ Divide both sides by 85. $24 = p \quad$ Simplify. Thus, 20.4 is 24% of 85.

Section 7.4   Applications of Percent

PERCENT INCREASE $\text{percent increase} = \dfrac{\text{amount of increase}}{\text{original amount}}$ **PERCENT DECREASE** $\text{percent decrease} = \dfrac{\text{amount of decrease}}{\text{original amount}}$	A town with a population of 16,480 decreased to 13,870 over a 12-year period. Find the percent decrease. Round to the nearest whole percent. $\text{amount of decrease} = 16{,}480 - 13{,}870$ $\qquad\qquad\qquad\quad = 2610$ $\text{percent decrease} = \dfrac{\text{amount of decrease}}{\text{original amount}}$ $\qquad\qquad\qquad\quad = \dfrac{2610}{16{,}480} \approx 0.16$ $\qquad\qquad\qquad\quad = 16\%$ The town's population decreased by about 16%.

DEFINITIONS AND CONCEPTS	EXAMPLES
Section 7.5 Percent and Problem Solving: Sales Tax, Commission, and Discount	

SALES TAX

$$\text{sales tax} = \text{sales tax rate} \cdot \text{purchase price}$$
$$\text{total price} = \text{purchase price} + \text{sales tax}$$

Find the sales tax and the total price of a purchase of $42.00 if the sales tax rate is 9%.

$$\boxed{\text{sales tax}} = \boxed{\text{sales tax rate}} \cdot \boxed{\text{purchase price}}$$
$$\downarrow \qquad\qquad \downarrow \qquad\qquad \downarrow$$
$$\text{sales tax} = \qquad 9\% \qquad \cdot \qquad \$42$$
$$= \quad 0.09 \cdot \$42$$
$$= \quad \$3.78$$

The total price is

$$\boxed{\text{total price}} = \boxed{\text{purchase price}} + \boxed{\text{sales tax}}$$
$$\downarrow \qquad\qquad \downarrow \qquad\qquad \downarrow$$
$$\text{total price} = \qquad \$42.00 \qquad + \qquad \$3.78$$
$$= \qquad \$45.78$$

The total price is $45.78.

COMMISSION

$$\text{commission} = \text{commission rate} \cdot \text{sales}$$

A salesperson earns a commission of 3%. Find the commission from sales of $12,500 worth of appliances.

$$\boxed{\text{commission}} = \boxed{\text{commission rate}} \cdot \boxed{\text{sales}}$$
$$\downarrow \qquad\qquad \downarrow \qquad\qquad \downarrow$$
$$\text{commission} = \qquad 3\% \qquad \cdot \$12,500$$
$$= \quad 0.03 \cdot 12,500$$
$$= \quad \$375$$

The commission is $375.

DISCOUNT AND SALE PRICE

$$\text{amount of discount} = \text{discount rate} \cdot \text{original price}$$
$$\text{sale price} = \text{original price} - \text{amount of discount}$$

A suit is priced at $320 and is on sale today for 25% off. What is the sale price?

$$\boxed{\begin{array}{c}\text{amount of}\\\text{discount}\end{array}} = \boxed{\text{discount rate}} \cdot \boxed{\text{original price}}$$
$$\downarrow \qquad\qquad \downarrow \qquad\qquad \downarrow$$
$$\begin{array}{c}\text{amount of}\\\text{discount}\end{array} = \qquad 25\% \qquad \cdot \qquad \$320$$
$$= \quad 0.25 \cdot 320$$
$$= \quad \$80$$

$$\boxed{\text{sale price}} = \boxed{\text{original price}} - \boxed{\begin{array}{c}\text{amount of}\\\text{discount}\end{array}}$$
$$\downarrow \qquad\qquad \downarrow \qquad\qquad \downarrow$$
$$\text{sale price} = \qquad \$320 \qquad - \qquad \$80$$
$$= \qquad \$240$$

The sale price is $240.

DEFINITIONS AND CONCEPTS	**EXAMPLES**

Section 7.6   Percent and Problem Solving: Interest

SIMPLE INTEREST

$$\text{interest} = \text{principal} \cdot \text{rate} \cdot \text{time}$$

where the rate is understood to be per year, unless told otherwise.

COMPOUND INTEREST FORMULA
The total amount A in an account is

$$A = P\left(1 + \frac{r}{n}\right)^{n \cdot t}$$

where P is the principal, r is the interest rate written as a decimal, t is time in years, and n is the number of times compounded per year.

Find the simple interest after 3 years on $800 at an interest rate of 5%.

interest	=	principal	·	rate	·	time
↓		↓		↓		↓
interest	=	$800	·	5%	·	3
	=	$800 · 0.05 · 3		Write 5% as 0.05.		
	=	$120		Multiply.		

The interest is $120.

$800 is invested at 5% compounded quarterly for 10 years. Find the total amount at the end of 10 years.

"Compounded quarterly" means $n = 4$. Also, $P = \$800$, $r = 5\% = 0.05$, and $t = 10$ years.

$$A = P\left(1 + \frac{r}{n}\right)^{n \cdot t}$$

$$= 800\left(1 + \frac{0.05}{4}\right)^{4 \cdot 10}$$

$$= 800(1.0125)^{40}$$

$$\approx 1314.90$$

The total amount is $1314.90.

 STUDY SKILLS BUILDER

Are You Prepared for a Test on Chapter 7?

Below I have listed some *common trouble areas* for students in Chapter 7. After studying for your test—but before taking your test—read these.

- Can you convert from percents to fractions or decimals and from fractions or decimals to percents?

  Percent to decimal: $7.5\% = 7.5(0.01) = 0.075$

  Percent to fraction: $11\% = 11 \cdot \dfrac{1}{100} = \dfrac{11}{100}$

  Decimal to percent: $0.36 = 0.36(100\%) = 36\%$

Fraction to percent: $\dfrac{6}{7} = \dfrac{6}{7} \cdot 100\% = \dfrac{6}{7} \cdot \dfrac{100}{1}\%$

$$= \dfrac{600}{7}\% = 85\dfrac{5}{7}\%$$

- Do you remember how to find percent increase or percent decrease? The number of CDs increased from 40 to 48. Find the percent increase.

  $$\dfrac{\text{percent}}{\text{increase}} = \dfrac{\text{increase}}{\text{original number}} = \dfrac{8}{40} = 0.20 = 20\%$$

CHAPTER REVIEW

(7.1) *Solve.*

1. In a survey of 100 adults, 37 preferred pepperoni on their pizzas. What percent preferred pepperoni?

2. A basketball player made 77 out 100 attempted free throws. What percent of free throws was made?

Write each percent as a decimal.

3. 26%

4. 75%

5. 3.5%

6. 1.5%

7. 275%

8. 400%

9. 47.85%

10. 85.34%

Write each decimal as a percent.

11. 1.6

12. 0.055

13. 0.076

14. 0.085

15. 0.71

16. 0.65

17. 6

18. 9

Write each percent as a fraction or mixed number in simplest form.

19. 7%

20. 15%

21. 25%

22. 8.5%

23. 10.2%

24. $16\frac{2}{3}\%$

25. $33\frac{1}{3}\%$

26. 110%

Write each fraction or mixed number as a percent.

27. $\frac{2}{5}$

28. $\frac{7}{10}$

29. $\frac{7}{12}$

30. $1\frac{2}{3}$

31. $1\frac{1}{4}$

32. $\frac{3}{5}$

33. $\frac{1}{16}$

34. $\frac{5}{8}$

(7.2) Translating *Translate each to an equation and solve.*

35. 1250 is 1.25% of what number?

36. What number is $33\frac{1}{3}$% of 24,000?

37. 124.2 is what percent of 540?

38. 22.9 is 20% of what number?

39. What number is 17% of 640?

40. 693 is what percent of 462?

(7.3) Translating *Translate each to a proportion and solve.*

41. 104.5 is 25% of what number?

42. 16.5 is 5.5% of what number?

43. What number is 30% of 532?

44. 63 is what percent of 35?

45. 93.5 is what percent of 85?

46. What number is 33% of 500?

(7.4) *Solve.*

47. In a survey of 2000 people, it was found that 1320 have a microwave oven. Find the percent of people who own microwaves.

48. Of the 12,360 freshmen entering County College, 2000 are enrolled in prealgebra. Find the percent of entering freshmen who are enrolled in prealgebra. Round to the nearest whole percent.

49. The number of violent crimes in a city decreased from 675 to 534. Find the percent decrease. Round to the nearest tenth of a percent.

50. The current charge for dumping waste in a local landfill is $16 per cubic foot. To cover new environmental costs, the charge will increase to $33 per cubic foot. Find the percent increase.

51. This year the fund drive for a charity collected $215,000. Next year, a 4% decrease is expected. Find how much is expected to be collected in next year's drive.

52. A local union negotiated a new contract that increases the hourly pay 15% over last year's pay. The old hourly rate was $11.50. Find the new hourly rate rounded to the nearest cent.

(7.5) *Solve.*

53. If the sales tax rate is 5.5%, what is the total amount charged for a $250 coat?

54. Find the sales tax paid on a $25.50 purchase if the sales tax rate is 4.5%.

55. Russ James is a sales representative for a chemical company and is paid a commission rate of 5% on all sales. Find his commission if he sold $100,000 worth of chemicals last month.

56. Carol Sell is a sales clerk in a clothing store. She receives a commission of 7.5% on all sales. Find her commission for the week if her sales for the week were $4005. Round to the nearest cent.

57. A $3000 mink coat is on sale for 30% off. Find the discount and the sale price.

58. A $90 calculator is on sale for 10% off. Find the discount and the sale price.

(7.6) *Solve.*

59. Find the simple interest due on $4000 loaned for 4 months at 12% interest.

60. Find the simple interest due on $6500 loaned for 3 months at 20%.

61. Find the total amount in an account if $5500 is compounded annually at 12% for 15 years.

62. Find the total amount in an account if $6000 is compounded semiannually at 11% for 10 years.

63. Find the total amount in an account if $100 is compounded quarterly at 12% for 5 years.

64. Find the total amount in an account if $1000 is compounded quarterly at 18% for 20 years.

Mixed Review

Write each percent as a decimal.

65. 3.8%

66. 124.5%

Write each decimal as a percent.

67. 0.54

68. 95.2

Write each percent as a fraction or mixed number in simplest form.

69. 47%

70. 5.6%

Write each fraction or mixed number as a percent.

71. $\dfrac{3}{8}$

72. $\dfrac{6}{5}$

Translating *Translate each into an equation and solve.*

73. 43 is 16% of what number?

74. 27.5 is what percent of 25?

75. What number is 36% of 1968?

76. 67 is what percent of 50?

Translate each into a proportion and solve.

77. 75 is what percent of 25?

78. What number is 16% of 240?

79. 28 is 5% of what number?

80. 52 is what percent of 16?

Solve.

81. The total number of cans in a soft drink machine is 300. If 78 soft drinks have been sold, find the percent of soft drink cans that have been sold.

82. A home valued at $96,950 last year has lost 7% of its value this year. Find the loss in value.

83. A dinette set sells for $568.00. If the sales tax rate is 8.75%, find the purchase price of the dinette set.

84. The original price of a video game is $23.00. It is on sale for 15% off. What is the amount of the discount?

85. A candy salesman makes a commission of $1.60 from each case of candy he sells. If a case of candy costs $12.80, what is his rate of commission?

86. Find the total amount due on a 6 month loan of $1400 at a simple interest rate of 13%.

87. Find the total amount due on a loan of $5,500 for 9 years at 12.5% simple interest.

7 CHAPTER TEST

 Use the Chapter Test Prep Video CD to see the fully worked-out solutions to any of the exercises you want to review.

Write each percent as a decimal.

1. 85% **2.** 500% **3.** 0.6%

Write each decimal as a percent.

4. 0.056 **5.** 6.1 **6.** 0.35

Write each percent as a fraction or a mixed number in simplest form.

7. 120% **8.** 38.5% **9.** 0.2%

Write each fraction or mixed number as a percent.

10. $\dfrac{11}{20}$ **11.** $\dfrac{3}{8}$ **12.** $1\dfrac{3}{4}$

13. Sales of bottled water have recently surged. Bottled water accounts for $\dfrac{1}{5}$ of the total noncarbonated beverage category in convenience stores. Write $\dfrac{1}{5}$ as a percent. (*Source:* Grocery Manufacturers of America)

14. In small firms in the United States, 64% of full-time employees receive medical insurance benefits. Write 64% as a fraction. (*Source:* U.S. Bureau of Labor Statistics)

Solve.

15. What number is 42% of 80?

16. 0.6% of what number is 7.5?

17. 567 is what percent of 756?

Answers
1.
2.
3.
4.
5.
6.
7.
8.
9.
10.
11.
12.
13.
14.
15.
16.
17.

18. _____

19. _____

20. _____

21. _____

22. _____

23. _____

24. _____

25. _____

26. _____

27. _____

28. _____

Solve. If necessary, round percents to the nearest tenth, dollar amounts to the nearest cent, and all other numbers to the nearest whole.

18. An alloy is 12% copper. How much copper is contained in 320 pounds of this alloy?

19. A farmer in Nebraska estimates that 20% of his potential crop, or $11,350, has been lost to a hard freeze. Find the total value of his potential crop.

20. If the local sales tax rate is 1.25%, find the total amount charged for a stereo system priced at $354.

21. A town's population increased from 25,200 to 26,460. Find the percent increase.

22. A $120 framed picture is on sale for 15% off. Find the discount and the sale price.

23. Randy Nguyen is paid a commission rate of 4% on all sales. Find Randy's commission if his sales were $9875.

24. A sales tax of $1.53 is added to an item's price of $152.99. Find the sales tax rate. Round to the nearest whole percent.

25. Find the simple interest earned on $2000 saved for $3\frac{1}{2}$ years at an interest rate of 9.25%.

26. $1365 is compounded annually at 8%. Find the total amount in the account after 5 years.

27. A couple borrowed $400 from a bank at 13.5% simple interest for 6 months for car repairs. Find the total amount due the bank at the end of the 6-month period.

28. The number of crimes reported in New York City was 125,587 in the first half of 2001 and 118,346 in the first half of 2002. Find the percent decrease in the number of crimes in New York City from 2001 to 2002. (*Source: Federal Bureau of Investigation. Uniform Crime Reports, January–June 2002*)

Chapters 1–7

1. Multiply: 236×86

2. Multiply: 409×76

3. Subtract: 7 from -3

4. Subtract: -2 from 8.

5. Solve: $x - 2 = -1$

6. Solve: $x + 4 = 3$

7. Solve: $3(2x - 6) + 6 = 0$

8. Solve: $5(x - 2) = 3x$

9. Write an equivalent fraction with the given denominator.

$$3 = \frac{}{7}$$

10. Write an equivalent fraction with the given denominator.

$$8 = \frac{}{5}$$

11. Simplify: $-\dfrac{10}{27}$

12. Simplify: $\dfrac{10y}{32}$

13. Divide: $-\dfrac{5}{16} \div -\dfrac{3}{4}$

14. Divide: $\dfrac{-2}{5} \div \dfrac{7}{10}$

15. Evaluate $y - x$ if $x = -\dfrac{3}{10}$ and $y = -\dfrac{8}{10}$.

16. Evaluate $2x + 3y$ if $x = \dfrac{2}{5}$ and $y = \dfrac{-1}{5}$.

17. Find: $-\dfrac{3}{4} - \dfrac{1}{14} + \dfrac{6}{7}$

18. Find: $\dfrac{2}{9} + \dfrac{7}{15} - \dfrac{1}{3}$

19. Simplify: $\dfrac{\frac{1}{2} + \frac{3}{8}}{\frac{3}{4} - \frac{1}{6}}$

20. Simplify: $\dfrac{\frac{2}{3} + \frac{1}{6}}{\frac{3}{4} - \frac{3}{5}}$

21. Solve: $\dfrac{x}{2} = \dfrac{x}{3} + \dfrac{1}{2}$

22. Solve: $\dfrac{x}{2} + \dfrac{1}{5} = 3 - \dfrac{x}{5}$

Answers

1. _____
2. _____
3. _____
4. _____
5. _____
6. _____
7. _____
8. _____
9. _____
10. _____
11. _____
12. _____
13. _____
14. _____
15. _____
16. _____
17. _____
18. _____
19. _____
20. _____
21. _____
22. _____

23. a. _____

　　b. _____

24. a. _____

　　b. _____

25. _____

26. _____

27. _____

28. _____

29. _____

30. _____

31. _____

32. _____

33. _____

34. _____

35. _____

36. _____

37. _____

38. _____

39. _____

40. _____

41. _____

42. _____

43. _____

44. _____

45. _____

46. _____

47. _____

48. _____

49. _____

50. _____

23. Write each mixed number as an improper fraction.

　a. $4\frac{2}{9}$

　b. $1\frac{8}{11}$

24. Write each mixed number as an improper fraction.

　a. $3\frac{2}{5}$

　b. $6\frac{2}{7}$

Write each decimal as a fraction or mixed number. Write your answer in simplest form.

25. 0.125

26. 0.85

27. -105.083

28. 17.015

29. Subtract:　$85 - 17.31$

30. Subtract:　$38 - 10.06$

Multiply.

31. 7.68×10

32. 12.483×100

33. $(-76.3)(1000)$

34. -853.75×10

35. Evaluate $x \div y$ for $x = 2.5$ and $y = 0.05$.

36. Is 470 a solution of the equation $\frac{x}{100} = 4.75$?

37. Find the median of the scores: 67, 91, 75, 86, 55, 91

38. Find the mean of 36, 40, 86, and 30.

39. Write the ratio of 2.5 to 3.15 as a fraction in simplest form.

40. Write the ratio of 5.8 to 7.6 as a fraction in simplest form.

41. A store charges $3.36 for a 16-ounce jar of picante sauce. What is the unit price in dollars per ounce?

42. Flooring tiles cost $90 for a box with 40 tiles. Each tile is 1 square foot. Find the unit price in dollars per square foot.

43. Is $\frac{4.1}{7} = \frac{2.9}{5}$ a true proportion?

44. Is $\frac{6.3}{9} = \frac{3.5}{5}$ a true proportion?

45. On a Chamber of Commerce map of Abita Springs, 5 miles corresponds to 2 inches. How many miles correspond to 7 inches?

46. A student doing math homework can complete 7 problems in about 6 minutes. At this rate, how many problems can be complete in 30 minutes?

Write each percent as a fraction or mixed number in simplest form.

47. 1.9%

48. 2.3%

49. $33\frac{1}{3}\%$

50. 108%

8

Graphing and Introduction to Statistics

We often need to make decisions based on known statistics or the probability of an event occurring. For example, we decide whether or not to bring an umbrella to work based on the probability of rain. We interpret numerical data, often presented in graph form, from newspaper and magazine articles and in television news reports. We can predict which football team will win based on the trend in its previous wins and losses. This chapter reviews presenting data in a usable form on a graph and the basic ideas of probability.

The graph below is a double line graph of U.S. spending on VHS and DVD home entertainment. A graph such as this helps us visually see comparisons and trends and may help us make predictions of future spending. In Section 8.4, Exercises 51 through 57, we use this graph for such applications.

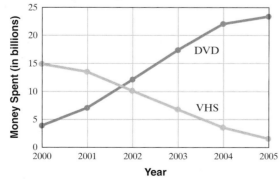

U.S. Spending on Home Entertainment

Source: Adams Media Research

8.1 READING PICTOGRAPHS, BAR GRAPHS, HISTOGRAMS, AND LINE GRAPHS

Often data is presented visually in a graph. In this section, we practice reading several kinds of graphs including pictographs, bar graphs, and line graphs.

Objective **A** Reading Pictographs

A **pictograph** such as the one below is a graph in which pictures or symbols are used. This type of graph contains a key that explains the meaning of the symbol used. An advantage of using a pictograph to display information is that comparisons can easily be made. A disadvantage of using a pictograph is that it is often hard to tell what fractional part of a symbol is shown. For example, in the pictograph below, Sweden shows a part of a symbol, but it's hard to read with any accuracy what fractional part of a symbol is shown.

PRACTICE PROBLEM 1

Use the pictograph shown in Example 1 to answer the following questions:

a. Approximate the amount of nuclear energy that was generated in Sweden.

b. Approximate the total nuclear energy generated in Sweden and Germany.

EXAMPLE 1 Calculating Nuclear Energy Generated

The following pictograph shows the approximate amount of nuclear energy generated by selected countries in the year 2005. Use this pictograph to answer the questions.

Nuclear Energy Generated by Selected Countries (2005)

a. Approximate the amount of nuclear energy that was generated in Russia.

b. Approximate how much more nuclear energy was generated in France than in Russia.

Solution:

a. Russia corresponds to 5 symbols, and each symbol represents 50 billion kilowatt-hours of energy. This means that Russia generated approximately $5 \cdot (50 \text{ billion})$ or 250 billion kilowatt-hours of energy.

b. France shows $3\frac{1}{2}$ more symbols than Russia. This means that France generated $3\frac{1}{2} \cdot (50 \text{ billion})$ or 175 billion more kilowatt-hours of nuclear energy than Russia.

Answers

1. a. 75 billion kilowatt-hours
b. 375 billion kilowatt-hours

■ **Work Practice Problem 1**

Objective B Reading and Constructing Bar Graphs

Another way to visually present data is with a **bar graph.** Bar graphs can appear with vertical bars or horizontal bars. Although we have studied bar graphs in previous sections, we now practice reading the height or length of the bars contained in a bar graph. An advantage to using bar graphs is that a scale is usually included for greater accuracy. Care must be taken when reading bar graphs, as well as other types of graphs—they may be misleading, as shown later in this section.

EXAMPLE 2 **Finding the Number of Endangered Species**

The following bar graph shows the number of endangered species in the United States in 2005. Use this graph to answer the questions.

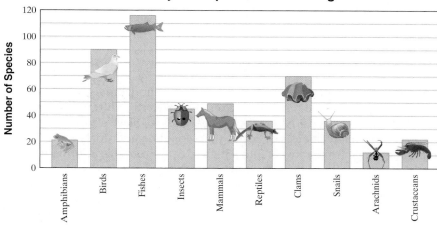

How Many U.S. Species Are Endangered?

Source: U.S. Fish and Wildlife Service

a. Approximate the number of endangered species that are reptiles.

b. Which category has the most endangered species?

Solution:

a. To approximate the number of endangered species that are reptiles, we go to the top of the bar that represents reptiles. From the top of this bar, we move horizontally to the left until the scale is reached. We read the height of the bar on the scale as approximately 36. There are approximately 36 reptile species that are endangered, as shown.

b. The most endangered species is represented by the tallest (longest) bar. The tallest bar corresponds to fishes.

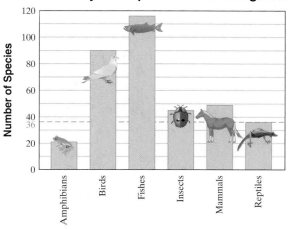

How Many U.S. Species Are Endangered?

Source: U.S. Fish and Wildlife Service

■ **Work Practice Problem 2**

PRACTICE PROBLEM 2

Use the bar graph in Example 2 to answer the following questions:

a. Approximate the number of endangered species that are insects.

b. Which category shows the fewest endangered species?

Answers

2. a. 45   **b.** arachnids

PRACTICE PROBLEM 3

Draw a vertical bar graph using the information in the table about electoral votes for selected states.

| Total Electoral Votes by Selected States ||
State	Electoral Votes
Texas	34
California	55
Florida	27
Nebraska	5
Indiana	11
Georgia	15

(*Source: World Almanac* 2006)

Next, we practice constructing a bar graph.

EXAMPLE 3 Draw a vertical bar graph using the information in the table below that gives the caffeine content of selected foods.

| Average Caffeine Content of Selected Foods ||||
Food	Milligrams	Food	Milligrams
Brewed coffee (percolator, 8 ounces)	124	Instant coffee (8 ounces)	104
Brewed decaffeinated coffee (8 ounces)	3	Brewed tea (U.S. brands, 8 ounces)	64
Coca-Cola Classic (8 ounces)	31	Mr. Pibb (8 ounces)	27
Dark chocolate (semisweet, $1\frac{1}{2}$ ounces)	30	Milk chocolate (8 ounces)	9

(*Sources:* International Food Information Council and the Coca-Cola Company)

Solution: We draw and label a vertical line and a horizontal line as shown next to the left. These lines are also called axes. We place the different food categories along the horizontal axis. Along the vertical axis, we place a scale.

There are many choices of scales that would be appropriate. Notice that the milligrams range from a low of 3 to a high of 124. From this information, we use a scale that starts at 0 and then shows multiples of 20 so that the scale is not too cluttered. The scale stops at 140, the smallest multiple of 20 that will allow all milligrams to be graphed. It may also be helpful to draw horizontal lines along the scale markings to help draw the vertical bars at the correct height. The finished bar graph is shown below.

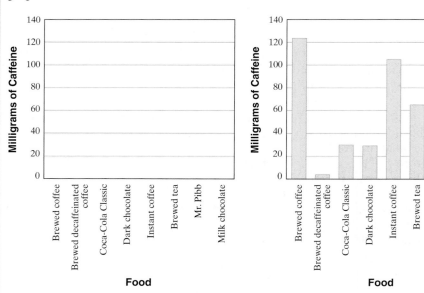

🔲 Work Practice Problem 3

As mentioned previously, graphs can be misleading. Both graphs on the next page show the same information, but with different scales. Special care should be taken when forming conclusions from the appearance of a graph.

Answer

3.

Notice the �$\succ$ symbol on each vertical scale on the graphs. This symbol alerts us that numbers are missing on that scale

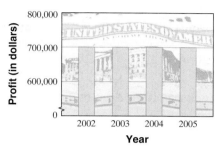

Are profits shown in the graphs above greatly increasing, or are they remaining about the same?

Objective C Reading and Constructing Histograms

Suppose that the test scores of 36 students are summarized in the table below:

Student Scores	Frequency (Number of Students)
40–49	1
50–59	3
60–69	2
70–79	10
80–89	12
90–99	8

The results in the table can be displayed in a histogram. A **histogram** is a special bar graph. The width of each bar represents a range of numbers called a **class interval.** The height of each bar corresponds to how many times a number in the class interval occurs and is called the **class frequency.** The bars in a histogram lie side by side with no space between them.

Reading a Histogram on Student Test Scores

Use the preceding histogram to determine how many students scored 50–59 on the test.

Solution: We find the bar representing 50–59. The height of this bar is 3, which means 3 students scored 50–59 on the test.

◻ **Work Practice Problem 4**

PRACTICE PROBLEM 4

Use the histogram on the left to determine how many students scored 80–89 on the test.

Answer
4. 12

PRACTICE PROBLEM 5

Use the histogram from Example 4 to determine how many students scored less than 80 on the test.

PRACTICE PROBLEM 6

Complete the frequency distribution table for the data below. Each number represents a credit card owner's unpaid balance for one month.

0	53	89	125
265	161	37	76
62	201	136	42

Class Intervals (Credit Card Balances)	Tally	Class Frequency (Number of Months)
$0–$49	___	___
$50–$99	___	___
$100–$149	___	___
$150–$199	___	___
$200–$249	___	___
$250–$299	___	___

PRACTICE PROBLEM 7

Construct a histogram from the frequency distribution table above.

EXAMPLE 5 Reading a Histogram on Student Test Scores

Use the preceding histogram to determine how many students scored 80 or above on the test.

Solution: We see that two different bars fit this description. There are 12 students who scored 80–89 and 8 students who scored 90–99. The sum of these two categories is 12 + 8 or 20 students. Thus, 20 students scored 80 or above on the test.

▣ **Work Practice Problem 5**

Now we will look at a way to construct histograms.

The daily high temperatures for 1 month in New Orleans, Louisiana, are recorded in the following list:

85°	90°	95°	89°	88°	94°
87°	90°	95°	92°	95°	94°
82°	92°	96°	91°	94°	92°
89°	89°	90°	93°	95°	91°
88°	90°	88°	86°	93°	89°

The data in this list have not been organized and can be hard to interpret. One way to organize the data is to place it in a **frequency distribution table.** We will do this in Example 6.

EXAMPLE 6 Completing a Frequency Distribution on Temperature

Complete the frequency distribution table for the preceding temperature data.

Solution: Go through the data and place a tally mark in the second column of the table next to the class interval. Then count the tally marks and write each total in the third column of the table.

Class Intervals (Temperatures)	Tally	Class Frequency (Number of Days)
82°–84°	I	1
85°–87°	III	3
88°–90°	⊮⊮I	11
91°–93°	⊮II	7
94°–96°	⊮III	8

▣ **Work Practice Problem 6**

EXAMPLE 7 Constructing a Histogram

Construct a histogram from the frequency distribution table in Example 6.

Solution:

▣ **Work Practice Problem 7**

Answers

5. 16

6.

	Tally	Class Frequency (Number of Months)
	III	3
	IIII	4
	II	2
	I	1
	I	1
	I	1

7.

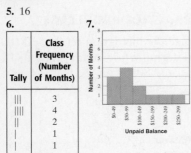

✔ **Concept Check** Which of the following sets of data is better suited to representation by a histogram? Explain.

Set 1		Set 2	
Grade on Final	# of Students	Section Number	Avg. Grade on Final
51–60	12	150	78
61–70	18	151	83
71–80	29	152	87
81–90	23	153	73
91–100	25		

Objective D Reading Line Graphs

Another common way to display information with a graph is by using a **line graph.** An advantage of a line graph is that it can be used to visualize relationships between two quantities. A line graph can also be very useful in showing a change over time.

EXAMPLE 8 Reading Temperatures from Line Graph

The following line graph shows the average daily temperature for each month for Omaha, Nebraska. Use this graph to answer the questions below.

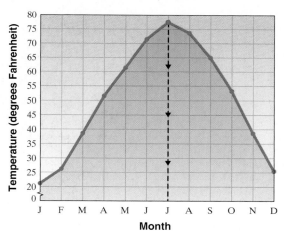

Average Daily Temperature for Omaha, Nebraska

Source: National Climatic Data Center

a. During what month is the average daily temperature the highest?

b. During what month, from July through December, is the average daily temperature 65°F?

c. During what months is the average daily temperature less than 30°F?

Solution:

a. The month with the highest temperature corresponds to the highest point. This is the red point shown on the graph above. We follow this highest point downward to the horizontal month scale and see that this point corresponds to July.

Continued on next page

PRACTICE PROBLEM 8

Use the temperature graph in Example 8 to answer the following questions:

a. During what month is the average daily temperature the lowest?

b. During what month is the average daily temperature 25°F?

c. During what months is the average daily temperature greater than 70°F?

Answers

8. a. January **b.** December
c. June, July, and August

✔ **Concept Check Answer**

Set 1; the grades are arranged in range of scores.

b. The months July through December correspond to the right side of the graph. We find the 65°F mark on the vertical temperature scale and move to the right until a point on the right side of the graph is reached. From that point, we move downward to the horizontal month scale and read the corresponding month. During the month of September, the average daily temperature is 65°F.

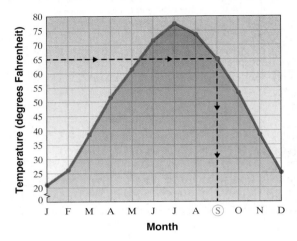

Source: National Climatic Data Center

c. To see what months the temperature is less than 30°F, we find what months correspond to points that fall below the 30°F mark on the vertical scale. These months are January, February, and December.

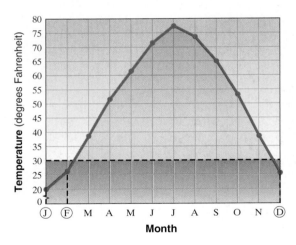

Source: National Climatic Data Center

■ **Work Practice Problem 8**

Vocabulary and Readiness Check

Fill in each blank with one of the choices below.

pictograph	bar	class frequency
histogram	line	class interval

1. A _____ graph presents data using vertical or horizontal bars.
2. A _____ is a graph in which pictures or symbols are used to visually present data.
3. A _____ graph displays information with a line that connects data points.
4. A _____ is a special bar graph in which the width of each bar represents a _____ and the height of each bar represents the _____ .

8.1 EXERCISE SET

Objective A *The following pictograph shows the annual apple production in 2005 by the top apple-producing states. Use this graph to answer Exercises 1 through 8. See Example 1. (Source: United States Apple Association)*

1. Which state produced the greatest quantity of apples?

2. In which of the states shown were the least amount of apples grown?

3. Approximate the number of bushels of apples grown in New York.

4. Approximate the number of bushels of apples grown in California.

5. Which two states together produced as many apples as New York?

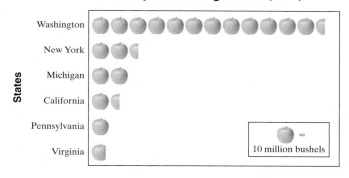

Annual Apple Production in Top Producing States (2005)

6. Washington produces about 53% of the apples grown in the United States. About how many bushels of apples were grown in the United States in 2005? Round to the nearest whole.

7. In 2001, New York produced about 23.8 million bushels of apples. What was the percent increase in apple production in New York from 2001 to 2005?

8. Which state produced about 20 million bushels of apples?

The following pictograph shows the average number of wildfires in the United States between 2000 and 2005. Use this graph to answer Exercises 9 through 16. See Example 1. (Source: National Interagency Fire Center)

9. Approximate the number of wildfires in 2002.

10. Which year, of the years shown, had the most wildfires?

11. Approximately how many wildfires were there in 2000?

12. In what years were the number of wildfires greater than 80,000?

Wildfires in the United States

13. What was the amount of decrease in wildfires from 2004 to 2005?

14. What was the amount of decrease in wildfires from 2000 to 2001?

15. What was the average annual number of wildfires from 2003 to 2005?

16. Describe a possible trend in wildfires based on this graph.

Objective $\boxed{B}$ *The National Weather Service has exacting definitions for hurricanes; they are tropical storms with winds in excess of 74 mph. The following bar graph shows the number of hurricanes, by month, that have made landfall on the mainland United States between 1851 and 2005. Use this graph to answer Exercises 17 through 22. See Example 2.* (*Source:* National Weather Service: National Hurricane Center)

17. In which month did the most hurricanes make landfall in the United States?

18. In which month did the fewest hurricanes make landfall in the United States?

19. Approximate the number of hurricanes that made landfall in the United States during the month of August.

20. Approximate the number of hurricanes that made landfall in the United States in September.

Hurricanes Making Landfall in the United States, by Month, 1851–2005

21. In 2005 alone, two hurricanes made landfall during the month of July. What fraction of all hurricanes that made landfall during July is this?

22. In 2005 alone, there were five hurricanes that made landfall on the United States mainland. If there have been 97 hurricanes to make landfall in the United States since 1851, approximately what percent of these arrived in 2005?

The following horizontal bar graph shows the 2005 population of the world's largest cities (including their suburbs). Use this graph to answer Exercises 23 through 28. See Example 2. (*Source:* CityPopulation)

23. Name the city with the largest population, and estimate its population.

24. Name the city whose population is between 16 million and 17 million, and estimate its population.

25. Name the city in the United States with the largest population, and estimate its population.

26. Name the two cities whose population is between 19 and 20 million.

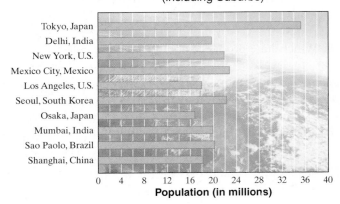

World's Largest Cities
(including Suburbs)

27. How much larger (in terms of population) is Seoul, South Korea, than São Paolo, Brazil?

28. How much larger (in terms of population) is Mexico City, Mexico, than Osaka, Japan?

Use the information given to draw a vertical bar graph. Clearly label the bars. See Example 3.

29.

Fiber Content of Selected Foods	
Food	**Grams of Total Fiber**
Kidney beans $\left(\frac{1}{2}c\right)$	4.5
Oatmeal, cooked $\left(\frac{3}{4}c\right)$	3.0
Peanut butter, chunky (2 tbsp)	1.5
Popcorn (1 c)	1.0
Potato, baked with skin (1 med)	4.0
Whole wheat bread (1 slice)	2.5
(*Sources:* American Dietetic Association and National Center for Nutrition and Dietetics)	

30.

U.S. Annual Food Sales	
Year	**Sales in Billions of Dollars**
1990	557
1995	655
2000	826
2003	947
(*Source:* U.S. Department of Agriculture)	

31.

Best-Selling Albums of All Time (U.S. Sales)	
Album	**Estimated Sales (in millions)**
Shania Twain: *Come on Over* (1997)	20
Pink Floyd: *The Wall* (1979)	23
Michael Jackson: *Thriller* (1982)	27
AC/DC: *Back in Black* (1980)	21
Billy Joel: *Greatest Hits Volumes I & II* (1985)	21
Eagles: *Their Greatest Hits* (1976)	28
Led Zeppelin: *Led Zeppelin IV* (1971)	22
(*Source:* Recording Industry Association of America)	

32.

Fuel Economy of Selected 2007 Vehicles in the United States*	
Vehicle (sales rank)	**Highway Fuel Economy* (in miles per gallon)**
Ford F-Series (1)	20
Chevrolet Silverado (2)	19
Dodge Nitro (3)	24
Toyota Camry (4)	34
Honda Accord (5)	34
Ford Explorer (6)	21
*Maximum fuel economy available among all model trims, 2007 models. (*Sources:* Internet research)	

Objective **C** *The following histogram shows the number of miles that each adult, from a survey of 100 adults, drives per week. Use this histogram to answer Exercises 33 through 42. See Examples 4 and 5.*

33. How many adults drive 100–149 miles per week?

34. How many adults drive 200–249 miles per week?

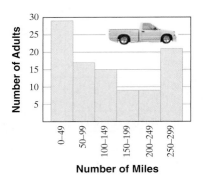

35. How many adults drive fewer than 150 miles per week?

36. How many adults drive 200 miles or more per week?

37. How many adults drive 100–199 miles per week?

38. How many adults drive 150–249 miles per week?

39. How many more adults drive 250–299 miles per week than 200–249 miles per week?

40. How many more adults drive 0–49 miles per week than 50–99 miles per week?

41. What is the ratio of adults who drive 150–199 miles per week to the total number of adults surveyed?

42. What is the ratio of adults who drive 50–99 miles per week to the total number of adults surveyed?

The following histogram shows the projected ages of householders for the year 2010. Use this histogram to answer Exercises 43 through 50. For Exercises 45 through 48, estimate to the nearest whole million. See Examples 4 and 5.

43. The most householders will be in what age range?

44. The least number of householders will be in what age range?

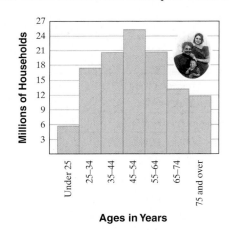

45. How many householders will be 55–64 years old?

46. How many householders will be 35–44 years old?

47. How many householders will be 44 years old or younger?

48. How many householders will be 55 years old or older?

49. How many more householders will be 45–54 years old than 55–64 years old?

50. Which age-interval will describe you in the year 2010?

Source: U.S. Bureau of the Census, *Current Population Reports*

The following list shows the golf scores for an amateur golfer. Use this list to complete the frequency distribution table to the right. See Example 6.

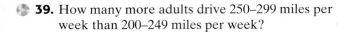

78	84	91	93	97
97	95	85	95	96
101	89	92	89	100

Class Intervals (Scores)	Tally	Class Frequency (Number of Games)
51. 70–79		
52. 80–89		
53. 90–99		
54. 100–109		

Twenty-five people in a survey were asked to give their current checking account balances. Use the balances shown in the following list to complete the frequency distribution table to the right. See Example 6.

$53	$105	$162	$443	$109
$468	$47	$259	$316	$228
$207	$357	$15	$301	$75
$86	$77	$512	$219	$100
$192	$288	$352	$166	$292

	Class Intervals (Account Balances)	Tally	Class Frequency (Number of People)
55.	$0–$99		
56.	$100–$199		
57.	$200–$299		
58.	$300–$399		
59.	$400–$499		
60.	$500–$599		

61. Use the frequency distribution table from Exercises 51 through 54 to construct a histogram. See Example 7.

Golf Scores

62. Use the frequency distribution table from Exercises 55 through 60 to construct a histogram. See Example 7.

Account Balances

Objective **D** *The following line graph shows the World Cup goals per game average during the years shown. Use this graph to answer Exercises 63 through 70. See Example 8.*

63. Find the average number of goals per game in 1994.

64. Find the average number of goals per game in 2006.

65. During what year shown was the average number of goals per game the highest?

66. During what year shown was the average number of goals per game the lowest?

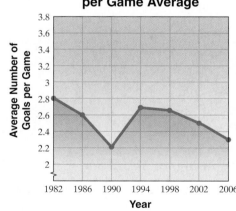

World Cup Goals per Game Average

Source: Soccer America Magazine

67. Between 2002 and 2006, did the average number of goals per game increase or decrease?

68. Between 1990 and 1994, did the average number of goals per game increase or decrease?

69. During what year(s) was the average goals per game less than 2.5?

70. During what year(s) was the average goals per game greater than 2.6?

Review

Find each percent. See Sections 7.2 and 7.3.

71. 30% of 12   **72.** 45% of 120   **73.** 10% of 62   **74.** 95% of 50

Write each fraction as a percent. See Section 7.1.

75. $\dfrac{1}{4}$ **76.** $\dfrac{2}{5}$ **77.** $\dfrac{17}{50}$ **78.** $\dfrac{9}{10}$

Concept Extensions

The following double-line graph shows temperature highs and lows for a week. Use this graph to answer Exercises 79 through 84.

79. What was the high temperature reading on Thursday?

80. What was the low temperature reading on Thursday?

81. What day was the temperature the lowest? What was this low temperature?

82. What day of the week was the temperature the highest? What was this high temperature?

83. On what day of the week was the difference between the high temperature and the low temperature the greatest? What was this difference in temperature?

84. On what day of the week was the difference between the high temperature and the low temperature the least? What was this difference in temperature?

 85. True or false? With a bar graph, the width of the bar is just as important as the height of the bar. Explain your answer.

STUDY SKILLS BUILDER

Are You Preparing for Your Final Exam?

To prepare for your final exam, try the following study techniques:

- Review the material that you will be responsible for on your exam. This includes material from your textbook, your notebook, and any handouts from your instructor.
- Review any formulas that you may need to memorize.
- Check to see if your instructor or mathematics department will be conducting a final exam review.
- Check with your instructor to see whether final exams from previous semesters/quarters are available to students for review.

- Use your previously taken exams as a practice final exam. To do so, rewrite the test questions in mixed order on blank sheets of paper. This will help you prepare for exam conditions.
- If you are unsure of a few concepts, see your instructor or visit a learning lab for assistance. Also, view the video segment of any troublesome sections.
- If you need further exercises to work, try the Cumulative Reviews at the end of the chapters.

Once again, good luck! I hope you are enjoying this textbook and your mathematics course.

8.2 READING CIRCLE GRAPHS

Objective **A** Reading Circle Graphs

In Exercise Set 7.1, the following **circle graph** was shown. This particular graph shows the favorite sport for 100 adults.

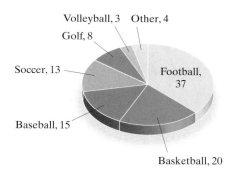

Each sector of the graph (shaped like a piece of pie) shows a category and the relative size of the category. In other words, the most popular sport is football, and it is represented by the largest sector.

EXAMPLE 1 Find the ratio of adults preferring basketball to total adults. Write the ratio as a fraction in simplest form.

Solution: The ratio is

$$\frac{\text{people preferring basketball}}{\text{total adults}} = \frac{20}{100} = \frac{1}{5}$$

Work Practice Problem 1

PRACTICE PROBLEM 1

Find the ratio of adults preferring golf to total adults. Write the ratio as a fraction in simplest form.

A circle graph is often used to show percents in different categories, with the whole circle representing 100%.

EXAMPLE 2 **Using a Circle Graph**

The following circle graph shows the percent of Americans with various numbers of working computers at home. Using the circle graph shown, determine the percent of Americans who have one or more working computers at home.

Number of Working Computers at Home

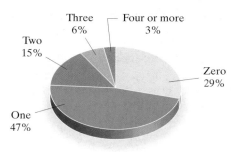

Source: UCLA Center for Communication Policy, 2003

Solution: To find this percent, we add the percents corresponding to one, two, three, and four or more working computers at home. The percent of Americans that have one or more working computers at home is

$$47\% + 15\% + 6\% + 3\% = 71\%$$

Work Practice Problem 2

PRACTICE PROBLEM 2

Using the circle graph shown in Example 2, determine the percent of Americans that have three or more working computers at home.

Answers

1. $\frac{2}{25}$ **2.** 9%

Helpful Hint

Since a circle graph represents a whole, the percents should add to 100% or 1. Notice this is true for Example 2.

EXAMPLE 3 Finding Percent of Population

In 2006, the population of the United States was approximately 299,800,000. Using the circle graph from Example 2, find the number of Americans that have no working computers at home.

Solution: We use the percent equation.

amount = percent · base

amount = 0.40 · 299,800,000

= 0.40(299,800,000) = 119,920,000

Thus, 119,920,000 Americans have no working computer at home.

▣ **Work Practice Problem 3**

✔**Concept Check** Can the following data be represented by a circle graph? Why or why not?

Responses to the Question, "In Which Activities Are You Involved?"	
Intramural sports	60%
On-campus job	42%
Fraternity/sorority	27%
Academic clubs	21%
Music programs	14%

Objective **B** **Drawing Circle Graphs**

To draw a circle graph, we use the fact that a whole circle contains 360° (degrees).

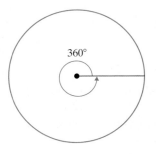

360°

EXAMPLE 4 **Drawing a Circle Graph for U.S. Armed Forces Personnel**

The following table shows the percent of U.S. armed forces personnel that are in each branch of service in 2006. (*Source:* U.S. Department of Defense)

Branch of Service	Percent
Army	43
Navy	22
Marine Corps	11
Air Force	22
Coast Guard	2

Draw a circle graph showing this data.

Solution: First we find the number of degrees in each sector representing each branch of service. Remember that the whole circle contains 360°. (We will round degrees to the nearest whole.)

Sector	Degrees in Each Sector
Army	$43\% \times 360° = 0.43 \times 360° = 154.8° \approx 155°$
Navy	$22\% \times 360° = 0.22 \times 360° = 79.2° \approx 79°$
Marine Corps	$11\% \times 360° = 0.11 \times 360° = 39.6° \approx 40°$
Air Force	$22\% \times 360° = 0.22 \times 360° = 79.2° = 79°$
Coast Guard	$2\% \times 360° = 0.02 \times 360° = 7.2° \approx 7°$

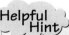
Helpful Hint

Check your calculations by finding the sum of the degrees.

$155° + 79° + 40° + 79° + 7° = 360°$

The sum should be 360°. (It may vary only slightly because of rounding.)

Next we draw a circle and mark its center. Then we draw a line from the center of the circle to the circle itself.

To construct the sectors, we will use a **protractor.** A protractor measures the number of degrees in an angle. We place the hole in the protractor over the center of the circle. Then we adjust the protractor so that 0° on the protractor is aligned with the line that we drew.

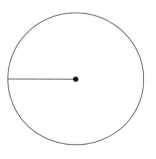

Continued on next page

PRACTICE PROBLEM 4

Use the data shown to draw a circle graph.

Freshmen	30%
Sophomores	27%
Juniors	25%
Seniors	18%

Answer

4.
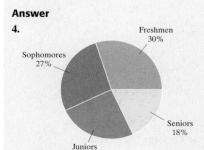

It makes no difference which sector we draw first. To construct the "Army" sector, we find 155° on the protractor and mark our circle. Then we remove the protractor and use this mark to draw a second line from the center to the circle itself.

To construct the "Navy" sector, we follow the same procedure as above, except that we line up 0° with the second line we drew and mark the protractor at 79°.

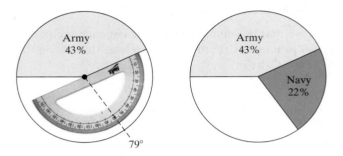

We continue in this manner until the circle graph is complete.

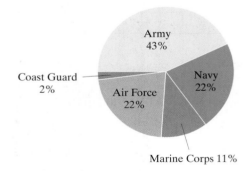

■ **Work Practice Problem 4**

✔**Concept Check** True or false? The larger a sector in a circle graph, the larger the percent of the total it represents. Explain your answer.

Vocabulary and Readiness Check

Use the choices below to fill in each blank.

> sector   circle   100   360

1. In a _____ graph, each section (shaped like a piece of pie) shows a category and the relative size of the category.
2. A circle graph contains pie-shaped sections, each called a _____.
3. The number of degrees in a whole circle is _____.
4. If a circle graph has percent labels, the percents should add up to _____.

8.2 EXERCISE SET

FOR EXTRA HELP

Student Solutions Manual   PH Math/Tutor Center   CD/Video for Review   MathXL®   MyMathLab

Objective A *The following circle graph is a result of surveying 700 college students. They were asked where they live while attending college. Use this graph to answer Exercises 1 through 6. Write all ratios as fractions in simplest form. See Example 1.*

1. Where do most of these college students live?

2. Besides the category "Other Arrangements," where do least of these college students live?

3. Find the ratio of students living in campus housing to total students.

4. Find the ratio of students living in off-campus rentals to total students.

5. Find the ratio of students living in campus housing to students living in a parent or guardian's home.

6. Find the ratio of students living in off-campus rentals to students living in a parent or guardian's home.

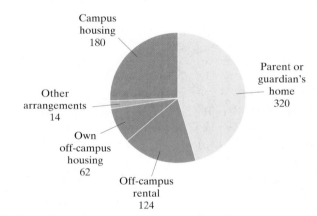

The following circle graph shows the percent of the land area of the continents of Earth. Use this graph for Exercises 7 through 14. See Examples 2 and 3.

7. Which continent is the largest?

8. Which continent is the smallest?

9. What percent of the land on Earth is accounted for by Asia and Europe together?

10. What percent of the land on Earth is accounted for by North and South America?

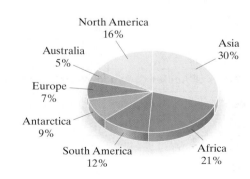

Source: National Geographic Society

The total amount of land on Earth is approximately 57,000,000 square miles. Use the graph to find the area of the continents given in Exercises 11 through 14.

11. Asia   12. South America   13. Australia   14. Europe

The following circle graph shows the percent of the types of books available at Midway Memorial Library. Use this graph for Exercises 15 through 24. See Examples 2 and 3.

15. What percent of books are classified as some type of fiction?

16. What percent of books are nonfiction or reference?

17. What is the second-largest category of books?

18. What is the third-largest category of books?

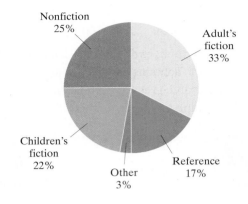

If this library has 125,600 books, find how many books are in each category given in Exercises 19 through 24.

19. Nonfiction   **20.** Reference   **21.** Children's fiction

22. Adult's fiction   **23.** Reference or other   **24.** Nonfiction or other

Objective **B** *Fill in the table. Round to the nearest degree. Then draw a circle graph to represent the information given in each table. (Remember: The total of "Degrees in Sector" column should equal 360° or very close to 360° because of rounding.) See Example 4.*

25.

U.S. Electricity Generation by Energy Source: 2006		
Energy Source	**Percent**	**Degrees in Sector**
Coal	50.6%	
Natural gas	15.8%	
Hydroelectric	8.2%	
Nuclear	20.6%	
Petroleum	1.6%	
Other	3.2%	
(*Source:* Energy Information Administration)		

26.

Types of Apples Grown in Washington State		
Type of Apple	**Percent**	**Degrees in Sector**
Red delicious	37%	
Golden delicious	13%	
Fuji	14%	
Gala	15%	
Granny Smith	12%	
Other varieties	6%	
Braeburn	3%	
(*Source:* U.S. Apple Association)		

27.

2006* Hybrid Sales by Make of Car		
Company	Percent	Degrees in Sector
Toyota	63%	
Honda	17%	
Lexus	10%	
Ford/Mercury	10%	
*Through six months of 2006		

28.

Number of Times the "Are We There Yet?" Question Is Asked to Parents During Road Trips:		
	Percent	Degrees in Sector
Never	20%	
Once	11%	
2–5 times	36%	
6–10 times	14%	
More than 10 times	19%	
(*Source:* KRC Research for Goodyear Tire & Rubber Co.)		

Review

Write the prime factorization of each number. See Section 4.2.

29. 20

30. 25

31. 40

32. 16

33. 85

34. 105

Concept Extensions

The following circle graph shows the relative sizes of the great oceans.

35. Without calculating, determine which ocean is the largest. How can you answer this question by looking at the circle graph?

36. Without calculating, determine which ocean is the smallest. How can you answer this question by looking at the circle graph?

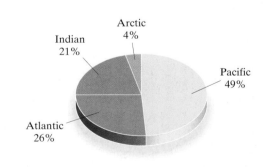

Source: Philip's World Atlas

These oceans together make up 264,489,800 square kilometers of the Earth's surface. Find the square kilometers for each ocean.

37. Pacific Ocean   **38.** Atlantic Ocean   **39.** Indian Ocean   **40.** Arctic Ocean

The following circle graph summarizes the results of a survey of 2800 Internet users who make purchases online. Use this graph for Exercises 41 through 45. Round to the nearest whole.

41. How many of the survey respondents said that they spend \$0–\$15 online each month?

42. How many of the survey repondents said that they spend \$15–\$175 online each month?

43. How many of the survey respondents said that they spend at least \$15 or over \$175 online each month?

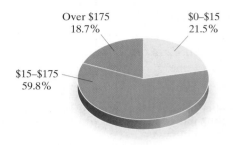

Online Spending per Month

Over \$175
18.7%

\$0–\$15
21.5%

\$15–\$175
59.8%

Source: UCLA Center for Communication Policy

44. Find the ratio of *percent* of respondents who spend \$0–\$15 online to *percent* of those who spend \$15–\$175. Write the ratio as a fraction with integers in the numerator and denominator.

45. Find the ratio of *number* of respondents who spend \$0–\$15 online to *number* of respondents who spend \$15–\$175 online. Write the ratio as a fraction with integers in the numerator and denominator.

See the Concept Checks in this section.

46. True or false? The smaller a sector in a circle graph, the smaller the percent of the total it represents. Explain why.

47. Can the data below be represented by a circle graph? Why or why not?

Responses to the Question, "In Which Activities Are You Involved?"	
Intramural sports	60%
On-campus job	42%
Fraternity/sorority	27%
Academic clubs	21%
Music programs	14%

8.3 THE RECTANGULAR COORDINATE SYSTEM AND PAIRED DATA

Objective **A** Plotting Points

In 8.1, we saw how bar and line graphs can be used to show relationships between items listed on the horizontal and vertical axes.

The table and line graph below both show the U.S. spending on DVD home entertainment for the years shown.

U.S. Spending on DVD Home Entertainment	
Year	Money Spent (in billions)
2000	$3.9
2001	$7.0
2002	$11.0
2003	$17.4
2004	$22.0
2005	$23.3

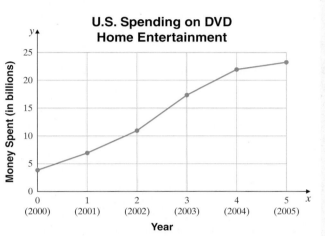

Both the table and the line graph show the relationship between two quantities (year and money), called **paired data.** These data are called paired data because each year corresponds to, or is paired with, a specific amount of money. In other words, the year 2005 is paired with $23.3 billion. Notice that the dots, or points, on the line graph visually show the same pairing. For the year 2005, the height of the dot is 23.3, representing $23.3 billion.

We can call the "year" axis the horizontal axis and the "money" axis the vertical axis.

In general, we can use this same horizontal and vertical axis idea to describe the location of points in a plane. A **plane** is a flat surface that extends indefinitely. Surfaces like a plane (except that they don't extend indefinitely) are a classroom floor, a blackboard or whiteboard, and this sheet of paper.

The system that we use to describe the location of points in a plane is called the **rectangular coordinate system.** It consists of two number lines, one horizontal and one vertical, intersecting at the point 0 on each number line. This point of intersection is called the **origin.** We call the horizontal number line the **x-axis** and the vertical number line the **y-axis.** Notice that the axes divide the plane into four regions, called **quadrants.** They are numbered as shown.

Every point in the rectangular coordinate system corresponds to an **ordered pair of numbers,** such as (3, 4). The first number, 3, of an ordered pair is associated with the *x*-axis and is called the *x*-**coordinate** or *x*-**value.** The second number, 4, is associated with the *y*-axis and is called the *y*-**coordinate** or *y*-**value.** To find the **single point** on the rectangular coordinate system corresponding to the ordered pair (3, 4), start at the origin. Move 3 units in the positive direction along the *x*-axis. From there, move 4 units in the positive direction parallel to the *y*-axis as shown to the right. This process of locating a point on the rectangular coordinate system is called **plotting the point.** Since the origin is located at 0 on the *x*-axis and 0 on the *y*-axis, the origin corresponds to the ordered pair (0, 0).

In general, to plot the ordered pair (x, y), start at the origin. Next,

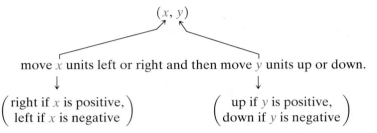

(x, y)

move x units left or right and then move y units up or down.

$$\left(\begin{array}{l} \text{right if } x \text{ is positive,} \\ \text{left if } x \text{ is negative} \end{array} \right) \qquad \left(\begin{array}{l} \text{up if } y \text{ is positive,} \\ \text{down if } y \text{ is negative} \end{array} \right)$$

Helpful Hint

Since the first number, or *x*-coordinate, of an ordered pair is associated with the *x*-axis, it tells how many units to move left or right. Similarly, the second number, or *y*-coordinate, tells how many units to move up or down.

To plot $(-1, 5)$, start at the origin. Move 1 unit left (because the *x*-coordinate is negative) and then 5 units up and draw a dot at that point. This dot is the graph of the point that corresponds to the ordered pair $(-1, 5)$.

To plot $(0, -3)$, start at the origin, move 0 units (left or right), then 3 units down, then draw a dot.

Below are some more plotted points with their corresponding ordered pairs. Another name for the rectangular coordinate system is the Cartesian coordinate system. This name comes from French scientist, philosopher, and mathematician René Descartes (1596–1650), who introduced the system.

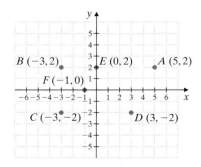

Helpful Hint

Remember that **each point** in the rectangular coordinate system corresponds to exactly **one ordered pair** and that **each ordered pair** corresponds to exactly **one point.**

EXAMPLE 1 Plot each point corresponding to the ordered pairs on the same set of axes.

$$(5, 4), (-2, 3), (-1, -2), (6, -3), (0, 2), (-4, 0), \left(3, \frac{1}{2}\right)$$

Solution:

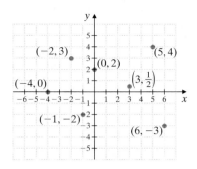

■ **Work Practice Problem 1**

EXAMPLE 2 Find the ordered pair corresponding to each point plotted below.

Solution:

Point A has coordinates $(-4, 2)$.
Point B has coordinates $(1, 2)$.
Point C has coordinates $(0, 1)$.
Point D has coordinates $(-3, 0)$.
Point E has coordinates $(5, -4)$.

■ **Work Practice Problem 2**

Have you noticed a pattern from the preceding examples? **If an ordered pair has a y-coordinate of 0, its graph lies on the x-axis. If an ordered pair has an x-coordinate of 0, its graph lies on the y-axis.**

Helpful Hint

     Order is the key word in ordered pair. The first value always corresponds to the x-value and the second value always corresponds to the y-value. For example, $(-2, 4)$ does not describe the same point as $(4, -2)$.

PRACTICE PROBLEM 1

Plot each point corresponding to the ordered pairs on the same set of axes. $(2, 3), (-4, 2)$ $(-6, -5)$ $(2, -2)$ $(5, 0)$ $(0, -3)$ $\left(-\frac{1}{2}, 4\right)$

PRACTICE PROBLEM 2

Find the ordered pair corresponding to each point plotted on the rectangular coordinate system.

Answers

1.

2. $A(5, 0); B(-5, 4); C(-3, -4);$
$D(0, -1); E(2, 1)$

Objective **B** Determining Whether an Ordered Pair Is a Solution

Let's see how we can use ordered pairs of numbers to record solutions of equations containing two variables.

Recall that an equation with one variable, such as $3 + x = 5$, has one solution—in this case, 2 because 2 is the only value that can be substituted for x so that the resulting equation $3 + 2 = 5$ is a true statement. We can graph this solution on a number line.

An equation with two variables such as $x + y = 7$ has many solutions. Each solution is a pair of numbers, one for each variable that makes the equation a true statement. For example, $x = 4$ and $y = 3$ is a solution of the equation $x + y = 7$ because, when x is replaced with 4 and y is replaced with 3, $x + y = 7$ becomes $4 + 3 = 7$, which is a true statement. We can write the solution $x = 4$ and $y = 3$ as the ordered pair $(4, 3)$. Study the chart below for more solutions.

Replacement Values	$x + y = 7$	True or False	Ordered-Pair Solution
$x = 4, y = 3$	$4 + 3 = 7$	True	$(4, 3)$
$x = 0, y = 7$	$0 + 7 = 7$	True	$(0, 7)$
$x = 7, y = 1$	$7 + 1 = 7$	False	No
$x = 11, y = -4$	$11 + (-4) = 7$	True	$(11, -4)$

In general, we say that an ordered pair of numbers is a solution of an equation if the equation is a true statement when the variables are replaced by the coordinates of the ordered pair.

Let's practice deciding whether an ordered pair of numbers is a solution of an equation in two variables.

PRACTICE PROBLEM 3

Is $(0, -4)$ a solution of the equation $x + 3y = -12$?

EXAMPLE 3 Is $(-1, 6)$ a solution of the equation $2x + y = 4$?

Solution: Replace x with -1 and y with 6 in the equation $2x + y = 4$ to see if the result is a true statement.

$2x + y = 4$ Original equation

$2(-1) + 6 \overset{?}{=} 4$ Replace x with -1 and y with 6.

$-2 + 6 \overset{?}{=} 4$ Multiply.

$4 = 4$ True

Since $4 = 4$ is true, $(-1, 6)$ is a solution of the equation $2x + y = 4$.

◻ **Work Practice Problem 3**

Answer

3. yes

Unlike equations with one variable, equations with two variables, such as $x + y = 7$, have so many solutions we cannot simply list them. Instead, to "see" all the solutions of a two-variable equation, we draw a graph of its solutions.

EXAMPLE 4 Each ordered pair listed is a solution of the equation $x + y = 7$. Plot them on the same set of axes.

a. $(3, 4)$ **b.** $(0, 7)$ **c.** $(-1, 8)$
d. $(7, 0)$ **e.** $(5, 2)$ **f.** $(4, 3)$

Solution:

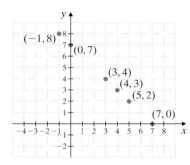

Work Practice Problem 4

Notice that the points in Example 4 all seem to lie on the same line. We will discuss this more in the next section.

An equation such as $x + y = 7$ is called a **linear or first-degree equation in two variables.** It is called a **first-degree equation** because the exponent on x and y is an understood 1. The equation is called an equation **in two variables** because it contains two different variables, x and y.

Objective C Completing Ordered-Pair Solutions

In the next section, we will graph linear equations in two variables. Before that, we need to practice finding ordered-pair solutions of these equations. If one coordinate of an ordered-pair solution is known, the other coordinate can be determined. To find the unknown coordinate, replace the appropriate variable with the known coordinate in the equation. Doing so results in an equation with one variable that we can solve.

EXAMPLE 5 Complete each ordered-pair solution of the equation $y = 2x$.

a. $(3, \ )$ **b.** $(\ , 0)$ **c.** $(-2, \ )$

Solution:

a. In the ordered pair $(3, \ )$, the x-value is 3. To find the corresponding y-value, let $x = 3$ in the equation $y = 2x$ and calculate the value of y.

$y = 2x$ Original equation
$y = 2(3)$ Replace x with 3.
$y = 6$ Multiply.

Check: To check, replace x with 3 and y with 6 in the original equation to see that a true statement results.

$y = 2x$ Original equation
$6 \overset{?}{=} 2(3)$ Let $x = 3$ and $y = 6$.
$6 = 6$ True

The ordered-pair solution is $(3, 6)$.

Continued on next page

PRACTICE PROBLEM 4

Each ordered pair listed is a solution of the equation $x - y = 5$. Plot them on the same set of axes.

a. $(6, 1)$ **b.** $(5, 0)$
c. $(0, -5)$ **d.** $(7, 2)$
e. $(-1, -6)$ **f.** $(2, -3)$

PRACTICE PROBLEM 5

Complete each ordered-pair solution of the equation $y = -5x$.

a. $(5, \ )$ **b.** $(\ , 0)$
c. $(-3, \ )$

Answers
4.

5. **a.** $(5, -25)$ **b.** $(0, 0)$ **c.** $(-3, 15)$

b. In the ordered pair (, 0), the y-value is 0. To find the corresponding x-value, replace y with 0 in the equation and solve for x.

$y = 2x$ Original equation   **Check:** $y = 2x$ Original equation

$0 = 2x$ Replace y with 0. $0 \overset{?}{=} 2 \cdot 0$ Let $x = 0$ and $y = 0$.

$\dfrac{0}{2} = \dfrac{2x}{2}$ Divide both sides by 2. $0 = 0$ True

$0 = x$ Simplify.

The ordered-pair solution is $(0, 0)$, the origin.

c. In the ordered pair $(-2, \ )$, the x-value is -2. To find the corresponding y-value, replace x with -2 in the equation and calculate the value for y.

$y = 2x$ Original equation   **Check:** $y = 2x$ Original equation

$y = 2(-2)$ Replace x with -2. $-4 \overset{?}{=} 2(-2)$ Let $x = -2$ and $y = -4$.

$y = -4$ Multiply. $-4 = -4$ True

The ordered-pair solution is $(-2, -4)$.

 Work Practice Problem 5

PRACTICE PROBLEM 6

Complete each ordered-pair solution of the equation $y = 5x + 2$.

a. $(0, \ )$

b. $(\ , -3)$

EXAMPLE 6 Complete each ordered-pair solution of the equation $y = 3x - 4$.

a. $(2, \ )$ **b.** $(\ , -16)$

Solution:

a. Replace x with 2 in the equation and calculate the value for y.

$y = 3x - 4$ Original equation

$y = 3 \cdot 2 - 4$ Replace x with 2.

$y = 2$ Simplify.

The ordered-pair solution is $(2, 2)$.

b. Replace y with -16 in the equation and solve for x.

$y = 3x - 4$ Original equation

$-16 = 3x - 4$ Replace y with -16.

$-16 + 4 = 3x - 4 + 4$ Add 4 to both sides.

$-12 = 3x$ Simplify.

$\dfrac{-12}{3} = \dfrac{3x}{3}$ Divide both sides by 3.

$-4 = x$ Simplify.

The ordered-pair solution is $(-4, -16)$.

 Work Practice Problem 6

Answers
6. a. $(0, 2)$ **b.** $(-1, -3)$

Vocabulary and Readiness Check

Use the words and phrases below to fill in each blank. The questions and statements below all have to do with the rectangular coordinate system.

plotting	x	one	$(0, 0)$	ordered pair of numbers
origin	y	four	plane	

1. In the ordered pair $(-1, 2)$, the _____-value is -1 and the _____-value is 2.

2. In the rectangular coordinate system, the point of intersection of the x-axis and the y-axis is called the _____.

3. The axes divide the plane into _____ regions, called quadrants.

4. Every ordered pair of numbers, such as $\left(\frac{1}{2}, -3\right)$, corresponds to how many points in the plane? _____

5. The process of locating a point on the rectangular coordinate system is called _____ the point.

6. The origin corresponds to the ordered pair _____.

7. A _____ is a flat surface that extends indefinitely in all directions.

8. Every point in the rectangular coordinate system corresponds to a(n) _____.

8.3 EXERCISE SET

FOR EXTRA HELP

Student Solutions Manual   PH Math/Tutor Center   CD/Video for Review   MathXL MathXL®   MyMathLab MyMathLab

Objective Ⓐ *Use the rectangular coordinate system below each exercise to plot the points corresponding to the ordered pairs. See Example 1.*

1. $(1, 3), (-2, 4), (0, 2), (-5, 0), (-3, -3), (5, -5)$

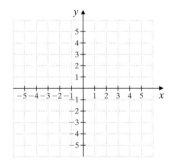

2. $(5, 2), (3, -4), (-1, -1), (0, -1), (4, 0), (-2, 4)$

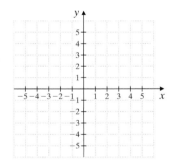

3. $\left(2\frac{1}{2}, 3\right), (0, -3), (-4, -6), \left(-1, 5\frac{1}{2}\right), (1, 0), (3, -5)$

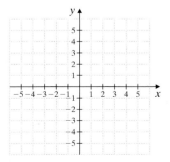

4. $\left(5, \frac{1}{2}\right), \left(-3\frac{1}{2}, 0\right), (-1, 4), (4, -1), (0, 2), (-5, -5)$

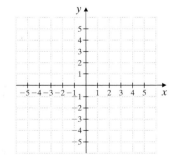

565

Find the x- and y-coordinates of each labeled point. See Example 2.

5.

6.

Objective **B** *Determine whether each ordered pair is a solution of the given linear equation. See Example 3.*

7. $(0, 0)$; $y = -20x$

8. $(1, 14)$; $x = 14y$

9. $(1, 2)$; $x - y = 3$

10. $(-1, 9)$; $x + y = 8$

11. $(-2, -3)$; $y = 2x + 1$

12. $(1, 1)$; $y = -x$

13. $(6, -2)$; $x = -3y$

14. $(-9, 1)$; $x = -9y$

15. $(5, 0)$; $3y + 2x = 10$

16. $(1, 1)$; $-5y + 4x = -1$

17. $(3, 1)$; $x - 5y = -1$

18. $(0, 2)$; $x - 7y = -15$

Use the rectangular coordinate system below each exercise to plot the three ordered-pair solutions of the given equation. See Example 4.

19. $2x + y = 5$; $(1, 3), (0, 5), (3, -1)$

20. $3x + y = 5$; $(1, 2), (0, 5), (2, -1)$

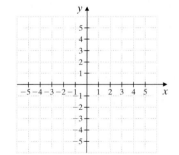

21. $x = 5y$; $(5, 1), (0, 0), (-5, -1)$

22. $y = -3x$; $(1, -3), (2, -6), (-1, 3)$

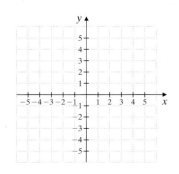

23. $y = -2x + 3$; $(4, -5)$, $(2, -1)$, $(0, 3)$

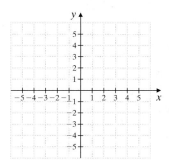

24. $y = 3x + 1$; $(0, 1)$, $(1, 4)$, $(-1, -2)$

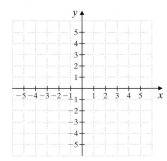

Objective **C** *Complete each ordered-pair solution of the given equations. See Examples 5 and 6.*

25. $y = -9x$; $(1, \quad)$, $(0, \quad)$, $(\quad, -18)$

26. $x = -7y$; $(\quad, 2)$, $(14, \quad)$, $(\quad, -1)$

27. $x - y = 14$; $(2, \quad)$, $(\quad, -8)$, $(0, \quad)$

28. $x - y = 8$; $(0, \quad)$, $(\quad, 0)$, $(5, \quad)$

29. $x + y = -2$; $(-2, \quad)$, $(1, \quad)$, $(\quad, 5)$

30. $x + y = -3$; $(\quad, 0)$, $(0, \quad)$, $(4, \quad)$

31. $y = x + 5$; $(1, \quad)$, $(\quad, 7)$, $(3, \quad)$

32. $x = y - 4$; $(\quad, -12)$, $(\quad, 3)$, $(100, \quad)$

33. $y = 3x - 5$; $(1, \quad)$, $(2, \quad)$, $(\quad, 4)$

34. $x = -12y + 5$; $(\quad, -1)$, $(\quad, 1)$, $(41, \quad)$

35. $y = -x$; $(\quad, 0)$, $(2, \quad)$, $(\quad, 2)$

36. $x = -y$; $(\quad, 0)$, $(3, \quad)$, $(\quad, -9)$

37. $x + 2y = -8$; $(4, \quad)$, $(\quad, -3)$, $(0, \quad)$

38. $-5x + y = 1$; $(0, \quad)$, $(-1, \quad)$, $(2, \quad)$

Review

Perform each indicated operation on decimals. See Sections 5.2 to 5.4.

39. $5.6 - 3.9$

40. $5 + 2.54 + 8.7$

41. $5.6 \cdot 3.9$

42. $0.56 \div 0.8$

43. $(0.236)(-100)$

44. $44.72 \div 100$

Concept Extensions

Recall that the axes divide the plane into four quadrants as shown. For Exercises 45 through 52, if a and b are both positive numbers, determine whether each statement is true or false.

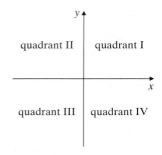

45. (a, b) lies in quadrant I.

46. $(-a, -b)$ lies in quadrant IV.

47. $(0, b)$ lies on the *y*-axis.

48. $(a, 0)$ lies on the *x*-axis.

49. $(0, -b)$ lies on the *x*-axis.

50. $(-a, 0)$ lies on the *y*-axis.

51. $(-a, b)$ lies in quadrant III.

52. $(a, -b)$ lies in quadrant IV.

53. Is the ordered pair $(4, -3)$ plotted to the left or right of the *y*-axis? Explain.

54. Is the ordered pair $(6, -2)$ plotted above or below the *x*-axis? Explain.

Plot the points $A(4, 3)$, $B(-2, 3)$, $C(-2, -1)$ and $D(4, -1)$ on a rectangular coordinate system.

55. Draw a line segment connecting A and B, B and C, C and D, and D and A. Name the figure drawn.

56. Find the length and the width of the figure.

57. Find the perimeter of the figure.

58. Find the area of the figure.

Reading Graphs

The following pictograph shows the six occupations with the largest estimated numerical increase in employment in the United States between 2004 and 2014. Use this graph to answer Exercises 1 through 4.

Jobs with Projected Highest Numerical Increase: 2004–2014

Jobs	
Retail Salespeople	🧍 (16)
Registered Nurses	🧍 (15)
Post-secondary Teachers	🧍 (11)
Customer Service Reps	🧍 (10)
Janitors and Cleaners	🧍 (9)
Waitstaff	🧍 (8)

🧍 = 50,000 workers

Source: Bureau of Labor Statistics

1. Approximate the increase in the number of customer service representatives from 2004 to 2014.

2. Approximate the increase in number of post-secondary teachers from 2004 to 2014.

3. Which occupation is expected to show the greatest increase in numbers of employees between the years shown?

4. Which of the listed occupations is expected to show the least increase in numbers of employees between the years shown?

The following bar graph shows the highest U.S. dams. Use this graph to answer Exercises 5 through 8.

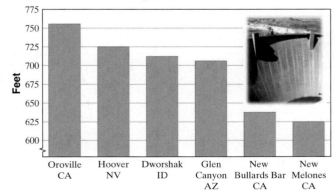

Highest U.S. Dams

Source: Committee on Register of Dams

5. Name the U.S. dam with the greatest height and estimate its height.

6. Name the U.S. dam whose height is between 625 and 650 feet and estimate its height.

7. Estimate how much higher the Hoover Dam is than the Glen Canyon Dam.

8. How many U.S. dams have heights over 700 feet?

9. _____

10. _____

11. _____

12. _____

13. _____

14. _____

15. _____

16. _____

17. see table

18. see table

19. see table

20. see table

21. see table

The following line graph shows the daily high temperatures for one week in Annapolis, Maryland. Use this graph to answer Exercises 9 through 12.

9. Name the day(s) of the week with the highest temperature and give that high temperature.

10. Name the day(s) of the week with the lowest temperature and give that low temperature.

11. On what days of the week was the temperature less than 90° Fahrenheit?

12. On what days of the week was the temperature greater than 90° Fahrenheit?

The following circle graph shows the type of beverage milk consumed in the United States. Use this graph for Exercises 13 through 16. If a store in Kerrville, Texas, sells 200 quart containers of milk per week, estimate how many quart containers are sold in each category below.

Types of Beverage Milk Consumed

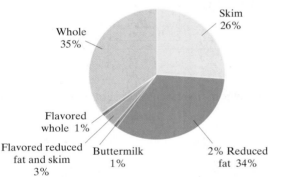

Source: U.S. Department of Agriculture

13. Whole milk

14. Skim milk

15. Buttermilk

16. Flavored reduced fat and skim milk

The following list shows weekly quiz scores for a student in prealgebra. Use this list to complete the frequency distribution table below.

50	80	71	83	86
67	89	93	88	97
	53	90		
75	80	78	93	99

	Class Intervals (Scores)	Tally	Class Frequency (Number of Quizzes)
17.	50–59		
18.	60–69		
19.	70–79		
20.	80–89		
21.	90–99		

22. Use the frequency distribution table from Exercises 17 through 21 to construct a histogram.

Number of Quizzes

2

1

Quiz Scores

23. Plot the points corresponding to the ordered pairs on the same set of axes. $(0, 2)$, $(-1, 4)$ $(2, 1)$ $(-3, 0)$ $(-3, -5)$ $(4, -1)$

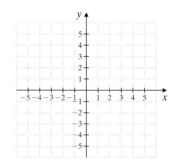

24. Determine whether the ordered pair $(1, 3)$ is a solution of the linear equation $x = 3y$.

25. Determine whether the ordered pair $(-2, -4)$ is a solution of the linear equation $x + y = -6$.

26. Complete each ordered pair solution of the equation $x - y = 6$. $(0, \ )$, $(\ , 0)$, $(2, \ )$

8.4 GRAPHING LINEAR EQUATIONS IN TWO VARIABLES

Now that we know how to plot points in a rectangular coordinate system and how to find ordered-pair solutions, we are ready to graph linear equations in two variables. First, we give a formal definition.

Linear Equation in Two Variables

A *linear equation in two variables* is an equation that can be written in the form

$$ax + by = c \qquad \text{Examples:} \begin{cases} 2x + 3y = 6 \\ 2x = 5 \\ y = 3 \end{cases}$$

where a, b, and c are numbers, and a and b are not both 0.

In the last section, we discovered that a linear equation in two variables has many solutions. For the linear equation $x + y = 7$, we listed, for example, the solutions $(3, 4)$, $(0, 7)$, $(-1, 8)$, $(7, 0)$, $(5, 2)$, and $(4, 3)$. Are these all the solutions? No. There are infinitely many solutions of the equation $x + y = 7$ since there are infinitely many pairs of numbers whose sum is 7. Every linear equation in two variables has infinitely many ordered-pair solutions. Since it is impossible to list every solution, we graph the solutions instead.

Objective **A** Graphing Linear Equations in Two Variables by Plotting Points

Fortunately, the pattern described by the solutions of a linear equation makes "seeing" the solutions possible by graphing. This is so because **all the solutions of a linear equation in two variables correspond to points on a single straight line.** If we plot just a few of these points and draw the straight line connecting them, we have a complete graph of all the solutions.

To graph the equation $x + y = 7$, then, we plot a few ordered-pair solutions, say $(3, 4)$, $(0, 7)$, and $(-1, 8)$.

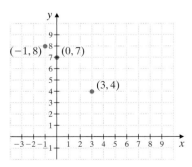

Now we connect these three points by drawing a straight line through them. The arrows at both ends of the line indicate that the line goes on forever in both directions.

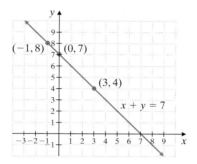

Every point on this line corresponds to an ordered-pair solution of the equation $x + y = 7$. Also, every ordered-pair solution of the equation $x + y = 7$ corresponds to a point on this line. In other words, this line is the graph of the equation $x + y = 7$. Although a line can be drawn using just two points, we will graph a third solution to check our work.

EXAMPLE 1 Graph the equation $y = 3x$ by plotting the following points that satisfy the equation and drawing a line through the points.

$$(1, 3), (0, 0), (-1, -3)$$

Solution: Plot the points and draw a line through them. The line is the graph of the linear equation $y = 3x$.

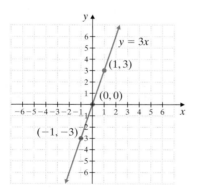

🖥 **Work Practice Problem 1**

Graphing Linear Equations and Finding Ordered-Pair Solutions

- To graph a linear equation in two variables, find three ordered-pair solutions, graph the solutions, and draw the line through the plotted points.
- To find an ordered-pair solution of an equation, choose either an x-value or y-value of the ordered pair and complete the ordered pair as we did in the previous section.

PRACTICE PROBLEM 1

Graph the equation $x - y = 1$ by plotting the following points that satisfy the equation and drawing a line through the points.

$$(3, 2), (0, -1), (1, 0)$$

Answer

1.

PRACTICE PROBLEM 2

Graph: $y - x = 4$

PRACTICE PROBLEM 3

Graph: $y = -3x + 2$

Answers

2.

3.

✔ **Concept Check Answer**

no

EXAMPLE 2 Graph: $x - y = 6$

Solution: Find any three ordered-pair solutions. For each solution, we choose a value for x or y, and replace x or y by its chosen value in the equation. Then we solve the resulting equation for the other variable.

For example, let $x = 3$.

$x - y = 6$	Original equation
$3 - y = 6$	Let $x = 3$.
$-y = 6 - 3$	Subtract 3 from both sides.
$-y = 3$	Simplify.
$\dfrac{-y}{-1} = \dfrac{3}{-1}$	Divide both sides by -1.
$y = -3$	Simplify.

The ordered pair is $(3, -3)$.

Also, let $y = 0$.

$x - y = 6$	Original equation
$x - 0 = 6$	Let $y = 0$.
$x = 6$	Simplify.

The ordered pair is $(6, 0)$.

If we let $x = 7$, and solve the resulting equation for y, we will see that $y = 1$. **A third ordered-pair solution is, then,** $(7, 1)$.

Plot the three ordered-pair solutions on a rectangular coordinate system and then draw a line through the three points. This line is the graph of $x - y = 6$. (For organization purposes, the ordered pairs are entered in the table below.)

x	y
3	−3
6	0
7	1

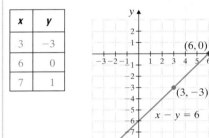

◻ **Work Practice Problem 2**

Helpful Hint
All three points should fall on the same straight line. If not, check your ordered-pair solutions for a mistake, since the graph of every linear equation is a line.

✔**Concept Check** In Example 2, is the point $(9, -3)$ on the graph of $x - y = 6$?

EXAMPLE 3 Graph: $y = 5x + 1$

Solution: For this equation, we will choose three x-values and find the corresponding y-values. We choose $x = 0$, $x = 1$, and $x = 2$. The first table on the next page shows our chosen x-values.

If $x = 0$, then
$y = 5x + 1$ becomes
$y = 5 \cdot 0 + 1$ or
$y = 1$

If $x = 1$, then
$y = 5x + 1$ becomes
$y = 5 \cdot 1 + 1$ or
$y = 6$

If $x = 2$, then
$y = 5x + 1$ becomes
$y = 5 \cdot 2 + 1$ or
$y = 11$

Continued on next page

Now we complete the table of values, plot the ordered-pair solutions, and graph the equation $y = 5x + 1$.

x	y
0	
1	
2	

x	y
0	1
1	6
2	11

⬛ **Work Practice Problem 3**

Next, we will graph a few special linear equations.

EXAMPLE 4 Graph: $y = 4$

Solution: The equation $y = 4$ can be written as $0x + y = 4$. When the equation is written in this form, notice that no matter what value we choose for x, y is always 4.

$$0 \cdot x + y = 4$$
$$0 \cdot (\text{any number}) + y = 4$$
$$0 + y = 4$$
$$y = 4$$

Fill in a table listing ordered-pair solutions of $y = 4$. Choose any three x-values. The y-values are always 4. Plot the ordered-pair solutions and graph $y = 4$.

x	y
0	4
1	4
−2	4

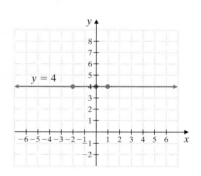

The graph is a horizontal line that crosses the y-axis at 4.

⬛ **Work Practice Problem 4**

PRACTICE PROBLEM 4

Graph: $y = -4$

Answer

4.

Horizontal Lines

If a is a number, then the graph of $y = a$ is a *horizontal line* that crosses the y-axis at a. For example, the graph of $y = 2$ is a horizontal line that crosses the y-axis at 2.

PRACTICE PROBLEM 5

Graph: $x = 5$

EXAMPLE 5 Graph: $x = -2$

Solution: The equation $x = -2$ can be written as $x + 0y = -2$. No matter what y-value we choose, x is always -2. Fill in a table listing ordered-pair solutions of $x = -2$. Choose any three y-values. The x-values are always -2. Plot the ordered-pair solutions and graph $x = -2$.

x	y
−2	3
−2	0
−2	−2

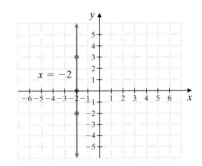

The graph is a vertical line that crosses the x-axis at -2.

■ **Work Practice Problem 5**

Vertical Lines

If a is a number, then the graph of $x = a$ is a *vertical line* that crosses the x-axis at a. For example, the graph of $x = -3$ is a vertical line that crosses the x-axis at -3.

Answer

5.

✔ Concept Check Answer

a. horizontal line   **b.** neither
c. vertical line   **d.** neither

✔ Concept Check   Determine whether the following equations represent vertical lines, horizontal lines, or neither.

a. $y = -6$ **b.** $x + y = 7$ **c.** $x = \dfrac{1}{2}$ **d.** $x = \dfrac{1}{2}y$

Vocabulary and Readiness Check

Use the choices below to fill in each blank.

horizontal   line

vertical   linear

1. A _____ equation in two variables can be written in the form $ax + by = c$.

2. The graph of the equation $x = 5$ is a _____ line.

3. The graph of the equation $y = -2$ is a _____ line.

4. The graph of the equation $3x - 2y = 6$ is a _____ that is not vertical or horizontal.

8.4 EXERCISE SET

FOR EXTRA HELP

Student Solutions Manual   PH Math/Tutor Center   CD/Video for Review   Math XL   MyMathLab

MathXL®   MyMathLab

Objective **A** *Graph each equation. See Examples 1 through 3.*

1. $x + y = 4$

2. $x + y = 2$

3. $x - y = -6$

4. $y - x = 6$

5. $y = 4x$

6. $x = 2y$

7. $y = 2x - 1$

8. $y = 2x + 5$

Graph each equation. See Examples 4 and 5.

9. $x = -3$

10. $y = 1$

 11. $y = -3$

12. $x = -7$

13. $x = 0$

14. $y = 0$

Mixed Practice *Graph each equation. See Examples 1 through 5.*

15. $y = -2x$

16. $x = y$

17. $y = -2$

18. $x = 1$

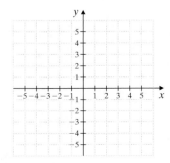

19. $x + 2y = 12$

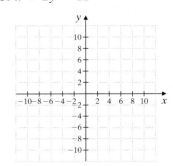

20. $3x - y = 3$

21. $x = 6$

22. $y = 7$

23. $y = x - 3$

24. $y = x - 1$

25. $x = y - 4$

26. $x = 5 - y$

27. $x + 3 = 0$

28. $y + 4 = 0$

29. $y = -\dfrac{1}{4}x$

30. $y = \dfrac{1}{4}x$

31. $y = \dfrac{1}{3}x$

32. $y = -\dfrac{1}{3}x$

33. $y = 4x + 2$

34. $y = -2x + 3$

35. $2x + 3y = 6$

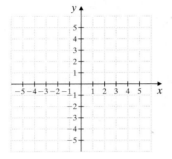

36. $5x - 2y = 10$

37. $x = -3.5$

38. $y = 5.5$

39. $3x - 4y = 24$

40. $4x + 2y = 16$

Review

Perform the indicated operations. See Sections 4.3 through 4.5, and 5.5.

41. $(0.5)\left(-\dfrac{1}{8}\right)$

42. $\dfrac{3}{4} - \dfrac{19}{20}$

43. $\dfrac{3}{4} \div \left(-\dfrac{19}{20}\right)$

44. $0.75 + \dfrac{3}{10}$

45. $\dfrac{2}{11} - \dfrac{x}{11}$

46. $\dfrac{2}{11}\left(-\dfrac{x}{11}\right)$

Concept Extensions

47. Fill in the table and use it to graph $y = |x|$ on the coordinate system.

x	y
−3	
−2	
−1	
0	
1	
2	
3	

 48. Graph: $15x - 18y = 270$

To do so, complete the table and find at least 2 additional ordered-pair solutions.

x	y
0	
	0

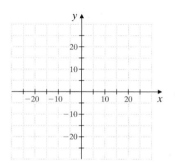

49. In your own words, explain how to graph a linear equation in two variables.

50. Suppose that a classmate tries to graph a linear equation in two variables and plots three ordered-pair solutions. If the three points do not all lie on the same line, what should the student do next?

The graph below is called a double line graph. Use this graph to answer Exercises 51 through 57.

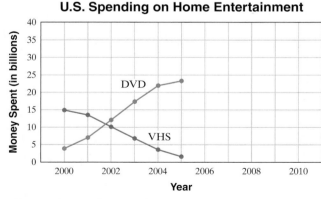

U.S. Spending on Home Entertainment

(*Source:* Adams Media Research)

51. Is U.S. spending on DVD home entertainment increasing or decreasing?

52. Is U.S. spending on VHS home entertainment increasing or decreasing?

53. Estimate the amount of money spent on DVD home entertainment in 2005.

54. Estimate the amount of money spent on VHS home entertainment in 2005.

55. Between what two years did the amount of spending on DVD entertainment equal the amount spent on VHS?

56. Describe any trends you see from this graph.

57. Place a straight edge (or ruler) on the DVD point for 2000 and for 2005. Draw a line using these two points and use this line to estimate between what two years this amount of money will be $40 billion.

8.5 COUNTING AND INTRODUCTION TO PROBABILITY

Objective **A** Using a Tree Diagram

In our daily conversations, we often talk about the likelihood or **probability** of a given result occurring. For example:

The *chance* of thundershowers is 70 percent.

What are the *odds* that the New Orleans Saints will go to the Super Bowl?

What is the *probability* that you will finish cleaning your room today?

Each of these chance happenings—thundershowers, the New Orleans Saints playing in the Super Bowl, and cleaning your room today—is called an **experiment.** The possible results of an experiment are called **outcomes.** For example, flipping a coin is an experiment, and the possible outcomes are heads (H) or tails (T).

One way to picture the outcomes of an experiment is to draw a **tree diagram.** Each outcome is shown on a separate branch. For example, the outcomes of flipping a coin are

Heads   Tails

PRACTICE PROBLEM 1

Draw a tree diagram for tossing a coin three times. Then use the diagram to find the number of possible outcomes.

EXAMPLE 1 Draw a tree diagram for tossing a coin twice. Then use the diagram to find the number of possible outcomes.

Solution: There are 4 possible outcomes when tossing a coin twice.

First Coin Toss	Second Coin Toss	Outcomes
H	H	H, H
	T	H, T
T	H	T, H
	T	T, T

Work Practice Problem 1

Answer

1.

8 outcomes

EXAMPLE 2 Draw a tree diagram for an experiment consisting of rolling a die and then tossing a coin. Then use the diagram to find the number of possible outcomes.

Die

Solution: Recall that a die has six sides and that each side represents a number, 1 through 6.

Roll a Die	Toss a coin	Outcomes
1	H	1, H
	T	1, T
2	H	2, H
	T	2, T
3	H	3, H
	T	3, T
4	H	4, H
	T	4, T
5	H	5, H
	T	5, T
6	H	6, H
	T	6, T

There are 12 possible outcomes for rolling a die and then tossing a coin.

Work Practice Problem 2

Any number of outcomes considered together are called an **event.** For example, when tossing a coin twice, H, H is an event. The event is tossing heads first and tossing heads second. Another event would be tossing tails first and then heads (T, H), and so on.

Objective B Finding the Probability of an Event

As we mentioned earlier, the **probability of an event is a measure of the chance or likelihood of it occurring.** For example, if a coin is tossed, what is the probability that heads occurs? Since one of two equally likely possible outcomes is heads, the probability is $\frac{1}{2}$.

The Probability of an Event

$$\text{probability of an event} = \frac{\text{number of ways that the event can occur}}{\text{number of possible outcomes}}$$

Note from the definition of probability that the probability of an event is always between 0 and 1, inclusive (i.e., including 0 and 1). A probability of 0 means that an event won't occur, and a probability of 1 means that an event is certain to occur.

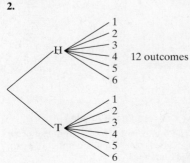

PRACTICE PROBLEM 3

If a coin is tossed three times, find the probability of tossing tails, then heads, then tails (T, H, T).

EXAMPLE 3 If a coin is tossed twice, find the probability of tossing heads on the first toss and then heads again on the second toss (H, H).

Solution: 1 way the event can occur

$$\underbrace{H, T, \quad \overbrace{H, H,}^{} \quad T, H, \quad T, T}_{4 \text{ possible outcomes}}$$

probability $= \dfrac{1}{4}$ Number of ways the event can occur
    Number of possible outcomes

The probability of tossing heads and then heads is $\dfrac{1}{4}$.

◻ **Work Practice Problem 3**

PRACTICE PROBLEM 4

If a die is rolled one time, find the probability of rolling a 2 or a 5.

EXAMPLE 4 If a die is rolled one time, find the probability of rolling a 3 or a 4.

Solution: Recall that there are 6 possible outcomes when rolling a die.

2 ways that the event can occur

possible outcomes: $\underbrace{1, \quad 2, \quad 3, \quad 4, \quad 5, \quad 6}_{6 \text{ possible outcomes}}$

probability of a 3 or a 4 $= \dfrac{2}{6}$ Number of ways the event can occur
    Number of possible outcomes

$= \dfrac{1}{3}$ Simplest form

◻ **Work Practice Problem 4**

✔ **Concept Check** Suppose you have calculated a probability of $\dfrac{11}{9}$. How do you know that you have made an error in your calculation?

PRACTICE PROBLEM 5

Use the diagram and information from Example 5 and find the probability of choosing a blue marble from the box.

EXAMPLE 5 Find the probability of choosing a red marble from a box containing 1 red, 1 yellow, and 2 blue marbles.

Solution: 1 way that event can occur

red   yellow   blue   blue

4 possible outcomes

probability $= \dfrac{1}{4}$

◻ **Work Practice Problem 5**

Answers

3. $\dfrac{1}{8}$ 4. $\dfrac{1}{3}$ 5. $\dfrac{1}{2}$

✔ **Concept Check Answer**

The number of ways an event can occur can't be larger than the number of possible outcomes.

Vocabulary and Readiness Check

Use the choices below to fill in each blank.

    0   probability   tree diagram

    1   outcome

1. A possible result of an experiment is called a(n) _____.

2. A _____ shows each outcome of an experiment as a separate branch.

3. The _____ of an event is a measure of the likelihood of it occurring.

4. _____ is calculated by the number of ways that the event can occur divided by the number of possible outcomes.

5. A probability of _____ means that an event won't occur.

6. A probability of _____ means that an event is certain to occur.

8.5 EXERCISE SET

FOR EXTRA HELP

Student Solutions Manual · PH Math/Tutor Center · CD/Video for Review · MathXL · MyMathLab

Objective A *Draw a tree diagram for each experiment. Then use the diagram to find the number of possible outcomes. See Examples 1 and 2.*

1. Choosing the letters in the word MATH, then a number (1, 2, or 3)

2. Choosing a number (1 or 2) and then a vowel (a, e, i, o, u)

3. Spinning Spinner A once

4. Spinning Spinner B once

Spinner A   Spinner B

5. Spinning Spinner B twice

6. Spinning Spinner A twice

7. Spinning Spinner A and then Spinner B

8. Spinning Spinner B and then Spinner A

9. Tossing a coin and then spinning Spinner B

10. Spinning Spinner A and then tossing a coin

Objective **B** *If a single die is tossed once, find the probability of each event. See Examples 3 through 5.*

11. A 5

12. A 9

13. A 1 or a 6

14. A 2 or a 3

15. An even number

16. An odd number

17. A number greater than 2

18. A number less than 6

Suppose the spinner shown is spun once. Find the probability of each event. See Examples 3 through 5.

19. The result of the spin is 2.

20. The result of the spin is 3.

21. The result of a spin is 1, 2, or 3.

22. The result of a spin is not 3.

23. The result of the spin is an odd number.

24. The result of the spin is an even number.

If a single choice is made from the bag of marbles shown, find the probability of each event. See Examples 3 through 5.

25. A red marble is chosen.

26. A blue marble is chosen.

27. A yellow marble is chosen.

28. A green marble is chosen.

29. A green or red marble is chosen.

30. A blue or yellow marble is chosen.

A new drug is being tested that is supposed to lower blood pressure. This drug was given to 200 people and the results are shown below.

Lower Blood Pressure	Higher Blood Pressure	Blood Pressure Not Changed
152	38	10

31. If a person is testing this drug, what is the probability that their blood pressure will be higher?

32. If a person is testing this drug, what is the probability that their blood pressure will be lower?

33. If a person is testing this drug, what is the probability that their blood pressure will not change?

34. What is the sum of the answers to Exercises 31, 32, and 33? In your own words, explain why.

Review

Perform each indicated operation. See Sections 4.3 and 4.5.

35. $\dfrac{1}{2} + \dfrac{1}{3}$ **36.** $\dfrac{7}{10} - \dfrac{2}{5}$ **37.** $\dfrac{1}{2} \cdot \dfrac{1}{3}$ **38.** $\dfrac{7}{10} \div \dfrac{2}{5}$ **39.** $5 \div \dfrac{3}{4}$ **40.** $\dfrac{3}{5} \cdot 10$

Concept Extensions

Recall that a deck of cards contains 52 cards. These cards consist of four suits (hearts, spades, clubs, and diamonds) of each of the following: 2, 3, 4, 5, 6, 7, 8, 9, 10, jack, queen, king, and ace. If a card is chosen from a deck of cards, find the probability of each event.

41. The king of hearts

42. The 10 of spades

43. A king

44. A 10

45. A heart

46. A club

47. A card in black ink

48. A queen or ace

Two dice are tossed. Find the probability of each sum of the dice. (Hint: Draw a tree diagram of the possibilities of two tosses of a die, and then find the sum of the numbers on each branch.)

49. A sum of 6

50. A sum of 10

51. A sum of 13

52. A sum of 2

Solve. See the Concept Check in this section.

53. In your own words, explain why the probability of an event cannot be greater than 1.

54. In your own words, explain when the probability of an event is 0.

CHAPTER 8   Group Activity

Scatter Diagrams
Sections 8.1–8.3

In Section 8.1, we learned about presenting data visually in a graph, such as a pictograph, bar graph, or line graph. Now we learn about another type of graph, based on ordered pairs, that can be used to present data.

Data that can be represented as an ordered pair is called paired data. Many types of data collected from the real world are paired data. For instance, the total amount of rainfall a location receives each year can be written as an ordered pair of the form (year, total rainfall in inches) and is paired data. The graph of paired data as points in the rectangular coordinate system is called a **scatter**

diagram, or scatter plot. Scatter diagrams can be used to look for patterns and trends in paired data.

For example, the data shown in the table is paired data that can be written as a set of ordered pairs of the form (year, number of restaurants in thousands), such as (2005, 25) and (2006, 27). A scatter diagram of the paired data is shown below. Notice that the horizontal axis is labeled "Year" to describe the x-coordinates in the ordered pairs. The vertical axis is labeled "Restaurants (in thousands)" to describe the y-coordinates in the ordered pairs.

Number of Subway Restaurants	
Year	Number of Restaurants (in thousands)
2002	18
2003	20
2004	21
2005	25
2006	27

(*Source:* Internet research Corporation)

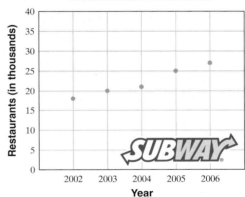

The table to the right gives the annual revenue of McDonald's corporation each year. Use the table to answer the following questions.

1. Write this paired data as a set of ordered pairs of the form (year, revenue in billions).
2. Create a scatter diagram of the paired data.
3. What trend in the paired data does the scatter diagram show?
4. Find or collect your own paired data and present it in a scatter diagram. What does the graph show?

Annual Revenue of McDonald's Corporation (in billions)	
Year	Annual Revenue of Company (in billions)
2001	$15
2002	$16
2003	$17
2004	$19
2005	$21

(*Source:* McDonald's Corporation)

Chapter 8 Vocabulary Check

Fill in each blank with one of the words or phrases listed below.

outcomes	quadrants	line	*y*	class frequency	probability
pictograph	plotting	origin	experiment	linear	class interval
histogram	bar	*x*	circle	tree diagram	

1. A _____ graph presents data using vertical or horizontal bars.

2. The possible results of an experiment are the _____.

3. A _____ is a graph in which pictures or symbols are used to visually present data.

4. A _____ graph displays information with a line that connects data points.

5. In the ordered pair (a, b) the _____-value is a and the _____-value is b.

6. A _____ is one way to picture and count outcomes.

7. An _____ is an activity being considered, such as tossing a coin or rolling a die.

8. In a _____ graph, each section (shaped like a piece of pie) shows a category and the relative size of the category.

9. The _____ of an event is $\dfrac{\text{number of ways that the event can occur}}{\text{number of possible outcomes}}$.

10. A _____ is a special bar graph in which the width of each bar represents a _____ and the height of each bar represents the _____.

11. The point of intersection of the *x*-axis and the *y*-axis in the rectangular coordinate system is called the _____.

12. The axes divide the plane into four regions, called _____.

13. The process of locating a point on the rectangular coordinate system is called _____ the point.

14. A _____ equation in two variables can be written in the form $ax + by = c$.

Helpful Hint

Are you preparing for your test? Don't forget to take the Chapter 8 Test on page 600. Then check your answers at the back of the text and use the Chapter Test Prep Video CD to see the fully worked-out solutions to any of the exercises you want to review.

8 Chapter Highlights

DEFINITIONS AND CONCEPTS	EXAMPLES
Section 8.1 Reading Pictographs, Bar Graphs, Histograms, and Line Graphs	

A **pictograph** is a graph in which pictures or symbols are used to visually present data.

A **line graph** displays information with a line that connects data points.

A **bar graph** presents data using vertical or horizontal bars.

The bar graph on the right shows the number of acres of wheat harvested in 1996 for leading states.

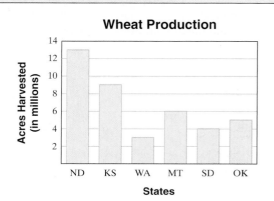

Wheat Production

Source: U.S. Department of Agriculture

DEFINITIONS AND CONCEPTS	**EXAMPLES**

Section 8.1   Reading Pictographs, Bar Graphs, Histograms, and Line Graphs (*continued*)

A **histogram** is a special bar graph in which the width of each bar represents a **class interval** and the height of each bar represents the **class frequency.** The histogram on the right shows student quiz scores.

1. Approximately how many acres of wheat were harvested in Kansas?

9,000,000 acres

2. About how many more acres of wheat were harvested in North Dakota than South Dakota?

$$\begin{array}{r} 13 \text{ million} \\ - \ 4 \text{ million} \\ \hline 9 \text{ million} \end{array} \quad \text{or } 9{,}000{,}000 \text{ acres}$$

1. How many students received a score of 6–10?

4 students

2. How many students received a score of 11–20?

9 + 13 = 22 students

Section 8.2   Reading Circle Graphs

In a **circle graph,** each section (shaped like a piece of pie) shows a category and the relative size of the category. The circle graph on the right classifies tornadoes by wind speed.

Tornado Wind Speeds

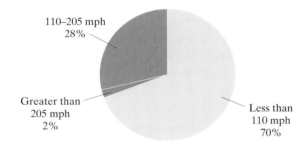

Source: National Oceanic and Atmospheric Administration

1. What percent of tornadoes have wind speeds of 110 mph or greater?

28% + 2% = 30%

2. If there were 1235 tornadoes in the United States in 1995, how many of these might we expect to have had wind speeds less than 110 mph? Find 70% of 1235.

$$70\%(1235) = 0.70(1235) = 864.5 \approx 865$$

Around 865 tornadoes would be expected to have had wind speeds of less than 110 mph.

Section 8.3   The Rectangular Coordinate System and Paired Data

The **rectangular coordinate system** consists of two number lines intersecting at the point 0 on each number line. The horizontal number line is called the **x-axis** and the vertical number line is called the **y-axis.**

Every point in the rectangular coordinate system corresponds to an **ordered pair of numbers** such as

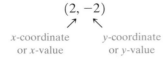

$$(2, -2)$$
 ↗ ↖
x-coordinate   *y*-coordinate
or *x*-value   or *y*-value

An ordered pair is a **solution** of an equation if the equation is a true statement when the variables are replaced by the coordinates of the ordered pair.

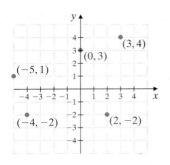

Is $(2, -1)$ a solution of $5x - y = 11$?

$$5x - y = 11$$
$$5(2) - (-1) \stackrel{?}{=} 11 \quad \text{Replace } x \text{ with 2 and } y \text{ with } -1.$$
$$10 + 1 \stackrel{?}{=} 11 \quad \text{Multiply.}$$
$$11 = 11 \quad \text{True}$$

Yes, $(2, -1)$ is a solution of $5x - y = 11$.

Section 8.4   Graphing Linear Equations in Two Variables

LINEAR EQUATIONS IN TWO VARIABLES

A linear equation in two variables is an equation that can be written in the form

$$ax + by = c$$

where *a*, *b*, and *c* are numbers, and *a* and *b* are not both 0.

To graph a linear equation in two variables, find three ordered-pair solutions and draw the line through the plotted points.

Graph: $y = 4x$

Let $x = 1$.   Let $x = -1$.   Let $x = 0$.
  $y = 4(1)$ $y = 4(-1)$ $y = 4(0)$
  $y = 4$ $y = -4$ $y = 0$

x	y
1	4
−1	−4
0	0

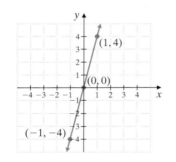

The graph of $y = a$ is a **horizontal line** that crosses the *y*-axis at *a*.

The graph of $x = a$ is a **vertical line** that crosses the *x*-axis at *a*.

An **experiment** is an activity being considered, such as tossing a coin or rolling a die. The possible results of an experiment are the **outcomes**. A **tree diagram** is one way to picture and count outcomes.

Draw a tree diagram for tossing a coin and then choosing a number from 1 to 4.

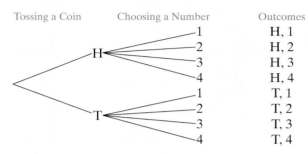

Any number of outcomes considered together is called an **event**. The **probability** of an event is a measure of the chance or likelihood of it occurring.

$$\text{probability of an event} = \frac{\text{number of ways that the event can occur}}{\text{number of possible outcomes}}$$

Find the probability of tossing a coin twice and tails occurring each time.

1 way the event can occur

$$\underbrace{(H,H), (H,T), (T,H), (T,T)}$$

4 possible outcomes

$$\text{probability} = \frac{1}{4}$$

8 CHAPTER REVIEW

(8.1) *The following pictograph shows the number of new homes constructed from June, 2005 to June, 2006, by region. Use this graph to answer Exercises 1 through 6.*

**New Home Construction
June 2005–June 2006**

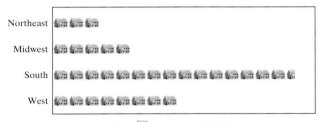

Each 🏠 represents 800,000 homes

Source: U.S. Census Bureau

1. How many housing starts were there in the Midwest during the given year?

2. How many housing starts were there in the Northeast during the given year?

3. Which region had the most housing starts?

4. Which region had the fewest housing starts?

5. Which region(s) had 6,400,000 or more housing starts?

6. Which region(s) has fewer than 6,400,000 housing starts?

The following bar graph shows the percent of persons age 25 or over who completed four or more years of college. Use this graph to answer Exercises 7 through 10.

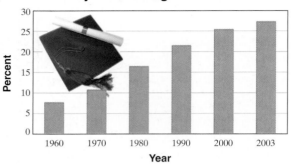

**Four or More Years of College
by Persons Age 25 or Over**

Source: U.S. Census Bureau

7. Approximate the percent of persons who completed four or more years of college in 1960.

8. What year shown had the greatest percent of persons completing four or more years of college?

9. What years shown had 15% or more of persons completing four or more years of college?

10. Describe any patterns you notice in this graph.

The following line graph shows the total number of Olympic Medal Events for Winter Olympics since 1992. Use this graph to answer Exercises 11 through 15.

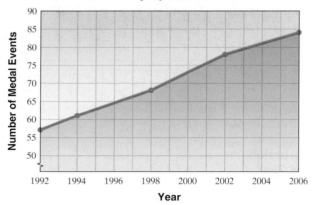

**Number of Olympic Medal Events
Winter Olympics: 1992–2006**

Source: International Olympic Committee

11. Approximate the number of medal events at the Winter Olympics of 1998.

12. Approximate the number of medal events at the Winter Olympics of 2002.

13. Approximate the number of medal events at the Winter Olympics of 2006.

14. How many more medal events were there at the Winter Olympics of 2006 than at the Winter Olympics of 2002?

15. How many more medal events were there at the Winter Olympics of 2006 than the Winter Olympics of 1992?

The following histogram shows the hours worked per week by the employees of Southern Star Furniture. Use this histogram to answer Exercises 16 through 19.

Southern Star Furniture

16. How many employees work 21–25 hours per week?

17. How many employees work 41–45 hours per week?

18. How many employees work 36 hours or more per week?

19. How many employees work 30 hours or less per week?

Following is a list of monthly record high temperatures for New Orleans, Louisiana. Use this list to complete the frequency distribution table below.

83	96	101	92
85	100	92	102
89	101	87	84

	Class Intervals (Temperatures)	Tally	Class Frequency (Number of Months)
20.	80°–89°	____	____
21.	90°–99°	____	____
22.	100°–109°	____	____

23. Use the table from Exercises 20–22 to draw a histogram.

(8.2) *The following circle graph shows a family's $4000 monthly budget. Use this graph to answer Exercises 24 through 30. Write all ratios as fractions in simplest form.*

24. What is the largest budget item?

25. What is the smallest budget item?

26. How much money is budgeted for the mortgage payment and utilities?

27. How much money is budgeted for savings and contributions?

28. Find the ratio of the mortage payment to the total monthly budget.

29. Find the ratio of food to the total monthly budget.

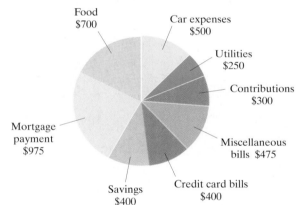

30. Find the ratio of car expenses to food.

The following circle graph shows the percent of states with various rural interstate highway speed limits in 2006. Use this graph to determine the number of states with each speed limit in Exercises 31 through 34. All states are included in this data.

Percent of States with Rural Interstate Highway Speed Limit

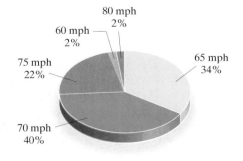

Source: Insurance Institute for Highway Safety

31. How many states have a rural interstate highway speed limit of 70 mph?

32. How many states have a rural interstate highway speed limit of 75 mph?

33. How many states have a rural interstate highway speed limit of 65 mph?

34. How many states have a rural interstate highway speed limit of 60 mph or 80 mph?

(8.3) *Complete and graph the ordered-pair solutions of each given equation.*

35. $x = -6y$; $(0, \quad), (\quad, -1), (-6, \quad)$

36. $y = 3x - 2$; $(0, \quad), (-1, \quad), (2, \quad)$

37. $x + y = -4$; $(-1, \quad), (\quad, 0), (-5, \quad)$

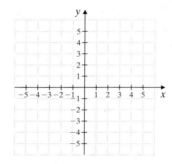

38. $x - y = 3$; $(4, \quad), (0, \quad), (\quad, 3)$

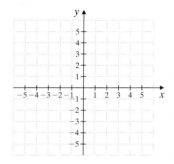

39. $y = 3x$; $(1, \quad), (-2, \quad), (\quad, 0)$

40. $x = y + 6$; $(1, \quad), (6, \quad), (\quad, -4)$

(8.4) *Graph each linear equation.*

41. $x = -5$

42. $y = 0$

43. $x + y = 11$

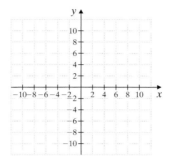

44. $x - y = 11$

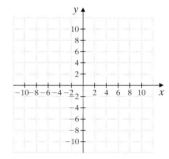

45. $y = 4x - 2$

46. $y = 5x$

47. $x = -2y$

48. $x + y = -1$

49. $2x - 3y = 12$

50. $x = \frac{1}{2}y$

(8.5) *Draw a tree diagram for each experiment. Then use the diagram to determine the number of outcomes.*

Spinner 1

Spinner 2

51. Tossing a coin and then spinning Spinner 1

52. Spinning Spinner 2 and then tossing a coin

53. Spinning Spinner 1 twice

54. Spinning Spinner 2 twice

55. Spinning Spinner 1 and then Spinner 2

Find the probability of each event.

Die

56. Rolling a 4 on a die

57. Rolling a 3 on a die

58. Spinning a 4 on Spinner 1

59. Spinning a 3 on Spinner 1

60. Spinning either a 1, 3, or 5 on Spinner 1

61. Spinning either a 2 or a 4 on Spinner 1

Mixed Review

Given a bag containing 2 red marbles, 2 blue marbles, 3 yellow marbles, and 1 green marble, find the following:

62. The probability of choosing a blue marble from the bag

63. The probability of choosing a yellow marble from the bag

64. The probability of choosing a red marble from the bag

65. The probability of choosing a green marble from the bag

Graph each equation.

66. $x = -4$

67. $y = 3$

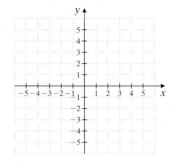

68. $x - 2y = 8$

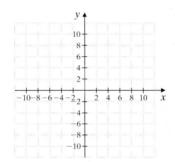

69. $2x + y = 6$

70. $x = y + 3$

71. $x + y = -4$

72. $y = \dfrac{3}{4}x$

73. $y = -\dfrac{3}{4}x$

CHAPTER TEST

 Remember to use the Chapter Test Prep Video CD to see the fully worked-out solutions to any of the exercises you want to review.

The following pictograph shows the money collected each week from a wrapping paper fundraiser. Use this graph to answer Exercises 1 through 3.

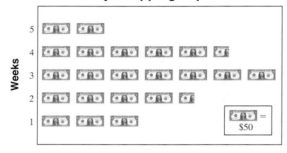

1. _____

2. _____

1. How much money was collected during the second week?

2. During which week was the most money collected? How much money was collected during that week?

3. What was the total money collected for the fundraiser?

The bar graph shows the normal monthly precipitation in centimeters for Chicago, Illinois. Use this graph to answer Exercises 4 through 6.

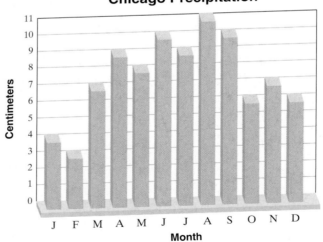

Source: U.S. National Oceanic and Atmospheric Administration, *Climatography of the United States*, No. 81

3. _____

4. _____

5. _____

4. During which month(s) does Chicago normally have more than 9 centimeters of precipitation?

5. During which month does Chicago normally have the least amount of precipitation? How much precipitation occurs during that month?

6. During which month(s) does 7 centimeters of precipitation normally occur?

7. Use the information in the table to draw a bar graph. Clearly label each bar.

Countries with the Highest Newspaper Cirulations	
Country	**Average Daily Circulation (in millions)**
Japan	72
U.S.	56
China	50
India	31
Germany	24
Russia	24
U.K.	19
(*Source:* World Association of Newspapers	

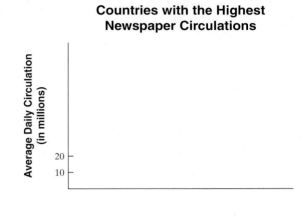

Countries with the Highest Newspaper Circulations

6. _____

7. see graph _____

8. _____

The following line graph shows the annual inflation rate in the United States for the years 1990–2003. Use this graph to answer Exercises 8 through 10.

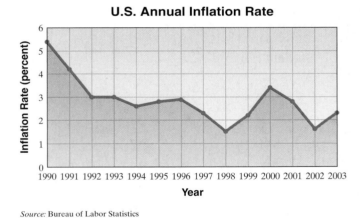

U.S. Annual Inflation Rate

Source: Bureau of Labor Statistics

8. Approximate the annual inflation rate in 2002.

9. During which of the years shown was the inflation rate greater than 3%?

10. During which sets of years was the inflation rate increasing?

9. _____

10. _____

The result of a survey of 200 people is shown in the following circle graph. Each person was asked to tell his or her favorite type of music. Use this graph to answer Exercises 11 and 12.

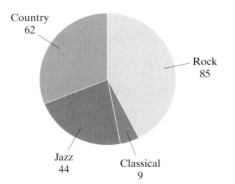

Country 62

Rock 85

Jazz 44

Classical 9

11. Find the ratio of those who prefer rock music to the total number surveyed.

12. Find the ratio of those who prefer country music to those who prefer jazz.

11. _____

12. _____

13. _____

The following circle graph shows the U.S. labor force employment by industry for 2000. There were approximately 132,000,000 people employed by these industries in the United States in 2000. Use the graph to find how many people were employed by the industries given in Exercises 13 and 14.

U.S. Labor Force Employment by Industry

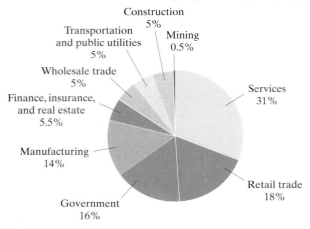

Source: Bureau of Labor Statistics

14. _____

13. Services

14. Government

A professor measures the heights of the students in her class. The results are shown in the following histogram. Use this histogram to answer Exercises 15 and 16.

15. _____

Student Heights

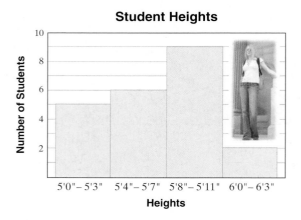

16. _____

15. How many students are 5′8″–5′11″ tall?

16. How many students are 5′7″ or shorter?

17. The history test scores of 25 students are shown below. Use these scores to complete the frequency distribution table.

70	86	81	65	92
43	72	85	69	97
82	51	75	50	68
88	83	85	77	99
77	63	59	84	90

Class Intervals (Scores)	Tally	Class Frequency (Number of Students)
40–49		
50–59		
60–69		
70–79		
80–89		
90–99		

17. see table

18. Use the results of Exercise 17 to draw a histogram.

Scores

Find the coordinates of each point in the graph below.

19. *A*   **20.** *B*   **21.** *C*   **22.** *D*

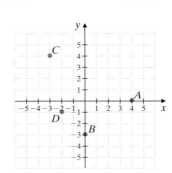

Complete and graph the ordered-pair solutions of each given equation.

23. $x = -6y$; $(0, \quad)$, $(\quad, 1)$, $(12, \quad)$ **24.** $y = 7x - 4$; $(2, \quad)$, $(-1, \quad)$, $(0, \quad)$

          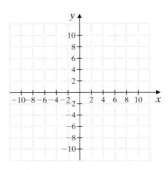

Graph each linear equation.

25. $y + x = -4$ **26.** $y = -4$

          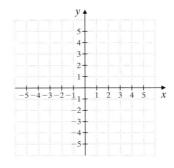

27. see graph

28. see graph

29. see graph

30. see graph

27. $y = 3x - 5$

28. $x = 5$

29. $y = -\dfrac{1}{2}x$

30. $3x - 2y = 12$

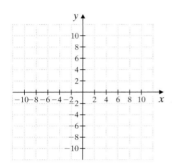

31. Draw a tree diagram for the experiment of spinning the spinner twice.

32. Draw a tree diagram for the experiment of tossing a coin twice.

Suppose that the numbers 1 through 10 are each written on the same size sheet of paper and placed in a bag. You then select one sheet of paper from the bag.

33. What is the probability of choosing a 6 from the bag?

34. What is the probability of choosing a 3 or a 4 from the bag?

31. _____

32. _____

33. _____

34. _____

1. Simplify: $4^3 + [3^2 - (10 \div 2)] - 7 \cdot 3$

2. $7^2 - [5^3 + (6 \div 3)] + 4 \cdot 2$

3. Evaluate $x - y$ for $x = -3$ and $y = 9$.

4. Evaluate $x - y$ for $x = 7$ and $y = -2$.

5. Solve: $3y - 7y = 12$

6. Solve: $2x - 6x = 24$

7. Solve: $\dfrac{x}{6} + 1 = \dfrac{4}{3}$

8. Solve: $\dfrac{7}{2} + \dfrac{a}{4} = 1$

9. Add: $2\dfrac{1}{3} + 5\dfrac{3}{8}$

10. Add: $3\dfrac{2}{5} + 4\dfrac{3}{4}$

11. Write 5.9 as a mixed number.

12. Write 2.8 as a mixed number.

13. Subtract: $3.5 - 0.068$

14. Subtract: $7.4 - 0.073$.

15. Multiply: 0.0531×16

16. Multiply: 0.147×0.2.

17. Divide: $-5.98 \div 115$

18. Divide: $27.88 \div 205$.

19. Simplify: $(-1.3)^2 + 2.4$

20. Simplify: $(-2.7)^2$

21. Write $\dfrac{1}{4}$ as a decimal.

22. Write $\dfrac{3}{8}$ as a decimal.

23. Solve: $5(x - 0.36) = -x + 2.4$

24. Solve: $4(0.35 - x) = x - 7$

25. Approximate $\sqrt{80}$ to the nearest thousandth.

26. Approximate $\sqrt{60}$ to the nearest thousandth.

Answers

1. _____
2. _____
3. _____
4. _____
5. _____
6. _____
7. _____
8. _____
9. _____
10. _____
11. _____
12. _____
13. _____
14. _____
15. _____
16. _____
17. _____
18. _____
19. _____
20. _____
21. _____
22. _____
23. _____
24. _____
25. _____
26. _____

606

27. _____

28. _____

29. _____

30. _____

31. _____

32. _____

33. _____

34. _____

35. _____

36. _____

37. _____

38. _____

39. _____

40. _____

41. _____

42. _____

43. _____

44. _____

45. _____

46. _____

47. _____

48. _____

49. _____

50. _____

27. Write the ratio of 21 to 29 using fractional notation.

28. Write the ratio of 7 to 15 using fractional notation.

29. Write as a unit rate: 337.5 miles every 15 gallons of gas.

30. Write as a unit rate: $1.59 for 3 ounces.

31. Solve $\dfrac{51}{34} = \dfrac{-3}{x}$ for x and then check.

32. Solve $\dfrac{8}{5} = \dfrac{x}{10}$ for x and then check.

△**33.** Find the ratio of corresponding sides for the similar triangles ABC and DEF.

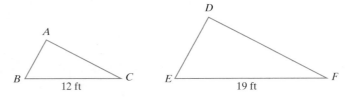

△**34.** Find the ratio of corresponding sides for the similar triangles GHJ and KLM.

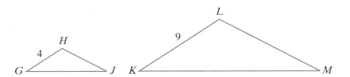

Write each percent as a decimal.

35. 4.6%

36. 32%

37. 0.74%

38. 2.7%

39. What number is 35% of 60?

40. What number is 40% of 36?

41. 20.8 is 40% of what number?

42. 9.5 is 25% of what number?

43. The sales tax on a $406 Sony flat screen digital 27-inch television is $34.51. Find the sales tax rate.

44. The sales tax on a $2.00 yo-yo is $0.13. Find the sales tax rate.

45. Four thousand dollars is invested at 5.3% compounded quarterly for 10 years. Find the total amount at the end of 10 years.

46. Linda Bonnett borrows $1600 for 1 year. If the interest is $128.60, find the monthly payment.

47. Find the median of the list of scores: 67, 91, 75, 86, 55, 91

48. Find the median of the list of numbers. 43, 46, 47, 50, 52, 83

49. If a die is rolled one time, find the probability of rolling a 3 or a 4.

50. If a die is rolled once, find the probability of rolling an even number.

9

Geometry and Measurement

The word *geometry* is formed from the Greek words *geo*, meaning "Earth," and *metron*, meaning "measure." Geometry literally means to measure the Earth. In this chapter we learn about various geometric figures and their properties, such as perimeter, area, and volume. Knowledge of geometry can help us solve practical problems in real-life situations. For instance, knowing certain measures of a circular swimming pool allows us to calculate how much water it can hold.

D id you know there are many more "grand canyons" in the United States in addition to the popular Grand Canyon of the Colorado River? One such canyon is called the Black Canyon of the Gunnison River. The walls of this canyon are deep and narrow, letting in only slanting rays of sunlight, thus the name "Black Canyon." The Gunnison River within the canyon is also unique in that it drops an average of 95 feet per mile (18 meters per kilometer). This is one of the greatest rates of fall for a river in North America. The diagram below shows canyon walls and river elevations for 12 miles (19 kilometers). In Section 9.4, Exercises 85 and 86, we explore the dimensions of some "grand canyons."

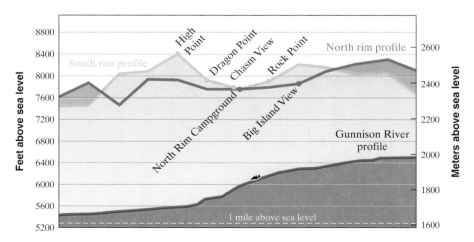

9.1 LINES AND ANGLES

Objective **A** Identifying Lines, Line Segments, Rays, and Angles

Let's begin with a review of two important concepts—space and plane.

Space extends in all directions indefinitely. Examples of objects in space are houses, grains of salt, bushes, your *Prealgebra* textbook, and you.

A **plane** is a flat surface that extends indefinitely. Surfaces like a plane are a classroom floor or a blackboard or whiteboard.

Plane

The most basic concept of geometry is the idea of a point in space. A **point** has no length, no width, and no height, but it does have location. We represent a point by a dot, and we usually label points with capital letters.

P

Point *P*

A **line** is a set of points extending indefinitely in two directions. A line has no width or height, but it does have length. We can name a line by any two of its points or by a single lowercase letter. A **line segment** is a piece of a line with two endpoints.

Line *AB*, $\overleftrightarrow{AB}$, or line *l**

Line segment *AB* or $\overline{AB}$

A **ray** is a part of a line with one endpoint. A ray extends indefinitely in one direction. An **angle** is made up of two rays that share the same endpoint. The common endpoint is called the **vertex.**

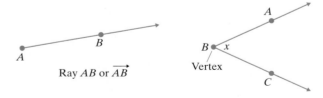

Ray *AB* or $\overrightarrow{AB}$

Vertex

The angle in the figure above can be named

$\angle ABC$ $\angle CBA$ $\angle B$ or $\angle x$
   ↑   ↑
The vertex is the
middle point.

Rays *BA* and *BC* are **sides** of the angle.

* Although line *l* is also line *BA* or $\overleftrightarrow{BA}$, we will use one order of points only to name a line or line segment.

Helpful Hint

Naming an Angle

When there is no confusion as to what angle is being named, you may use the vertex alone.

Name of ∠B is all right.
There is no confusion. ∠B means ∠1.

Name of ∠B is *not* all right.
There is confusion. Does ∠B mean
∠1, ∠2, ∠3, or ∠4?

EXAMPLE 1 Identify each figure as a line, a ray, a line segment, or an angle. Then name the figure using the given points.

a.

b.

c.

d.

Solution:

Figure (a) extends indefinitely in two directions. It is line CD or $\overleftrightarrow{CD}$.
Figure (b) has two endpoints. It is line segment EF or $\overline{EF}$.
Figure (c) has two rays with a common endpoint. It is $\angle MNO$, $\angle ONM$, or $\angle N$.
Figure (d) is part of a line with one endpoint. It is ray PT or $\overrightarrow{PT}$.

▪ **Work Practice Problem 1**

EXAMPLE 2 List other ways to name $\angle y$.

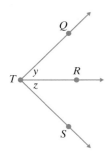

Solution: Two other ways to name $\angle y$ are $\angle QTR$ and $\angle RTQ$. We may *not* use the vertex alone to name this angle because three different angles have T as their vertex.

▪ **Work Practice Problem 2**

PRACTICE PROBLEM 1

Identify each figure as a line, a ray, a line segment, or an angle. Then name the figure using the given points.

PRACTICE PROBLEM 2

Use the figure in Example 2 to list other ways to name $\angle z$.

Answers

1. **a.** ray; ray AB or $\overrightarrow{AB}$
 b. line segment; line segment RS or $\overline{RS}$
 c. line; line EF or $\overleftrightarrow{EF}$
 d. angle; $\angle TVH$, or $\angle HVT$ or $\angle V$
2. $\angle RTS$, $\angle STR$

Objective B Classifying Angles as Acute, Right, Obtuse, or Straight

An angle can be measured in **degrees.** The symbol for degrees is a small, raised circle, °. There are 360° in a full revolution, or a full circle.

$\frac{1}{2}$ of a revolution measures $\frac{1}{2}(360°) = 180°$. An angle that measures 180° is called a **straight angle.**

∠RST is a straight angle.

$\frac{1}{4}$ of a revolution measures $\frac{1}{4}(360°) = 90°$. An angle that measures 90° is called a **right angle.** The symbol ∟ is used to denote a right angle.

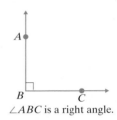

∠ABC is a right angle.

An angle whose measure is between 0° and 90° is called an **acute angle.**

Acute angles

An angle whose measure is between 90° and 180° is called an **obtuse angle.**

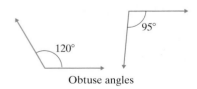

Obtuse angles

PRACTICE PROBLEM 3

Classify each angle as acute, right, obtuse, or straight.

a.
b.

c.
d.

EXAMPLE 3 Classify each angle as acute, right, obtuse, or straight.

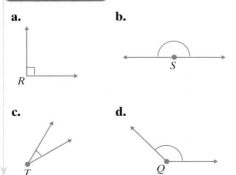

a.
b.

c.
d.

Answers

3. **a.** obtuse **b.** straight **c.** acute **d.** right

Solution:

a. $\angle R$ is a right angle, denoted by $\llcorner$. It measures 90°.
b. $\angle S$ is a straight angle. It measures 180°.
c. $\angle T$ is an acute angle. It measures between 0° and 90°.
d. $\angle Q$ is an obtuse angle. It measures between 90° and 180°.

▨ **Work Practice Problem 3**

Let's look at $\angle B$ below, whose measure is 62°.

There is a shorthand notation for writing the measure of this angle. To write "The measure of $\angle B$ is 62°," we can write,

$$\downarrow \quad \downarrow \;\; \downarrow \;\; \downarrow$$

$$m\angle B = 62°.$$

By the way, note that $\angle B$ is an acute angle because $m\angle B$ is between 0° and 90°.

Objective ⬤ Identifying Complementary and Supplementary Angles

Two angles that have a sum of 90° are called **complementary angles.** We say that each angle is the **complement** of the other.

$\angle R$ and $\angle S$ are complementary angles because

$$m\angle R + m\angle S = 60° + 30° = 90°$$

Complementary angles
$60° + 30° = 90°$

Two angles that have a sum of 180° are called **supplementary angles.** We say that each angle is the **supplement** of the other.

$\angle M$ and $\angle N$ are supplementary angles because

$$m\angle M + m\angle N = 125° + 55° = 180°$$

Supplementary angles
$125° + 55° = 180°$

EXAMPLE 4 Find the complement of a 48° angle.

Solution: Two angles that have a sum of 90° are complementary. This means that the complement of an angle that measures 48° is an angle that measures $90° - 48° = 42°$.

▨ **Work Practice Problem 4**

PRACTICE PROBLEM 4

Find the complement of a 29° angle.

Answer
4. 61°

PRACTICE PROBLEM 5

Find the supplement of a 67° angle.

EXAMPLE 5 Find the supplement of a 107° angle.

Solution: Two angles that have a sum of 180° are supplementary. This means that the supplement of an angle that measures 107° is an angle that measures 180° − 107° = 73°.

◼ **Work Practice Problem 5**

✔**Concept Check**   True or false? The supplement of a 48° angle is 42°. Explain.

Objective ⒟ Finding Measures of Angles

Measures of angles can be added or subtracted to find measures of related angles.

PRACTICE PROBLEM 6

a. Find the measure of ∠y.

b. Find the measure of ∠x.

c. Classify ∠x and ∠y as acute, obtuse, or right angles.

EXAMPLE 6 Find the measure of ∠x. Then classify ∠x as an acute, obtuse, or right angle.

Solution: $m\angle x = m\angle QTS - m\angle RTS$
$$= 87° - 52°$$
$$= 35°$$

Thus, the measure of ∠x ($m\angle x$) is 35°.
   Since ∠x measures between 0° and 90°, it is an acute angle.

◼ **Work Practice Problem 6**

Two lines in a plane can be either parallel or intersecting. **Parallel lines** never meet. **Intersecting lines** meet at a point. The symbol ‖ is used to indicate "is parallel to." For example, in the figure $p\|q$.

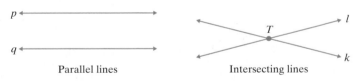

Parallel lines   Intersecting lines

Some intersecting lines are perpendicular. Two lines are **perpendicular** if they form right angles when they intersect. The symbol ⊥ is used to denote "is perpendicular to." For example, in the figure below, $n \perp m$.

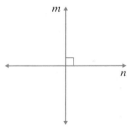

Perpendicular lines

When two lines intersect, four angles are formed. Two angles that are opposite each other are called **vertical angles.** Vertical angles have the same measure.
   Two angles that share a common side are called **adjacent angles.** Adjacent angles formed by intersecting lines are supplementary. That is, they have a sum of 180°.

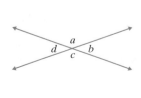

Vertical angles:
∠a and ∠c
∠d and ∠b

Adjacent angles:
∠a and ∠b
∠b and ∠c
∠c and ∠d
∠d and ∠a

Answers

5. 113° **6. a.** 44° **b.** 28° **c.** acute

✔ **Concept Check Answer**

false; the *complement* of a 48° angle is 42°; the *supplement* of a 48° angle is 132°

Here are a few real-life examples of the lines we just discussed.

Parallel lines

Vertical angles

Perpendicular lines

EXAMPLE 7 Find the measure of $\angle x$, $\angle y$, and $\angle z$ if the measure of $\angle t$ is 42°.

Solution: Since $\angle t$ and $\angle x$ are vertical angles, they have the same measure, so $\angle x$ measures 42°.

Since $\angle t$ and $\angle y$ are adjacent angles, their measures have a sum of 180°. So $\angle y$ measures $180° - 42° = 138°$.

Since $\angle y$ and $\angle z$ are vertical angles, they have the same measure. So $\angle z$ measures 138°.

🔲 **Work Practice Problem 7**

A line that intersects two or more lines at different points is called a **transversal.** Line l is a transversal that intersects lines m and n. The eight angles formed have special names. Some of these names are:

Corresponding angles: $\angle a$ and $\angle e$, $\angle c$ and $\angle g$, $\angle b$ and $\angle f$, $\angle d$ and $\angle h$

Alternate interior angles: $\angle c$ and $\angle f$, $\angle d$ and $\angle e$

When two lines cut by a transversal are *parallel,* the following statement is true:

Parallel Lines Cut by a Transversal

If two parallel lines are cut by a transversal, then the measures of **corresponding angles are equal** and the measures of the **alternate interior angles are equal.**

EXAMPLE 8 Given that $m \| n$ and that the measure of $\angle w$ is 100°, find the measures of $\angle x$, $\angle y$, and $\angle z$.

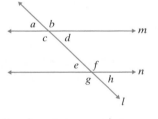

Solution:

$m\angle x = 100°$. $\angle x$ and $\angle w$ are vertical angles.

$m\angle z = 100°$. $\angle x$ and $\angle z$ are corresponding angles.

$m\angle y = 180° - 100° = 80°$. $\angle z$ and $\angle y$ are supplementary angles.

🔲 **Work Practice Problem 8**

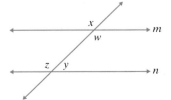

Vocabulary and Readiness Check

Use the choices below to fill in each blank.

acute	straight	degrees	adjacent	parallel	intersecting
obtuse	space	plane	point	vertical	vertex
right	angle	ray	line	perpendicular	transversal

1. $A(n)$ _____ is a flat surface that extends indefinitely.
2. $A(n)$ _____ has no length, no width, and no height.
3. _____ extends in all directions indefinitely.
4. $A(n)$ _____ is a set of points extending indefinitely in two directions.
5. $A(n)$ _____ is part of a line with one endpoint.
6. $A(n)$ _____ is made up of two rays that share a common endpoint. The common endpoint is called the _____.
7. $A(n)$ _____ angle measures 180°.
8. $A(n)$ _____ angle measures 90°.
9. $A(n)$ _____ angle measures between 0° and 90°.
10. $A(n)$ _____ angle measures between 90° and 180°.
11. _____ lines never meet and _____ lines meet at a point.
12. Two intersecting lines are _____ if they form right angles when they intersect.
13. An angle can be measured in _____.
14. A line that intersects two or more lines at different points is called a _____.
15. When two lines intersect, four angles are formed, called _____ angles.
16. Two angles that share a common side are called _____ angles.

9.1 EXERCISE SET

FOR EXTRA HELP

Student Solutions Manual · PH Math/Tutor Center · CD/Video for Review · MathXL MathXL® · MyMathLab

Objective A *Identify each figure as a line, a ray, a line segment, or an angle. Then name the figure using the given points. See Examples 1 and 2.*

1.

2.

3.

4.

5.

6.

7.

8.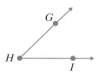

List two other ways to name each angle. See Example 2.

9. ∠x

10. ∠w

11. ∠z

12. ∠y

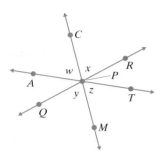

Objective **B** *Classify each angle as acute, right, obtuse, or straight. See Example 3.*

13.

14.

15.

16.

17.

18.

19.

20.

Objective **C** *Find each complementary or supplementary angle as indicated. See Examples 4 and 5.*

21. Find the complement of a 23° angle.

22. Find the complement of an 77° angle.

23. Find the supplement of a 17° angle.

24. Find the supplement of an 77° angle.

25. Find the complement of a 58° angle.

26. Find the complement of a 22° angle.

27. Find the supplement of a 150° angle.

28. Find the supplement of a 130° angle.

29. Identify the pairs of complementary angles.

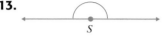

30. Identify the pairs of complementary angles.

31. Identify the pairs of supplementary angles.

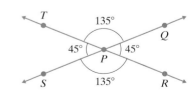

32. Identify the pairs of supplementary angles.

Objective **D** *Find the measure of ∠x in each figure. See Example 6.*

33.

34.

35.

36.
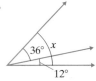

Find the measures of angles x, y, and z in each figure. See Examples 7 and 8.

37.

38.

39.

40.

41. *m ∥ n*

42. *m ∥ n*

43. *m ∥ n*

44. *m ∥ n*
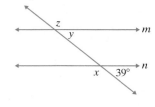

Objectives **A** **D** **Mixed Practice** *Find two other ways of naming each angle. See Example 2.*

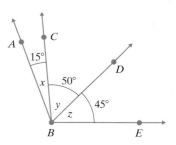

45. ∠x

46. ∠y

47. ∠z

48. ∠ABE (just name one other way)

Find the measure of each angle in the figure above.

49. ∠ABC

50. ∠EBD

51. ∠CBD

52. ∠CBA

53. ∠DBA

54. ∠EBC

55. ∠CBE

56. ∠ABE

Review

Perform each indicated operation. See Sections 4.3, 4.5, and 4.7.

57. $\dfrac{7}{8} + \dfrac{1}{4}$

58. $\dfrac{7}{8} - \dfrac{1}{4}$

59. $\dfrac{7}{8} \cdot \dfrac{1}{4}$

60. $\dfrac{7}{8} \div \dfrac{1}{4}$

61. $3\dfrac{1}{3} - 2\dfrac{1}{2}$

62. $3\dfrac{1}{3} + 2\dfrac{1}{2}$

63. $3\dfrac{1}{3} \div 2\dfrac{1}{2}$

64. $3\dfrac{1}{3} \cdot 2\dfrac{1}{2}$

Concept Extensions

65. The angle between the two walls of the Vietnam Veterans Memorial in Washington, D.C., is 125.2°. Find the supplement of this angle. (*Source:* National Park Service)

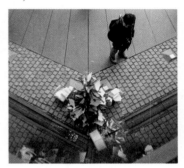

66. The faces of Khafre's Pyramid at Giza, Egypt, are inclined at an angle of 53.13°. Find the complement of this angle. (*Source:* PBS *NOVA* Online)

Answer true or false for Exercises 67 through 70. See the Concept Check in this section. If false, explain why.

67. The complement of a 100° angle is an 80° angle.

68. It is possible to find the complement of a 120° angle.

69. It is possible to find the supplement of a 120° angle.

70. The supplement of a 5° angle is a 175° angle.

71. If lines m and n are parallel, find the measures of angles a through e.

72. In your own words, describe how to find the complement and the supplement of a given angle.

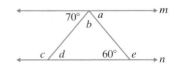

73. Can two supplementary angles both be acute? Explain why or why not.

74. Find two complementary angles with the same measure.

75. Is the figure below possible? Why or why not?

76. Below is a rectangle. List which segments, if extended, would be parallel lines.

9.2 PERIMETER

Objective A Using Formulas to Find Perimeters

Recall from Section 1.3 that the perimeter of a polygon is the distance around the polygon. This means that the perimeter of a polygon is the sum of the lengths of its sides.

EXAMPLE 1 Find the perimeter of the rectangle below.

5 inches

9 inches

Solution:

perimeter = 9 inches + 9 inches + 5 inches + 5 inches

= 28 inches

■ Work Practice Problem 1

PRACTICE PROBLEM 1

a. Find the perimeter of the rectangle.

18 meters

10 meters

b. Find the perimeter of the rectangular lot shown below:

50 feet

125 feet

Notice that the perimeter of the rectangle in Example 1 can be written as $2 \cdot (9 \text{ inches}) + 2 \cdot (5 \text{ inches})$.

↑ length   ↑ width

In general, we can say that the perimeter of a rectangle is always

$2 \cdot \text{length} + 2 \cdot \text{width}$

As we have just seen, the perimeter of some special figures such as rectangles form patterns. These patterns are given as **formulas.** The formula for the perimeter of a rectangle is shown next:

Perimeter of a Rectangle

Perimeter = $2 \cdot$ **length** + $2 \cdot$ **width**

In symbols, this can be written as

$$P = 2l + 2w$$

length

width   width

length

PRACTICE PROBLEM 2

Find the perimeter of a rectangle with a length of 32 centimeters and a width of 15 centimeters.

EXAMPLE 2 Find the perimeter of a rectangle with a length of 11 inches and a width of 3 inches.

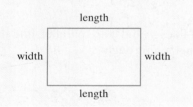

11 in.

3 in.

Solution: We use the formula for perimeter and replace the letters by their known lengths.

$P = 2l + 2w$

$= 2 \cdot 11 \text{ in.} + 2 \cdot 3 \text{ in.}$ Replace l with 11 in. and w with 3 in.

$= 22 \text{ in.} + 6 \text{ in.}$

$= 28 \text{ in.}$

The perimeter is 28 inches.

■ Work Practice Problem 2

Answers

1. a. 56 m **b.** 350 ft **2.** 94 cm

Recall that a square is a special rectangle with all four sides the same length. The formula for the perimeter of a square is shown next:

Perimeter of a Square

Perimeter = **side** + **side** + **side** + **side**
= 4 · **side**

In symbols,

$P = 4s$

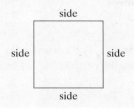

EXAMPLE 3 Finding the Perimeter of a Field

How much fencing is needed to enclose a square field 50 yards on a side?

Solution: To find the amount of fencing needed, we find the distance around, or perimeter. The formula for the perimeter of a square is $P = 4 \cdot s$. We use this formula and replace s by 50 yards.

$P = 4 \cdot s$
$= 4 \cdot 50 \text{ yd}$
$= 200 \text{ yd}$

The amount of fencing needed is 200 yards.

■ **Work Practice Problem 3**

The formula for the perimeter of a triangle with sides of lengths a, b, and c is given next:

Perimeter of a Triangle

Perimeter = side **a** + side **b** + side **c**

In symbols,

$P = a + b + c$

side a
side b
side c

PRACTICE PROBLEM 3

Find the perimeter of a square tabletop if each side is 4 feet long.

4 feet
4 feet

PRACTICE PROBLEM 4

Find the perimeter of a triangle if the sides are 6 centimeters, 10 centimeters, and 8 centimeters in length.

EXAMPLE 4 Find the perimeter of a triangle if the sides are 3 inches, 7 inches, and 6 inches.

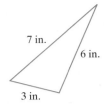

Solution: The formula for the perimeter is $P = a + b + c$, where a, b, and c are the lengths of the sides. Thus,

$$P = a + b + c$$
$$= 3 \text{ in.} + 7 \text{ in.} + 6 \text{ in.}$$
$$= 16 \text{ in.}$$

The perimeter of the triangle is 16 inches.

▢ **Work Practice Problem 4**

The method for finding the perimeter of any polygon is given next:

Perimeter of a Polygon

The perimeter of a polygon is the sum of the lengths of its sides.

PRACTICE PROBLEM 5

Find the perimeter of the trapezoid shown.

EXAMPLE 5 Find the perimeter of the trapezoid shown below:

Solution: To find the perimeter, we find the sum of the lengths of its sides.

perimeter = 3 cm + 2 cm + 6 cm + 3 cm = 14 cm

The perimeter is 14 centimeters.

▢ **Work Practice Problem 5**

PRACTICE PROBLEM 6

Find the perimeter of the room shown.

EXAMPLE 6 Finding the Perimeter of a Room

Find the perimeter of the room shown below:

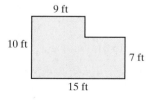

Solution: To find the perimeter of the room, we first need to find the lengths of all sides of the room.

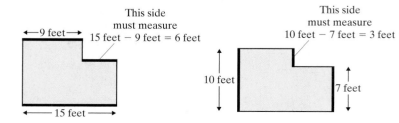

Answers

4. 24 cm   **5.** 23 km   **6.** 102 m

Now that we know the measures of all sides of the room, we can add the measures to find the perimeter.

perimeter = 10 ft + 9 ft + 3 ft + 6 ft + 7 ft + 15 ft

= 50 ft

The perimeter of the room is 50 feet.

📖 **Work Practice Problem 6**

EXAMPLE 7 **Calculating the Cost of Wallpaper Border**

A rectangular room measures 10 feet by 12 feet. Find the cost to hang a wallpaper border on the walls close to the ceiling if the cost of the wallpaper border is $1.09 per foot.

Solution: First we find the perimeter of the room.

$P = 2 \cdot l + 2 \cdot w$

$= 2 \cdot 12 \text{ ft} + 2 \cdot 10 \text{ ft}$ Replace *l* with 12 feet and *w* with 10 feet.

$= 24 \text{ ft} + 20 \text{ ft}$

$= 44 \text{ ft}$

The cost of the wallpaper is

cost = $1.09 \cdot 44$ ft = 47.96

The cost of the wallpaper is $47.96.

📖 **Work Practice Problem 7**

PRACTICE PROBLEM 7

A rectangular lot measures 60 feet by 120 feet. Find the cost to install fencing around the lot if the cost of fencing is $1.90 per foot.

Objective B Using Formulas to Find Circumferences

Recall from Section 5.3 that the distance around a circle is called the **circumference.** This distance depends on the radius or the diameter of the circle.

The formulas for circumference are shown next:

Circumference of a Circle

Circumference = $2 \cdot \pi \cdot$ radius or Circumference = $\pi \cdot$ diameter

In symbols,

$C = 2\pi r$ or $C = \pi d$

where $\pi \approx 3.14$ or $\pi \approx \dfrac{22}{7}$.

To better understand circumference and π(pi), try the following experiment. Take any can and measure its circumference and its diameter.

The can in the figure above has a circumference of 23.5 centimeters and a diameter of 7.5 centimeters. Now divide the circumference by the diameter.

$$\frac{\text{circumference}}{\text{diameter}} = \frac{23.5 \text{ cm}}{7.5 \text{ cm}} \approx 3.13$$

Try this with other sizes of cylinders and circles—you should always get a number close to 3.1. The exact ratio of circumference to diameter is π. (Recall that $\pi \approx 3.14$ or $\approx \frac{22}{7}$.)

PRACTICE PROBLEM 8

An irrigation device waters a circular region with a diameter of 20 yards. Find the exact circumference of the watered region, then use $\pi \approx 3.14$ to give an approximation.

EXAMPLE 8 **Finding Circumference of a Circular Spa**

A homeowner plans to install a border of new tiling around the circumference of her circular spa. If her spa has a diameter of 14 feet, find its exact circumference. Then use the approximation 3.14 for π to approximate the circumference.

14 feet

Solution: Because we are given the diameter, we use the formula $C = \pi \cdot d$.

$C = \pi \cdot d$

$\quad = \pi \cdot 14 \text{ ft}$ Replace d with 14 feet.

$\quad = 14\pi \text{ ft}$

The circumference of the spa is *exactly* 14π feet. By replacing π with the *approximation* 3.14, we find that the circumference is *approximately* 14 feet · 3.14 = 43.96 feet.

◻ **Work Practice Problem 8**

✔**Concept Check** The distance around which figure is greater: a square with side length 5 inches or a circle with radius 3 inches?

Answer

8. exactly 20π yd ≈ 62.8 yd

✔ **Concept Check Answer**

a square with side length 5 in.

Vocabulary and Readiness Check

Use the choices below to fill in each blank.

circumference	radius	π	$\frac{22}{7}$
diameter	perimeter	3.14	

1. The _____ of a polygon is the sum of the lengths of its sides.

2. The distance around a circle is called the _____ .

3. The exact ratio of circumference to diameter is _____ .

4. The diameter of a circle is double its _____ .

5. Both _____ and _____ are approximations for π.

6. The radius of a circle is half its _____ .

9.2 EXERCISE SET

FOR EXTRA HELP

Student Solutions Manual   PH Math/Tutor Center   CD/Video for Review   Math XL MathXL®   MyMathLab MyMathLab

Objective A *Find the perimeter of each figure. See Examples 1 through 6.*

1.
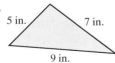
15 ft | Rectangle
17 ft

2.
Rectangle | 14 m
5 m

3.

Parallelogram   25 cm
35 cm

4.

Parallelogram
3 yd
2 yd

5.

5 in.   7 in.
9 in.

6.

5 units   11 units
10 units

7.

10 ft   8 ft
7 ft   8 ft
15 ft

8.

10 m
4 m
13 m   10 m
20 m   9 m

Find the perimeter of each regular polygon. (The sides of a regular polygon have the same length.)

9.

14 inches

10.

50 m

11.

31 cm

12.

15 yd

Solve. See Examples 1 through 7.

13. A polygon has sides of length 5 feet, 3 feet, 2 feet, 7 feet, and 4 feet. Find its perimeter.

14. A triangle has sides of length 8 inches, 12 inches, and 10 inches. Find its perimeter.

15. A line marking machine lays down lime powder to mark the base path on a new baseball diamond. If a baseball diamond has 90 feet between each base, how many feet of lime will be deposited on the diamond?

16. Find how much fencing is needed to enclose a rectangular rose garden 85 feet by 15 feet.

17. If a football field is 53 yards wide and 120 yards long, what is the perimeter?

18. A stop sign has eight equal sides of length 12 inches. Find its perimeter.

19. A metal strip is being installed around a workbench that is 8 feet long and 3 feet wide. Find how much stripping is needed.

20. Find how much fencing is needed to enclose a rectangular garden 70 feet by 21 feet.

21. If the stripping in Exercise 19 costs $2.50 per foot, find the total cost of the stripping.

22. If the fencing in Exercise 20 costs $2 per foot, find the total cost of the fencing.

23. A regular octagon has a side length of 9 inches. Find its perimeter.

24. A regular pentagon has a side length of 14 meters. Find its perimeter.

25. Find the perimeter of the top of a square compact disc case if the length of one side is 7 inches.

26. Find the perimeter of a square ceramic tile with a side of length 3 inches.

27. A rectangular room measures 10 feet by 11 feet. Find the cost of installing a strip of wallpaper around the room if the wallpaper costs $0.86 per foot.

28. A rectangular house measures 85 feet by 70 feet. Find the cost of installing gutters around the house if the cost is $2.36 per foot.

Find the perimeter of each figure. See Example 6.

29.

30.

31.

32.
16 cm
2 cm
11 cm
4 cm
3 cm
9 cm

33.
18 cm
5 cm
9 cm
14 cm

34.
22 km
12 km
5 km
6 km

Objective B *Find the circumference of each circle. Give the exact circumference and then an approximation. Use* $\pi \approx 3.14$. *See Example 8.*

35.
17 cm

36.
2.5 in.

37.
8 mi

38.
50 ft

39.
26 m

40.
10 yd

41. Wyley Robinson just bought a trampoline for his children to use. The trampoline has a diameter of 15 feet. If Wyley wishes to buy netting to go around the outside of the trampoline, how many feet of netting does he need?

42. The largest round barn in the world is located at the Marshfield Fairgrounds in Wisconsin. The barn has a diameter of 150 ft. What is the circumference of the barn? (*Source: The Milwaukee Journal Sentinel*)

43. Meteor Crater, near Winslow, Arizona, is 4000 feet in diameter. Approximate the distance around the crater. Use 3.14 for π. (*Source: The Handy Science Answer Book*)

44. The largest pearl, the *Pearl of Lao-tze*, has a diameter of $5\frac{1}{2}$ inches. Approximate the distance around the pearl. Use $\frac{22}{7}$ for π. (*Source: The Guinness World Records*)

Objectives **A** **B** **Mixed Practice** *Find the distance around each figure. For circles, give the exact circumference and then an approximation. Use $\pi \approx 3.14$.*

45.

9 mi
4.7 mi
6 mi
11 mi

46.
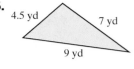
4.5 yd
7 yd
9 yd

47.

14 cm

48.

11 m

49.

Regular Pentagon
8 mm

50.

Regular Parallelogram
19 km

51.

7 ft
8 ft
22 ft
20 ft

52.

44 mi
40 mi
9 mi

Review

Simplify. See Section 1.7.

53. $5 + 6 \cdot 3$

54. $25 - 3 \cdot 7$

55. $(20 - 16) \div 4$

56. $6 \cdot (8 + 2)$

57. $(18 + 8) - (12 + 4)$

58. $72 \div (2 \cdot 6)$

59. $(72 \div 2) \cdot 6$

60. $4^1 \cdot (2^3 - 8)$

Concept Extensions

There are a number of factors that determine the dimensions of a rectangular soccer field. Use the table below to answer Exercises 61 and 62.

Soccer Field Width and Length		
Age	**Width Min–Max**	**Length Min–Max**
Under 6/7:	15–20 yards	25–30 yards
Under 8:	20–25 yards	30–40 yards
Under 9:	30–35 yards	40–50 yards
Under 10:	40–50 yards	60–70 yards
Under 11:	40–50 yards	70–80 yards
Under 12:	40–55 yards	100–105 yards
Under 13:	50–60 yards	100–110 yards
International:	70–80 yards	110–120 yards

61. a. Find the minimum length and width of a soccer field for 8-year-old children. (Carefully consider the age.)
b. Find the perimeter of this field.

62. a. Find the maximum length and width of a soccer field for 12-year-old children.
b. Find the perimeter of this field.

Solve. See the Concept Check in this section. Choose the figure that has greater distance around.

63. a. A square with side length 3 inches
b. A circle with diameter 4 inches

64. a. A circle with diameter 7 inches
b. A square with side length 7 inches

65. a. Find the circumference of each circle. Approximate the circumference by using 3.14 for π.

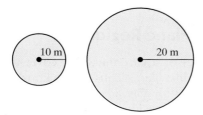

b. If the radius of a circle is doubled, is its corresponding circumference doubled?

66. a. Find the circumference of each circle. Approximate the circumference by using 3.14 for π.

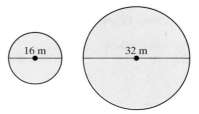

b. If the diameter of a circle is doubled, is its corresponding circumference doubled?

67. Find the perimeter of the skating rink.

68. In your own words, explain how to find the perimeter of any polygon.

Find the perimeter. Round your results to the nearest tenth.

69.

70.

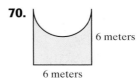

9.3 AREA, VOLUME, AND SURFACE AREA

Objective **A** Finding Area of Plane Regions

Recall that area measures the number of square units that cover the surface of a plane region; that is, a region that lies in a plane. Thus far, we know how to find the area of a rectangle and a square. These formulas, as well as formulas for finding the areas of other common geometric figures, are given next.

Area Formulas of Common Geometric Figures

Geometric Figure	**Area Formula**
RECTANGLE 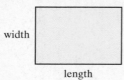	Area of a rectangle: **Area = length · width** $A = lw$
SQUARE	Area of a square: **Area = side · side** $A = s \cdot s = s^2$
TRIANGLE 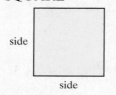	Area of a triangle: **Area** $= \dfrac{1}{2} \cdot$ **base · height** $A = \dfrac{1}{2}bh$
PARALLELOGRAM	Area of a parallelogram: **Area = base · height** $A = bh$
TRAPEZOID	Area of a trapezoid: **Area** $= \dfrac{1}{2} \cdot$ (one **b**ase + other **b**ase) · **height** $A = \dfrac{1}{2}(b + B)h$

Use these formulas for the following examples.

> **Helpful Hint**
>
> Area is always measured in square units.

EXAMPLE 1 Find the area of the triangle.

Solution: $A = \dfrac{1}{2}bh$

$= \dfrac{1}{2} \cdot 14 \text{ cm} \cdot 8 \text{ cm}$ Replace b, base, with 14 cm and h, height, with 8 cm.

$= \dfrac{\overset{1}{\cancel{2}} \cdot 7 \cdot 8}{\underset{1}{\cancel{2}}}$ square centimeters   Write 14 as $2 \cdot 7$.

$= 56$ square centimeters

The area is 56 square centimeters.

◼ **Work Practice Problem 1**

EXAMPLE 2 Find the area of the parallelogram.

Solution: $A = bh$

$= 3.4 \text{ miles} \cdot 1.5 \text{ miles}$ Replace b, base, with 3.4 miles and h, height, with 1.5 miles.

$= 5.1$ square miles

The area is 5.1 square miles.

◼ **Work Practice Problem 2**

> **Helpful Hint**
>
> When finding the area of figures, check to make sure that all measurements are in the same units before calculations are made.

EXAMPLE 3 Find the area of the figure.

Continued on next page

Continued on next page

PRACTICE PROBLEM 1

Find the area of the triangle.

PRACTICE PROBLEM 2

Find the area of the trapezoid.

PRACTICE PROBLEM 3

Find the area of the figure.

Answers

1. $49\dfrac{1}{2}$ sq in.   **2.** 48.8 sq yd

3. 396 sq m

Solution: Split the figure into two rectangles. To find the area of the figure, we find the sum of the areas of the two rectangles.

area of Rectangle 1 = lw

$= 8 \text{ feet} \cdot 4 \text{ feet}$

$= 32 \text{ square feet}$

Notice that the length of Rectangle 2 is 12 feet − 4 feet or 8 feet.

area of Rectangle 2 = lw

$= 8 \text{ feet} \cdot 5 \text{ feet}$

$= 40 \text{ square feet}$

area of the figure = area of Rectangle 1 + area of Rectangle 2

$= 32 \text{ square feet} + 40 \text{ square feet}$

$= 72 \text{ square feet}$

🔲 **Work Practice Problem 3**

Helpful Hint

The figure in Example 3 could also be split into two rectangles as shown.

To better understand the formula for area of a circle, try the following. Cut a circle into many pieces, as shown.

The circumference of a circle is $2\pi r$. This means that the circumference of half a circle is half of $2\pi r$, or πr.

Then unfold the two halves of the circle and place them together, as shown.

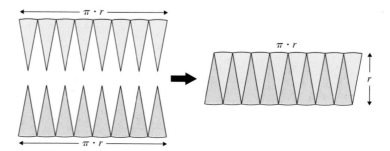

The figure on the right is almost a parallelogram with a base of πr and a height of r. The area is

$A =$ base $\cdot$ height

$= (\pi r) \cdot r$

$= \pi r^2$

This is the formula for area of a circle.

Area Formula of a Circle

Circle

radius

Area of a circle

Area $= \pi \cdot (\text{radius})^2$

$A = \pi r^2$

(A fraction approximation for π is $\dfrac{22}{7}$.)

(A decimal approximation for π is 3.14.)

EXAMPLE 4 Find the area of a circle with a radius of 3 feet. Find the exact area and an approximation. Use 3.14 as an approximation for π.

3 ft

Solution: We let $r = 3$ feet and use the formula

$A = \pi r^2$

$= \pi \cdot (3 \text{ feet})^2$ Replace r with 3 feet.

$= 9 \cdot \pi$ square feet   Replace $(3 \text{ feet})^2$ with 9 sq ft.

To approximate this area, we substitute 3.14 for π.

$9 \cdot \pi$ square feet $\approx 9 \cdot 3.14$ square feet

$= 28.26$ square feet

The *exact* area of the circle is 9π square feet, which is *approximately* 28.26 square feet.

Work Practice Problem 4

✔ **Concept Check** Use estimation to decide which figure would have a larger area: a circle of diameter 10 in. or a square 10 in. long on each side.

PRACTICE PROBLEM 4

Find the area of the given circle. Find the exact area and an approximation. Use 3.14 as an approximation for π.

7 cm

Answer

4. 49π sq cm ≈ 153.86 sq cm

✔ Concept Check Answer

A square 10 in. long on each side would have a larger area.

Objective **B** Finding Volume and Surface Area of Solids

A convex solid is a set of points, S, not all in one plane, such that for any two points A and B in S, all points between A and B are also in S. In this section, we will find the volume and surface area of special types of solids called polyhedrons. A solid formed by the intersection of a finite number of planes is called a **polyhedron.** The box below is an example of a polyhedron.

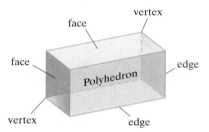

Each of the plane regions of the polyhedron is called a **face** of the polyhedron. If the intersection of two faces is a line segment, this line segment is an **edge** of the polyhedron. The intersections of the edges are the **vertices** of the polyhedron.

Volume is a measure of the space of a region. The volume of a box or can, for example, is the amount of space inside. Volume can be used to describe the amount of juice in a pitcher or the amount of concrete needed to pour a foundation for a house.

The volume of a solid is the number of **cubic units** in the solid. A cubic centimeter and a cubic inch are illustrated.

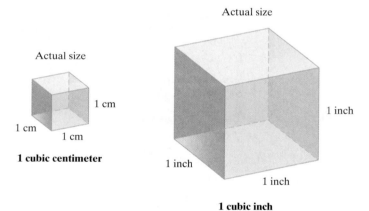

The **surface area** of a polyhedron is the sum of the areas of the faces of the polyhedron. For example, each face of the cube to the left above has an area of 1 square centimeter. Since there are 6 faces of the cube, the sum of the areas of the faces is 6 square centimeters. Surface area can be used to describe the amount of material needed to cover a solid. Surface area is measured in square units.

Formulas for finding the volumes, V, and surface areas, SA, of some common solids are given next.

Volume and Surface Area Formulas of Common Solids	
Solid	**Formulas**
RECTANGULAR SOLID height width length	$V = lwh$ $SA = 2lh + 2wh + 2lw$ where h = height, w = width, l = length
CUBE side side side	$V = s^3$ $SA = 6s^2$ where s = side
SPHERE radius	$V = \dfrac{4}{3}\pi r^3$ $SA = 4\pi r^2$ where r = radius
CIRCULAR CYLINDER height radius	$V = \pi r^2 h$ $SA = 2\pi rh + 2\pi r^2$ where h = height, r = radius
CONE height radius	$V = \dfrac{1}{3}\pi r^2 h$ $SA = \pi r\sqrt{r^2 + h^2} + \pi r^2$ where h = height, r = radius
SQUARE-BASED PYRAMID height side	$V = \dfrac{1}{3}s^2 h$ $SA = B + \dfrac{1}{2}pl$ where B = area of base; p = perimeter of base, h = height, s = side, l = slant height

PRACTICE PROBLEM 5

Find the volume and surface area of a rectangular box that is 7 feet long, 3 feet wide, and 4 feet deep.

EXAMPLE 5 Find the volume and surface area of a rectangular box that is 12 inches long, 6 inches wide, and 3 inches high.

3 in.

6 in.   12 in.

Solution: Let $h = 3$ in., $l = 12$ in., and $w = 6$ in.

$$V = lwh$$

$$V = 12 \text{ inches} \cdot 6 \text{ inches} \cdot 3 \text{ inches} = 216 \text{ cubic inches}$$

The volume of the rectangular box is 216 cubic inches.

$$SA = 2lh + 2wh + 2lw$$

$$= 2(12 \text{ in.})(3 \text{ in.}) + 2(6 \text{ in.})(3 \text{ in.}) + 2(12 \text{ in.})(6 \text{ in.})$$

$$= 72 \text{ sq in.} + 36 \text{ sq in.} + 144 \text{ sq in.}$$

$$= 252 \text{ sq in.}$$

The surface area of rectangular box is 252 square inches.

Work Practice Problem 5

✔ **Concept Check** Juan is calculating the volume of the following rectangular solid. Find the error in his calculation.

$$\text{Volume} = l + w + h$$

$$= 14 \text{ cm} + 8 \text{ cm} + 5 \text{ cm}$$

$$= 27 \text{ cu cm}$$

5 cm

8 cm   14 cm

PRACTICE PROBLEM 6

Find the volume and surface area of a ball of radius $\frac{1}{2}$ centimeter. Give the exact volume and surface area. Then use $\frac{22}{7}$ for π.

EXAMPLE 6 Find the volume and surface area of a ball of radius 2 inches. Give the exact volume and surface area. Then use the approximation $\frac{22}{7}$ for π.

2 in.

Answers

5. $V = 84$ cu ft; $SA = 122$ sq ft

6. $V = \frac{1}{6}\pi$ cu cm $\approx \frac{11}{21}$ cu cm;

$SA = \pi$ sq cm $\approx 3\frac{1}{7}$ sq cm

✔ **Concept Check Answer**

Volume $= l \cdot w \cdot h$

$= 14 \cdot 8 \cdot 5$

$= 560$ cu cm

Solution:

$$V = \frac{4}{3}\pi r^3 \qquad \text{Formula for volume of a sphere}$$

$$V = \frac{4}{3} \cdot \pi (2 \text{ in.})^3 \qquad \text{Let } r = 2 \text{ inches.}$$

$$= \frac{32}{3}\pi \text{ cu in.} \qquad \text{Exact volume}$$

$$\approx \frac{32}{3} \cdot \frac{22}{7} \text{ cu in.} \qquad \text{Approximate } \pi \text{ with } \frac{22}{7}.$$

$$= \frac{704}{21} \text{ or } 33\frac{11}{21} \text{ cu in.} \qquad \text{Approximate volume}$$

The volume of the sphere is exactly $\dfrac{32}{3}\pi$ cubic inches or approximately $33\dfrac{11}{21}$ cubic inches.

$$SA = 4\pi r^2 \qquad \text{Formula for surface area}$$
$$SA = 4 \cdot \pi (2\,\text{in.})^2 \qquad \text{Let } r = 2 \text{ inches.}$$
$$= 16\pi \text{ sq in.} \qquad \text{Exact surface area}$$
$$\approx 16 \cdot \dfrac{22}{7} \text{ sq in.} \qquad \text{Approximate } \pi \text{ with } \dfrac{22}{7}.$$
$$= \dfrac{352}{7} \text{ or } 50\dfrac{2}{7} \text{ sq in.} \qquad \text{Approximate surface area}$$

The surface area of the sphere is exactly 16π square inches or approximately $50\dfrac{2}{7}$ square inches.

◻ **Work Practice Problem 6**

EXAMPLE 7 Approximate the volume of a can that has a $3\dfrac{1}{2}$-inch radius and a height of 6 inches. Use $\dfrac{22}{7}$ for π. Give an exact volume and an approximate volume.

$3\dfrac{1}{2}$ in.

6 in.

Solution: Using the formula for a circular cylinder, we have

$$V = \pi \cdot r^2 \cdot h \qquad \qquad 3\dfrac{1}{2} = \dfrac{7}{2}$$

$$= \pi \cdot \left(\dfrac{7}{2}\,\text{in.}\right)^2 \cdot 6\,\text{in.}$$

$$= \pi \cdot \dfrac{49}{4} \text{ sq in.} \cdot 6\,\text{in.}$$

$$= \dfrac{\pi \cdot 49 \cdot \overset{1}{\cancel{2}} \cdot 3}{\underset{1}{\cancel{2}} \cdot 2} \text{ cu in.}$$

$$= 73\dfrac{1}{2}\pi \text{ cu in. or } 73.5\pi \text{ cu in.}$$

This is the exact volume. To approximate the volume, use the approximation $\dfrac{22}{7}$ for π.

$$V = 73\dfrac{1}{2}\pi \text{ or } \dfrac{147}{2} \cdot \dfrac{22}{7} \text{ cu in.} \qquad \text{Replace } \pi \text{ with } \dfrac{22}{7}.$$

$$= \dfrac{21 \cdot \overset{1}{\cancel{7}} \cdot \overset{1}{\cancel{2}} \cdot 11}{\underset{1}{\cancel{2}} \cdot \underset{1}{\cancel{7}}} \text{ cu in.}$$

$$= 231 \text{ cubic in.}$$

The volume is approximately 231 cubic inches.

◻ **Work Practice Problem 7**

PRACTICE PROBLEM 7

Approximate the volume of a cylinder of radius 5 inches and height 9 inches. Use 3.14 for π. Give an exact answer and an approximate answer.

Answer
7. 225π cu in. ≈ 706.5 cu in.

PRACTICE PROBLEM 8

Find the volume of a square-based pyramid that has a 3-meter side and a height of 5.1 meters.

EXAMPLE 8 Approximate the volume of a cone that has a height of 14 centimeters and a radius of 3 centimeters. Use 3.14 for π. Give an exact answer and an approximate answer.

Solution: Using the formula for volume of a cone, we have

$$V = \frac{1}{3} \cdot \pi \cdot r^2 \cdot h$$

$$= \frac{1}{3} \cdot \pi \cdot (3 \text{ cm})^2 \cdot 14 \text{ cm} \quad \text{Replace } r \text{ with 3 cm and } h \text{ with 14 cm.}$$

$$= 42\pi \text{ cu cm}$$

Thus, 42π cubic centimeters is the exact volume. To approximate the volume, use the approximation 3.14 for π.

$$V \approx 42 \cdot 3.14 \text{ cu cm} \quad \text{Replace } \pi \text{ with 3.14.}$$

$$= 131.88 \text{ cu cm}$$

The volume is approximately 131.88 cubic centimeters.

⬛ **Work Practice Problem 8**

Answer

8. 15.3 cu m

Vocabulary and Readiness Check

Use the choices below to fill in each blank. Some choices may be used more than once.

area surface area cubic

volume square

1. The _____ of a polyhedron is the sum of the areas of its faces.
2. The measure of the space of a solid is its _____.
3. _____ measures the amount of surface enclosed by a region.
4. Volume is measured in _____ units.
5. Area is measured in _____ units.
6. Surface area is measured in _____ units.

9.3 EXERCISE SET

FOR EXTRA HELP

Student Solutions Manual   PH Math/Tutor Center   CD/Video for Review   Math XL   MyMathLab
MathXL®   MyMathLab

Objective A *Find the area of each geometric figure. If the figure is a circle, give an exact area and then use the given* **approximation** *for π to approximate the area. See Examples 1 through 4.*

1.

2 m | Rectangle
3.5 m

2.

1.2 ft | Rectangle
3.5 ft

3.

3 yd
$6\frac{1}{2}$ yd

4.
5 ft
$4\frac{1}{2}$ ft

5.
6 yd
5 yd

6.
5 ft   7 ft

7. Use 3.14 for π.

$d = 3$ in.

8. Use $\frac{22}{7}$ for π.

$r = 5$ cm

9.
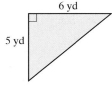
Parallelogram
5.25 ft
7 ft

10.

Parallelogram
4.25 cm
3 cm

11.

5 m
Trapezoid
4 m
9 m

12.

Trapezoid
6 in.
5 in. $8\frac{1}{2}$ in.

13.

4 yd

4 yd   Trapezoid

7 yd

14.

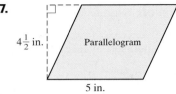

10 ft

3 ft   Trapezoid

5 ft

15.

7 ft

Parallelogram

$5\frac{1}{4}$ ft

16.

Parallelogram

$4\frac{1}{4}$ cm

3 cm

17.

$4\frac{1}{2}$ in.   Parallelogram

5 in.

18.

4 m

6 m

Parallelogram

19.

2 cm

$1\frac{1}{2}$ cm $1\frac{1}{2}$ cm

3 cm

7 cm

20.

6 km

4 km

5 km

10 km

21.

5 mi

10 mi

3 mi

17 mi

22.

25 cm

15 cm   12 cm

5 cm

23.

5 cm

3 cm

24.

4 in.

5 in.

25. Use $\frac{22}{7}$ for π.

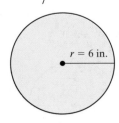

$r = 6$ in.

26. Use 3.14 for π.

$d = 5$ m

Objective **B** *Find the volume and surface area of each solid. See Examples 5 through 8. For formulas containing π, give an exact answer and then approximate using $\frac{22}{7}$ for π.*

27.

3 in.

4 in.   6 in.

28.

4 cm

4 cm   8 cm

29.

8 cm

8 cm

8 cm   8 cm

30.

11 mi

11 mi   11 mi

31. For surface area, use $\pi = 3.14$.

3 yd

2 yd

32. For surface area, use $\pi = 3.14$.

$1\frac{3}{4}$ in.

9 in.

33.

10 in.

34.

3 mi

35. Find the volume only.

2 in.

9 in.

36. Find the volume only.

10 ft

6 ft

37. Find the volume only.

9 cm

5 cm

38. Find the volume only.

15 m

7 m

Objectives Ⓐ Ⓑ **Mixed Practice** *Solve. See Examples 1 through 8.*

39. Find the volume of a cube with edges of $1\frac{1}{3}$ inches.

$1\frac{1}{3}$ inches

40. A water storage tank is in the shape of a cone with the pointed end down. If the radius is 14 ft and the depth of the tank is 15 ft, approximate the volume of the tank in cubic feet. Use $\frac{22}{7}$ for π.

14 ft

15 ft

41. Find the volume and surface area of a rectangular box 2 ft by 1.4 ft by 3 ft.

42. Find the volume and surface area of a box in the shape of a cube that is 5 ft on each side.

43. The world's largest flag measures 505 feet by 225 feet. It's the U.S. "Super flag" owned by "Ski" Demski of Long Beach, California. Find its area. (*Source: Guinness World Records,* 2005)

44. The longest illuminated sign is in Ramat Gan, Israel, and measures 197 feet by 66 feet. Find its area. (*Source: The Guinness Book of World Records*)

225 feet

505 feet

45. A drapery panel measures 6 ft by 7 ft. Find how many square feet of material are needed for *four* panels.

46. A page in this book measures 27.5 cm by 21.8 cm. Find its area.

47. A paperweight is in the shape of a square-based pyramid 20 centimeters tall. If an edge of the base is 12 centimeters, find the volume of the paperweight.

48. A birdbath is made in the shape of a hemisphere (half-sphere). If its radius is 10 inches, approximate the volume. Use $\frac{22}{7}$ for π.

10 in.

49. Find how many square feet of land are in the following plot:

50. For Gerald Gomez to determine how much grass seed he needs to buy, he must know the size of his yard. Use the drawing to determine how many square feet are in his yard.

90 feet

80 feet

140 feet

96 feet

48 feet

48 feet

24 feet

132 feet

51. Find the exact volume and surface area of a sphere with a radius of 7 inches.

52. A tank is in the shape of a cylinder 8 feet tall and 3 feet in radius. Find the exact volume and surface area of the tank.

53. The shaded part of the roof shown is in the shape of a trapezoid and needs to be shingled. The number of packages of shingles to buy depends on the area. Use the dimensions given to find the area of the shaded part of the roof to the nearest whole square foot.

54. The end of the building shaded in the drawing is to be bricked. The number of bricks to buy depends on the area.

a. Find the area.

b. If the side area of each brick (including mortar room) is $\frac{1}{6}$ square ft, find the number of bricks needed to buy.

55. A snow globe has a diameter of 6 inches. Find its exact volume. Then approximate its volume using 3.14 for π.

56. Find the exact volume of a waffle ice cream cone with a 3-in. diameter and a height of 7 inches.

57. Paul Revere's Pizza in the USA will bake and deliver a round pizza with a 4-foot diameter. This pizza is called the "Ultimate Party Pizza" and its current price is $99.99. Find the exact area of the top of the pizza and an approximation. Use 3.14 as an approximation for π.

58. The face of a circular watch has a diameter of 2 centimeters. What is its area? Find the exact area and an approximation. Use 3.14 as an approximation for π.

59. Zorbing is an extreme sport invented by two New Zealanders who joke that they were looking for a way to walk on water. A Zorb is a large sphere inside a second sphere with the space between the spheres pumped full of air. There is a tunnel-like opening so a person can crawl into the inner sphere. You are strapped in and sent down a zorbing hill. A standard zorb is approximately 3 m in diameter. Find the exact volume of a zorb, and approximate the volume using 3.14 for π.

60. Mount Fuji, in Japan, is considered the most beautiful composite volcano in the world. The mountain is in the shape of a cone whose height is about 3.5 kilometers and whose base radius is about 3 kilometers. Approximate the volume of Mt. Fuji in cubic kilometers. Use $\frac{22}{7}$ for π.

61. A $10\frac{1}{2}$-foot by 16-foot concrete wall is to be built using concrete blocks. Find the area of the wall.

62. The floor of Terry's attic is 24 feet by 35 feet. Find how many square feet of insulation are needed to cover the attic floor.

63. Find the volume of a pyramid with a square base 5 inches on a side and a height of $1\frac{3}{10}$ inches.

64. Approximate to the nearest hundredth the volume of a sphere with a radius of 2 centimeters. Use 3.14 for π.

Review

Evaluate. See Section 1.7.

65. 5^2

66. 7^2

67. 3^2

68. 20^2

69. $1^2 + 2^2$

70. $5^2 + 3^2$

71. $4^2 + 2^2$

72. $1^2 + 6^2$

Concept Extensions

Given the following situations, tell whether you are more likely to be concerned with area or perimeter.

73. ordering fencing to fence a yard

74. ordering grass seed to plant in a yard

75. buying carpet to install in a room

76. buying gutters to install on a house

77. ordering paint to paint a wall

78. ordering baseboards to install a room

79. buying a wallpaper border to go on the walls around a room

80. buying fertilizer for your yard

81. A pizza restaurant recently advertised two specials. The first special was a 12-inch pizza for $10. The second special was two 8-inch pizzas for $9. Determine the better buy. (*Hint:* First compare the areas of the two specials and then find a price per square inch for both specials.)

82. Find the approximate area of the state of Utah.

83. The centerpiece of the New England Aquarium in Boston is the Caribbean Coral Reef tank. The centerpiece of this four-story giant tank is a human-made coral reef with over 300 species of coral displayed. This giant reef, with its many caves, is the perfect backdrop for sharks, sea turtles, barracudas, moray eels, and hundreds of smaller tropical fish. The radius of the tank is 16.3 feet and its height is 32 feet. What is the volume of the Caribbean Coral Reef Tank? Use $\pi = 3.14$ and round to the nearest tenth of a cubic foot. (*Source:* New England Aquarium)

84. The Great Pyramid of Khufu at Giza is the largest of the ancient Egyptian pyramids. Its original height was 146.5 meters. The length of each side of its square base was originally 230 meters. Find the volume of the Great Pyramid of Khufu as it was originally built. Round to the nearest whole cubic meter. (*Source:* PBS *NOVA* Online)

85. Can you compute the volume of a rectangle? Why or why not?

86. In your own words, explain why perimeter is measured in units and area is measured in square units. (*Hint:* See Section 1.5 for an introduction on the meaning of area.)

87. Find the area of the shaded region. Use the approximation 3.14 for π.

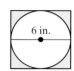
6 in.

88. The largest pumpkin pie was made and served in Windsor, California. The pie had a diameter of 72 inches. Find the exact area of the top of the pie and an approximation. Use $\pi \approx 3.14$. (*Source: Guinness Book of World Records*)

Find the area of each figure. If needed, use $\pi \approx 3.14$ and round results to the nearest tenth.

89. Find the skating area.

ROYALS
5 m
22 m

90.

←5 feet→
7 feet

91. Do two rectangles with the same perimeter have the same area? To see, find the perimeter and the area of each rectangle.

6 in.
8 in.
3 in.
11 in.

📖 **STUDY SKILLS BUILDER**

How Well Do You Know Your Textbook?

Let's check to see whether you are familiar with your textbook yet. Remember, for help, see Section 1.1 in this text.

1. What does the 🔘 icon mean?

2. What does the ✎ icon mean?

3. What does the △ icon mean?

4. Where can you find a review for each chapter? What answers to this review can be found in the back of your text?

5. Each chapter contains an overview of the chapter along with examples. What is this feature called?

6. Each chapter contains a review of vocabulary. What is this feature called?

7. There are free CDs in your text. What content is contained on these CDs?

8. What is the location of the section that is entirely devoted to study skills?

9. There are Practice Problems that are contained in the margin of the text. What are they and how can they be used?

Geometry Concepts

1. _____

2. _____

3. _____

4. _____

5. _____

6. _____

7. _____

8. _____

9. _____

10. _____

11. _____

12. _____

13. _____

14. _____

15. _____

16. _____

△ **1.** Find the supplement and the complement of a 27° angle.

Find the measures of angles x, y, and z in each figure.

2.

105°
x z
y

3. $m \| n$

x / 52° → m

y
z → n

4. Find the measure of ∠x.

x

38°

5. Find the diameter.

2.3 in.

6. Find the radius.

$8\frac{1}{2}$ in.

For Exercises 7 through 11, find the perimeter (or circumference) and area of each figure. For the circle give an exact circumference and area. Then use $\pi \approx 3.14$ to approximate each. Don't forget to attach correct units.

7.

Square 5 m

8.

4 ft
3 ft   5 ft

9.

5 cm

10.

11 mi

Parallelogram 5 mi
4 mi

11.

8 cm
3 cm
7 cm
17 cm

12. The smallest cathedral is in Highlandville, Missouri. The rectangular floor of the cathedral measures 14 feet by 17 feet. Find its perimeter and its area. (*Source: The Guinness Book of World Records*)

Find the volume of each solid. Don't forget to attach correct units. For Exercises 13 and 14, find the surface area, also.

13. A cube with edges of 4 inches each.

14. A rectangular box 2 feet by 3 feet by 5.1 feet.

15. A pyramid with a square base 10 centimeters on a side and a height of 12 centimeters.

16. A sphere with a diameter of 3 miles. Give the exact volume and then use $\pi \approx \frac{22}{7}$ to approximate.

9.4 LINEAR MEASUREMENT

Objective A Defining and Converting U.S. System Units of Length

In the United States, two systems of measurement are commonly used. They are the **United States (U.S.), or English, measurement system** and the **metric system.** The U.S. measurement system is familiar to most Americans. Units such as feet, miles, ounces, and gallons are used. However, the metric system is also commonly used in fields such as medicine, sports, international marketing, and certain physical sciences. We are accustomed to buying 2-liter bottles of soft drinks, watching televised coverage of the 100-meter dash at the Olympic Games, or taking a 200-milligram dose of pain reliever.

The U.S. system of measurement uses the **inch, foot, yard,** and **mile** to measure **length.** The following is a summary of equivalencies between units of length:

Objectives

A Define U.S. Units of Length and Convert from One Unit to Another.

B Use Mixed Units of Length.

C Perform Arithmetic Operations on U.S. Units of Length.

D Define Metric Units of Length and Convert from One Unit to Another.

E Perform Arithmetic Operations on Metric Units of Length.

U.S. Units of Length

$$12 \text{ inches (in.)} = 1 \text{ foot (ft)}$$
$$3 \text{ feet} = 1 \text{ yard (yd)}$$
$$36 \text{ inches} = 1 \text{ yard}$$
$$5280 \text{ feet} = 1 \text{ mile (mi)}$$

To convert from one unit of length to another, we will use **unit fractions.** We define a unit fraction to be a fraction that is equivalent to 1. Examples of unit fractions are as follows:

Unit Fractions

$$\frac{12 \text{ in.}}{1 \text{ ft}} = 1 \text{ or } \frac{1 \text{ ft}}{12 \text{ in.}} = 1 \text{ (since 12 in. = 1 ft)}$$

$$\frac{3 \text{ ft}}{1 \text{ yd}} = 1 \text{ or } \frac{1 \text{ yd}}{3 \text{ ft}} = 1 \text{ (since 3 ft = 1 yd)}$$

$$\frac{5280 \text{ ft}}{1 \text{ mi}} = 1 \text{ or } \frac{1 \text{ mi}}{5280 \text{ ft}} = 1 \text{ (since 5280 ft = 1 mi)}$$

Remember that multiplying a number by 1 does not change the value of the number.

EXAMPLE 1 Convert 8 feet to inches.

Solution: We multiply 8 feet by a unit fraction that uses the equality 12 inches = 1 foot. The unit fraction should be in the form $\frac{\text{units to convert to}}{\text{original units}}$ or in this case $\frac{12 \text{ inches}}{1 \text{ foot}}$. We do this so that like units will divide out to 1, as shown.

$$8 \text{ ft} = \frac{8 \text{ ft}}{1} \cdot 1 \qquad \text{Multiply by 1 in the form of } \frac{12 \text{ in.}}{1 \text{ ft}}.$$

$$= \frac{8 \text{ ft}}{1} \cdot \frac{12 \text{ in.}}{1 \text{ ft}}$$

$$= 8 \cdot 12 \text{ in.}$$

$$= 96 \text{ in.} \qquad \text{Multiply.}$$

PRACTICE PROBLEM 1

Convert 6 feet to inches.

Answer
1. 72 in.

Continued on next page

645

Thus, 8 ft = 96 in., as shown in the diagram:

8 feet = 96 inches

■ **Work Practice Problem 1**

PRACTICE PROBLEM 2

Convert 8 yards to feet.

EXAMPLE 2 Convert 7 feet to yards.

Solution: We multiply by a unit fraction that compares 1 yard to 3 feet.

$$7 \text{ ft} = \frac{7 \text{ ft}}{1} \cdot 1$$

$$= \frac{7 \text{ ft}}{1} \cdot \frac{1 \text{ yd}}{3 \text{ ft}} \quad \leftarrow \text{Units to convert to}$$
$$\qquad\qquad\qquad\quad \leftarrow \text{Original units}$$

$$= \frac{7}{3} \text{ yd}$$

$$= 2\frac{1}{3} \text{ yd} \qquad \text{Divide.}$$

Thus, 7 ft = $2\frac{1}{3}$ yd, as shown in the diagram.

7 feet = $2\frac{1}{3}$ yards

■ **Work Practice Problem 2**

PRACTICE PROBLEM 3

Suppose the bill in the photo measures 18 inches. Convert 18 inches to feet, using decimals.

EXAMPLE 3 **Finding Length of Pelican's Bill**

The Australian pelican has the longest bill, measuring from 13 to 18.5 inches long. The pelican in the photo has a 15-inch bill. Convert 15 inches to feet, using decimals in your final answer.

Solution:

$$15 \text{ in.} = \frac{15 \text{ in.}}{1} \cdot \frac{1 \text{ ft}}{12 \text{ in.}} \quad \leftarrow \text{Units to convert to}$$
$$\qquad\qquad\qquad\qquad\qquad \leftarrow \text{Original units}$$

$$= \frac{15}{12} \text{ ft}$$

$$= \frac{5}{4} \text{ ft} \qquad \text{Simplify } \frac{15}{12}.$$

$$= 1.25 \text{ ft} \qquad \text{Divide.}$$

Thus, 15 in. = 1.25 ft, as shown in the diagram.

15 inches = 1.25 ft

1 ft

$\frac{1}{4}$ or 0.25 ft

Answers

2. 24 ft   **3.** 1.5 ft

■ **Work Practice Problem 3**

Objective B Using Mixed U.S. System Units of Length

Sometimes it is more meaningful to express a measurement of length with mixed units such as 1 ft and 5 in. We usually condense this and write 1 ft 5 in.

In Example 2, we found that 7 feet was the same as $2\frac{1}{3}$ yards. The measurement can also be written as a mixture of yards and feet. That is,

7 ft = _____ yd _____ ft

Because 3 ft = 1 yd, we divide 3 into 7 to see how many whole yards are in 7 feet. The quotient is the number of yards, and the remainder is the number of feet.

$$\begin{array}{r} 2 \text{ yd } 1 \text{ ft} \\ 3\overline{)7} \\ -6 \\ \hline 1 \end{array}$$

Thus, 7 ft = 2 yd 1 ft, as seen in the diagram:

EXAMPLE 4 Convert: 134 in. = _____ ft _____ in.

Solution: Because 12 in. = 1 ft, we divide 12 into 134. The quotient is the number of feet. The remainder is the number of inches. To see why we divide 12 into 134, notice that

$$134 \text{ in.} = \frac{134 \text{ in.}}{1} \cdot \frac{1 \text{ ft}}{12 \text{ in.}} = \frac{134}{12} \text{ ft}$$

$$\begin{array}{r} 11 \text{ ft } 2 \text{ in.} \\ 12\overline{)134} \\ -12 \\ \hline 14 \\ -12 \\ \hline 2 \end{array}$$

Thus, 134 in. = 11 ft 2 in.

🔲 **Work Practice Problem 4**

EXAMPLE 5 Convert 3 feet 7 inches to inches.

Solution: First, we convert 3 feet to inches. Then we add 7 inches.

$$3 \text{ ft} = \frac{3 \text{ ft}}{1} \cdot \frac{12 \text{ in.}}{1 \text{ ft}} = 36 \text{ in.}$$

Then

$$3 \text{ ft } 7 \text{ in.} = 36 \text{ in.} + 7 \text{ in.} = 43 \text{ in.}$$

🔲 **Work Practice Problem 5**

PRACTICE PROBLEM 4
Convert:
68 in. = _____ ft _____ in.

PRACTICE PROBLEM 5
Convert 5 yards 2 feet to feet.

Answers
4. 5 ft 8 in. **5.** 17 ft

Objective C Performing Operations on U.S. System Units of Length

Finding sums or differences of measurements often involves converting units, as shown in the next example. Just remember that, as usual, only like units can be added or subtracted.

PRACTICE PROBLEM 6

Add 4 ft 8 in. to 8 ft 11 in.

EXAMPLE 6 Add 3 ft 2 in. and 5 ft 11 in.

Solution: To add, we line up the similar units.

$$
\begin{array}{r}
3\text{ ft }\ \ 2\text{ in.}\\
+\ 5\text{ ft }11\text{ in.}\\
\hline
8\text{ ft }13\text{ in.}
\end{array}
$$

Since 13 inches is the same as 1 ft 1 in., we have

$$
8\text{ ft }13\text{ in.} = 8\text{ ft} + \overbrace{1\text{ ft }\ 1\text{ in.}}
$$
$$
= 9\text{ ft }1\text{ in.}
$$

■ **Work Practice Problem 6**

✔ **Concept Check**   How could you estimate the following sum?

$$
\begin{array}{r}
7\text{ yd }\ \ 4\text{ in.}\\
+\ 3\text{ yd }27\text{ in.}\\
\hline
\end{array}
$$

PRACTICE PROBLEM 7

Multiply 4 ft 7 in. by 4.

EXAMPLE 7 Multiply 8 ft 9 in. by 3.

Solution: By the distributive property, we multiply 8 ft by 3 and 9 in. by 3.

$$
\begin{array}{r}
8\text{ ft }\ \ 9\text{ in.}\\
\times \qquad\quad 3\\
\hline
24\text{ ft }27\text{ in.}
\end{array}
$$

Since 27 in. is the same as 2 ft 3 in., we simplify the product as

$$
24\text{ ft }27\text{ in.} = 24\text{ ft} + 2\text{ ft }3\text{ in.}
$$
$$
= 26\text{ ft }3\text{ in.}
$$

We divide in a similar manner as above.

■ **Work Practice Problem 7**

PRACTICE PROBLEM 8

A carpenter cuts 1 ft 9 in. from a board of length 5 ft 8 in. Find the remaining length of the board.

EXAMPLE 8 Finding the Length of a Piece of Rope

A rope of length 6 yd 1 ft has 2 yd 2 ft cut from one end. Find the length of the remaining rope.

Solution: Subtract 2 yd 2 ft from 6 yd 1 ft.

$$
\begin{array}{lr}
\text{beginning length} \rightarrow & 6\text{ yd }1\text{ ft}\\
-\qquad\text{amount cut} \rightarrow & -\ 2\text{ yd }2\text{ ft}\\
\hline
\text{remaining length} &
\end{array}
$$

We cannot subtract 2 ft from 1 ft, so we borrow 1 yd from the 6 yd. One yard is converted to 3 ft and combined with the 1 ft already there.

Answers

6. 13 ft 7 in.   **7.** 18 ft 4 in.
8. 3 ft 11 in.

✔**Concept Check Answer**

round each to the nearest yard:
7 yd + 4 yd = 11 yd

Borrow 1 yd = 3 ft

5 yd + ⟨1 yd⟩ ⟨3 ft⟩

$$
\begin{array}{rclr}
\cancel{6\ \text{yd}}\ 1\ \text{ft} & = & 5\ \text{yd}\ 4\ \text{ft} \\
-\ 2\ \text{yd}\ 2\ \text{ft} & = & -\ 2\ \text{yd}\ 2\ \text{ft} \\
\hline
 & & 3\ \text{yd}\ 2\ \text{ft}
\end{array}
$$

The remaining rope is 3 yd 2 ft long.

🔲 **Work Practice Problem 8**

Objective Ⓓ Defining and Converting Metric System Units of Length

The basic unit of length in the metric system is the **meter.** A meter is slightly longer than a yard. It is approximately 39.37 inches long. Recall that a yard is 36 inches long.

1 yard = 36 inches

1 meter ≈ 39.37 inches

All units of length in the metric system are based on the meter. The following is a summary of the prefixes used in the metric system. Also shown are equivalencies between units of length. Like the decimal system, the metric system uses powers of 10 to define units.

Metric Unit of Length
1 **kilo**meter (km) = 1000 meters (m)
1 **hecto**meter (hm) = 100 m
1 **deka**meter (dam) = 10 m
1 meter (m) = 1 m
1 **deci**meter (dm) = 1/10 m or 0.1 m
1 **centi**meter (cm) = 1/100 m or 0.01 m
1 **milli**meter (mm) = 1/1000 m or 0.001 m

The figure below will help you with decimeters, centimeters, and millimeters.

1 decimeter = $\frac{1}{10}$ meter　　1 centimeter = $\frac{1}{100}$ meter　　1 millimeter = $\frac{1}{1000}$ meter

> **Helpful Hint**
>
> Study the figure above for other equivalencies between metric units of length.
>
> 10 decimeters = 1 meter　　10 millimeters = 1 centimeter
> 100 centimeters = 1 meter　　10 centimeters = 1 decimeter
> 1000 millimeters = 1 meter

These same prefixes are used in the metric system for mass and capacity. The most commonly used measurements of length in the metric system are the **meter, millimeter, centimeter,** and **kilometer.**

✔**Concept Check**   Is this statement reasonable? "The screen of a home television set has a 30-meter diagonal." Why or why not?

Being comfortable with the metric units of length means gaining a "feeling" for metric lengths, just as you have a "feeling" for the length of an inch, a foot, and a mile. To help you accomplish this, study the following examples:

A millimeter is about the thickness of a large paper clip.

A centimeter is about the width of a large paper clip.

A meter is slightly longer than a yard.

A kilometer is about two-thirds of a mile.

The width of this book is approximately 21.5 centimeters.

The distance between New York and Philadelphia is about 160 kilometers.

1.7 meters

7 millimeters

19 centimeters

$2\frac{1}{2}$ centimeters is about 1 inch.

As with the U.S. system of measurement, unit fractions may be used to convert from one unit of length to another. For example, let's convert 1200 meters to kilometers. To do so, we will multiply by 1 in the form of the unit fraction

$$\frac{1 \text{ km}}{1000 \text{ m}} \quad \begin{array}{l} \leftarrow \text{Units to convert to} \\ \leftarrow \text{Original units} \end{array}$$

Unit fraction

$$1200 \text{ m} = \frac{1200 \text{ m}}{1} \cdot 1 = \frac{1200 \ \cancel{m}}{1} \cdot \frac{\overbrace{1 \text{ km}}}{1000 \ \cancel{m}} = \frac{1200 \text{ km}}{1000} = 1.2 \text{ km}$$

The metric system does, however, have a distinct advantage over the U.S. system of measurement: The ease of converting from one unit of length to another. Since all units of length are powers of 10 of the meter, converting from one unit of length to another is as simple as moving the decimal point. Listing units of length in order from largest to smallest helps to keep track of how many places to move the decimal point when converting.

Let's again convert 1200 meters to kilometers. This time, to convert from meters to kilometers, we move along the chart shown 3 units to the left, from meters to kilometers. This means that we move the decimal point 3 places to the left.

km	hm	dam	**m**	dm	cm	mm

3 units to the left

Thus, 1200 m = 1.2 km as shown in the diagram.

✔**Concept Check Answer**

no; answers may vary

EXAMPLE 9 Convert 2.3 m to centimeters.

Solution: First we will convert by using a unit fraction.

Unit fraction

$$2.3 \text{ m} = \frac{2.3 \text{ m}}{1} \cdot \frac{\overbrace{100 \text{ cm}}}{1 \text{ m}} = 230 \text{ cm}$$

Now we will convert by listing the units of length in order from left to right and moving from meters to centimeters.

km   hm   dam   m   dm   cm   mm

2 units to the right

2.30 m = 230. cm

2 places to the right

With either method, we get 230 cm.

■ **Work Practice Problem 9**

EXAMPLE 10 Convert 450,000 mm to meters.

Solution: We list the units of length in order from left to right and move from millimeters to meters.

km   hm   dam   m   dm   cm   mm

3 units to the left

Thus, move the decimal point 3 places to the left.

450,000 mm = 450.000 m or 450 m

■ **Work Practice Problem 10**

✔ **Concept Check** What is wrong with the following conversion of 150 cm to meters?

150.00 cm = 15,000 m

Objective E Performing Operations on Metric System Units of Length

To add, subtract, multiply, or divide with metric measurements of length, we write all numbers using the same unit of length and then add, subtract, multiply, or divide as with decimals.

EXAMPLE 11 Subtract 430 m from 1.3 km.

Solution: First we convert both measurements to kilometers or both to meters.

430 m = 0.43 km   or   1.3 km = 1300 m

$$\begin{array}{r} 1.30 \text{ km} \\ - 0.43 \text{ km} \\ \hline 0.87 \text{ km} \end{array} \qquad \begin{array}{r} 1300 \text{ m} \\ - 430 \text{ m} \\ \hline 870 \text{ m} \end{array}$$

The difference is 0.87 km or 870 m.

■ **Work Practice Problem 11**

PRACTICE PROBLEM 9
Convert 2.5 m to millimeters.

PRACTICE PROBLEM 10
Convert 3500 m to kilometers.

PRACTICE PROBLEM 11
Subtract 640 m from 2.1 km.

Answers
9. 2500 mm **10.** 3.5 km
11. 1.46 km or 1460 m

✔ **Concept Check Answer**
decimal point should be moved two places to the left: 1.5 m

PRACTICE PROBLEM 12

Multiply 18.3 hm by 5.

EXAMPLE 12 Multiply 5.7 mm by 4.

Solution: Here we simply multiply the two numbers. Note that the unit of measurement remains the same.

$$\begin{array}{r} 5.7 \text{ mm} \\ \times \quad 4 \\ \hline 22.8 \text{ mm} \end{array}$$

📖 **Work Practice Problem 12**

PRACTICE PROBLEM 13

Doris Blackwell is knitting a scarf that is currently 0.8 meter long. If she knits an additional 45 centimeters, how long will the scarf be?

EXAMPLE 13 **Finding a Person's Height**

Fritz Martinson was 1.2 meters tall on his last birthday. Since then, he has grown 14 centimeters. Find his current height in meters.

Solution:

original height	→	1.20 m	
+ height grown	→	+ 0.14 m	(Since 14 cm = 0.14 m)
current height		1.34 m	

Fritz is now 1.34 meters tall.

📖 **Work Practice Problem 13**

Answers

12. 91.5 hm **13.** 125 cm or 1.25 m

Vocabulary and Readiness Check

Use the choices below to fill in each blank. Some choices may be used more than once.

inches   yard   unit fraction
feet   meter

1. The basic unit of length in the metric system is the _____.

2. The expression $\dfrac{1 \text{ foot}}{12 \text{ inches}}$ is an example of a _____.

3. A meter is slightly longer than a _____.

4. One foot equals 12 _____.

5. One yard equals 3 _____.

6. One yard equals 36 _____.

7. One mile equals 5280 _____.

9.4 EXERCISE SET

FOR EXTRA HELP

Student Solutions Manual   PH Math/Tutor Center   CD/Video for Review   MathXL MathXL®   MyMathLab MyMathLab

Objective A *Convert each measurement as indicated. See Examples 1 through 3.*

1. 60 in. to feet

2. 84 in. to feet

3. 12 yd to feet

4. 18 yd to feet

5. 42,240 ft to miles

6. 36,960 ft to miles

7. $8\frac{1}{2}$ ft to inches

8. $12\frac{1}{2}$ ft to inches

9. 10 ft to yards

10. 25 ft to yards

11. 6.4 mi to feet

12. 3.8 mi to feet

13. 162 in. to yd (Write answer as a decimal.)

14. 7216 yd to mi (Write answer as a decimal.)

15. 3 in. to ft (Write answer as a decimal.)

16. 129 in. to ft (Write answer as a decimal.)

Objective B *Convert each measurement as indicated. See Examples 4 and 5.*

17. 40 ft = _____ yd _____ ft

18. 100 ft = _____ yd _____ ft

19. 85 in. = _____ ft _____ in.

20. 59 in. = _____ ft _____ in.

21. 10,000 ft = _____ mi _____ ft

22. 25,000 ft = _____ mi _____ ft

23. 5 ft 2 in. = _____ in.

24. 4 ft 11 in. = _____ in.

25. 8 yd 2 ft = _____ ft

26. 4 yd 1 ft = _____ ft

27. 2 yd 1 ft = _____ in.

28. 1 yd 2 ft = _____ in.

Objective C *Perform each indicated operation. Simplify the result if possible. See Examples 6 through 8.*

29. 3 ft 10 in. + 7 ft 4 in.

30. 12 ft 7 in. + 9 ft 11 in.

31. 12 yd 2 ft + 9 yd 2 ft

32. 16 yd 2 ft + 8 yd 2 ft

33. 22 ft 8 in. − 16 ft 3 in.

34. 15 ft 5 in. − 8 ft 2 in.

35. 18 ft 3 in. − 10 ft 9 in.

36. 14 ft 8 in. − 3 ft 11 in.

37. 28 ft 8 in. ÷ 2

38. 34 ft 6 in. ÷ 2

39. 16 yd 2 ft × 5

40. 15 yd 1 ft × 8

Objective D *Convert as indicated. See Examples 9 and 10.*

41. 60 m to centimeters

42. 46 m to centimeters

43. 40 mm to centimeters

44. 14 mm to centimeters

45. 500 m to kilometers

46. 400 m to kilometers

47. 1700 mm to meters

48. 6400 mm to meters

49. 1500 cm to meters

50. 6400 cm to meters

51. 0.42 km to centimeters

52. 0.95 km to centimeters

53. 7 km to meters

54. 5 km to meters

55. 8.3 cm to millimeters

56. 4.6 cm to millimeters

57. 20.1 mm to decimeters

58. 140.2 mm to decimeters

59. 0.04 m to millimeters

60. 0.2 m to millimeters

Objective E *Perform each indicated operation. See Examples 11 through 13.*

61. 8.6 m + 0.34 m

62. 14.1 cm + 3.96 cm

63. 2.9 m + 40 mm

64. 30 cm + 8.9 m

65. 24.8 mm − 1.19 cm

66. 45.3 m − 2.16 dam

67. 15 km − 2360 m

68. 14 cm − 15 mm

69. 18.3 m × 3

70. 14.1 m × 4

71. 6.2 km ÷ 4

72. 9.6 m ÷ 5

Objectives **A** **C** **D** **E** **Mixed Practice** *Solve. Remember to insert units when writing your answers. For Exercises 73 through 82, complete the charts. See Examples 1 through 13.*

		Yards	Feet	Inches
73.	Chrysler Building in New York City		1046	
74.	4-story building			792
75.	Python length		35	
76.	Ostrich height			108

Complete the chart.

		Meters	Millimeters	Kilometers	Centimeters
77.	Length of elephant	5			
78.	Height of grizzly bear	3			
79.	Tennis ball diameter				6.5
80.	Golf ball diameter				4.6
81.	Distance from London to Paris			342	
82.	Distance from Houston to Dallas			396	

83. The National Zoo maintains a small patch of bamboo, which it grows as a food supply for its pandas. Two weeks ago, the bamboo was 6 ft 10 in. tall. Since then, the bamboo has grown 3 ft 8 in. How tall is the bamboo now?

84. While exploring in the Marianas Trench, a submarine probe was lowered to a point 1 mile 1400 feet below the ocean's surface. Later it was lowered an additional 1 mile 4000 feet below this point. How far is the probe below the surface of the Pacific?

85. At its deepest point, the Grand Canyon of the Colorado River in Arizona is about 6000 ft. The Grand Canyon of the Yellowstone River, which is in Yellowstone National Park in Wyoming, is at most 900 feet deep. How much deeper is the Grand Canyon of the Colorado River than the Grand Canyon of the Yellowstone River? (*Source:* National Park Service)

86. The Grand Canyon of the Gunnison River, in Colorado, is often called the Black Canyon of the Gunnison because it is so steep that light rarely penetrates the depth of the canyon. The Black Canyon of the Gunnison is only 1150 ft wide at its narrowest point. At its narrowest, the Grand Canyon of the Yellowstone is $\frac{1}{2}$ mile wide. Find the difference in width between the Grand Canyon of the Yellowstone and the Black Canyon of the Gunnison. (*Note:* Notice that the dimensions are different.) (*Source:* National Park Service)

87. The tallest man in the world is recorded as Robert Pershing Wadlow of Alton, Illinois. Born is 1918, he measured 8 ft 11 in. at his tallest. The shortest man in the world is Gul Mohammed of India, who measures 22.5 in. How many times taller than Gul is Robert? Round to one decimal place. (*Source:* Guinness World Records)

88. A 3.4-m rope is attached to a 5.8-m rope. However, when the ropes are tied, 8 cm of length is lost to form the knot. What is the length of the tied ropes?

89. The ice on a pond is 5.33 cm thick. For safe skating, the owner of the pond insists that it must be 80 mm thick. How much thicker must the ice be before skating is allowed?

90. The sediment on the bottom of the Towamencin Creek is normally 14 cm thick, but the recent flood washed away 22 mm of sediment. How thick is it now?

91. The Amana Corporation stacks up its microwave ovens in a distribution warehouse. Each stack is 1 ft 9 in. wide. How far from the wall would 9 of these stacks extend?

92. The highway commission is installing concrete sound barriers along a highway. Each barrier is 1 yd 2 ft long. Find the total length of 25 barriers placed end to end.

1 ft 9 in.

1 yd 2ft

93. A logging firm needs to cut a 67-m-long redwood log into 20 equal pieces before loading it onto a truck for shipment. How long will each piece be?

94. An 18.3-m-tall flagpole is mounted on a 65-cm-high pedestal. How far is the top of the flagpole from the ground?

95. The world's longest Coca-Cola truck is in Sweden and is 79 feet long. How many *yards* long are 4 of these trucks? (*Source: Coca-Cola Today*)

△ **96.** The world's largest Coca-Cola sign is in Arica, Chile. It is in the shape of a rectangle whose length is $133\frac{1}{3}$ yards and whose width is 131 feet. Find the area of the sign in square feet. (*Source: Coca-Cola Today*) (*Hint:* Recall that area of a rectangle is the product: length times width.)

Review

Write each decimal as a fraction and each fraction as a decimal. See Section 5.1.

97. 0.21   **98.** 0.86   **99.** $\frac{13}{100}$ **100.** $\frac{47}{100}$ **101.** $\frac{1}{4}$ **102.** $\frac{3}{20}$

Concept Extensions

Determine whether the measurement in each statement is reasonable.

103. The width of a twin-size bed is 20 meters.

104. A window measures 1 meter by 0.5 meter.

105. A drinking glass is made of glass 2 millimeters thick.

106. A paper clip is 4 kilometers long.

107. The distance across the Colorado River is 50 kilometers.

108. A model's hair is 30 centimeters long.

Estimate each sum or difference. See the first Concept Check in this section.

109.   5 yd   2 in.
     + 7 yd 30 in.

110.   45 ft   1 in.
     − 10 ft 11 in.

111. Using a unit other than the foot, write a length that is equivalent to 4 feet. (*Hint:* There are many possibilities.)

112. Using a unit other than the meter, write a length that is equivalent to 7 meters. (*Hint:* There are many possibilities.)

113. To convert from meters to centimeters, the decimal point is moved two places to the right. Explain how this relates to the fact that the prefix *centi* means $\frac{1}{100}$.

114. Explain why conversions in the metric system are easier to make than conversions in the U.S. system of measurement.

115. An advertisement sign outside Fenway Park in Boston measures 18.3 m by 18.3 m. What is the area of this sign?

A Define U.S. Units of Weight and Convert from One Unit to Another.

B Perform Arithmetic Operations on Units of Weight.

C Define Metric Units of Mass and Convert from One Unit to Another.

D Perform Arithmetic Operations on Units of Mass.

9.5 WEIGHT AND MASS

Objective **A** Defining and Converting U.S. System Units of Weight

Whenever we talk about how heavy an object is, we are concerned with the object's **weight.** We discuss weight when we refer to a 12-ounce box of Rice Krispies, a 15-pound tabby cat, or a barge hauling 24 tons of garbage.

12 ounces

15 pounds

24 tons of garbage

The most common units of weight in the U.S. measurement system are the **ounce,** the **pound,** and the **ton.** The following is a summary of equivalencies between units of weight:

U.S. Units of Weight	Unit Fractions
16 ounces (oz) = 1 pound (lb)	$\dfrac{16\ oz}{1\ lb} = \dfrac{1\ lb}{16\ oz} = 1$
2000 pounds = 1 ton	$\dfrac{2000\ lb}{1\ ton} = \dfrac{1\ ton}{2000\ lb} = 1$

✔ **Concept Check** If you were describing the weight of a fully-loaded semi-trailer, which type of unit would you use: ounce, pound, or ton? Why?

Unit fractions that equal 1 are used to convert between units of weight in the U.S. system. When converting using unit fractions, recall that the numerator of a unit fraction should contain the units we are converting to and the denominator should contain the original units.

PRACTICE PROBLEM 1

Convert 6500 pounds to tons.

Answer

1. $3\dfrac{1}{4}$ tons

✔ **Concept Check Answer**

ton

EXAMPLE 1 Convert 9000 pounds to tons.

Solution: We multiply 9000 lb by a unit fraction that uses the equality

2000 pounds = 1 ton.

Remember, the unit fraction should be $\dfrac{\text{units to convert to}}{\text{original units}}$ or $\dfrac{1\ ton}{2000\ lb}$.

$$9000 \text{ lb} = \frac{9000 \text{ lb}}{1} \cdot 1 = \frac{9000 \text{ lb}}{1} \cdot \frac{1 \text{ ton}}{2000 \text{ lb}} = \frac{9000 \text{ tons}}{2000} = \frac{9}{2} \text{ tons or } 4\frac{1}{2} \text{ tons}$$

2000 lb   2000 lb   2000 lb   2000 lb   1000 lb

1 ton   1 ton   1 ton   1 ton $\frac{1}{2}$ ton

$9000 \text{ lb} = 4\frac{1}{2} \text{ tons}$

🔲 **Work Practice Problem 1**

EXAMPLE 2 Convert 3 pounds to ounces.

Solution: We multiply by the unit fraction $\frac{16 \text{ oz}}{1 \text{ lb}}$ to convert from pounds to ounces.

$$3 \text{ lb} = \frac{3 \text{ lb}}{1} \cdot 1 = \frac{3 \text{ lb}}{1} \cdot \frac{16 \text{ oz}}{1 \text{ lb}} = 3 \cdot 16 \text{ oz} = 48 \text{ oz}$$

1 pound   1 pound   1 pound

3 lb = 48 oz

16 ounces   16 ounces   16 ounces

🔲 **Work Practice Problem 2**

As with length, it is sometimes useful to simplify a measurement of weight by writing it in terms of mixed units.

EXAMPLE 3 Convert: 33 ounces = _____ lb _____ oz

Solution: Because 16 oz = 1 lb, divide 16 into 33 to see how many pounds are in 33 ounces. The quotient is the number of pounds, and the remainder is the number of ounces. To see why we divide 16 into 33, notice that

$$33 \text{ oz} = 33 \text{ oz} \cdot \frac{1 \text{ lb}}{16 \text{ oz}} = \frac{33}{16} \text{ lb}$$

```
        2 lb 1 oz
  16)33
    -32
      1
```

Thus, 33 ounces is the same as 2 lb 1 oz.

16 ounces   16 ounces   1 ounce

33 oz = 2 lb 1 oz

1 pound   1 pound   1 ounce

🔲 **Work Practice Problem 3**

PRACTICE PROBLEM 2

Convert 72 ounces to pounds.

PRACTICE PROBLEM 3

Convert:
47 ounces = _____ lb _____ oz

Objective B Performing Operations on U.S. System Units of Weight

Performing arithmetic operations on units of weight works the same way as performing arithmetic operations on units of length.

PRACTICE PROBLEM 4

Subtract 5 tons 1200 lb from 8 tons 100 lb.

EXAMPLE 4 Subtract 3 tons 1350 lb from 8 tons 1000 lb.

Solution: To subtract, we line up similar units.

$$\begin{array}{r} 8 \text{ tons } 1000 \text{ lb} \\ - 3 \text{ tons } 1350 \text{ lb} \\ \hline \end{array}$$

Since we cannot subtract 1350 lb from 1000 lb, we borrow 1 ton from the 8 tons. To do so, we write 1 ton as 2000 lb and combine it with the 1000 lb.

7 tons + (1 ton) 2000 lb

$$\begin{array}{rcl} 8 \text{ tons } 1000 \text{ lb} & = & 7 \text{ tons } 3000 \text{ lb} \\ - 3 \text{ tons } 1350 \text{ lb} & = & - 3 \text{ tons } 1350 \text{ lb} \\ \hline & & 4 \text{ tons } 1650 \text{ lb} \end{array}$$

To check, see that the sum of 4 tons 1650 lb and 3 tons 1350 lb is 8 tons 1000 lb.

🔲 **Work Practice Problem 4**

PRACTICE PROBLEM 5

Divide 5 lb 8 oz by 4.

EXAMPLE 5 Divide 9 lb 6 oz by 2.

Solution: We divide each of the units by 2.

$$\begin{array}{r} 4 \text{ lb} \quad\; 11 \text{ oz} \\ 2 \overline{)\; 9 \text{ lb} \quad\;\; 6 \text{ oz}} \\ -8 \quad\quad\quad \\ \hline 1 \text{ lb} = 16 \text{ oz} \\ \hline 22 \text{ oz} \end{array}$$ Divide 2 into 22 oz to get 11 oz.

To check, multiply 4 pounds 11 ounces by 2. The result is 9 pounds 6 ounces.

🔲 **Work Practice Problem 5**

PRACTICE PROBLEM 6

A 5-lb 14-oz batch of cookies is packed into a 6-oz container before it is mailed. Find the total weight.

EXAMPLE 6 Finding the Weight of a Child

Bryan weighed 8 lb 8 oz at birth. By the time he was 1 year old, he had gained 11 lb 14 oz. Find his weight at age 1 year.

Solution:

$$\begin{array}{rcl} \text{birth weight} & \rightarrow & 8 \text{ lb} \;\; 8 \text{ oz} \\ + \text{ weight gained} & \rightarrow & + 11 \text{ lb } 14 \text{ oz} \\ \hline \text{total weight} & \rightarrow & 19 \text{ lb } 22 \text{ oz} \end{array}$$

Since 22 oz equals 1 lb 6 oz,

$$\begin{aligned} 19 \text{ lb } 22 \text{ oz} &= 19 \text{ lb} + 1 \text{ lb } 6 \text{ oz} \\ &= 20 \text{ lb } 6 \text{ oz} \end{aligned}$$

Bryan weighed 20 lb 6 oz on his first birthday.

🔲 **Work Practice Problem 6**

Answers

4. 2 tons 900 lb   **5.** 1 lb 6 oz
6. 6 lb 4 oz

Objective C Defining and Converting Metric System Units of Mass

In scientific and technical areas, a careful distinction is made between **weight** and **mass. Weight** is really a measure of the pull of gravity. The farther from Earth an object gets, the less it weighs. However, **mass** is a measure of the amount of substance in the object and does not change. Astronauts orbiting Earth weigh much less than they weigh on Earth, but they have the same mass in orbit as they do on Earth. Here on Earth weight and mass are the same, so either term may be used.

The basic unit of mass in the metric system is the **gram.** It is defined as the mass of water contained in a cube 1 centimeter (cm) on each side.

The following examples may help you get a feeling for metric masses:

A tablet contains 200 milligrams of ibuprofen.

A large paper clip weighs approximately 1 gram.

A box of crackers weighs 453 grams.

A kilogram is slightly over 2 pounds. An adult woman may weigh 60 kilograms.

The prefixes for units of mass in the metric system are the same as for units of length, as shown in the following table:

Metric Unit of Mass
1 **kilo**gram (kg) = 1000 grams (g)
1 **hecto**gram (hg) = 100 g
1 **deka**gram (dag) = 10 g
1 gram (g) = 1 g
1 **deci**gram (dg) = 1/10 g or 0.1 g
1 **centi**gram (cg) = 1/100 g or 0.01 g
1 **milli**gram (mg) = 1/1000 g or 0.001 g

✔ *Concept Check*   True or false? A decigram is larger than a dekagram. Explain.

The **milligram,** the **gram,** and the **kilogram** are the three most commonly used units of mass in the metric system.

As with lengths, all units of mass are powers of 10 of the gram, so converting from one unit of mass to another involves moving only the decimal point. To convert

✔ **Concept Check Answer**

false

from one unit of mass to another in the metric system, list the units of mass in order from largest to smallest.

Let's convert 4300 milligrams to grams. To convert from milligrams to grams, we move along the list 3 units to the left.

kg   hg   dag   **g**   dg   cg   **mg**

3 units to the left

This means that we move the decimal point 3 places to the left to convert from milligrams to grams.

4300 mg = 4.3 g

Don't forget, the same conversion can be done with unit fractions.

$$4300 \text{ mg} = \frac{4300 \text{ mg}}{1} \cdot 1 = \frac{4300 \text{ mg}}{1} \cdot \frac{0.001 \text{ g}}{1 \text{ mg}}$$

$$= 4300 \cdot 0.001 \text{ g}$$

$$= 4.3 \text{ g} \quad \text{To multiply by 0.001, move the decimal point 3 places to the left.}$$

To see that this is reasonable, study the diagram:

Thus, 4300 mg = 4.3 g

PRACTICE PROBLEM 7

Convert 3.41 g to milligrams.

EXAMPLE 7 Convert 3.2 kg to grams.

Solution: First we convert by using a unit fraction.

Unit fraction

$$3.2 \text{ kg} = 3.2 \text{ kg} \cdot 1 = 3.2 \text{ kg} \cdot \frac{1000 \text{ g}}{1 \text{ kg}} = 3200 \text{ g}$$

Now let's list the units of mass in order from left to right and move from kilograms to grams.

kg   hg   dag   g   dg   cg   mg

3 units to the right

3.200 kg = 3200. g

3 places to the right

Answer

7. 3410 mg

📖 **Work Practice Problem 7**

EXAMPLE 8 Convert 2.35 cg to grams.

Solution: We list the units of mass in a chart and move from centigrams to grams.

kg   hg   dag   g   dg   cg   mg

2 units to the left

02.35 cg = 0.0235 g

2 places to the left

◼ **Work Practice Problem 8**

Convert 56.2 cg to grams.

Objective D Performing Operations on Metric System Units of Mass

Arithmetic operations can be performed with metric units of mass just as we performed operations with metric units of length. We convert each number to the same unit of mass and add, subtract, multiply, or divide as with decimals.

EXAMPLE 9 Subtract 5.4 dg from 1.6 g.

Solution: We convert both numbers to decigrams or to grams before subtracting.

5.4 dg = 0.54 g	or	1.6 g = 16 dg
1.60 g		16.0 dg
− 0.54 g		− 5.4 dg
1.06 g		10.6 dg

The difference is 1.06 g or 10.6 dg.

◼ **Work Practice Problem 9**

PRACTICE PROBLEM 9

Subtract 3.1 dg from 2.5 g.

EXAMPLE 10 Calculating Allowable Weight in an Elevator

An elevator has a weight limit of 1400 kg. A sign posted in the elevator indicates that the maximum capacity of the elevator is 17 persons. What is the average allowable weight for each passenger, rounded to the nearest kilogram?

Solution: To solve, notice that the total weight of 1400 kilograms ÷ 17 = average weight

$$
\begin{array}{r}
82.3 \text{ kg} \approx 82 \text{ kg} \\
17\overline{)\ 1400.0 \text{ kg}} \\
-136 \\
\hline
40 \\
-34 \\
\hline
60 \\
-51 \\
\hline
9
\end{array}
$$

Each passenger can weigh an average of 82 kg. (Recall that a kilogram is slightly over 2 pounds, so 82 kilograms is over 164 pounds.)

◼ **Work Practice Problem 10**

PRACTICE PROBLEM 10

Twenty-four bags of cement weigh a total of 550 kg. Find the average weight of 1 bag, rounded to the nearest kilogram.

Answers
8. 0.562 g **9.** 2.19 g or 21.9 dg
10. 23 kg

Vocabulary and Readiness Check

Use the choices below to fill in each blank.

    mass   weight   gram

1. _____ is a measure of the amount of substance in an object. This measure does not change.

2. _____ is the measure of the pull of gravity.

3. The basic unit of mass in the metric system is the _____.

Fill in these blanks with the correct number. Choices for these blanks are not shown in the list of terms above.

4. One pound equals _____ ounces.

5. One ton equals _____ pounds.

9.5 EXERCISE SET

FOR EXTRA HELP

Student Solutions Manual   PH Math/Tutor Center   CD/Video for Review   MathXL®   MyMathLab

Objective A *Convert as indicated. See Examples 1 through 3.*

1. 2 pounds to ounces

2. 5 pounds to ounces

3. 5 tons to pounds

4. 7 tons to pounds

5. 18,000 pounds to tons

6. 28,000 pounds to tons

7. 60 ounces to pounds

8. 90 ounces to pounds

9. 3500 pounds to tons

10. 11,000 pounds to tons

11. 12.75 pounds to ounces

12. 9.5 pounds to ounces

13. 4.9 tons to pounds

14. 8.3 tons to pounds

15. $4\frac{3}{4}$ pounds to ounces

16. $9\frac{1}{8}$ pounds to ounces

17. 2950 pounds to the nearest tenth of a ton

18. 51 ounces to the nearest tenth of a pound

19. $\frac{4}{5}$ oz to pounds

20. $\frac{1}{4}$ oz to pounds

21. $5\frac{3}{4}$ lb to ounces

22. $2\frac{1}{4}$ lb to ounces

23. 10 lb 1 oz to ounces

24. 7 lb 6 oz to ounces

25. 89 oz to _____ lb _____ oz

26. 100 oz = _____ lb _____ oz

Objective **B** *Perform each indicated operation. See Examples 4 through 6.*

27. 34 lb 12 oz + 18 lb 14 oz   **28.** 6 lb 10 oz + 10 lb 8 oz   **29.** 3 tons 1820 lb + 4 tons 930 lb

30. 1 ton 1140 lb + 5 tons 1200 lb   **31.** 5 tons 1050 lb − 2 tons 875 lb   **32.** 4 tons 850 lb − 1 ton 260 lb

33. 12 lb 4 oz − 3 lb 9 oz   **34.** 45 lb 6 oz − 26 lb 10 oz   **35.** 5 lb 3 oz × 6

36. 2 lb 5 oz × 5   **37.** 6 tons 1500 lb ÷ 5   **38.** 5 tons 400 lb ÷ 4

Objective **C** *Convert as indicated. See Examples 7 and 8.*

39. 500 g to kilograms   **40.** 820 g to kilograms   **41.** 4 g to milligrams   **42.** 9 g to milligrams

43. 25 kg to grams   **44.** 18 kg to grams   **45.** 48 mg to grams   **46.** 112 mg to grams

47. 6.3 g to kilograms   **48.** 4.9 g to kilograms   **49.** 15.14 g to milligrams   **50.** 16.23 g to milligrams

51. 6.25 kg to grams   **52.** 3.16 kg to grams   **53.** 35 hg to centigrams   **54.** 4.26 cg to dekagrams

Objective **D** *Perform each indicated operation. See Examples 9 and 10.*

55. 3.8 mg + 9.7 mg   **56.** 41.6 g + 9.8 g   **57.** 205 mg + 5.61 g   **58.** 2.1 g + 153 mg

59. 9 g − 7150 mg   **60.** 4 kg − 2410 g   **61.** 1.61 kg − 250 g   **62.** 6.13 g − 418 mg

63. 5.2 kg × 2.6   **64.** 4.8 kg × 9.3   **65.** 17 kg ÷ 8   **66.** 8.25 g ÷ 6

Objectives A B C D **Mixed Practice** *Solve. Remember to insert units when writing your answers. For Exercises 67 through 74, complete the chart. See Examples 1 through 10.*

	Object	Tons	Pounds	Ounces
67.	Statue of Liberty—weight of copper sheeting	100		
68.	Statue of Liberty—weight of steel	125		
69.	A 12-inch cube of osmium (heaviest metal)		1,345	
70.	A 12-inch cube of lithium (lightest metal)		32	

	Object	Grams	Kilograms	Milligrams	Centigrams
71.	Capsule of Amoxicillin (antibiotic)			500	
72.	Tablet of Topamax (epilepsy and migraine uses)			25	
73.	A six-year-old boy		21		
74.	A golf ball	45			

75. A can of 7-Up weighs 336 grams. Find the weight in kilograms of 24 cans.

76. Guy Green normally weighs 73 kg, but he lost 2800 grams after being sick with the flu. Find Guy's new weight.

77. Sudafed is a decongestant that comes in two strengths. Regular strength contains 60 mg of medication. Extra strength contains 0.09 g of medication. How much extra medication is in the extra-strength tablet?

78. A small can of Planters sunflower seeds weighs 177 g. If each can contains 6 servings, find the weight of one serving.

79. Doris Johnson has two open containers of Uncle Ben's rice. If she combines 1 lb 10 oz from one container with 3 lb 14 oz from the other container, how much total rice does she have?

80. Dru Mizel maintains the records of the amount of coal delivered to his department in the steel mill. In January, 3 tons 1500 lb were delivered. In February, 2 tons 1200 lb were delivered. Find the total amount delivered in these two months.

81. Carla Hamtini was amazed when she grew a 28 lb 10 oz zucchini in her garden, but later she learned that the heaviest zucchini ever grown weighed 64 lb 8 oz in Llanharry, Wales, by B. Lavery in 1990. How far below the record weight was Carla's zucchini? (*Source: Guinness World Records*)

82. The heaviest baby born in good health weighed an incredible 22 lb 8 oz. He was born in Italy in September, 1955. How much heavier is this than a 7 lb 12 oz baby? (*Source: Guinness World Records*)

83. The smallest baby born in good health weighed only 8.6 ounces, less than a can of soda. She was born in Chicago in December, 2004. How much lighter was she than an average baby, who weighs about 7 lb 8 ounces?

84. A large bottle of Hire's Root Beer weighs 1900 grams. If a carton contains 6 large bottles of root beer, find the weight in kilograms of 5 cartons.

85. Three milligrams of preservatives are added to a 0.5-kg box of dried fruit. How many milligrams of preservatives are in 3 cartons of dried fruit if each carton contains 16 boxes?

86. One box of Swiss Miss Cocoa Mix weighs 0.385 kg, but 39 grams of this weight is the packaging. Find the actual weight of the cocoa in 8 boxes.

87. A carton of 12 boxes of Quaker Oats Oatmeal weighs 6.432 kg. Each box includes 26 grams of packaging material. What is the actual weight of the oatmeal in the carton?

88. The supermarket prepares hamburger in 85-gram market packages. When Leo Gonzalas gets home, he divides the package in half before refrigerating the meat. How much will each package weigh?

89. The Shop 'n Bag supermarket chain ships hamburger meat by placing 10 packages of hamburger in a box, with each package weighing 3 lb 4 oz. How much will 4 boxes of hamburger weigh?

90. The Quaker Oats Company ships its 1-lb 2-oz boxes of oatmeal in cartons containing 12 boxes of oatmeal. How much will 3 such cartons weigh?

91. A carton of Del Monte Pineapple weighs 55 lb 4 oz, but 2 lb 8 oz of this weight is due to packaging. Subtract the weight of the packaging to find the actual weight of the pineapple in 4 cartons.

92. The Hormel Corporation ships cartons of canned ham weighing 43 lb 2 oz each. Of this weight, 3 lb 4 oz is due to packaging. Find the actual weight of the ham found in 3 cartons.

Review

Write each fraction as a decimal. See Section 5.5.

93. $\dfrac{4}{25}$

94. $\dfrac{3}{5}$

95. $\dfrac{7}{8}$

96. $\dfrac{3}{16}$

Concept Extensions

Determine whether the measurement in each statement is reasonable.

97. The doctor prescribed a pill containing 2 kg of medication.

98. A full-grown cat weighs approximately 15 g.

99. A bag of flour weighs 4.5 kg.

100. A staple weighs 15 mg.

101. A professor weighs less than 150 g.

102. A car weighs 2000 mg.

103. Use a unit other than centigram and write a mass that is equivalent to 25 centigrams. (*Hint:* There are many possibilities.)

104. Use a unit other than pound and write a weight that is equivalent to 4000 pounds. (*Hint:* There are many possibilities.)

True or False? See the Concept Check in this section.

105. A kilogram is larger than a gram.

106. A decigram is larger than a milligram.

107. Why is the decimal point moved to the right when grams are converted to milligrams?

108. To change 8 pounds to ounces, multiply by 16. Why is this the correct procedure?

 STUDY SKILLS BUILDER

How Are Your Homework Assignments Going?

Remember that it is important to keep up with homework. Why? Many concepts in mathematics build on each other. Often, your understanding of a day's lecture depends on an understanding of the previous day's material.

To complete a homework assignment, remember these 4 things:

- Attempt all of it.
- Check it.
- Correct it.
- If needed, ask questions about it.

Take a moment and review your completed homework assignments. Answer the exercises below based on this review.

1. Approximate the fraction of your homework you have attempted.

2. Approximate the fraction of your homework you have checked (if possible).

3. If you are able to check your homework, have you corrected it when errors have been found?

4. When working homework, if you do not understand a concept, what do you personally do?

9.6 CAPACITY

A Define U.S. Units of Capacity and Convert from One Unit to Another.

B Perform Arithmetic Operations on U.S. Units of Capacity.

C Define Metric Units of Capacity and Convert from One Unit to Another.

D Perform Arithmetic Operations on Metric Units of Capacity.

Objective **A** Defining and Converting U.S. System Units of Capacity

Units of **capacity** are generally used to measure liquids. The number of gallons of gasoline needed to fill a gas tank in a car, the number of cups of water needed in a bread recipe, and the number of quarts of milk sold each day at a supermarket are all examples of using units of capacity. The following summary shows equivalencies between units of capacity:

U.S. Units of Capacity

$$8 \text{ fluid ounces (fl oz)} = 1 \text{ cup (c)}$$
$$2 \text{ cups} = 1 \text{ pint (pt)}$$
$$2 \text{ pints} = 1 \text{ quart (qt)}$$
$$4 \text{ quarts} = 1 \text{ gallon (gal)}$$

Just as with units of length and weight, we can form unit fractions to convert between different units of capacity. For instance,

$$\frac{2 \text{ c}}{1 \text{ pt}} = \frac{1 \text{ pt}}{2 \text{ c}} = 1 \quad \text{and} \quad \frac{2 \text{ pt}}{1 \text{ qt}} = \frac{1 \text{ qt}}{2 \text{ pt}} = 1$$

EXAMPLE 1 Convert 9 quarts to gallons.

Solution: We multiply by the unit fraction $\dfrac{1 \text{ gal}}{4 \text{ qt}}$.

$$9 \text{ qt} = \frac{9 \text{ qt}}{1} \cdot 1$$

$$= \frac{9 \text{ q\!t}}{1} \cdot \frac{1 \text{ gal}}{4 \text{ q\!t}}$$

$$= \frac{9 \text{ gal}}{4}$$

$$= 2\frac{1}{4} \text{ gal}$$

Thus, 9 quarts is the same as $2\dfrac{1}{4}$ gallons, as shown in the diagram:

1 gallon   1 gallon $\frac{1}{4}$ gallon

9 quarts $= 2\frac{1}{4}$ gal

◼ **Work Practice Problem 1**

Answer

1. $21\dfrac{1}{2}$ qt

669

PRACTICE PROBLEM 2

Convert 26 quarts to cups.

EXAMPLE 2 Convert 14 cups to quarts.

Solution: Our equivalency table contains no direct conversion from cups to quarts. However, from this table we know that

$$1 \text{ qt} = 2 \text{ pt} = \frac{2 \text{ pt}}{1} \cdot 1 = \frac{2 \text{ pt}}{1} \cdot \frac{2 \text{ c}}{1 \text{ pt}} = 4 \text{ c}$$

so 1 qt = 4 c. Now we have the unit fraction $\frac{1 \text{ qt}}{4 \text{ c}}$. Thus,

$$14 \text{ c} = \frac{14 \text{ c}}{1} \cdot 1 = \frac{14 \text{ c}}{1} \cdot \frac{1 \text{ qt}}{4 \text{ c}} = \frac{14 \text{ qt}}{4} = \frac{7}{2} \text{ qt} \quad \text{or} \quad 3\frac{1}{2} \text{ qt}$$

1 quart + 1 quart + 1 quart + $\frac{1}{2}$ quart   14 cups = $3\frac{1}{2}$ qt

Work Practice Problem 2

✔**Concept Check** If 50 cups are converted to quarts, will the equivalent number of quarts be less than or greater than 50? Explain.

Objective B Performing Operations on U.S. System Units of Capacity

As is true of units of length and weight, units of capacity can be added, subtracted, multiplied, and divided.

PRACTICE PROBLEM 3

Subtract 2 qt from 1 gal 1 qt.

EXAMPLE 3 Subtract 3 qt from 4 gal 2 qt.

Solution: To subtract, we line up similar units.

```
  4 gal 2 qt
−       3 qt
```

We cannot subtract 3 qt from 2 qt. We need to borrow 1 gallon from the 4 gallons, convert it to 4 quarts, and then combine it with the 2 quarts.

3 gal + (1 gal) 4 qt

```
  4 gal 2 qt   =    3 gal 6 qt
−       3 qt   =  −       3 qt
                   3 gal 3 qt
```

To check, see that the sum of 3 gal 3 qt and 3 qt is 4 gal 2 qt.

Work Practice Problem 3

Answers
2. 104 c   **3.** 3 qt

✔ **Concept Check Answer**
less than 50

EXAMPLE 4 **Finding the Amount of Water in an Aquarium**

An aquarium contains 6 gal 3 qt of water. If 2 gal 2 qt of water is added, what is the total amount of water in the aquarium?

Solution:

		beginning water	→	6 gal 3 qt
+		water added	→	+ 2 gal 2 qt
		total water	→	8 gal 5 qt

Since 5 qt = 1 gal 1 qt, we have

$$= \overbrace{8 \text{ gal}}^{8 \text{ gal}} + \overbrace{1 \text{ gal } 1 \text{ qt}}^{5 \text{ qt}}$$
$$= 9 \text{ gal } 1 \text{ qt}$$

The total amount of water is 9 gal 1 qt.

■ **Work Practice Problem 4**

Objective **C** **Defining and Converting Metric System Units of Capacity**

Thus far, we know that the basic unit of length in the metric system is the meter and that the basic unit of mass in the metric system is the gram. What is the basic unit of capacity? The **liter.** By definition, a **liter** is the capacity or volume of a cube measuring 10 centimeters on each side.

The following examples may help you get a feeling for metric capacities:
One liter of liquid is slightly more than one quart.
Many soft drinks are packaged in 2-liter bottles.

The metric system was designed to be a consistent system. Once again, the prefixes for metric units of capacity are the same as for metric units of length and mass, as summarized in the following table:

Metric Unit of Capacity
1 **kilo**liter (kl) = 1000 liters (L)
1 **hecto**liter (hl) = 100 L
1 **deka**liter (dal) = 10 L
1 liter (L) = 1 L
1 **deci**liter (dl) = 1/10 L or 0.1 L
1 **centi**liter (cl) = 1/100 L or 0.01 L
1 **milli**liter (ml) = 1/1000 L or 0.001 L

The **milliliter** and the **liter** are the two most commonly used metric units of capacity.

Converting from one unit of capacity to another involves multiplying by powers of 10 or moving the decimal point to the left or to the right. Listing units of capacity in order from largest to smallest helps to keep track of how many places to move the decimal point when converting.

Let's convert 2.6 liters to milliliters. To convert from liters to milliliters, we move along the chart 3 units to the right.

kl   hl   dal   **L**   dl   cl   **ml**

3 units to the right

PRACTICE PROBLEM 4

A large oil drum contains 15 gal 3 qt of oil. How much will be in the drum if an additional 4 gal 3 qt of oil is poured into it?

10 cm

10 cm

10 cm

1 liter   1 quart

Answer
4. 20 gal 2 qt

This means that we move the decimal point 3 places to the right to convert from liters to milliliters.

$$2.600 \, \text{L} = 2600. \, \text{ml}$$

This same conversion can be done with unit fractions.

$$2.6 \, \text{L} = \frac{2.6 \, \text{L}}{1} \cdot 1$$

$$= \frac{2.6 \, \cancel{\text{L}}}{1} \cdot \frac{1000 \, \text{ml}}{1 \, \cancel{\text{L}}}$$

$$= 2.6 \cdot 1000 \, \text{ml}$$

$$= 2600 \, \text{ml} \quad \text{To multiply by 1000, move the decimal point 3 places to the right.}$$

To visualize the result, study the diagram below:

1000 ml   1000 ml   600 ml   =   2600 ml

Thus, 2.6 L = 2600 ml.

PRACTICE PROBLEM 5

Convert 2100 ml to liters.

EXAMPLE 5 Convert 3210 ml to liters.

Solution: Let's use the unit fraction method first.

Unit fraction

$$3210 \, \text{ml} = \frac{3210 \, \text{ml}}{1} \cdot 1 = 3210 \, \cancel{\text{ml}} \cdot \frac{1 \, \text{L}}{1000 \, \cancel{\text{ml}}} = 3.21 \, \text{L}$$

Now let's list the unit measures in order from left to right and move from milliliters to liters.

kl   hl   dal   L   dl   cl   ml

3 units to the left

3210 ml = 3.210 L, the same results as before and shown below in the diagram.

3 places to the left

1 L   1 L   1 L   0.210 L = 3.210 L

■ **Work Practice Problem 5**

Answer

5. 2.1 L

EXAMPLE 6 Convert 0.185 dl to milliliters.

Solution: We list the unit measures in order from left to right and move from deciliters to milliliters.

kl   hl   dal   L   dl   cl   ml

2 units to the right

0.185 dl = 18.5 ml

2 places to the right

■ Work Practice Problem 6

PRACTICE PROBLEM 6
Convert 2.13 dal to liters.

Objective D Performing Operations on Metric System Units of Capacity

As was true for length and weight, arithmetic operations involving metric units of capacity can also be performed. Make sure that the metric units of capacity are the same before adding, subtracting, multiplying, or dividing.

EXAMPLE 7 Add 2400 ml to 8.9 L.

Solution: We must convert both to liters or both to milliliters before adding the capacities together.

2400 ml = 2.4 L   or   8.9 L = 8900 ml

2.4 L   2400 ml
+ 8.9 L   + 8900 ml
11.3 L   11,300 ml

The total is 11.3 L or 11,300 ml. They both represent the same capacity.

■ Work Practice Problem 7

✔ Concept Check How could you estimate the following operation? Subtract 950 ml from 7.5 L.

PRACTICE PROBLEM 7
Add 1250 ml to 2.9 L.

EXAMPLE 8 Finding the Amount of Medication a Person Has Received

A patient hooked up to an IV unit in the hospital is to receive 12.5 ml of medication every hour. How much medication does the patient receive in 3.5 hours?

Solution: We multiply 12.5 ml by 3.5.

medication per hour → 12.5 ml
× hours → × 3.5
total medication   625
              3750
            43.75 ml

The patient receives 43.75 ml of medication.

■ Work Practice Problem 8

PRACTICE PROBLEM 8
If 28.6 L of water can be pumped every minute, how much water can be pumped in 85 minutes?

Answers
6. 21.3 L **7.** 4150 ml or 4.15 L
8. 2431 L

✔ Concept Check Answer
950 ml = 0.95 L; round 0.95 to 1;
7.5 − 1 = 6.5 L

Vocabulary and Readiness Check

Use the choices below to fill in each blank. Some choices may be used more than once.

cups   pints   liter

quarts   fluid ounces   capacity

1. Units of _____ are generally used to measure liquids.

2. The basic unit of capacity in the metric system is the _____.

3. One cup equals 8 _____.

4. One quart equals 2 _____.

5. One pint equals 2 _____.

6. One quart equals 4 _____.

7. One gallon equals 4 _____.

9.6 EXERCISE SET

FOR EXTRA HELP

Student Solutions Manual   PH Math/Tutor Center   CD/Video for Review   Math XL MathXL®   MyMathLab MyMathLab

Objective A *Convert each measurement as indicated. See Examples 1 and 2.*

1. 32 fluid ounces to cups

2. 16 quarts to gallons

3. 8 quarts to pints

4. 9 pints to quarts

5. 14 quarts to gallons

6. 11 cups to pints

 7. 80 fluid ounces to pints

8. 18 pints to gallons

9. 2 quarts to cups

10. 3 pints to fluid ounces

11. 120 fluid ounces to quarts

12. 20 cups to gallons

13. 42 cups to quarts

14. 7 quarts to cups

15. $4\frac{1}{2}$ pints to cups

16. $6\frac{1}{2}$ gallons to quarts

17. 5 gal 3 qt to quarts

18. 4 gal 1 qt to quarts

19. $\frac{1}{2}$ cup to pint

20. $\frac{1}{2}$ pint to quarts

21. 58 qt = _____ gal _____ qt

22. 70 qt = _____ gal _____ qt

23. 39 pt = _____ gal _____ qt _____ pt

24. 29 pt = _____ gal _____ qt _____ pt

25. $2\frac{3}{4}$ gallons to pints

26. $3\frac{1}{4}$ quarts to cups

Objective B *Perform each indicated operation. See Examples 3 and 4.*

27. 5 gal 3 qt + 7 gal 3 qt

28. 2 gal 2 qt + 9 gal 3 qt

29. 1 c 5 fl oz + 2 c 7 fl oz

30. 2 c 3 fl oz + 2 c 6 fl oz

31. 3 gal − 1 gal 3 qt

32. 2 pt − 1 pt 1 c

33. 3 gal 1 qt − 1 qt 1 pt

34. 3 qt 1 c − 1 c 4 fl oz

35. 8 gal 2 qt × 2

36. 6 gal 1 pt × 2

37. 9 gal 2 qt ÷ 2

38. 5 gal 6 fl oz ÷ 2

Objective **C** *Convert as indicated. See Examples 5 and 6.*

39. 5 L to milliliters

40. 8 L to milliliters

41. 0.16 L to kiloliters

42. 0.127 L to kiloliters

43. 5600 ml to liters

44. 1500 ml to liters

45. 3.2 L to centiliters

46. 1.7 L to centiliters

47. 410 L to kiloliters

48. 250 L to kiloliters

49. 64 ml to liters

50. 39 ml to liters

51. 0.16 kl to liters

52. 0.48 kl to liters

53. 3.6 L to milliliters

54. 1.9 L to milliliters

Objective **D** *Perform each indicated operation. See Examples 7 and 8.*

55. 3.4 L + 15.9 L

56. 18.5 L + 4.6 L

57. 2700 ml + 1.8 L

58. 4.6 L + 1600 ml

59. 8.6 L − 190 ml

60. 4.8 L − 283 ml

61. 17,500 ml − 0.9 L

62. 6850 ml − 0.3 L

63. 480 ml × 8

64. 290 ml × 6

65. 81.2 L ÷ 0.5

66. 5.4 L ÷ 3.6

Objectives **A** **B** **C** **D** **Mixed Practice** *Solve. Remember to insert units when writing your answers. For Exercises 67 through 70, complete the chart.*

	Capacity	Cups	Gallons	Quarts	Pints
67.	An average-size bath of water		21		
68.	A dairy cow's daily milk yield				38
69.	Your kidneys filter about this amount of blood every minute	4			
70.	The amount of water needed in a punch recipe	2			

71. Mike Schaferkotter drank 410 ml of Mountain Dew from a 2-liter bottle. How much Mountain Dew remains in the bottle?

72. The Werners' Volvo has a 54.5-L gas tank. Only 3.8 liters of gasoline still remain in the tank. How much is needed to fill it?

73. Margie Phitts added 354 ml of Prestone dry gas to the 18.6 L of gasoline in her car's tank. Find the total amount of gasoline in the tank.

74. Chris Peckaitis wishes to share a 2-L bottle of Coca Cola equally with 7 of his friends. How much will each person get?

75. A garden tool engine requires a 30 to 1 gas to oil mixture. This means that $\frac{1}{30}$ of a gallon of oil should be mixed with 1 gallon of gas. Convert $\frac{1}{30}$ gallon to ounces. Round to the nearest tenth.

76. Henning's Supermarket sells homemade soup in 1 qt 1 pt containers. How much soup is contained in three such containers?

77. Can 5 pt 1 c of fruit punch and 2 pt 1 c of ginger ale be poured into a 1-gal container without it overflowing?

78. Three cups of prepared Jell-O are poured into 6 dessert dishes. How many fluid ounces of Jell-O are in each dish?

79. Stanley Fisher paid $14 to fill his car with 44.3 liters of gasoline. Find the price per liter of gasoline to the nearest thousandth of a dollar.

80. A student carelessly misread the scale on a cylinder in the chemistry lab and added 40 cl of water to a mixture instead of 40 ml. Find the excess amount of water.

Review

Write each fraction in simplest form. See Section 4.2.

81. $\frac{20}{25}$ **82.** $\frac{75}{100}$ **83.** $\frac{27}{45}$ **84.** $\frac{56}{60}$ **85.** $\frac{72}{80}$ **86.** $\frac{18}{20}$

Concept Extensions

Determine whether the measurement in each statement is reasonable.

87. Clair took a dose of 2 L of cough medicine to cure her cough.

88. John drank 250 ml of milk for lunch.

89. Jeannie likes to relax in a tub filled with 3000 ml of hot water.

90. Sarah pumped 20 L of gasoline into her car yesterday.

Solve. See the Concept Checks in this section.

91. If 70 pints are converted to gallons, will the equivalent number of gallons be less than or greater than 70? Explain why.

92. If 30 gallons are converted to quarts, will the equivalent number of quarts be less than or greater than 30? Explain why.

93. Explain how to estimate the following operation: Add 986 ml to 6.9 L.

94. Find the number of fluid ounces in 1 gallon.

95. Explain how to borrow in order to subtract 1 gal 2 qt from 3 gal 1 qt.

A cubic centimeter (cc) is the amount of space that a volume of 1 ml occupies. Because of this, we will say that 1 cc = 1 ml.

A common syringe is one with a capacity of 3 cc. Use the diagram and give the measurement indicated by each arrow.

96. A **97.** B **98.** C **99.** D

In order to measure small dosages, such as for insulin, u-100 syringes are used. For these syringes, 1 cc has been divided into 100 equal units (u). Use the diagram and give the measurement indicated by each arrow in units (u) and then cubic centimeters. Use 100 u = 1 cc and round to the nearest hundredth.

100. A **101.** B

102. C **103.** D

9.7 TEMPERATURE AND CONVERSIONS BETWEEN THE U.S. AND METRIC SYSTEMS

Objective

A Convert Between the U.S. and Metric Systems.

B Convert Temperatures from Degrees Celsius to Degrees Fahrenheit.

C Convert Temperatures from Degrees Fahrenheit to Degrees Celsius.

Objective **A** Converting Between the U.S. and Metric Systems

The metric system probably had its beginnings in France in the 1600s, but it was the Metric Act of 1866 that made the use of this system legal (but not mandatory) in the United States. Other laws have followed that allow for a slow, but deliberate, transfer to the modernized metric system. In April, 2001, for example, the U.S. Stock Exchanges completed their change to decimal trading instead of fractions. By the end of 2009, all products sold in Europe (with some exceptions) will be required to have only metric units on their labels. (*Source:* U.S. Metric Association and National Institute of Standards and Technology)

You may be surprised at the number of everyday items we use that are already manufactured in metric units. We easily recognize 1L and 2L soda bottles, but what about the following?

Pencil leads (0.5 mm or 0.7 mm)
Camera film (35 mm)
Sporting events (5 km or 10 km races)
Medicines (500 mg capsules)
Labels on retail goods (dual-labeled since 1994)

Since the United States has not completely converted to the metric system, we need to practice converting from one system to the other. Below is a table of mostly approximate conversions.

Length:		**Capacity:**		**Weight** (mass):	
metric	U.S. System	metric	U.S. System	metric	U.S. System
1 m	≈ 1.09 yd	1 L	≈ 1.06 qt	1 kg	≈ 2.20 lb
1 m	≈ 3.28 ft	1 L	≈ 0.26 gal	1 g	≈ 0.04 oz
1 km	≈ 0.62 mi	3.79 L	≈ 1 gal	0.45 kg	≈ 1 lb
2.54 cm	= 1 in.	0.95 L	≈ 1 qt	28.35 g	≈ 1 oz
0.30 m	≈ 1 ft	29.57 ml	≈ 1 fl oz		
1.61 km	≈ 1 mi				

There are many ways to perform these metric to U.S. conversions. We will do so by using unit fractions.

1 yard
1 meter

1 quart   1 liter

1 pound   1 kilogram

EXAMPLE 1 Compact Discs

Standard-sized compact discs are 12 centimeters in diameter. Convert this length to inches. Round the result to two decimal places. (*Source:* usByte.com)

Solution: From our length conversion table, we know that 2.54 cm = 1 in. This fact gives us two unit fractions: $\frac{2.54 \text{ cm}}{1 \text{ in.}}$ and $\frac{1 \text{ in.}}{2.54 \text{ cm}}$. We use the unit fraction with cm in the denominator so that these units divide out.

1.5 cm

12 cm

$$12 \text{ cm} = \frac{12 \text{ cm}}{1} \cdot 1 = \frac{12 \text{ cm}}{1} \cdot \overbrace{\frac{1 \text{ in.}}{2.54 \text{ cm}}}^{\text{Unit fraction}} \quad \begin{array}{l} \leftarrow \text{Units to convert to} \\ \leftarrow \text{Original units} \end{array}$$

$$= \frac{12 \text{ in.}}{2.54}$$

$$\approx 4.72 \text{ in.} \quad \text{Divide.}$$

PRACTICE PROBLEM 1

The center hole of a standard-sized compact disc is 1.5 centimeters in diameter. Convert this length to inches. Round the result to 2 decimal places.

Answer
1. 0.59 in.

Continued on next page

Thus, the diameter of a standard compact disc is exactly 12 cm or approximately 4.72 inches. For a dimension this size, you can use a ruler to check. Another method is to approximate. Our result, 4.72 in., is close to 5 inches. Since 1 in. is about 2.5 cm, then 5 in. is about 5(2.5 cm) = 12.5 cm, which is close to 12 cm.

⬛ **Work Practice Problem 1**

PRACTICE PROBLEM 2

A full-grown human heart weighs about 8 ounces. Convert this weight to grams. If necessary, round your result to the nearest tenth of a gram.

EXAMPLE 2 **Liver**

The liver is your largest internal organ. It weighs about 3.5 pounds in a grown man. Convert this weight to kilograms. Round to the nearest tenth. (*Source: Some Body!* by Dr. Pete Rowan)

Solution: $3.5 \text{ lb} \approx \dfrac{3.5 \text{ lb}}{1} \cdot \overbrace{\dfrac{0.45 \text{ kg}}{1 \text{ lb}}}^{\text{Unit fraction}} = 3.5(0.45 \text{ kg}) \approx 1.6 \text{ kg}$

Thus 3.5 pounds are approximately 1.6 kilograms. From the table of conversions, we know that 1 kg ≈ 2.2 lb. So that 0.5 kg ≈ 1.1 lb and after adding, we have 1.5 kg ≈ 3.3 lb. Our result is reasonable.

⬛ **Work Practice Problem 2**

PRACTICE PROBLEM 3

Convert 237 ml to fluid ounces. Round to the nearest whole fluid ounce.

EXAMPLE 3 **Postage Stamp**

Australia converted to the metric system in 1973. In that year, four postage stamps were issued to publicize this conversion. One such stamp is shown. Let's check the mathematics on the stamp by converting 7 fluid ounces to milliliters. Round to the nearest hundred.

Solution: $7 \text{ fl oz} \approx \dfrac{7 \text{ fl oz}}{1} \cdot \overbrace{\dfrac{29.57 \text{ ml}}{1 \text{ fl oz}}}^{\text{Unit fraction}} = 7(29.57 \text{ ml}) = 206.99 \text{ ml}$

Rounded to the nearest hundred, 7 fl oz ≈ 200 ml.

⬛ **Work Practice Problem 3**

Now that we have practiced converting between two measurement systems, let's practice converting between two temperature scales.

Temperature When Gabriel Fahrenheit and Anders Celsius independently established units for temperature scales, each based his unit on the heat of water the moment it boils compared to the moment it freezes. One degree Celsius is $\dfrac{1}{100}$ of the difference in heat. One degree Fahrenheit is $\dfrac{1}{180}$ of the difference in heat. Celsius arbitrarily labeled the temperature at the freezing point at 0°C, making the boiling point 100°C; Fahrenheit labeled the freezing point 32°F, making the boiling point 212°F. Water boils at 212°F or 100°C.

By comparing the two scales in the figure, we see that a 20°C day is as warm as a 68°F day. Similarly, a sweltering 104°F day in the Mojave desert corresponds to a 40°C day.

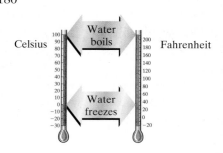

Answers

2. 226.8 g   **3.** 8 fl oz

✔ **Concept Check** Which of the following statements is correct? Explain.

a. 6°C is below the freezing point of water.

b. 6°F is below the freezing point of water.

Objective B Converting Degrees Celsius to Degrees Fahrenheit

To convert from Celsius temperatures to Fahrenheit temperatures, see the box below. In this box, we use the symbol F to represent degrees Fahrenheit and the symbol C to represent degrees Celsius.

Converting Celsius to Fahrenheit

$$F = \frac{9}{5}C + 32 \quad \text{or} \quad F = 1.8C + 32$$

(To convert to Fahrenheit temperature, multiply the Celsius temperature by $\frac{9}{5}$ or 1.8, and then add 32.)

EXAMPLE 4 Convert 15°C to degrees Fahrenheit.

Solution: $F = \frac{9}{5}C + 32$

$\qquad = \frac{9}{5} \cdot 15 + 32 \qquad$ Replace C with 15.

$\qquad = 27 + 32 \qquad\qquad$ Simplify.

$\qquad = 59 \qquad\qquad\quad$ Add.

Thus, 15°C is equivalent to 59°F.

🔲 **Work Practice Problem 4**

PRACTICE PROBLEM 4

Convert 60°C to degrees Fahrenheit.

EXAMPLE 5 Convert 29°C to degrees Fahrenheit.

Solution: $F = 1.8\,C + 32$

$\qquad = 1.8 \cdot 29 + 32 \qquad$ Replace C with 29.

$\qquad = 52.2 + 32 \qquad\quad$ Multiply 1.8 by 29.

$\qquad = 84.2 \qquad\qquad\quad$ Add.

Therefore, 29°C is the same as 84.2°F.

🔲 **Work Practice Problem 5**

PRACTICE PROBLEM 5

Convert 32°C to degrees Fahrenheit.

Objective C Converting Degrees Fahrenheit to Degrees Celsius

To convert from Fahrenheit temperatures to Celsius temperatures, see the box below. The symbol C represents degrees Celsius and the symbol F represents degrees Fahrenheit.

Converting Fahrenheit to Celsius

$$C = \frac{5}{9}(F - 32)$$

(To convert to Celsius temperature, subtract 32 from the Fahrenheit temperature, and then multiply by $\frac{5}{9}$.)

Answers

4. 140°F   **5.** 89.6°F

✔ **Concept Check Answer**

a. false   **b.** true

PRACTICE PROBLEM 6

Convert 68°F to degrees Celsius.

EXAMPLE 6 Convert 59°F to degrees Celsius.

Solution: We evaluate the formula $C = \frac{5}{9}(F - 32)$ when F is 59.

$$C = \frac{5}{9}(F - 32)$$

$$= \frac{5}{9} \cdot (59 - 32) \quad \text{Replace F with 59.}$$

$$= \frac{5}{9} \cdot (27) \quad\quad\quad \text{Subtract inside parentheses.}$$

$$= 15 \quad\quad\quad\quad\quad \text{Multiply.}$$

Therefore, 59°F is the same temperature as 15°C.

◾ **Work Practice Problem 6**

PRACTICE PROBLEM 7

Convert 113°F to degrees Celsius. If necessary, round to the nearest tenth of a degree.

EXAMPLE 7 Convert 114°F to degrees Celsius. If necessary, round to the nearest tenth of a degree.

Solution: $C = \frac{5}{9}(F - 32)$

$$= \frac{5}{9}(114 - 32) \quad \text{Replace F with 114.}$$

$$= \frac{5}{9} \cdot (82) \quad\quad\quad \text{Subtract inside parentheses.}$$

$$\approx 45.6 \quad\quad\quad\quad \text{Multiply.}$$

Therefore, 114°F is approximately 45.6°C.

◾ **Work Practice Problem 7**

PRACTICE PROBLEM 8

During a bout with the flu, Albert's temperature reaches 102.8°F. What is his temperature measured in degrees Celsius? Round to the nearest tenth of a degree.

EXAMPLE 8 **Body Temperature**

Normal body temperature is 98.6°F. What is this temperature in degrees Celsius?

Solution: We evaluate the formula $C = \frac{5}{9}(F - 32)$ when F is 98.6.

$$C = \frac{5}{9}(F - 32)$$

$$= \frac{5}{9}(98.6 - 32) \quad \text{Replace F with 98.6.}$$

$$= \frac{5}{9} \cdot (66.6) \quad\quad\quad \text{Subtract inside parentheses.}$$

$$= 37 \quad\quad\quad\quad\quad\quad \text{Multiply.}$$

Therefore, normal body temperature is 37°C.

◾ **Work Practice Problem 8**

Answers

6. 20°C   **7.** 45°C   **8.** 39.3°C

✔**Concept Check Answer**

She used the conversion for Celsius to Fahrenheit instead of Fahrenheit to Celsius.

✔**Concept Check** Clarissa must convert 40°F to degrees Celsius. What is wrong with her work shown below?

$$F = 1.8 \cdot C + 32$$
$$F = 1.8 \cdot 40 + 32$$
$$F = 72 + 32$$
$$F = 104$$

Note: Because approximations are used, your answers may vary slightly from the answers given in the back of the book.

Objective A *Convert as indicated. If necessary, round answers to two decimal places. See Examples 1 through 3.*

1. 756 milliliters to fluid ounces

2. 18 liters to quarts

3. 86 inches to centimeters

4. 86 miles to kilometers

5. 1000 grams to ounces

6. 100 kilograms to pounds

7. 93 kilometers to miles

8. 9.8 meters to feet

9. 14.5 liters to gallons

10. 150 milliliters to fluid ounces

11. 30 pounds to kilograms

12. 15 ounces to grams

Fill in the chart. Give exact answers or round to 1 decimal place.

		Meters	Yards	Centimeters	Feet	Inches
13.	The Height of a Woman				5	
14.	Statue of Liberty Length of Nose	1.37				
15.	Leaning Tower of Pisa		60			
16.	Blue Whale		36			

Solve. If necessary, round answers to two decimal places. See Examples 1 through 3.

17. The balance beam for female gymnasts is 10 centimeters wide. Convert this width to inches.

18. In men's gymnastics, the rings are 250 centimeters from the floor. Convert this height to inches, then to feet.

19. In many states, the maximum speed limit for recreational vehicles is 50 miles per hour. Convert this to kilometers per hour.

20. The speed limit is 70 miles per hour. Convert this to kilometers per hour.

21. Ibuprofen comes in 200 milligram tablets. Convert this to ounces. (Round your answer to this exercise to 3 decimal places.)

22. Vitamin C tablets come in 500 milligram caplets. Convert this to ounces.

23. A stone is a unit in the British customary system. Use the conversion: 14 pounds = 1 stone to check the equivalencies in this 1973 Australian stamp. Is 100 kilograms approximately 15 stone 10 pounds?

24. Convert 5 feet 11 inches to centimeters and check the conversion on this 1973 Australian stamp. Is it correct?

25. The Monarch butterfly migrates annually between the northern United States and central Mexico. The trip is about 4500 km long. Convert this to miles.

26. There is a species of African termite that builds nests up to 18 ft high. Convert this to meters.

27. A $3\frac{1}{2}$-inch diskette is not really $3\frac{1}{2}$ inches. To find its actual width, convert this measurement to centimeters, then to millimeters. Round the result to the nearest ten.

28. The average two-year-old is 84 centimeters tall. Convert this to feet and inches.

29. For an average adult, the weight of a right lung is greater than the weight of a left lung. If the right lung weighs 1.5 pounds and the left lung weighs 1.25 pounds, find the difference in grams. (*Source: Some Body!*)

30. The skin of an average adult weighs 9 pounds and is the heaviest organ. Find the weight in grams. (*Source: Some Body!*)

31. A fast sneeze has been clocked at about 167 kilometers per hour. Convert this to miles per hour. Round to the nearest whole.

32. A Boeing 747 has a cruising speed of about 980 kilometers per hour. Convert this to miles per hour. Round to the nearest whole.

33. The General Sherman giant sequoia tree has a diameter of about 8 meters at its base. Convert this to feet. (*Source: Fantastic Book of Comparisions*)

34. The largest crater on the near side of the moon is Billy Crater. It has a diameter of 303 kilometers. Convert this to miles. (*Source: Fantastic Book of Comparisions*)

35. The total length of the track on a CD is about 4.5 kilometers. Convert this to miles. Round to the nearest whole mile.

36. The distance between Mackinaw City, Michigan, and Cheyenne, Wyoming, is 2079 kilometers. Convert this to miles. Round to the nearest whole mile.

37. A doctor orders a dosage of 5 ml of medicine every 4 hours for 1 week. How many fluid ounces of medicine should be purchased? Round up to the next whole fluid ounce.

38. A doctor orders a dosage of 12 ml of medicine every 6 hours for 10 days. How many fluid ounces of medicine should be purchased? Round up to the next whole fluid ounce.

Without actually converting, choose the most reasonable answer.

39. This math book has a height of about _____.

   a. 28 mm   **b.** 28 cm
   c. 28 m   **d.** 28 km

40. A mile is _____ a kilometer.

   a. shorter than   **b.** longer than
   c. the same length as

41. A liter has _____ capacity than a quart.

   a. less   **b.** greater
   c. the same

42. A foot is _____ a meter.

   a. shorter than   **b.** longer than
   c. the same length as

43. A kilogram weighs _____ a pound.

    a. the same as   **b.** less than
    c. greater than

44. A football field is 100 yards, which is about _____.

    a. 9 m   **b.** 90 m
    c. 900 m   **d.** 9000 m

45. An $8\frac{1}{2}$ ounce glass of water has a capacity of about _____.

    a. 250 L   **b.** 25 L
    c. 2.5 L   **d.** 250 ml

46. A 5-gallon gasoline can has a capacity of about _____.

    a. 19 L   **b.** 1.9 L
    c. 19 ml   **d.** 1.9 ml

47. The weight of an average man is about _____.

    a. 700 kg   **b.** 7 kg
    c. 0.7 kg   **d.** 70 kg

48. The weight of a pill is about _____.

    a. 200 kg   **b.** 20 kg
    c. 2 kg   **d.** 200 mg

Objectives **B** **C** *Convert as indicated. When necessary, round to the nearest tenth of a degree. See Examples 4 through 8.*

49. 77°F to degrees Celsius

50. 86°F to degrees Celsius

51. 104°F to degrees Celsius

52. 140°F to degrees Celsius

53. 50°C to degrees Fahrenheit

54. 80°C to degrees Fahrenheit

55. 115°C to degrees Fahrenheit

56. 225°C to degrees Fahrenheit

57. 20°F to degrees Celsius

58. 26°F to degrees Celsius

59. 142.1°F to degrees Celsius

60. 43.4°F to degrees Celsius

61. 92°C to degrees Fahrenheit

62. 75°C to degrees Fahrenheit

63. 12.4°C to degrees Fahrenheit

64. 48.6°C to degrees Fahrenheit

65. The hottest temperature ever recorded in the United States, in Death Valley, was 134°F. Convert this temperature to degrees Celsius. (*Source:* National Climatic Data Center)

66. The hottest temperature ever recorded in the United States in January was 95°F in Los Angeles. Convert this temperature to degrees Celsius. (*Source:* National Climatic Data Center)

67. A weather forecaster in Caracas predicts a high temperature of 27°C. Find this measurement in degrees Fahrenheit.

68. While driving to work, Alan Olda notices a temperature of 18°C flash on the local bank's temperature display. Find the corresponding temperature in degrees Fahrenheit.

69. At Mack Trucks' headquarters, the room temperature is to be set at 70°F, but the thermostat is calibrated in degrees Celsius. Find the temperature to be set.

70. The computer room at Merck, Sharp, and Dohm is normally cooled to 66°F. Find the corresponding temperature in degrees Celsius.

71. In a European cookbook, a recipe requires the ingredients for caramels to be heated to 118°C, but the cook has access only to a Fahrenheit thermometer. Find the temperature in degrees Fahrenheit that should be used to make the caramels.

72. The ingredients for divinity should be heated to 127°C, but the candy thermometer that Myung Kim has is calibrated to degrees Fahrenheit. Find how hot he should heat the ingredients.

73. The temperature of Earth's core is estimated to be 4000°C. Find the corresponding temperature in degrees Fahrenheit.

74. In 2005, the average temperature of the Earth's surface was 58.1°F, the second warmest in recorded history. Convert this temperature to degrees Celsius.

Review

Perform the indicated operations. See Section 1.7.

75. $6 \cdot 4 + 5 \div 1$ **76.** $10 \div 2 + 9(8)$ **77.** $3[(1 + 5) \cdot (8 - 6)]$ **78.** $5[(18 - 8) - 9]$

Concept Extensions

Determine whether the measurement in each statement is reasonable.

79. A 72°F room feels comfortable.

80. Water heated to 110°F will boil.

81. Josiah has a fever if a thermometer shows his temperature to be 40°F.

82. An air temperature of 20°F on a Vermont ski slope can be expected in the winter.

83. When the temperature is 30°C outside, an overcoat is needed.

84. An air-conditioned room at 60°C feels quite chilly.

85. Barbara has a fever when a thermometer records her temperature at 40°C.

86. Water cooled to 32°C will freeze.

Body surface area (BSA) is often used to calculate dosages for some drugs. BSA is calculated in square meters using a person's weight and height.

$$ \text{BSA} = \sqrt{\frac{(\text{weight in kg}) \times (\text{height in cm})}{3600}} $$

For Exercises 87 through 92, calculate the BSA for each person. Round to the nearest hundredth. You will need to use the square root key on your calculator.

87. An adult whose height is 182 cm and weight is 90 kg.

88. An adult whose height is 157 cm and weight is 63 kg.

89. A child whose height is 40 in. and weight is 50 kg. (*Hint:* Don't forget to first convert inches to centimeters.)

90. A child whose height is 26 in. and weight is 13 kg.

91. An adult whose height is 60 in. and weight is 150 lb.

92. An adult whose height is 69 in. and weight is 172 lb.

93. On February 17, 1995, in the Tokamak Fusion Test Reactor at Princeton University, the highest temperature produced in a laboratory was achieved. This temperature was 918,000,000°F. Convert this temperature to degrees Celsius. Round your answer to the nearest ten million degrees. (*Source: Guinness World Records*)

94. The hottest-burning substance known is carbon subnitride. Its flame at one atmospheric pressure reaches 9010°F. Convert this temperature to degrees Celsius. (*Source: Guinness World Records*)

95. In your own words, describe how to convert from degrees Celsius to degrees Fahrenheit.

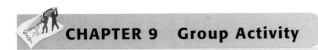

CHAPTER 9   Group Activity

Map Reading

Sections 9.1 and 9.4

Materials:

- ruler
- string
- calculator

This activity may be completed by working in groups or individually.

Investigate the route you would take from Santa Rosa, New Mexico, to San Antonio, New Mexico. Use the map in the figure to answer the following questions. You may find that using string to match the roads on the map is useful when measuring distances.

1. How many miles is it from Santa Rosa to San Antonio via Interstate 40 and Interstate 25? Convert this distance to kilometers.

2. How many miles is it from Santa Rosa to San Antonio via U.S. 54 and U.S. 380? Convert this distance to kilometers.

3. Assume that the speed limit on Interstates 40 and 25 is 65 miles per hour. How long would the trip take if you took this route and traveled 65 miles per hour the entire trip?

4. At what average speed would you have to travel on the U.S. routes to make the trip from Santa Rosa to San Antonio in the same amount of time that it would take on the interstate routes? Do you think this speed is reasonable on this route? Explain your reasoning.

5. Discuss in general the factors that might affect your decision among the different routes.

6. Explain which route you would choose in this case and why.

Chapter 9 Vocabulary Check

Fill in each blank with one of the words or phrases listed below.

transversal	line segment	obtuse	straight	adjacent	right	volume	area
acute	perimeter	vertical	supplementary	ray	angle	line	complementary
vertex	mass	unit fractions	gram	weight	meter	liter	surface area

1. _____ is a measure of the pull of gravity.
2. _____ is a measure of the amount of substance in an object. This measure does not change.
3. The basic unit of length in the metric system is the _____.
4. To convert from one unit of length to another, _____ may be used.
5. A _____ is the basic unit of mass in the metric system.
6. The _____ is the basic unit of capacity in the metric system.
7. A _____ is a piece of a line with two endpoints.
8. Two angles that have a sum of 90° are called _____ angles.
9. A _____ is a set of points extending indefinitely in two directions.
10. The _____ of a polygon is the distance around the polygon.
11. An _____ is made up of two rays that share the same endpoint. The common endpoint is called the _____.
12. _____ measures the amount of surface of a region.
13. A _____ is a part of a line with one endpoint. A ray extends indefinitely in one direction.
14. A line that intersects two or more lines at different points is called a _____.
15. An angle that measures 180° is called a _____ angle.
16. The measure of the space of a solid is called its _____.
17. When two lines intersect, four angles are formed. Two of these angles that are opposite each other are called _____ angles.
18. Two of the angles from #17 that share a common side are called _____ angles.
19. An angle whose measure is between 90° and 180° is called an _____ angle.
20. An angle that measures 90° is called a _____ angle.
21. An angle whose measure is between 0° and 90° is called an _____ angle.
22. Two angles that have a sum of 180° are called _____ angles.
23. The _____ of a polyhedron is the sum of the areas of the faces of the polyhedrons.

STUDY SKILLS BUILDER

Are You Prepared for a Test on Chapter 9?

Below I have listed some common trouble areas for topics covered in Chapter 9. After studying for your test—but before taking your test—read these.

- Don't forget the difference between complementary and supplementary angles.

  Complementary angles have a sum of 90°.

  Supplementary angles have a sum of 180°.

  The complement of a 15° angle measures $90° - 15° = 75°$.

  The supplement of a 15° angle measures $180° - 15° = 165°$.

- Remember:

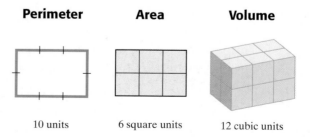

Perimeter	Area	Volume
10 units	6 square units	12 cubic units

Remember: This is simply a checklist of common trouble areas. For a review of Chapter 9, see the Highlights and Chapter Review at the end of this chapter.

Helpful Hint

Are you preparing for your test? Don't forget to take the Chapter 9 Test on page 697. Then check your answers at the back of the text and use the Chapter Test Prep Video CD to see the fully worked-out solutions to any of the exercises you want to review.

9 Chapter Highlights

DEFINITIONS AND CONCEPTS	**EXAMPLES**

Section 9.1   Lines and Angles

A **line** is a set of points extending indefinitely in two directions. A line has no width or height, but it does have length. We name a line by any two of its points.

A **line segment** is a piece of a line with two endpoints.

A **ray** is a part of a line with one endpoint. A ray extends indefinitely in one direction.

An **angle** is made up of two rays that share the same endpoint. The common endpoint is called the **vertex.**

Line AB or $\overleftrightarrow{AB}$

Line segment AB or $\overline{AB}$

Ray AB or $\overrightarrow{AB}$

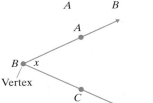

Section 9.2   Perimeter

PERIMETER FORMULAS

Rectangle: $P = 2l + 2w$

Square: $P = 4s$

Triangle: $P = a + b + c$

Circumference of a Circle: $C = 2\pi r$ or $C = \pi d$

where $\pi \approx 3.14$ or $\pi \approx \dfrac{22}{7}$

Find the perimeter of the rectangle.

$$P = 2l + 2w$$
$$= 2 \cdot 28 \text{ meters} + 2 \cdot 15 \text{ meters}$$
$$= 56 \text{ meters} + 30 \text{ meters}$$
$$= 86 \text{ meters}$$

The perimeter is 86 meters.

Section 9.3   Area, Volume, and Surface Area

AREA FORMULAS

Rectangle: $A = lw$

Square: $A = s^2$

Triangle: $A = \dfrac{1}{2}bh$

Parallelogram: $A = bh$

Trapezoid: $A = \dfrac{1}{2}(b + B)h$

Circle: $A = \pi r^2$

Find the area of the square.

$$A = s^2$$
$$= (8 \text{ centimeters})^2$$
$$= 64 \text{ square centimeters}$$

The area of the square is 64 square centimeters.

DEFINITIONS AND CONCEPTS	**EXAMPLES**

Section 9.3　Area, Volume, and Surface Area (*continued*)

VOLUME FORMULAS

Surface Area Formulas: See page 633.

Rectangular Solid: $V = lwh$

Cube: $V = s^3$

Sphere: $V = \dfrac{4}{3}\pi r^3$

Right Circular Cylinder: $V = \pi r^2 h$

Cone: $V = \dfrac{1}{3}\pi r^2 h$

Square-Based Pyramid: $V = \dfrac{1}{3}s^2 h$

Find the volume of the sphere. Use $\dfrac{22}{7}$ for π.

4 in.

$$V = \frac{4}{3}\pi r^3$$

$$\approx \frac{4}{3} \cdot \frac{22}{7} \cdot (4 \text{ inches})^3$$

$$= \frac{4 \cdot 22 \cdot 64}{3 \cdot 7} \text{ cubic inches}$$

$$= \frac{5632}{21} \quad \text{or} \quad 268\frac{4}{21} \text{ cubic inches}$$

Section 9.4　Linear Measurement

To convert from one unit of length to another, multiply by a **unit fraction** in the form

$$\frac{\text{units to convert to}}{\text{original units}}$$

LENGTH: U.S. SYSTEM OF MEASUREMENT

$$12 \text{ inches (in.)} = 1 \text{ foot (ft)}$$
$$3 \text{ feet} = 1 \text{ yard (yd)}$$
$$5280 \text{ feet} = 1 \text{ mile (mi)}$$

The basic unit of length in the metric system is the **meter.** A meter is slightly longer than a yard.

LENGTH: METRIC SYSTEM OF MEASUREMENT

Metric Unit of Length
1 **kilo**meter (km) = 1000 meters (m)
1 **hecto**meter (hm) = 100 m
1 **deka**meter (dam) = 10 m
1 meter (m) = 1 m
1 **deci**meter (dm) = 1/10 m or 0.1 m
1 **centi**meter (cm) = 1/100 m or 0.01 m
1 **milli**meter (mm) = 1/1000 m or 0.001 m

$$\frac{12 \text{ inches}}{1 \text{ foot}}, \frac{1 \text{ foot}}{12 \text{ inches}}, \frac{3 \text{ feet}}{1 \text{ yard}}$$

Convert 6 feet to inches.

$$6 \text{ ft} = \frac{6 \text{ ft}}{1} \cdot 1$$

$$= \frac{6 \text{ ft}}{1} \cdot \frac{12 \text{ in.}}{1 \text{ ft}} \quad \begin{matrix} \leftarrow \text{ units to convert to} \\ \leftarrow \text{ original units} \end{matrix}$$

$$= 6 \cdot 12 \text{ in.}$$

$$= 72 \text{ in.}$$

Convert 3650 centimeters to meters.

$$3650 \text{ cm} = 3650 \text{ cm} \cdot 1$$

$$= \frac{3650 \text{ cm}}{1} \cdot \frac{0.01 \text{ m}}{1 \text{ cm}} = 36.5 \text{ m}$$

or

km　hm　dam　m　dm　cm　mm

2 units to the left

$$3650 \text{ cm} = 36.5 \text{ m}$$

2 places to the left

DEFINITIONS AND CONCEPTS	**EXAMPLES**

Section 9.5   Weight and Mass

Weight is really a measure of the pull of gravity. **Mass** is a measure of the amount of substance in an object and does not change.

Convert 5 pounds to ounces.

$$5 \text{ lb} = 5 \text{ lb} \cdot 1 = \frac{5 \cancel{\text{lb}}}{1} \cdot \frac{16 \text{ oz}}{1 \cancel{\text{lb}}} = 80 \text{ oz}$$

WEIGHT: U.S. SYSTEM OF MEASUREMENT

     16 ounces (oz) = 1 pound (1b)

     2000 pounds = 1 ton

A **gram** is the basic unit of mass in the metric system. It is the mass of water contained in a cube 1 centimeter on each side. A paper clip weighs about 1 gram.

Convert 260 grams to kilograms.

$$260 \text{ g} = \frac{260 \text{ g}}{1} \cdot 1 = \frac{260 \cancel{\text{g}}}{1} \cdot \frac{1 \text{ kg}}{1000 \cancel{\text{g}}} = 0.26 \text{ kg}$$

or

MASS: METRIC SYSTEM OF MEASUREMENT

Metric Unit of Mass
1 kilogram (kg) = 1000 grams (g)
1 hectogram (hg) = 100 g
1 dekagram (dag) = 10 g
1 gram (g) = 1 g
1 decigram (dg) = 1/10 g or 0.1 g
1 centigram (cg) = 1/100 g or 0.01 g
1 milligram (mg) = 1/1000 g or 0.001 g

kg   hg   dag   g   dg   cg   mg

3 units to the left

260 g = 0.260 kg

3 places to the left

Section 9.6   Capacity

CAPACITY: U.S. SYSTEM OF MEASUREMENT

     8 fluid ounces (fl oz) = 1 cup (c)

           2 cups = 1 pint (pt)

           2 pints = 1 quart (qt)

        4 quarts = 1 gallon (gal)

Convert 5 pints to gallons.

$$1 \text{ gal} = 4 \text{ qt} = 8 \text{ pt}$$

$$5 \text{ pt} = 5 \text{ pt} \cdot 1 = \frac{5 \cancel{\text{pt}}}{1} \cdot \frac{1 \text{ gal}}{8 \cancel{\text{pt}}} = \frac{5}{8} \text{ gal}$$

The **liter** is the basic unit of capacity in the metric system. It is the capacity or volume of a cube measuring 10 centimeters on each side. A liter of liquid is slightly more than 1 quart.

Convert 1.5 liters to milliliters.

$$1.5 \text{ L} = \frac{1.5 \text{ L}}{1} \cdot 1 = \frac{1.5 \cancel{\text{L}}}{1} \cdot \frac{1000 \text{ ml}}{1 \cancel{\text{L}}} = 1500 \text{ ml}$$

or

CAPACITY: METRIC SYSTEM OF MEASUREMENT

Metric Unit of Capacity
1 kiloliter (kl) = 1000 liters (L)
1 hectoliter (hl) = 100 L
1 dekaliter (dal) = 10 L
1 liter (L) = 1 L
1 deciliter (dl) = 1/10 L or 0.1 L
1 centiliter (cl) = 1/100 L or 0.01 L
1 milliliter (ml) = 1/1000 L or 0.001 L

kl   hl   dal   L   dl   cl   ml

3 units to the right

1.500 L = 1500 ml

3 places to the right

DEFINITIONS AND CONCEPTS	EXAMPLES

Section 9.7   Temperature and Conversions Between the U.S. and Metric Systems

To convert between systems, use approximate unit fractions.

Convert 7 feet to meters.

$$7 \text{ ft} \approx \frac{7 \text{ ft}}{1} \cdot \frac{0.30 \text{ m}}{1 \text{ ft}} = 2.1 \text{ m}$$

Convert 8 liters to quarts.

$$8 \text{ L} \approx \frac{8 \text{ L}}{1} \cdot \frac{1.06 \text{ qt}}{1 \text{ L}} = 8.48 \text{ qt}$$

Convert 363 grams to ounces.

$$363 \text{ g} \approx \frac{363 \text{ g}}{1} \cdot \frac{0.04 \text{ oz}}{1 \text{ g}} = 14.52 \text{ oz}$$

CELSIUS TO FAHRENHEIT

$$F = \frac{9}{5}C + 32 \quad \text{or} \quad F = 1.8C + 32$$

Convert 35°C to degrees Fahrenheit.

$$F = \frac{9}{5} \cdot 35 + 32 = 63 + 32 = 95$$

$$35°C = 95°F$$

FAHRENHEIT TO CELSIUS

$$C = \frac{5}{9}(F - 32)$$

Convert 50°F to degrees Celsius.

$$C = \frac{5}{9} \cdot (50 - 32) = \frac{5}{9} \cdot (18) = 10$$

$$50°F = 10°C$$

9 CHAPTER REVIEW

(9.1) *Classify each angle as acute, right, obtuse, or straight.*

1.

2.

3.

C

4.

5. Find the complement of a 25° angle.

6. Find the supplement of a 105° angle.

Find the measure of angle x in each figure.

7.

8.

9.

10.

11. Identify the pairs of supplementary angles.

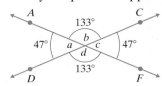

12. Identify the pairs of complementary angles.

Find the measures of angles x, y, and z in each figure.

13.

14.

15. Given that $m \parallel n$.

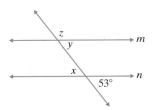

16. Given that $m \parallel n$.

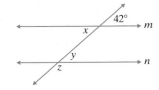

(9.2) *Find the perimeter of each figure.*

17.

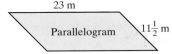

23 m

Parallelogram $11\frac{1}{2}$ m

18.

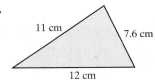

11 cm

7.6 cm

12 cm

19.

7 m

8 m

5 m

10 m

20.

5 ft

4 ft

11 ft   3 ft

22 ft

Solve.

21. Find the perimeter of a rectangular sign that measures 6 feet by 10 feet.

22. Find the perimeter of a town square that measures 110 feet on a side.

Find the circumference of each circle. Use $\pi \approx 3.14$.

23.

1.7 in.

24.

5 yd

(9.3) *Find the area of each figure. For the circles, find the exact area and then use* $\pi \approx 3.14$ *to approximate the area.*

25.

12 ft
10 ft
36 ft

26.

14 m
20 m

27.

15 cm
40 cm

28.

9 yd
21 yd

29.

7 ft

30.

Square   9.1 m

31.

34 in.
7 in.

32.

64 cm
26 cm
32 cm

33.

4 m
3 m
12 m
13 m

34. The amount of sealer necessary to seal a driveway depends on the area. Find the area of a rectangular driveway 36 feet by 12 feet.

35. Find how much carpet is necessary to cover the floor of the room shown.

10 feet
13 feet

Find the volume and surface area of the solids in Exercises 36 and 37. For Exercises 38 and 39, give an exact volume and an approximation.

36.
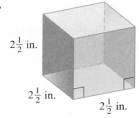
$2\frac{1}{2}$ in.
$2\frac{1}{2}$ in.
$2\frac{1}{2}$ in.

37.

6 ft
2 ft
7 ft

38. Use $\pi \approx 3.14$.

50 cm
20 cm

39. Use $\pi \approx \dfrac{22}{7}$.
$\frac{1}{2}$ km

40. Find the volume of a pyramid with a square base 2 feet on a side and a height of 2 feet.

41. Approximate the volume of a tin can 8 inches high and 3.5 inches in radius. Use 3.14 for π.

42. A chest has 3 drawers. If each drawer has inside measurements of $2\frac{1}{2}$ feet by $1\frac{1}{2}$ feet by $\frac{2}{3}$ foot, find the total volume of the 3 drawers.

43. A cylindrical canister for a shop vacuum is 2 feet tall and 1 foot in *diameter*. Find its exact volume.

(9.4) *Convert.*

44. 108 in. to feet

45. 72 ft to yards

46. 1.5 mi to feet

47. $\frac{1}{2}$ yd to inches

48. 52 ft = _____ yd _____ ft

49. 46 in. = _____ ft _____ in.

50. 42 m to centimeters

51. 82 cm to millimeters

52. 12.18 mm to meters

53. 2.31 m to kilometers

Perform each indicated operation.

54. 4 yd 2 ft + 16 yd 2 ft

55. 7 ft 4 in. ÷ 2

56. 8 cm + 15 mm

57. 4 m − 126 cm

Solve.

58. A bolt of cloth contains 333 yd 1 ft of cotton ticking. Find the amount of material that remains after 163 yd 2 ft is removed from the bolt.

59. The student activities club is sponsoring a walk for hunger, and all students who participate will receive a sash with the name of the school to wear on the walk. If each sash requires 5 ft 2 in. of material and there are 50 students participating in the walk, how much material will the student activities club need?

60. The trip from Philadelphia to Washington, D.C., is 217 km each way. Four friends agree to share the driving equally. How far must each drive on this round-trip vacation?

△ **61.** The college has ordered that NO SMOKING signs be placed above the doorway of each classroom. Each sign is 0.8 m long and 30 cm wide. Find the area of each sign. (*Hint:* Recall that the area of a rectangle = width · length.)

0.8 meter

30 centimeters

(9.5) *Convert.*

62. 66 oz to pounds

63. 2.3 tons to pounds

64. 52 oz = _____ lb _____ oz.

65. 10,300 lb = _____ tons _____ lb

66. 27 mg to grams

67. 40 kg to grams

68. 2.1 hg to dekagrams

69. 0.03 mg to decigrams

Perform each indicated operation.

70. 6 lb 5 oz − 2 lb 12 oz

71. 8 lb 6 oz × 4

72. 4.3 mg × 5

73. 4.8 kg − 4200 g

Solve.

74. Donshay Berry ordered 1 lb 12 oz of soft-center candies and 2 lb 8 oz of chewy-center candies for his party. Find the total weight of the candy ordered.

75. Four local townships jointly purchase 38 tons 300 lb of cinders to spread on their roads during an ice storm. Determine the weight of the cinders each township receives if they share the purchase equally.

(9.6) *Convert.*

76. 28 pints to quarts

77. 40 fluid ounces to cups

78. 3 qt 1 pt to pints

79. 18 quarts to cups

80. 9 pt = _____ qt _____ pt

81. 15 qt = _____ gal _____ qt

82. 3.8 L to milliliters

83. 14 hl to kiloliters

84. 30.6 L to centiliters

Perform each indicated operation.

85. 1 qt 1 pt + 3 qt 1 pt

86. 3 gal 2 qt × 2

87. 0.946 L − 210 ml

88. 6.1 L + 9400 ml

Solve.

89. Carlos Perez prepares 4 gal 2 qt of iced tea for a block party. During the first 30 minutes of the party, 1 gal 3 qt of the tea is consumed. How much iced tea remains?

90. A recipe for soup stock calls for 1 c 4 fl oz of beef broth. How much should be used if the recipe is cut in half?

91. Each bottle of Kiwi liquid shoe polish holds 85 ml of the polish. Find the number of liters of shoe polish contained in 8 boxes if each box contains 16 bottles.

92. Ivan Miller wants to pour three separate containers of saline solution into a single vat with a capacity of 10 liters. Will 6 liters of solution in the first container combined with 1300 milliliters in the second container and 2.6 liters in the third container fit into the larger vat?

(9.7) *Note: Because approximations are used in this section, your answers may vary slightly from the answers given in the back of the book.*

Convert as indicated. If necessary, round to two decimal places.

93. 7 meters to feet

94. 11.5 yards to meters

95. 17.5 liters to gallons

96. 7.8 liters to quarts

97. 15 ounces to grams

98. 23 pounds to kilograms

99. A compact disc is 1.2 mm thick. Find the height (in inches) of 50 discs.

100. If a person weighs 82 kilograms, how many pounds is this?

Convert. Round to the nearest tenth of a degree, if necessary.

101. 42°C to degrees Fahrenheit

102. 160°C to degrees Fahrenheit

103. 41.3°F to degrees Celsius

104. 80°F to degrees Celsius

Solve. Round to the nearest tenth of a degree, if necessary.

105. A sharp dip in the jet stream caused the temperature in New Orleans to drop to 35°F. Find the corresponding temperature in degrees Celsius.

106. The recipe for meat loaf calls for a 165°C oven. Find the setting used if the oven has a Fahrenheit thermometer.

Mixed Review

Find the following.

107. Find the supplement of a 72° angle.

108. Find the complement of a 1° angle.

Find the measure of angle x in each figure.

109.

x

$85°$

110.

x

m

$m \parallel n$

n

$123°$

Find the perimeter of each figure.

111.

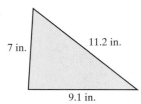

7 in.   11.2 in.

9.1 in.

112.

22 ft   11 ft

15 ft

42 ft

40 ft

Find the area of each figure. For the circle, find the exact area and then use $\pi = 3.14$ to approximate the area.

113.

43 m

42 m

13 m

14 m

114.

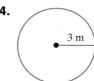

3 m

Find the volume of each solid.

115. Give an approximation using $\dfrac{22}{7}$ for π.

$5\frac{1}{4}$ in.

12 in.

116. Find the surface area, also.

5 in.

4 in.   7 in.

Convert the following.

117. 6.25 ft to inches

118. 8200 lb = _____ tons _____ lb

119. 5 m to centimeters

120. 286 mm to kilometers

121. 1400 mg to grams

122. 6.75 gallons to quarts

123. 86°C to degrees Fahrenheit

124. 51.8°F to degrees Celsius

Perform the indicated operations and simplify.

125. 9.3 km − 183 m

126. 35 L + 700 ml

127. 3 gal 3 qt + 4 gal 2 qt

128. 3.2 kg × 4

CHAPTER TEST

Use the Chapter Test Prep Video CD to see the fully worked-out solutions to any of the exercises you want to review.

△ **1.** Find the complement of a 78° angle.   △ **2.** Find the supplement of a 124° angle.

△ **3.** Find the measure of ∠x.

Find the measure of x, y, and z in each figure.

△ **4.**

△ **5.**

Find the unknown diameter or radius as indicated.

△ **6.**

$d = ?$
$r = 3.1$ m

△ **7.**

$r = ?$
$d = 20$ in.

Find the perimeter (or circumference) and area of each figure. For the circle, give the exact value and then use $\pi \approx 3.14$.

△ **8.**

9 in.

△ **9.**

Rectangle   5.3 yd

7 yd

△ **10.**

6 in.

11 in.

7 in.

23 in.

11. _____

12. _____

13. _____

14. _____

15. _____

16. _____

17. _____

18. _____

19. _____

20. _____

21. _____

22. _____

23. _____

24. _____

25. _____

26. _____

27. _____

28. _____

29. _____

30. _____

31. _____

32. _____

33. _____

34. _____

35. _____

36. _____

37. _____

Find the volume of each solid. For the cylinder, use $\pi \approx \dfrac{22}{7}$.

△ **11.**

5 in.

2 in.

△ **12.**

2 ft

3 ft   5 ft

Solve.

△ **13.** Find the perimeter of a square frame with a side length of 4 inches.

4 in.

△ **14.** How much soil is needed to fill a rectangular hole 3 feet by 3 feet by 2 feet?

15. Find how much baseboard is needed to go around a rectangular room that measures 18 feet by 13 feet. If baseboard costs $1.87 per foot, also calculate the total cost needed for materials.

Convert.

16. 280 in. = _____ ft _____ in.

17. $2\dfrac{1}{2}$ gal to quarts

18. 30 oz to pounds

19. 2.8 tons to pounds

20. 38 pt to gallons

21. 40 mg to grams

22. 2.4 kg to grams

23. 3.6 cm to millimeters

24. 4.3 dg to grams

25. 0.83 L to milliliters

Perform each indicated operation.

26. 3 qt 1 pt + 2 qt 1 pt

27. 8 lb 6 oz − 4 lb 9 oz

28. 2 ft 9 in. × 3

29. 5 gal 2 qt ÷ 2

30. 8 cm − 14 mm

31. 1.8 km + 456 m

Convert. Round to the nearest tenth of a degree, if necessary.

32. 84°F to degrees Celsius

33. 12.6°C to degrees Fahrenheit

34. The sugar maples in front of Bette MacMillan's house are 8.4 meters tall. Because they interfere with the phone lines, the telephone company plans to remove the top third of the trees. How tall will the maples be after they are shortened?

35. A total of 15 gal 1 qt of oil has been removed from a 20-gallon drum. How much oil still remains in the container?

36. The engineer in charge of bridge construction said that the span of a certain bridge would be 88 m. But the actual construction required it to be 340 cm longer. Find the span of the bridge, in meters.

37. If 2 ft 9 in. of material is used to manufacture one scarf, how much material is needed for 6 scarves?

1. Solve: $3a - 6 = a + 4$

2. Solve: $2x + 1 = 3x - 5$.

3. Evaluate:

    a. $\left(\dfrac{2}{5}\right)^4$ **b.** $\left(-\dfrac{1}{4}\right)^2$

4. Evaluate:

    a. $\left(-\dfrac{1}{3}\right)^3$ **b.** $\left(\dfrac{3}{7}\right)^2$.

5. Add: $2\dfrac{4}{5} + 5 + 1\dfrac{1}{2}$

6. Add: $2\dfrac{1}{3} + 4\dfrac{2}{5} + 3$.

7. Simplify by combining like terms:

    $11.1x - 6.3 + 8.9x - 4.6$

8. Simplify by combining like terms:

    $2.5y + 3.7 - 1.3y - 1.9$

9. Simplify: $\dfrac{5.68 + (0.9)^2 \div 100}{0.2}$

10. Simplify: $\dfrac{0.12 + 0.96}{0.5}$

11. Insert $<, >$, or $=$ to form a true statement.

    $0.\overline{7}$ $\dfrac{7}{9}$

12. Insert $<, >$, or $=$ to form a true statement.

    0.43 $\dfrac{2}{5}$

13. Solve: $0.5y + 2.3 = 1.65$

14. Solve: $0.4x - 9.3 = 2.7$

△**15.** An inner-city park is in the shape of a square that measures 300 feet on a side. Find the length of the diagonal of the park rounded to the nearest whole foot.

△**16.** A rectangular field is 200 feet by 125 feet. Find the length of the diagonal of the field, rounded to the nearest whole foot.

Answers

1. _____

2. _____

3. a. _____

   b. _____

4. a. _____

   b. _____

5. _____

6. _____

7. _____

8. _____

9. _____

10. _____

11. _____

12. _____

13. _____

14. _____

15. _____

16. _____

17. a. _____

b. _____

18. a. _____

b. _____

19. _____

20. _____

21. _____

22. _____

23. _____

24. _____

25. _____

26. _____

27. _____

28. _____

29. _____

30. _____

31. _____

32. _____

△ **17.** Given the rectangle shown:

7 feet

5 feet

a. Find the ratio of its width to its length.
b. Find the ratio of its length to its perimeter.

△ **18.** A square is 9 inches by 9 inches.

9 inches

Square   9 inches

a. Find the ratio of a side to its perimeter.
b. Find the ratio of its perimeter to its area.

19. Write the rate as a fraction in simplest form: $2160 for 12 weeks.

20. Write the rate as a fraction in simplest form: 8 chaperones for 40 students

21. Solve: $\dfrac{1.6}{1.1} = \dfrac{x}{0.3}$

Round the solution to the nearest hundredth.

22. Solve: $\dfrac{2.4}{3.5} = \dfrac{0.7}{x}$

Round the solution to the nearest hundredth.

23. The standard dose of an antibiotic is 4 cc (cubic centimeters) for every 25 pounds (lb) of body weight. At this rate, find the standard dose for a 140-lb woman.

24. A recipe that makes 2 pie crusts calls for 3 cups of flour. How much flour is needed to make 5 pie crusts?

25. In a survey of 100 people, 17 people drive blue cars. What percent of people drive blue cars?

26. Of 100 shoppers surveyed at a mall, 38 paid for their purchases using only cash. What percent of shoppers used only cash to pay for their purchases?

27. 13 is $6\frac{1}{2}$% of what number?

28. 54 is $4\frac{1}{2}$% of what number?

29. What number is 30% of 9?

30. What number is 42% of 30?

31. The number of applications for a mathematics scholarship at Yale increased from 34 to 45 in one year. What is the percent increase? Round to the nearest whole percent.

32. The price of a gallon of paint rose from $15 to $19. Find the percent increase, rounded to the nearest whole percent.

33. Find the sales tax and the total price on a purchase of an $85.50 atlas in a city where the sales tax rate is 7.5%.

34. A sofa has a purchase price of $375. If the sales tax rate is 8%, find the amount of sales tax and the total cost of the sofa.

35. Find the ordered pair corresponding to each point plotted on the rectangular coordinate system.

36. Find the ordered pair corresponding to each point plotted on the rectangular coordinate system.

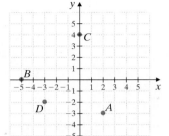

37. Graph $y = 4$.

38. Graph $y = -2$.

39. Find the median of the list of numbers: 25, 54, 56, 57, 60, 71, 98

40. Find the median in the list of scores: 60, 95, 89, 72, 83

41. Find the probability of choosing a red marble from a box containing 1 red, 1 yellow, and 2 blue marbles.

42. Find the probability of choosing a nickel at random in a coin purse that contains 2 pennies, 2 nickels, and 3 quarters.

43. Find the complement of a 48° angle.

44. Find the supplement of a 137° angle.

45. Convert 8 feet to inches.

46. Convert 7 yards to feet.

47. Subtract 3 tons 1350 lb from 8 tons 1000 lb.

48. Add 8 lb 15 oz to 9 lb 3 oz.

49. Convert 59°F to degrees Celsius.

50. Convert 86°F to degrees Celsius.

33. _____

34. _____

35. _____

36. _____

37. see graph

38. see graph

39. _____

40. _____

41. _____

42. _____

43. _____

44. _____

45. _____

46. _____

47. _____

48. _____

49. _____

50. _____

10

Exponents and Polynomials

Recall that an exponent is a shorthand way of representing repeated multiplication. In this chapter, we learn more about exponents and a special type of expression containing exponents, called a *polynomial*. Studying polynomials is a major part of algebra. Polynomials are also useful for modeling many real-world situations. This chapter serves as an introduction to polynomials and some operations that can be performed on them.

Once upon a time, your telephone and your computer were wired into your house, and you had to be there to use them. That is no longer true. With wireless technology, you are no longer restricted to working only at school, at the office, or at home. You can also be reached anywhere and anytime. You may or may not like this new advancement in technology, but it appears to be here to stay. Below is a graph showing the growth of wireless subscribers. In Exercises 61 and 62 of Section 10.1, you will explore the growth of cellular telephone and wireless computer technology.

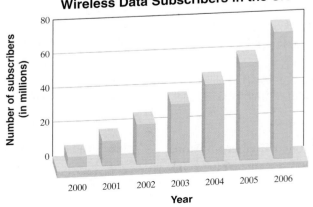

Wireless Data Subscribers in the U.S.

Source: Gartner Dataquest

10.1 ADDING AND SUBTRACTING POLYNOMIALS

Objectives

A Add Polynomials.

B Subtract Polynomials.

C Evaluate Polynomials at Given Replacement Values.

Before we add and subtract polynomials, let's first review some definitions presented in Section 3.1. Recall that the *addends* of an algebraic expression are the *terms* of the expression.

Expression

$3x + 5$ — 2 terms

$7y^2 + (-6y) + 4$ — 3 terms

Also, recall that *like terms* can be added or subtracted by using the distributive property. For example,

$7x + 3x = (7 + 3)x = 10x$

Objective **A** Adding Polynomials

Some terms are also **monomials.** A term is a monomial if the term contains only whole-number exponents and no variable in the denominator.

Monomials	Not Monomials	
$3x^2$	$\dfrac{2}{y}$	Variable in denominator
$-\dfrac{1}{2}a^2bc^3$	$-2x^{-5}$	Not a whole number exponent
7		

A monomial or a sum and/or difference of monomials is called a **polynomial.**

Polynomial

A **polynomial** is a monomial or a sum and/or difference of monomials.

Examples of Polynomials

$5x^3 - 6x^2 + 2x + 10, \quad -1.2y^3 + 0.7y, \quad z, \quad \dfrac{1}{3}r - \dfrac{1}{2}, \quad 0$

Some polynomials are given special names depending on their number of terms.

Types of Polynomials

A **monomial** is a polynomial with exactly one term.
A **binomial** is a polynomial with exactly two terms.
A **trinomial** is a polynomial with exactly three terms.

The next page contains examples of monomials, binomials, and trinomials. Each of these examples is also a polynomial.

Polynomials			
Monomials	**Binomials**	**Trinomials**	**More Than Three Terms**
z	$x + 2$	$x^2 - 2x + 1$	$5x^3 - 6x^2 + 2x - 10$
4	$\dfrac{1}{3}r - \dfrac{1}{2}$	$y^5 + 3y^2 - 1.7$	$t^7 - t^5 + t^3 - t + 1$
$0.2x^2$ $\uparrow$ 1 term	$-1.2y^3 + 0.7y$ $\uparrow$ 2 terms	$-a^3 + 2a^2 - 5a$ $\uparrow$ 3 terms	$z^8 - z^4 + 3z^2 - 2z$

To add polynomials, we use the commutative and associative properties to rearrange and group like terms. Then, we combine like terms.

Adding Polynomials

To add polynomials, combine like terms.

PRACTICE PROBLEM 1

Add:

$(3y + 7) + (-9y - 14)$

EXAMPLE 1 Add: $(3x - 1) + (-6x + 2)$

Solution:

$$
\begin{aligned}
(3x - 1) + (-6x + 2) &= (3x - 6x) + (-1 + 2) &&\text{Group like terms.}\\
&= (-3x) + (1) &&\text{Combine like terms.}\\
&= -3x + 1
\end{aligned}
$$

▢ **Work Practice Problem 1**

PRACTICE PROBLEM 2

Add:

$(x^2 - 4x - 3) + (5x^2 - 6x)$

EXAMPLE 2 Add: $(9y^2 - 6y) + (7y^2 + 10y + 2)$

Solution:

$$
\begin{aligned}
(9y^2 - 6y) + (7y^2 + 10y + 2) &= 9y^2 + 7y^2 - 6y + 10y + 2 &&\text{Group like terms.}\\
&= 16y^2 + 4y + 2
\end{aligned}
$$

▢ **Work Practice Problem 2**

PRACTICE PROBLEM 3

Find the sum of
$(-z^2 - 4.2z + 11)$ and
$(9z^2 - 1.9z + 6.3)$.

EXAMPLE 3 Find the sum of $(-y^2 + 2y + 1.7)$ and $(12y^2 - 6y - 3.6)$.

Solution: Recall that "sum" means addition.

$$
\begin{aligned}
&(-y^2 + 2y + 1.7) + (12y^2 - 6y - 3.6)\\
&= \underbrace{-y^2 + 12y^2}_{} + \underbrace{2y - 6y}_{} + \underbrace{1.7 - 3.6}_{} &&\text{Group like terms.}\\
&= 11y^2 - 4y - 1.9 &&\text{Combine like terms.}
\end{aligned}
$$

▢ **Work Practice Problem 3**

Polynomials can also be added vertically. To do this, line up like terms underneath one another. Let's vertically add the polynomials in Example 3.

PRACTICE PROBLEM 4

Add vertically:

$(x^2 - x + 1.1)$
$+ (-8x^2 - x - 6.7)$

EXAMPLE 4 Find the sum of $(-y^2 + 2y + 1.7)$ and $(12y^2 - 6y - 3.6)$. Use a vertical format.

Solution: Line up like terms underneath one another.

$$
\begin{array}{r}
-y^2 + 2y + 1.7\\
+12y^2 - 6y - 3.6\\
\hline
11y^2 - 4y - 1.9
\end{array}
$$

▢ **Work Practice Problem 4**

Answers

1. $-6y - 7$ **2.** $6x^2 - 10x - 3$
3. $8z^2 - 6.1z + 17.3$
4. $-7x^2 - 2x - 5.6$

Notice that we are finding the same sum in Example 4 as in Example 3. Of course, the results are the same.

Objective **B** Subtracting Polynomials

To subtract one polynomial from another, recall how we subtract numbers. Recall from Section 2.3 that to subtract a number, we add its opposite: $a - b = a + (-b)$.

For example,

$$7 - 10 = 7 + (-10)$$
$$= -3$$

To subtract a polynomial, we also add its opposite. Just as the opposite of 3 is -3, the opposite of $(2x^2 - 5x + 1)$ is $-(2x^2 - 5x + 1)$. Let's practice simplifying the opposite of a polynomial.

EXAMPLE 5 Simplify: $-(2x^2 - 5x + 1)$

Solution: Rewrite $-(2x^2 - 5x + 1)$ as $-1(2x^2 - 5x + 1)$ and use the distributive property.

$$-(2x^2 - 5x + 1) = -1(2x^2 - 5x + 1)$$
$$= -1(2x^2) + (-1)(-5x) + (-1)(1)$$
$$= -2x^2 + 5x - 1$$

Work Practice Problem 5

PRACTICE PROBLEM 5

Simplify: $-(3y^2 + y - 2)$

Notice the result of Example 5.

$$-(2x^2 - 5x + 1) = -2x^2 + 5x - 1$$

This means that **the opposite of a polynomial can be found by changing the signs of the terms of the polynomial.** This leads to the following.

Subtracting Polynomials

To subtract polynomials, change the signs of the terms of the polynomial being subtracted, then add.

EXAMPLE 6 Subtract: $(5a + 7) - (2a - 10)$

Solution:

$$(5a + 7) - (2a - 10) = (5a + 7) + (-2a + 10) \quad \text{Add the opposite of } 2a - 10.$$
$$= 5a - 2a + 7 + 10 \quad \text{Group like terms.}$$
$$= 3a + 17$$

Work Practice Problem 6

PRACTICE PROBLEM 6

Subtract: $(9b + 8) - (11b - 20)$

EXAMPLE 7 Subtract: $(8x^2 - 4x + 1) - (10x^2 + 4)$

Solution:

$$(8x^2 - 4x + 1) - (10x^2 + 4) = (8x^2 - 4x + 1) + (-10x^2 - 4) \quad \begin{array}{l}\text{Add the opposite} \\ \text{of } 10x^2 + 4.\end{array}$$
$$= 8x^2 - 10x^2 - 4x + 1 - 4 \quad \text{Group like terms.}$$
$$= -2x^2 - 4x - 3$$

Work Practice Problem 7

PRACTICE PROBLEM 7

Subtract:
$(11x^2 + 7x + 2) - (15x^2 + 4x)$

Answers

5. $-3y^2 - y + 2$ **6.** $-2b + 28$
7. $-4x^2 + 3x + 2$

PRACTICE PROBLEM 8

Subtract $(-7y^2 + y - 4)$ from $(-3y^2 + 5y)$.

EXAMPLE 8 Subtract $(-6z^2 - 2z + 13)$ from $(4z^2 - 20z)$.

Solution: Be careful when arranging the polynomials in this example.

$$(4z^2 - 20z) - (-6z^2 - 2z + 13) = (4z^2 - 20z) + (6z^2 + 2z - 13)$$
$$= 4z^2 + 6z^2 - 20z + 2z - 13 \quad \text{Group like terms.}$$
$$= 10z^2 - 18z - 13$$

Work Practice Problem 8

✔**Concept Check** Find and explain the error in the following subtraction.

$$(3x^2 + 4) - (x^2 - 3x) = (3x^2 + 4) + (-x^2 - 3x)$$
$$= 3x^2 - x^2 - 3x + 4$$
$$= 2x^2 - 3x + 4$$

Just as with adding polynomials, we can subtract polynomials using a vertical format. Let's subtract the polynomials in Example 8 using a vertical format.

PRACTICE PROBLEM 9

Subtract $(3x^2 - 12x)$ from $(-4x^2 + 20x + 17)$. Use a vertical format.

EXAMPLE 9 Subtract $(-6z^2 - 2z + 13)$ from $(4z^2 - 20z)$. Use a vertical format.

Solution: Line up like terms underneath one another.

$$\begin{array}{r} 4z^2 - 20z \\ -(-6z^2 - 2z + 13) \end{array} \qquad \begin{array}{r} 4z^2 - 20z \\ +6z^2 + 2z - 13 \\ \hline 10z^2 - 18z - 13 \end{array}$$

Work Practice Problem 9

Notice that the answers to Examples 8 and 9 are the same regardless of which format is used.

Objective C **Evaluating Polynomials**

Polynomials have different values depending on the replacement values for the variables.

PRACTICE PROBLEM 10

Find the value of the polynomial $2y^3 + y^2 - 6$ when $y = 3$.

EXAMPLE 10 Find the value of the polynomial $3t^3 - 2t + 5$ when $t = 1$.

Solution: Replace t with 1 and simplify.

$$3t^3 - 2t + 5 = 3(1)^3 - 2(1) + 5 \quad \text{Let } t = 1.$$
$$= 3(1) - 2(1) + 5 \quad (1)^3 = 1.$$
$$= 3 - 2 + 5$$
$$= 6$$

The value of $3t^3 - 2t + 5$ when $t = 1$ is 6.

Work Practice Problem 10

Many real-world applications are modeled by polynomials.

Answers

8. $4y^2 + 4y + 4$
9. $-7x^2 + 32x + 17$ 10. 57

✔ **Concept Check Answer**

$(3x^2 + 4) - (x^2 - 3x)$
$= (3x^2 + 4) + (-x^2 + 3x)$
$= 3x^2 - x^2 + 3x + 4$
$= 2x^2 + 3x + 4$

EXAMPLE 11 **Finding the Height of an Object**

An object is dropped from the top of an 800-foot-tall building. Its height at time t seconds is given by the polynomial $-16t^2 + 800$. Find the height of the object when $t = 1$ second and when $t = 3$ seconds.

800 feet

Solution: To find each height, we evaluate the polynomial when $t = 1$ and when $t = 3$.

$$-16t^2 + 800 = -16(1)^2 + 800$$
$$= -16 + 800$$
$$= 784$$

The height of the object at 1 second is 784 feet.

$$-16t^2 + 800 = -16(3)^2 + 800$$
$$= -16(9) + 800$$
$$= -144 + 800$$
$$= 656$$

$t = 1$
$t = 3$
800 feet
784 feet
656 feet

The height of the object at 3 seconds is 656 feet.

🔲 **Work Practice Problem 11**

PRACTICE PROBLEM 11

An object is dropped from the top of a 530-foot cliff. Its height in feet at time t seconds is given by the polynomial $-16t^2 + 530$. Find the height of the object when $t = 1$ second and when $t = 4$ seconds.

Helpful Hint Don't forget to insert units, if appropriate.

Answer
11. 514 feet; 274 feet

Vocabulary and Readiness Check

Use the choices below to fill in each blank.

trinomial	binomial	add
terms	monomial	subtract

1. The addends of an algebraic expression are the _____ of the expression.

2. A polynomial with exactly one term is called a _____.

3. A polynomial with exactly two terms is called a _____.

4. A polynomial with exactly three terms is called a _____.

5. To _____ polynomials, combine like terms.

6. To _____ polynomials, change the sign of the terms of the polynomial being subtracted; then add.

10.1 EXERCISE SET

FOR EXTRA HELP

Student Solutions Manual · PH Math/Tutor Center · CD/Video for Review · MathXL® · MyMathLab

Objective A *Add the polynomials. See Examples 1 through 4.*

 1. $(2x + 3) + (-7x - 27)$

2. $(9y - 16) + (-43y + 16)$

3. $(-3z^2 + 5z - 5) + (-8z^2 - 8z + 4)$

4. $(8a^2 + 5a - 9) + (5a^2 - 11a + 6)$

5. $(12y - 20) + (9y^2 + 13y - 20)$

6. $(5x^2 - 6) + (-3x^2 + 17x - 2)$

 7. $(4.3a^4 + 5) + (-8.6a^4 - 2a^2 + 4)$

8. $(-12.7z^3 - 14z) + (-8.9z^3 + 12z + 2)$

Objective B *Simplify. See Example 5.*

9. $-(9x - 16)$

10. $-(4y - 12)$

11. $-(-3z^2 + z - 7)$

12. $-(-2x^2 - x + 1)$

Subtract the polynomials. See Examples 6 through 9.

13. $(8a - 5) - (3a + 8)$

14. $(3b + 5) - (-2b + 9)$

15. $(3x^2 - 2x + 1) - (5x^2 - 6x)$

16. $(-9z^2 + 6z + 2) - (3z^2 + 1)$

17. $(10y^2 - 7) - (20y^3 - 2y^2 - 3)$

18. $(11x^3 + 15x - 9) - (-x^3 + 10x^2 - 9)$

19. Subtract $(9x^2 + 3x - 4)$ from $(2x + 12)$.

20. Subtract $(4a^2 + 6a + 1)$ from $(-7a + 7)$.

 21. Subtract $(5y^2 + 4y - 6)$ from $(13y^2 - 6y - 14)$.

22. Subtract $(16x^2 - x + 1)$ from $(12x^2 - 3x - 12)$.

708

Objectives Ⓐ Ⓑ **Mixed Practice** *Perform each indicated operation. See Examples 1 through 9.*

23. $(25x - 5) + (-20x - 7)$

24. $(14x + 2) + (-7x - 1)$

25. $(4y + 4) - (3y + 8)$

26. $(6z - 3) - (8z + 5)$

27. $(3x^2 + 3x - 4) + (-8x^2 + 9)$

28. $(-2a^2 - 5a) + (6a^2 - 2a + 9)$

29. $(5x + 4.5) + (-x - 8.6)$

30. $(20x - 0.8) + (x + 1.2)$

31. $(a - 5) - (-3a + 2)$

32. $(t + 9) - (-2t + 6)$

33. $(21y - 4.6) - (36y - 8.2)$

34. $(8.6x + 4) - (9.7x - 93)$

35. $(18t^2 - 4t + 2) - (-t^2 + 7t - 1)$

36. $(35x^2 + x - 5) - (17x^2 - x + 5)$

37. $(2b^3 + 5b^2 - 5b - 8) + (8b^2 + 9b + 6)$

38. $(3z^2 - 8z + 5) + (-3z^3 - 5z^2 - 2z - 4)$

39. Add $(6x^2 - 7)$ and $(-11x^2 - 11x + 20)$.

40. Add $(-2x^2 + 3x)$ and $(9x^2 - x + 14)$.

41. Subtract $\left(3z - \dfrac{3}{7}\right)$ from $\left(3z + \dfrac{6}{7}\right)$.

42. Subtract $\left(8y^2 - \dfrac{7}{10}y\right)$ from $\left(-5y^2 + \dfrac{3}{10}y\right)$.

Objective Ⓒ *Find the value of each polynomial when $x = 2$. See Examples 10 and 11.*

43. $-2x + 9$

44. $-5x - 7$

45. $x^2 - 6x + 3$

46. $5x^2 + 4x - 100$

47. $\dfrac{3x^2}{2} - 14$

48. $\dfrac{7x^3}{14} - x + 5$

Find the value of each polynomial when $x = 5$. See Examples 10 and 11.

49. $2x + 10$

50. $-5x - 6$

51. x^2

52. x^3

53. $2x^2 + 4x - 20$

54. $4x^2 - 5x + 10$

Solve. See Example 11.
The distance in feet traveled by a free-falling object in t seconds is given by the polynomial $16t^2$. Use this polynomial for Exercises 55 and 56.

55. Find the distance traveled by an object that falls for 6 seconds.

56. It takes 8 seconds for a hard hat to fall from the top of a building. How high is the building?

Office Supplies, Inc. manufactures office products. The company determines that the total cost for manufacturing x file cabinets is given by the polynomial $3000 + 20x$. Use this polynomial for Exercises 57 and 58.

57. Find the total cost to manufacture 10 file cabinets.

58. Find the total cost to manufacture 100 file cabinets.

An object is dropped from the deck of the Royal Gorge Bridge, which stretches across Royal Gorge at a height of 1053 feet above the Arkansas River. The height of the object above the river after t seconds is given by the polynomial $1053 - 16t^2$. Use this polynomial for Exercises 59 and 60. (Source: Royal Gorge Bridge Co.)

59. How far above the river is an object that has been falling for 3 seconds?

60. How far above the river is an object that has been falling for 8 seconds?

61. Cellular telephone subscriptions are increasing every year. The number of cellular subscribers (in millions of subscribers) in a certain year can be modeled by the equation $17x^2 + 210x + 700$, where x stands for the number of years since 2000. That means that in 2000, there were approximately 700 million subscribers. If the rate of growth continues as it has, how many cellular subscribers should we expect in the year 2007? (*Source: Statistical Abstract of the United States*)

62. Now that WiFi is the big new thing, many more people in America are browsing the Internet using wireless technology. The number of wireless Internet users (in millions) can be modeled by the equation $3.4x^2 + 1.7x + 1.9$, where x stands for the number of years since 2000. If wireless Internet use continues to grow at this rate, how many wireless Internet users should we expect in the year 2007? (*Source: Statistical Abstract of the United States*)

Review

Evaluate. See Sections 1.7 and 2.4.

63. 3^4

64. $(-2)^5$

65. $(-5)^2$

66. 4^3

Write using exponential notation. See Section 1.7.

67. $x \cdot x \cdot x$

68. $y \cdot y \cdot y \cdot y \cdot y$

69. $2 \cdot 2 \cdot a \cdot a \cdot a \cdot a$

70. $5 \cdot 5 \cdot 5 \cdot b \cdot b$

Concept Extensions

Find the perimeter of each figure.

△ **71.**

$(5x - 10)$ inches $(2x + 1)$ inches

$(x + 11)$ inches

△ **72.**

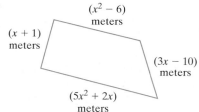

$(x^2 - 6)$ meters

$(x + 1)$ meters

$(3x - 10)$ meters

$(5x^2 + 2x)$ meters

Given the lengths in the figure below, we find the unknown length by subtracting. Use the information to find the unknown lengths in Exercises 73 and 74.

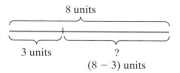

8 units

3 units   ?
$(8 - 3)$ units

73.

$(7x - 10)$ units

$(3x + 5)$ units   ? units

74.

$(x^2 - 7x + 6)$ units

$(x^2 + 2)$ units   ? units

Fill in the blanks.

75. $(3x^2 + \underline{\quad} x - \underline{\quad}) + (\underline{\quad} x^2 - 6x + 2) = 5x^2 + 14x - 4$

76. $(\underline{\quad} y^2 + 4y - 3) + (8y^2 - \underline{\quad} y + \underline{\quad}) = 9y^2 + 2y + 7$

77. Find the value of $7a^4 - 6a^2 + 2a - 1$ when $a = 1.2$.

78. Find the value of $3b^3 + 4b^2 - 100$ when $b = -2.5$.

79. For Exercises 59 and 60, the polynomial $1053 - 16t^2$ was used to give the height of an object above the river after t seconds. Find the height when $t = 8$ seconds and $t = 9$ seconds. Explain what happened and why.

10.2 MULTIPLICATION PROPERTIES OF EXPONENTS

Objective **A** Using the Product Rule

Recall from Section 1.8 that an exponent has the same meaning whether the base is a number or a variable. For example,

$$5^3 = \underbrace{5 \cdot 5 \cdot 5}_{3 \text{ factors of } 5} \quad \text{and} \quad x^3 = \underbrace{x \cdot x \cdot x}_{3 \text{ factors of } x}$$

We can use this definition of an exponent to discover properties that will help us to simplify products and powers of exponential expressions.

For example, let's use the definition of an exponent to find the product of x^3 and x^4.

$$x^3 \cdot x^4 = (x \cdot x \cdot x)(x \cdot x \cdot x \cdot x)$$
$$= \underbrace{x \cdot x \cdot x \cdot x \cdot x \cdot x \cdot x}_{7 \text{ factors of } x}$$
$$= x^7$$

Notice that the result is the same if we add the exponents.

$$x^3 \cdot x^4 = x^{3+4} = x^7$$

This suggests the following product rule or property for exponents.

Product Property for Exponents

If m and n are positive integers and a is a real number, then

$$a^m \cdot a^n = a^{m+n}$$

In other words, to multiply two exponential expressions with the same base, keep the base and add the exponents.

PRACTICE PROBLEM 1

Multiply: $z^5 \cdot z^6$

EXAMPLE 1 Multiply: $y^7 \cdot y^2$

Solution: $y^7 \cdot y^2 = y^{7+2}$ Use the product property for exponents.
$$= y^9 \qquad \text{Simplify.}$$

▪ **Work Practice Problem 1**

PRACTICE PROBLEM 2

Multiply: $8y^5 \cdot 4y^9$

EXAMPLE 2 Multiply: $3x^5 \cdot 6x^3$

Solution: $3x^5 \cdot 6x^3 = (3 \cdot 6)(x^5 \cdot x^3)$ Apply the commutative and associative properties.
$$= 18x^{5+3} \qquad \text{Use the product property for exponents.}$$
$$= 18x^8 \qquad \text{Simplify.}$$

▪ **Work Practice Problem 2**

PRACTICE PROBLEM 3

Multiply: $(-4r^6s^2)(-3r^2s^5)$

EXAMPLE 3 Multiply: $(-2a^4b^{10})(9a^5b^3)$

Solution: Use properties of multiplication to group numbers and like variables together.

$$(-2a^4b^{10})(9a^5b^3) = (-2 \cdot 9)(a^4 \cdot a^5)(b^{10} \cdot b^3)$$
$$= -18a^{4+5}b^{10+3}$$
$$= -18a^9b^{13}$$

▪ **Work Practice Problem 3**

Answers

1. z^{11} **2.** $32y^{14}$ **3.** $12r^8s^7$

EXAMPLE 4 Multiply: $2x^3 \cdot 3x \cdot 5x^6$

Solution: First notice the factor $3x$. Since there is one factor of x in $3x$, it can also be written as $3x^1$.

$$2x^3 \cdot 3x^1 \cdot 5x^6 = (2 \cdot 3 \cdot 5)(x^3 \cdot x^1 \cdot x^6)$$
$$= 30x^{10}$$

> **Helpful Hint**
> Don't forget that if an exponent is not written, it is assumed to be 1.

■ **Work Practice Problem 4**

PRACTICE PROBLEM 4

Multiply: $11y^5 \cdot 3y^2 \cdot y$. (Recall that $y = y^1$.)

> **Helpful Hint**
> These examples will remind you of the difference between adding and multiplying terms.

Addition	Multiplication
$5x^3 + 3x^3 = (5 + 3)x^3 = 8x^3$	$(5x^3)(3x^3) = 5 \cdot 3 \cdot x^3 \cdot x^3 = 15x^{3+3} = 15x^6$
$7x + 4x^2 = 7x + 4x^2$	$(7x)(4x^2) = 7 \cdot 4 \cdot x \cdot x^2 = 28x^{1+2} = 28x^3$

Objective B Using the Power Rule

Next suppose that we want to simplify an exponential expression raised to a power. To see how we simplify $(x^2)^3$, we again use the definition of an exponent.

$$(x^2)^3 = \underbrace{(x^2) \cdot (x^2) \cdot (x^2)}_{\text{3 factors of } x^2} \quad \text{Apply the definition of an exponent.}$$

$$= x^{2+2+2} \qquad \text{Use the product property for exponents.}$$
$$= x^6 \qquad \text{Simplify.}$$

Notice the result is exactly the same if we multiply the exponents.

$$(x^2)^3 = x^{2 \cdot 3} = x^6$$

This suggests the following power rule or property for exponents.

Power Property for Exponents

If m and n are positive integers and a is a real number, then

$$(a^m)^n = a^{m \cdot n}$$

In other words, to raise a power to a power, keep the base and multiply the exponents.

> **Helpful Hint**
> Take a moment to make sure that you understand when to apply the product rule and when to apply the power rule.

Product Property → Add Exponents	Power Property → Multiply Exponents
$x^5 \cdot x^7 = x^{5+7} = x^{12}$	$(x^5)^7 = x^{5 \cdot 7} = x^{35}$
$y^6 \cdot y^2 = y^{6+2} = y^8$	$(y^6)^2 = y^{6 \cdot 2} = y^{12}$

Answer
4. $33y^8$

PRACTICE PROBLEM 5

Simplify: $(z^3)^6$

EXAMPLE 5 Simplify: $(y^8)^2$

Solution:

$$(y^8)^2 = y^{8 \cdot 2} \quad \text{Use the power property.}$$
$$= y^{16}$$

◾ **Work Practice Problem 5**

PRACTICE PROBLEM 6

Simplify: $(z^4)^5 \cdot (z^3)^7$

EXAMPLE 6 Simplify: $(a^3)^4 \cdot (a^2)^9$

Solution:

$$(a^3)^4 \cdot (a^2)^9 = a^{12} \cdot a^{18} \quad \text{Use the power property.}$$
$$= a^{12+18} \quad \text{Use the product property.}$$
$$= a^{30} \quad \text{Simplify.}$$

◾ **Work Practice Problem 6**

Objective C Using the Power of a Product Rule

Next, let's simplify the power of a product.

$$(xy)^3 = xy \cdot xy \cdot xy \quad \text{Apply the definition of an exponent.}$$
$$= (x \cdot x \cdot x)(y \cdot y \cdot y) \quad \text{Group like bases.}$$
$$= x^3 y^3 \quad \text{Simplify.}$$

Notice that the power of a product can be written as the product of powers. This leads to the following power of a product rule or property.

Power of a Product Property for Exponents

If n is a positive integer and a and b are real numbers, then

$$(ab)^n = a^n b^n$$

In other words, to raise a product to a power, raise each factor to the power.

✔**Concept Check** Which property is needed to simplify $(x^6)^3$? Explain.

a. Product property for exponents
b. Power property for exponents
c. Power of a product property for exponents

PRACTICE PROBLEM 7

Simplify: $(3b)^4$

EXAMPLE 7 Simplify: $(5t)^3$

Solution:

$$(5t)^3 = 5^3 t^3 \quad \text{Apply the power of a product property.}$$
$$= 125t^3 \quad \text{Write } 5^3 \text{ as } 125.$$

◾ **Work Practice Problem 7**

Answers

5. z^{18} **6.** z^{41} **7.** $81b^4$

✔ **Concept Check Answer**

b

EXAMPLE 8 Simplify: $(2a^5b^3)^3$

Solution:

$$(2a^5b^3)^3 = 2^3(a^5)^3(b^3)^3 \quad \text{Apply the power of a product property.}$$
$$= 8a^{15}b^9 \quad \text{Apply the power property.}$$

Work Practice Problem 8

PRACTICE PROBLEM 8
Simplify: $(4x^2y^6)^3$

EXAMPLE 9 Simplify: $(3y^4z^2)^4(2y^3z^5)^5$

Solution:

$$(3y^4z^2)^4(2y^3z^5)^5 = 3^4(y^4)^4(z^2)^4 \cdot 2^5(y^3)^5(z^5)^5 \quad \text{Apply the power of a product property.}$$
$$= 81y^{16}z^8 \cdot 32y^{15}z^{25} \quad \text{Apply the power property.}$$
$$= (81 \cdot 32)(y^{16} \cdot y^{15})(z^8 \cdot z^{25}) \quad \text{Group like bases.}$$
$$= 2592y^{31}z^{33} \quad \text{Apply the product property.}$$

Work Practice Problem 9

PRACTICE PROBLEM 9
Simplify: $(2x^2y^4)^4(3x^6y^9)^2$

Answers
8. $64x^6y^{18}$ **9.** $144x^{20}y^{34}$

Vocabulary and Readiness Check

Use the choices below to fill in each blank. Not all choices will be used.

add   multiply   exponent $6x^2$

subtract   divide $36x$ $36x^2$

1. In $7x^2$, the 2 is called the _____ .

2. To simplify $x^4 \cdot x^3$, we _____ the exponents.

3. To simplify $(x^4)^3$, we _____ the exponents.

4. The expression $(6x)^2$ simplifies to _____ .

10.2 EXERCISE SET

FOR EXTRA HELP

Student Solutions Manual   PH Math/Tutor Center   CD/Video for Review   Math XL   MathXL®   MyMathLab   MyMathLab

Objective A *Multiply. See Examples 1 through 4.*

 1. $x^5 \cdot x^9$ **2.** $y^4 \cdot y^7$ **3.** $a^3 \cdot a$ **4.** $b \cdot b^4$

5. $3z^3 \cdot 5z^2$ **6.** $8r^2 \cdot 2r^{15}$ **7.** $-4x \cdot 10x$ **8.** $-9y \cdot 3y$

9. $2x \cdot 3x \cdot 7x$ **10.** $4y \cdot 3y \cdot 5y$ **11.** $a \cdot 4a^{11} \cdot 3a^5$ **12.** $b \cdot 7b^{10} \cdot 5b^8$

13. $(-5x^2y^3)(-5x^4y)$ **14.** $(-2xy^4)(-6x^3y^7)$ **15.** $(7ab)(4a^4b^5)$ **16.** $(3a^3b^6)(12a^2b^9)$

Objectives B C **Mixed Practice** *Simplify. See Examples 5 through 9.*

 17. $(x^5)^3$ **18.** $(y^4)^7$ **19.** $(z^3)^{10}$ **20.** $(a^9)^3$

21. $(b^7)^6(b^2)^{10}$ **22.** $(x^2)^9 \cdot (x^5)^3$ **23.** $(3a)^4$ **24.** $(2y)^5$

25. $(a^{11}b^8)^3$ **26.** $(x^7y^4)^8$ **27.** $(10x^5y^3)^3$ **28.** $(8a^5b^7)^2$

29. $(-3y)(2y^7)^3$ **30.** $(-2x)(5x^2)^4$ **31.** $(4xy)^3(2x^3y^5)^2$ **32.** $(2xy)^4(3x^4y^3)^3$

Review

Multiply. See Section 3.1.

33. $7(x - 3)$ **34.** $4(y + 2)$ **35.** $-2(3a + 2b)$

36. $-3(8r + 3s)$ **37.** $9(x + 2y - 3)$ **38.** $5(a + 7b - 3)$

716

Concept Extensions

Find the area of each figure.

△ **39.**

Square $4x^6$ inches

△ **40.**

$9y^2$ centimeters

Rectangle $9y$ centimeters

△ **41.**

$4ab$ meters

$6a^3b^4$ meters

△ **42.**

$30y^{12}$ feet Parallelogram

$50y^{15}$ feet

(*Hint:* Area = base · height)

Multiply and simplify.

🖩 **43.** $(14a^7b^6)^3(9a^6b^3)^4$

🖩 **44.** $(5x^{14}y^6)^7(3x^{20}y^{19})^5$

45. $(8.1x^{10})^5$

46. $(4.6a^{14})^4$

47. In your own words, explain why $x^2 \cdot x^3 = x^5$ and $(x^2)^3 = x^6$.

48. $(x^{90}y^{72})^3$

49. $(a^{20}b^{10}c^5)^5(a^9b^{12})^3$

INTEGRATED REVIEW

Sections 10.1–10.2

Operations on Polynomials

1.
2.
3.
4.
5.
6.
7.
8.
9.
10.
11.
12.
13.
14.
15.
16.
17.
18.
19.
20.
21.
22.
23.
24.
25.
26.
27.
28.
29.
30.

Add or subtract the polynomials as indicated.

1. $(3x + 5) + (-x - 8)$　　　　　**2.** $(15y - 7) + (5y - 4)$

3. $(7x + 1) - (-3x - 2)$　　　　　**4.** $(14y - 6) - (19y - 2)$

5. $(a^4 + 5a) + (3a^4 - 3a^2 - 4a)$　　　**6.** $(2a^3 - 6a^2 + 11) - (6a^3 + 6a^2 + 11)$

7. $(4.5x^2 + 8.1x) + (2.8x^2 - 12.3x - 5.3)$　**8.** $(1.2y^2 - 3.6y) + (0.6y^2 + 1.2y - 5.6)$

9. Subtract $(2x - 6)$ from $(8x + 1)$.　　**10.** Subtract $(3x^2 - x + 2)$ from $(5x^2 + 2x - 10)$.

Find the value of each polynomial when $x = 3$.

11. $2x - 7$　　　　　　　　　　**12.** $x^2 + 5x + 2$

Multiply.

13. $x^9 \cdot x^{11}$　　　　**14.** $x^5 \cdot x^5$　　　　**15.** $y^3 \cdot y$

16. $a \cdot a^{10}$　　　　**17.** $(x^7)^{11}$　　　　**18.** $(x^6)^6$

19. $(x^3)^4 \cdot (x^5)^6$　**20.** $(y^2)^9 \cdot (y^3)^3$　**21.** $(5x)^3$

22. $(2y)^5$　　　　　　　　**23.** $(-6xy^2)(2xy^5)$

24. $(-4a^2b^3)(-3ab)$　　　　**25.** $(y^{11}z^{13})^3$

26. $(a^5b^{12})^4$　　　　　　　**27.** $(10x^2y)^2(3y)$

28. $(8y^3z)^2(2z^5)$　　　　　**29.** $(2a^5b)^4(3a^9b^4)^2$

30. $(5x^4y^6)^3(x^2y^2)^5$

10.3 MULTIPLYING POLYNOMIALS

Objectives

A Multiply a Monomial and Any Polynomial.

B Multiply Two Binomials.

C Square a Binomial.

D Use the FOIL Order to Multiply Binomials.

E Multiply Any Two Polynomials.

Objective **A** Multiplying a Monomial and a Polynomial

Recall from Section 10.1 that a polynomial that consists of one term is called a **monomial.** For example, $5x$ is a monomial. To multiply a monomial and any polynomial, we use the distributive property

$$a(b + c) = a \cdot b + a \cdot c$$

and apply properties of exponents.

EXAMPLE 1 Multiply: $5x(3x^2 + 2)$

Solution:

$$5x(3x^2 + 2) = 5x \cdot 3x^2 + 5x \cdot 2 \quad \text{Apply the distributive property.}$$
$$= 15x^3 + 10x$$

▪ **Work Practice Problem 1**

PRACTICE PROBLEM 1

Multiply: $4y(8y^2 + 5)$

EXAMPLE 2 Multiply: $2z(4z^2 + 6z - 9)$

Solution:

$$2z(4z^2 + 6z - 9) = 2z \cdot 4z^2 + 2z \cdot 6z + 2z(-9)$$
$$= 8z^3 + 12z^2 - 18z$$

▪ **Work Practice Problem 2**

PRACTICE PROBLEM 2

Multiply: $3r(8r^2 - r + 11)$

To visualize multiplication by a monomial, let's look at two ways we can represent the area of the same rectangle.

Method 1:   The width of the rectangle is x and its length is $x + 3$. One way to calculate the area of the rectangle is

area = width · length
   = $x(x + 3)$

Method 2:   Another way to calculate the area of the rectangle is to find the sum of the areas of the smaller figures.

area = $x^2 + 3x$

Since the areas must be equal, we have that

$$x(x + 3) = x^2 + 3x \quad \text{As expected by the distributive property}$$

Answers

1. $32y^3 + 20y$ **2.** $24r^3 - 3r^2 + 33r$

Objective B Multiplying Binomials

Recall also from Section 10.1 that a polynomial that consists of exactly two terms is called a **binomial.** To multiply two binomials, we use a version of the distributive property:

$$(b + c)a = b \cdot a + c \cdot a$$

PRACTICE PROBLEM 3
Multiply: $(b + 3)(b + 5)$

EXAMPLE 3   Multiply: $(x + 2)(x + 3)$

Solution:

$$
\begin{aligned}
(x + 2)(x + 3) &= x(x + 3) + 2(x + 3) &&\text{Apply the distributive property.}\\
&= x \cdot x + x \cdot 3 + 2 \cdot x + 2 \cdot 3 &&\text{Apply the distributive property.}\\
&= x^2 + 3x + 2x + 6 &&\text{Multiply.}\\
&= x^2 + 5x + 6 &&\text{Combine like terms.}
\end{aligned}
$$

Work Practice Problem 3

PRACTICE PROBLEM 4
Multiply: $(7x - 1)(5x + 4)$

EXAMPLE 4   Multiply: $(4y + 9)(3y - 2)$

Solution:

$$
\begin{aligned}
(4y + 9)(3y - 2) &= 4y(3y - 2) + 9(3y - 2) &&\text{Apply the distributive property.}\\
&= 4y \cdot 3y + 4y(-2) + 9 \cdot 3y + 9(-2) &&\text{Apply the distributive property.}\\
&= 12y^2 - 8y + 27y - 18 &&\text{Multiply.}\\
&= 12y^2 + 19y - 18 &&\text{Combine like terms.}
\end{aligned}
$$

Work Practice Problem 4

Objective C Squaring a Binomial

Raising a binomial to the power of 2 is also called squaring a binomial. To square a binomial, we use the definition of an exponent, and then multiply.

PRACTICE PROBLEM 5
Multiply: $(6y - 1)^2$

EXAMPLE 5   Multiply: $(2x + 1)^2$

Solution:

$$
\begin{aligned}
(2x + 1)^2 &= (2x + 1)(2x + 1) &&\text{Apply the definition of an exponent.}\\
&= 2x(2x + 1) + 1(2x + 1) &&\text{Apply the distributive property.}\\
&= 2x \cdot 2x + 2x \cdot 1 + 1 \cdot 2x + 1 \cdot 1 &&\text{Apply the distributive property.}\\
&= 4x^2 + 2x + 2x + 1 &&\text{Multiply.}\\
&= 4x^2 + 4x + 1 &&\text{Combine like terms.}
\end{aligned}
$$

Work Practice Problem 5

✔ **Concept Check**   Correct and explain the error:

$$(x + 5)^2 = x^2 + 25$$

Answers
3. $b^2 + 8b + 15$ **4.** $35x^2 + 23x - 4$
5. $36y^2 - 12y + 1$

✔ **Concept Check Answer**
$$
\begin{aligned}
(x + 5)^2 &= (x + 5)(x + 5)\\
&= x^2 + 10x + 25
\end{aligned}
$$

Objective D Using the FOIL Order to Multiply Binomials

Recall from Example 3 that

$$(x + 2)(x + 3) = x \cdot x + x \cdot 3 + 2 \cdot x + 2 \cdot 3$$
$$= x^2 + 5x + 6$$

One way to remember these products $x \cdot x$, $x \cdot 3$, $2 \cdot x$, and $2 \cdot 3$ is to use a special order for multiplying binomials, called the FOIL order. Of course, the product is the same no matter what order or method you choose to use.

FOIL stands for the products of the First terms, Outer terms, Inner terms, then Last terms. For example,

$$(x + 2)(x + 3) = x \cdot x + x \cdot 3 + 2 \cdot x + 2 \cdot 3 = x^2 + 3x + 2x + 6$$
$$= x^2 + 5x + 6$$

F   O   I   L

> **Helpful Hint** The product is the same no matter what order or method you choose to use.

EXAMPLES Use the FOIL order to multiply.

6. $(3x - 6)(2x + 1) = 3x \cdot 2x + 3x \cdot 1 + (-6)(2x) + (-6)(1)$

F   O   I   L

$$= 6x^2 + 3x - 12x - 6 \quad \text{Multiply.}$$
$$= 6x^2 - 9x - 6 \quad \text{Combine like terms.}$$

7. $(3x - 5)^2 = (3x - 5)(3x - 5)$

F   O   I   L

$$= 3x \cdot 3x + 3x(-5) + (-5)(3x) + (-5)(-5)$$
$$= 9x^2 - 15x - 15x + 25 \quad \text{Multiply.}$$
$$= 9x^2 - 30x + 25 \quad \text{Combine like terms.}$$

☐ **Work Practice Problems 6–7**

PRACTICE PROBLEMS

Use the FOIL order to multiply.
6. $(10x - 7)(2x + 3)$
7. $(3x + 2)^2$

> **Helpful Hint** Remember that the FOIL order can only be used to multiply **two binomials.**

Objective E Multiplying Polynomials

In Section 10.1, we learned that a polynomial that consists of exactly three terms is called a **trinomial.** Next, we multiply a binomial by a trinomial.

EXAMPLE 8 Multiply: $(3a + 2)(a^2 - 6a + 3)$

Solution: Use the distributive property to multiply $3a$ by the trinomial $(a^2 - 6a + 3)$ and then 2 by the trinomial.

$$(3a + 2)(a^2 - 6a + 3) = 3a(a^2 - 6a + 3) + 2(a^2 - 6a + 3) \quad \text{Apply the distributive property.}$$
$$= 3a \cdot a^2 + 3a(-6a) + 3a \cdot 3 + \quad \text{Apply the distributive property.}$$
$$2 \cdot a^2 + 2(-6a) + 2 \cdot 3$$
$$= 3a^3 - 18a^2 + 9a + 2a^2 - 12a + 6 \quad \text{Multiply.}$$
$$= 3a^3 - 16a^2 - 3a + 6 \quad \text{Combine like terms.}$$

☐ **Work Practice Problem 8**

PRACTICE PROBLEM 8

Multiply: $(2x + 5)(x^2 + 4x - 1)$

Answers
6. $20x^2 + 16x - 21$ **7.** $9x^2 + 12x + 4$
8. $2x^3 + 13x^2 + 18x - 5$

In general, we have the following.

To Multiply Two Polynomials

Multiply each term of the first polynomial by each term of the second polynomial, and then combine like terms.

A convenient method of multiplying polynomials is to use a vertical format similar to multiplying real numbers.

✔**Concept Check**   True or false? When a trinomial is multiplied by a trinomial, the result will have at most nine terms. Explain.

PRACTICE PROBLEM 9

Multiply $(x^2 + 4x - 1)$ and $(2x + 5)$ vertically.

EXAMPLE 9   Find the product of $(a^2 - 6a + 3)$ and $(3a + 2)$ vertically.

Solution:

$$
\begin{array}{r}
a^2 - 6a + 3 \\
\times \qquad 3a + 2 \\
\hline
2a^2 - 12a + 6 \\
3a^3 - 18a^2 + 9a \\
\hline
3a^3 - 16a^2 - 3a + 6
\end{array}
$$

Multiply $a^2 - 6a + 3$ by 2.

Multiply $a^2 - 6a + 3$ by $3a$. Line up like terms.

Combine like terms

Notice that this example is the same as Example 8 and of course the products are the same.

◼ **Work Practice Problem 9**

10.3 EXERCISE SET

Objective A *Multiply. See Examples 1 and 2.*

1. $3x(9x^2 - 3)$

2. $4y(10y^3 + 2y)$

3. $-3a(2a^2 - 3a - 5)$

4. $-4b(-2b^2 - 5b + 8)$

5. $7x^2(6x^2 - 5x + 7)$

6. $6z^2(-3z^2 - z + 4)$

Objectives B C D Mixed Practice *Multiply. See Examples 3 through 7.*

7. $(x + 3)(x + 10)$

8. $(y + 5)(y + 9)$

9. $(2x - 6)(x + 4)$

10. $(7z + 1)(z - 6)$

11. $(6a + 4)^2$

12. $(8b - 3)^2$

Objective E *Multiply. See Examples 8 and 9.*

13. $(a + 6)(a^2 - 6a + 3)$

14. $(y + 4)(y^2 + 8y - 2)$

15. $(4x - 5)(2x^2 + 3x - 10)$

16. $(9z - 2)(2z^2 + z + 1)$

17. $(x^3 + 2x + x^2)(3x + 1 + x^2)$

18. $(y^2 - 2y + 5)(y^3 + 2 + y)$

Objectives A B C D E Mixed Practice *Multiply. See Examples 1 through 9.*

19. $10r(-3r + 2)$

20. $5x(4x^2 + 5)$

21. $-2y^2(3y + y^2 - 6)$

22. $3z^3(4z^4 - 2z + z^3)$

23. $(x + 2)(x + 12)$

24. $(y + 7)(y - 7)$

25. $(2a + 3)(2a - 3)$

26. $(6s + 1)(3s - 1)$

27. $(x + 5)^2$

28. $(x + 3)^2$

29. $\left(b + \dfrac{3}{5}\right)\left(b + \dfrac{4}{5}\right)$

30. $\left(a - \dfrac{7}{10}\right)\left(a + \dfrac{3}{10}\right)$

31. $(6x + 1)(x^2 + 4x + 1)$

32. $(9y - 1)(y^2 + 3y - 5)$

33. $(7x + 5)^2$

34. $(5x + 9)^2$

35. $(2x - 1)^2$

36. $(4a - 3)^2$

37. $(2x^2 - 3)(4x^3 + 2x - 3)$ **38.** $(3y^2 + 2)(5y^2 - y + 2)$ **39.** $(x^3 + x^2 + x)(x^2 + x + 1)$

40. $(a^4 + a^2 + 1)(a^4 + a^2 - 1)$ **41.** $(2z^2 - z + 1)(5z^2 + z - 2)$ **42.** $(2b^2 - 4b + 3)(b^2 - b + 2)$

Review

Write each number as a product of prime numbers. See Section 4.2.

43. 50   **44.** 48   **45.** 72

46. 36   **47.** 200   **48.** 300

Concept Extensions

Find the area of each figure.

 49.

$(y - 6)$ feet

$(y^2 + 3y + 2)$ feet

△ **50.**

Square $(2x + 11)$ centimeters

Find the area of the shaded figure. To do so, subtract the area of the smaller square from the area of the larger geometric figure.

△ **51.**

Square

$(x^2 - 1)$ meters

x meters

△ **52.**

$(3x + 5)$ miles

$2x$ miles

$(3x - 5)$ miles

53. Suppose that a classmate asked you why $(2x + 1)^2$ is **not** $4x^2 + 1$. Write down your response to this classmate.

10.4 INTRODUCTION TO FACTORING POLYNOMIALS

Objectives

A Find the Greatest Common Factor of a List of Integers.

B Find the Greatest Common Factor of a List of Terms.

C Factor the Greatest Common Factor from the Terms of a Polynomial.

Recall from Section 2.4 that when an integer is written as the product of two or more integers, each of these integers is called a *factor* of the product. This is true of polynomials also. When a polynomial is written as the product of two or more other polynomials, each of these polynomials is called a factor of the product.

factor · factor	=	product
$-2 \cdot 4$	=	-8
$x^3 \cdot x^7$	=	x^{10}
$5(x + 2)$	=	$5x + 10$

The process of writing a polynomial as a product is called **factoring.** Notice that factoring is the reverse process of multiplying.

$$\underbrace{5x + 10}_{\text{multiplying}} = \overbrace{5(x + 2)}^{\text{factoring}}$$

Objective **A** Finding the GCF of a List of Integers

Before we factor polynomials, let's practice finding the greatest common factor of a list of integers. The **greatest common factor (GCF)** of a list of integers is the largest integer that is a factor of all the integers in the list. For example,

the GCF of 30 and 18 is 6

because 6 is the largest integer that is a factor of both 30 and 18.

If the GCF cannot be found by inspection, the following steps can be used.

To Find the GCF of a List of Integers

Step 1: Write each number as a product of prime numbers.

Step 2: Identify the common prime factors.

Step 3: The product of all common prime factors found in *Step 2* is the greatest common factor. If there are no common prime factors, the greatest common factor is 1.

✔**Concept Check**   Which of the following is the prime factorization of 36?

a. $4 \cdot 9$ **b.** $2 \cdot 2 \cdot 3 \cdot 3$ **c.** $6 \cdot 6$

Recall from Section 4.2 that a prime number is a whole number other than 1, whose only factors are 1 and itself.

✔ Concept Check Answer
b

PRACTICE PROBLEM 1

Find the GCF of 42 and 28.

EXAMPLE 1 Find the GCF of 12 and 20.

Solution:

Step 1: Write each number as a product of primes.

$12 = 2 \cdot 2 \cdot 3$
$20 = 2 \cdot 2 \cdot 5$

Step 2: $12 = 2 \cdot 2 \cdot 3$
$20 = 2 \cdot 2 \cdot 5$

$\downarrow \quad \downarrow$

$2 \cdot 2$ Identify the common factors.

Step 3: The GCF is $2 \cdot 2 = 4$

■ **Work Practice Problem 1**

Objective B Finding the GCF of a List of Terms

How do we find the GCF of a list of variables raised to powers? For example, what is the GCF of y^3, y^5, and y^{10}? Notice that each variable term contains a factor of y^3 and no higher power of y is a factor of each term.

$y^3 = y^3$
$y^5 = y^3 \cdot y^2$ Recall the product property for exponents.
$y^{10} = y^3 \cdot y^7$

The GCF of y^3, y^5, and y^{10} is y^3. From this example, we can see that **the GCF of a list of variables raised to powers is the variable raised to the smallest exponent in the list.**

PRACTICE PROBLEM 2

Find the GCF of z^7, z^8, and z.

EXAMPLE 2 Find the GCF of x^{11}, x^4, and x^6.

Solution: The GCF is x^4 since 4 is the smallest exponent to which x is raised.

■ **Work Practice Problem 2**

In general, **the GCF of a list of terms is the product of all common factors.**

PRACTICE PROBLEM 3

Find the GCF of $6a^4$, $3a^5$, and $15a^2$.

EXAMPLE 3 Find the GCF of $4x^3$, $12x$, and $10x^5$.

Solution:

The GCF of 4, 12, and 10 is 2.
The GCF of x^3, x^1, and x^5 is x^1.
Thus, the GCF of $4x^3$, $12x$, and $10x^5$ is $2x^1$ or $2x$.

■ **Work Practice Problem 3**

Helpful Hint

If you ever have trouble finding the GCF, remember that you can always use the method below.

Example 3:

$4x^3 = 2 \cdot 2 \cdot x \cdot x \cdot x$
$12x = 2 \cdot 2 \cdot 3 \cdot x$
$10x^5 = 2 \cdot 5 \cdot x \cdot x \cdot x \cdot x \cdot x$
GCF $= 2 \cdot x$ or $2x$

Answers
1. 14 **2.** z **3.** $3a^2$

Objective C Factoring Out the GCF

Next, we practice factoring a polynomial by factoring the GCF from its terms. To do so, we write each term of the polynomial as a product of the GCF and another factor, and then apply the distributive property.

EXAMPLE 4 Factor: $7x^3 + 14x^2$

Solution: The GCF of $7x^3$ and $14x^2$ is $7x^2$.

$$7x^3 + 14x^2 = 7x^2 \cdot x^1 + 7x^2 \cdot 2$$
$$= 7x^2(x + 2) \qquad \text{Apply the distributive property.}$$

■ **Work Practice Problem 4**

PRACTICE PROBLEM 4

Factor: $10y^7 + 5y^9$

Notice in Example 4 that we factored $7x^3 + 14x^2$ by writing it as the product $7x^2(x + 2)$. Also notice that to check factoring, we multiply

$$7x^2(x + 2) = 7x^2 \cdot x + 7x^2 \cdot 2$$
$$= 7x^3 + 14x^2$$

which is the original binomial.

EXAMPLE 5 Factor: $6x^2 - 24x + 6$

Solution: The GCF of the terms is 6.

$$6x^2 - 24x + 6 = 6 \cdot x^2 - 6 \cdot 4x + 6 \cdot 1$$
$$= 6(x^2 - 4x + 1)$$

■ **Work Practice Problem 5**

PRACTICE PROBLEM 5

Factor: $4z^2 - 12z + 2$

Helpful Hint
Don't forget to include the term 1.

Helpful Hint
A common mistake in the example above is to forget to write down the term of 1. Remember to mentally check by multiplying.

$$6(x^2 - 4x) = 6x^2 - 24x \qquad \text{Not the original trinomial}$$

$$6(x^2 - 4x + 1) = 6x^2 - 24x + 6 \qquad \text{The original trinomial}$$

EXAMPLE 6 Factor: $-2a + 20b - 4b^2$

Solution:

$$-2a + 20b - 4b^2 = 2 \cdot -a + 2 \cdot 10b - 2 \cdot 2b^2$$
$$= 2(-a + 10b - 2b^2)$$

When the coefficient of the first term is a negative number, we often factor out a negative common factor.

$$-2a + 20b - 4b^2 = (-2)(a) + (-2)(-10b) + (-2)(2b^2)$$
$$= -2(a - 10b + 2b^2)$$

Both $2(-a + 10b - 2b^2)$ and $-2(a - 10b + 2b^2)$ are factorizations of $-2a + 20b - 4b^2$.

■ **Work Practice Problem 6**

✔ Concept Check Check both factorizations given in Example 6.

PRACTICE PROBLEM 6

Factor: $-3y^2 - 9y + 15x^2$

Answers
4. $5y^7(2 + y^2)$ **5.** $2(2z^2 - 6z + 1)$
6. $-3(y^2 + 3y - 5x^2)$ or
$3(-y^2 - 3y + 5x^2)$

✔ Concept Check Answer
answers may vary

Vocabulary and Readiness Check

Use the choices below to fill in each blank. Not all choices will be used.

factoring   smallest   factor

product   largest   greatest common factor (GCF)

1. In $-3 \cdot x^4 = -3x^4$, the -3 and the x^4 are each called a _____ and $-3x^4$ is called a _____.

2. The _____ of a list of integers is the largest integer that is a factor of all integers in the list.

3. The GCF of a list of variables raised to powers is the variable raised to the _____ exponent in the list.

4. _____ is the process of writing an expression as a product.

10.4 EXERCISE SET

FOR EXTRA HELP

Student Solutions Manual   PH Math/Tutor Center   CD/Video for Review   MathXL®   MyMathLab

Objective A *Find the greatest common factor of each list of numbers. See Example 1.*

1. 48 and 15

2. 36 and 20

3. 60 and 72

4. 96 and 45

5. 12, 20, and 36

6. 18, 24, and 60

7. 8, 32, and 100

8. 30, 50, and 200

Objective B *Find the greatest common factor of each list of terms. See Examples 2 and 3.*

9. y^7, y^2, y^{10}

10. x^3, x, x^5

11. a^5, a^5, a^5

12. b^6, b^6, b^4

13. x^3y^2, xy^2, x^4y^2

14. a^5b^3, a^5b^2, a^5b

15. $3x^4, 5x^7, 10x$

16. $9z^6, 4z^5, 2z^3$

17. $2z^3, 14z^5, 18z^3$

18. $6y^7, 9y^6, 15y^5$

Objective C *Factor. Check by multiplying. See Examples 4 through 6.*

19. $3y^2 + 18y$

20. $2x^2 + 18x$

21. $10a^6 - 5a^8$

22. $21y^5 + y^{10}$

23. $4x^3 + 12x^2 + 20x$

24. $9b^3 - 54b^2 + 9b$

25. $z^7 - 6z^5$

26. $y^{10} + 4y^5$

27. $-35 + 14y - 7y^2$

28. $-20x + 4x^2 - 2$

29. $12a^5 - 36a^6$

30. $25z^3 - 20z^2$

Review

Solve. See Sections 7.1–7.3.

31. Find 30% of 120.

32. Find 45% of 265.

33. Write 80% as a fraction in simplified form.

34. Write 65% as a fraction in simplified form.

35. Write $\frac{3}{8}$ as a percent.

36. Write $\frac{3}{4}$ as a percent.

Concept Extensions

△ **37.** The area of the larger rectangle below is $x(x + 2)$. Find another expression for the area by writing the sum of the areas of the smaller rectangles.

△ **38.** Write an expression for the area of the largest rectangle in two different ways.

39. In your own words, define the greatest common factor of a list of numbers.

40. Suppose that a classmate asks you why $4x^2 + 6x + 2$ does not factor as $2(2x^2 + 3x)$. Write down your response to this classmate.

41. Evaluate $(xy + z)^x$ for $x = 2$ and $z = 7$. Then multiply and simplify.

CHAPTER 10   Group Activity

Business Analysis

This activity may be completed by working in groups or individually.

Suppose you own a small business that manufactures specialized iPod covers. You need to decide how many covers to make. The more covers you make, the lower the price you will have to charge to sell them all. Naturally, each cover you make costs you money because you must buy the materials to make each cover. The following table summarizes some factors you must consider in deciding how many covers to make, along with algebraic representations of those factors.

Number of Covers	Description	Algebraic Expression
	Unknown	x
Total manufacturing expenses	This is the total amount that it will cost to manufacture all the iPod covers. It will cost $100 to buy special equipment to manufacture the covers in addition to materials costing $0.50 per cover	$100 + 0.50x$
Price charged per cover	For each additional cover produced, the price that must be charged per cover decreases from $40 by an additional $0.05	$40 - 0.05x$

1. *Revenue* is the amount of money collected from selling the iPod covers. Revenue can be found by multiplying the price charged per cover by the number of covers sold. Use the algebraic expressions given in the table above to find a polynomial that represents the revenue from sales of covers. Then write this polynomial in the Polynomial column next to "Revenue" in the table to the right.

2. *Profit* is the amount of money you make from selling the iPod covers after deducting the expenses for making the covers. Profit can be found by subtracting total manufacturing expenses from revenue. Find a polynomial that represents the profit from the sales of covers. Then write this polynomial in the Polynomial column next to "Profit" in the table to the right.

3. Complete the following table by evaluating each polynomial for each of the numbers of covers given in the table.

	Polynomial	Number of Covers, x				
		200	300	400	500	600
Revenue						
Total manufacturing expenses	$100 + 0.50x$					
Profit						

4. Study the table. Which number of covers will give you the largest profit from making and selling iPod covers?

Chapter 10 Vocabulary Check

Fill in each blank with one of the words or phrases listed below.

trinomial   monomial   greatest common factor   binomial

exponent   factoring   polynomials   FOIL

1. _____ is the process of writing an expression as a product.
2. The _____ of a list of terms is the product of all common factors.
3. The _____ method may be used when multiplying two binomials.
4. A polynomial with exactly 3 terms is called a _____.
5. A polynomial with exactly 2 terms is called a _____.
6. A polynomial with exactly 1 term is called a _____.
7. Monomials, binomials, and trinomials are all examples of _____.
8. In $5x^3$, the 3 is called an _____.

> **Helpful Hint**
>
> Are you preparing for your test? Don't forget to take the Chapter 10 Test on page 735. Then check your answers at the back of the text and use the Chapter Test Prep Video CD to see the fully worked-out solutions to any of the exercises you want to review.

10 Chapter Highlights

DEFINITIONS AND CONCEPTS	EXAMPLES

Section 10.1   Adding and Subtracting Polynomials

A **monomial** is a term that contains whole-number exponents and no variable in the denominator.	Monomial: $-2x^2y^3$
A **polynomial** is a monomial or a sum or difference of monomials.	Polynomials $$5x^2 - 6x + 2, \quad -\frac{9}{10}y, \quad 7$$
A **binomial** is a polynomial with two terms.	Binomial: $5x - y$
A **trinomial** is a polynomial with three terms.	Trinomial: $7z^3 + 0.5z + 1$
To add polynomials, combine like terms.	Add: $(7z^2 - 6z + 2) + (5z^2 - 4z + 5)$ $(7z^2 - 6z + 2) + (5z^2 - 4z + 5)$ $= \underline{7z^2 + 5z^2} \quad \underline{- 6z - 4z} \quad \underline{+ 2 + 5}$ Group like terms. $= 12z^2 - 10z + 7$ Combine like terms.
To subtract polynomials, change the signs of the terms being subtracted, then add.	Subtract: $(20x - 6) - (30x - 6)$ $(20x - 6) - (30x - 6)$ $= (20x - 6) + (-30x + 6)$ $= \underline{20x - 30x} \quad \underline{- 6 + 6}$ Group like terms. $= -10x$ Combine like terms.

DEFINITIONS AND CONCEPTS	**EXAMPLES**

Section 10.2   Multiplication Properties of Exponents

Product property for exponents

$a^m \cdot a^n = a^{m+n}$

Power property for exponents

$(a^m)^n = a^{m \cdot n}$

Power of a product property for exponents

$(ab)^n = a^n b^n$

$x^3 \cdot x^{11} = x^{3+11} = x^{14}$

$(y^5)^3 = y^{5 \cdot 3} = y^{15}$

$(2z^5)^4 = 2^4(z^5)^4 = 16z^{20}$

Section 10.3   Multiplying Polynomials

To multiply two polynomials, multiply each term of the first polynomial by each term of the second polynomial, and then combine like terms.

$(x + 2)(x^2 + 5x - 1)$

$= x(x^2 + 5x - 1) + 2(x^2 + 5x - 1)$
$= x \cdot x^2 + x \cdot 5x + x(-1) + 2 \cdot x^2 + 2 \cdot 5x + 2(-1)$
$= x^3 + 5x^2 - x + 2x^2 + 10x - 2$
$= x^3 + 7x^2 + 9x - 2$

Section 10.4   Introduction to Factoring Polynomials

TO FIND THE GREATEST COMMON FACTOR OF A LIST OF INTEGERS

Step 1. Write each number as a product of prime numbers.

Step 2. Identify the common prime factors.

Step 3. The product of all common prime factors found in *Step 2* is the greatest common factor. If there are no common prime factors, the greatest common factor is 1.

The **GCF of a list of variables** raised to powers is the variable raised to the smallest exponent in the list.

The **GCF of a list of terms** is the product of all common factors.

To factor the GCF from the terms of a polynomial, write each term as a product of the GCF and another factor, then apply the distributive property.

Find the GCF of 18 and 30.

$18 = 2 \cdot 3 \cdot 3$
$30 = 2 \cdot 3 \cdot 5$

The GCF is $2 \cdot 3$ or 6.

The GCF of x^6, x^8, and x^3 is x^3.

Find the GCF of $6y^3$, $12y$, and $4y^7$.
  The GCF of 6, 12, and 4 is 2.
  The GCF of y^3, y, and y^7 is y.
The GCF of $6y^3$, $12y$, and $4y^7$ is $2y$.

Factor $4y^6 + 6y^5$.
The GCF of $4y^6$ and $6y^5$ is $2y^5$.

$4y^6 + 6y^5 = 2y^5 \cdot 2y + 2y^5 \cdot 3$
$= 2y^5(2y + 3)$

10 CHAPTER REVIEW

(10.1) *Perform each indicated operation.*

1. $(2b + 7) + (8b - 10)$

2. $(7s - 6) + (14s - 9)$

3. $(3x + 0.2) - (4x - 2.6)$

4. $(10y - 6) - (11y + 6)$

5. $(4z^2 + 6z - 1) + (5z - 5)$

6. $(17a^3 + 11a^2 + a) + (14a^2 - a)$

7. $\left(9y^2 - y + \dfrac{1}{2}\right) - \left(20y^2 - \dfrac{1}{4}\right)$

8. Subtract $(x - 2)$ from $(x^2 - 6x + 1)$.

Find the value of each polynomial when $x = 3$.

9. $5x^2$

10. $2 - 7x$

△ **11.** Find the perimeter of the given rectangle.

$(3x + 16)$ feet

$(10x - 2)$ feet

12. Find the perimeter of a square whose side length is $(4x^2 + 1)$ meters.

(10.2) *Multiply and simplify.*

13. $x^{10} \cdot x^{14}$

14. $y \cdot y^6$

15. $4z^2 \cdot 6z^5$

16. $(-3x^2y)(5xy^4)$

17. $(a^5)^7$

18. $(x^2)^4 \cdot (x^{10})^2$

19. $(9b)^2$

20. $(a^4b^2c)^5$

21. $(7x)(2x^5)^3$

22. $(3x^6y^5)^3(2x^6y^5)^2$

△ **23.** Find the area of the square.

$(9a^7)$ miles

24. Find the area of a rectangle whose length is $3x^4$ inches and whose width is $9x$ inches.

(10.3) *Multiply.*

25. $2a(5a^2 - 6)$

26. $-3y^2(y^2 - 2y + 1)$

27. $(x + 2)(x + 6)$

28. $(3x - 1)(5x - 9)$

29. $(y - 5)^2$

30. $(7a + 1)^2$

31. $(x + 1)(x^2 - 2x + 3)$

32. $(4y^2 - 3)(2y^2 + y + 1)$

33. $(3z^2 + 2z + 1)(z^2 + z + 1)$

△ **34.** Find the area of the given rectangle.

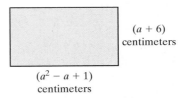

$(a + 6)$ centimeters

$(a^2 - a + 1)$ centimeters

(10.4) *Find the greatest common factor (GCF) of each list.*

35. 20 and 35

36. 12 and 32

37. 24, 30, and 60

38. 10, 20, and 25

39. x^3, x^2, x^{10}

40. y^{10}, y^7, y^7

41. xy^2, xy, x^3y^3

42. a^5b^4, a^6b^3, a^7b^2

43. $5a^3, 10a, 20a^4$

44. $12y^2z, 20y^2z, 24y^5z$

Factor out the GCF.

45. $2x^2 + 12x$

46. $6a^2 - 12a$

47. $6y^4 - y^6$

48. $7x^2 - 14x + 7$

49. $5a^7 - a^4 + a^3$

50. $10y^6 - 10y$

Mixed Review

Perform the indicated operations.

51. $(z^2 - 5z + 8) + (6z - 4)$

52. $(8y - 5) - (12y - 3)$

53. $x^5 \cdot x^{16}$

54. $y^8 \cdot y$

55. $(a^3b^5c)^6$

56. $(9x^2) \cdot (3x^2)^2$

57. $3a(4a^3 - 5)$

58. $(x + 4)(x + 5)$

59. $(3x + 4)^2$

60. $(6z + 5)(z - 2)$

Find the greatest common factor (GCF) of each list.

61. 28, 32, and 40

62. $5z^5, 12z^8, 3z^4$

Factor out the GCF.

63. $z^9 - 4z^7$

64. $x^{12} + 6x^5$

65. $15a^4 + 45a^5$

66. $16z^5 - 24z^8$

 STUDY SKILLS BUILDER

Are You Prepared for Your Final Exam?

To prepare for your final exam, try the following study techniques:

- Review the material that you will be responsible for on your exam. This includes material from your textbook, your notebook, and any handouts from your instructor.
- Review any formulas that you may need to memorize.
- Check to see if your instructor or mathematics department will be conducting a final exam review.
- Check with your instructor to see whether final exams from previous semesters/quarters are available to students for review.

- Use your previously taken exams as a practice final exam. To do so, rewrite the test questions in mixed order on blank sheets of paper. This will help you prepare for exam conditions.
- If you are unsure of a few concepts, see your instructor or visit a learning lab for assistance. Also, view the video segment of any troublesome sections.
- If you need further exercises to work, try the Cumulative Reviews at the end of the chapters.

Once again, good luck! I hope you have enjoyed this textbook and your mathematics course.

10 CHAPTER TEST

Use the Chapter Test Prep Video CD to see the fully worked-out solutions to any of the exercises you want to review.

Add or subtract as indicated.

1. $(11x - 3) + (4x - 1)$

2. $(11x - 3) - (4x - 1)$

3. $(1.3y^2 + 5y) + (2.1y^2 - 3y - 3)$

4. Subtract $(8a^2 + a)$ from $(6a^2 + 2a + 1)$.

5. Find the value of $x^2 - 6x + 1$ when $x = 8$.

Multiply and simplify.

6. $y^3 \cdot y^{11}$

7. $(y^3)^{11}$

8. $(2x^2)^4$

9. $(6a^3)(-2a^7)$

10. $(p^6)^7(p^2)^6$

11. $(3a^4b)^2(2ba^4)^3$

12. $5x(2x^2 + 1.3)$

13. $-2y(y^3 + 6y^2 - 4)$

14. $(x - 3)(x + 2)$

15. $(5x + 2)^2$

16. $(a + 2)(a^2 - 2a + 4)$

△**17.** Find the area and the perimeter of the parallelogram. (*Hint:* $A = b \cdot h$.)

$(x + 7)$ in. $2x$ in.
$(5x - 2)$ in.

Find the greatest common factor of each list.

18. 45 and 60

19. $6y^3, 9y^5, 18y^4$

Factor out the GCF.

20. $3y^2 - 15y$

21. $10a^2 + 12a$

22. $6x^2 - 12x - 30$

23. $7x^6 - 6x^4 + x^3$

Answers

1. _____

2. _____

3. _____

4. _____

5. _____

6. _____

7. _____

8. _____

9. _____

10. _____

11. _____

12. _____

13. _____

14. _____

15. _____

16. _____

17. _____

18. _____

19. _____

20. _____

21. _____

22. _____

23. _____

Answers

1. _____

2. _____

3. _____

4. _____

5. _____

6. _____

7. _____

8. _____

9. _____

10. _____

11. _____

12. _____

13. _____

14. _____

15. _____

16. _____

17. _____

18. _____

19. _____

20. _____

△ **1.** The state of Colorado is in the shape of a rectangle whose length is about 380 miles and whose width is about 280 miles. Find its area.

2. In a pecan orchard, there are 21 trees in each row and 7 rows of trees. How many pecan trees are there?

3. Add: $1 + (-10) + (-8) + 9$

4. Add: $-2 + (-7) + 3 + (-4)$

Subtract.

5. $8 - 15$

6. $4 - 7$

7. $-4 - (-5)$

8. $3 - (-2)$

9. Solve: $7x = 6x + 4$

10. Solve: $4x = -2 + 3x$

11. Solve: $17 - 7x + 3 = -3x + 21 - 3x$

12. $20 - 6x + 4 = -2x + 18 + 2x$

13. $\dfrac{2x}{15} + \dfrac{3x}{10}$

14. $\dfrac{5}{7y} - \dfrac{9}{14y}$

15. Round 736.2359 to the nearest tenth.

16. Round 328.174 to the nearest tenth.

17. Add: $23.85 + 1.604$

18. Add: $12.762 + 4.29$

19. Is -9 a solution of the equation $3.7y = -3.33$?

20. Is 6 a solution of the equation $2.8x = 16.8$?

Divide.

21. $\dfrac{786.1}{1000}$

22. $\dfrac{818}{1000}$

23. $-\dfrac{0.12}{10}$

24. $\dfrac{5.03}{100}$

25. Evaluate $-2x + 5$ for $x = 3.8$.

26. Evaluate $6x - 1$ for $x = -2.1$.

27. Write $\dfrac{22}{7}$ as a decimal. Round to the nearest hundreth.

28. Write $\dfrac{37}{19}$ as a decimal. Round to the nearest thousandth.

29. Find: $\sqrt{\dfrac{1}{36}}$

30. Find: $\sqrt{\dfrac{4}{25}}$

31. Mel Wagstaff is a 6-foot-tall park ranger who needs to know the height of a particular tree. He notices that when the shadow of the tree is 69 feet long, his own shadow is 9 feet long. Find the height of the tree.

32. Phoebe, a very intelligent dog, wants to estimate the height of a fire hydrant. She notices that when her shadow is 2 feet long, the shadow of the hydrant is 6 feet long. Find the height of the hydrant if Phoebe is 1 foot tall.

33. Translate to an equation: 1.2 is 30% of what number?

34. Translate to an equation: 9 is 45% of what number?

35. What percent of 50 is 8?

36. What percent of 16 is 4?

37. Mr. Percy, the principal at Slidell High School, counted 31 freshmen absent during a particular day. If this is 4% of the total number of freshmen, how many freshmen are there at Slidell High School?

38. Two percent of the apples in a shipment are rotten. If there are 29 rotten apples, how many apples are in the shipment?

21. _____

22. _____

23. _____

24. _____

25. _____

26. _____

27. _____

28. _____

29. _____

30. _____

31. _____

32. _____

33. _____

34. _____

35. _____

36. _____

37. _____

38. _____

39. _____

40. _____

41. _____

42. _____

43. _____

44. _____

45. _____

46. _____

47. _____

48. _____

49. _____

50. _____

39. Ivan Borski borrowed $2400 at 10% simple interest for 8 months to buy a used Toyota Corolla. Find the simple interest he paid.

40. Find the amount of simple interest earned on a $1000 CD for 10 months at an interest rate of 3%.

41. Using the circle graph shown, determine the percent of Americans that have one or more working computers at home.

42. Using the circle graph in Exercise 41, find the percent of Americans that have fewer than 3 working computers at home.

Number of Working Computers at Home

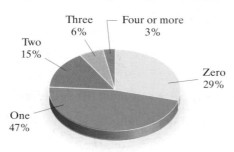

Source: UCLA Center for Communication Policy, 2003

△ **43.** Find the perimeter of a rectangle with a length of 11 inches and a width of 3 inches.

△ **44.** Find the perimeter of a triangular yard whose sides are 6 feet, 8 feet, and 11 feet.

△ **45.** Find the area of the parallelogram.

△ **46.** Find the area of the triangle.

47. Subtract 3 tons 1350 lb from 8 tons 1000 lb.

48. Multiply 5 tons 700 lb by 3.

49. Convert 3210 ml to L.

50. Convert 4321 cl to L.

A Tables

A.1 TABLE OF GEOMETRIC FIGURES

Plane Figures Have Length and Width but No Thickness or Depth		
Name	**Description**	**Figure**
Polygon	Union of three or more coplanar line segments that intersect with each other only at each endpoint, with each endpoint shared by two segments.	
Triangle	Polygon with three sides (sum of measures of three angles is 180°).	
Scalene Triangle	Triangle with no sides of equal length.	
Isosceles Triangle	Triangle with two sides of equal length.	
Equilateral Triangle	Triangle with all sides of equal length.	
Right Triangle	Triangle that contains a right angle.	leg hypotenuse leg
Quadrilateral	Polygon with four sides (sum of measures of four angles is 360°).	
Trapezoid	Quadrilateral with exactly one pair of opposite sides parallel.	base leg parallel sides leg base
Isosceles Trapezoid	Trapezoid with legs of equal length.	
Parallelogram	Quadrilateral with both pairs of opposite sides parallel.	
Rhombus	Parallelogram with all sides of equal length.	

(*Continued*)

Plane Figures Have Length and Width but No Thickness or Depth (continued)

Name	Description	Figure
Rectangle	Parallelogram with four right angles.	
Square	Rectangle with all sides of equal length.	
Circle	All points in a plane the same distance from a fixed point called the **center.**	radius, center, diameter

Solid Figures Have Length, Width, and Height or Depth

Name	Description	Figure
Rectangular Solid	A solid with six sides, all of which are rectangles.	
Cube	A rectangular solid whose six sides are squares.	
Sphere	All points the same distance from a fixed point, called the **center.**	radius, center
Right Circular Cylinder	A cylinder having two circular bases that are perpendicular to its altitude.	
Right Circular Cone	A cone with a circular base that is perpendicular to its altitude.	

Percent	Decimal	Fraction
1%	0.01	$\frac{1}{100}$
5%	0.05	$\frac{1}{20}$
10%	0.1	$\frac{1}{10}$
12.5% or $12\frac{1}{2}$%	0.125	$\frac{1}{8}$
$16.\overline{6}$% or $16\frac{2}{3}$%	$0.1\overline{6}$	$\frac{1}{6}$
20%	0.2	$\frac{1}{5}$
25%	0.25	$\frac{1}{4}$
30%	0.3	$\frac{3}{10}$
$33.\overline{3}$% or $33\frac{1}{3}$%	$0.\overline{3}$	$\frac{1}{3}$
37.5% or $37\frac{1}{2}$%	0.375	$\frac{3}{8}$
40%	0.4	$\frac{2}{5}$
50%	0.5	$\frac{1}{2}$
60%	0.6	$\frac{3}{5}$
62.5% or $62\frac{1}{2}$%	0.625	$\frac{5}{8}$
$66.\overline{6}$% or $66\frac{2}{3}$%	$0.\overline{6}$	$\frac{2}{3}$
70%	0.7	$\frac{7}{10}$
75%	0.75	$\frac{3}{4}$
80%	0.8	$\frac{4}{5}$
$83.\overline{3}$% or $83\frac{1}{3}$%	$0.8\overline{3}$	$\frac{5}{6}$
87.5% or $87\frac{1}{2}$%	0.875	$\frac{7}{8}$
90%	0.9	$\frac{9}{10}$
100%	1.0	1
110%	1.1	$1\frac{1}{10}$
125%	1.25	$1\frac{1}{4}$
$133.\overline{3}$% or $133\frac{1}{3}$%	$1.\overline{3}$	$1\frac{1}{3}$
150%	1.5	$1\frac{1}{2}$
$166.\overline{6}$% or $166\frac{2}{3}$%	$1.\overline{6}$	$1\frac{2}{3}$
175%	1.75	$1\frac{3}{4}$
200%	2.0	2

A.3 TABLE ON FINDING COMMON PERCENTS OF A NUMBER

Common Percent Equivalences*	Shortcut Method for Finding Percent	Example
$1\% = 0.01 \left(\text{or } \frac{1}{100}\right)$	To find 1% of a number, multiply by 0.01. To do so, move the decimal point 2 places to the left.	1% of 210 is 2.10 or 2.1. 1% of 1500 is 15. 1% of 8.6 is 0.086.
$10\% = 0.1 \left(\text{or } \frac{1}{10}\right)$	To find 10% of a number, multiply by 0.1, or move the decimal point of the number one place to the left.	10% of 140 is 14. 10% of 30 is 3. 10% of 17.6 is 1.76.
$25\% = \frac{1}{4}$	To find 25% of a number, find $\frac{1}{4}$ of the number, or divide the number by 4.	25% of 20 is $\frac{20}{4}$ or 5. 25% of 8 is 2. 25% of 10 is $\frac{10}{4}$ or $2\frac{1}{2}$.
$50\% = \frac{1}{2}$	To find 50% of a number, find $\frac{1}{2}$ of the number, or divide the number by 2.	50% of 64 is $\frac{64}{2}$ or 32. 50% of 1000 is 500. 50% of 9 is $\frac{9}{2}$ or $4\frac{1}{2}$.
$100\% = 1$	To find 100% of a number, multiply the number by 1. In other words, 100% of a number is the number.	100% of 98 is 98. 100% of 1407 is 1407. 100% of 18.4 is 18.4.
$200\% = 2$	To find 200% of a number, multiply the number by 2.	200% of 31 is 31 · 2 or 62. 200% of 750 is 1500. 200% of 6.5 is 13.

*See Appendix A.2.

A.4 TABLE OF SQUARES AND SQUARE ROOTS

n	n^2	$\sqrt{n}$	n	n^2	$\sqrt{n}$
1	1	1.000	51	2601	7.141
2	4	1.414	52	2704	7.211
3	9	1.732	53	2809	7.280
4	16	2.000	54	2916	7.348
5	25	2.236	55	3025	7.416
6	36	2.449	56	3136	7.483
7	49	2.646	57	3249	7.550
8	64	2.828	58	3364	7.616
9	81	3.000	59	3481	7.681
10	100	3.162	60	3600	7.746
11	121	3.317	61	3721	7.810
12	144	3.464	62	3844	7.874
13	169	3.606	63	3969	7.937
14	196	3.742	64	4096	8.000
15	225	3.873	65	4225	8.062
16	256	4.000	66	4356	8.124
17	289	4.123	67	4489	8.185
18	324	4.243	68	4624	8.246
19	361	4.359	69	4761	8.307
20	400	4.472	70	4900	8.367
21	441	4.583	71	5041	8.426
22	484	4.690	72	5184	8.485
23	529	4.796	73	5329	8.544
24	576	4.899	74	5476	8.602
25	625	5.000	75	5625	8.660
26	676	5.099	76	5776	8.718
27	729	5.196	77	5929	8.775
28	784	5.292	78	6084	8.832
29	841	5.385	79	6241	8.888
30	900	5.477	80	6400	8.944
31	961	5.568	81	6561	9.000
32	1024	5.657	82	6724	9.055
33	1089	5.745	83	6889	9.110
34	1156	5.831	84	7056	9.165
35	1225	5.916	85	7225	9.220
36	1296	6.000	86	7396	9.274
37	1369	6.083	87	7569	9.327
38	1444	6.164	88	7744	9.381
39	1521	6.245	89	7921	9.434
40	1600	6.325	90	8100	9.487
41	1681	6.403	91	8281	9.539
42	1764	6.481	92	8464	9.592
43	1849	6.557	93	8649	9.644
44	1936	6.633	94	8836	9.695
45	2025	6.708	95	9025	9.747
46	2116	6.782	96	9216	9.798
47	2209	6.856	97	9409	9.849
48	2304	6.928	98	9604	9.899
49	2401	7.000	99	9801	9.950
50	2500	7.071	100	10,000	10.000

B The Bigger Picture

To see the sections that formed this outline, see Sections 1.6, 1.7, 2.4, 3.3, 4.3, 4.7, 5.4, 6.2, 6.4, 7.2, and 7.3.

I. Operations on Sets of Numbers

A. Whole Numbers

1. Add or Subtract:

$$\begin{array}{r} 14 \\ +39 \\ \hline 53 \end{array} \qquad \begin{array}{r} 300 \\ -27 \\ \hline 273 \end{array}$$

2. Multiply or Divide:

$$\begin{array}{r} 238 \\ \times 47 \\ \hline 1666 \\ 9520 \\ \hline 11{,}186 \end{array} \qquad \begin{array}{r} 127 \text{ R2} \\ 7{\overline{)891}} \\ -7 \\ \hline 19 \\ -14 \\ \hline 51 \\ -49 \\ \hline 2 \end{array}$$

3. Exponent:

4 factors of 3

$$3^4 = \overbrace{3 \cdot 3 \cdot 3 \cdot 3} = 81$$

4. Order of Operations:

$24 \div 3 \cdot 2 - (2 + 8) = 24 \div 3 \cdot 2 - (10)$ Simplify within parentheses.

$\qquad\qquad\qquad\qquad = 8 \cdot 2 - 10$ Multiply or divide from left to right.

$\qquad\qquad\qquad\qquad = 16 - 10$ Multiply or divide from left to right.

$\qquad\qquad\qquad\qquad = 6$ Add or subtract from left to right.

5. Square Root:

$\sqrt{25} = 5$ *because* $5 \cdot 5 = 25$ and 5 is a positive number.

B. Integers

1. Add: $-5 + (-2) = -7$ Adding like signs
Add absolute value. Attach the common sign.

$\qquad\qquad -5 + 2 = -3$ Adding unlike signs
Subtract absolute values. Attach the sign of the number with the larger absolute value.

2. Subtract: Add the first number to the opposite of the second number.

$7 - 10 = 7 + (-10) = -3$

3. Multiply or Divide: Multiply or divide as usual. If the signs of the two numbers are the same, the answer is positive. If the signs of the two numbers are different, the answer is negative.

$-5 \cdot 5 = -25, \quad \dfrac{-32}{-8} = 4$

C. Fractions

1. Simplify: Factor the numerator and denominator. Then divide out factors of 1 by dividing out common factors in the numerator and denominator.

Simplify: $\dfrac{20}{28} = \dfrac{4 \cdot 5}{4 \cdot 7} = \dfrac{5}{7}$

2. **Multiply:** Numerator times numerator over denominator times denominator.

$$\frac{5}{9} \cdot \frac{2}{7} = \frac{10}{63}$$

3. **Divide:** First fraction times the reciprocal of the second fraction.

$$\frac{2}{11} \div \frac{3}{4} = \frac{2}{11} \cdot \frac{4}{3} = \frac{8}{33}$$

4. **Add or Subtract:** Must have same denominators. If not, find the LCD, and write each fraction as an equivalent fraction with the LCD as denominator.

$$\frac{2}{5} + \frac{1}{15} = \frac{2}{5} \cdot \frac{3}{3} + \frac{1}{15} = \frac{6}{15} + \frac{1}{15} = \frac{7}{15}$$

D. Decimals

1. **Add or Subtract:** Line up decimal points.

$$\begin{array}{r} 1.27 \\ + \ 0.6 \\ \hline 1.87 \end{array}$$

2. **Multiply:**

$$\begin{array}{r} 2.56 \quad \text{2 decimal places} \\ \times \ 3.2 \quad \text{1 decimal place} \\ \hline 512 \quad {}^{2 + 1 = 3} \\ 7680 \\ \hline 8.192 \quad \text{3 decimal places} \end{array}$$

3. **Divide:** $8\overline{)5.6}^{\,0.7} \qquad 0.6\overline{)0.786}^{\,1.31}$

II. Solving Equations

A. Equations in General: Simplify both sides of the equation by removing parentheses and adding any like terms. Then use the addition property to write variable terms on one side, constants (or numbers) on the other side. Then use the multiplication property to solve for the variable by dividing both sides of the equation by the coefficient of the variable.

Solve: $2(x - 5) = 80$

$2x - 10 = 80$	Use the distributive property.
$2x - 10 + 10 = 80 + 10$	Add 10 to both sides.
$2x = 90$	Simplify.
$\dfrac{2x}{2} = \dfrac{90}{2}$	Divide both sides by 2.
$x = 45$	Simplify.

B. Proportions: Set cross products equal to each other. Then solve.

$$\frac{14}{3} = \frac{2}{n}, \ or \ 14 \cdot n = 3 \cdot 2, \ or \ 14 \cdot n = 6, \ or \ n = \frac{6}{14} = \frac{3}{7}$$

C. Percent Problems

1. **Solved by Equations:** Remember that "of" means multiplication and "is" means equals.

   "12% of some number is 6" translates to

   $$12\% \cdot n = 6 \ or \ 0.12 \cdot n = 6 \ or \ n = \frac{6}{0.12} \ or \ n = 50$$

2. **Solved by Proportions:** Remember that percent, p, is identified by % or percent, base, b, usually appears after "of" and amount, a, is the part compared to the whole.

   "12% of some number is 6" translates to

   $$\frac{6}{b} = \frac{12}{100} \ or \ 6 \cdot 100 = b \cdot 12 \ or \ \frac{600}{12} = b \ or \ 50 = b$$

C Quotient Rule and Negative Exponents

Objective **A** Using the Quotient Rule and Defining the Zero Exponent

Let's study pattern for simplifying exponential expressions that involves quotients.

$$\frac{x^5}{x^3} = \frac{x \cdot x \cdot x \cdot x \cdot x}{x \cdot x \cdot x}$$

$$= \frac{\boxed{x} \cdot \boxed{x} \cdot \boxed{x} \cdot x \cdot x}{\boxed{x} \cdot \boxed{x} \cdot \boxed{x}}$$

$$= 1 \cdot 1 \cdot 1 \cdot x \cdot x$$

$$= x \cdot x$$

$$= x^2$$

Notice that the result is exactly the same if we subtract exponents of the common bases.

$$\frac{x^5}{x^3} = x^{5-3} = x^2$$

The following rule states this result in a general way.

Quotient Rule for Exponents

If m and n are positive integers and a is a real number, then

$$\frac{a^m}{a^n} = a^{m-n}, \quad a \neq 0$$

For example,

$$\frac{x^6}{x^2} = x^{6-2} = x^4, \quad x \neq 0$$

In other words, to divide one exponential expression by another with a common base, we keep the base and subtract the exponents.

PRACTICE PROBLEMS 1–3

Simplify each quotient.

1. $\dfrac{y^{10}}{y^6}$ **2.** $\dfrac{5^{11}}{5^8}$ **3.** $\dfrac{12a^4b^{11}}{ab}$

EXAMPLES Simplify each quotient.

1. $\dfrac{x^5}{x^2} = x^{5-2} = x^3$ Use the quotient rule.

2. $\dfrac{4^7}{4^3} = 4^{7-3} = 4^4 = 256$ Use the quotient rule.

3. $\dfrac{2x^5y^2}{xy} = 2 \cdot \dfrac{x^5}{x^1} \cdot \dfrac{y^2}{y^1}$

$\qquad = 2 \cdot (x^{5-1}) \cdot (y^{2-1})$ Use the quotient rule.

$\qquad = 2x^4y^1 \quad \text{or} \quad 2x^4y$

Work Practice Problems 1–3

Answers

1. y^4 **2.** 125 **3.** $12a^3b^{10}$

Let's now give meaning to an expression such as x^0. To do so, we will simplify $\dfrac{x^3}{x^3}$ in two ways and compare the results.

$$\dfrac{x^3}{x^3} = x^{3-3} = x^0 \qquad \text{Apply the quotient rule.}$$

$$\dfrac{x^3}{x^3} = \dfrac{\boxed{x} \cdot \boxed{x} \cdot \boxed{x}}{\boxed{x} \cdot \boxed{x} \cdot \boxed{x}} = 1 \qquad \text{Apply the fundamental principle for fractions.}$$

Since $\dfrac{x^3}{x^3} = x^0$ and $\dfrac{x^3}{x^3} = 1$, we define that $x^0 = 1$ as long as x is not 0.

Zero Exponent

$a^0 = 1$, as long as a is not 0.   For example, $5^0 = 1$.

In other words, a base raised to the 0 power is 1, as long as the base is not 0.

EXAMPLES   Simplify each expression.

4. $3^0 = 1$

5. $(-4)^0 = 1$

6. $-4^0 = -1 \cdot 4^0 = -1 \cdot 1 = -1$

7. $5x^0 = 5 \cdot x^0 = 5 \cdot 1 = 5$

■ **Work Practice Problems 4–7**

PRACTICE PROBLEMS 4–7

Simplify each expression.
4. 6^0 **5.** $(-8)^0$ **6.** -8^0
7. $7y^0$

Objective **B** **Simplifying Expressions Containing Negative Exponents**

Our work with exponential expressions so far has been limited to exponents that are positive integers or 0. Here we will also give meaning to an expression like x^{-3}.

Suppose that we wish to simplify the expression $\dfrac{x^2}{x^5}$. If we use the quotient rule for exponents, we subtract exponents:

$$\dfrac{x^2}{x^5} = x^{2-5} = x^{-3}, \quad x \neq 0$$

But what does x^{-3} mean? Let's simplify $\dfrac{x^2}{x^5}$ using the definition of a^n.

$$\dfrac{x^2}{x^5} = \dfrac{x \cdot x}{x \cdot x \cdot x \cdot x \cdot x}$$

$$= \dfrac{\boxed{x} \cdot \boxed{x}}{\boxed{x} \cdot \boxed{x} \cdot x \cdot x \cdot x} \qquad \begin{array}{l}\text{Divide numerator and denominator by common factors by}\\ \text{applying the fundamental principle for fractions.}\end{array}$$

$$= \dfrac{1}{x^3}$$

If the quotient rule is to hold true for negative exponents, then x^{-3} must equal $\dfrac{1}{x^3}$.

Answers
4. 1 **5.** 1 **6.** -1 **7.** 7

From this example, we state the definition for negative exponents.

Negative Exponents

If a is a real number other than 0 and n is an integer, then

$$a^{-n} = \frac{1}{a^n}$$

For example,

$$x^{-3} = \frac{1}{x^3}$$

PRACTICE PROBLEMS 8–10

Simplify by writing each expression with positive exponents only.

8. 5^{-2} **9.** $5x^{-2}$ **10.** $4^{-1} + 3^{-1}$

> **Helpful Hint**
> Don't forget that since there are no parentheses, only x is the base for the exponent -3.

EXAMPLES   Simplify by writing each expression with positive exponents only.

8. $3^{-2} = \frac{1}{3^2} = \frac{1}{9}$ Use the definition of negative exponents.

9. $2x^{-3} = 2^1 \cdot \frac{1}{x^3} = \frac{2^1}{x^3}$ or $\frac{2}{x^3}$ Use the definition of negative exponents.

10. $2^{-1} + 4^{-1} = \frac{1}{2} + \frac{1}{4} = \frac{2}{4} + \frac{1}{4} = \frac{3}{4}$

■ **Work Practice Problems 8–10**

> **Helpful Hint**
> A negative exponent *does not affect* the sign of its base.
> Remember: Another way to write a^{-n} is to take its reciprocal and change the sign of its exponent: $a^{-n} = \frac{1}{a^n}$. For example,
>
> $$x^{-2} = \frac{1}{x^2}, \qquad 2^{-3} = \frac{1}{2^3} \text{ or } \frac{1}{8}$$
> $$\frac{1}{y^{-4}} = \frac{1}{\frac{1}{y^4}} = y^4, \qquad \frac{1}{5^{-2}} = 5^2 \text{ or } 25$$

From the preceding Helpful Hint, we know that $x^{-2} = \frac{1}{x^2}$ and $\frac{1}{y^{-4}} = y^4$. We can use this to include another statement in our definition of negative exponents.

Negative Exponents

If a is a real number other than 0 and n is an integer, then

$$a^{-n} = \frac{1}{a^n} \quad \text{and} \quad \frac{1}{a^{-n}} = a^n$$

Answers

8. $\frac{1}{25}$ **9.** $\frac{5}{x^2}$ **10.** $\frac{7}{12}$

EXAMPLES Simplify each expression. Write each result using positive exponents only.

11. $\left(\dfrac{2}{x}\right)^{-3} = \dfrac{2^{-3}}{x^{-3}} = \dfrac{2^{-3}}{1} \cdot \dfrac{1}{x^{-3}} = \dfrac{1}{2^3} \cdot \dfrac{x^3}{1} = \dfrac{x^3}{2^3} = \dfrac{x^3}{8}$ Use the negative exponents rule.

12. $\dfrac{y}{y^{-2}} = \dfrac{y^1}{y^{-2}} = y^{1-(-2)} = y^3$ Use the quotient rule.

13. $\dfrac{x^{-5}}{x^7} = x^{-5-7} = x^{-12} = \dfrac{1}{x^{12}}$

🔲 **Work Practice Problems 11–13**

Objective **C** Simplifying Exponential Expressions

All the previously stated rules for exponents apply for negative exponents also. Here is a summary of the rules and definitions for exponents. Notice that there is a power of a quotient rule.

Summary of Exponent Rules

If m and n are integers and a, b, and c are real numbers, then

Product rule for exponents:	$a^m \cdot a^n = a^{m+n}$
Power rule for exponents:	$(a^m)^n = a^{m \cdot n}$
Power of a product:	$(ab)^n = a^n b^n$
Power of a quotient:	$\left(\dfrac{a}{c}\right)^n = \dfrac{a^n}{c^n}, \quad c \neq 0$
Quotient rule for exponents:	$\dfrac{a^m}{a^n} = a^{m-n}, \quad a \neq 0$
Zero exponent:	$a^0 = 1, \quad a \neq 0$
Negative exponent:	$a^{-n} = \dfrac{1}{a^n}, \quad a \neq 0$

EXAMPLES Simplify each expression. Write each result using positive exponents only.

14. $x^{-3} \cdot x^2 \cdot x^{-7} = x^{-3+2} \cdot x^{-7}$ Use the product rule.
$= x^{-1} \cdot x^{-7}$
$= x^{-1+(-7)}$ Use the product rule.
$= x^{-8}$
$= \dfrac{1}{x^8}$ Use the definition of negative exponent.

15. $(x^4 y^{-9})(x^{-1} y^{11}) = x^{4+(-1)} \cdot y^{-9+11}$ Use the product rule.
$= x^3 y^2$ Simplify.

16. $(4m^2 n^5)(5m^6 n^{-8}) = 4 \cdot 5 \cdot m^{2+6} \cdot n^{5+(-8)}$ Use the product rule.
$= 20 \cdot m^8 \cdot n^{-3}$ Simplify.
$= \dfrac{20m^8}{n^3}$ Use the definition of a negative exponent.

🔲 **Work Practice Problems 14–16**

PRACTICE PROBLEMS 11–13

Simplify each expression. Write each result using positive exponents only.

11. $\left(\dfrac{6}{7}\right)^{-2}$ **12.** $\dfrac{x}{x^{-4}}$ **13.** $\dfrac{y^{-4}}{y^6}$

PRACTICE PROBLEMS 14–16

Simplify each expression. Write each result using positive exponents only.

14. $y^{-6} \cdot y^3 \cdot y^{-4}$
15. $(a^6 b^{-4})(a^{-3} b^8)$
16. $(3y^9 z^{10})(2y^3 z^{-12})$

Answers

11. $\dfrac{49}{36}$ **12.** x^5 **13.** $\dfrac{1}{y^{10}}$

14. $\dfrac{1}{y^7}$ **15.** $a^3 b^4$ **16.** $\dfrac{6y^{12}}{z^2}$

Student Solutions Manual PH Math/Tutor Center CD/Video for Review MathXL® MyMathLab

Objective A *Use the quotient rule and simplify each expression. See Examples 1 through 3.*

1. $\dfrac{x^3}{x}$

2. $\dfrac{y^{10}}{y^9}$

3. $\dfrac{9^8}{9^6}$

4. $\dfrac{5^7}{5^4}$

5. $\dfrac{p^7 q^{20}}{pq^{15}}$

6. $\dfrac{x^8 y^6}{xy^5}$

7. $\dfrac{7x^2 y^6}{14x^2 y^3}$

8. $\dfrac{9a^4 b^7}{27ab^2}$

Simplify each expression. See Examples 4 through 7.

9. 7^0

10. 23^0

11. $2x^0$

12. $4y^0$

13. -7^0

14. -2^0

15. $(-7)^0$

16. $(-2)^0$

Objective B *Simplify each expression. Write each result using positive exponents only. See Examples 8 through 13.*

17. 4^{-3}

18. 6^{-2}

19. $7x^{-3}$

20. $5y^{-4}$

21. $3^{-1} + 2^{-1}$

22. $4^{-1} + 4^{-2}$

23. $\dfrac{1}{p^{-3}}$

24. $\dfrac{1}{q^{-5}}$

25. $\dfrac{x^{-2}}{x}$

26. $\dfrac{y}{y^{-3}}$

27. $\dfrac{z^{-4}}{z^{-7}}$

28. $\dfrac{x^{-4}}{x^{-1}}$

29. $3^{-2} + 3^{-1}$

30. $4^{-2} - 4^{-3}$

31. $\left(\dfrac{5}{y}\right)^{-2}$

32. $\left(\dfrac{3}{x}\right)^{-3}$

33. $\dfrac{1}{p^{-4}}$

34. $\dfrac{1}{y^{-6}}$

Objective C *Simplify each expression. Write each result using positive exponents only. See Examples 14 through 16.*

35. $a^2 \cdot a^{-9} \cdot a^{13}$

36. $z^4 \cdot z^{-5} \cdot z^3$

37. $(x^8 y^{-6})(x^{-2} y^{12})$

38. $(a^{-20} b^8)(a^{22} b^{-4})$

39. $x^{-7} \cdot x^{-8} \cdot x^4$

40. $y^{-6} \cdot y^{-3} \cdot y^2$

41. $(5x^{-7})(3x^4)$

42. $(4x^9)(6x^{-13})$

43. $y^5 \cdot y^{-7} \cdot y^{-10}$

44. $x^8 \cdot x^{-11} \cdot x^{-2}$

45. $(8m^5 n^{-1})(7m^2 n^{-4})$

46. $(2x^{10} y^{-3})(9x^4 y^{-7})$

Objectives A B C Mixed Practice *Simplify each expression. Write each result using positive exponents only. See Examples 1 through 16.*

47. $\dfrac{x^{15}}{x^8}$

48. $\dfrac{y^{19}}{y^{10}}$

49. $\dfrac{a^9 b^{14}}{ab}$

50. $\dfrac{x^{11} y^7}{xy}$

51. $\dfrac{x^3}{x^9}$

52. $\dfrac{z^4}{z^{12}}$

53. $3z^0$

54. $5y^0$

55. 5^{-3}

56. 7^{-2}

57. $8x^{-9}$

58. $9y^{-7}$

59. $5^{-1} + 10^{-1}$

60. $7^{-1} + 14^{-1}$

61. $\dfrac{z^{-8}}{z^{-1}}$

62. $\dfrac{r^{-15}}{r^{-4}}$

63. $x^{-7} \cdot x^5 \cdot x^{-7}$

64. $y^{-9} \cdot y^6 \cdot y^{-9}$

65. $(a^{-2} b^3)(a^{10} b^{-11})$

66. $(x^{-4} y^5)(x^{13} y^{-14})$

67. $(3x^{20} y^{-1})(10x^{-11} y^{-5})$

68. $(4m^{16} n^{-3})(11m^{-6} n^{-3})$

Appendix

D Scientific Notation

Objective A Writing Numbers in Scientific Notation

Both very large and very small numbers frequently occur in many fields of science. For example, the distance between the Sun and the dwarf planet Pluto is approximately 5,906,000,000 kilometers, and the mass of a proton is approximately 0.0000000000000000000000165 gram. It can be tedious to write these numbers in this standard decimal notation, so **scientific notation** is used as a convenient shorthand for expressing very large and very small numbers.

Objectives

A Write Numbers in Scientific Notation.

B Convert Numbers in Scientific Notation to Standard Form.

C Perform Operations on Numbers Written in Scientific Notation.

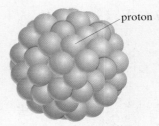

5,906,000,000 kilometers

Pluto

proton

Mass of proton is approximately 0.0000000000000000000000165 gram

---

Scientific Notation

A positive number is written in scientific notation if it is written as the product of a number a, where $1 \leq a < 10$, and an integer power r of 10: $a \times 10^r$.

The following numbers are written in scientific notation. The $\times$ sign for multiplication is used as part of the notation.

(Mass of a proton) 1.65×10^{-24} 5.906×10^9 (Distance between the Sun and Pluto)

The following steps are useful when writing numbers in scientific notation.

To Write a Number in Scientific Notation

Step 1: Move the decimal point in the original number so that the new number has a value between 1 and 10.

Step 2: Count the number of decimal places the decimal point is moved in Step 1. If the original number is 10 or greater, the count is positive. If the original number is less than 1, the count is negative.

Step 3: Multiply the new number in Step 1 by 10 raised to an exponent equal to the count found in Step 2.

PRACTICE PROBLEM 1

Write each number in scientific notation.

a. 760,000   **b.** 0.00035

EXAMPLE 1   Write each number in scientific notation.

a. 367,000,000   **b.** 0.000003

Solution: **a.** Move each decimal point until the number is between 1 and 10.

$$367,000,000. = 3.67 \times 10^8$$
8 places

The original number is 10 or greater, so the exponent is positive.

b. $0.000003 = 3.0 \times 10^{-6}$
6 places

The original number is less than 1, so the exponent is negative.

💻 **Work Practice Problem 1**

Objective B Converting Numbers to Standard Form

A number written in scientific notation can be rewritten in standard form. For example, to write 8.63×10^3 in standard form, recall that $10^3 = 1000$.

$$8.63 \times 10^3 = 8.63(1000) = 8630$$

Notice that the exponent on the 10 is positive 3, and we moved the decimal point 3 places to the right.

To write 7.29×10^{-3} in standard form, recall that $10^{-3} = \dfrac{1}{10^3} = \dfrac{1}{1000}$.

$$7.29 \times 10^{-3} = 7.29\left(\frac{1}{1000}\right) = \frac{7.29}{1000} = 0.00729$$

The exponent on the 10 is negative 3, and we moved the decimal to the left 3 places.

In general, **to write a scientific notation number in standard form,** move the decimal point the same number of places as the exponent on 10. If the exponent is positive, move the decimal point to the right; if the exponent is negative, move the decimal point to the left.

PRACTICE PROBLEM 2

Write the numbers in standard notation, without exponents.

a. 9.062×10^{-4}

b. 8.002×10^6

EXAMPLE 2   Write each number in standard notation, without exponents.

a. 1.02×10^5 **b.** 7.358×10^{-3}

Solution:

a. Move the decimal point 5 places to the right.

$$1.02 \times 10^5 = 102,000.$$

b. Move the decimal point 3 places to the left.

$$7.358 \times 10^{-3} = 0.007358$$

💻 **Work Practice Problem 2**

Objective C Performing Operations on Numbers Written in Scientific Notation

Performing operations on numbers written in scientific notation makes use of the rules and definitions for exponents.

Answers

1. a. 7.6×10^5 **b.** 3.5×10^{-4}

2. a. 0.0009062   **b.** 8,002,000

EXAMPLE 3 Perform each indicated operation. Write each result in standard decimal notation.

a. $(8 \times 10^{-6})(7 \times 10^{3})$

b. $\dfrac{12 \times 10^{2}}{6 \times 10^{-3}}$

Solution:

a. $(8 \times 10^{-6})(7 \times 10^{3}) = 8 \cdot 7 \cdot 10^{-6} \cdot 10^{3}$
$$= 56 \times 10^{-3}$$
$$= 0.056$$

b. $\dfrac{12 \times 10^{2}}{6 \times 10^{-3}} = \dfrac{12}{6} \times 10^{2-(-3)} = 2 \times 10^{5} = 200{,}000$

▨ **Work Practice Problem 3**

PRACTICE PROBLEM 3

Perform each indicated operation. Write each result in standard decimal notation.

a. $(8 \times 10^{7})(3 \times 10^{-9})$

b. $\dfrac{8 \times 10^{4}}{2 \times 10^{-3}}$

Answers
3. a. 0.24　**b.** 40,000,000,

Objective A *Write each number in scientific notation. See Example 1.*

1. 78,000

2. 9,300,000,000

3. 0.00000167

4. 0.00000017

5. 0.00635

6. 0.00194

7. 1,160,000

8. 700,000

9. At this writing, the world's largest optical telescopes are the twin Keck Telescopes located near the summit of Mauna Kea in Hawaii. The elevation of the Keck Telescopes is about 13,600 feet above sea level. Write 13,600 in scientific notation. (*Source:* W. M. Keck Observatory)

10. After more than 30 years, the *Pioneer 10* spacecraft sent its last signal to Earth. Launched on March 2, 1972, it became the first spacecraft to leave our solar system. When it transmitted its last signal, in January 2003, it was approximately 8,000,000,000 miles from Earth. Write 8,000,000,000 in scientific notation. (*Source:* NASA Ames Research Center)

Objective B *Write each number in standard notation. See Example 2.*

11. 8.673×10^{-10}

12. 9.056×10^{-4}

13. 3.3×10^{-2}

14. 4.8×10^{-6}

15. 2.032×10^{4}

16. 9.07×10^{10}

17. Each second, the Sun converts 7.0×10^{8} tons of hydrogen into helium and energy in the form of gamma rays. Write this number in standard notation. (*Source:* Students for the Exploration and Development of Space)

18. In chemistry, Avogadro's number is the number of atoms in one mole of an element. Avogadro's number is $6.02214199 \times 10^{23}$. Write this number in standard notation. (*Source:* National Institute of Standards and Technology)

Objectives C D Mixed Practice *See Examples 1 and 2. Below are some interesting facts about the Internet. If a number is written in standard form, write it in scientific notation. If a number is written in scientific notation, write it in standard form.*

19. The total number of Internet users is 940,000,000.

20. The total number of Internet hosts (sites) is 233,000,000.

21. In a recent year, the revenue generated by the Internet was 1.23×10^{12} dollars.

22. The estimated number of e-mail boxes is 1.2×10^{9}.

23. Junk e-mail (SPAM) costs consumers and businesses an estimated $23,000,000,000.

24. The weather channel has the most visited news/information Web sites with 30,135,000 visitors in a month.

Objective C *Evaluate each expression using exponential rules. Write each result in standard notation. See Example 3.*

25. $(1.2 \times 10^{-3})(3 \times 10^{-2})$

26. $(2.5 \times 10^{6})(2 \times 10^{-6})$

27. $(4 \times 10^{-10})(7 \times 10^{-9})$

28. $(5 \times 10^{6})(4 \times 10^{-8})$

29. $\dfrac{8 \times 10^{-1}}{16 \times 10^{5}}$

30. $\dfrac{25 \times 10^{-4}}{5 \times 10^{-9}}$

31. $\dfrac{1.4 \times 10^{-2}}{7 \times 10^{-8}}$

32. $\dfrac{0.4 \times 10^{5}}{0.2 \times 10^{11}}$

33. Although the actual amount varies by season and time of day, the average volume of water that flows over Niagara Falls (the American and Canadian falls combined) each second is 7.5×10^{5} gallons. How much water flows over Niagara Falls in an hour? Write the result in scientific notation. (*Hint:* 1 hour equals 3600 seconds) (*Source:* niagarafallslive.com)

34. A beam of light travels 9.460×10^{12} kilometers per year. How far does light travel in 10,000 years? Write the result in scientific notation.

Appendix

E Geometric Formulas

Rectangle

Perimeter: $P = 2l + 2w$
Area: $A = lw$

Square

Perimeter: $P = 4s$
Area: $A = s^2$

Triangle

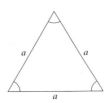

Perimeter: $P = a + b + c$
Area: $A = \frac{1}{2}bh$

Sum of Angles of Triangle

$A + B + C = 180°$
The sum of the measures of the three angles is 180°.

Pythagorean Theorem (for right triangles)

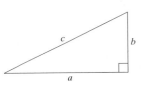

$a^2 + b^2 = c^2$
One 90° (right) angle

Isosceles Triangle

Triangle has:
two equal sides and
two equal angles.

Equilateral Triangle

Triangle has:
three equal sides and
three equal angles.
Measure of each angle is 60°.

Trapezoid

Perimeter:
$P = a + b + c + B$
Area: $A = \frac{1}{2}h(B + b)$

Parallelogram

Perimeter: $P = 2a + 2b$
Area: $A = bh$

Circle

Circumference: $C = \pi d$
$C = 2\pi r$
Area: $A = \pi r^2$

Rectangular Solid

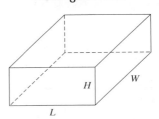

Volume: $V = LWH$
Surface Area:
$S = 2LW + 2HL + 2HW$

Cube

Volume: $V = s^3$
Surface Area: $S = 6s^2$

Cone

Volume: $V = \frac{1}{3}\pi r^2 h$

Right Circular Cylinder

Volume: $V = \pi r^2 h$
Surface Area:
$S = 2\pi r^2 + 2\pi rh$

Sphere

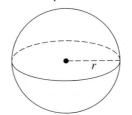

Volume: $V = \frac{4}{3}\pi r^3$
Surface Area: $S = 4\pi r^2$

Square-Based Pyramid

Volume: $V = \frac{1}{3} \cdot s^2 \cdot h$

ANSWERS TO SELECTED EXERCISES

CHAPTER 1   The Whole Numbers

Vocabulary and Readiness Check   **1.** whole   **3.** words   **5.** period

Exercise Set 1.2   **1.** tens   **3.** thousands   **5.** hundred-thousands   **7.** millions   **9.** three hundred fifty-four
11. eight thousand, two hundred seventy-nine   **13.** twenty-six thousand, nine hundred ninety
15. two million, three hundred eighty-eight thousand   **17.** twenty-four million, three hundred fifty thousand, one hundred eighty-five
19. sixty-five thousand, seven hundred seventy-three   **21.** one thousand, six hundred seventy-nine   **23.** two million, eight hundred thousand
25. eleven thousand, two hundred thirty-nine   **27.** two hundred two thousand, seven hundred   **29.** 6587   **31.** 59,800   **33.** 13,601,011
35. 7,000,017   **37.** 260,997   **39.** 353   **41.** 484,235   **43.** 54,500,000   **45.** 714   **47.** $200 + 9$ **49.** $3000 + 400 + 70$
51. $80,000 + 700 + 70 + 4$ **53.** $60,000 + 6000 + 40 + 9$ **55.** $30,000,000 + 9,000,000 + 600,000 + 80,000$
57. 5532: five thousand, five hundred thirty-two   **59.** $5000 + 400 + 90 + 2$ **61.** Mt. Washington   **63.** Chihuahua
65. Labrador retriever; one hundred forty-six thousand, seven hundred fourteen   **67.** 25 pounds   **69.** 9861   **71.** no; one hundred five
73. answers may vary   **75.** 135,500,000,000,000

Calculator Explorations   **1.** 134   **3.** 340   **5.** 2834   **7.** 770   **9.** 109   **11.** 8978

Vocabulary and Readiness Check   **1.** number   **3.** 0   **5.** minuend, subtrahend, difference   **7.** order, commutative

Exercise Set 1.3   **1.** 36   **3.** 292   **5.** 49   **7.** 5399   **9.** 209,078   **11.** 25   **13.** 212   **15.** 11,926   **17.** 16,717   **19.** 35,901
21. 632,389   **23.** 25   **25.** 600   **27.** 288   **29.** 168   **31.** 5723   **33.** 504   **35.** 79   **37.** 32,711   **39.** 5041   **41.** 31,213
43. 1034   **45.** 9   **47.** 8518   **49.** 22,876   **51.** 25 ft   **53.** 24 in.   **55.** 29 in.   **57.** 44 m   **59.** 2093   **61.** 266   **63.** 20
65. 544   **67.** 72   **69.** 88   **71.** 2028 thousand   **73.** $619   **75.** 264,000 sq mi   **77.** 283,000 sq mi   **79.** 340 ft   **81.** 264 pages
83. 813,109   **85.** 100 dB   **87.** 58 dB   **89.** 9042   **91.** 124 ft   **93.** California   **95.** 421 stores   **97.** Florida and Georgia
99. 5894 mi   **101.** minuend: 48; subtrahend: 1   **103.** minuend: 70; subtrahend: 7   **105.** answers may vary   **107.** correct
109. incorrect; 530   **111.** incorrect; 685   **113.** correct   **115.** $\begin{array}{r} 52\underline{6}9 \\ -\ 2385 \\ \hline 28\underline{8}4 \end{array}$ **117.** answers may vary   **119.** no; 1089 more pages

Vocabulary and Readiness Check   **1.** graph   **3.** 70; 60

Exercise Set 1.4   **1.** 420   **3.** 640   **5.** 2800   **7.** 500   **9.** 21,000   **11.** 34,000   **13.** 328,500   **15.** 36,000   **17.** 39,990
19. 30,000,000   **21.** 5280; 5300; 5000   **23.** 9440; 9400; 9000   **25.** 14,880; 14,900; 15,000   **27.** 19,000   **29.** 38,000 points
31. $68,000,000,000   **33.** $3,000,000   **35.** US 208,000,000; India 69,000,000   **37.** 130   **39.** 80   **41.** 5700
43. 300   **45.** 11,400   **47.** incorrect   **49.** correct   **51.** correct   **53.** $3400   **55.** 900 mi   **57.** 6000 ft   **59.** 400,000 people
61. 3,000 children   **63.** $339,000,000; $340,000,000; $300,000,000   **65.** $179,000,000; $180,000,000; $200,000,000   **67.** 5723, for example
69. a. 8550   **b.** 8649   **71.** answers may vary   **73.** 140 m

Calculator Explorations   **1.** 3456   **3.** 15,322   **5.** 272,291

Vocabulary and Readiness Check   **1.** 0   **3.** product; factor   **5.** grouping; associative   **7.** length

Exercise Set 1.5   **1.** 24   **3.** 0   **5.** 0   **7.** 87   **9.** $6 \cdot 3 + 6 \cdot 8$ **11.** $4 \cdot 3 + 4 \cdot 9$ **13.** $20 \cdot 14 + 20 \cdot 6$ **15.** 512   **17.** 3678
19. 1662   **21.** 6444   **23.** 1157   **25.** 24,418   **27.** 24,786   **29.** 15,600   **31.** 0   **33.** 6400   **35.** 48,126   **37.** 142,506
39. 2,369,826   **41.** 64,790   **43.** 3,949,935   **45.** 63 sq m   **47.** 680 sq ft   **49.** 240,000   **51.** 300,000   **53.** c   **55.** c   **57.** 880
59. 4200   **61.** 4480   **63.** 375 cal   **65.** $3290   **67. a.** 20   **b.** 100   **c.** 2000 lb   **69.** 8800 sq ft   **71.** 56,000 sq ft   **73.** 5828 pixels
75. 2100 characters   **77.** 1280 cal   **79.** $10, $60; $10, $200; $12, $36; $12, $36; $372   **81.** 21,700,000 qt   **83.** 135   **85.** 2144
87. 23   **89.** 15   **91.** $5 \cdot 6$ or $6 \cdot 5$ **93. a.** $5 + 5 + 5$ or $3 + 3 + 3 + 3 + 3$ **b.** answers may vary   **95.** $\begin{array}{r} 203 \\ \times\ 14 \\ \hline 812 \\ 2030 \\ \hline 2842 \end{array}$ **97.** $\begin{array}{r} 42 \\ \times\ 93 \\ \hline \end{array}$

99. answers may vary   **101.** 506 windows

Calculator Explorations   **1.** 53   **3.** 62   **5.** 261   **7.** 0

Vocabulary and Readiness Check   **1.** quotient; dividend; divisor   **3.** 1   **5.** undefined

Exercise Set 1.6   **1.** 6   **3.** 12   **5.** 0   **7.** 31   **9.** 1   **11.** 8   **13.** undefined   **15.** 1   **17.** 0   **19.** 9   **21.** 29   **23.** 74
25. 338   **27.** undefined   **29.** 9   **31.** 25   **33.** 68 R 3   **35.** 236 R 5   **37.** 38 R 1   **39.** 326 R 4   **41.** 13   **43.** 49   **45.** 97 R 8
47. 209 R 11   **49.** 506   **51.** 202 R 7   **53.** 54   **55.** 99 R 100   **57.** 202 R 15   **59.** 579 R 72   **61.** 17   **63.** 511 R 3
65. 2132 R 32   **67.** 6080   **69.** 23 R 2   **71.** 5 R 25   **73.** 20 R 2   **75.** 33 students   **77.** 165 lb   **79.** 310 yd   **81.** 89 bridges
83. 11 light poles   **85.** 5 mi   **87.** 1760 yd   **89.** 20   **91.** 387   **93.** 79   **95.** 74°   **97.** 9278   **99.** 15,288   **101.** 679
103. undefined   **105.** 9 R 12   **107.** c   **109.** b   **111.** $732,000,000   **113.** increase; answers may vary   **115.** no; answers may vary
117. answers may vary

121   **2.** 43   **3.** 3198   **4.** 89 R 11   **5.** 0   **6.** 0   **7.** 26   **8.** undefined   **9.** 82   **10.** 1844

..ew   **1.** 194   **2.** 6555   **3.** 4524   **4.** 562   **5.** 67   **6.** undefined   **7.** 1   **8.** 5   **9.** 0   **10.** 0   **11.** 0   **12.** 3
14. 9   **15.** 138   **16.** 276   **17.** 1169   **18.** 9826   **19.** 182 R 4   **20.** 79,317   **21.** 1099 R 2   **22.** 111 R 1   **23.** 663 R 24
..6 R 60   **25.** 1037   **26.** 9899   **27.** 30,603   **28.** 47,500   **29.** 71   **30.** 558   **31.** 6 R 8   **32.** 53   **33.** 183   **34.** 231
..5. 9740; 9700; 10,000   **36.** 1400; 1000   **37.** 20,800; 20,800; 21,000   **38.** 432,200; 432,200; 432,000
39. perimeter: 24 ft; area: 36 sq ft   **40.** perimeter: 42 in.; area: 98 sq in.   **41.** 28 mi   **42.** 26 m   **43.** 24   **44.** 124
45. Lake Pontchartrain Bridge: 2175 ft   **46.** $6246

Calculator Explorations   1. 729   **3.** 1024   **5.** 2048   **7.** 2526   **9.** 4295   **11.** 8

Vocabulary and Readiness Check   1. base; exponent   **3.** addition   **5.** division

Exercise Set 1.7   1. 4^3 **3.** 7^6 **5.** 12^3 **7.** $6^2 \cdot 5^3$ **9.** $9 \cdot 8^2$ **11.** $3 \cdot 2^4$ **13.** $3 \cdot 2^4 \cdot 5^5$ **15.** 64   **17.** 125   **19.** 32   **21.** 1
23. 7   **25.** 128   **27.** 256   **29.** 256   **31.** 729   **33.** 144   **35.** 100   **37.** 20   **39.** 729   **41.** 192   **43.** 162   **45.** 21   **47.** 7
49. 5   **51.** 16   **53.** 46   **55.** 8   **57.** 64   **59.** 83   **61.** 2   **63.** 48   **65.** 4   **67.** undefined   **69.** 59   **71.** 52   **73.** 44
75. 12   **77.** 21   **79.** 3   **81.** 43   **83.** 8   **85.** 16   **87.** 49 sq m   **89.** 529 sq mi   **91.** true   **93.** false   **95.** $(2 + 3) \cdot 6 - 2$
97. $24 \div (3 \cdot 2) + 2 \cdot 5$ **99.** 1260 ft   **101.** 6,384,814   **103.** answers may vary; $(20 - 10) \cdot 5 \div 25 + 3$

The Bigger Picture   1. 216   **2.** 48   **3.** 200   **4.** 15   **5.** 37   **6.** 200   **7.** 799   **8.** 2160   **9.** 10 R 34   **10.** 27

Vocabulary and Readiness Check   1. expression   **3.** expression; variables   **5.** equation

Exercise Set 1.8   1. 28; 14; 147; 3   **3.** 152; 152; 0; undefined   **5.** 57; 55; 56; 56   **7.** 9   **9.** 8   **11.** 6   **13.** 5   **15.** 117   **17.** 94
19. 5   **21.** 34   **23.** 20   **25.** 4   **27.** 4   **29.** 0   **31.** 33   **33.** 125   **35.** 121   **37.** 100   **39.** 60   **41.** 4   **43.** 16; 64; 144; 256
45. yes   **47.** no   **49.** no   **51.** yes   **53.** no   **55.** yes   **57.** 12   **59.** 6   **61.** 4   **63.** none   **65.** 11   **67.** $x + 8$
69. $x + 8$ **71.** $20 - x$ **73.** $512x$ **75.** $\dfrac{6}{x}$ **77.** $5x + (17 + x)$ **79.** $5x$ **81.** $11 - x$ **83.** $x - 5$ **85.** $6 \div x$ or $\dfrac{6}{x}$
87. $50 - 8x$ **89.** 274,657   **91.** 777   **93.** $5x$ **95.** As t gets larger $16t^2$ gets larger.

Chapter 1 Vocabulary Check   1. whole numbers   **2.** perimeter   **3.** place value   **4.** exponent   **5.** area   **6.** digits   **7.** variable
8. equation   **9.** expression   **10.** solution   **11.** set   **12.** sum   **13.** divisor   **14.** dividend   **15.** quotient   **16.** factor
17. product   **18.** minuend   **19.** subtrahend   **20.** difference   **21.** addend

Chapter Review   1. tens   **2.** ten-millions   **3.** seven thousand, six hundred forty
4. forty-six million, two hundred thousand, one hundred twenty   **5.** $3000 + 100 + 50 + 8$
6. $400,000,000 + 3,000,000 + 200,000 + 20,000 + 5000$ **7.** 81,900   **8.** 6,304,000,000   **9.** 1,953,631   **10.** 3,844,829
11. San Jose, CA   **12.** New York, NY   **13.** 67   **14.** 67   **15.** 65   **16.** 304   **17.** 449   **18.** 840   **19.** 3914   **20.** 7908
21. 4211   **22.** 1967   **23.** 1334   **24.** 886   **25.** 17,897   **26.** 34,658   **27.** 7523 mi   **28.** $197,699   **29.** 216 ft   **30.** 66 km
31. 111,863   **32.** 54,269   **33.** May   **34.** August   **35.** $110   **36.** $240   **37.** 40   **38.** 50   **39.** 880   **40.** 500   **41.** 3800
42. 58,000   **43.** 40,000,000   **44.** 800,000   **45.** 7300   **46.** 4100   **47.** 2700 mi   **48.** 300,000   **49.** 2208   **50.** 1396   **51.** 2280
52. 2898   **53.** 560   **54.** 900   **55.** 0   **56.** 0   **57.** 16,994   **58.** 8954   **59.** 113,634   **60.** 44,763   **61.** 411,426   **62.** 636,314
63. 1500   **64.** 4920   **65.** $898   **66.** $96,400   **67.** 91 sq mi   **68.** 500 sq cm   **69.** 7   **70.** 4   **71.** 5 R 2   **72.** 4 R 2
73. undefined   **74.** 0   **75.** 33 R 2   **76.** 19 R 7   **77.** 24 R 2   **78.** 35 R 15   **79.** 506 R 10   **80.** 907 R 40   **81.** 2793 R 140
82. 2012 R 60   **83.** 18 R 2   **84.** 21 R 2   **85.** 27 boxes   **86.** 13 miles   **87.** 51   **88.** 59   **89.** 64   **90.** 125   **91.** 405
92. 400   **93.** 16   **94.** 10   **95.** 15   **96.** 7   **97.** 12   **98.** 9   **99.** 42   **100.** 33   **101.** 9   **102.** 2   **103.** 6   **104.** 29
105. 40   **106.** 72   **107.** 5   **108.** 7   **109.** 49 sq m   **110.** 9 sq in.   **111.** 5   **112.** 17   **113.** undefined   **114.** 0   **115.** 121
116. 2   **117.** 4   **118.** 20   **119.** $x - 5$ **120.** $x + 7$ **121.** $10 \div x$ **122.** $5x$ **123.** yes   **124.** no   **125.** no   **126.** yes
127. 11   **128.** 175   **129.** 14   **130.** none   **131.** 417   **132.** 682   **133.** 2196   **134.** 2516   **135.** 1101   **136.** 1411
137. 458 R 8   **138.** 237 R 1   **139.** 70,848   **140.** 95,832   **141.** 1644   **142.** 8481   **143.** 840   **144.** 300,000   **145.** 12   **146.** 6
147. no   **148.** yes   **149.** 53 full boxes with 18 left over   **150.** $86

Chapter Test   1. eighty-two thousand, four hundred twenty-six   **2.** 402,550   **3.** 141   **4.** 113   **5.** 14,880   **6.** 766 R 42   **7.** 200
8. 98   **9.** 0   **10.** undefined   **11.** 33   **12.** 21   **13.** 48   **14.** 36   **15.** 5,698,000   **16.** 82   **17.** 52,000   **18.** 13,700
19. 1600   **20.** 92   **21.** 122   **22.** 1605   **23.** 7 R 2   **24.** $17   **25.** $126   **26.** 360 cal   **27.** $7905   **28.** 20 cm; 25 sq cm
29. 60 yd; 200 sq yd   **30.** 30   **31.** 1   **32. a.** $\dfrac{x}{17}$ **b.** $2x - 20$ **33.** yes   **34.** 10

CHAPTER 2   Integers and Introduction to Solving Equations

Vocabulary and Readiness Check   1. integers   **3.** inequality symbols   **5.** is less than; is greater than   **7.** absolute value

Exercise Set 2.1   1. -1235 **3.** $+14,433$ **5.** $+118$ **7.** $-13,000$ **9.** $-10,458$ million   **11.** $-160, -147$; Guillermo   **13.** -81
15.   **17.**   **19.**
21.   **23.** >   **25.** <   **27.** >   **29.** <   **31.** 5   **33.** 8   **35.** 0   **37.** 55   **39.** -5
41. 4   **43.** -23 **45.** 85   **47.** 7   **49.** -20 **51.** -3 **53.** 43   **55.** 15   **57.** 33   **59.** 6   **61.** -2 **63.** 32   **65.** -7
67. <   **69.** <   **71.** =   **73.** <   **75.** >   **77.** <   **79.** >   **81.** <   **83.** 31; -31 **85.** 28; 28   **87.** Caspian Sea

89. Lake Superior **91.** Mars **93.** Uranus **95.** 13 **97.** 35 **99.** 360 **101.** $-|-8|, -|3|, 2^2, -(-5)$ **103.** $-|-6|, -|1|, |-1|, -(-6)$ **105.** $-10, -|-9|, -(-2), |-12|, 5^2$ **107.** a, d **109.** 8 **111.** false **113.** true **115.** false **117.** answers may vary **119.** no; answers may vary

Calculator Explorations 1. -159 **3.** 44 **5.** $-894{,}855$

Vocabulary and Readiness Check 1. 0 **3.** a

Exercise Set 2.2

1. **3.** **5.** **7.** 67

9. -10 **11.** 0 **13.** 4 **15.** -6 **17.** -2 **19.** -9 **21.** -24 **23.** -840 **25.** 7 **27.** -3 **29.** -30 **31.** 40 **33.** -20 **35.** -125 **37.** -7 **39.** -246 **41.** 16 **43.** 13 **45.** -28 **47.** -11 **49.** 20 **51.** -34 **53.** -1 **55.** 0 **57.** -42 **59.** -70 **61.** 3 **63.** -21 **65.** 19 **67.** -10 **69.** $0 + (-215) + (-16) = -231$; 231 ft below the surface **71.** Sorenstam: -1; Hurst: $+3$ **73.** \$65,000,000 **75.** \$40,000,000 **77.** 2°C **79.** \$13,609 **81.** -47°F **83.** -7679 m **85.** 44 **87.** 141 **89.** answers may vary **91.** -3 **93.** 10 **95.** true **97.** false **99.** answers may vary

Vocabulary and Readiness Check 1. b **3.** d

Exercise Set 2.3 1. 0 **3.** 3 **5.** -5 **7.** 22 **9.** 3 **11.** -20 **13.** -12 **15.** -13 **17.** 508 **19.** -14 **21.** -4 **23.** -12 **25.** -42 **27.** -19 **29.** 14 **31.** -56 **33.** -5 **35.** -145 **37.** -37 **39.** 3 **41.** 1 **43.** -1 **45.** -31 **47.** 44 **49.** -32 **51.** -9 **53.** 14 **55.** -11 **57.** 31 **59.** 12 **61.** 20 **63.** 15°F **65.** 4°F **67.** 265°F **69.** \$49 **71.** -10°C **73.** 154 ft **75.** 69 ft **77.** 652 ft **79.** 144 ft **81.** 1197°F **83.** $-\$782$ billion **85.** $-5 + x$ **87.** $-20 - x$ **89.** 5 **91.** 1058 **93.** answers may vary **95.** 16 **97.** -20 **99.** -4 **101.** 0 **103.** -14 **105.** false **107.** answers may vary

Vocabulary and Readiness Check 1. negative **3.** positive **5.** 0 **7.** undefined

Exercise Set 2.4 1. 12 **3.** -36 **5.** -81 **7.** 0 **9.** 48 **11.** -12 **13.** 80 **15.** 0 **17.** -15 **19.** -9 **21.** -27 **23.** -36 **25.** -64 **27.** -8 **29.** -5 **31.** 7 **33.** 0 **35.** undefined **37.** -14 **39.** 0 **41.** -15 **43.** -63 **45.** 42 **47.** -24 **49.** 49 **51.** -5 **53.** -9 **55.** -6 **57.** 120 **59.** -1080 **61.** 0 **63.** -6 **65.** -7 **67.** 3 **69.** -1 **71.** -32 **73.** 180 **75.** 1 **77.** -30 **79.** -1104 **81.** -2870 **83.** -56 **85.** -18 **87.** 35 **89.** -1 **91.** undefined **93.** 6 **95.** 16; 4 **97.** 0; 0 **99.** -6 **101.** 252 **103.** $-71 \cdot x$ or $-71x$ **105.** $-16 - x$ **107.** $-29 + x$ **109.** $\dfrac{x}{-33}$ or $x \div (-33)$ **111.** $3 \cdot (-4) = -12$ yd; a loss of 12 yd **113.** $5 \cdot (-20) = -100$; a depth of 100 feet **115.** -210°C **117.** -189°C **119.** $-\$1828$ million **121. a.** -576 million or 576 million less **b.** -144 million or 144 million less **123.** 109 **125.** 8 **127.** -19 **129.** -28 **131.** -8 **133.** negative **135.** $(-5)^{17}, (-2)^{17}, (-2)^{12}, (-5)^{12}$ **137.** answers may vary

The Bigger Picture 1. 5 **2.** 55 **3.** -6 **4.** -12 **5.** -12 **6.** 81 **7.** -81 **8.** -34 **9.** 10 **10.** -30 **11.** 30 **12.** -11 **13.** -30 **14.** -80 **15.** 15 **16.** 45

Integrated Review 1. $-50; +122$ or 122 **2.** [number line with points at $-3, -1, 1, 3$] **3.** $>$ **4.** $<$ **5.** $<$ **6.** $>$ **7.** 3 **8.** -4 **9.** 9 **10.** 5 **11.** -11 **12.** 3 **13.** -64 **14.** 0 **15.** 12 **16.** -20 **17.** -48 **18.** -9 **19.** 10 **20.** -2 **21.** 106 **22.** -3 **23.** 0 **24.** 4 **25.** 42 **26.** 6 **27.** 19 **28.** -900 **29.** -12 **30.** -19 **31.** undefined **32.** 0 **33.** -4 **34.** -44 **35.** 125 **36.** 20 **37.** $\dfrac{x}{-17}$ or $x \div (-17)$ **38.** $-3 + x$ **39.** $x - (-18)$ **40.** $-7 \cdot x$ or $-7x$ **41.** 9 **42.** -15 **43.** 27 **44.** 33 **45.** -15 **46.** -4

Calculator Explorations 1. 48 **3.** -258

Vocabulary and Readiness Check 1. division **3.** average **5.** subtraction

Exercise Set 2.5 1. -125 **3.** -64 **5.** 32 **7.** -8 **9.** -11 **11.** -43 **13.** -8 **15.** 17 **17.** -1 **19.** 4 **21.** -3 **23.** 16 **25.** 13 **27.** -77 **29.** 80 **31.** 256 **33.** 53 **35.** 4 **37.** -64 **39.** 4 **41.** 16 **43.** -27 **45.** 34 **47.** 65 **49.** -7 **51.** 36 **53.** -117 **55.** 30 **57.** -3 **59.** -30 **61.** 1 **63.** -12 **65.** 0 **67.** -20 **69.** 9 **71.** -16 **73.** -128 **75.** 1 **77.** -50 **79.** -2 **81.** -19 **83.** 28 **85.** 2 **87.** no; answers may vary **89.** 4050 **91.** 45 **93.** 32 in. **95.** 30 ft **97.** $2 \cdot (7 - 5) \cdot 3$ **99.** $-6 \cdot (10 - 4)$ **101.** answers may vary **103.** answers may vary **105.** 20,736 **107.** 8900 **109.** 9

Vocabulary and Readiness Check 1. expression **3.** equation; expression **5.** solution **7.** addition

Exercise Set 2.6 1. yes **3.** no **5.** yes **7.** yes **9.** 18 **11.** -12 **13.** 9 **15.** -17 **17.** 4 **19.** -4 **21.** -14 **23.** -17 **25.** 0 **27.** 1 **29.** -7 **31.** -50 **33.** -25 **35.** 36 **37.** 21 **39.** 12 **41.** -80 **43.** -2 **45.** $x - (-2)$ **47.** $-6 \cdot x$ or $-6x$ **49.** $-15 + x$ **51.** $-8 \div x$ or $\dfrac{-8}{x}$ **53.** 41,574 **55.** -409 **57.** answers may vary **59.** answers may vary

Chapter 2 Vocabulary Check 1. opposites **2.** absolute value **3.** integers **4.** negative **5.** positive **6.** inequality symbols **7.** solution **8.** average **9.** expression **10.** equation **11.** is less than; is greater than **12.** addition **13.** multiplication

Chapter Review   **1.** -1572 **2.** $+11,239$ **3.** (number line: points at $-5, -3, 0, 3, 7$; labeled $-7\ -6\ -5\ -4\ -3\ -2\ -1\ 0\ 1\ 2\ 3\ 4\ 5\ 6\ 7$)

4. (number line: points at $-6, -2, 0, 5$; labeled $-7\ -6\ -5\ -4\ -3\ -2\ -1\ 0\ 1\ 2\ 3\ 4\ 5\ 6\ 7$)   **5.** 11 **6.** 0 **7.** -8 **8.** 9 **9.** -16 **10.** 2 **11.** $>$ **12.** $<$

13. $>$ **14.** $>$ **15.** 18 **16.** -42 **17.** false   **18.** true   **19.** true   **20.** true   **21.** 2 **22.** 3 **23.** -5 **24.** -10 **25.** Elevator D   **26.** Elevator B   **27.** 2 **28.** 14 **29.** 4 **30.** 17 **31.** -23 **32.** -22 **33.** -21 **34.** -70 **35.** 0 **36.** 0 **37.** -151 **38.** -606 **39.** $-20°C$ **40.** -150 ft   **41.** -7 **42.** $+1$ **43.** 8 **44.** -16 **45.** -24 **46.** -10 **47.** 20 **48.** 8 **49.** 0 **50.** -32 **51.** 0 **52.** 7 **53.** -10 **54.** -27 **55.** 692 ft   **56.** -25 **57.** -14 or 14 ft below ground   **58.** 82 ft   **59.** true   **60.** false   **61.** 21 **62.** -18 **63.** -64 **64.** 60 **65.** 25 **66.** -1 **67.** 0 **68.** 24 **69.** -5 **70.** 3 **71.** 0 **72.** undefined   **73.** -20 **74.** -9 **75.** 38 **76.** -5 **77.** $(-5)(2) = -10$ **78.** $(-50)(4) = -200$ **79.** $-1024 \div 4 = -256$ **80.** $-45 \div 9 = -5$ **81.** 49 **82.** -49 **83.** 0 **84.** -8 **85.** -16 **86.** 35 **87.** -32 **88.** -8 **89.** 7 **90.** -14 **91.** 39 **92.** -117 **93.** -2 **94.** -12 **95.** -3 **96.** -35 **97.** -5 **98.** 5 **99.** -1 **100.** -7 **101.** no   **102.** yes   **103.** -13 **104.** -20 **105.** -3 **106.** -9 **107.** -13 **108.** -31 **109.** 44 **110.** -26 **111.** -19 **112.** 38 **113.** 6 **114.** -5 **115.** -15 **116.** -19 **117.** 48 **118.** -21 **119.** 21 **120.** -5 **121.** $-\$9$ **122.** $6°C$ **123.** $13,118$ ft   **124.** $-27°C$ **125.** 2 **126.** 3 **127.** -5 **128.** -25 **129.** -20 **130.** 17 **131.** -21 **132.** -17 **133.** 12 **134.** -9 **135.** -200 **136.** 3

Chapter Test   **1.** 3 **2.** -6 **3.** -100 **4.** 4 **5.** -30 **6.** 12 **7.** 65 **8.** 5 **9.** 12 **10.** -6 **11.** 50 **12.** -2 **13.** -11 **14.** -46 **15.** -117 **16.** 3456 **17.** 28 **18.** -213 **19.** -2 **20.** 2 **21.** -5 **22.** -32 **23.** -17 **24.** 1 **25.** -1 **26.** 88 ft below sea level   **27.** 45 **28.** $31,642$ **29.** 3820 ft below sea level   **30.** -4 **31. a.** $17x$ **b.** $20 - 2x$ **32.** 5 **33.** -28 **34.** -20 **35.** -4

Cumulative Review   **1.** hundred-thousands; Sec. 1.2, Ex. 1   **2.** hundreds; Sec. 1.2   **3.** thousands; Sec. 1.2, Ex. 2   **4.** thousands; Sec. 1.2   **5.** ten-millions; Sec. 1.2, Ex. 3   **6.** hundred-thousands; Sec. 1.2   **7. a.** $<$ **b.** $>$ **c.** $>$; Sec. 2.1, Ex. 3   **8. a.** $>$ **b.** $<$ **c.** $>$; Sec. 2.1   **9.** 39; Sec. 1.3, Ex. 3   **10.** 39; Sec. 1.3   **11.** 7321; Sec. 1.3, Ex. 6   **12.** 3013; Sec. 1.3   **13.** 36,184 mi; Sec. 1.3, Ex. 11   **14.** $\$525$; Sec. 1.3   **15.** 570; Sec. 1.4, Ex. 1   **16.** 600; Sec. 1.4   **17.** 1800; Sec. 1.4, Ex. 5   **18.** 5000; Sec. 1.4   **19. a.** $5 \cdot 6 + 5 \cdot 5$ **b.** $20 \cdot 4 + 20 \cdot 7$ **c.** $2 \cdot 7 + 2 \cdot 9$; Sec. 1.5, Ex. 2   **20. a.** $5 \cdot 2 + 5 \cdot 12$ **b.** $9 \cdot 3 + 9 \cdot 6$ **c.** $4 \cdot 8 + 4 \cdot 1$; Sec. 1.5   **21.** 78,875; Sec. 1.5, Ex. 5   **22.** 31,096; Sec. 1.5   **23. a.** 6   **b.** 8   **c.** 7; Sec. 1.6, Ex. 1   **24. a.** 7   **b.** 8   **c.** 12; Sec. 1.6   **25.** 741; Sec. 1.6, Ex. 4   **26.** 456; Sec. 1.6   **27.** 12 cards each; 10 cards left over; Sec. 1.6, Ex. 11   **28.** $\$9$; Sec. 1.6   **29.** 81; Sec. 1.7, Ex. 5   **30.** 125; Sec. 1.7   **31.** 6; Sec. 1.7, Ex. 6   **32.** 4; Sec. 1.7   **33.** 180; Sec. 1.7, Ex. 8   **34.** 56; Sec. 1.7   **35.** 2; Sec. 1.7, Ex. 13   **36.** 5; Sec. 1.7   **37.** 14; Sec. 1.8, Ex. 1   **38.** 14; Sec. 1.8   **39. a.** 9   **b.** 3   **c.** 0; Sec. 2.1, Ex. 4   **40. a.** 4   **b.** 7; Sec. 2.1   **41.** 3; Sec. 2.2, Ex. 7   **42.** 5; Sec. 2.2   **43.** 22; Sec. 2.3, Ex. 12   **44.** 5; Sec. 2.3   **45.** -21; Sec. 2.4, Ex. 1   **46.** -10; Sec. 2.4   **47.** 0; Sec. 2.4, Ex. 3   **48.** -54; Sec. 2.4   **49.** -16; Sec. 2.5, Ex. 8   **50.** -27; Sec. 2.5

CHAPTER 3   Solving Equations and Problem Solving

Vocabulary and Readiness Check   **1.** expression; term   **3.** combine like terms   **5.** variable; constant   **7.** associative   **9.** numerical coefficient

Exercise Set 3.1   **1.** $8x$ **3.** $-n$ **5.** $-2c$ **7.** $-6x$ **9.** $12a - 5$ **11.** $42x$ **13.** $-33y$ **15.** $72a$ **17.** $2y + 6$ **19.** $3a - 18$ **21.** $-12x - 28$ **23.** $2x + 1$ **25.** $15c + 3$ **27.** $-21n + 20$ **29.** $7w + 15$ **31.** $11x - 8$ **33.** $-2y + 16$ **35.** $-2y$ **37.** $-7z$ **39.** $8d - 3c$ **41.** $6y - 14$ **43.** $-q$ **45.** $2x + 22$ **47.** $-3x - 35$ **49.** $-3z - 15$ **51.** $-6x + 6$ **53.** $3x - 30$ **55.** $-r + 8$ **57.** $-7n + 3$ **59.** $9z - 14$ **61.** -6 **63.** $-4x + 10$ **65.** $2xy + 20$ **67.** $7a + 12$ **69.** $3y + 5$ **71.** $(14y + 22)$ m   **73.** $(11a + 12)$ ft   **75.** $(-25x + 55)$ in.   **77.** $36y$ sq in.   **79.** $(32x - 64)$ sq km   **81.** $(60y + 20)$ sq mi   **83.** 2000 sq ft   **85.** 64 ft   **87.** $(3x + 6)$ ft   **89.** -3 **91.** 8 **93.** 0 **95.** incorrect; $15x - 10$ **97.** correct   **99.** distributive   **101.** associative   **103.** $(20x + 16)$ sq mi   **105.** $4824q + 12,274$ **107.** answers may vary   **109.** answers may vary

Vocabulary and Readiness Check   **1.** equivalent   **3.** simplifying   **5.** addition

Exercise Set 3.2   **1.** 6 **3.** 8 **5.** -4 **7.** 6 **9.** -1 **11.** 18 **13.** -8 **15.** -50 **17.** 3 **19.** -22 **21.** -6 **23.** 24 **25.** -30 **27.** -4 **29.** 4 **31.** 3 **33.** -3 **35.** 1 **37.** 5 **39.** -9 **41.** -4 **43.** -3 **45.** -1 **47.** 0 **49.** 3 **51.** -6 **53.** -35 **55.** 10 **57.** -2 **59.** 0 **61.** 28 **63.** -5 **65.** -28 **67.** -28 **69.** 5 **71.** -4 **73.** $-7 + x$ **75.** $x - 11$ **77.** $-13x$ **79.** $\dfrac{x}{-12}$ or $-\dfrac{x}{12}$ **81.** $-11x + 5$ **83.** $-10 - 7x$ **85.** $4x + 7$ **87.** $2x - 17$ **89.** $-6(x + 15)$ **91.** $\dfrac{45}{-5x}$ **93.** $\dfrac{17}{x} + (-15)$ **95.** 2005 **97.** $35,000$ **99.** answers may vary   **101.** no; answers may vary   **103.** answers may vary   **105.** $67,896$ **107.** -48 **109.** 42

Integrated Review   **1.** expression   **2.** equation   **3.** equation   **4.** expression   **5.** simplify   **6.** solve   **7.** $8x$ **8.** $-4y$ **9.** $-2a - 2$ **10.** $5a - 26$ **11.** $-8x - 14$ **12.** $-6x + 30$ **13.** $5y - 10$ **14.** $15x - 31$ **15.** $(12x - 6)$ sq m   **16.** $(2x + 9)$ ft   **17.** -4 **18.** -3 **19.** -10 **20.** 6 **21.** -15 **22.** -120 **23.** -5 **24.** -13 **25.** -24 **26.** -54 **27.** 12 **28.** -42 **29.** 2 **30.** 2 **31.** -3 **32.** 5 **33.** -5 **34.** 6 **35.** $x - 10$ **36.** $-20 + x$ **37.** $10x$ **38.** $\dfrac{10}{x}$ **39.** $-2x + 5$ **40.** $-4(x - 1)$

Calculator Explorations   **1.** yes   **3.** no   **5.** yes

Vocabulary and Readiness Check   **1.** $3x - 9 + x - 16; 5(2x + 6) - 1 = 39$ **3.** addition   **5.** distributive

Exercise Set 3.3 **1.** -12 **3.** -3 **5.** 1 **7.** -45 **9.** -9 **11.** 6 **13.** 8 **15.** 5 **17.** 0 **19.** -5 **21.** -22 **23.** 6 **25.** -11
27. -7 **29.** -5 **31.** 270 **33.** 5 **35.** 3 **37.** 9 **39.** -6 **41.** 11 **43.** 3 **45.** 4 **47.** -4 **49.** 3 **51.** -1 **53.** -4

55. -5 **57.** 0 **59.** 4 **61.** 1 **63.** -30 **65.** $-42 + 16 = -26$ **67.** $-5(-29) = 145$ **69.** $3(-14 - 2) = -48$ **71.** $\dfrac{100}{2(50)} = 1$

73. 97 million returns **75.** 20 million returns **77.** 33 **79.** -37 **81.** b **83.** a **85.** $6x - 10 = 5x - 7$
$6x = 5x + 3$
$x = 3$

87. -81 **89.** 891 **91.** no; answers may vary

The Bigger Picture **1.** -5 **2.** 11 **3.** -7 **4.** 103 **5.** 75 **6.** -11 **7.** -6 **8.** 9

Exercise Set 3.4 **1.** $-5 + x = -7$ **3.** $3x = 27$ **5.** $-20 - x = 104$ **7.** $2x = 108$ **9.** $5(-3 + x) = -20$ **11.** $9 + 3x = 33; 8$
13. $3 + 4 + x = 16; 9$ **15.** $x - 3 = \dfrac{10}{5}; 5$ **17.** $30 - x = 3(x + 6); 3$ **19.** $5x - 40 = x + 8; 12$ **21.** $3(x - 5) = \dfrac{108}{12}; 8$
23. $4x = 30 - 2x; 5$ **25.** California: 55 votes; Florida: 27 **27.** falcon: 185 mph; pheasant: 37 mph **29.** India: 8407; US: 5758
31. Xbox: \$420; games: \$140 **33.** 5470 mi **35.** Michigan Stadium: 107,501; Neyland Stadium: 102,854
37. China: 140 million; Spain: 70 million **39.** Spain: 6609 cars per day; Germany: 13,218 cars per day **41.** \$225 **43.** 78 **45.** 590
47. 1000 **49.** 3000 **51.** answers may vary **53.** \$8250 **55.** \$5

Chapter 3 Vocabulary Check **1.** simplified; combined **2.** like **3.** variable **4.** algebraic expression **5.** terms
6. numerical coefficient **7.** evaluating the expression **8.** constant **9.** equation **10.** solution **11.** distributive
12. multiplication **13.** addition

Chapter Review **1.** $10y - 15$ **2.** $-6y - 10$ **3.** $-6a - 7$ **4.** $-8y + 2$ **5.** $2x + 10$ **6.** $-3y - 24$ **7.** $11x - 12$
8. $-4m - 12$ **9.** $-5a + 4$ **10.** $12y - 9$ **11.** $16y - 5$ **12.** $x - 2$ **13.** $(4x + 6)$yd **14.** $20y$ m **15.** $(6x - 3)$ sq yd
16. $(45x + 8)$ sq cm **17.** -2 **18.** 10 **19.** -7 **20.** 5 **21.** -12 **22.** 45 **23.** -6 **24.** -1 **25.** -25 **26.** -8 **27.** -2
28. -2 **29.** -8 **30.** -45 **31.** 5 **32.** -5 **33.** -63 **34.** -15 **35.** 5 **36.** 12 **37.** -6 **38.** 4 **39.** $-5x$ **40.** $x - 3$
41. $-5 + x$ **42.** $\dfrac{-2}{x}$ **43.** $2x + 11$ **44.** $-5x - 50$ **45.** $\dfrac{70}{x + 6}$ **46.** $2(x - 13)$ **47.** 21 **48.** -10 **49.** 2 **50.** 2 **51.** 11
52. -5 **53.** -15 **54.** 10 **55.** -2 **56.** -6 **57.** -1 **58.** 1 **59.** 0 **60.** 20 **61.** $20 - (-8) = 28$
62. $5[2 + (-6)] = -20$ **63.** $\dfrac{-75}{5 + 20} = -3$ **64.** $-2 - 19 = -21$ **65.** $2x - 8 = 40$ **66.** $6x = x + 2$ **67.** $\dfrac{x}{2} - 12 = 10$
68. $x - 3 = \dfrac{x}{4}$ **69.** 5 **70.** -16 **71.** 2386 votes **72.** 84 DVDs **73.** $-11x$ **74.** $-35x$ **75.** $22x - 19$ **76.** $-9x - 32$
77. -1 **78.** -25 **79.** 13 **80.** -6 **81.** -22 **82.** -6 **83.** -15 **84.** 18 **85.** -5 **86.** 2 **87.** 11 **88.** -1 **89.** 0
90. -6 **91.** 5 **92.** 1 **93.** Hawaii: 4309 mi; Delaware: 5894 mi **94.** North Dakota: 86,782 mi; South Dakota: 83,688 mi

Chapter Test **1.** $-5x + 5$ **2.** $-6y - 14$ **3.** $-8z - 20$ **4.** $(15x + 15)$ in. **5.** $(12x - 4)$ sq m **6.** -6 **7.** -6 **8.** 24 **9.** -2
10. 6 **11.** 3 **12.** -2 **13.** 0 **14.** 4 **15.** $-23 + x$ **16.** $-2 - 3x$ **17.** $2 \cdot 5 + (-15) = -5$ **18.** $3x + 6 = -30$ **19.** -2
20. 8 free throws **21.** 244 women

Cumulative Review **1.** three hundred eight million, sixty-three thousand, five hundred fifty-seven; Sec. 1.2, Ex. 7
2. two hundred seventy-six thousand, four; Sec. 1.2 **3.** 13 in.; Sec. 1.3, Ex. 9 **4.** 18 in.; Sec. 1.3 **5.** 726; Sec. 1.3, Ex. 8 **6.** 9585; Sec. 1.3
7. 249,000; Sec. 1.4, Ex. 3 **8.** 844,000; Sec. 1.4 **9.** 200; Sec. 1.5, Ex. 3a **10.** 29,230; Sec. 1.5 **11.** 208; Sec. 1.6, Ex. 5 **12.** 86; Sec. 1.6
13. 7; Sec. 1.7, Ex. 9 **14.** 35; Sec. 1.7 **15.** 26; Sec. 1.8, Ex. 4 **16.** 10; Sec. 1.8 **17.** 20 is a solution.; Sec. 1.8, Ex. 7
18. a. $<$; **b.** $>$, Sec. 2.1 **19.** 3; Sec. 2.2, Ex. 1 **20.** -7; Sec. 2.2 **21.** -25; Sec. 2.2, Ex. 5 **22.** -4; Sec. 2.2 **23.** 3; Sec. 2.2, Ex. 7
24. 17; Sec. 2.2 **25.** -14; Sec. 2.3, Ex. 2 **26.** -5; Sec. 2.3 **27.** 11; Sec. 2.3, Ex. 3 **28.** 29; Sec. 2.3 **29.** -4; Sec. 2.3, Ex. 4
30. -3; Sec. 2.4 **31.** -2; Sec. 2.4, Ex. 10 **32.** 6; Sec. 2.4 **33.** 5; Sec. 2.4, Ex. 11 **34.** -13; Sec. 2.4 **35.** -16; Sec. 2.4, Ex. 12
36. -10; Sec. 2.4 **37.** 9; Sec. 2.5, Ex. 1 **38.** -32; Sec. 2.5 **39.** -9; Sec. 2.5, Ex. 2 **40.** 25; Sec. 2.5 **41.** $6y + 2$; Sec. 3.1, Ex. 2
42. $3x + 9$; Sec. 3.1 **43.** not a solution; Sec. 2.6, Ex. 1 **44.** solution; Sec. 2.6 **45.** 3; Sec. 2.6, Ex. 7 **46.** -5; Sec. 2.6
47. 12; Sec. 3.2, Ex. 7 **48.** -2; Sec. 3.2 **49.** software: \$420; computer system: \$1680; Sec. 3.4, Ex. 4 **50.** 11; Sec. 3.4

CHAPTER 4   Fractions and Mixed Numbers

Vocabulary and Readiness Check **1.** fraction; denominator; numerator **3.** improper; proper; mixed number

Exercise Set 4.1 **1.** numerator: 1; denominator: 2; proper **3.** numerator: 10; denominator: 3; improper **5.** numerator: 15; denominator: 15;
improper **7.** $\dfrac{1}{3}$ **9. a.** $\dfrac{11}{4}$ **b.** $2\dfrac{3}{4}$ **11. a.** $\dfrac{23}{6}$ **b.** $3\dfrac{5}{6}$ **13.** $\dfrac{7}{12}$ **15.** $\dfrac{3}{7}$ **17.** $\dfrac{4}{9}$ **19. a.** $\dfrac{4}{3}$ **b.** $1\dfrac{1}{3}$ **21. a.** $\dfrac{11}{2}$ **b.** $-5\dfrac{1}{2}$

23. $\dfrac{1}{6}$ **25.** $\dfrac{5}{8}$ **27.** **29.** **31.** **33.** $\dfrac{42}{131}$

35. a. 89 **b.** $\dfrac{89}{131}$ **37.** $\dfrac{8}{43}$ **39.** $\dfrac{15}{28}$ of the tropical storms **41.** $\dfrac{11}{31}$ of the month **43.** $\dfrac{10}{31}$ of the class **45. a.** $\dfrac{33}{50}$ of the states

b. 17 states **c.** $\dfrac{17}{50}$ of the states **47. a.** $\dfrac{21}{50}$ **b.** 29 **c.** $\dfrac{29}{50}$ **49.**

51.

0    $\frac{4}{7}$ 1

53.

0   1 $\frac{8}{5}$ 2

55.

0   1   2 $\frac{7}{3}$ 3

57. 1

59. -5 **61.** 0   **63.** 1   **65.** undefined   **67.** 3   **69.** $\frac{7}{3}$ **71.** $\frac{18}{5}$ **73.** $\frac{53}{8}$ **75.** $\frac{83}{7}$ **77.** $\frac{187}{20}$ **79.** $\frac{500}{3}$ **81.** $3\frac{2}{5}$ **83.** $4\frac{5}{8}$

85. $3\frac{2}{15}$ **87.** 15   **89.** $1\frac{7}{175}$ **91.** $6\frac{65}{112}$ **93.** 9   **95.** 125   **97.** $\frac{-11}{2}; \frac{11}{-2}$ **99.** $\frac{13}{-15}; -\frac{13}{15}$ **101.** answers may vary

103.

105. 6   **107.** $\frac{57}{171}$ of the licensees   **109.** $\frac{5}{13}$ of the training centers

Calculator Explorations   **1.** $\frac{4}{7}$ **3.** $\frac{20}{27}$ **5.** $\frac{8}{15}$ **7.** $\frac{2}{9}$

Vocabulary and Readiness Check   **1.** prime factorization   **3.** prime   **5.** equivalent

Exercise Set 4.2   **1.** $2^2 \cdot 5$ **3.** $2^4 \cdot 3$ **5.** 3^4 **7.** $2 \cdot 3^4$ **9.** $2 \cdot 5 \cdot 11$ **11.** $5 \cdot 17$ **13.** $2^4 \cdot 3 \cdot 5$ **15.** $2^2 \cdot 3^2 \cdot 23$ **17.** $\frac{1}{4}$

19. $\frac{2x}{21}$ **21.** $\frac{7}{8}$ **23.** $\frac{2}{3}$ **25.** $\frac{7}{10}$ **27.** $-\frac{7}{9}$ **29.** $\frac{5x}{6}$ **31.** $\frac{27}{64}$ **33.** $\frac{5x}{8}$ **35.** $-\frac{5}{8}$ **37.** $\frac{3x^2y}{2}$ **39.** $\frac{3}{4}$ **41.** $\frac{5}{8z}$ **43.** $\frac{3}{14}$

45. $-\frac{11}{17y}$ **47.** $\frac{7}{8}$ **49.** $14a^2$ **51.** equivalent   **53.** not equivalent   **55.** equivalent   **57.** equivalent   **59.** not equivalent

61. not equivalent   **63.** $\frac{1}{4}$ of a shift   **65.** $\frac{1}{2}$ mi   **67. a.** $\frac{3}{10}$ **b.** 35 states   **c.** $\frac{7}{10}$ **69.** $\frac{5}{12}$ of the wall   **71. a.** 28   **b.** $\frac{14}{25}$

73. $\frac{291}{449}$ of individuals   **75.** -3 **77.** -14 **79.** answers may vary   **81.** $\frac{3}{5}$ **83.** $\frac{9}{25}$ **85.** $\frac{1}{25}$ **87.** $2^2 \cdot 3^5 \cdot 5 \cdot 7$

89. answers may vary   **91.** no; answers may vary   **93.** $\frac{2}{25}$ **95.** answers may vary   **97.** $2235, 105, 900, 1470$ **99.** 15; answers may vary

Vocabulary and Readiness Check   **1.** $\frac{a \cdot c}{b \cdot d}$ **3.** $\frac{2 \cdot 2 \cdot 2}{7}; \frac{2}{7} \cdot \frac{2}{7} \cdot \frac{2}{7}$ **5.** $\frac{a \cdot d}{b \cdot c}$

Exercise Set 4.3   **1.** $\frac{18}{77}$ **3.** $-\frac{5}{28}$ **5.** $\frac{1}{15}$ **7.** $\frac{18x}{55}$ **9.** $\frac{3a^2}{4}$ **11.** $\frac{x^2}{y}$ **13.** 0   **15.** $-\frac{17}{25}$ **17.** $\frac{1}{56}$ **19.** $\frac{1}{125}$ **21.** $\frac{4}{9}$

23. $-\frac{4}{27}$ **25.** $\frac{4}{5}$ **27.** $-\frac{1}{6}$ **29.** $-\frac{16}{9x}$ **31.** $\frac{121y}{60}$ **33.** $-\frac{1}{6}$ **35.** $\frac{x}{25}$ **37.** $\frac{10}{27}$ **39.** $\frac{18x^2}{35}$ **41.** $\frac{10}{9}$ **43.** $-\frac{1}{4}$ **45.** $\frac{9}{16}$

47. xy^2 **49.** $\frac{77}{2}$ **51.** $-\frac{36}{x}$ **53.** $\frac{3}{49}$ **55.** $-\frac{19y}{7}$ **57.** $\frac{4}{11}$ **59.** $\frac{8}{3}$ **61.** $\frac{15x}{4}$ **63.** $\frac{8}{9}$ **65.** $\frac{1}{60}$ **67.** b **69.** $\frac{2}{5}$ **71.** $-\frac{5}{3}$

73. a. $\frac{1}{3}$ **b.** $\frac{12}{25}$ **75. a.** $-\frac{36}{55}$ **b.** $-\frac{44}{45}$ **77.** yes   **79.** no   **81.** 50   **83.** 20   **85.** 128   **87.** 868 mi   **89.** $\frac{3}{16}$ in.   **91.** $1838

93. 20 contestants   **95.** $\frac{1}{14}$ sq ft   **97.** 3840 mi   **99.** 2400 mi   **101.** 201   **103.** 196   **105.** answers may vary   **107.** 5

109. 34,897,200 people   **111.** 37

The Bigger Picture   **1.** $\frac{16}{27}$ **2.** $-\frac{3}{4}$ **3.** $\frac{4}{3}$ **4.** $\frac{12}{5}$ **5.** 8   **6.** 72   **7.** $\frac{1}{24}$ **8.** 40   **9.** 35   **10.** 24   **11.** 0   **12.** -6

Vocabulary and Readiness Check   **1.** like; unlike   **3.** $-\frac{a}{b}$ **5.** least common denominator (LCD)

Exercise Set 4.4   **1.** $\frac{7}{11}$ **3.** $\frac{2}{3}$ **5.** $-\frac{1}{4}$ **7.** $-\frac{1}{2}$ **9.** $\frac{2}{3x}$ **11.** $-\frac{x}{9}$ **13.** $\frac{6}{11}$ **15.** $\frac{3}{4}$ **17.** $-\frac{3}{y}$ **19.** $-\frac{19}{33}$ **21.** $-\frac{1}{3}$ **23.** $\frac{7a-3}{4}$

25. $\frac{9}{10}$ **27.** $-\frac{13x}{14}$ **29.** $\frac{9x+1}{15}$ **31.** $-\frac{x}{2}$ **33.** $-\frac{2}{3}$ **35.** $\frac{3x}{4}$ **37.** $\frac{5}{4}$ **39.** $\frac{2}{5}$ **41.** 1 in.   **43.** 2 m   **45.** $\frac{7}{25}$ **47.** $\frac{1}{50}$

49. $\frac{7}{10}$ mi   **51.** $\frac{2}{5}$ **53.** 45   **55.** 72   **57.** 150   **59.** $24x$ **61.** 126   **63.** 168   **65.** 14   **67.** 20   **69.** 25   **71.** $56x$ **73.** $8y$

75. $20a$ **77.** $\frac{89}{100}, \frac{90}{100}, \frac{96}{100}, \frac{92}{100}, \frac{68}{100}, \frac{91}{100}, \frac{80}{100}, \frac{69}{100}, \frac{8}{100}, \frac{70}{100}$ **79.** India   **81.** 9   **83.** 125   **85.** 49   **87.** 24   **89.** $\frac{2}{7} + \frac{9}{7} = \frac{11}{7}$

91. answers may vary   **93.** 1; answers may vary   **95.** $814x$ **97.** answers may vary   **99.** a, b, and d

Calculator Explorations   **1.** $\frac{37}{80}$ **3.** $\frac{95}{72}$ **5.** $\frac{394}{323}$

Vocabulary and Readiness Check   **1.** equivalent; least common denominator   **3.** $\frac{4}{24}, \frac{15}{24}, \frac{19}{24}$ **5.** expression; equation

Exercise Set 4.5   **1.** $\frac{5}{6}$ **3.** $\frac{1}{6}$ **5.** $-\frac{4}{33}$ **7.** $-\frac{3}{14}$ **9.** $\frac{3x}{5}$ **11.** $\frac{24-y}{12}$ **13.** $\frac{11}{36}$ **15.** $-\frac{44}{7}$ **17.** $\frac{89a}{99}$ **19.** $\frac{4y-1}{6}$ **21.** $\frac{x+6}{2x}$

23. $-\frac{8}{33}$ **25.** $\frac{3}{14}$ **27.** $\frac{11y-10}{35}$ **29.** $-\frac{11}{36}$ **31.** $\frac{1}{20}$ **33.** $\frac{33}{56}$ **35.** $\frac{17}{16}$ **37.** $-\frac{2}{9}$ **39.** $-\frac{11}{30}$ **41.** $\frac{15+11y}{33}$ **43.** $-\frac{53}{42}$

45. $\frac{7x}{8}$ **47.** $\frac{11}{18}$ **49.** $\frac{44a}{39}$ **51.** $-\frac{11}{60}$ **53.** $\frac{5y+9}{9y}$ **55.** $\frac{19}{20}$ **57.** $\frac{56}{45}$ **59.** $\frac{40+9x}{72x}$ **61.** $-\frac{5}{24}$ **63.** $\frac{37x-20}{56}$ **65.** <

67. >   **69.** >   **71.** $\frac{13}{12}$ **73.** $\frac{1}{4}$ **75.** $\frac{11}{6}$ **77.** $\frac{34}{15}$ or $2\frac{4}{15}$ cm   **79.** $\frac{17}{10}$ or $1\frac{7}{10}$ m   **81.** $x+\frac{1}{2}$ **83.** $\frac{3}{8}-x$ **85.** $\frac{5}{8}$ in.

87. $\frac{7}{100}$ mph   **89.** $\frac{47}{32}$ in.   **91.** $\frac{49}{100}$ of students   **93.** $\frac{19}{25}$ **95.** $\frac{11}{50}$ **97.** $\frac{21}{25}$ **99.** -20 **101.** -6 **103.** $\frac{3}{5}+\frac{4}{5}=\frac{7}{5}$ **105.** $\frac{223}{540}$

107. $\frac{49}{44}$ **109.** answers may vary   **111.** standard mail

Integrated Review   **1.** $\frac{3}{7}$ **2.** $\frac{5}{4}$ or $1\frac{1}{4}$ **3.** $\frac{73}{85}$ **4.**   **5.** -1 **6.** 17

7. 0   **8.** undefined   **9.** $5 \cdot 13$ **10.** $2 \cdot 5 \cdot 7$ **11.** $3^2 \cdot 5 \cdot 7$ **12.** $3^2 \cdot 7^2$ **13.** $\frac{1}{7}$ **14.** $\frac{6}{5}$ **15.** $-\frac{14}{15}$ **16.** $-\frac{9}{10}$ **17.** $\frac{2x}{5}$

18. $\frac{3}{8y}$ **19.** $\frac{11z^2}{14}$ **20.** $\frac{7}{11ab^2}$ **21.** not equivalent   **22.** equivalent   **23. a.** $\frac{1}{25}$ **b.** 48 **c.** $\frac{24}{25}$ **24. a.** $\frac{1}{5}$ **b.** 368 **c.** $\frac{4}{5}$ **25.** 30

26. 14   **27.** 90   **28.** $\frac{28}{36}$ **29.** $\frac{55}{75}$ **30.** $\frac{40}{48}$ **31.** $\frac{4}{5}$ **32.** $-\frac{2}{5}$ **33.** $\frac{3}{25}$ **34.** $\frac{1}{3}$ **35.** $\frac{4}{5}$ **36.** $\frac{5}{9}$ **37.** $-\frac{1}{6y}$ **38.** $\frac{3x}{2}$

39. $\frac{1}{18}$ **40.** $\frac{4}{21}$ **41.** $-\frac{7}{48z}$ **42.** $-\frac{9}{50}$ **43.** $\frac{37}{40}$ **44.** $\frac{11}{36}$ **45.** $\frac{11}{18}$ **46.** $\frac{25y+12}{50}$ **47.** $\frac{2}{3} \cdot x$ or $\frac{2}{3}x$ **48.** $x \div \left(-\frac{1}{5}\right)$

49. $-\frac{8}{9}-x$ **50.** $\frac{6}{11}+x$ **51.** 1020   **52.** 24 lots   **53.** $\frac{3}{4}$ ft

Vocabulary and Readiness Check   **1.** complex   **3.** division   **5.** addition

Exercise Set 4.6   **1.** $\frac{1}{6}$ **3.** $\frac{7}{3}$ **5.** $\frac{x}{6}$ **7.** $\frac{23}{22}$ **9.** $\frac{2x}{13}$ **11.** $\frac{17}{60}$ **13.** $\frac{5}{8}$ **15.** $\frac{35}{9}$ **17.** $-\frac{17}{45}$ **19.** $\frac{11}{8}$ **21.** $\frac{29}{10}$ **23.** $\frac{27}{32}$

25. $\frac{1}{100}$ **27.** $\frac{9}{64}$ **29.** $\frac{7}{6}$ **31.** $-\frac{2}{5}$ **33.** $-\frac{2}{9}$ **35.** $\frac{5a}{2}$ **37.** $\frac{7}{2}$ **39.** $\frac{9}{20}$ **41.** $-\frac{13}{2}$ **43.** $\frac{9}{25}$ **45.** $-\frac{5}{32}$ **47.** 1   **49.** $-\frac{2}{5}$

51. $-\frac{11}{40}$ **53.** $\frac{x+6}{16}$ **55.** $\frac{7}{2}$ or $3\frac{1}{2}$ **57.** $\frac{49}{6}$ or $8\frac{1}{6}$ **59.** no; answers may vary   **61.** $\frac{5}{8}$ **63.** $\frac{11}{56}$ **65.** halfway between a and b
67. false   **69.** false   **71.** true   **73.** no; answers may vary   **75.** addition, multiplication, subtraction, division   **77.** subtraction,

multiplication, division, addition   **79.** $\frac{17}{7}$ **81.** $\frac{18}{5}$

Calculator Explorations   **1.** $\frac{280}{11}$ **3.** $\frac{3776}{35}$ **5.** $26\frac{1}{14}$ **7.** $92\frac{3}{10}$

Vocabulary and Readiness Check   **1.** mixed number   **3.** round

Exercise Set 4.7   **1.**   **3.**   **5.** b   **7.** a

9. $\frac{8}{21}$ **11.** $4\frac{3}{8}$ **13.** $\frac{77}{10}$ or $7\frac{7}{10}$; 8   **15.** $\frac{836}{35}$ or $23\frac{31}{35}$; 24   **17.** $\frac{25}{2}$ or $12\frac{1}{2}$ **19.** $5\frac{1}{2}$ **21.** $18\frac{2}{3}$ **23.** a   **25.** b   **27.** $6\frac{2}{3}$; 7

29. $13\frac{11}{14}$; 14   **31.** $17\frac{7}{25}$ **33.** $47\frac{53}{84}$ **35.** $25\frac{5}{14}$ **37.** $13\frac{13}{24}$ **39.** $2\frac{3}{5}$; 3   **41.** $7\frac{5}{14}$; 7   **43.** $\frac{24}{25}$ **45.** $3\frac{5}{9}$ **47.** $15\frac{3}{4}$

49. 4   **51.** $5\frac{11}{14}$ **53.** $6\frac{2}{9}$ **55.** $\frac{25}{33}$ **57.** $35\frac{13}{18}$ **59.** $2\frac{1}{2}$ **61.** $72\frac{19}{30}$ **63.** $\frac{11}{14}$ **65.** $5\frac{4}{7}$ **67.** $13\frac{16}{33}$ **69.** $-5\frac{2}{7}-x$ **71.** $1\frac{9}{10} \cdot x$

73. $9\frac{2}{5}$ in.   **75.** $7\frac{13}{20}$ in.   **77.** $\frac{7}{2}$ or $3\frac{1}{2}$ sq yd   **79.** $\frac{15}{16}$ sq in.   **81.** $21\frac{5}{24}$ m   **83.** no; she will be $\frac{1}{12}$ ft short   **85.** $3\frac{3}{16}$ mi   **87.** $4\frac{2}{3}$ m

89. $9\frac{7}{12}$ min   **91.** $1\frac{4}{5}$ min   **93.** $-10\frac{3}{25}$ **95.** $-24\frac{7}{8}$ **97.** $-13\frac{59}{60}$ **99.** $-\frac{1}{2}$ **101.** $-1\frac{23}{24}$ **103.** $\frac{73}{1000}$ **105.** $1x$ or x **107.** $1a$ or a

109. a, b, c   **111.** Incorrect; to divide mixed numbers, first write each mixed number as an improper fraction.   **113.** answers may vary

115. Incorrect; $3\frac{2}{3} \cdot 1\frac{1}{7} = \frac{11}{3} \cdot \frac{8}{7} = \frac{88}{21}$ or $4\frac{4}{21}$ **117.** answers may vary

The Bigger Picture   **1.** $\frac{5}{17}$ **2.** $\frac{4}{5}$ **3.** $\frac{29}{30}$ **4.** $\frac{1}{24}$ **5.** $1\frac{33}{40}$ **6.** $\frac{27}{64}$ **7.** $12\frac{3}{7}$ **8.** $9\frac{13}{24}$ **9.** $\frac{16}{33}$ **10.** $\frac{34}{27}$ or $1\frac{7}{27}$

Exercise Set 4.8   **1.** $-\frac{2}{3}$ **3.** $\frac{1}{13}$ **5.** $\frac{4}{5}$ **7.** $\frac{11}{12}$ **9.** $-\frac{7}{10}$ **11.** $\frac{11}{16}$ **13.** $\frac{11}{18}$ **15.** $\frac{2}{7}$ **17.** 12   **19.** -27 **21.** $\frac{27}{8}$ **23.** $\frac{1}{21}$

25. $\frac{2}{11}$ **27.** $-\frac{3}{10}$ **29.** 1   **31.** $\frac{15}{2}$ **33.** -1 **35.** -15 **37.** $\frac{3x-28}{21}$ **39.** $\frac{y+10}{2}$ **41.** $\frac{7x}{15}$ **43.** $\frac{4}{3}$ **45.** 2   **47.** $\frac{21}{10}$

49. $-\dfrac{1}{14}$ **51.** -3 **53.** $-\dfrac{1}{9}$ **55.** 50 **57.** $-\dfrac{3}{7}$ **59.** 4 **61.** $\dfrac{3}{5}$ **63.** $-\dfrac{1}{24}$ **65.** $-\dfrac{5}{14}$ **67.** 4 **69.** -36 **71.** $57{,}200$

73. 330 **75.** answers may vary   **77.** -2

Chapter 4 Vocabulary Check   1. reciprocals   **2.** composite number   **3.** equivalent   **4.** improper fraction   **5.** prime number   **6.** simplest form   **7.** proper fraction   **8.** mixed number   **9.** numerator; denominator   **10.** prime factorization   **11.** undefined   **12.** 0 **13.** like   **14.** least common denominator   **15.** complex fraction   **16.** cross products

Chapter Review   1. $\dfrac{2}{6}$ **2.** $\dfrac{4}{7}$ **3.** $\dfrac{7}{3}$ or $2\dfrac{1}{3}$ **4.** $\dfrac{13}{4}$ or $3\dfrac{1}{4}$ **5.** $\dfrac{11}{12}$ **6.** $\dfrac{108}{131}$ **7.** -1 **8.** 1 **9.** 0 **10.** undefined

11. $\dfrac{7}{9}$ **12.** $\dfrac{4}{7}$ **13.** $1\dfrac{5}{4}$

14. $\dfrac{7}{5}$ **15.** $3\dfrac{3}{4}$ **16.** 3 **17.** $\dfrac{11}{5}$ **18.** $\dfrac{35}{9}$ **19.** $\dfrac{3}{7}$ **20.** $\dfrac{5}{9}$ **21.** $-\dfrac{1}{3x}$ **22.** $-\dfrac{y^2}{2}$ **23.** $\dfrac{29}{32c}$

24. $\dfrac{18z}{23}$ **25.** $\dfrac{5x}{3y^2}$ **26.** $\dfrac{7b}{5c^2}$ **27.** $\dfrac{2}{3}$ of a foot   **28.** $\dfrac{3}{5}$ of the cars   **29.** no   **30.** yes   **31.** $\dfrac{3}{10}$ **32.** $-\dfrac{5}{14}$ **33.** $\dfrac{9}{x^2}$ **34.** $\dfrac{y}{2}$

35. $-\dfrac{1}{27}$ **36.** $\dfrac{25}{144}$ **37.** -2 **38.** $\dfrac{15}{4}$ **39.** $\dfrac{27}{2}$ **40.** $-\dfrac{5}{6y}$ **41.** $\dfrac{12}{7}$ **42.** $-\dfrac{63}{10}$ **43.** $\dfrac{77}{48}$ sq ft   **44.** $\dfrac{4}{9}$ sq m   **45.** $\dfrac{10}{11}$ **46.** $\dfrac{2}{3}$

47. $-\dfrac{1}{3}$ **48.** $\dfrac{4x}{5}$ **49.** $\dfrac{4y-3}{21}$ **50.** $-\dfrac{1}{15}$ **51.** $3x$ **52.** 24 **53.** 20 **54.** 35 **55.** $49a$ **56.** $45b$ **57.** 40 **58.** 10

59. $\dfrac{3}{4}$ of his homework   **60.** $\dfrac{3}{2}$ mi   **61.** $\dfrac{11}{18}$ **62.** $\dfrac{7}{26}$ **63.** $-\dfrac{1}{12}$ **64.** $-\dfrac{5}{12}$ **65.** $\dfrac{25x+2}{55}$ **66.** $\dfrac{4+3b}{15}$ **67.** $\dfrac{7y}{36}$ **68.** $\dfrac{11x}{18}$

69. $\dfrac{4y+45}{9y}$ **70.** $-\dfrac{15}{14}$ **71.** $\dfrac{91}{150}$ **72.** $\dfrac{5}{18}$ **73.** $\dfrac{19}{9}$ m   **74.** $\dfrac{3}{2}$ ft   **75.** $\dfrac{21}{50}$ of the donors   **76.** $\dfrac{1}{4}$ yd   **77.** $\dfrac{4x}{7}$ **78.** $\dfrac{3y}{11}$ **79.** -2

80. $-7y$ **81.** $\dfrac{15}{4}$ **82.** $-\dfrac{5}{24}$ **83.** $\dfrac{8}{13}$ **84.** $-\dfrac{1}{27}$ **85.** $\dfrac{29}{110}$ **86.** $-\dfrac{1}{7}$ **87.** $20\dfrac{7}{24}$ **88.** $2\dfrac{51}{55}$; 3 **89.** $\dfrac{26}{5}$ or $5\dfrac{1}{5}$; 6 **90.** $\dfrac{21}{4}$ or $5\dfrac{1}{4}$

91. 22 mi   **92.** $\dfrac{110}{3}$ or $36\dfrac{2}{3}$ g   **93.** each measurement is $4\dfrac{1}{4}$ in.   **94.** $\dfrac{7}{10}$ yd   **95.** $-27\dfrac{5}{14}$ **96.** $1\dfrac{5}{27}$ **97.** $-3\dfrac{15}{16}$ **98.** $-\dfrac{33}{40}$ **99.** $\dfrac{5}{6}$

100. $-\dfrac{9}{10}$ **101.** -10 **102.** -6 **103.** 15 **104.** 1 **105.** 5 **106.** -4 **107.** $\dfrac{1}{4}$ **108.** $\dfrac{y^2}{2x}$ **109.** $\dfrac{1}{5}$ **110.** $-\dfrac{7}{12x}$ **111.** $\dfrac{11x}{12}$

112. $-\dfrac{23}{55}$ **113.** $-\dfrac{32}{5}$ or $-6\dfrac{2}{5}$ **114.** $12\dfrac{3}{8}$; 13 **115.** $2\dfrac{19}{35}$; 2 **116.** $\dfrac{22}{7}$ or $3\dfrac{1}{7}$ **117.** $-\dfrac{1}{12}$ **118.** $-\dfrac{9}{14}$ **119.** $-\dfrac{4}{9}$ **120.** -19

121. $44\dfrac{1}{2}$ yd   **122.** $\dfrac{81}{2}$ or $40\dfrac{1}{2}$ sq ft

Chapter Test   1. $\dfrac{7}{16}$ **2.** $\dfrac{23}{3}$ **3.** $18\dfrac{4}{3}$ **4.** $\dfrac{4}{35}$ **5.** $-\dfrac{3x}{5}$ **6.** not equivalent   **7.** equivalent   **8.** $2^2 \cdot 3 \cdot 7$ **9.** $3^2 \cdot 5 \cdot 11$

10. $\dfrac{4}{3}$ **11.** $-\dfrac{4}{3}$ **12.** $\dfrac{8x}{9}$ **13.** $\dfrac{x-21}{7x}$ **14.** y^2 **15.** $\dfrac{16}{45}$ **16.** $\dfrac{9a+4}{10}$ **17.** $-\dfrac{2}{3y}$ **18.** $\dfrac{1}{a^2}$ **19.** $\dfrac{3}{4}$ **20.** $14\dfrac{1}{40}$ **21.** $16\dfrac{8}{11}$

22. $\dfrac{64}{3}$ or $21\dfrac{1}{3}$ **23.** $\dfrac{45}{2}$ or $22\dfrac{1}{2}$ **24.** $-\dfrac{5}{3}$ or $-1\dfrac{2}{3}$ **25.** $\dfrac{9}{16}$ **26.** $\dfrac{3}{8}$ **27.** $\dfrac{11}{12}$ **28.** $\dfrac{3}{4x}$ **29.** $\dfrac{76}{21}$ **30.** -2 **31.** -4 **32.** 1

33. $\dfrac{5}{2}$ **34.** $\dfrac{4}{31}$ **35.** $3\dfrac{3}{4}$ ft   **36.** $\dfrac{23}{50}$ **37.** $\dfrac{13}{50}$ **38.** $\$2820$ **39.** perimeter: $3\dfrac{1}{3}$ ft; area: $\dfrac{2}{3}$ sq ft   **40.** 24 mi

Cumulative Review   1. five hundred forty-six; Sec. 1.2, Ex. 5   **2.** one hundred fifteen; Sec. 1.2   **3.** twenty-seven thousand, thirty-four (Sec 1.2, Ex. 6   **4.** six thousand five hundred seventy-three; Sec. 1.2   **5.** 759; Sec. 1.3, Ex. 1   **6.** 631; Sec. 1.3   **7.** 514; Sec. 1.3, Ex. 7   **8.** 933; Sec. 1.3   **9.** $278{,}000$; Sec. 1.4, Ex. 2   **10.** 1440; Sec. 1.4   **11.** $57{,}600$ megabytes; Sec. 1.5, Ex. 7   **12.** 1305 mi; Sec. 1.5   **13.** 7089 R 5; Sec. 1.6, Ex. 7   **14.** 379 R 10; Sec. 1.6   **15.** 7^3; Sec. 1.7, Ex. 1   **16.** 7^2; Sec. 1.7   **17.** $3^4 \cdot 9^3$; Sec. 1.7, Ex. 4   **18.** $9^4 \cdot 5^2$; Sec 1.7   **19.** 6; Sec. 1.8, Ex. 2   **20.** 52; Sec. 1.8   **21.** -6824; Sec. 2.1, Ex. 1   **22.** -21; Sec. 2.1   **23.** -4; Sec. 2.2, Ex. 3   **24.** 5; Sec. 2.2   **25.** 3; Sec. 2.3, Ex. 9   **26.** 10; Sec. 2.3   **27.** 25; Sec. 2.4, Ex. 8   **28.** -16; Sec. 2.4   **29.** -16; Sec. 2.5, Ex. 8   **30.** 25; Sec. 2.5   **31.** $6y + 2$; Sec. 3.1, Ex. 2   **32.** $6x + 9$; Sec. 3.1   **33.** -14; Sec. 3.2, Ex. 4   **34.** -18; Sec. 3.2   **35.** -1; Sec. 3.3, Ex. 2   **36.** -11; Sec. 3.3   **37.** $\dfrac{2}{5}$, Sec. 4.1, Ex. 3   **38.** $2^2 \cdot 3 \cdot 13$, Sec. 4.2   **39. a.** $\dfrac{38}{9}$ **b.** $\dfrac{19}{11}$, Sec. 4.1, Ex. 20

40. $\dfrac{39}{5}$, Sec. 4.1   **41.** $\dfrac{7x}{11}$, Sec. 4.2, Ex. 5   **42.** $\dfrac{2}{3y}$, Sec. 4.2   **43.** $\dfrac{35}{12}$ or $2\dfrac{11}{12}$, Sec. 4.7, Ex. 2   **44.** $\dfrac{8}{3}$ or $2\dfrac{2}{3}$, Sec. 4.7

45. $\dfrac{5}{12}$, Sec. 4.3, Ex. 11   **46.** $\dfrac{11}{56}$, Sec. 4.7

CHAPTER 5   Decimals

Vocabulary and Readiness Check   **1.** words; standard form   **3.** decimals   **5.** tenths; tens

Exercise Set 5.1   **1.** five and sixty-two hundredths   **3.** sixteen and twenty-three hundredths   **5.** negative two hundred five thousandths
7. one hundred sixty-seven and nine thousandths   **9.** three thousand and four hundredths   **11.** one hundred five and six tenths
13. two and forty-three hundredths

15.

| Preprinted Name | Current date |
| Preprinted Address | DATE |

PAY TO THE
ORDER OF ___R.W. Financial___ $ [321.42]
Three hundred twenty-one and 42/100   DOLLARS

FOR _____   _Signature_

17.

| Preprinted Name | Current date |
| Preprinted Address | DATE |

PAY TO THE
ORDER OF ___Bell South___ $ [59.68]
Fifty-nine and 68/100   DOLLARS

FOR _____   _Signature_

19. 2.8   **21.** 9.08   **23.** −705.625   **25.** 0.0046   **27.** $\frac{7}{10}$ **29.** $\frac{27}{100}$ **31.** $\frac{2}{5}$ **33.** $5\frac{2}{5}$ **35.** $-\frac{29}{500}$ **37.** $7\frac{1}{125}$ **39.** $15\frac{401}{500}$
41. $\frac{601}{2000}$ **43.** $0.8; \frac{8}{10}$ or $\frac{4}{5}$ **45.** seventy-seven thousandths; $\frac{77}{1000}$ **47.** <   **49.** <   **51.** <   **53.** =   **55.** <   **57.** >   **59.** <
61. >   **63.** 0.6   **65.** 98,210   **67.** −0.23   **69.** 0.594   **71.** 3.1   **73.** 3.142   **75.** $27   **77.** $0.20   **79.** 10.4 cm   **81.** 1.73 hr
83. $48   **85.** 24.623 hr   **87.** 5766   **89.** 35   **91.** b   **93.** a   **95.** answers may vary   **97.** 7.12   **99.** $\frac{26,849,576}{100,000,000,000}$
101. answers may vary   **103.** answers may vary   **105.** 0.0612, 0.0586   **107.** 0.01, 0.0839, 0.09, 0.1   **109.** $20
Calculator Explorations   **1.** 328.742   **3.** 5.2414   **5.** 865.392

Vocabulary and Readiness Check   **1.** last   **3.** like   **5.** vertically

Exercise Set 5.2   **1.** 7.8   **3.** 10.35   **5.** 27.0578   **7.** −8.57   **9.** 10.33   **11.** 465.56; $\begin{array}{r}230\\+230\\\hline 460\end{array}$ **13.** 115.123; $\begin{array}{r}100\\6\\+\ 9\\\hline 115\end{array}$ **15.** 50.409

17. 4.4   **19.** 15.3   **21.** 598.23   **23.** 1.83; 6 − 4 = 2   **25.** 876.6; $\begin{array}{r}1000\\-\ 100\\\hline 900\end{array}$ **27.** 194.4   **29.** −6.32   **31.** −6.15   **33.** 3.1   **35.** 2.9988

37. 16.3   **39.** 3.1   **41.** −5.62   **43.** 363.36   **45.** −549.8   **47.** 861.6   **49.** 115.123   **51.** 0.088   **53.** −180.44   **55.** −1.1
57. 3.81   **59.** 3.39   **61.** 1.61   **63.** no   **65.** yes   **67.** no   **69.** $6.9x + 6.9$ **71.** $-10.97 + 3.47y$ **73.** $0.17   **75.** 28.56 m
77. 11.2 in.   **79.** $7.52   **81.** 285.8 mph   **83.** $3.4 billion   **85.** $1498.49   **87.** 240.8 in.   **89.** 67.44 ft   **91.** 18.193 mph

93. Switzerland   **95.** 8.1 lb   **97.**

Country	Pounds of Chocolate per Person
Switzerland	22.0
Norway	16.0
Germany	15.8
United Kingdom	14.5
Belgium	13.9

99. 138   **101.** $\frac{1}{125}$ **103.** 6.08 in.   **105.** $1.20
107. 1 nickel, 1 dime, and 2 pennies; 3 nickels and 2 pennies; 1 dime and 7 pennies   **109.** answers may vary
111. 0.777 mi   **113.** $-109.544x + 15.604$

Vocabulary and Readiness Check   **1.** sum   **3.** right; zeros   **5.** circumference

Exercise Set 5.3   **1.** 1.36   **3.** 0.6   **5.** −17.595   **7.** 55.008   **9.** 28.56; 7 × 4 = 28   **11.** 0.1041   **13.** 8.23854; $\begin{array}{r}1\\\times 8\\\hline 8\end{array}$ **15.** 11.2746

17. 65   **19.** 0.83   **21.** −7093   **23.** 709.3   **25.** 0.0983   **27.** 0.02523   **29.** 0.0492   **31.** 14,790   **33.** 1.29   **35.** −9.3762
37. 0.5623   **39.** 36.024   **41.** 1,500,000,000   **43.** 49,800,000   **45.** −0.6   **47.** 17.3   **49.** no   **51.** yes   **53.** 10π cm ≈ 31.4 cm
55. 18.2π yd ≈ 57.148 yd   **57.** 24.8 g   **59.** 7.2 sq in.   **61.** 250π ft ≈ 785 ft   **63.** 135π m ≈ 423.9 m   **65.** 64.9605 in.   **67.** $715.20
69. a. 62.8 m and 125.6 m   **b.** yes   **71.** $340   **73.** 77,800.5 yen   **75.** 715.59 euros   **77.** 8   **79.** −9   **81.** 3.64   **83.** 3.56
85. −0.1105   **87.** 3,831,600 mi   **89.** answers may vary   **91.** answers may vary

Calculator Explorations   **1.** not reasonable   **3.** reasonable

Vocabulary and Readiness Check   **1.** quotient; divisor; dividend   **3.** left; zeros

Exercise Set 5.4   **1.** 4.6   **3.** 0.094   **5.** 300   **7.** 5.8   **9.** 6.6; $6\overline{)36}$ **11.** 0.413   **13.** −600   **15.** 7   **17.** 4.8   **19.** 2100

21. 5.8   **23.** 5.5   **25.** 9.8; $7\overline{)70}$ **27.** 9.6   **29.** 45   **31.** 54.592   **33.** 0.0055   **35.** 23.87   **37.** 114.0   **39.** 0.83397   **41.** 2.687
43. −0.0129   **45.** 12.6   **47.** 1.31   **49.** 0.045625   **51.** 0.413   **53.** −7   **55.** −4.8   **57.** 2100   **59.** 30   **61.** −58,000
63. −0.69   **65.** 0.024   **67.** 65   **69.** −5.65   **71.** −7.0625   **73.** yes   **75.** no   **77.** 11 qt   **79.** 5.1 m   **81.** 11.4 boxes

83. 24 tsp   **85.** 8 days   **87.** 248.1 mi   **89.** $5109.62 per hr   **91.** $\frac{21}{50}$ **93.** $-\frac{1}{10}$ **95.** 4.26   **97.** 1.578   **99.** −26.66

101. 904.29   **103.** c   **105.** b   **107.** 85.5   **109.** 8.6 ft   **111.** answers may vary   **113.** 65.2−82.6 knots   **115.** 319.64 m

The Bigger Picture   **1.** 22.172   **2.** 3.951   **3.** 9133.2   **4.** −6.8   **5.** 1.404   **6.** 8.66   **7.** 0.051   **8.** 2.14   **9.** $\frac{2}{15}$ **10.** $-\frac{16}{75}$

Integrated Review   **1.** 2.57   **2.** 4.05   **3.** 8.9   **4.** 3.5   **5.** 0.16   **6.** 0.24   **7.** 0.27   **8.** 0.52   **9.** −4.8   **10.** 6.09   **11.** 75.56
12. 289.12   **13.** −24.974   **14.** −43.875   **15.** −8.6   **16.** 5.4   **17.** −280   **18.** 1600   **19.** 224.938   **20.** 145.079   **21.** 0.56
22. −0.63   **23.** 27.6092   **24.** 145.6312   **25.** 5.4   **26.** −17.74   **27.** −414.44   **28.** −1295.03   **29.** −34   **30.** −28   **31.** 116.81
32. 18.79   **33.** 156.2   **34.** 1.562   **35.** 25.62   **36.** 5.62   **37.** 200 mi   **38.** $0.90   **39.** $24.88 billion or $24,880,000,000

Vocabulary and Readiness Check   **1.** False   **3.** False

Exercise Set 5.5   **1.** 0.2   **3.** 0.68   **5.** 0.75   **7.** −0.08   **9.** 2.25   **11.** $0.91\overline{6}$ **13.** 0.425   **15.** 0.45   **17.** $-0.\overline{3}$ **19.** 0.4375
21. $0.\overline{63}$ **23.** 5.85   **25.** 0.624   **27.** −0.33   **29.** 0.44   **31.** 0.6   **33.** 0.62   **35.** 0.71   **37.** 0.02   **39.** <   **41.** =   **43.** <

45. <   **47.** <   **49.** >   **51.** <   **53.** <   **55.** 0.32, 0.34, 0.35   **57.** 0.49, 0.491, 0.498   **59.** 5.23, $\frac{42}{8}$, 5.34   **61.** 0.612, $\frac{5}{8}$, 0.649

63. 0.59   **65.** −3   **67.** 5.29   **69.** 9.24   **71.** 0.2025   **73.** −1.29   **75.** −15.4   **77.** −3.7   **79.** 25.65 sq in.   **81.** 0.248 sq yd

83. 5.76   **85.** 5.7   **87.** 3.6   **89.** $\frac{77}{50}$ **91.** $\frac{5}{2}$ **93.** = 1   **95.** > 1   **97.** < 1   **99.** 0.189   **101.** 6000 stations

103. answers may vary

Exercise Set 5.6   **1.** 5.9   **3.** −0.43   **5.** 10.2   **7.** −4.5   **9.** 4   **11.** 0.45   **13.** 4.2   **15.** −4   **17.** 1.8   **19.** 10   **21.** 7.6

23. 60   **25.** −0.07   **27.** 20   **29.** 0.0148   **31.** −8.13   **33.** 1.5   **35.** −1   **37.** −7   **39.** 7   **41.** 53.2   **43.** $3x − 16$ **45.** $\frac{3}{5x}$

47. $\frac{32}{55}$ **49.** 3.7   **51.** $6x − 0.61$ **53.** $-2y + 6.8$ **55.** 9.1   **57.** −3   **59.** $-4z + 16.67$ **61.** 15.7   **63.** 5.85

65. $-2.1 − 10.1$ **67.** answers may vary   **69.** 7.683   **71.** 4.683

Vocabulary and Readiness Check   **1.** average   **3.** mean (or average)   **5.** grade point average

Exercise Set 5.7   **1.** mean: 21; median: 23; no mode   **3.** mean: 8.1; median: 8.2; mode: 8.2   **5.** mean: 0.5; median: 0.5; mode: 0.2 and 0.5
7. mean: 370.9; median: 313.5; no mode   **9.** 1495 ft   **11.** 1415 ft   **13.** answers may vary   **15.** 2.79   **17.** 3.64   **19.** 6.8   **21.** 6.9

23. 88.5   **25.** 73   **27.** 70 and 71   **29.** 9 rates   **31.** $\frac{1}{3}$ **33.** $\frac{3}{5y}$ **35.** $\frac{11}{15}$ **37.** 35, 35, 37, 43   **39.** yes; answers may vary

Chapter 5 Vocabulary Check   **1.** decimal   **2.** numerator; denominator   **3.** vertically   **4.** and   **5.** sum   **6.** mode
7. circumference   **8.** median; mean   **9.** mean   **10.** standard form

Chapter Review   **1.** tenths   **2.** hundred-thousandths   **3.** negative twenty-three and forty-five hundredths   **4.** three hundred forty-five
hundred-thousandths   **5.** one hundred nine and twenty-three hundredths   **6.** two hundred and thirty-two millionths   **7.** 8.06   **8.** −503.102

9. 16,025.0014   **10.** 14.011   **11.** $\frac{4}{25}$ **12.** $-12\frac{23}{1000}$ **13.** 0.00231   **14.** 25.25   **15.** >   **16.** =   **17.** <   **18.** >   **19.** 0.6

20. 0.94   **21.** −42.90   **22.** 16.349   **23.** 887,000,000   **24.** 600,000   **25.** 18.1   **26.** 5.1   **27.** −7.28   **28.** −12.04   **29.** 320.312

30. 148.74236   **31.** 1.7   **32.** 2.49   **33.** −1324.5   **34.** −10.136   **35.** 65.02   **36.** 199.99802   **37.** 52.6 mi   **38.** −5.7   **39.** 22.2 in.
40. 38.9 ft   **41.** 72   **42.** 9345   **43.** −78.246   **44.** 73,246.446   **45.** 14π m ≈ 43.96 m   **46.** 20π in. ≈ 62.8 in.   **47.** 0.0877

48. 15.825   **49.** 70   **50.** −0.21   **51.** 8.059   **52.** 30.4   **53.** 0.02365   **54.** −9.3   **55.** 7.3 m   **56.** 45 mo   **57.** 0.8   **58.** −0.923
59. $2.\overline{3}$ or 2.333   **60.** $0.21\overline{6}$ or 0.217   **61.** =   **62.** <   **63.** <   **64.** <   **65.** 0.832, 0.837, 0.839   **66.** $\frac{5}{8}$, 0.626, 0.685

67. 0.42, $\frac{3}{7}$, 0.43   **68.** $\frac{19}{12}$, 1.63, $\frac{18}{11}$ **69.** −11.94   **70.** 3.89   **71.** −129   **72.** 0.81   **73.** 55   **74.** 7.26

75. 6.9 sq ft   **76.** 5.46 sq in.   **77.** 0.3   **78.** 92.81   **79.** 8.6   **80.** −80   **81.** 1.98   **82.** −1.5   **83.** −20   **84.** 1

85. mean: 17.8; median: 14; no mode   **86.** mean: 58.1; median: 60; mode: 45 and 86   **87.** mean: 24,500; median: 20,000; mode: 20,000
88. mean: 447.3; median: 420; mode: 400   **89.** 3.25   **90.** 2.57   **91.** two hundred and thirty-two ten-thousandths   **92.** −16.09

93. $\frac{847}{10,000}$ **94.** 0.75, $\frac{6}{7}$, $\frac{8}{9}$ **95.** −0.07   **96.** 0.1125   **97.** 51.057   **98.** >   **99.** <   **100.** 86.91   **101.** 3.115   **102.** $123.00

103. $3646.00   **104.** −1.7   **105.** 5.26   **106.** −12.76   **107.** −14.907   **108.** 8.128   **109.** −7.245   **110.** 4900   **111.** 23.904
112. 9600 sq ft   **113.** no   **114.** 0.1024   **115.** 3.6   **116.** mean: 74.4; median: 73; mode: none   **117.** mean: 619.17; median: 647.5; mode: 327

Chapter Test   **1.** forty-five and ninety-two thousandths   **2.** 3000.059   **3.** 17.595   **4.** −51.20   **5.** −20.42   **6.** 40.902   **7.** 0.037

8. 34.9   **9.** 0.862   **10.** <   **11.** <   **12.** $\frac{69}{200}$ **13.** $-24\frac{73}{100}$ **14.** −0.5   **15.** 0.941   **16.** 1.93   **17.** −6.2   **18.** $0.5x - 13.4$

19. −3   **20.** 3.7   **21.** mean: 38.4; median: 42; no mode   **22.** mean: 12.625; median: 12.5; mode: 12 and 16   **23.** 3.07   **24.** 4,583,000,000

25. 2.31 sq mi   **26.** 18π mi, 56.52 mi   **27. a.** 9904 sq ft   **b.** 198.08 oz   **28.** 54 mi

Cumulative Review   **1.** seventy-two; Sec. 1.2, Ex. 4   **2.** one hundred seven; Sec. 1.2   **3.** five hundred forty-six; Sec. 1.2, Ex. 5
4. five thousand, twenty-six; Sec. 1.2   **5.** 759; Sec. 1.3, Ex. 1   **6.** 19 in.; Sec. 1.3   **7.** 514; Sec. 1.3, Ex. 7   **8.** 121 R 1; Sec. 1.6
9. 278,000; Sec. 1.4, Ex. 2   **10.** 1, 2, 3, 5, 6, 10, 15, 30; Sec. 4.2   **11.** 20,296; Sec. 1.5, Ex. 4   **12.** 0; Sec. 1.5
13. a. 7   **b.** 12   **c.** 1   **d.** 1   **e.** 20   **f.** 1; Sec. 1.6, Ex. 2   **14.** 25; Sec. 1.6   **15.** 7; Sec. 1.7, Ex. 9   **16.** 49; Sec. 1.7
17. 81; Sec. 1.7, Ex. 5   **18.** 125; Sec. 1.7   **19.** 81; Sec. 1.7, Ex. 7   **20.** 1000; Sec. 1.7   **21.** 2; Sec. 1.8, Ex. 3   **22.** 6; Sec. 1.8
23. a. −13   **b.** 2   **c.** 0; Sec. 2.1, Ex. 5   **24. a.** 7   **b.** −4   **c.** 1; Sec. 2.1   **25.** −23; Sec. 2.2, Ex. 4   **26.** −22; Sec. 2.2
27. 180; Sec. 1.7, Ex. 8   **28.** 32; Sec. 1.7   **29.** −49; Sec. 2.4, Ex. 9   **30.** −32; Sec. 2.4   **31.** 25; Sec. 2.4, Ex. 8   **32.** −9; Sec. 2.4

33. $\frac{4}{3}$ or $1\frac{1}{3}$; Sec. 4.1, Ex. 10   **34.** $\frac{11}{4}$ or $2\frac{3}{4}$; Sec. 4.1   **35.** $\frac{11}{4}$ or $2\frac{3}{4}$; Sec. 4.1, Ex. 11   **36.** $\frac{14}{3}$ or $4\frac{2}{3}$; Sec. 4.1   **37.** $2^2 \cdot 3^2 \cdot 7$; Sec. 4.2, Ex. 3

38. 62; Sec. 1.3   **39.** $-\frac{36}{13}$; Sec. 4.2, Ex. 8   **40.** $\frac{79}{8}$; Sec. 4.1   **41.** equivalent; Sec. 4.2, Ex. 10   **42.** >, Sec. 4.5

43. $\frac{10}{33}$; Sec. 4.3, Ex. 1   **44.** $\frac{3}{2}$ or $1\frac{1}{2}$; Sec. 4.7   **45.** $\frac{1}{8}$; Sec. 4.3, Ex. 2   **46.** 37; Sec. 4.7   **47.** −24; Sec. 3.2, Ex. 3   **48.** −8; Sec. 3.2

49. 829.6561; Sec. 5.2, Ex. 2   **50.** 230.8628; Sec. 5.2   **51.** 18.408; Sec. 5.3, Ex. 1   **52.** 28.251; Sec. 5.3

Chapter 6   Ratio, Proportion, and Triangle Applications

Vocabulary and Readiness Check   **1.** unit   **3.** division   **5.** numerator; denominator   **7.** false

Exercise Set 6.1   **1.** $\frac{2}{3}$ **3.** $\frac{77}{100}$ **5.** $\frac{463}{821}$ **7.** $\frac{3}{8}$ **9.** $\frac{8}{25}$ **11.** $\frac{12}{7}$ **13.** $\frac{2}{7}$ **15.** $\frac{4}{1}$ **17.** $\frac{13}{51}$ **19.** $\frac{5}{11}$ **21.** $\frac{5}{4}$ **23.** $\frac{15}{1}$ **25.** $\frac{1}{3}$

27. $\frac{9}{19}$ **29.** $\frac{7}{6}$ **31.** $\frac{1\,\text{shrub}}{3\,\text{ft}}$ **33.** $\frac{3\,\text{returns}}{20\,\text{sales}}$ **35.** $\frac{3\,\text{laser printers}}{14\,\text{computers}}$ **37.** $\frac{9\,\text{gal}}{2\,\text{acres}}$ **39.** 110 cal/oz   **41.** 75 riders/car

43. 90 wingbeats/sec   **45.** $50,000/yr   **47.** 1,369,022 voters/senator   **49.** 300 good/defective   **51.** $65,000 house

53. a. 31.25 computer boards/hr   **b.** 33.5 computer boards/hr   **c.** Suellen   **55. a.** ≈27.6 miles/gal   **b.** ≈29.2 miles/gal   **c.** the truck

57. $11.50 per compact disc   **59.** $0.17 per banana   **61.** 8 oz: $0.149 per oz; 12 oz: $0.133 per oz; 12 oz
63. 16 oz: $0.106 per oz; 6 oz: $0.115 per oz; 16 oz   **65.** 12 oz: $0.191 per oz; 8 oz: $0.186 per oz; 8 oz
67. 100: $0.006 per napkin; 180: $0.005 per napkin; 180 napkins   **69.** 2.3   **71.** 0.15   **73.** no; answers may vary   **75.** no; $\frac{2}{5}$

77. yes, the machine should be repaired   **79.** 257; 19.2; 352; 22.3; 347; 21.6   **81.** 544 students/school   **83.** answers may vary

85. no; answers may vary   **87. a.** $\frac{2}{5}$ **b.** $\frac{2}{3}$ **c.** no; answers may vary

Vocabulary and Readiness Check   **1.** proportion; ratio   **3.** true

Exercise Set 6.2   **1.** $\frac{10\,\text{diamonds}}{6\,\text{opals}} = \frac{5\,\text{diamonds}}{3\,\text{opals}}$ **3.** $\frac{20\,\text{students}}{5\,\text{microscopes}} = \frac{4\,\text{students}}{1\,\text{microscope}}$ **5.** $\frac{6\,\text{eagles}}{58\,\text{sparrows}} = \frac{3\,\text{eagles}}{29\,\text{sparrows}}$

7. $\frac{2\frac{1}{4}\,\text{cups flour}}{24\,\text{cookies}} = \frac{6\frac{3}{4}\,\text{cups flour}}{72\,\text{cookies}}$ **9.** $\frac{22\,\text{vanilla wafers}}{1\,\text{cup cookie crumbs}} = \frac{55\,\text{vanilla wafers}}{2.5\,\text{cups cookie crumbs}}$ **11.** true   **13.** false   **15.** true   **17.** true

19. false   **21.** true   **23.** true   **25.** false   **27.** true   **29.** $\frac{10}{15} = \frac{4}{6}$; true   **31.** $\frac{11}{4} = \frac{5}{2}$; false   **33.** $\frac{0.15}{3} = \frac{0.35}{7}$; true   **35.** $\frac{\frac{2}{3}}{\frac{1}{5}} = \frac{\frac{2}{5}}{\frac{1}{9}}$; false

37. 3   **39.** −9   **41.** 5   **43.** 3.2   **45.** 38.4   **47.** 25   **49.** 0.0025   **51.** 1   **53.** $\frac{9}{20}$ **55.** 12   **57.** $\frac{3}{4}$ **59.** $\frac{35}{18}$

61. 14.9   **63.** 0.07   **65.** 1.9   **67.** 3.163   **69.** <   **71.** >   **73.** $\frac{3}{5}$ **75.** $\frac{2x}{7}$ **77.** $\frac{9}{3} = \frac{15}{5}; \frac{5}{15} = \frac{3}{9}; \frac{15}{9} = \frac{5}{3}$

79. $\frac{6}{1} = \frac{18}{3}; \frac{3}{18} = \frac{1}{6}; \frac{18}{6} = \frac{3}{1}$ **81.** $\frac{d}{b} = \frac{c}{a}; \frac{a}{c} = \frac{b}{d}; \frac{b}{a} = \frac{d}{c}$ **83.** answers may vary   **85.** 0   **87.** 1400   **89.** 252.5

The Bigger Picture   **1.** $\frac{1}{4}$ **2.** $\frac{7}{200}$ **3.** $\frac{7}{2}$ or $3\frac{1}{2}$ **4.** $\frac{9}{20}$ **5.** 7.62   **6.** 0.152   **7.** 8   **8.** $\frac{5}{2}$ or $2\frac{1}{2}$ **9.** 0.004   **10.** $\frac{35}{12}$ or $2\frac{11}{12}$

Integrated Review   **1.** $\frac{9}{10}$ **2.** $\frac{9}{25}$ **3.** $\frac{47}{50}$ **4.** $\frac{8}{23}$ **5.** $\frac{173}{139}$ **6.** $\frac{6}{7}$ **7.** $\frac{7}{26}$ **8.** $\frac{20}{33}$ **9.** $\frac{2}{3}$ **10.** $\frac{1}{8}$ **11.** $\frac{2}{3}$ **12. a.** 10   **b.** $\frac{1}{2}$

13. $\frac{1\,\text{professor}}{5\,\text{graduate assistants}}$ **14.** $\frac{3\,\text{lights}}{10\,\text{ft}}$ **15.** $\frac{2\,\text{Senators}}{1\,\text{state}}$ **16.** $\frac{1\,\text{teacher}}{28\,\text{students}}$ **17.** $\frac{7\,\text{inches}}{1\,\text{second}}$ **18.** $\frac{\$8}{1\,\text{hour}}$

19. $\dfrac{19\,\text{households with computers}}{25\,\text{households}}$ **20.** $\dfrac{269\,\text{electoral votes}}{25\,\text{states}}$ **21.** 140 ft/sec **22.** 65 mi/hr **23.** 21 employees/fax line

24. 17 phone calls/teenager **25.** 26 mi/gal **26.** 16 teachers/computer **27.** 6.5 books/student **28.** 165 lb/adult
29. 8 lb: $0.27 per lb; 18 lb: $0.277 per lb; 8 lb **30.** 100: $0.020 per plate; 500: $0.018 per plate; 500 paper plates
31. 3 packs: $0.797 per pack; 8 packs: $0.749 per pack; 8 packs **32.** 4: $0.923 per battery; 10: $0.989 per battery; 4 batteries **33.** no **34.** yes
35. 24 **26.** 32.5 **27.** $2.\overline{72}$ or $2\dfrac{8}{11}$ **28.** 18

Exercise Set 6.3 **1.** 360 baskets **3.** 165 min **5.** 630 applications **7.** 23 ft **9.** 270 sq ft **11.** 25 gal **13.** 450 km **15.** 16 bags

17. 15 hits **19.** 27 people **21.** 18 applications **23.** 5 weeks **25.** $10\dfrac{2}{3}$ servings **27.** 120 calories **29. a.** 18 tsp **b.** 6 tbsp

31. 6 people **33.** 112 ft; 11-in. difference **35.** 102.9 mg **37.** 1248 ft; coincidentally, this is the actual height of the Empire State Building

39. 434 emergency room visits **41.** 28 workers **43.** 2.4 c **45. a.** 0.1 gal **b.** 13 fl oz **47. a.** 2062.5 mg **b.** no **49.** $2^3 \cdot 5^2$ **51.** 2^5

53. 0.8 ml **55.** 1.25 ml **57.** $11 \approx 12$ or 1 dozen; $1.5 \times 8 = 12$; 12 cups of milk **59.** $4\dfrac{2}{3}$ ft **61.** answers may vary

Calculator Explorations **1.** 32 **3.** 3.873 **5.** 9.849

Vocabulary and Readiness Check **1.** $10; -10$ **3.** radical **5.** perfect squares **7.** $c^2; b^2$

Exercise Set 6.4 **1.** 2 **3.** 11 **5.** $\dfrac{1}{9}$ **7.** $\dfrac{4}{8} = \dfrac{1}{2}$ **9.** 1.732 **11.** 3.873 **13.** 5.568 **15.** 5.099 **17.** 16 **19.** 9.592 **21.** $\dfrac{7}{12}$

23. 8.426 **25.** 13 in. **27.** 6.633 cm **29.** 52.802 m **31.** 117 mm **33.** 5 **35.** 12 **37.** 17.205 **39.** 44.822 **41.** 42.426

43. 1.732 **45.** 8.5 **47.** 141.42 yd **49.** 25.0 ft **51.** 340 ft **53.** $\dfrac{5}{6}$ **55.** $\dfrac{x}{30}$ **57.** $\dfrac{3x-5}{9}$ **59.** $\dfrac{9x}{64}$ **61.** 6, 7 **63.** 10, 11

65. answers may vary **67.** no

The Bigger Picture **1.** 9 **2.** $\dfrac{2}{7}$ **3.** 15 **4.** 8 **5.** 24 **6.** 33 **7.** 28 **8.** 42 **9.** $\dfrac{3}{2}$ or $1\dfrac{1}{2}$ **10.** $\dfrac{17}{16}$ or $1\dfrac{1}{16}$

Vocabulary and Readiness Check **1.** false **3.** true **5.** false

Exercise Set 6.5 **1.** congruent **3.** not congruent **5.** congruent **7.** congruent **9.** $\dfrac{2}{1}$ **11.** $\dfrac{3}{2}$ **13.** 4.5 **15.** 6 **17.** 5

19. 13.5 **21.** 17.5 **23.** 10 **25.** 28.125 **27.** 10 **29.** 500 ft **31.** approx 560 ft high **33.** 14.4 ft **35.** 52 neon tetras

37. 381 ft **39.** 4.01 **41.** -1.23 **43.** $3\dfrac{8}{9}$ in.; no **45.** 32.7 **47.** 200 ft, 300 ft, 425 ft

Chapter 6 Vocabulary Check **1.** ratio **2.** proportion **3.** unit rate **4.** unit price **5.** rate **6.** cross products **7.** equal
8. not equal **9.** congruent **10.** similar **11.** leg **12.** leg **13.** hypotenuse **14.** right **15.** Pythagorean

Chapter Review **1.** $\dfrac{23}{37}$ **2.** $\dfrac{11}{13}$ **3.** $\dfrac{17}{35}$ **4.** $\dfrac{18}{35}$ **5.** $\dfrac{9}{4}$ **6.** $\dfrac{2}{13}$ **7.** $\dfrac{5\,\text{people}}{2\,\text{pets}}$ **8.** $\dfrac{5\,\text{pages}}{2\,\text{min}}$ **9.** 52 mi/hr **10.** 15 ft/sec

11. $1.74/diskette **12.** 13 bushels/tree **13.** 8 oz: $0.124 per oz; 12 oz: $0.141 per oz; 8-oz size **14.** 18 oz: $0.083; 28 oz: $0.085; 18-oz size

15. $\dfrac{24\,\text{uniforms}}{8\,\text{players}} = \dfrac{3\,\text{uniforms}}{1\,\text{player}}$ **16.** $\dfrac{12\,\text{tires}}{3\,\text{cars}} = \dfrac{4\,\text{tires}}{1\,\text{car}}$ **17.** no **18.** yes **19.** 5 **20.** 21 **21.** -5.625 **22.** 60 **23.** 0.94

24. $1\dfrac{1}{8}$ **25.** 0.63 **26.** 30.9 **27.** 14 **28.** 35 **29.** 8 bags **30.** 16 bags **31.** $213\dfrac{1}{3}$ mi **32.** 9.6 in. **33.** 8 **34.** 12

35. 3.464 **36.** 3.873 **37.** 0 **38.** 1 **39.** 7.071 **40.** 8.062 **41.** $\dfrac{2}{5}$ **42.** $\dfrac{1}{10}$ **43.** 13 **44.** 29 **45.** 10.7 **46.** 86.6

47. 28.28 cm **48.** 88.2 ft **49.** congruent: ASA **50.** not congruent **51.** $13\dfrac{1}{3}$ **52.** 7.4 **53.** approximately 33 ft

54. $x = \dfrac{5}{6}$ in.; $y = 2\dfrac{1}{6}$ in. **55.** $\dfrac{3}{5}$ **56.** $\dfrac{1}{27}$ **57.** $\dfrac{1\,\text{teacher}}{9\,\text{students}}$ **58.** $\dfrac{1\,\text{nurse}}{4\,\text{patients}}$ **59.** 34 mi/hr **60.** 2 gal/cow **61.** $\dfrac{29}{256}$

62. 166 steps/min **63.** 4 oz: $1.235; 8 oz $1.248; 4-oz size **64.** 12 oz: $0.054; 64 oz: $0.047; 64-oz size

65. $\dfrac{2\,\text{cups cookie dough}}{30\,\text{cookies}} = \dfrac{4\,\text{cups cookie dough}}{60\,\text{cookies}}$ **66.** $\dfrac{5\,\text{nickels}}{3\,\text{dollars}} = \dfrac{20\,\text{nickels}}{12\,\text{dollars}}$ **67.** 1.6 **68.** 25 **69.** 3.75 **70.** $\dfrac{2}{5}$ **71.** 6

72. $\dfrac{4}{9}$ **73.** 10.247 **74.** 5.657 **75.** 86.6 **76.** 20.8 **77.** 12 **78.** $6\dfrac{1}{2}$

Chapter Test **1.** $\dfrac{9}{13}$ **2.** $\dfrac{3\,\text{in.}}{10\,\text{days}}$ **3.** $\dfrac{43}{50}$ **4.** $\dfrac{47}{78}$ **5.** $\dfrac{138}{77}$ **6.** 81.25 km/hr **7.** 28 students/teacher **8.** 9 in./sec

9. 8 oz: $0.149 per oz; 12 oz: $0.158 per oz; 8-oz size **10.** 16 oz: $0.0931 per oz; 24 oz: $0.0996 per oz; 16-oz size **11.** true **12.** false

13. 5 **14.** $4\dfrac{4}{11}$ **15.** $\dfrac{7}{3}$ **16.** 8 **17.** $49\dfrac{1}{2}$ ft **18.** $3\dfrac{3}{4}$ hr **19.** $53\dfrac{1}{3}$ g **20.** 7 **21.** 12.530 **22.** $\dfrac{4}{5}$

23. 5.66 cm **24.** 7.5 **25.** 69 ft

Cumulative Review **1. a.** 3 **b.** 15 **c.** 0 **d.** 70; Sec. 1.3, Ex. 5 **2. a.** 0 **b.** 20 **c.** 0 **d.** 20; Sec. 1.5 **3.** 249,000; Sec. 1.4, Ex. 3
4. 249,000; Sec. 1.4 **5. a.** 200 **b.** 1230; Sec. 1.5, Ex. 3 **6.** 373 R 24; Sec. 1.6 **7.** −8, Sec. 2.2, Ex. 15 **8.** −84, Sec. 2.4

9. $2^4 \cdot 5$, Sec. 4.2, Ex. 2 **10.** 8, Sec. 1.7 **11.** $\frac{3}{5}$, Sec. 4.2, Ex. 4 **12.** 243; Sec. 1.7 **13.** $\frac{2}{5}$, Sec. 4.3, Ex. 6 **14.** $\frac{123}{8}$ or $15\frac{3}{8}$, Sec. 4.7

15. 2; Sec. 4.4, Ex. 3 **16.** $\frac{4}{5}$, Sec. 4.4 **17.** 14; Sec. 4.4, Ex. 12 **18.** $\frac{49}{50}$, Sec. 4.5 **19.** $\frac{15}{20}$, Sec. 4.4, Ex. 17 **20.** yes; Sec. 4.2

21. $-\frac{8}{33}$; Sec. 4.5, Ex. 4 **22.** $7\frac{47}{72}$; Sec. 4.7 **23.** $\frac{1}{6}$ hr; Sec. 4.5, Ex. 11 **24.** 27; Sec. 1.7 **25.** $7\frac{17}{24}$, Sec. 4.7, Ex. 9 **26.** $\frac{16}{27}$; Sec. 4.6

27. $<$, Sec. 4.5, Ex. 7 **28.** −8, Sec. 3.2 **29.** −3, Sec. 3.2, Ex. 1 **30.** 0; Sec. 3.3 **31.** 5; Sec. 3.3, Ex. 1 **32.** 5; Sec. 3.3
33. 2; Sec. 3.3, Ex. 4 **34.** 0.075; Sec. 5.1 **35.** 736.2; Sec. 5.1, Ex. 15 **36.** 736.236; Sec. 5.1 **37.** 25.454; Sec. 5.2, Ex. 1 **38.** 681.24; Sec. 5.2

39. 0.8496; Sec. 5.3, Ex. 2 **40.** 0.375; Sec. 5.5 **41.** −0.052, Sec. 5.4, Ex. 3 **42.** $\frac{79}{10}$, Sec. 5.5 **43.** −3.7, Sec. 5.5, Ex. 12 **44.** 3; Sec. 6.2

45. $\frac{4}{9}, \frac{9}{20}$, 0.456; Sec. 5.5, Ex. 10 **46.** 140 m/sec; Sec. 6.1 **47.** $\frac{3}{2}$, Sec. 6.1, Ex. 2 **48.** $\frac{1}{3}$, Sec. 6.1 **49.** $\frac{50}{63}$, Sec. 6.1, Ex. 3 **50.** $\frac{1}{10}$, Sec. 6.1

CHAPTER 7 Percent

Vocabulary and Readiness Check **1.** percent **3.** percent **5.** 0.01

Exercise Set 7.1 **1.** 96% **3.** football; 37% **5.** 50% **7.** 0.41 **9.** 0.06 **11.** 1.00 or 1 **13.** 0.736 **15.** 0.028 **17.** 0.006

19. 3.00 or 3 **21.** 0.3258 **23.** $\frac{2}{25}$ **25.** $\frac{1}{25}$ **27.** $\frac{9}{200}$ **29.** $\frac{7}{4}$ or $1\frac{3}{4}$ **31.** $\frac{1}{16}$ **33.** $\frac{31}{300}$ **35.** $\frac{179}{800}$ **37.** 0.6% **39.** 22%

41. 530% **43.** 5.6% **45.** 22.28% **47.** 300% **49.** 70% **51.** 70% **53.** 80% **55.** 68% **57.** $37\frac{1}{2}$% **59.** $33\frac{1}{3}$%

61. 450% **63.** 190% **65.** 81.82% **67.** 26.67% **69.** $0.6; \frac{3}{5}; 23\frac{1}{2}\%; \frac{47}{200}; 80\%; 0.8; 0.333\overline{3}; \frac{1}{3}; 87.5\%; 0.875; 0.075; \frac{3}{40}$

71. $2; 2; 280\%; 2\frac{4}{5}; 7.05; 7\frac{1}{20}; 454\%; 4.54$ **73.** $0.38; \frac{19}{50}$ **75.** $0.252; \frac{63}{250}$ **77.** $0.322; \frac{161}{500}$ **79.** $0.004; \frac{1}{250}$ **81.** $0.12; \frac{3}{25}$ **83.** 44.2%

85. 0.7% **87.** $\frac{39}{1000}$ **89.** $\frac{11}{36}$ **91.** $1\frac{5}{6}$ **93. a.** 52.9% **b.** 52.86% **95.** b, d **97.** 4% **99.** 75% **101.** greater

103. 0.266; 26.6% **105.** network systems and data communication analysts **107.** 0.49 **109.** answers may vary

Vocabulary and Readiness Check **1.** is **3.** amount; base; percent **5.** greater

Exercise Set 7.2 **1.** $18\% \cdot 81 = x$ **3.** $20\% \cdot x = 105$ **5.** $0.6 = 40\% \cdot x$ **7.** $x \cdot 80 = 3.8$ **9.** $x = 9\% \cdot 43$ **11.** $x \cdot 250 = 150$
13. 3.5 **15.** 28.7 **17.** 10 **19.** 600 **21.** 110% **23.** 34% **25.** 1 **27.** 645 **29.** 500 **31.** 5.16% **33.** 25.2
35. 35% **37.** 35 **39.** 0.624 **41.** 0.5% **43.** 145 **45.** 63% **47.** 4% **49.** 30 **51.** $3\frac{7}{11}$ **53.** $\frac{17}{12} = \frac{x}{20}$ **55.** $\frac{8}{9} = \frac{14}{x}$
57. c **59.** b **61.** What number equals thirty-three and one-third percent of twenty-four? **63.** a **65.** c **67.** a **69.** a
71. answers may vary **73.** 686.625 **75.** 12,285

The Bigger Picture **1.** $\frac{19}{45}$ **2.** 63 **3.** −0.021 **4.** 8 **5.** 48 **6.** 1.6 **7.** 250% **8.** 45.6 or $45\frac{3}{5}$ **9.** 28% **10.** 180

Vocabulary and Readiness Check **1.** amount; base; percent **3.** amount

Exercise Set 7.3 **1.** $\frac{a}{45} = \frac{98}{100}$ **3.** $\frac{a}{150} = \frac{4}{100}$ **5.** $\frac{14.3}{b} = \frac{26}{100}$ **7.** $\frac{84}{b} = \frac{35}{100}$ **9.** $\frac{70}{400} = \frac{p}{100}$ **11.** $\frac{8.2}{82} = \frac{p}{100}$ **13.** 26 **15.** 18.9

17. 600 **19.** 10 **21.** 120% **23.** 28% **25.** 37 **27.** 1.68 **29.** 1000 **31.** 210% **33.** 55.18 **35.** 45 **37.** 75 **39.** 0.864

41. 0.5% **43.** 140 **45.** 9.6 **47.** 113% **49.** $-\frac{7}{8}$ **51.** $3\frac{2}{15}$ **53.** 0.7 **55.** 2.19 **57.** answers may vary **59.** no; $a = 16$

61. no; 300 or 300% **63.** answers may vary **65.** 80.0%

The Bigger Picture **1.** $\frac{19}{45}$ **2.** 63 **3.** −0.021 **4.** 8 **5.** 48 **6.** 1.6 **7.** 250% **8.** $45\frac{3}{5}$ or 45.6 **9.** 28% **10.** 180

Integrated Review **1.** 94% **2.** 17% **3.** 37.5% **4.** 350% **5.** 470% **6.** 800% **7.** 45% **8.** 106% **9.** 675%
10. 325% **11.** 2% **12.** 6% **13.** 0.71 **14.** 0.31 **15.** 0.03 **16.** 0.04 **17.** 2.24 **18.** 7 **19.** 0.029 **20.** 0.066 **21.** $0.07; \frac{7}{100}$
22. $0.05; \frac{1}{20}$ **23.** $0.068; \frac{17}{250}$ **24.** $0.1125; \frac{9}{80}$ **25.** $0.74; \frac{37}{50}$ **26.** $0.45; \frac{9}{20}$ **27.** $0.163; \frac{49}{300}$ **28.** $0.127; \frac{19}{150}$ **29.** 13.5 **30.** 100
31. 350 **32.** 120% **33.** 28% **34.** 76 **35.** 34 **36.** 130% **37.** 46% **38.** 37.8 **39.** 150 **40.** 62

Exercise Set 7.4 **1.** 8.8 pounds **3.** 1600 bolts **5.** 14% **7.** 545 **9.** 5.5% **11.** 496 chairs; 5704 chairs
13. 93,870 physician assistants **15.** 50% **17.** 12.5% **19.** 29.2% **21.** $175,000 **23.** 31.2 hr **25.** $867.87; $20,153.87 **27.** 2224
29. 28 million; 63 million **31.** 30; 60% **33.** 52; 80% **35.** 2; 25% **37.** 120; 75% **39.** 44% **41.** 1.3% **43.** 155.2% **45.** 2%

47. 5.6% **49.** 20.1% **51.** 51.9% **53.** 92.7% **55.** 4.56 **57.** 11.18 **59.** $-\frac{19}{24}$ **61.** $\frac{28}{39}$

63. The increased number is double the original number. **65.** percent increase $= \frac{30}{150} = 20\%$ **67.** False; the percents are different.

Vocabulary and Readiness Check   **1.** sales tax   **3.** commission   **5.** sale price

Exercise Set 7.5   **1.** $7.50   **3.** $858.93   **5.** 7%   **7. a.** $120   **b.** $130.20   **9.** $177; $1917   **11.** $485   **13.** 6%   **15.** $16.10; $246.10   **17.** $53,176.04   **19.** 14%   **21.** $4888.50   **23.** $185,500   **25.** $8.90; $80.10   **27.** $98.25; $98.25   **29.** $143.50; $266.50   **31.** $3255; $18,445   **33.** $45; $255   **35.** $27.45; $332.45   **37.** $3.08; $59.08   **39.** $7074   **41.** 8%   **43.** 1200   **45.** 132   **47.** 16   **49.** d   **51.** $4.00; $6.00; $8.00   **53.** $7.20; $10.80; $14.40   **55.** a discount of 60% is better; answers may vary   **57.** $26,838.45

Calculator Explorations   **1.** $936.31   **3.** $9674.77   **5.** $634.49

Vocabulary and Readiness Check   **1.** simple   **3.** compound   **5.** total amount

Exercise Set 7.6   **1.** $32   **3.** $73.60   **5.** $750   **7.** $33.75   **9.** $700   **11.** $101,562.50   **13.** $5562.50   **15.** $14,280   **17.** $46,815.37   **19.** $2327.14   **21.** $58,163.65   **23.** $2915.75   **25.** $2938.66   **27.** $2971.89   **29.** 32 yd   **31.** 35 m   **33.** $\frac{9x}{20}$ **35.** $-\frac{131}{225}$ **37.** answers may vary   **39.** answers may vary

Chapter 7 Vocabulary Check   **1.** of   **2.** is   **3.** percent   **4.** compound interest   **5.** $\frac{\text{amount}}{\text{base}}$ **6.** 100%   **7.** 0.01   **8.** $\frac{1}{100}$ **9.** base; amount   **10.** percent of decrease   **11.** percent of increase   **12.** sales tax   **13.** total price   **14.** commission   **15.** amount of discount   **16.** sale price

Chapter Review   **1.** 37%   **2.** 77%   **3.** 0.26   **4.** 0.75   **5.** 0.035   **6.** 0.015   **7.** 2.75   **8.** 4.00 or 4   **9.** 0.4785   **10.** 0.8534   **11.** 160%   **12.** 5.5%   **13.** 7.6%   **14.** 8.5%   **15.** 71%   **16.** 65%   **17.** 600%   **18.** 900%   **19.** $\frac{7}{100}$ **20.** $\frac{3}{20}$ **21.** $\frac{1}{4}$ **22.** $\frac{17}{200}$ **23.** $\frac{51}{500}$ **24.** $\frac{1}{6}$ **25.** $\frac{1}{3}$ **26.** $1\frac{1}{10}$ **27.** 40%   **28.** 70%   **29.** $58\frac{1}{3}$%   **30.** $166\frac{2}{3}$%   **31.** 125%   **32.** 60%   **33.** 6.25%   **34.** 62.5%   **35.** 100,000   **36.** 8000   **37.** 23%   **38.** 114.5   **39.** 108.8   **40.** 150%   **41.** 418   **42.** 300   **43.** 159.6   **44.** 180%   **45.** 110%   **46.** 165   **47.** 66%   **48.** 16%   **49.** 20.9%   **50.** 106.25%   **51.** $206,400   **52.** $13.23   **53.** $263.75   **54.** $1.15   **55.** $5000   **56.** $300.38   **57.** discount: $900; sale price: $2100   **58.** discount: $9; sale price: $81   **59.** $160   **60.** $325   **61.** $30,104.61   **62.** $17,506.54   **63.** $180.61   **64.** $33,830.10   **65.** 0.038   **66.** 1.245   **67.** 54%   **68.** 9520%   **69.** $\frac{47}{100}$ **70.** $\frac{7}{125}$ **71.** $37\frac{1}{2}$%   **72.** 120%   **73.** 268.75   **74.** 110%   **75.** 708.48   **76.** 134%   **77.** 300%   **78.** 38.4   **79.** 560   **80.** 325%   **81.** 26%   **82.** $6786.50   **83.** $617.70   **84.** $3.45   **85.** 12.5%   **86.** $1491   **87.** $11,687.50

Chapter Test   **1.** 0.85   **2.** 5   **3.** 0.006   **4.** 5.6%   **5.** 610%   **6.** 35%   **7.** $1\frac{1}{5}$ **8.** $\frac{77}{200}$ **9.** $\frac{1}{500}$ **10.** 55%   **11.** 37.5%   **12.** 175%   **13.** 20%   **14.** $\frac{16}{25}$ **15.** 33.6   **16.** 1250   **17.** 75%   **18.** 38.4 lb   **19.** $56,750   **20.** $358.43   **21.** 5%   **22.** discount: $18; sale price: $102   **23.** $395   **24.** 1%   **25.** $647.50   **26.** $2005.63   **27.** $427   **28.** 5.8%

Cumulative Review   **1.** 20,296; Sec. 1.5, Ex. 4   **2.** 31,084; Sec. 1.5   **3.** −10; Sec. 2.3, Ex. 8   **4.** 10; Sec. 2.3   **5.** 1; Sec. 2.6, Ex. 2   **6.** −1; Sec. 2.6   **7.** 2; Sec. 3.3, Ex. 4   **8.** 5; Sec. 3.3   **9.** $\frac{21}{7}$; Sec. 4.4, Ex. 20   **10.** $\frac{40}{5}$; Sec. 4.4   **11.** $-\frac{10}{27}$; Sec. 4.2, Ex. 6   **12.** $\frac{5y}{16}$; Sec. 4.2   **13.** $\frac{5}{12}$; Sec. 4.3, Ex. 13   **14.** $\frac{-4}{7}$; Sec. 4.3   **15.** $-\frac{1}{2}$; Sec. 4.4, Ex. 9   **16.** $\frac{1}{5}$; Sec. 4.4   **17.** $\frac{1}{28}$; Sec. 4.5, Ex. 5   **18.** $\frac{16}{45}$; Sec. 4.5   **19.** $\frac{3}{2}$; Sec. 4.6, Ex. 2   **20.** $\frac{50}{9}$; Sec. 4.6   **21.** 3; Sec. 4.8, Ex. 9   **22.** 4; Sec. 4.8   **23. a.** $\frac{38}{9}$ **b.** $\frac{19}{11}$; Sec. 4.1, Ex. 20   **24. a.** $\frac{17}{5}$ **b.** $\frac{44}{7}$; Sec. 4.1   **25.** $\frac{1}{8}$; Sec. 5.1, Ex. 9   **26.** $\frac{17}{20}$; Sec. 5.1   **27.** $-105\frac{83}{1000}$; Sec. 5.1, Ex. 11   **28.** $17\frac{3}{200}$; Sec. 5.1   **29.** 67.69; Sec. 5.2, Ex. 6   **30.** 27.94; Sec. 5.2   **31.** 76.8; Sec. 5.3, Ex. 5   **32.** 1248.3; Sec. 5.3   **33.** −76,300; Sec. 5.3, Ex. 7   **34.** −8537.5; Sec. 5.3   **35.** 50; Sec. 5.4, Ex. 10   **36.** no; Sec. 5.4   **37.** 80.5; Sec. 5.7, Ex. 4   **38.** 48; Sec. 5.7   **39.** $\frac{50}{63}$; Sec. 6.1, Ex. 3   **40.** $\frac{29}{38}$; Sec. 6.1   **41.** $0.21/oz; Sec. 6.1, Ex. 11   **42.** $2.25 per sq ft; Sec. 6.1   **43.** no; Sec. 6.2, Ex. 3   **44.** yes; Sec. 6.2   **45.** 17.5 mi; Sec. 6.3, Ex. 1   **46.** 35; Sec. 6.3   **47.** $\frac{19}{1000}$; Sec. 7.1, Ex. 9   **48.** $\frac{23}{1000}$; Sec. 7.1   **49.** $\frac{1}{3}$; Sec. 7.1, Ex. 11   **50.** $1\frac{2}{25}$; Sec. 7.1

CHAPTER 8   Graphing and Introduction to Statistics

Vocabulary and Readiness Check   **1.** bar   **3.** line

Exercise Set 8.1   **1.** Washington   **3.** 25 million bushels   **5.** California and Pennsylvania   **7.** about 5% increase   **9.** 90,000

11. 120,000   **13.** 12,000   **15.** 76,000 wildfires/year   **17.** September   **19.** 27   **21.** $\frac{1}{3}$ **23.** Tokyo, Japan; about 35.2 million or

35,200,000   **25.** New York; 21.9 million, or 21,900,000   **27.** approximately 2 million

29.
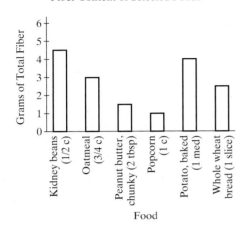
Fiber Content of Selected Foods

31.
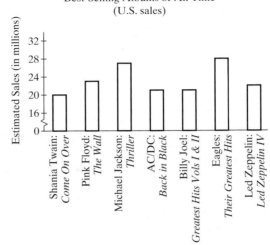
Best-Selling Albums of All Time
(U.S. sales)

33. 15 adults   **35.** 61 adults   **37.** 24 adults   **39.** 12 adults   **41.** $\frac{9}{100}$ **43.** 45–54   **45.** 21 million householders

47. 44 million householders   **49.** 4 million householders   **51.** |; 1   **53.** ⊥⊥⊥⊥ |||; 8   **55.** ⊥⊥⊥⊥ |; 6   **57.** ⊥⊥⊥⊥ |; 6   **59.** ||; 2

61.

Golf Scores

63. 2.7 goals   **65.** 1982   **67.** decrease   **69.** 1990, 2006   **71.** 3.6   **73.** 6.2   **75.** 25%   **77.** 34%

79. 83°F   **81.** Sunday; 68°F   **83.** Tuesday; 13°F   **85.** answers may vary

Vocabulary and Readiness Check   **1.** circle   **3.** 360

Exercise Set 8.2   **1.** parent or guardian's home   **3.** $\frac{9}{35}$ **5.** $\frac{9}{16}$ **7.** Asia   **9.** 37%   **11.** 17,100,000 sq mi   **13.** 2,850,000 sq mi

15. 55%   **17.** nonfiction   **19.** 31,400 books   **21.** 27,632 books   **23.** 25,120 books

25.

Coal

Petroleum
Other
Nuclear
Hydroelectric
Natural Gas

27.

Toyota

Honda
Ford/mercury
Lexus

29. $2^2 \times 5$ **31.** $2^3 \times 5$ **33.** 5×17

35. answers may vary   **37.** 129,600,002 sq km

39. 55,542,858 sq km   **41.** 602 respondents

43. 2198 respondents   **45.** $\frac{301}{837}$

47. no; answers may vary

Vocabulary and Readiness Check   **1.** $x; y$ **3.** four   **5.** plotting   **7.** plane

Exercise Set 8.3

1.

3.

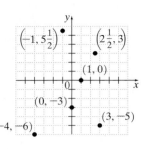

5. $A(0, 0); B\left(3\frac{1}{2}, 0\right); C(3, 2); D(-1, 3); E(-2, -2); F(0, -1); G(2, -1)$

7. yes   **9.** no   **11.** yes   **13.** yes   **15.** yes   **17.** no

19.

21.

23.

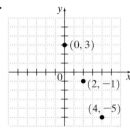

25. $(1, -9), (0, 0), (2, -18)$
27. $(2, -12), (6, -8), (0, -14)$
29. $(-2, 0), (1, -3), (-7, 5)$
31. $(1, 6), (2, 7), (3, 8)$
33. $(1, -2), (2, 1), (3, 4)$
35. $(0, 0), (2, -2), (-2, 2)$
37. $(4, -6), (-2, -3), (0, -4)$
39. 1.7   **41.** 21.84   **43.** -23.6
45. true   **47.** true   **49.** false
51. false   **53.** right   **55.** rectangle
57. 20 units

Integrated Review   **1.** 500,000   **2.** 550,000   **3.** Retail salespeople   **4.** Waitstaff   **5.** Oroville Dam; 755 ft
6. New Bullards Bar Dam; 635 ft   **7.** 15 ft   **8.** 4 dams   **9.** Thursday and Saturday; 100°F   **10.** Monday; 82°F
11. Sunday, Monday, and Tuesday   **12.** Wednesday, Thursday, Friday, and Saturday   **13.** 70 qt containers   **14.** 52 qt containers
15. 2 qt containers   **16.** 6 qt containers   **17.** ‖; 2   **18.** |; 1   **19.** ‖‖; 3   **20.** ⦀ |; 6   **21.** ⦀; 5
24. no   **25.** yes   **26.** $(0, -6), (6, 0), (2, -4)$

22.

23.

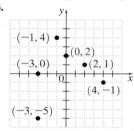

Vocabulary and Readiness Check   **1.** linear   **3.** horizontal

Exercise Set 8.4   **1.**

3.

5.

7.

9.

11.

13.

15.

$y = -2x$

17.

$y = -2$

19.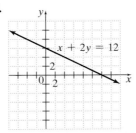

$x + 2y = 12$

21.

$x = 6$

23.

$y = x - 3$

25.

$x = y - 4$

27.

$x + 3 = 0$

29.

$y = -\frac{1}{4}x$

31.

$y = \frac{1}{3}x$

33.

$y = 4x + 2$

35.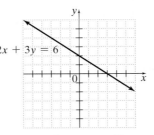

$2x + 3y = 6$

37.

$x = -3.5$

39.

$3x - 4y = 24$

41. $-\dfrac{1}{16}$ **43.** $-\dfrac{15}{19}$ **45.** $\dfrac{2 - x}{11}$

47.

$y = |x|$

49. answers may vary **51.** increasing **53.** $23.3 billion **55.** 2001–2002 **57.** 2009–2010

Vocabulary and Readiness Check **1.** outcome **3.** probability **5.** 0

Exercise Set 8.5 **1.**

M — 1, 2, 3
A — 1, 2, 3
T — 1, 2, 3
H — 1, 2, 3

12 outcomes

3.

Red
Blue
Yellow

3 outcomes

5.

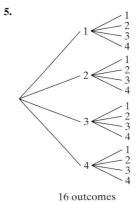

1 — 1, 2, 3, 4
2 — 1, 2, 3, 4
3 — 1, 2, 3, 4
4 — 1, 2, 3, 4

16 outcomes

7.

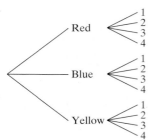

12 outcomes

9.

8 outcomes

11. $\frac{1}{6}$ **13.** $\frac{1}{3}$ **15.** $\frac{1}{2}$ **17.** $\frac{2}{3}$ **19.** $\frac{1}{3}$ **21.** 1   **23.** $\frac{2}{3}$ **25.** $\frac{1}{7}$ **27.** $\frac{2}{7}$ **29.** $\frac{4}{7}$ **31.** $\frac{19}{100}$ **33.** $\frac{1}{20}$ **35.** $\frac{5}{6}$ **37.** $\frac{1}{6}$

39. $\frac{20}{3}$ or $6\frac{2}{3}$ **41.** $\frac{1}{52}$ **43.** $\frac{1}{13}$ **45.** $\frac{1}{4}$ **47.** $\frac{1}{2}$ **49.** $\frac{5}{36}$ **51.** 0   **53.** answers may vary

Chapter 8 Vocabulary Check   **1.** bar   **2.** outcomes   **3.** pictograph   **4.** line   **5.** x; y **6.** tree diagram   **7.** experiment
8. circle   **9.** probability   **10.** histogram; class interval; class frequency   **11.** origin   **12.** quadrants   **13.** plotting   **14.** linear

Chapter Review   **1.** 4,000,000   **2.** 2,400,000   **3.** South   **4.** Northeast   **5.** South, West   **6.** Northeast, Midwest   **7.** 7.5%
8. 2003   **9.** 1980, 1990, 2000, 2003   **10.** answers may vary   **11.** 68   **12.** 78   **13.** 84   **14.** 6   **15.** 27   **16.** 4 employees
17. 1 employee   **18.** 9 employees   **19.** 18 employees   **20.** ⦀; 5   **21.** ⦀; 3   **22.** ⦀; 4

23.

Temperatures

24. mortgage payment   **25.** utilities   **26.** $1225   **27.** $700   **28.** $\frac{39}{160}$ **29.** $\frac{7}{40}$ **30.** $\frac{5}{7}$ **31.** 20 states
32. 11 states   **33.** 17 states   **34.** 2 states

35. $(0, 0)$, $(6, -1)$, $(-6, 1)$

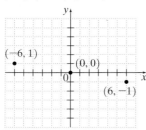

36. $(0, -2)$, $(-1, -5)$, $(2, 4)$

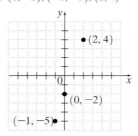

37. $(-1, -3)$, $(-4, 0)$, $(-5, 1)$

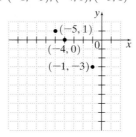

38. $(4, 1)$, $(0, -3)$, $(6, 3)$

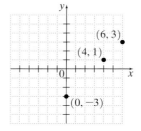

39. $(1, 3)$, $(-2, -6)$, $(0, 0)$

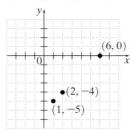

40. $(1, -5)$, $(6, 0)$, $(2, -4)$

41.

42.

43.

44.

45.

46.

47.

48.

49.

50.

51.

10 outcomes

52.

4 outcomes

53.

25 outcomes

54.

4 outcomes

55.

10 outcomes

56. $\dfrac{1}{6}$ **57.** $\dfrac{1}{6}$ **58.** $\dfrac{1}{5}$ **59.** $\dfrac{1}{5}$

60. $\dfrac{3}{5}$ **61.** $\dfrac{2}{5}$ **62.** $\dfrac{1}{4}$ **63.** $\dfrac{3}{8}$

64. $\dfrac{1}{4}$ **65.** $\dfrac{1}{8}$

66.

67.

68.

69.

70.

71.

72.

73.

Chapter Test   **1.** $225   **2.** 3rd week; $350   **3.** $1100   **4.** June, August, September   **5.** February; 3 cm   **6.** March and November

7.

Countries with the Highest Newspaper Circulations

8. 1.5%   **9.** 1990, 1991, 2000
10. 1994–1995, 1995–1996, 1998–1999, 1999–2000, 2002–2003   **11.** $\dfrac{17}{40}$ **12.** $\dfrac{31}{22}$

13. 40,920,000 people   **14.** 21,120,000 people   **15.** 9 students
16. 11 students   **17.** |; 1; |||; 3; ||||; 4; ||| |; 5; ||| |||; 8; ||||; 4

18.

19. $(4, 0)$ **20.** $(0, -3)$ **21.** $(-3, 4)$ **22.** $(-2, -1)$

23. $(0, 0), (-6, 1), (12, -2)$

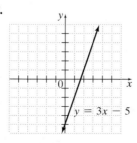

24. $(2, 10), (-1, -11), (0, -4)$

25.

26.

27.

28.

29.

30.

31.

32.

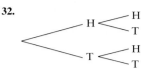

33. $\frac{1}{10}$ **34.** $\frac{1}{5}$

Cumulative Review **1.** 47; Sec. 1.7, Ex. 12 **2.** -70; Sec. 1.7 **3.** -12; Sec. 2.3, Ex. 11 **4.** 9; Sec. 2.3 **5.** -3; Sec. 3.2, Ex. 2

6. -6; Sec. 3.2 **7.** 2; Sec. 4.8, Ex. 7 **8.** -10; Sec. 4.8 **9.** $7\frac{17}{24}$; Sec. 4.7, Ex. 9 **10.** $8\frac{3}{20}$; Sec. 4.7 **11.** $5\frac{9}{10}$; Sec. 5.1, Ex. 8

12. $2\frac{4}{5}$; Sec. 5.1 **13.** 3.432; Sec. 5.2, Ex. 5 **14.** 7.327; Sec. 5.2 **15.** 0.8496; Sec. 5.3, Ex. 2 **16.** 0.0294; Sec. 5.3 **17.** -0.052; Sec. 5.4, Ex. 3

18. 0.136; Sec. 5.4 **19.** 4.09; Sec. 5.5, Ex. 13 **20.** 7.29; Sec. 5.5 **21.** 0.25; Sec. 5.5, Ex. 1 **22.** 0.375; Sec. 5.5 **23.** 0.7; Sec. 5.6, Ex. 5

24. 1.68; Sec. 5.6 **25.** 8.944; Sec. 6.4, Ex. 7b **26.** 7.746; Sec. 6.4 **27.** $\frac{21}{29}$; Sec. 6.1, Ex. 1 **28.** $\frac{7}{15}$; Sec. 6.1 **29.** 22.5 mi/gal; Sec. 6.1, Ex. 10

30. \$0.53 per oz **31.** -2; Sec. 6.2, Ex. 6 **32.** 16; Sec. 6.2 **33.** $\frac{12}{19}$; Sec. 6.5, Ex. 2 **34.** $\frac{4}{9}$; Sec. 6.5 **35.** 0.046; Sec. 7.1, Ex. 4

36. 0.32; Sec. 7.1 **37.** 0.0074; Sec. 7.1, Ex. 6 **38.** 0.027; Sec. 7.1 **39.** 21; Sec. 7.2, Ex. 7 **40.** 14.4; Sec. 7.2 **41.** 52; Sec. 7.3, Ex. 9

42. 38; Sec. 7.3 **43.** 8.5%; Sec. 7.5, Ex. 2 **44.** 6.5%; Sec. 7.5 **45.** \$6772.12; Sec. 7.6, Ex. 5 **46.** \$144.05; Sec. 7.6 **47.** 80.5; Sec. 5.7, Ex. 4

48. 48.5; Sec. 5.7 **49.** $\frac{1}{3}$; Sec. 8.5, Ex. 4 **50.** $\frac{1}{2}$; Sec. 8.5

CHAPTER 9 Geometry and Measurement

Vocabulary and Readiness Check **1.** plane **3.** space **5.** ray **7.** straight **9.** acute **11.** parallel; intersecting
13. degrees **15.** vertical

Exercise Set 9.1 **1.** line; line CD or line l or $\overleftrightarrow{CD}$ **3.** line segment; line segment MN or $\overline{MN}$ **5.** angle; $\angle GHI$ or $\angle IHG$ or $\angle H$
7. ray; ray UW or $\overrightarrow{UW}$ **9.** $\angle CPR$, $\angle RPC$ **11.** $\angle TPM$, $\angle MPT$ **13.** straight **15.** right **17.** obtuse **19.** acute **21.** $67°$
23. $163°$ **25.** $32°$ **27.** $30°$ **29.** $\angle MNP$ and $\angle RNO$; $\angle PNQ$ and $\angle QNR$ **31.** $\angle SPT$ and $\angle TPQ$; $\angle SPR$ and $\angle RPQ$; $\angle SPT$ and
$\angle SPR$; $\angle TPQ$ and $\angle QPR$ **33.** $27°$ **35.** $132°$ **37.** $x = 30°$, $y = 150°$, $z = 30°$ **39.** $m\angle x = 77°$; $m\angle y = 103°$; $m\angle z = 77°$
41. $m\angle x = 100°$; $m\angle y = 80°$; $m\angle z = 100°$ **43.** $m\angle x = 134°$; $m\angle y = 46°$; $m\angle z = 134°$ **45.** $\angle ABC$ or $\angle CBA$ **47.** $\angle DBE$ or $\angle EBD$
49. $15°$ **51.** $50°$ **53.** $65°$ **55.** $95°$ **57.** $\frac{9}{8}$ or $1\frac{1}{8}$ **59.** $\frac{7}{32}$ **61.** $\frac{5}{6}$ **63.** $\frac{4}{3}$ or $1\frac{1}{3}$ **65.** $54.8°$ **67.** false; answers may vary
69. true **71.** $m\angle a = 60°$; $m\angle b = 50°$; $m\angle c = 110°$; $m\angle d = 70°$; $m\angle e = 120°$ **73.** no; answers may vary **75.** no; answers may vary

Vocabulary and Readiness Check **1.** perimeter **3.** π **5.** $\frac{22}{7}$ (or 3.14); $3.14\left(\text{or } \frac{22}{7}\right)$

Exercise Set 9.2 **1.** 64 ft **3.** 120 cm **5.** 21 in. **7.** 48 ft **9.** 42 in. **11.** 155 cm **13.** 21 ft **15.** 360 ft **17.** 346 yd
19. 22 ft **21.** \$55 **23.** 72 in. **25.** 28 in. **27.** \$36.12 **29.** 96 m **31.** 66 ft **33.** 74 cm **35.** 17π cm; 53.38 cm
37. 16π mi; 50.24 mi **39.** 26π m; 81.64 m **41.** $C = 30\pi$ ft; 94.2 ft **43.** 12,560 ft **45.** 30.7 mi **47.** $14\pi \approx 43.96$ cm
49. 40 mm **51.** 84 ft **53.** 23 **55.** 1 **57.** 10 **59.** 216 **61. a.** width: 30 yd; length: 40 yd **b.** 140 yd **63.** b
65. a. 62.8 m; 125.6 m **b.** yes **67.** $44 + 10\pi \approx 75.4$ m **69.** 26.9 ft

Vocabulary and Readiness Check **1.** surface area **3.** area **5.** square

Exercise Set 9.3 **1.** 7 sq m **3.** $9\frac{3}{4}$ sq yd **5.** 15 sq yd **7.** 2.25π sq in. ≈ 7.065 sq in. **9.** 36.75 sq ft **11.** 28 sq m **13.** 22 sq yd
15. $36\frac{3}{4}$ sq ft **17.** $22\frac{1}{2}$ sq in. **19.** 25 sq cm **21.** 86 sq mi **23.** 24 sq cm **25.** 36π sq in. $\approx 113\frac{1}{7}$ sq in. **27.** $V = 72$ cu in.;

$SA = 108$ sq in. **29.** $V = 512$ cu cm; $SA = 384$ sq cm **31.** $V = 4\pi$ cu yd $\approx 12\frac{4}{7}$ cu yd; $SA = (2\sqrt{13}\pi + 4\pi)$ sq yd ≈ 35.20 sq yd

33. $V = \frac{500}{3}\pi$ cu in. $\approx 523\frac{17}{21}$ cu in.; $SA = 100\pi$ sq in. $\approx 314\frac{2}{7}$ sq in. **35.** $V = 9\pi$ cu in. $\approx 28\frac{2}{7}$ cu in. **37.** $V = 75$ cu cm

39. $2\frac{10}{27}$ cu in. **41.** $V = 8.4$ cu ft; $SA = 26$ sq ft **43.** 113,625 sq ft **45.** 168 sq ft **47.** 960 cu cm **49.** 9200 sq ft

51. $V = \frac{1372}{3}\pi$ cu in. or $457\frac{1}{3}\pi$ cu in.; $SA = 196\pi$ sq in. **53.** 381 sq ft **55.** $V = 36\pi$ cu in. ≈ 113.04 cu in. **57.** 4π sq ft ≈ 12.56 sq ft

59. $V = 4.5\pi$ cu m; 14.13 cu m **61.** 168 sq ft **63.** $10\frac{5}{6}$ cu in. **65.** 25 **67.** 9 **69.** 5 **71.** 20 **73.** perimeter **75.** area

77. area **79.** perimeter **81.** 12-in. pizza **83.** 26,696.5 cu ft **85.** no; answers may vary **87.** 7.74 sq in. **89.** 298.5 sq m

91. no; answers may vary

Integrated Review 1. $153°; 63°$ **2.** $m\angle x = 75°; m\angle y = 105°; m\angle z = 75°$ **3.** $m\angle x = 128°; m\angle y = 52°; m\angle z = 128°$ **4.** $m\angle x = 52°$

5. 4.6 in. **6.** $4\frac{1}{4}$ in. **7.** 20 m; 25 sq m **8.** 12 ft; 6 sq ft **9.** 10π cm ≈ 31.4 cm; 25π sq cm ≈ 78.5 sq cm **10.** 32 mi; 44 sq mi

11. 54 cm; 143 sq cm **12.** 62 ft; 238 sq ft **13.** $V = 64$ cu in.; $SA = 96$ sq in. **14.** $V = 30.6$ cu ft; $SA = 63$ sq ft

15. $V = 400$ cu cm; **16.** $V = 4\frac{1}{2}\pi$ cu mi $\approx 14\frac{1}{7}$ cu mi

Vocabulary and Readiness Check 1. meter **3.** yard **5.** feet **7.** feet

Exercise Set 9.4 1. 5 ft **3.** 36 ft **5.** 8 mi **7.** 102 in. **9.** $3\frac{1}{3}$ yd **11.** 33,792 ft **13.** 4.5 yd **15.** 0.25 ft **17.** 13 yd 1 ft

19. 7 ft 1 in. **21.** 1 mi 4720 ft **23.** 62 in. **25.** 26 ft **27.** 84 in. **29.** 11 ft 2 in. **31.** 22 yd 1 ft **33.** 6 ft 5 in. **35.** 7 ft 6 in.

37. 14 ft 4 in. **39.** 83 yd 1 ft **41.** 6000 cm **43.** 4.0 cm **45.** 0.5 km **47.** 1.7 m **49.** 15 m **51.** 42,000 cm **53.** 7000 m

55. 83 mm **57.** 0.201 dm **59.** 40 mm **61.** 8.94 m **63.** 2.94 m or 2940 mm **65.** 1.29 cm or 12.9 mm **67.** 12.640 km or 12,640 m

69. 54.9 m **71.** 1.55 km **73.** $348\frac{2}{3}$; 12,552 **75.** $11\frac{2}{3}$; 420 **77.** 5000; 0.005; 500 **79.** 0.065; 65; 0.000065

81. 342,000; 342,000,000; 34,200,000 **83.** 10 ft 6 in. **85.** 5100 ft **87.** 4.8 times **89.** 26.7 mm **91.** 15 ft 9 in. **93.** 3.35 m

95. $105\frac{1}{3}$ yd **97.** $\frac{21}{100}$ **99.** 0.13 **101.** 0.25 **103.** no **105.** yes **107.** no **109.** Estimate: 13 yd

111. answers may vary; for example, $1\frac{1}{3}$ yd or 48 in. **113.** answers may vary **115.** 334.89 sq m

Vocabulary and Readiness Check 1. mass **3.** gram **5.** 2000

Exercise Set 9.5 1. 32 oz **3.** 10,000 lb **5.** 9 tons **7.** $3\frac{3}{4}$ lb **9.** $1\frac{3}{4}$ tons **11.** 204 oz **13.** 9800 lb **15.** 76 oz **17.** 1.5 tons

19. $\frac{1}{20}$ lb **21.** 92 oz **23.** 161 oz **25.** 5 lb 9 oz **27.** 53 lb 10 oz **29.** 8 tons 750 lb **31.** 3 tons 175 lb **33.** 8 lb 11 oz

35. 31 lb 2 oz **37.** 1 ton 700 lb **39.** 0.5 kg **41.** 4000 mg **43.** 25,000 g **45.** 0.048 g **47.** 0.0063 kg **49.** 15,140 mg

51. 6250 g **53.** 350,000 cg **55.** 13.5 mg **57.** 5.815 g or 5815 mg **59.** 1850 mg or 1.850 g **61.** 1360 g or 1.360 kg **63.** 13.52 kg

65. 2.125 kg **67.** 200,000; 3,200,000 **69.** $\frac{269}{400}$ or 0.6725; 21,520 **71.** 0.5; 0.0005; 50 **73.** 21,000; 21,000,000; 2,100,000

75. 8.064 kg **77.** 30 mg **79.** 5 lb 8 oz **81.** 35 lb 14 oz **83.** 6 lb 15.4 oz **85.** 144 mg **87.** 6.12 kg **89.** 130 lb **91.** 211 lb

93. 0.16 **95.** 0.875 **97.** no **99.** yes **101.** no **103.** answers may vary **105.** true **107.** answers may vary

Vocabulary and Readiness Check 1. capacity **3.** fluid ounces **5.** cups **7.** quarts

Exercise Set 9.6 1. 4 c **3.** 16 pt **5.** $3\frac{1}{2}$ gal **7.** 5 pt **9.** 8 c **11.** $3\frac{3}{4}$ qt **13.** $10\frac{1}{2}$ qt **15.** 9 c **17.** 23 qt **19.** $\frac{1}{4}$ pt

21. 14 gal 2 qt **23.** 4 gal 3 qt 1 pt **25.** 22 pt **27.** 13 gal 2 qt **29.** 4 c 4 fl oz **31.** 1 gal 1 qt **33.** 2 gal 3 qt 1 pt **35.** 17 gal
37. 4 gal 3 qt **39.** 5000 ml **41.** 0.00016 kl **43.** 5.6 L **45.** 320 cl **47.** 0.41 kl **49.** 0.064 L **51.** 160 L **53.** 3600 ml
55. 19.3 L **57.** 4.5 L or 4500 ml **59.** 8410 ml or 8.41 L **61.** 16,600 ml or 16.6 L **63.** 3840 ml **65.** 162.4 L **67.** 336; 84; 168
69. $\frac{1}{4}$; 1; 2 **71.** 1.59 L **73.** 18.954 L **75.** 4.3 oz **77.** yes **79.** $0.316 **81.** $\frac{4}{5}$ **83.** $\frac{3}{5}$ **85.** $\frac{9}{10}$ **87.** no **89.** no
91. answers may vary **93.** answers may vary **95.** answers may vary **97.** 1.5 cc **99.** 2.7 cc
101. 54 u or 0.54 cc **103.** 86 u or 0.86 cc

Exercise Set 9.7 1. 25.57 fl oz **3.** 218.44 cm **5.** 40 oz **7.** 57.66 mi **9.** 3.77 gal **11.** 13.5 kg **13.** 1.5; $1\frac{2}{3}$; 150; 60

15. 54.9; 5486.4; 180; 2160 **17.** 3.94 in. **19.** 80.5 kph **21.** 0.007 oz **23.** yes **25.** 2795 mi **27.** 90 mm **29.** 112.5 g
31. 104 mph **33.** 26.24 ft **35.** 3 mi **37.** 8 fl oz **39.** b **41.** b **43.** c **45.** d **47.** d **49.** 25°C **51.** 40°C
53. 122°F **55.** 239°F **57.** −6.7°C **59.** 61.2°C **61.** 197.6°F **63.** 54.3°F **65.** 56.7°C **67.** 80.6°F **69.** 21.1°C
71. 244.4°F **73.** 7232°F **75.** 29 **77.** 36 **79.** yes **81.** no **83.** no **85.** yes **87.** 2.13 sq m **89.** 1.19 sq m
91. 1.69 sq m **93.** 510,000,000°C **95.** answers may vary

Chapter 9 Vocabulary Check 1. weight **2.** mass **3.** meter **4.** unit fractions **5.** gram **6.** liter **7.** line segment
8. complementary **9.** line **10.** perimeter **11.** angle; vertex **12.** area **13.** ray **14.** transversal **15.** straight **16.** volume
17. vertical **18.** adjacent **19.** obtuse **20.** right **21.** acute **22.** supplementary **23.** surface area

Chapter Review 1. right **2.** straight **3.** acute **4.** obtuse **5.** 65° **6.** 75° **7.** 58° **8.** 98° **9.** 90° **10.** 25°
11. $\angle a$ and $\angle b$; $\angle b$ and $\angle c$; $\angle c$ and $\angle d$; $\angle d$ and $\angle a$ **12.** $\angle x$ and $\angle w$; $\angle y$ and $\angle z$ **13.** $m\angle x = 100°; m\angle y = 80°; m\angle z = 80°$
14. $m\angle x = 155°; m\angle y = 155°; m\angle z = 25°$ **15.** $m\angle x = 53°; m\angle y = 53°; m\angle z = 127°$ **16.** $m\angle x = 42°; m\angle y = 42°; m\angle z = 138°$
17. 69 m **18.** 30.6 cm **19.** 36 m **20.** 90 ft **21.** 32 ft **22.** 440 ft **23.** 5.338 in. **24.** 31.4 yd **25.** 240 sq ft
26. 140 sq m **27.** 600 sq cm **28.** 189 sq yd **29.** 49π sq ft ≈ 153.86 sq ft **30.** 82.81 sq m **31.** 119 sq in. **32.** 1248 sq cm

33. 144 sq m   **34.** 432 sq ft   **35.** 130 sq ft   **36.** $V = 15\frac{5}{8}$ cu in.; $SA = 37\frac{1}{2}$ sq in.   **37.** $V = 84$ cu ft; $SA = 136$ sq ft

38. $V = 20{,}000\pi$ cu cm $\approx 62{,}800$ cu cm   **39.** $V = \frac{1}{6}\pi$ cu km $\approx \frac{11}{21}$ cu km   **40.** $2\frac{2}{3}$ cu ft   **41.** 307.72 cu in.   **42.** $7\frac{1}{2}$ cu ft

43. 0.5π cu ft or $\frac{1}{2}\pi$ cu ft   **44.** 9 ft   **45.** 24 yd   **46.** 7920 ft   **47.** 18 in.   **48.** 17 yd 1 ft   **49.** 3 ft 10 in.   **50.** 4200 cm

51. 820 mm   **52.** 0.01218 m   **53.** 0.00231 km   **54.** 21 yd 1 ft   **55.** 3 ft 8 in.   **56.** 9.5 cm or 95 mm   **57.** 2.74 m or 274 cm
58. 169 yd 2 ft   **59.** 258 ft 4 in.   **60.** 108.5 km   **61.** 0.24 sq m   **62.** 4.125 lb   **63.** 4600 lb   **64.** 3 lb 4 oz   **65.** 5 tons 300 lb
66. 0.027 g   **67.** 40,000 g   **68.** 21 dag   **69.** 0.0003 dg   **70.** 3 lb 9 oz   **71.** 33 lb 8 oz   **72.** 21.5 mg   **73.** 0.6 kg or 600 g
74. 4 lb 4 oz   **75.** 9 tons 1075 lb   **76.** 14 qt   **77.** 5 c   **78.** 7 pt   **79.** 72 c   **80.** 4 qt 1 pt   **81.** 3 gal 3 qt   **82.** 3800 ml
83. 1.4 kl   **84.** 3060 cl   **85.** 1 gal 1 qt   **86.** 7 gal   **87.** 736 ml or 0.736 L   **88.** 15.5 L or 15,500 ml   **89.** 2 gal 3 qt   **90.** 6 fl oz
91. 10.88 L   **92.** yes   **93.** 22.96 ft   **94.** 10.55 m   **95.** 4.55 gal   **96.** 8.27 qt   **97.** 425.25 g   **98.** 10.35 kg   **99.** 2.36 in.
100. 180.4 lb   **101.** 107.6°F   **102.** 320°F   **103.** 5.2°C   **104.** 26.7°C   **105.** 1.7°C   **106.** 329°F   **107.** 108°   **108.** 89°
109. 95°   **110.** 57°   **111.** 27.3 in.   **112.** 194 ft   **113.** 1624 sq m   **114.** 9π sq m ≈ 28.26 sq m   **115.** $346\frac{1}{2}$ cu in.
116. $V = 140$ cu in.; $SA = 166$ sq in.   **117.** 75 in.   **118.** 4 tons 200 lb   **119.** 500 cm   **120.** 0.000286 km   **121.** 1.4 g
122. 27 qt   **123.** 186.8°F   **124.** 11°C   **125.** 9117 m or 9.117 km   **126.** 35.7 L or 35,700 ml   **127.** 8 gal 1 qt   **128.** 12.8 kg

Chapter Test 1. 12°   **2.** 56°   **3.** 50°   **4.** $\angle x = 118°$; $\angle y = 62°$; $\angle z = 118°$ **5.** $\angle x = 73°$; $\angle y = 73°$; $\angle z = 73°$
6. 6.2 m   **7.** 10 in.   **8.** circumference $= 18\pi \approx 56.52$ in.; area $= 81\pi \approx 254.34$ sq in.   **9.** perimeter $= 24.6$ yd; area $= 37.1$ sq yd

10. perimeter $= 68$ in.; area $= 185$ sq in.   **11.** $62\frac{6}{7}$ cu in.   **12.** 30 cu ft   **13.** 16 in.   **14.** 18 cu ft   **15.** 62 ft; $115.94   **16.** 23 ft 4 in.

17. 10 qt   **18.** 1.875 lb   **19.** 5600 lb   **20.** $4\frac{3}{4}$ gal   **21.** 0.04 g   **22.** 2400 g   **23.** 36 mm   **24.** 0.43 g   **25.** 830 ml   **26.** 1 gal 2 qt

27. 3 lb 13 oz   **28.** 8 ft 3 in.   **29.** 2 gal 3 qt   **30.** 66 mm or 6.6 cm   **31.** 2.256 km or 2256 m   **32.** 28.9°C   **33.** 54.7°F
34. 5.6 m   **35.** 4 gal 3 qt   **36.** 91.4 m   **37.** 16 ft 6 in.

Cumulative Review 1. 5; Sec. 3.3, Ex. 1   **2.** 6; Sec. 3.3   **3. a.** $\frac{16}{625}$ **b.** $\frac{1}{16}$; Sec. 4.3, Ex. 10   **4. a.** $-\frac{1}{27}$ **b.** $\frac{9}{49}$; Sec. 4.3

5. $9\frac{3}{10}$; Sec. 4.7, Ex. 11   **6.** $9\frac{11}{15}$; Sec. 4.7   **7.** $20x - 10.9$; Sec. 5.2, Ex. 13   **8.** $1.2y + 1.8$; Sec. 5.2   **9.** 28.4405; Sec. 5.5, Ex. 14

10. 2.16; Sec. 5.5   **11.** $=$; Sec. 5.5, Ex. 9   **12.** $>$; Sec. 5.5   **13.** -1.3; Sec. 5.6, Ex. 6   **14.** 30; Sec. 5.6   **15.** 424 ft; Sec. 6.4, Ex. 11
16. 236 ft; Sec. 6.4   **17. a.** $\frac{5}{7}$ **b.** $\frac{7}{24}$; Sec. 6.1, Ex. 6   **18. a.** $\frac{1}{4}$ **b.** $\frac{4}{9}$; Sec. 6.1   **19.** $\frac{180 \text{ dollars}}{1 \text{ week}}$; Sec. 6.1, Ex. 7   **20.** $\frac{1 \text{ chaperone}}{5 \text{ students}}$; Sec. 6.1
21. 0.44; Sec. 6.2; Ex. 10   **22.** 1.02; Sec. 6.2   **23.** 22.4 cc; Sec. 6.3, Ex. 2   **24.** 7.5 cups; Sec. 6.3   **25.** 17%; Sec. 7.1, Ex. 1
26. 38%; Sec. 7.1   **27.** 200; Sec. 7.2, Ex. 10   **28.** 1200; Sec. 7.2   **29.** 2.7; Sec. 7.3, Ex. 7   **30.** 12.6; Sec. 7.3   **31.** 32%; Sec. 7.4, Ex. 5
32. 27%; Sec. 7.4   **33.** sales tax: $6.41, total price: $91.91; Sec. 7.5, Ex. 1   **34.** sales tax: $30; total price: $405; Sec. 7.5
35. $A(-4, 2)$, $B(1, 2)$, $C(0, 1)$, $D(-3, 0)$, $E(5, -4)$; Sec. 8.3, Ex. 2   **36.** $A(2, -3)$, $B(-5, 0)$, $C(0, 4)$, $D(-3, -2)$; Sec. 8.3
37.

; Sec. 8.4, Ex. 4   **38.**

; Sec. 8.4   **39.** 57; Sec. 5.7, Ex. 3

40. 83; Sec. 5.7   **41.** $\frac{1}{4}$; Sec. 8.5, Ex. 5   **42.** $\frac{2}{7}$; Sec. 8.5   **43.** 42°; Sec. 9.1, Ex. 4   **44.** 43°; Sec. 9.1   **45.** 96 in.; Sec. 9.4, Ex. 1
46. 21 feet; Sec. 9.4   **47.** 4 tons 1650 lb; Sec. 9.5, Ex. 4   **48.** 18 lb 2 oz; Sec. 9.5   **49.** 15°C; Sec. 9.7, Ex. 6   **50.** 30°C; Sec. 9.7

CHAPTER 10   Exponents and Polynomials

Vocabulary and Readiness Check 1. terms   **3.** binomial   **5.** add

Exercise Set 10.1 1. $-5x - 24$ **3.** $-11z^2 - 3z - 1$ **5.** $9y^2 + 25y - 40$ **7.** $-4.3a^4 - 2a^2 + 9$ **9.** $-9x + 16$ **11.** $3z^2 - z + 7$
13. $5a - 13$ **15.** $-2x^2 + 4x + 1$ **17.** $-20y^3 + 12y^2 - 4$ **19.** $-9x^2 - x + 16$ **21.** $8y^2 - 10y - 8$ **23.** $5x - 12$ **25.** $y - 4$
27. $-5x^2 + 3x + 5$ **29.** $4x - 4.1$ **31.** $4a - 7$ **33.** $-15y + 3.6$ **35.** $19t^2 - 11t + 3$ **37.** $2b^3 + 13b^2 + 4b - 2$
39. $-5x^2 - 11x + 13$ **41.** $\frac{9}{7}$ **43.** 5   **45.** -5 **47.** -8 **49.** 20   **51.** 25   **53.** 50   **55.** 576 ft   **57.** $3200   **59.** 909 ft
61. 3003 million   **63.** 81   **65.** 25   **67.** x^3 **69.** $2^2 a^4$ **71.** $(8x + 2)$ in.   **73.** $(4x - 15)$ units   **75.** 20; 6; 2   **77.** 7.2752
79. 29 ft; -243 ft; answers may vary

Vocabulary and Readiness Check 1. exponent   **3.** multiply

Exercise Set 10.2   **1.** x^{14} **3.** a^4 **5.** $15z^5$ **7.** $-40x^2$ **9.** $42x^3$ **11.** $12a^{17}$ **13.** $25x^6y^4$ **15.** $28a^5b^6$ **17.** x^{15} **19.** z^{30}
21. b^{62} **23.** $81a^4$ **25.** $a^{33}b^{24}$ **27.** $1000x^{15}y^9$ **29.** $-24y^{22}$ **31.** $256x^9y^{13}$ **33.** $7x - 21$ **35.** $-6a - 4b$ **37.** $9x + 18y - 27$
39. $16x^{12}$ sq in.   **41.** $12a^4b^5$ sq m   **43.** $18{,}003{,}384a^{45}b^{30}$ **45.** $34{,}867.84401x^{50}$ **47.** answers may vary   **49.** $a^{127}b^{86}c^{25}$

Integrated Review   **1.** $2x - 3$ **2.** $20y - 11$ **3.** $10x + 3$ **4.** $-5y - 4$ **5.** $4a^4 - 3a^2 + a$ **6.** $-4a^3 - 12a^2$ **7.** $7.3x^2 - 4.2x - 5.3$
8. $1.8y^2 - 2.4y - 5.6$ **9.** $6x + 7$ **10.** $2x^2 + 3x - 12$ **11.** -1 **12.** 26 **13.** x^{20} **14.** x^{10} **15.** y^4 **16.** a^{11} **17.** x^{77}
18. x^{36} **19.** x^{42} **20.** y^{27} **21.** $125x^3$ **22.** $32y^5$ **23.** $-12x^2y^7$ **24.** $12a^3b^4$ **25.** $y^{33}z^{39}$ **26.** $a^{20}b^{48}$ **27.** $300x^4y^3$ **28.** $128y^6z^7$
29. $144a^{38}b^{12}$ **30.** $125x^{22}y^{28}$

Exercise Set 10.3   **1.** $27x^3 - 9x$ **3.** $-6a^3 + 9a^2 + 15a$ **5.** $42x^4 - 35x^3 + 49x^2$ **7.** $x^2 + 13x + 30$ **9.** $2x^2 + 2x - 24$
11. $36a^2 + 48a + 16$ **13.** $a^3 - 33a + 18$ **15.** $8x^3 + 2x^2 - 55x + 50$ **17.** $x^5 + 4x^4 + 6x^3 + 7x^2 + 2x$ **19.** $-30r^2 + 20r$
21. $-6y^3 - 2y^4 + 12y^2$ **23.** $x^2 + 14x + 24$ **25.** $4a^2 - 9$ **27.** $x^2 + 10x + 25$ **29.** $b^2 + \dfrac{7}{5}b + \dfrac{12}{25}$ **31.** $6x^3 + 25x^2 + 10x + 1$
33. $49x^2 + 70x + 25$ **35.** $4x^2 - 4x + 1$ **37.** $8x^5 - 8x^3 - 6x^2 - 6x + 9$ **39.** $x^5 + 2x^4 + 3x^3 + 2x^2 + x$ **41.** $10z^4 - 3z^3 + 3z - 1$
43. $2 \cdot 5^2$ **45.** $2^3 \cdot 3^2$ **47.** $2^3 \cdot 5^2$ **49.** $(y^3 - 3y^2 - 16y - 12)$ sq ft   **51.** $(x^4 - 3x^2 + 1)$ sq m   **53.** answers may vary

Vocabulary and Readiness Check   **1.** factor; product   **3.** smallest

Exercise Set 10.4   **1.** 3 **3.** 12 **5.** 4 **7.** 4 **9.** y^2 **11.** a^5 **13.** xy^2 **15.** x **17.** $2z^3$ **19.** $3y(y + 6)$ **21.** $5a^6(2 - a^2)$
23. $4x(x^2 + 3x + 5)$ **25.** $z^5(z^2 - 6)$ **27.** $-7(5 - 2y + y^2)$ or $7(-5 + 2y - y^2)$ **29.** $12a^5(1 - 3a)$ **31.** 36 **33.** $\dfrac{4}{5}$ **35.** 37.5%
37. $x^2 + 2x$ **39.** answers may vary   **41.** $4y^2 + 28y + 49$

Chapter 10 Vocabulary Check   **1.** factoring   **2.** greatest common factor   **3.** FOIL   **4.** trinomial   **5.** binomial   **6.** monomial
7. polynomials   **8.** exponent

Chapter Review   **1.** $10b - 3$ **2.** $21s - 15$ **3.** $-x + 2.8$ **4.** $-y - 12$ **5.** $4z^2 + 11z - 6$ **6.** $17a^3 + 25a^2$ **7.** $-11y^2 - y + \dfrac{3}{4}$
8. $x^2 - 7x + 3$ **9.** 45 **10.** -19 **11.** $(26x + 28)$ ft   **12.** $(16x^2 + 4)$ m   **13.** x^{24} **14.** y^7 **15.** $24z^7$ **16.** $-15x^3y^5$ **17.** a^{35}
18. x^{28} **19.** $81b^2$ **20.** $a^{20}b^{10}c^5$ **21.** $56x^{16}$ **22.** $108x^{30}y^{25}$ **23.** $81a^{14}$ sq mi   **24.** $27x^5$ sq in.   **25.** $10a^3 - 12a$
26. $-3y^4 + 6y^3 - 3y^2$ **27.** $x^2 + 8x + 12$ **28.** $15x^2 - 32x + 9$ **29.** $y^2 - 10y + 25$ **30.** $49a^2 + 14a + 1$ **31.** $x^3 - x^2 + x + 3$
32. $8y^4 + 4y^3 - 2y^2 - 3y - 3$ **33.** $3z^4 + 5z^3 + 6z^2 + 3z + 1$ **34.** $(a^3 + 5a^2 - 5a + 6)$ sq cm   **35.** 5 **36.** 4 **37.** 6 **38.** 5
39. x^2 **40.** y^7 **41.** xy **42.** a^5b^2 **43.** $5a$ **44.** $4y^2z$ **45.** $2x(x + 6)$ **46.** $6a(a - 2)$ **47.** $y^4(6 - y^2)$ **48.** $7(x^2 - 2x + 1)$
49. $a^3(5a^4 - a + 1)$ **50.** $10y(y^5 - 1)$ **51.** $z^2 + z + 4$ **52.** $-4y - 2$ **53.** x^{21} **54.** y^9 **55.** $a^{18}b^{30}c^6$ **56.** $81x^6$
57. $12a^4 - 15a$ **58.** $x^2 + 9x + 20$ **59.** $9x^2 + 24x + 9$ **60.** $6z^2 - 7z - 10$ **61.** 4 **62.** z^4 **63.** $z^7(z^2 - 4)$ **64.** $x^5(x^7 + 6)$
65. $15a^4(1 + 3a)$ **66.** $8z^5(2 - 3z^3)$

Chapter Test   **1.** $15x - 4$ **2.** $7x - 2$ **3.** $3.4y^2 + 2y - 3$ **4.** $-2a^2 + a + 1$ **5.** 17 **6.** y^{14} **7.** y^{33} **8.** $16x^8$ **9.** $-12a^{10}$
10. p^{54} **11.** $72a^{20}b^5$ **12.** $10x^3 + 6.5x$ **13.** $-2y^4 - 12y^3 + 8y$ **14.** $x^2 - x - 6$ **15.** $25x^2 + 20x + 4$ **16.** $a^3 + 8$
17. perimeter: $(14x - 4)$ in.; area: $(5x^2 + 33x - 14)$ sq in.   **18.** 15 **19.** $3y^3$ **20.** $3y(y - 5)$ **21.** $2a(5a + 6)$ **22.** $6(x^2 - 2x - 5)$
23. $x^3(7x^3 - 6x + 1)$

Cumulative Review   **1.** $106{,}400$ sq mi; Sec. 1.5, Ex. 6   **2.** 147 trees; Sec. 1.5   **3.** -8; Sec. 2.2, Ex. 15   **4.** -10; Sec. 2.2
5. -7; Sec. 2.3, Ex. 6   **6.** -3; Sec. 2.3   **7.** 1; Sec. 2.3, Ex. 7   **8.** 5; Sec. 2.3   **9.** 4; Sec. 2.6, Ex. 4   **10.** -2; Sec. 2.6
11. -1; Sec. 3.3, Ex. 2   **12.** 1; Sec. 3.3   **13.** $\dfrac{13x}{30}$; Sec. 4.5, Ex. 2   **14.** $\dfrac{1}{14y}$; Sec. 4.5   **15.** 736.2; Sec. 5.1, Ex. 15   **16.** 328.2; Sec. 5.1
17. 25.454; Sec. 5.2, Ex. 1   **18.** 17.052; Sec. 5.2   **19.** no; Sec. 5.3, Ex. 13   **20.** yes; Sec. 5.3   **21.** 0.7861; Sec. 5.4, Ex. 8   **22.** 0.818; Sec. 5.4
23. -0.012; Sec. 5.4, Ex. 9   **24.** 0.0503; Sec. 5.4   **25.** -2.6; Sec. 5.5, Ex. 16   **26.** -13.6; Sec. 5.5   **27.** 3.14; Sec. 5.5, Ex. 4
28. 1.947; Sec. 5.5   **29.** $\dfrac{1}{6}$; Sec. 6.4, Ex. 5   **30.** $\dfrac{2}{5}$; Sec. 6.4   **31.** 46 ft; Sec. 6.5, Ex. 4   **32.** 3 ft; Sec. 6.5   **33.** $1.2 = 30\% \cdot x$; Sec. 7.2, Ex. 2
34. $9 = 45\% \cdot x$; Sec. 7.2   **35.** 16%; Sec. 7.3, Ex. 10   **36.** 25%; Sec. 7.3   **37.** 775 freshmen; Sec. 7.4, Ex. 3   **38.** 1450 apples; Sec. 7.4
39. $\$160$; Sec. 7.6, Ex. 2   **40.** $\$25$; Sec. 7.6   **41.** 71%; Sec. 8.2, Ex. 2   **42.** 91%; Sec. 8.2   **43.** 28 in.; Sec. 9.2, Ex. 2   **44.** 25 ft; Sec. 9.2
45. 5.1 sq mi; Sec. 9.3, Ex. 2   **46.** 68 sq in.; Sec. 9.3   **47.** 4 tons 1650 lb; Sec. 9.5, Ex. 4   **48.** 16 tons 100 lb; Sec. 9.5
49. 3.21 L; Sec. 9.6, Ex. 5   **50.** 43.21 L; Sec. 9.6

APPENDICES

Appendix C Exercise Set   **1.** x^2 **3.** 81 **5.** p^6q^5 **7.** $\dfrac{y^3}{2}$ **9.** 1 **11.** 2 **13.** -1 **15.** 1 **17.** $\dfrac{1}{64}$ **19.** $\dfrac{7}{x^3}$ **21.** $\dfrac{5}{6}$ **23.** p^3
25. $\dfrac{1}{x^3}$ **27.** z^3 **29.** $\dfrac{4}{9}$ **31.** $\dfrac{y^2}{25}$ **33.** p^4 **35.** a^6 **37.** x^6y^6 **39.** $\dfrac{1}{x^{11}}$ **41.** $\dfrac{15}{x^3}$ **43.** $\dfrac{1}{y^{12}}$ **45.** $\dfrac{56\,m^7}{n^5}$ **47.** x^7 **49.** a^8b^{13}
51. $\dfrac{1}{x^6}$ **53.** 3 **55.** $\dfrac{1}{125}$ **57.** $\dfrac{8}{x^9}$ **59.** $\dfrac{3}{10}$ **61.** $\dfrac{1}{z^7}$ **63.** $\dfrac{1}{x^9}$ **65.** $\dfrac{a^8}{b^8}$ **67.** $\dfrac{30x^9}{y^6}$

Appendix D Exercise Set   **1.** 7.8×10^4 **3.** 1.67×10^{-6} **5.** 6.35×10^{-3} **7.** 1.16×10^6 **9.** 1.36×10^4 **11.** 0.0000000008673
13. 0.033 **15.** $20{,}320$ **17.** $700{,}000{,}000$ **19.** 9.4×10^8 **21.** $1{,}230{,}000{,}000{,}000$ **23.** 2.3×10^{10} **25.** 0.000036
27. 0.0000000000000000000028 **29.** 0.0000005 **31.** $200{,}000$ **33.** 2.7×10^9 gal

SOLUTIONS TO SELECTED EXERCISES

CHAPTER 1

Exercise Set 1.2

1. The place value of the 5 in 657 is tens.

5. The place value of the 5 in 43,526,000 is hundred-thousands.

9. 354 is written as three hundred fifty-four.

13. 26,990 is written as twenty-six thousand, nine hundred ninety.

17. 24,350,185 is written as twenty-four million, three hundred fifty thousand, one hundred eighty-five.

21. 1679 is written as one thousand, six hundred seventy-nine.

25. 11,239 is written as eleven thousand, two hundred thirty-nine.

29. Six thousand, five hundred eighty-seven in standard form is 6587.

33. Thirteen million, six hundred one thousand, eleven in standard form is 13,601,011.

37. Two hundred sixty thousand, nine hundred ninety-seven in standard form is 260,997.

41. Four hundred eighty-four thousand, two hundred thirty-five in standard form is 484,235.

45. Seven hundred fourteen in standard form is 714.

49. 3470 = 3000 + 400 + 70

53. 66,049 = 60,000 + 6000 + 40 + 9

57. The elevation of Mt. Clay is 5532 which is five thousand, five hundred thirty-two in words.

61. Mt. Washington is the tallest mountain in New England.

65. Labrador retrievers have the most registrations; 146,714 is written as one hundred forty-six thousand, seven hundred fourteen.

69. The largest number is 9861.

73. answers may vary

Exercise Set 1.3

1.
$$\begin{array}{r} 14 \\ + 22 \\ \hline 36 \end{array}$$

5.
$$\begin{array}{r} 12 \\ 13 \\ + 24 \\ \hline 49 \end{array}$$

9.
$$\begin{array}{r} \overset{1\ \ 1\ 1}{22{,}781} \\ + 186{,}297 \\ \hline 209{,}078 \end{array}$$

13.
$$\begin{array}{r} \overset{2\ 2}{81} \\ 17 \\ 23 \\ 79 \\ + 12 \\ \hline 212 \end{array}$$

17.
$$\begin{array}{r} \overset{1\ 1\ \ 1}{6\ 820} \\ 4\ 271 \\ + 5\ 626 \\ \hline 16{,}717 \end{array}$$

21.
$$\begin{array}{r} \overset{1\ 2\ 2\ \ 2\ 1}{121{,}742} \\ 57{,}279 \\ 26{,}586 \\ + 426{,}782 \\ \hline 632{,}389 \end{array}$$

25.
$$\begin{array}{r} 749 \\ - 149 \\ \hline 600 \end{array}$$
Check:
$$\begin{array}{r} 600 \\ + 149 \\ \hline 749 \end{array}$$

29.
$$\begin{array}{r} 600 \\ - 432 \\ \hline 168 \end{array}$$
Check:
$$\begin{array}{r} \overset{1\ 1}{168} \\ + 432 \\ \hline 600 \end{array}$$

33.
$$\begin{array}{r} 533 \\ - 29 \\ \hline 504 \end{array}$$
Check:
$$\begin{array}{r} \overset{1}{504} \\ + 29 \\ \hline 533 \end{array}$$

37.
$$\begin{array}{r} 50{,}000 \\ - 17{,}289 \\ \hline 32{,}711 \end{array}$$
Check:
$$\begin{array}{r} \overset{1\ 1\ \ 1\ 1}{32{,}711} \\ + 17{,}289 \\ \hline 50{,}000 \end{array}$$

41.
$$\begin{array}{r} 51{,}111 \\ - 19{,}898 \\ \hline 31{,}213 \end{array}$$
Check:
$$\begin{array}{r} \overset{1\ 1\ \ 1\ 1}{31{,}213} \\ + 19{,}898 \\ \hline 51{,}111 \end{array}$$

45.
$$\begin{array}{r} 76 \\ - 67 \\ \hline 9 \end{array}$$

49.
$$\begin{array}{r} \overset{1\ 1\ \ 1}{10{,}962} \\ 4851 \\ + 7063 \\ \hline 22{,}876 \end{array}$$

53. Opposite sides of a rectangle have the same length.
4 + 8 + 4 + 8 = 12 + 12 = 24
The perimeter is 24 inches.

57. The unknown vertical side has length 12 − 5 = 7 meters.
The unknown horizontal side has length 10 − 5 = 5 meters.
10 + 12 + 5 + 7 + 5 + 5 = 44
The perimeter is 44 meters.

61. "Find the total" indicates addition.

$$\begin{array}{r} \scriptstyle 13 \\ 76 \\ 39 \\ 8 \\ 17 \\ +\ 126 \\ \hline 266 \end{array}$$

The total of 76, 39, 8, 17, and 126 is 266.

65. "Increased by" indicates addition.

$$\begin{array}{r} \scriptstyle 1 \\ 452 \\ +\ 92 \\ \hline 544 \end{array}$$

452 increased by 92 is 544.

69. "Subtracted from" indicates subtraction.

$$\begin{array}{r} 100 \\ -\ 12 \\ \hline 88 \end{array}$$

12 subtracted from 100 is 88.

73. Subtract the cost of the DVD player from the amount in her savings account.

$$\begin{array}{r} 914 \\ -\ 295 \\ \hline 619 \end{array}$$

She will have $619 left.

77.

$$\begin{array}{r} 530{,}000 \\ -\ 247{,}000 \\ \hline 283{,}000 \end{array}$$

The Missouri sub-basin drains 283,000 square miles more than the Arkansas Red-White sub-basin.

81.

$$\begin{array}{r} 503 \\ -\ 239 \\ \hline 264 \end{array}$$

She must read 264 more pages.

85. Live rock music has a decibel level of 100 dB.

89.

$$\begin{array}{r} \scriptstyle 1\ 1\ 1 \\ 5696 \\ +\ 3346 \\ \hline 9042 \end{array}$$

There were 9042 stores worldwide.

93. California has the most Target stores.

97. Florida and Georgia:

$$\begin{array}{r} \scriptstyle 1 \\ 95 \\ +\ 44 \\ \hline 139 \end{array}$$

Michigan and Ohio:

$$\begin{array}{r} \scriptstyle 1\ 1 \\ 53 \\ +\ 49 \\ \hline 102 \end{array}$$

Florida and Georgia have more Target stores.

101. The minuend is 48 and the subtrahend is 1.

105. answers may vary

109.

$$\begin{array}{r} \scriptstyle 22 \\ 14 \\ 173 \\ 86 \\ +\ 257 \\ \hline 530 \end{array}$$

The given sum is incorrect, the correct sum is 530.

113. The given difference is correct.

117. answers may vary

Exercise Set 1.4

1. To round 423 to the nearest ten, observe that the digit in the ones place is 3. Since this digit is less than 5, we do not add 1 to the digit in the tens place. The number 423 rounded to the nearest ten is 420.

5. To round 2791 to the nearest hundred, observe that the digit in the tens place is 9. Since this digit is at least 5, we add 1 to the digit in the hundreds place. The number 2791 rounded to the nearest hundred is 2800.

9. To round 21,094 to the nearest thousand, observe that the digit in the hundreds place is 0. Since this digit is less than 5, we do not add 1 to the digit in the thousands place. The number 21,094 rounded to the nearest thousand is 21,000.

13. To round 328,495 to the nearest hundred, observe that the digit in the tens place is 9. Since this digit is at least 5, we add 1 to the digit in the hundreds place. The number 328,495 rounded to the nearest hundred is 328,500.

17. To round 39,994 to the nearest ten, observe that the digit in the ones place is 4. Since this digit is less than 5, we do not add 1 to the digit in the tens place. The number 39,994 rounded to the nearest ten is 39,990.

21. Estimate 5281 to a given place value by rounding it to that place value. 5281 rounded to the tens place is 5280, to the hundreds place is 5300, and to the thousands place is 5000.

25. Estimate 14,876 to a given place value by rounding it to that place value. 14,876 rounded to the tens place is 14,880, to the hundreds place is 14,900, and to the thousands place is 15,000.

29. To round 38,387 to the nearest thousand, observe that the digit in the hundreds place is 3. Since this digit is less than 5, we do not add 1 to the digit in the thousands place. Therefore, 38,387 points rounded to the nearest thousand is 38,000 points.

33. To round 3,023,894 to the nearest hundred-thousand, observe that the digit in the ten-thousands place is 2. Since this digit is less than 5, we do not add 1 to the digit in the hundred-thousands place. Therefore, $3,023,894 rounded to the nearest hundred-thousand is $3,000,000.

37.

39	rounds to	40
45	rounds to	50
22	rounds to	20
+ 17	rounds to	+ 20
		130

41.

1913	rounds to	1900
1886	rounds to	1900
+ 1925	rounds to	+ 1900
		5700

45.

3995	rounds to	4000
2549	rounds to	2500
+ 4944	rounds to	+ 4900
		11,400

49. $229 + 443 + 606$ is approximately $230 + 440 + 600 = 1270$.

The answer of 1278 is correct.

53.

899	rounds to	900
1499	rounds to	1500
+ 999	rounds to	+ 1000
		3400

The total cost is approximately $3400.

57.

20,320	rounds to	20,000
− 14,410	rounds to	− 14,000
		6000

The difference in elevation is approximately 6000 feet.

61.

909,608	rounds to	910,000
− 906,993	rounds to	− 907,000
		3000

The decrease in enrollment is approximately 3000 children.

65. $179 million is $179,000,000 in standard form.

$179,000,000 rounded to the nearest ten-million is $180,000,000.

$179,000,000 rounded to the nearest hundred-million is $200,000,000.

69. a. The smallest possible number that rounds to 8600 is 8550.

b. The largest possible number that rounds to 8600 is 8649.

73. 54 rounds to 50

17 rounds to 20

$50 + 20 + 50 + 20 = 140$

The perimeter is approximately 140 meters.

Exercise Set 1.5

1. $1 \cdot 24 = 24$

5. $8 \cdot 0 \cdot 9 = 0$

9. $6(3 + 8) = 6 \cdot 3 + 6 \cdot 8$

13. $20(14 + 6) = 20 \cdot 14 + 20 \cdot 6$

17.

613
× 6
3678

21.

1074
× 6
6444

25.

421
× 58
3 368
21 050
24,418

29.

780
× 20
15,600

33. $(640)(1)(10) = (640)(10) = 6400$

37.

609
× 234
2 436
18 270
121 800
142,506

41.

589
× 110
5 890
58 900
64,790

45. Area = (length)(width)

= (9 meters)(7 meters)

= 63 square meters

49.

576	rounds to	600
× 354	rounds to	× 400
		240,000

53. 38×42 is approximately 40×40, which is 1600. The best estimate is c.

57. $80 \times 11 = (8 \times 10) \times 11$

$= 8 \times (10 \times 11)$

$= 8 \times 110$

$= 880$

61.

2240
× 2
4480

65.

94
× 35
470
2820
3290

The total cost is $3290.

69. Area = (length)(width)

= (110 feet)(80 feet)

= 8800 square feet

The area is 8800 square feet.

73.

94
× 62
188
5640
5828

There are 5828 pixels on the screen.

77.
$$\begin{array}{r} 160 \\ \times\ 8 \\ \hline 1280 \end{array}$$

There are 1280 calories in 8 ounces.

81. There are 31 days in March.

$$\begin{aligned} 31 \times 700{,}000 &= 31 \times 7 \times 100{,}000 \\ &= 217 \times 100{,}000 \\ &= 21{,}700{,}000 \end{aligned}$$

They would use 21,700,000 quarts in March.

85.
$$\begin{array}{r} 134 \\ \times\ 16 \\ \hline 804 \\ 1340 \\ \hline 2144 \end{array}$$

89.
$$\begin{array}{r} 19 \\ -\ 4 \\ \hline 15 \end{array}$$

The difference of 19 and 4 is 15.

93. a. $3 \cdot 5 = 5 + 5 + 5 = 3 + 3 + 3 + 3 + 3$

b. answers may vary

97. $42 \times 3 = 126$

$42 \times 9 = 378$

The problem is $\begin{array}{r} 42 \\ \times\ 93 \end{array}$

101. On a side with 7 windows per row, there are $7 \times 23 = 161$ windows. On a side with 4 windows per row, there are $4 \times 23 = 92$ windows. $161 + 161 + 92 + 92 = 506$

There are 506 windows on the building.

Exercise Set 1.6

1. $54 \div 9 = 6$

5. $0 \div 8 = 0$

9. $\dfrac{18}{18} = 1$

13. $26 \div 0$ is undefined.

17. $0 \div 14 = 0$

21.
$$\begin{array}{r} 29 \\ 3\overline{)\ 87} \\ -\ 6 \\ \hline 27 \\ -\ 27 \\ \hline 0 \end{array}$$

Check: $3 \cdot 29 = 87$

25.
$$\begin{array}{r} 338 \\ 3\overline{)1014} \\ -\ 9 \\ \hline 11 \\ -\ 9 \\ \hline 24 \\ -\ 24 \\ \hline 0 \end{array}$$

Check: $3 \cdot 338 = 1014$

29.
$$\begin{array}{r} 9 \\ 7\overline{)\ 63} \\ -\ 63 \\ \hline 0 \end{array}$$

Check: $7 \cdot 9 = 63$

33.
$$\begin{array}{r} 68\ \text{R}\ 3 \\ 7\overline{)\ 479} \\ -\ 42 \\ \hline 59 \\ -\ 56 \\ \hline 3 \end{array}$$

Check: $7 \cdot 68 + 3 = 479$

37.
$$\begin{array}{r} 38\ \text{R}\ 1 \\ 8\overline{)\ 305} \\ -\ 24 \\ \hline 65 \\ -\ 64 \\ \hline 1 \end{array}$$

Check: $8 \cdot 38 + 1 = 305$

41.
$$\begin{array}{r} 13 \\ 55\overline{)\ 715} \\ -\ 55 \\ \hline 165 \\ -\ 165 \\ \hline 0 \end{array}$$

Check: $55 \cdot 13 = 715$

45.
$$\begin{array}{r} 97\ \text{R}\ 8 \\ 97\overline{)\ 9417} \\ -\ 873 \\ \hline 687 \\ -\ 679 \\ \hline 8 \end{array}$$

$97 \cdot 97 + 8 = 9417$

49.
$$
\begin{array}{r}
506\\
13)\overline{6578}\\
-65\\
\overline{07}\\
-0\\
\overline{78}\\
-78\\
\overline{0}
\end{array}
$$
Check: $13 \cdot 506 = 6578$

53.
$$
\begin{array}{r}
54\\
236)\overline{12744}\\
-1180\\
\overline{944}\\
-944\\
\overline{0}
\end{array}
$$
Check: $236 \cdot 54 = 12{,}744$

57.
$$
\begin{array}{r}
202\ R\ 15\\
102)\overline{20619}\\
-204\\
\overline{21}\\
-0\\
\overline{219}\\
-204\\
\overline{15}
\end{array}
$$
Check: $102 \cdot 202 + 15 = 20{,}619$

61.
$$
\begin{array}{r}
17\\
7)\overline{119}\\
-7\\
\overline{49}\\
-49\\
\overline{0}
\end{array}
$$

65.
$$
\begin{array}{r}
2132\ R\ 32\\
40)\overline{85312}\\
-80\\
\overline{53}\\
-40\\
\overline{131}\\
-120\\
\overline{112}\\
-80\\
\overline{32}
\end{array}
$$

69.
$$
\begin{array}{r}
23\ R\ 2\\
5)\overline{117}\\
-10\\
\overline{17}\\
-15\\
\overline{2}
\end{array}
$$
The quotient is 23 R 2.

73.
$$
\begin{array}{r}
20\ R\ 2\\
3)\overline{62}\\
-6\\
\overline{02}\\
-0\\
\overline{2}
\end{array}
$$
The quotient is 20 R 2.

77.
$$
\begin{array}{r}
165\\
318)\overline{52470}\\
-318\\
\overline{2067}\\
-1908\\
\overline{1590}\\
-1590\\
\overline{0}
\end{array}
$$
The person weighs 165 pounds on Earth.

81.
$$
\begin{array}{r}
88\ R\ 1\\
3)\overline{265}\\
-24\\
\overline{25}\\
-24\\
\overline{1}
\end{array}
$$
There are 88 bridges every 3 miles over the 265 miles, plus the first bridge for a total of 89 bridges.

85.
$$
\begin{array}{r}
5\\
5280)\overline{26400}\\
-26400\\
\overline{0}
\end{array}
$$
Broad Peak is 5 miles tall.

89.
$$
\begin{array}{r}
\overset{2}{1}0\\
24\\
35\\
22\\
17\\
+12\\
\overline{120}
\end{array}
\qquad
\begin{array}{r}
20\\
6)\overline{120}\\
-12\\
\overline{0}
\end{array}
$$
Average $= \dfrac{120}{6} = 20$

93.
$$
\begin{array}{r}
\overset{2}{8}6\\
79\\
81\\
69\\
+80\\
\overline{395}
\end{array}
\qquad
\begin{array}{r}
79\\
5)\overline{395}\\
-35\\
\overline{45}\\
-45\\
\overline{0}
\end{array}
$$
Average $= \dfrac{395}{5} = 79$

97.
$$\begin{array}{r} {}^{111}82 \\ 463 \\ 29 \\ + \ 8704 \\ \hline 9278 \end{array}$$

101.
$$\begin{array}{r} 722 \\ - \ 43 \\ \hline 679 \end{array}$$

105. $24\overline{)\ 228}$ 9 R 12
$$\begin{array}{r} - \ 216 \\ \hline 12 \end{array}$$

109. 200 divided by 20 is $200 \div 20$, which is choice b.

113. The average will increase; answers may vary.

117. answers may vary

Exercise Set 1.7

1. $4 \cdot 4 \cdot 4 = 4^3$

5. $12 \cdot 12 \cdot 12 = 12^3$

9. $9 \cdot 8 \cdot 8 = 9 \cdot 8^2$

13. $3 \cdot 2 \cdot 2 \cdot 2 \cdot 2 \cdot 5 \cdot 5 \cdot 5 \cdot 5 = 3 \cdot 2^4 \cdot 5^5$

17. $5^3 = 5 \cdot 5 \cdot 5 = 125$

21. $1^{10} = 1 \cdot 1 \cdot 1 \cdot 1 \cdot 1 \cdot 1 \cdot 1 \cdot 1 \cdot 1 \cdot 1 = 1$

25. $2^7 = 2 \cdot 2 \cdot 2 \cdot 2 \cdot 2 \cdot 2 \cdot 2 = 128$

29. $4^4 = 4 \cdot 4 \cdot 4 \cdot 4 = 256$

33. $12^2 = 12 \cdot 12 = 144$

37. $20^1 = 20$

41. $3 \cdot 2^6 = 3 \cdot 2 \cdot 2 \cdot 2 \cdot 2 \cdot 2 \cdot 2 = 192$

45. $15 + 3 \cdot 2 = 15 + 6 = 21$

49. $32 \div 4 - 3 = 8 - 3 = 5$

53. $6 \cdot 5 + 8 \cdot 2 = 30 + 16 = 46$

57. $(7 + 5^2) \div 4 \cdot 2^3 = (7 + 25) \div 4 \cdot 2^3$
$$\begin{aligned} &= 32 \div 4 \cdot 2^3 \\ &= 32 \div 4 \cdot 8 \\ &= 8 \cdot 8 \\ &= 64 \end{aligned}$$

61. $\dfrac{18 + 6}{2^4 - 2^2} = \dfrac{24}{16 - 4} = \dfrac{24}{12} = 2$

65. $\dfrac{7(9 - 6) + 3}{3^2 - 3} = \dfrac{7(3) + 3}{9 - 3} = \dfrac{21 + 3}{6} = \dfrac{24}{6} = 4$

69. $2^4 \cdot 4 - (25 \div 5) = 2^4 \cdot 4 - 5$
$$\begin{aligned} &= 16 \cdot 4 - 5 \\ &= 64 - 5 \\ &= 59 \end{aligned}$$

73. $(7 \cdot 5) + [9 \div (3 \div 3)] = (7 \cdot 5) + [9 \div (1)]$
$$\begin{aligned} &= 35 + 9 \\ &= 44 \end{aligned}$$

77. $\dfrac{9^2 + 2^2 - 1^2}{8 \div 2 \cdot 3 \cdot 1 \div 3} = \dfrac{81 + 4 - 1}{4 \cdot 3 \cdot 1 \div 3}$
$$= \dfrac{85 - 1}{12 \cdot 1 \div 3}$$
$$= \dfrac{84}{12 \div 3}$$
$$= \dfrac{84}{4}$$
$$= 21$$

81. $9 \div 3 + 5^2 \cdot 2 - 10 = 9 \div 3 + 25 \cdot 2 - 10$
$$\begin{aligned} &= 3 + 50 - 10 \\ &= 53 - 10 \\ &= 43 \end{aligned}$$

85. $7^2 - \{18 - [40 \div (5 \cdot 1) + 2] + 5^2\}$
$$\begin{aligned} &= 7^2 - \{18 - [40 \div (5) + 2] + 5^2\} \\ &= 7^2 - \{18 - [8 + 2] + 5^2\} \\ &= 7^2 - \{18 - [10] + 5^2\} \\ &= 7^2 - \{18 - 10 + 25\} \\ &= 7^2 - \{8 + 25\} \\ &= 7^2 - \{33\} \\ &= 49 - 33 \\ &= 16 \end{aligned}$$

89. Area of a square $= (\text{side})^2$
$$\begin{aligned} &= (23 \text{ miles})^2 \\ &= 529 \text{ square miles} \end{aligned}$$

93. $2^5 = 2 \cdot 2 \cdot 2 \cdot 2 \cdot 2$

The statement is false.

97. $24 \div (3 \cdot 2) + 2 \cdot 5 = 24 \div 6 + 2 \cdot 5 = 4 + 10 = 14$

101. $(7 + 2^4)^5 - (3^5 - 2^4)^2 = (7 + 16)^5 - (243 - 16)^2$
$$\begin{aligned} &= 23^5 - 227^2 \\ &= 6{,}436{,}343 - 51{,}529 \\ &= 6{,}384{,}814 \end{aligned}$$

Exercise Set 1.8

1.

a	b	$a + b$	$a - b$	$a \cdot b$	$a \div b$
21	7	$21 + 7 = 28$	$21 - 7 = 14$	$21 \cdot 7 = 147$	$21 \div 7 = 3$

5.

a	b	$a + b$	$a - b$	$a \cdot b$	$a \div b$
56	1	$56 + 1 = 57$	$56 - 1 = 55$	$56 \cdot 1 = 56$	$56 \div 1 = 56$

9. $3xz - 5x = 3(2)(3) - 5(2) = 18 - 10 = 8$

13. $4x - z = 4(2) - 3 = 8 - 3 = 5$

17. $2xy^2 - 6 = 2(2)(5)^2 - 6$
$$\begin{aligned} &= 2 \cdot 2 \cdot 25 - 6 \\ &= 100 - 6 \\ &= 94 \end{aligned}$$

21. $x^5 + (y - z) = 2^5 + (5 - 3)$
$$= 2^5 + 2$$
$$= 32 + 2$$
$$= 34$$

25. $\dfrac{2y - 2}{x} = \dfrac{2(5) - 2}{2} = \dfrac{10 - 2}{2} = \dfrac{8}{2} = 4$

29. $\dfrac{5x}{y} - \dfrac{10}{y} = \dfrac{5(2)}{5} - \dfrac{10}{5} = \dfrac{10}{5} - 2 = 2 - 2 = 0$

33. $(4y - 5z)^3 = (4 \cdot 5 - 5 \cdot 3)^3$
$$= (20 - 15)^3$$
$$= (5)^3$$
$$= 125$$

37. $2y(4z - x) = 2 \cdot 5(4 \cdot 3 - 2)$
$$= 10(12 - 2)$$
$$= 10(10)$$
$$= 100$$

41. $\dfrac{7x + 2y}{3x} = \dfrac{7(2) + 2(5)}{3(2)} = \dfrac{14 + 10}{6} = \dfrac{24}{6} = 4$

45. Let n be 10.
$$n - 8 = 2$$
$$10 - 8 \stackrel{?}{=} 2$$
$$2 = 2 \quad \text{True}$$
Yes, 10 is a solution.

49. Let n be 7.
$$3n - 5 = 10$$
$$3(7) - 5 \stackrel{?}{=} 10$$
$$21 - 5 \stackrel{?}{=} 10$$
$$16 = 10 \quad \text{False}$$
No, 7 is not a solution.

53. Let x be 0.
$$5x + 3 = 4x + 13$$
$$5(0) + 3 \stackrel{?}{=} 4(0) + 13$$
$$0 + 3 \stackrel{?}{=} 0 + 13$$
$$3 = 13 \quad \text{False}$$
No, 0 is not a solution.

57. $n - 2 = 10$
Let n be 10.
$$10 - 2 \stackrel{?}{=} 10$$
$$8 = 10 \quad \text{False}$$
Let n be 12.
$$12 - 2 \stackrel{?}{=} 10$$
$$10 = 10 \quad \text{True}$$
Let n be 14.
$$14 - 2 \stackrel{?}{=} 10$$
$$12 = 10 \quad \text{False}$$
12 is a solution.

61. $6n + 2 = 26$
Let n be 0.
$$6(0) + 2 \stackrel{?}{=} 26$$
$$0 + 2 \stackrel{?}{=} 26$$
$$2 = 26 \quad \text{False}$$

Let n be 2.
$$6(2) + 2 \stackrel{?}{=} 26$$
$$12 + 2 \stackrel{?}{=} 26$$
$$14 = 26 \quad \text{False}$$
Let n be 4.
$$6(4) + 2 \stackrel{?}{=} 26$$
$$24 + 2 \stackrel{?}{=} 26$$
$$26 = 26 \quad \text{True}$$
4 is a solution.

65. $7x - 9 = 5x + 13$
Let x be 3.
$$7(3) - 9 \stackrel{?}{=} 5(3) + 13$$
$$21 - 9 \stackrel{?}{=} 15 + 13$$
$$12 = 28 \quad \text{False}$$
Let x be 7.
$$7(7) - 9 \stackrel{?}{=} 5(7) + 13$$
$$49 - 9 \stackrel{?}{=} 35 + 13$$
$$40 = 48 \quad \text{False}$$
Let x be 11.
$$7(11) - 9 \stackrel{?}{=} 5(11) + 13$$
$$77 - 9 \stackrel{?}{=} 55 + 13$$
$$68 = 68 \quad \text{True}$$
11 is a solution.

69. The total of a number and 8 is $x + 8$.

73. The product of 512 and a number is $512x$.

77. The sum of seventeen and a number added to the product of five and the number is $5x + (17 + x)$.

81. A number subtracted from 11 is $11 - x$.

85. 6 divided by a number is $6 \div x$ or $\dfrac{6}{x}$.

89. $x^4 - y^2 = 23^4 - 72^2$
$$= 279,841 - 5184$$
$$= 274,657$$

93. $5x$ is the largest; answers may vary.

Chapter Test

1. 82,426 in words is eighty-two thousand, four hundred twenty-six.

5. $\overset{21}{496}$
$\underline{\times \ 30}$
$14,880$

9. $0 \div 49 = 0$

13. $6^1 \cdot 2^3 = 6 \cdot 8 = 48$

17. To round 52,369 to the nearest thousand, observe that the digit in the hundreds place is 3. Since this digit is less than 5, we do not add 1 to the digit in the thousands place. The number 52,369 rounded to the nearest thousand is 52,000.

21. "sum" indicates addition
$\overset{1}{15}$
$\underline{+ 107}$
122
The sum of 15 and 107 is 122.

25. "How much more" indicates subtraction

$$725$$
$$\underline{-599}$$
$$126$$

The higher-priced refrigerator is $126 more expensive.

29. Opposite sides of a rectangle have the same length.

Perimeter $= 20 + 10 + 20 + 10 = 60$

The perimeter is 60 yards.

Area $=$ (length)(width)

$\quad\quad = (20 \text{ yards})(10 \text{ yards})$

$\quad\quad = 200$ square yards.

The area is 200 square yards.

33. Let n be 6 in $5n - 11 = 19$.

$5 \cdot 6 - 11 \overset{?}{=} 19$ Let $n = 6$.

$30 - 11 \overset{?}{=} 19$ Multiply.

$19 = 19$ True.

Since $19 = 19$ is a true statement, 6 is a solution of $5n - 11 = 19$.

CHAPTER 2

Exercise Set 2.1

1. If 0 represents ground level, then 1235 feet underground is -1235.

5. If 0 represents zero degrees Fahrenheit, then 118 degrees above zero is $+118$.

9. If 0 represents a loss of $0, then a loss of $10,458 is $-10,458$.

13. If 0 represents a loss of 0%, then an 81 percent loss is -81.

17.

21.

25. $-7 < -5$ since -7 is to the left of -5 on a number line.

29. $-26 < 26$ since -26 is to the left of 26 on a number line.

33. $|-8| = 8$ since -8 is 8 units from 0 on a number line.

37. $|-55| = 55$ since -55 is 55 units from 0 on a number line.

41. The opposite of negative 4 is 4.
$-(-4) = 4$

45. The opposite of negative 85 is 85.
$-(-85) = 85$

49. $-|20| = -20$

53. $-(-43) = 43$

57. $-(-33) = 33$

61. $-|-2| = -2$

65. $-|7| = -7$

69. $|-8| = 8$
$|-11| = 11$
Since $8 < 11, |-8| < |-11|$.

73. $-|-12| = -12$
$-(-12) = 12$
Since $-12 < 12, -|-12| < -(-12)$.

77. $|0| = 0$
$|-9| = 9$
Since $0 < 9, |0| < |-9|$.

81. $-(-12) = 12$
$-(-18) = 18$
Since $12 < 18, -(-12) < -(-18)$.

85. If the opposite of a number is -28, then the number is 28, and its absolute value is 28.

89. The tallest bar on the graph corresponds to Lake Superior, so Lake Superior has the highest elevation.

93. The number on the graph closest to $-300°$F is $-323°$F, which corresponds to the planet Uranus.

97. 15
$\underline{+20}$
35

101. $2^2 = 4, -|3| = -3, -(-5) = 5,$ and $-|-8| = -8$, so the numbers in order from least to greatest are $-|-8|, -|3|, 2^2, -(-5)$.

105. $-(-2) = 2, 5^2 = 25, -10 = -10, -|-9| = -9,$ and $|-12| = 12$, so the numbers in order from least to greatest are $-10, -|-9|, -(-2), |-12|, 5^2$.

109. $-(-|-8|) = -(-8) = 8$

113. True; a positive number will always be to the right of a negative number on a number line.

117. answers may vary

Exercise Set 2.2

1.

5.

9. $|-8| + |-2| = 8 + 2 = 10$
The common sign is negative, so $(-8) + (-2) = -10$.

13. $|6| - |-2| = 6 - 2 = 4$
$6 > 2$, so the answer is positive.
$6 + (-2) = 4$

17. $|-5| - |3| = 5 - 3 = 2$
$5 > 3$, so the answer is negative.
$3 + (-5) = -2$

21. $|-12| + |-12| = 12 + 12 = 24$
The common sign is negative, so $-12 + (-12) = -24$.

25. $|12| - |-5| = 12 - 5 = 7$

$12 > 5$, so the answer is positive.

$12 + (-5) = 7$

29. $|-56| - |26| = 56 - 26 = 30$

$56 > 26$, so the answer is negative.

$-56 + 26 = -30$

33. $|-144| - |124| = 144 - 124 = 20$

$144 > 124$, so the answer is negative.

$124 + (-144) = -20$

37. $-4 + 2 + (-5) = -2 + (-5) = -7$

41. $12 + (-4) + (-4) + 12 = 8 + (-4) + 12$

$= 4 + 12$

$= 16$

45. $-6 + (-15) + (-7) = -21 + (-7) = -28$

49. $5 + (-2) + 17 = 3 + 17 = 20$

53. $3 + 14 + (-18) = 17 + (-18) = -1$

57. $-13 + 8 + (-10) + (-27) = -5 + (-10) + (-27)$

$= -15 + (-27)$

$= -42$

61. $3x + y = 3(2) + (-3) = 6 + (-3) = 3$

65. The sum of -6 and 25 is $-6 + 25 = 19$.

69. $0 + (-215) + (-16) = -215 + (-16) = -231$
The diver's final depth is 231 feet below the surface.

73. The bar for 2002 has a height of 65, so the net income in 2002 was $65,000,000.

77. The starting temperature was $-10°C$ and it rose $12°C$.

$-10 + 12 = 2$

The temperature at 11 p.m. was $2°C$.

81. $-55 + 8 = -47$

West Virginia's record low temperature is $-47°F$.

85. $44 - 0 = 44$

89. answers may vary

93. $-4 + 14 = 10$

97. False; for example, $4 + (-2) = 2 > 0$.

Exercise Set 2.3

1. $-8 - (-8) = -8 + 8 = 0$

5. $3 - 8 = 3 + (-8) = -5$

9. $-4 - (-7) = -4 + 7 = 3$

13. $3 - 15 = 3 + (-15) = -12$

17. $478 - (-30) = 478 + 30 = 508$

21. $-7 - (-3) = -7 + 3 = -4$

25. $-25 - 17 = -25 + (-17) = -42$

29. $2 - (-12) = 2 + 12 = 14$

33. $8 - 13 = 8 + (-13) = -5$

37. $30 - 67 = 30 + (-67) = -37$

41. $13 - 5 - 7 = 13 + (-5) + (-7) = 8 + (-7) = 1$

45. $-11 + (-6) - 14 = -11 + (-6) + (-14)$

$= -17 + (-14)$

$= -31$

49. $-(-5) - 21 + (-16) = 5 + (-21) + (-16)$

$= -16 + (-16)$

$= -32$

53. $-3 + 4 - (-23) - 10 = -3 + 4 + 23 + (-10)$

$= 1 + 23 + (-10)$

$= 24 + (-10)$

$= 14$

57. $x - y = 8 - (-23) = 8 + 23 = 31$

61. $2x - y = 2(1) - (-18) = 2 + 18 = 20$

65. The two months with the lowest temperatures are January, $-10°F$, and December, $-6°F$.

$-6 - (-10) = -6 + 10 = 4$

The difference is $4°F$.

69. $93 - 18 - 26 = 93 + (-18) + (-26) = 49$

She owes $49 on her account.

73. $-282 - (-436) = -282 + 436 = 154$

The difference in elevation is 154 feet.

77. $600 - (-52) = 600 + 52 = 652$

The difference in elevation is 652 feet.

81. $867 - (-330) = 867 + 330 = 1197$

The difference in temperatures is $1197°F$.

85. The sum of -5 and a number is $-5 + x$.

89. $\frac{100}{20} = 5$

93. answers may vary

97. $10 - 30 = 10 + (-30) = -20$

101. $|-5| - |5| = 5 - 5 = 0$

105. $|-8 - 3| = |-8 + (-3)| = |-11| = 11$

$8 - 3 = 8 + (-3) = 5$

Since $11 \neq 5$, the statement is false.

Exercise Set 2.4

1. $-6(-2) = 12$

5. $9(-9) = -81$

9. $6(-2)(-4) = -12(-4) = 48$

13. $-4(4)(-5) = -16(-5) = 80$

17. $-5(3)(-1)(-1) = -15(-1)(-1) = 15(-1) = -15$

21. $(-3)^3 = (-3)(-3)(-3) = 9(-3) = -27$

25. $(-4)^3 = (-4)(-4)(-4) = 16(-4) = -64$

29. $\dfrac{-30}{6} = -5$

33. $\dfrac{0}{-21} = 0$

37. $\dfrac{56}{-4} = -14$

41. $-5(3) = -15$

45. $-7(-6) = 42$

49. $(-7)^2 = (-7)(-7) = 49$

53. $-\dfrac{72}{8} = -9$

57. $4(-10)(-3) = -40(-3) = 120$

61. $3 \cdot (-8) \cdot 0 = -24 \cdot 0 = 0$

65. $280 \div (-40) = \dfrac{280}{-40} = -7$

69. $-1^4 = -(1 \cdot 1 \cdot 1 \cdot 1) = -1$

73. $-2(3)(5)(-6) = -6(5)(-6) = -30(-6) = 180$

77. $-2(-3)(-5) = 6(-5) = -30$

81.
$$\begin{array}{r} 35 \\ \times\, 82 \\ \hline 70 \\ 2800 \\ \hline 2870 \end{array}$$
$35 \cdot (-82) = -2870$

85. $ab = 9(-2) = -18$

89. $\dfrac{x}{y} = \dfrac{5}{-5} = -1$

93. $\dfrac{x}{y} = \dfrac{-36}{-6} = 6$

97. $xy = 0(-8) = 0$

$\dfrac{x}{y} = \dfrac{0}{-8} = 0$

101.
$$\begin{array}{r} 42 \\ \times\, 6 \\ \hline 252 \end{array}$$
$-42(-6) = 252$

105. Subtract a number from -16 is $-16 - x$.

109. Divide a number by -33 is $\dfrac{x}{-33}$ or $x \div (-33)$.

113. Each move of 20 feet down is represented by -20.

$5 \cdot (-20) = -100$

The diver is at a depth of 100 feet.

117. $-3 \cdot 63 = -189$

The melting point of argon is $-189°C$.

121. a. $626 - 50 = 576$

576 million fewer VHS cassettes were shipped to retailers in 2005 than in 2001. This is a change of -576 million.

b. This is a period of 4 years.

$\dfrac{-576}{4} = -144$

The average change was -144 VHS cassettes per year.

125. $12 \div 4 - 2 + 7 = 3 - 2 + 7 = 1 + 7 = 8$

129. $-8 - 20 = -8 + (-20) = -28$

133. The product of an odd number of negative numbers is negative, so the product of seven negative numbers is negative.

137. answers may vary

Exercise Set 2.5

1. $(-5)^3 = (-5)(-5)(-5) = 25(-5) = -125$

5. $8 \cdot 2^2 = 8 \cdot 4 = 32$

9. $7 + 3(-6) = 7 + (-18) = -11$

13. $(-10) + 4 \div 2 = (-10) + 2 = -8$

17. $\dfrac{16 - 13}{-3} = \dfrac{3}{-3} = -1$

21. $5(-3) - (-12) = -15 + 12 = -3$

25. $8 \cdot 6 - 3 \cdot 5 + (-20) = 48 - 3 \cdot 5 + (-20)$
$$= 48 - 15 + (-20)$$
$$= 33 + (-20)$$
$$= 13$$

29. $|7 + 3| \cdot 2^3 = |10| \cdot 2^3 = 10 \cdot 2^3 = 10 \cdot 8 = 80$

33. $7^2 - (4 - 2^3) = 7^2 - (4 - 8)$
$$= 7^2 - (-4)$$
$$= 49 + 4$$
$$= 53$$

37. $-(-2)^6 = -64$

41. $|8 - 24| \cdot (-2) \div (-2) = |-16| \cdot (-2) \div (-2)$
$$= 16 \cdot (-2) \div (-2)$$
$$= -32 \div (-2)$$
$$= 16$$

45. $5(5 - 2) + (-5)^2 - 6 = 5(3) + (-5)^2 - 6$
$$= 5(3) + 25 - 6$$
$$= 15 + 25 - 6$$
$$= 40 - 6$$
$$= 34$$

49. $(-36 \div 6) - (4 \div 4) = -6 - 1 = -7$

53. $2(8 - 10)^2 - 5(1 - 6)^2 = 2(-2)^2 - 5(-5)^2$
$$= 2(4) - 5(25)$$
$$= 8 - 125$$
$$= -117$$

57. $\dfrac{(-7)(-3) - (4)(3)}{3[7 \div (3 - 10)]} = \dfrac{21 - 12}{3[7 \div (-7)]}$

$$= \frac{9}{3(-1)}$$

$$= \frac{9}{-3}$$

$$= -3$$

61. $x + y + z = -2 + 4 + (-1) = 2 + (-1) = 1$

65. $x^2 - y = (-2)^2 - 4 = 4 - 4 = 0$

69. $x^2 = (-3)^2 = 9$

73. $2z^3 = 2(-4)^3 = 2(-64) = -128$

77. $2x^3 - z = 2(-3)^3 - (-4)$

$$= 2(-27) + 4$$

$$= -54 + 4$$

$$= -50$$

81. average $= \dfrac{-17 + (-26) + (-20) + (-13)}{4}$

$$= \frac{-76}{4}$$

$$= -19$$

85. average $= \dfrac{-13 + (-2) + 0 + 5 + 7 + 15}{6}$

$$= \frac{12}{6}$$

$$= 2$$

The average of the scores is 2.

89. $\begin{array}{r} 45 \\ \times\ 90 \\ \hline 4050 \end{array}$

93. $8 + 8 + 8 + 8 = 32$

The perimeter is 32 inches.

97. $2 \cdot (7 - 5) \cdot 3 = 2 \cdot 2 \cdot 3 = 4 \cdot 3 = 12$

101. answers may vary

105. $(-12)^4 = (-12)(-12)(-12)(-12) = 20{,}736$

109. $(xy + z)^x = [2(-5) + 7]^2$

$$= [-10 + 7]^2$$

$$= [-3]^2$$

$$= 9$$

Exercise Set 2.6

1. $x - 8 = -2$

$6 - 8 \overset{?}{=} -2$

$-2 = -2$ True

Since $-2 = -2$ is true, 6 is a solution of the equation.

5. $-9f = 64 - f$

$-9(-8) \overset{?}{=} 64 - (-8)$

$72 \overset{?}{=} 64 + 8$

$72 = 72$ True

Since $72 = 72$ is true, -8 is a solution of the equation.

9. $a + 5 = 23$

$a + 5 - 5 = 23 - 5$

$a = 18$

Check: $a + 5 = 23$

$18 + 5 \overset{?}{=} 23$

$23 = 23$ True

The solution is 18.

13. $7 = y - 2$

$7 + 2 = y - 2 + 2$

$9 = y$

Check: $7 = y - 2$

$7 \overset{?}{=} 9 - 2$

$7 = 7$ True

The solution is 9.

17. $5x = 20$

$\dfrac{5x}{5} = \dfrac{20}{5}$

$\dfrac{5}{5} \cdot x = \dfrac{20}{5}$

$x = 4$

Check: $5x = 20$

$5(4) \overset{?}{=} 20$

$20 = 20$ True

The solution is 4.

21. $\dfrac{n}{7} = -2$

$7 \cdot \dfrac{n}{7} = 7 \cdot (-2)$

$\dfrac{7}{7} \cdot n = 7 \cdot (-2)$

$n = -14$

Check: $\dfrac{n}{7} = -2$

$\dfrac{-14}{7} \overset{?}{=} -2$

$-2 = -2$ True

The solution is -14.

25. $-4y = 0$

$$\frac{-4y}{-4} = \frac{0}{-4}$$

$$\frac{-4}{-4} \cdot y = \frac{0}{-4}$$

$$y = 0$$

Check: $-4y = 0$

$$-4(0) \overset{?}{=} 0$$

$$0 = 0 \quad \text{True}$$

The solution is 0.

29. $5x = -35$

$$\frac{5x}{5} = \frac{-35}{5}$$

$$\frac{5}{5} \cdot x = \frac{-35}{5}$$

$$x = -7$$

The solution is -7.

33. $-15 = y + 10$

$$-15 - 10 = y + 10 - 10$$

$$-25 = y$$

The solution is -25.

37. $11n = 10n + 21$

$$11n - 10n = 10n - 10n + 21$$

$$1n = 21$$

$$n = 21$$

The solution is 21.

41. $\frac{n}{4} = -20$

$$4 \cdot \frac{n}{4} = 4 \cdot (-20)$$

$$\frac{4}{4} \cdot n = 4 \cdot (-20)$$

$$n = -80$$

The solution is -80.

45. A number decreased by -2 is $x - (-2)$.

49. The sum of -15 and a number is $-15 + x$.

53. $n - 42{,}860 = -1286$

$$n - 42{,}860 + 42{,}860 = -1286 + 42{,}860$$

$$n = 41{,}574$$

The solution is 41,574.

57. answers may vary

Chapter Test

1. $-5 + 8 = 3$

5. $(-18) + (-12) = -30$

9. $|-25| + (-13) = 25 + (-13) = 12$

13. $(-8) + 9 \div (-3) = -8 + (-3) = -11$

17. $-(-7)^2 \div 7 \cdot (-4) = -49 \div 7 \cdot (-4) = -7 \cdot (-4) = 28$

21. $\dfrac{|25 - 30|^2}{2(-6) + 7} = \dfrac{|-5|^2}{-12 + 7} = \dfrac{(5)^2}{-5} = \dfrac{25}{-5} = -5$

25. $\dfrac{3z}{2y} = \dfrac{3(2)}{2(-3)} = \dfrac{6}{-6} = -1$

29. Subtract the depth of the lake from the elevation of the surface.
$1495 - 5315 = 1495 + (-5315) = -3820$

The deepest point on the lake is 3820 feet below sea level.

33. $\dfrac{n}{-7} = 4$

$$-7 \cdot \frac{n}{-7} = -7 \cdot 4$$

$$\frac{-7}{-7} \cdot n = -7 \cdot 4$$

$$n = -28$$

CHAPTER 3

Exercise Set 3.1

1. $3x + 5x = (3 + 5)x = 8x$

5. $4c + c - 7c = (4 + 1 - 7)c = -2c$

9. $3a + 2a + 7a - 5 = (3 + 2 + 7)a - 5 = 12a - 5$

13. $-3(11y) = (-3 \cdot 11)y = -33y$

17. $2(y + 3) = 2 \cdot y + 2 \cdot 3 = 2y + 6$

21. $-4(3x + 7) = -4 \cdot 3x + (-4) \cdot 7 = -12x - 28$

25. $8 + 5(3c - 1) = 8 + 5 \cdot 3c + 5 \cdot (-1)$
$$= 8 + 15c - 5$$
$$= 15c + 8 - 5$$
$$= 15c + 3$$

29. $3 + 6(w + 2) + w = 3 + 6 \cdot w + 6 \cdot 2 + w$
$$= 3 + 6w + 12 + w$$
$$= 6w + w + 3 + 12$$
$$= 7w + 15$$

33. $-(2y - 6) + 10 = -1(2y - 6) + 10$
$$= -1 \cdot 2y + (-1) \cdot (-6) + 10$$
$$= -2y + 6 + 10$$
$$= -2y + 16$$

37. $z - 8z = (1 - 8)z = -7z$

41. $2y - 6 + 4y - 8 = 2y + 4y - 6 - 8$
$$= 6y - 14$$

45. $2(x + 1) + 20 = 2 \cdot x + 2 \cdot 1 + 20$
$$= 2x + 2 + 20$$
$$= 2x + 22$$

49. $-5(z + 3) + 2z = -5 \cdot z + (-5) \cdot 3 + 2z$
$$= -5z - 15 + 2z$$
$$= -5z + 2z - 15$$
$$= -3z - 15$$

53. $-7(x + 5) + 5(2x + 1)$

$= -7 \cdot x + (-7) \cdot 5 + 5 \cdot 2x + 5 \cdot 1$

$= -7x - 35 + 10x + 5$

$= -7x + 10x - 35 + 5$

$= 3x - 30$

57. $-3(n - 1) - 4n = -3 \cdot n - (-3) \cdot 1 - 4n$

$= -3n + 3 - 4n$

$= -3n - 4n + 3$

$= -7n + 3$

61. $6(2x - 1) - 12x = 6 \cdot 2x - 6 \cdot 1 - 12x$

$= 12x - 6 - 12x$

$= 12x - 12x - 6$

$= -6$

65. $-(4xy - 10) + 2(3xy + 5)$

$= -1(4xy - 10) + 2 \cdot 3xy + 2 \cdot 5$

$= -1 \cdot 4xy - (-1) \cdot 10 + 6xy + 10$

$= -4xy + 10 + 6xy + 10$

$= -4xy + 6xy + 10 + 10$

$= 2xy + 20$

69. $5y - 2(y - 1) + 3 = 5y - 2y - 2(-1) + 3$

$= 3y + 2 + 3$

$= 3y + 5$

73. $2a + 2a + 6 + 5a + 6 + 2a$

$= 2a + 2a + 5a + 2a + 6 + 6$

$= (2 + 2 + 5 + 2)a + 6 + 6$

$= 11a + 12$

The perimeter is $(11a + 12)$ feet.

77. Area $= $ (length) $\cdot$ (width)

$= (4y) \cdot (9)$

$= (4 \cdot 9)y$

$= 36y$

The area is $36y$ square inches.

81. Area $= $ (length) $\cdot$ (width)

$= (3y + 1) \cdot (20)$

$= 3y \cdot 20 + 1 \cdot 20$

$= (3 \cdot 20)y + 20$

$= 60y + 20$

The area is $(60y + 20)$ square miles.

85. Perimeter $= 2 \cdot$ (length) $+ 2 \cdot$ (width)

$= 2 \cdot (18) + 2 \cdot (14)$

$= 36 + 28$

$= 64$

The perimeter is 64 feet.

89. $-13 + 10 = -3$

93. $-4 + 4 = 0$

97. $7x - (x + 2) = 7x + (-1) \cdot (x + 2)$

$= 7x + (-1) \cdot x + (-1) \cdot 2$

$= 7x - x - 2$

The expressions are equivalent.

101. The order of the terms is not changed, but the grouping is. This is the associative property of addition.

105. $9684q - 686 - 4860q + 12,960$

$= 9684q - 4860q - 686 + 12,960$

$= 4824q + 12,274$

109. answers may vary

Exercise Set 3.2

1. $x - 3 = -1 + 4$

$x - 3 = 3$

$x - 3 + 3 = 3 + 3$

$x = 6$

5. $2w - 12w = 40$

$-10w = 40$

$\dfrac{-10w}{-10} = \dfrac{40}{-10}$

$w = -4$

9. $2z = 12 - 14$

$2z = -2$

$\dfrac{2z}{2} = \dfrac{-2}{2}$

$z = -1$

13. $-3x - 3x = 50 - 2$

$-6x = 48$

$\dfrac{-6x}{-6} = \dfrac{48}{-6}$

$x = -8$

17. $7x + 7 - 6x = 10$

$7x - 6x + 7 = 10$

$x + 7 = 10$

$x + 7 - 7 = 10 - 7$

$x = 3$

21. $2(5x - 3) = 11x$

$2 \cdot 5x - 2 \cdot 3 = 11x$

$10x - 6 = 11x$

$10x - 10x - 6 = 11x - 10x$

$-6 = x$

25. $21y = 5(4y - 6)$

$21y = 5 \cdot 4y - 5 \cdot 6$

$21y = 20y - 30$

$21y - 20y = 20y - 20y - 30$

$y = -30$

29.
$$2x - 8 = 0$$
$$2x - 8 + 8 = 0 + 8$$
$$2x = 8$$
$$\frac{2x}{2} = \frac{8}{2}$$
$$x = 4$$

33.
$$-7 = 2x - 1$$
$$-7 + 1 = 2x - 1 + 1$$
$$-6 = 2x$$
$$\frac{-6}{2} = \frac{2x}{2}$$
$$-3 = x$$

37.
$$11(x - 6) = -4 - 7$$
$$11 \cdot x - 11 \cdot 6 = -11$$
$$11x - 66 = -11$$
$$11x - 66 + 66 = -11 + 66$$
$$11x = 55$$
$$\frac{11x}{11} = \frac{55}{11}$$
$$x = 5$$

41.
$$y - 20 = 6y$$
$$y - y - 20 = 6y - y$$
$$-20 = 5y$$
$$\frac{-20}{5} = \frac{5y}{5}$$
$$-4 = y$$

45.
$$-2 - 3 = -4 + x$$
$$-5 = -4 + x$$
$$-5 + 4 = -4 + 4 + x$$
$$-1 = x$$

49.
$$3w - 12w = -27$$
$$-9w = -27$$
$$\frac{-9w}{-9} = \frac{-27}{-9}$$
$$w = 3$$

53.
$$18 - 11 = \frac{x}{-5}$$
$$7 = \frac{x}{-5}$$
$$-5 \cdot 7 = -5 \cdot \frac{x}{-5}$$
$$-35 = x$$

57.
$$10 = 7t - 12t$$
$$10 = -5t$$
$$\frac{10}{-5} = \frac{-5t}{-5}$$
$$-2 = t$$

61.
$$50y = 7(7y + 4)$$
$$50y = 7 \cdot 7y + 7 \cdot 4$$
$$50y = 49y + 28$$
$$50y - 49y = 49y - 49y + 28$$
$$y = 28$$

65.
$$7x + 14 - 6x = -4 - 10$$
$$7x - 6x + 14 = -14$$
$$x + 14 = -14$$
$$x + 14 - 14 = -14 - 14$$
$$x = -28$$

69.
$$23x + 8 - 25x = 7 - 9$$
$$23x - 25x + 8 = -2$$
$$-2x + 8 = -2$$
$$-2x + 8 - 8 = -2 - 8$$
$$-2x = -10$$
$$\frac{-2x}{-2} = \frac{-10}{-2}$$
$$x = 5$$

73. The sum of –7 and a number is $-7 + x$.

77. The product of –13 and a number is $-13x$.

81. The product of –11 and a number, increased by 5 is $-11x + 5$.

85. Seven added to the product of 4 and a number is $4x + 7$.

89. The product of –6 and the sum of a number and 15 is $-6(x + 15)$.

93. The quotient of seventeen and a number, increased by –15
is $\frac{17}{x} + (-15)$.

97. From the length of the bar, there were approximately 35,000 trumpeter swans in 2005.

101. no; answers may vary

105.
$$\frac{y}{72} = -86 - (-1029)$$
$$\frac{y}{72} = -86 + 1029$$
$$\frac{y}{72} = 943$$
$$72 \cdot \frac{y}{72} = 72 \cdot 943$$
$$y = 67{,}896$$

109.
$$|-13| + 3^2 = 100y - |-20| - 99y$$
$$13 + 3^2 = 100y - 20 - 99y$$
$$13 + 9 = 100y - 99y - 20$$
$$22 = y - 20$$
$$22 + 20 = y - 20 + 20$$
$$42 = y$$

Exercise Set 3.3

1.
$$3x - 7 = 4x + 5$$
$$3x - 3x - 7 = 4x - 3x + 5$$
$$-7 = x + 5$$
$$-7 - 5 = x + 5 - 5$$
$$-12 = x$$

5.
$$19 - 3x = 14 + 2x$$
$$19 - 3x + 3x = 14 + 2x + 3x$$
$$19 = 14 + 5x$$
$$19 - 14 = 14 - 14 + 5x$$
$$5 = 5x$$
$$\frac{5}{5} = \frac{5x}{5}$$
$$1 = x$$

9.
$$x + 20 + 2x = -10 - 2x - 15$$
$$x + 2x + 20 = -10 - 15 - 2x$$
$$3x + 20 = -25 - 2x$$
$$3x + 2x + 20 = -25 - 2x + 2x$$
$$5x + 20 = -25$$
$$5x + 20 - 20 = -25 - 20$$
$$5x = -45$$
$$\frac{5x}{5} = \frac{-45}{5}$$
$$x = -9$$

13.
$$35 - 17 = 3(x - 2)$$
$$18 = 3x - 6$$
$$18 + 6 = 3x - 6 + 6$$
$$24 = 3x$$
$$\frac{24}{3} = \frac{3x}{3}$$
$$8 = x$$

17.
$$2(y - 3) = y - 6$$
$$2y - 6 = y - 6$$
$$2y - y - 6 = y - y - 6$$
$$y - 6 = -6$$
$$y - 6 + 6 = -6 + 6$$
$$y = 0$$

21.
$$2t - 1 = 3(t + 7)$$
$$2t - 1 = 3t + 21$$
$$2t - 2t - 1 = 3t - 2t + 21$$
$$-1 = t + 21$$
$$-1 - 21 = t + 21 - 21$$
$$-22 = t$$

25.
$$-4x = 44$$
$$\frac{-4x}{-4} = \frac{44}{-4}$$
$$x = -11$$

29.
$$8 - b = 13$$
$$8 - 8 - b = 13 - 8$$
$$-b = 5$$
$$-1 \cdot (-b) = -1 \cdot 5$$
$$b = -5$$

33.
$$3r + 4 = 19$$
$$3r + 4 - 4 = 19 - 4$$
$$3r = 15$$
$$\frac{3r}{3} = \frac{15}{3}$$
$$r = 5$$

37.
$$8y - 13y = -20 - 25$$
$$-5y = -45$$
$$\frac{-5y}{-5} = \frac{-45}{-5}$$
$$y = 9$$

41.
$$-4 + 12 = 16x - 3 - 15x$$
$$8 = x - 3$$
$$8 + 3 = x - 3 + 3$$
$$11 = x$$

45.
$$4x + 3 = 2x + 11$$
$$4x - 2x + 3 = 2x - 2x + 11$$
$$2x + 3 = 11$$
$$2x + 3 - 3 = 11 - 3$$
$$2x = 8$$
$$\frac{2x}{2} = \frac{8}{2}$$
$$x = 4$$

49.
$$-8n + 1 = -6n - 5$$
$$-8n + 8n + 1 = -6n + 8n - 5$$
$$1 = 2n - 5$$
$$1 + 5 = 2n - 5 + 5$$
$$6 = 2n$$
$$\frac{6}{2} = \frac{2n}{2}$$
$$3 = n$$

53.
$$9a + 29 + 7 = 0$$
$$9a + 36 = 0$$
$$9a + 36 - 36 = 0 - 36$$
$$9a = -36$$
$$\frac{9a}{9} = \frac{-36}{9}$$
$$a = -4$$

57.
$$12 + 5t = 6(t + 2)$$
$$12 + 5t = 6t + 12$$
$$12 + 5t - 5t = 6t - 5t + 12$$
$$12 = t + 12$$
$$12 - 12 = t + 12 - 12$$
$$0 = t$$

61.
$$10 + 5(z - 2) = 4z + 1$$
$$10 + 5z - 10 = 4z + 1$$
$$5z = 4z + 1$$
$$5z - 4z = 4z - 4z + 1$$
$$z = 1$$

65. The sum of -42 and 16 is -26 translates to $-42 + 16 = -26$.

69. Three times the difference of -14 and 2 amounts to -48 translates to $3(-14 - 2) = -48$.

73. From the height of the bar, approximately 97 million returns are expected to be filed electronically in 2010.

77.
$$
\begin{aligned}
x^3 - 2xy &= 3^3 - 2(3)(-1) \\
&= 27 - 2(3)(-1) \\
&= 27 - (-6) \\
&= 27 + 6 \\
&= 33
\end{aligned}
$$

81. The first step in solving $2x - 5 = -7$ is to add 5 to both sides, which is choice b.

85. The error is in the second line.
$$
\begin{aligned}
2(3x - 5) &= 5x - 7 \\
6x - 10 &= 5x - 7 \\
6x - 10 + 10 &= 5x - 7 + 10 \\
6x &= 5x + 3 \\
6x - 5x &= 5x + 3 - 5x \\
x &= 3
\end{aligned}
$$

89.
$$
\begin{aligned}
x + 45^2 &= 54^2 \\
x + 2025 &= 2916 \\
x + 2025 - 2025 &= 2916 - 2025 \\
x &= 891
\end{aligned}
$$

Exercise Set 3.4

1. A number added to -5 is -7 is $-5 + x = -7$.

5. A number subtracted from -20 amounts to 104 is $-20 - x = 104$.

9. The product of 5 and the sum of -3 and a number is -20 is $5(-3 + x) = -20$.

13. The sum of 3, 4, and a number amounts to 16 is
$$
\begin{aligned}
3 + 4 + x &= 16 \\
7 + x &= 16 \\
x + 7 - 7 &= 16 - 7 \\
x &= 9
\end{aligned}
$$

17. Thirty less a number is equal to the product of 3 and the sum of the number and 6 is
$$
\begin{aligned}
30 - x &= 3(x + 6) \\
30 - x &= 3x + 18 \\
30 - 30 - x &= 3x + 18 - 30 \\
-x &= 3x - 12 \\
-x - 3x &= 3x - 3x - 12 \\
-4x &= -12 \\
\frac{-4x}{-4} &= \frac{-12}{-4} \\
x &= 3
\end{aligned}
$$

21. Three times the difference of some number and 5 amounts to the quotient of 108 and 12 is
$$
\begin{aligned}
3(x - 5) &= \frac{108}{12} \\
3x - 15 &= 9 \\
3x - 15 + 15 &= 9 + 15 \\
3x &= 24 \\
\frac{3x}{3} &= \frac{24}{3} \\
x &= 8
\end{aligned}
$$

25. Let x be the number of electoral votes California has. Since Florida has 28 fewer electoral votes than California, Florida has $x - 28$ electoral votes. The total of these votes is 82, so the sum of x and $x - 28$ is 82.
$$
\begin{aligned}
x + x - 28 &= 82 \\
2x - 28 &= 82 \\
2x - 28 + 28 &= 82 + 28 \\
2x &= 110 \\
\frac{2x}{2} &= \frac{110}{2} \\
x &= 55
\end{aligned}
$$

California has 55 electoral votes and Florida has $55 - 28 = 27$ votes.

29. Let x be the number of universities in the United States. Since India has 2649 more universities than the United States, India has $x + 2649$ universities. Since their combined total is 14,165, the sum of x and $x + 2649$ is 14,165.
$$
\begin{aligned}
x + x + 2649 &= 14{,}165 \\
2x + 2649 &= 14{,}165 \\
2x + 2649 - 2649 &= 14{,}165 - 2649 \\
2x &= 11{,}516 \\
\frac{2x}{2} &= \frac{11{,}516}{2} \\
x &= 5758
\end{aligned}
$$

Thus, the United States has 5758 universities and India has $5758 + 2649 = 8407$ universities.

33. Let x be the distance from Los Angeles to Tokyo. Since the distance from New York to London is 2001 miles less than the distance from Los Angeles to Tokyo, the distance from New York to London is $x - 2001$. Since the total of the two distances is 8939, the sum of x and $x - 2001$ is 8939.
$$
\begin{aligned}
x + x - 2001 &= 8939 \\
2x - 2001 &= 8939 \\
2x - 2001 + 2001 &= 8939 + 2001 \\
2x &= 10{,}940 \\
\frac{2x}{2} &= \frac{10{,}940}{2} \\
x &= 5470
\end{aligned}
$$

The distance from Los Angeles to Tokyo is 5470 miles.

37. Let x be the number of tourists projected to visit Spain in 2020. Since the number of tourists projected to visit China in 2020 is twice the number projected for Spain, the number projected for China is $2x$. Since the total number of tourists projected for the two countries is 210 million, the sum of x and $2x$ is 210 million.

$$x + 2x = 210$$
$$3x = 210$$
$$\frac{3x}{3} = \frac{210}{3}$$
$$x = 70$$
$$2x = 2(70) = 140$$

70 million tourists are projected to visit Spain in 2020; 140 million are projected to visit China in 2020.

41. Let x be the amount Anthony received for the accessories. Since he received five times as much for the bike, he received $5x$ for the bike. Since he received a total of $270 for the bike and accessories, the sum of x and $5x$ is 270.

$$x + 5x = 270$$
$$6x = 270$$
$$\frac{6x}{6} = \frac{270}{6}$$
$$x = 45$$

Thus, Anthony received $5 \cdot \$45 = \$225$ for the bike.

45. 586 rounded to the nearest ten is 590.

49. 2986 rounded to the nearest thousand is 3000.

53. Use $P = A + C$, where $P = 165{,}000$ and $A = 156{,}750$.
$$P = A + C$$
$$165{,}000 = 156{,}750 + C$$
$$165{,}000 - 156{,}750 = 156{,}750 - 156{,}750 + C$$
$$8250 = C$$
The agent will receive $8250.

Chapter Test

1. $7x - 5 - 12x + 10 = 7x - 12x - 5 + 10$
$$= (7 - 12)x - 5 + 10$$
$$= -5x + 5$$

5. Area $= $ (length) $\cdot$ (width)
$$= 4 \cdot (3x - 1)$$
$$= 4 \cdot 3x + 4 \cdot (-1)$$
$$= 12x - 4$$
The area is $(12x - 4)$ square meters.

9. $-4x + 7 = 15$
$$-4x + 7 - 7 = 15 - 7$$
$$-4x = 8$$
$$\frac{-4x}{-4} = \frac{8}{-4}$$
$$x = -2$$

13. $4(5x + 3) = 2(7x + 6)$
$$20x + 12 = 14x + 12$$
$$20x + 12 - 14x = 14x + 12 - 14x$$
$$6x + 12 = 12$$
$$6x + 12 - 12 = 12 - 12$$
$$6x = 0$$
$$\frac{6x}{6} = \frac{0}{6}$$
$$x = 0$$

17. The sum of twice 5 and -15 is -5 translates to $2(5) + (-15) = -5$.

21. Let x be the number of women runners entered in the race. Since the number of men entered in the race is 112 more than the number of women, the number of men is $x + 112$. Since the total number of runners in the race is 600, the sum of x and $x + 112$ is 600.

$$x + x + 112 = 600$$
$$2x + 112 = 600$$
$$2x + 112 - 112 = 600 - 112$$
$$2x = 488$$
$$\frac{2x}{2} = \frac{488}{2}$$
$$x = 244$$

244 women entered the race.

CHAPTER 4

Exercise Set 4.1

1. In the fraction $\frac{1}{2}$, the numerator is 1 and the denominator is 2. Since $1 < 2$, the fraction is proper.

5. In the fraction $\frac{15}{15}$, the numerator is 15 and the denominator is 15. Since $15 \geq 15$, the fraction is improper.

9. Each part is $\frac{1}{4}$ of a whole and there are 11 parts shaded, or 2 wholes and 3 more parts.
a. $\frac{11}{4}$
b. $2\frac{3}{4}$

13. 7 out of 12 equal parts are shaded: $\frac{7}{12}$.

17. 4 out of 9 equal parts are shaded: $\frac{4}{9}$.

21. Each part is $\frac{1}{2}$ of a whole and there are 11 parts shaded, or 5 wholes and 1 more part.

   a. $\frac{11}{2}$

   b. $5\frac{1}{2}$

25. 5 of 8 equal parts are shaded: $\frac{5}{8}$.

29.

33. freshmen $\rightarrow \underline{42}$

   students $\rightarrow 131$

   $\frac{42}{131}$ of the students are freshmen.

37. born in Virginia $\rightarrow \underline{8}$

   U.S. presidents $\rightarrow 43$

   $\frac{8}{43}$ of U.S. presidents were born in Virginia.

41. 11 of 31 days of March is $\frac{11}{31}$ of the month.

45. There are 50 states total. 33 states contain federal Indian reservations.

   a. $\frac{33}{50}$ of the states contain Indian reservations.

   b. $50 - 33 = 17$

   17 states do not contain Indian reservations.

   c. $\frac{17}{50}$ of the states do not contain Indian reservations.

49.
$$\begin{array}{c} \xleftarrow{\quad\bullet\quad\;\;\;\;\;\;\;\;\;\;} \\ 0 \quad \frac{1}{4} \quad\quad\quad 1 \end{array}$$

53.
$$\begin{array}{c} \xleftarrow{\;\;\;\;\;\;\;\;\;\;\;\bullet\;\;} \\ 0 \quad\quad 1 \quad \frac{8}{5} \quad 2 \end{array}$$

57. $\frac{12}{12} = 12 \div 12 = 1$

61. $\frac{0}{-2} = 0 \div -2 = 0$

65. $\frac{-9}{0} = -9 \div 0 =$ undefined

69. $2\frac{1}{3} = \frac{3 \cdot 2 + 1}{3} = \frac{6 + 1}{3} = \frac{7}{3}$

73. $6\frac{5}{8} = \frac{8 \cdot 6 + 5}{8} = \frac{48 + 5}{8} = \frac{53}{8}$

77. $9\frac{7}{20} = \frac{20 \cdot 9 + 7}{20} = \frac{180 + 7}{20} = \frac{187}{20}$

81.
$$5\overline{)17}$$
$$\underline{15}$$
$$2$$
   $\frac{17}{5} = 3\frac{2}{5}$

85.
$$15\overline{)47}$$
$$\underline{45}$$
$$2$$
   $\frac{47}{15} = 3\frac{2}{15}$

89.
$$175\overline{)182}$$
$$\underline{175}$$
$$7$$
   $\frac{182}{175} = 1\frac{7}{175}$

93. $3^2 = 3 \cdot 3 = 9$

97. $-\frac{11}{2} = \frac{-11}{2} = \frac{11}{-2}$

101. answers may vary

105. 11 is close to 12, so $\frac{11}{12}$ is close to 1. $5\frac{11}{12}$ rounded to the nearest whole number is 6.

109. $5 + 3 + 2 + 1 + 1 + 1 = 13$

   Five of the 13 USMC training centers are in California: $\frac{5}{13}$.

Exercise Set 4.2

1. $20 = 2 \cdot 10$
$$\downarrow \;\downarrow\searrow$$
$$2 \cdot 2 \cdot 5 = 2^2 \cdot 5$$

5. $81 = 9 \quad \cdot \quad 9$
$$\swarrow\searrow \quad \swarrow\searrow$$
$$3 \cdot 3 \cdot 3 \cdot 3 = 3^4$$

9. $110 = 2 \cdot 55$
$$\downarrow \;\downarrow\searrow$$
$$2 \cdot 5 \cdot 11 = 2 \cdot 5 \cdot 11$$

13. $240 = 2 \cdot 120$
$$\downarrow\;\;\downarrow\searrow$$
$$2 \cdot 2 \cdot 60$$
$$\downarrow\;\downarrow\;\downarrow\searrow$$
$$2 \cdot 2 \cdot 2 \cdot 30$$
$$\downarrow\;\downarrow\;\downarrow\;\downarrow\searrow$$
$$2 \cdot 2 \cdot 2 \cdot 2 \cdot 15$$
$$\downarrow\;\downarrow\;\downarrow\;\downarrow\;\downarrow\searrow$$
$$2 \cdot 2 \cdot 2 \cdot 2 \cdot 3 \cdot 5 = 2^4 \cdot 3 \cdot 5$$

17. $\frac{3}{12} = \frac{3 \cdot 1}{3 \cdot 4} = \frac{1}{4}$

21. $\frac{14}{16} = \frac{2 \cdot 7}{2 \cdot 8} = \frac{7}{8}$

25. $\frac{35a}{50a} = \frac{5 \cdot 7 \cdot a}{5 \cdot 10 \cdot a} = \frac{7}{10}$

29. $\frac{30x^2}{36x} = \frac{5 \cdot 6 \cdot x \cdot x}{6 \cdot 6 \cdot x} = \frac{5 \cdot x}{6} = \frac{5x}{6}$

33. $\frac{25xy}{40y} = \frac{5 \cdot 5 \cdot x \cdot y}{5 \cdot 8 \cdot y} = \frac{5 \cdot x}{8} = \frac{5x}{8}$

37. $\frac{36x^3y^2}{24xy} = \frac{3 \cdot 12 \cdot x \cdot x^2 \cdot y \cdot y}{2 \cdot 12 \cdot x \cdot y} = \frac{3 \cdot x^2 \cdot y}{2} = \frac{3x^2y}{2}$

41. $\frac{40xy}{64xyz} = \frac{5 \cdot 8 \cdot x \cdot y}{8 \cdot 8 \cdot x \cdot y \cdot z} = \frac{5}{8 \cdot z} = \frac{5}{8z}$

45. $-\frac{55}{85y} = -\frac{5 \cdot 11}{5 \cdot 17 \cdot y} = -\frac{11}{17 \cdot y} = -\frac{11}{17y}$

49. $\frac{224a^3b^4c^2}{16ab^4c^2} = \frac{14 \cdot 16 \cdot a \cdot a^2 \cdot b^4 \cdot c^2}{1 \cdot 16 \cdot a \cdot b^4 \cdot c^2}$
$$= \frac{14 \cdot a^2}{1}$$
$$= 14a^2$$

53. Not equivalent, since the cross products are not equal: $7 \cdot 8 = 56$ and $5 \cdot 11 = 55$.

57. Equivalent, since the cross products are equal: $3 \cdot 18 = 54$ and $9 \cdot 6 = 54$.

61. Not equivalent, since the cross products are not equal: $8 \cdot 24 = 192$ and $12 \cdot 18 = 216$.

65. $\dfrac{2640 \text{ feet}}{5280 \text{ feet}} = \dfrac{2640 \cdot 1}{2640 \cdot 2} = \dfrac{1}{2}$

2640 feet represents $\dfrac{1}{2}$ of a mile.

69. $\dfrac{10 \text{ inches}}{24 \text{ inches}} = \dfrac{2 \cdot 5}{2 \cdot 12} = \dfrac{5}{12}$

$\dfrac{5}{12}$ of the wall is concrete.

73. $\dfrac{291 \text{ individuals}}{449 \text{ individuals}} = \dfrac{291}{449}$

$\dfrac{291}{449}$ of the individuals who had flown in space were American.

77. $2y = 2(-7) = -14$

81. $\dfrac{3975}{6625} = \dfrac{3 \cdot 1325}{5 \cdot 1325} = \dfrac{3}{5}$

85. $3 + 1 = 4$ blood donors having an AB blood type.

$\dfrac{4 \text{ donors}}{100 \text{ donors}} = \dfrac{4 \cdot 1}{4 \cdot 25} = \dfrac{1}{25}$

$\dfrac{1}{25}$ of blood donors have an AB blood type.

89. answers may vary

93. The piece representing biological sciences is labeled $\dfrac{2}{25}$, so $\dfrac{2}{25}$ of entering college freshmen plan to major in the biological sciences.

97. $1235, 2235, 85, 105, 900,$ and 1470 are divisible by 5 because each number ends with a 0 or 5. $8691, 786, 2235, 105, 222, 900,$ and 1470 are divisible by 3 because the sum of each number's digits is divisible by 3.

$2235, 105, 900,$ and 1470 are divisible by both 3 and 5.

Exercise Set 4.3

1. $\dfrac{6}{11} \cdot \dfrac{3}{7} = \dfrac{6 \cdot 3}{11 \cdot 7} = \dfrac{18}{77}$

5. $-\dfrac{1}{2} \cdot -\dfrac{2}{15} = \dfrac{1 \cdot 2}{2 \cdot 15} = \dfrac{1}{15}$

9. $3a^2 \cdot \dfrac{1}{4} = \dfrac{3a^2}{1} \cdot \dfrac{1}{4} = \dfrac{3a^2 \cdot 1}{1 \cdot 4} = \dfrac{3a^2}{4}$

13. $0 \cdot \dfrac{8}{9} = 0$

17. $\dfrac{11}{20} \cdot \dfrac{1}{7} \cdot \dfrac{5}{22} = \dfrac{11 \cdot 1 \cdot 5}{20 \cdot 7 \cdot 22}$

$= \dfrac{11 \cdot 1 \cdot 5}{5 \cdot 2 \cdot 2 \cdot 7 \cdot 11 \cdot 2}$

$= \dfrac{1}{2 \cdot 2 \cdot 7 \cdot 2}$

$= \dfrac{1}{56}$

21. $\left(-\dfrac{2}{3}\right)^2 = -\dfrac{2}{3} \cdot -\dfrac{2}{3} = \dfrac{2 \cdot 2}{3 \cdot 3} = \dfrac{4}{9}$

25. $\dfrac{2}{3} \div \dfrac{5}{6} = \dfrac{2}{3} \cdot \dfrac{6}{5} = \dfrac{2 \cdot 6}{3 \cdot 5} = \dfrac{2 \cdot 2 \cdot 3}{3 \cdot 5} = \dfrac{2 \cdot 2}{5} = \dfrac{4}{5}$

29. $-\dfrac{8}{9} \div \dfrac{x}{2} = -\dfrac{8}{9} \cdot \dfrac{2}{x} = -\dfrac{8 \cdot 2}{9 \cdot x} = -\dfrac{16}{9x}$

33. $-\dfrac{2}{3} \div 4 = -\dfrac{2}{3} \div \dfrac{4}{1}$

$= -\dfrac{2}{3} \cdot \dfrac{1}{4}$

$= -\dfrac{2 \cdot 1}{3 \cdot 4}$

$= -\dfrac{2 \cdot 1}{3 \cdot 2 \cdot 2}$

$= -\dfrac{1}{3 \cdot 2}$

$= -\dfrac{1}{6}$

37. $\dfrac{2}{3} \cdot \dfrac{5}{9} = \dfrac{2 \cdot 5}{3 \cdot 9} = \dfrac{10}{27}$

41. $\dfrac{16}{27y} \div \dfrac{8}{15y} = \dfrac{16}{27y} \cdot \dfrac{15y}{8}$

$= \dfrac{16 \cdot 15y}{27y \cdot 8}$

$= \dfrac{8 \cdot 2 \cdot 3 \cdot 5 \cdot y}{3 \cdot 9 \cdot y \cdot 8}$

$= \dfrac{2 \cdot 5}{9}$

$= \dfrac{10}{9}$

45. $\left(-\dfrac{3}{4}\right)^2 = -\dfrac{3}{4} \cdot -\dfrac{3}{4} = \dfrac{3 \cdot 3}{4 \cdot 4} = \dfrac{9}{16}$

49. $7 \div \dfrac{2}{11} = \dfrac{7}{1} \div \dfrac{2}{11} = \dfrac{7}{1} \cdot \dfrac{11}{2} = \dfrac{7 \cdot 11}{1 \cdot 2} = \dfrac{77}{2}$

53. $\left(\dfrac{2}{7} \div \dfrac{7}{2}\right) \cdot \dfrac{3}{4} = \left(\dfrac{2}{7} \cdot \dfrac{2}{7}\right) \cdot \dfrac{3}{4}$

$= \dfrac{2 \cdot 2 \cdot 3}{7 \cdot 7 \cdot 4}$

$= \dfrac{2 \cdot 2 \cdot 3}{7 \cdot 7 \cdot 2 \cdot 2}$

$= \dfrac{3}{7 \cdot 7}$

$= \dfrac{3}{49}$

57. $-\dfrac{2}{3} \cdot -\dfrac{6}{11} = \dfrac{2 \cdot 6}{3 \cdot 11} = \dfrac{2 \cdot 2 \cdot 3}{3 \cdot 11} = \dfrac{2 \cdot 2}{11} = \dfrac{4}{11}$

61. $\dfrac{21x^2}{10y} \div \dfrac{14x}{25y} = \dfrac{21x^2}{10y} \cdot \dfrac{25y}{14x}$

$= \dfrac{21x^2 \cdot 25y}{10y \cdot 14x}$

$= \dfrac{3 \cdot 7 \cdot x \cdot x \cdot 5 \cdot 5 \cdot y}{2 \cdot 5 \cdot y \cdot 2 \cdot 7 \cdot x}$

$= \dfrac{3 \cdot x \cdot 5}{2 \cdot 2}$

$= \dfrac{15x}{4}$

65. $\dfrac{a^3}{2} \div 30a^3 = \dfrac{a^3}{2} \div \dfrac{30a^3}{1}$

$\qquad = \dfrac{a^3}{2} \cdot \dfrac{1}{30a^3}$

$\qquad = \dfrac{a^3 \cdot 1}{2 \cdot 30a^3}$

$\qquad = \dfrac{1}{2 \cdot 30}$

$\qquad = \dfrac{1}{60}$

69. $\left(\dfrac{1}{2} \cdot \dfrac{2}{3}\right) \div \dfrac{5}{6} = \left(\dfrac{1 \cdot 2}{2 \cdot 3}\right) \div \dfrac{5}{6}$

$\qquad = \dfrac{1}{3} \div \dfrac{5}{6}$

$\qquad = \dfrac{1}{3} \cdot \dfrac{6}{5}$

$\qquad = \dfrac{1 \cdot 6}{3 \cdot 5}$

$\qquad = \dfrac{1 \cdot 2 \cdot 3}{3 \cdot 5}$

$\qquad = \dfrac{1 \cdot 2}{5}$

$\qquad = \dfrac{2}{5}$

73. **a.** $xy = \dfrac{2}{5} \cdot \dfrac{5}{6} = \dfrac{2 \cdot 5}{5 \cdot 6} = \dfrac{2 \cdot 5}{5 \cdot 2 \cdot 3} = \dfrac{1}{3}$

b. $x \div y = \dfrac{2}{5} \div \dfrac{5}{6} = \dfrac{2}{5} \cdot \dfrac{6}{5} = \dfrac{2 \cdot 6}{5 \cdot 5} = \dfrac{12}{25}$

77. $\qquad 3x = -\dfrac{5}{6}$

$\qquad 3\left(-\dfrac{5}{18}\right) \overset{?}{=} -\dfrac{5}{6}$

$\qquad \dfrac{3}{1} \cdot -\dfrac{5}{18} \overset{?}{=} -\dfrac{5}{6}$

$\qquad -\dfrac{3 \cdot 5}{1 \cdot 18} \overset{?}{=} -\dfrac{5}{6}$

$\qquad -\dfrac{3 \cdot 5}{3 \cdot 6} \overset{?}{=} -\dfrac{5}{6}$

$\qquad -\dfrac{5}{6} = -\dfrac{5}{6}$ True

81. $\dfrac{1}{4}$ of $200 = \dfrac{1}{4} \cdot 200$

$\qquad = \dfrac{1}{4} \cdot \dfrac{200}{1}$

$\qquad = \dfrac{1 \cdot 200}{4 \cdot 1}$

$\qquad = \dfrac{1 \cdot 4 \cdot 50}{4 \cdot 1}$

$\qquad = \dfrac{50}{1}$

$\qquad = 50$

85. $\dfrac{4}{25}$ of $800 = \dfrac{4}{25} \cdot 800$

$\qquad = \dfrac{4}{25} \cdot \dfrac{800}{1}$

$\qquad = \dfrac{4 \cdot 800}{25 \cdot 1}$

$\qquad = \dfrac{4 \cdot 25 \cdot 32}{25 \cdot 1}$

$\qquad = \dfrac{4 \cdot 32}{1}$

$\qquad = 128$

128 of the students would be expected to major in business.

89. $\dfrac{1}{2}$ of $\dfrac{3}{8} = \dfrac{1}{2} \cdot \dfrac{3}{8} = \dfrac{1 \cdot 3}{2 \cdot 8} = \dfrac{3}{16}$

The radius of the circle is $\dfrac{3}{16}$ inch.

93. $\dfrac{1}{9}$ of $180 = \dfrac{1}{9} \cdot 180$

$\qquad = \dfrac{1}{9} \cdot \dfrac{180}{1}$

$\qquad = \dfrac{1 \cdot 180}{9 \cdot 1}$

$\qquad = \dfrac{1 \cdot 9 \cdot 20}{9 \cdot 1}$

$\qquad = \dfrac{20}{1}$

$\qquad = 20$

20 contestants participated more than once.

97. $\dfrac{8}{25} \cdot 12{,}000 = \dfrac{8}{25} \cdot \dfrac{12{,}000}{1}$

$\qquad = \dfrac{8 \cdot 25 \cdot 480}{25 \cdot 1}$

$\qquad = \dfrac{8 \cdot 480}{1}$

$\qquad = 3840$

The family drove 3840 miles for work.

101. $\quad 27$

$\qquad 76$

$\underline{+\ 98}$

$\quad 201$

105. answers may vary

109. $\dfrac{3}{25}$ of $290{,}810{,}000 = \dfrac{3}{25} \cdot 290{,}810{,}000$

$\qquad = \dfrac{3}{25} \cdot \dfrac{290{,}810{,}000}{1}$

$\qquad = \dfrac{3 \cdot 290{,}810{,}000}{25 \cdot 1}$

$\qquad = \dfrac{3 \cdot 25 \cdot 11{,}632{,}400}{25 \cdot 1}$

$\qquad = \dfrac{3 \cdot 11{,}632{,}400}{1}$

$\qquad = 34{,}897{,}200$

The population of California is approximately 34,897,200.

Exercise Set 4.4

1. $\dfrac{5}{11} + \dfrac{2}{11} = \dfrac{5+2}{11} = \dfrac{7}{11}$

5. $-\dfrac{6}{20} + \dfrac{1}{20} = \dfrac{-6+1}{20} = \dfrac{-5}{20} = -\dfrac{1 \cdot 5}{4 \cdot 5} = -\dfrac{1}{4}$

9. $\dfrac{2}{9x} + \dfrac{4}{9x} = \dfrac{2+4}{9x} = \dfrac{6}{9x} = \dfrac{2 \cdot 3}{3 \cdot 3 \cdot x} = \dfrac{2}{3x}$

13. $\dfrac{10}{11} - \dfrac{4}{11} = \dfrac{10-4}{11} = \dfrac{6}{11}$

17. $\dfrac{1}{y} - \dfrac{4}{y} = \dfrac{1-4}{y} = \dfrac{-3}{y} = -\dfrac{3}{y}$

21. $\dfrac{20}{21} - \dfrac{10}{21} - \dfrac{17}{21} = \dfrac{20-10-17}{21}$

$= \dfrac{-7}{21}$

$= -\dfrac{1 \cdot 7}{3 \cdot 7} = -\dfrac{1}{3}$

25. $-\dfrac{9}{100} + \dfrac{99}{100} = \dfrac{-9+99}{100} = \dfrac{90}{100} = \dfrac{9 \cdot 10}{10 \cdot 10} = \dfrac{9}{10}$

29. $\dfrac{9x}{15} + \dfrac{1}{15} = \dfrac{9x+1}{15}$

33. $\dfrac{9}{12} - \dfrac{7}{12} - \dfrac{10}{12} = \dfrac{9-7-10}{12} = \dfrac{-8}{12} = -\dfrac{2 \cdot 4}{3 \cdot 4} = -\dfrac{2}{3}$

37. $x + y = \dfrac{3}{4} + \dfrac{2}{4} = \dfrac{3+2}{4} = \dfrac{5}{4}$

41. $\dfrac{4}{20} + \dfrac{7}{20} + \dfrac{9}{20} = \dfrac{4+7+9}{20} = \dfrac{20}{20} = 1$

The perimeter is 1 inch.

45. North America takes up $\dfrac{16}{100}$ of the world's land area, while South America takes up $\dfrac{12}{100}$ of the land area.

$\dfrac{16}{100} + \dfrac{12}{100} = \dfrac{16+12}{100} = \dfrac{28}{100} = \dfrac{4 \cdot 7}{4 \cdot 25} = \dfrac{7}{25}$

$\dfrac{7}{25}$ of the world's land area is within North America and South America.

49. To find the remaining amount of track to be inspected, subtract the $\dfrac{5}{20}$ mile that has already been inspected from the $\dfrac{19}{20}$ mile total that must be inspected.

$\dfrac{19}{20} - \dfrac{5}{20} = \dfrac{19-5}{20} = \dfrac{14}{20} = \dfrac{2 \cdot 7}{2 \cdot 10} = \dfrac{7}{10}$

$\dfrac{7}{10}$ of a mile of track remains to be inspected.

53. Multiples of 9: 9, 18, 27, 36, $\boxed{45}$, 54

Multiples of 15: 15, 30, $\boxed{45}$, 60

LCD: 45

57. $6 = \boxed{2} \cdot 3$

$15 = \boxed{3} \cdot 5$

$25 = \boxed{5 \cdot 5}$

LCD $= 2 \cdot 3 \cdot 5 \cdot 5 = 150$

61. $18 = \boxed{2} \cdot \boxed{3 \cdot 3}$

$21 = 3 \cdot \boxed{7}$

LCD $= 2 \cdot 3 \cdot 3 \cdot 7 = 126$

65. $\dfrac{2}{3} = \dfrac{2 \cdot 7}{3 \cdot 7} = \dfrac{14}{21}$

69. $\dfrac{1}{2} = \dfrac{1 \cdot 25}{2 \cdot 25} = \dfrac{25}{50}$

73. $\dfrac{2y}{3} = \dfrac{2y \cdot 4}{3 \cdot 4} = \dfrac{8y}{12}$

77. Denmark: $\dfrac{89}{100} = \dfrac{89}{100}$

Finland: $\dfrac{9}{10} = \dfrac{9 \cdot 10}{10 \cdot 10} = \dfrac{90}{100}$

Israel: $\dfrac{24}{25} = \dfrac{24 \cdot 4}{25 \cdot 4} = \dfrac{96}{100}$

Spain: $\dfrac{23}{25} = \dfrac{23 \cdot 4}{25 \cdot 4} = \dfrac{92}{100}$

Japan: $\dfrac{17}{25} = \dfrac{17 \cdot 4}{25 \cdot 4} = \dfrac{68}{100}$

Norway: $\dfrac{91}{100} = \dfrac{91}{100}$

Singapore: $\dfrac{4}{5} = \dfrac{4 \cdot 20}{5 \cdot 20} = \dfrac{80}{100}$

South Korea: $\dfrac{69}{100} = \dfrac{69}{100}$

India: $\dfrac{2}{25} = \dfrac{2 \cdot 4}{25 \cdot 4} = \dfrac{8}{100}$

United States: $\dfrac{7}{10} = \dfrac{7 \cdot 10}{10 \cdot 10} = \dfrac{70}{100}$

81. $3^2 = 3 \cdot 3 = 9$

85. $7^2 = 7 \cdot 7 = 49$

89. $\dfrac{2}{7} + \dfrac{9}{7} = \dfrac{2+9}{7} = \dfrac{11}{7}$

93. $\dfrac{16}{100} + \dfrac{12}{100} + \dfrac{7}{100} + \dfrac{20}{100} + \dfrac{30}{100} + \dfrac{6}{100} + \dfrac{9}{100}$

$= \dfrac{16 + 12 + 7 + 20 + 30 + 6 + 9}{100}$

$= \dfrac{100}{100}$

$= 1$

97. answers may vary

Exercise Set 4.5

1. The LCD of 3 and 6 is 6.

$$\frac{2}{3} + \frac{1}{6} = \frac{2 \cdot 2}{3 \cdot 2} + \frac{1}{6} = \frac{4}{6} + \frac{1}{6} = \frac{5}{6}$$

5. The LCD of 11 and 33 is 33.

$$-\frac{2}{11} + \frac{2}{33} = -\frac{2 \cdot 3}{11 \cdot 3} + \frac{2}{33} = -\frac{6}{33} + \frac{2}{33}$$
$$= -\frac{4}{33}$$

9. The LCD of 35 and 7 is 35.

$$\frac{11x}{35} + \frac{2x}{7} = \frac{11x}{35} + \frac{2x \cdot 5}{7 \cdot 5}$$
$$= \frac{11x}{35} + \frac{10x}{35}$$
$$= \frac{21x}{35}$$
$$= \frac{3 \cdot 7 \cdot x}{5 \cdot 7}$$
$$= \frac{3x}{5}$$

13. The LCD of 12 and 9 is 36.

$$\frac{5}{12} - \frac{1}{9} = \frac{5 \cdot 3}{12 \cdot 3} - \frac{1 \cdot 4}{9 \cdot 4} = \frac{15}{36} - \frac{4}{36} = \frac{11}{36}$$

17. The LCD of 11 and 9 is 99.

$$\frac{5a}{11} + \frac{4a}{9} = \frac{5a \cdot 9}{11 \cdot 9} + \frac{4a \cdot 11}{9 \cdot 11}$$
$$= \frac{45a}{99} + \frac{44a}{99}$$
$$= \frac{89a}{99}$$

21. The LCD of 2 and x is $2x$.

$$\frac{1}{2} + \frac{3}{x} = \frac{1 \cdot x}{2 \cdot x} + \frac{3 \cdot 2}{x \cdot 2} = \frac{x}{2x} + \frac{6}{2x} = \frac{x + 6}{2x}$$

25. The LCD of 14 and 7 is 14.

$$\frac{9}{14} - \frac{3}{7} = \frac{9}{14} - \frac{3 \cdot 2}{7 \cdot 2} = \frac{9}{14} - \frac{6}{14} = \frac{3}{14}$$

29. The LCD of 9 and 12 is 36.

$$\frac{1}{9} - \frac{5}{12} = \frac{1 \cdot 4}{9 \cdot 4} - \frac{5 \cdot 3}{12 \cdot 3} = \frac{4}{36} - \frac{15}{36} = -\frac{11}{36}$$

33. The LCD of 7 and 8 is 56.

$$\frac{5}{7} - \frac{1}{8} = \frac{5 \cdot 8}{7 \cdot 8} - \frac{1 \cdot 7}{8 \cdot 7} = \frac{40}{56} - \frac{7}{56} = \frac{33}{56}$$

37. $\dfrac{3}{9} - \dfrac{5}{9} = \dfrac{3 - 5}{9} = -\dfrac{2}{9}$

41. The LCD of 11 and 3 is 33.

$$\frac{5}{11} + \frac{y}{3} = \frac{5 \cdot 3}{11 \cdot 3} + \frac{y \cdot 11}{3 \cdot 11}$$
$$= \frac{15}{33} + \frac{11y}{33}$$
$$= \frac{15 + 11y}{33}$$

45. The LCD of 2, 4, and 16 is 16.

$$\frac{x}{2} + \frac{x}{4} + \frac{2x}{16} = \frac{x \cdot 8}{2 \cdot 8} + \frac{x \cdot 4}{4 \cdot 4} + \frac{2x}{16}$$
$$= \frac{8x}{16} + \frac{4x}{16} + \frac{2x}{16}$$
$$= \frac{14x}{16}$$
$$= \frac{2 \cdot 7x}{2 \cdot 8}$$
$$= \frac{7x}{8}$$

49. The LCD of 3 and 13 is 39.

$$\frac{2a}{3} + \frac{6a}{13} = \frac{2a \cdot 13}{3 \cdot 13} + \frac{6a \cdot 3}{13 \cdot 3}$$
$$= \frac{26a}{39} + \frac{18a}{39}$$
$$= \frac{44a}{39}$$

53. The LCD of 9 and y is $9y$.

$$\frac{5}{9} + \frac{1}{y} = \frac{5 \cdot y}{9 \cdot y} + \frac{1 \cdot 9}{y \cdot 9}$$
$$= \frac{5y}{9y} + \frac{9}{9y}$$
$$= \frac{5y + 9}{9y}$$

57. The LCD of 5 and 9 is 45.

$$\frac{4}{5} + \frac{4}{9} = \frac{4 \cdot 9}{5 \cdot 9} + \frac{4 \cdot 5}{9 \cdot 5} = \frac{36}{45} + \frac{20}{45} = \frac{56}{45}$$

61. The LCD of 12, 24, and 6 is 24.

$$-\frac{9}{12} + \frac{17}{24} - \frac{1}{6} = -\frac{9 \cdot 2}{12 \cdot 2} + \frac{17}{24} - \frac{1 \cdot 4}{6 \cdot 4}$$
$$= -\frac{18}{24} + \frac{17}{24} - \frac{4}{24}$$
$$= -\frac{5}{24}$$

65. The LCD of 7 and 10 is 70. Write each fraction as an equivalent fraction with a denominator of 70.

$$\frac{2}{7} = \frac{2 \cdot 10}{7 \cdot 10} = \frac{20}{70}$$
$$\frac{3}{10} = \frac{3 \cdot 7}{10 \cdot 7} = \frac{21}{70}$$

Since $20 < 21, \dfrac{20}{70} < \dfrac{21}{70}$, so $\dfrac{2}{7} < \dfrac{3}{10}$.

69. The LCD of 4 and 14 is 28. Write each fraction as an equivalent fraction with a denominator of 28.

$$-\frac{3}{4} = -\frac{3 \cdot 7}{4 \cdot 7} = -\frac{21}{28}$$
$$-\frac{11}{14} = -\frac{11 \cdot 2}{14 \cdot 2} = -\frac{22}{28}$$

Since $-21 > -22, -\dfrac{21}{28} > -\dfrac{22}{28}$, so $-\dfrac{3}{4} > -\dfrac{11}{14}$.

73. $xy = \dfrac{1}{3} \cdot \dfrac{3}{4} = \dfrac{1 \cdot 3}{3 \cdot 4} = \dfrac{1}{4}$

77. The LCD of 3 and 5 is 15.

$$\frac{4}{5} + \frac{1}{3} + \frac{4}{5} + \frac{1}{3} = \frac{4}{5}\cdot\frac{3}{3} + \frac{1}{3}\cdot\frac{5}{5} + \frac{4}{5}\cdot\frac{3}{3} + \frac{1}{3}\cdot\frac{5}{5}$$
$$= \frac{12}{15} + \frac{5}{15} + \frac{12}{15} + \frac{5}{15}$$
$$= \frac{34}{15}$$

The perimeter is $\frac{34}{15}$ or $2\frac{4}{15}$ centimeters.

81. The sum of a number and $\frac{1}{2}$ translates as $x + \frac{1}{2}$.

85. $1 - \frac{3}{16} - \frac{3}{16} = \frac{16}{16} - \frac{3}{16} - \frac{3}{16} = \frac{10}{16} = \frac{5}{8}$

The inner diameter is $\frac{5}{8}$ inch.

89. The LCD of 2, 16, and 32 is 32.

$$\frac{1}{2} + \frac{11}{16} + \frac{9}{32} = \frac{1}{2}\cdot\frac{16}{16} + \frac{11}{16}\cdot\frac{2}{2} + \frac{9}{32}$$
$$= \frac{16}{32} + \frac{22}{32} + \frac{9}{32}$$
$$= \frac{47}{32}$$

The total length is $\frac{47}{32}$ inches.

93. The LCD of 50 and 2 is 50.

$$\frac{13}{50} + \frac{1}{2} = \frac{13}{50} + \frac{1}{2}\cdot\frac{25}{25}$$
$$= \frac{13}{50} + \frac{25}{50}$$
$$= \frac{38}{50}$$
$$= \frac{2\cdot19}{2\cdot25}$$
$$= \frac{19}{25}$$

The Pacific and Atlantic Oceans account for $\frac{19}{25}$ of the world's water surface.

97. $1 - \frac{4}{25} = \frac{25}{25} - \frac{4}{25} = \frac{21}{25}$

$\frac{21}{25}$ of the beginning freshmen do not plan to major in business.

101. $(8-6)\cdot(4-7) = 2\cdot(-3) = -6$

105. The LCD of 3, 4, and 540 is 540.

$$\frac{2}{3} - \frac{1}{4} - \frac{2}{540} = \frac{2}{3}\cdot\frac{180}{180} - \frac{1}{4}\cdot\frac{135}{135} - \frac{2}{540}$$
$$= \frac{360}{540} - \frac{135}{540} - \frac{2}{540}$$
$$= \frac{225}{540} - \frac{2}{540}$$
$$= \frac{223}{540}$$

109. answers may vary

Exercise Set 4.6

1. $\dfrac{\frac{1}{8}}{\frac{3}{4}} = \frac{1}{8} \div \frac{3}{4} = \frac{1}{8}\cdot\frac{4}{3} = \frac{1\cdot4}{8\cdot3} = \frac{1\cdot4}{2\cdot4\cdot3} = \frac{1}{6}$

5. $\dfrac{\frac{2x}{27}}{\frac{4}{9}} = \frac{2x}{27} \div \frac{4}{9} = \frac{2x}{27}\cdot\frac{9}{4} = \frac{2x\cdot9}{27\cdot4} = \frac{2\cdot x\cdot9}{3\cdot9\cdot2\cdot2} = \frac{x}{6}$

9. $\dfrac{\frac{3x}{4}}{5 - \frac{1}{8}} = \dfrac{8\cdot\left(\frac{3x}{4}\right)}{8\cdot\left(5 - \frac{1}{8}\right)}$
$$= \frac{6x}{8\cdot5 - 8\cdot\frac{1}{8}}$$
$$= \frac{6x}{40 - 1}$$
$$= \frac{6x}{39}$$
$$= \frac{2\cdot3\cdot x}{3\cdot13}$$
$$= \frac{2x}{13}$$

13. $\frac{5}{6} \div \frac{1}{3}\cdot\frac{1}{4} = \frac{5}{6}\cdot\frac{3}{1}\cdot\frac{1}{4}$
$$= \frac{5\cdot3\cdot1}{2\cdot3\cdot1\cdot2\cdot2}$$
$$= \frac{5}{2\cdot2\cdot2}$$
$$= \frac{5}{8}$$

17. $\left(\frac{2}{9} + \frac{4}{9}\right)\left(\frac{1}{3} - \frac{9}{10}\right) = \left(\frac{2+4}{9}\right)\left(\frac{1}{3}\cdot\frac{10}{10} - \frac{9}{10}\cdot\frac{3}{3}\right)$
$$= \left(\frac{6}{9}\right)\left(\frac{10}{30} - \frac{27}{30}\right)$$
$$= \left(\frac{6}{9}\right)\left(\frac{-17}{30}\right)$$
$$= -\frac{6\cdot17}{9\cdot30}$$
$$= -\frac{3\cdot2\cdot17}{3\cdot3\cdot5\cdot2\cdot3}$$
$$= -\frac{17}{3\cdot3\cdot5}$$
$$= -\frac{17}{45}$$

21. $2 \cdot \left(\dfrac{1}{4} + \dfrac{1}{5}\right) + 2 = 2 \cdot \left(\dfrac{1}{4} \cdot \dfrac{5}{5} + \dfrac{1}{5} \cdot \dfrac{4}{4}\right) + 2$

$= 2 \cdot \left(\dfrac{5}{20} + \dfrac{4}{20}\right) + 2$

$= 2 \cdot \left(\dfrac{9}{20}\right) + 2$

$= \dfrac{2}{1} \cdot \dfrac{9}{20} + 2$

$= \dfrac{2 \cdot 9}{1 \cdot 2 \cdot 10} + 2$

$= \dfrac{9}{10} + \dfrac{2}{1} \cdot \dfrac{10}{10}$

$= \dfrac{9}{10} + \dfrac{20}{10}$

$= \dfrac{29}{10}$

25. $\left(\dfrac{2}{5} - \dfrac{3}{10}\right)^2 = \left(\dfrac{2}{5} \cdot \dfrac{2}{2} - \dfrac{3}{10}\right)^2$

$= \left(\dfrac{4}{10} - \dfrac{3}{10}\right)^2$

$= \left(\dfrac{4-3}{10}\right)^2$

$= \left(\dfrac{1}{10}\right)^2$

$= \dfrac{1}{10} \cdot \dfrac{1}{10}$

$= \dfrac{1}{100}$

29. $5y - z = 5\left(\dfrac{2}{5}\right) - \dfrac{5}{6}$

$= 2 - \dfrac{5}{6}$

$= \dfrac{2}{1} \cdot \dfrac{6}{6} - \dfrac{5}{6}$

$= \dfrac{12}{6} - \dfrac{5}{6}$

$= \dfrac{7}{6}$

33. $x^2 - yz = \left(-\dfrac{1}{3}\right)^2 - \left(\dfrac{2}{5}\right)\left(\dfrac{5}{6}\right)$

$= \left(-\dfrac{1}{3}\right)\left(-\dfrac{1}{3}\right) - \dfrac{2 \cdot 5}{5 \cdot 2 \cdot 3}$

$= \dfrac{1}{9} - \dfrac{1}{3}$

$= \dfrac{1}{9} - \dfrac{1}{3} \cdot \dfrac{3}{3}$

$= \dfrac{1}{9} - \dfrac{3}{9}$

$= \dfrac{1-3}{9}$

$= \dfrac{-2}{9}$

$= -\dfrac{2}{9}$

37. $\left(\dfrac{3}{2}\right)^3 + \left(\dfrac{1}{2}\right)^3 = \dfrac{27}{8} + \dfrac{1}{8} = \dfrac{28}{8} = \dfrac{4 \cdot 7}{4 \cdot 2} = \dfrac{7}{2}$

41. $\dfrac{2 + \frac{1}{6}}{1 - \frac{4}{3}} = \dfrac{6(2 + \frac{1}{6})}{6(1 - \frac{4}{3})}$

$= \dfrac{6 \cdot 2 + 6 \cdot \frac{1}{6}}{6 \cdot 1 - 6 \cdot \frac{4}{3}}$

$= \dfrac{12 + 1}{6 - 8}$

$= \dfrac{13}{-2}$

$= -\dfrac{13}{2}$

45. $\left(\dfrac{3}{4} - 1\right)\left(\dfrac{1}{8} + \dfrac{1}{2}\right) = \left(\dfrac{3}{4} - \dfrac{4}{4}\right)\left(\dfrac{1}{8} + \dfrac{4}{8}\right)$

$= \left(-\dfrac{1}{4}\right)\left(\dfrac{5}{8}\right)$

$= -\dfrac{1 \cdot 5}{4 \cdot 8}$

$= -\dfrac{5}{32}$

49. $\dfrac{\frac{1}{3} - \frac{5}{6}}{\frac{3}{4} + \frac{1}{2}} = \dfrac{12(\frac{1}{3} - \frac{5}{6})}{12(\frac{3}{4} + \frac{1}{2})}$

$= \dfrac{12 \cdot \frac{1}{3} - 12 \cdot \frac{5}{6}}{12 \cdot \frac{3}{4} + 12 \cdot \frac{1}{2}}$

$= \dfrac{4 - 10}{9 + 6}$

$= \dfrac{-6}{15}$

$= -\dfrac{2 \cdot 3}{3 \cdot 5}$

$= -\dfrac{2}{5}$

53. $\dfrac{\frac{x}{3} + 2}{5 + \frac{1}{3}} = \dfrac{3(\frac{x}{3} + 2)}{3(5 + \frac{1}{3})}$

$= \dfrac{3 \cdot \frac{x}{3} + 3 \cdot 2}{3 \cdot 5 + 3 \cdot \frac{1}{3}}$

$= \dfrac{x + 6}{15 + 1}$

$= \dfrac{x + 6}{16}$

57. $9 - \dfrac{5}{6} = \dfrac{9}{1} \cdot \dfrac{6}{6} - \dfrac{5}{6} = \dfrac{54}{6} - \dfrac{5}{6} = \dfrac{49}{6}$ or $8\dfrac{1}{6}$

61. $\dfrac{\frac{1}{2} + \frac{3}{4}}{2} = \dfrac{4(\frac{1}{2} + \frac{3}{4})}{4(2)}$

$= \dfrac{4 \cdot \frac{1}{2} + 4 \cdot \frac{3}{4}}{4 \cdot 2}$

$= \dfrac{2 + 3}{8}$

$= \dfrac{5}{8}$

65. The average of a and b should be halfway between a and b.

69. False, if the absolute value of the negative fraction is greater than the absolute value of the positive fraction.

73. No; answers may vary.

77. Subtraction, multiplication, division, addition

81. $\dfrac{\frac{9}{14}}{x+y} = \dfrac{\frac{9}{14}}{\frac{3}{4}+\left(-\frac{4}{7}\right)}$

$= \dfrac{28\left(\frac{9}{14}\right)}{28\left(\frac{3}{4}-\frac{4}{7}\right)}$

$= \dfrac{18}{28\cdot\frac{3}{4}-28\cdot\frac{4}{7}}$

$= \dfrac{18}{21-16}$

$= \dfrac{18}{5}$

Exercise Set 4.7

1.

5. $2\frac{11}{12}$ rounds to 3.

$1\frac{1}{4}$ rounds to 1.

$3\cdot 1 = 3$

The best estimate is b.

9. $2\frac{2}{3}\cdot\frac{1}{7} = \frac{8}{3}\cdot\frac{1}{7} = \frac{8}{21}$

13. Exact: $2\frac{1}{5}\cdot 3\frac{1}{2} = \frac{11}{5}\cdot\frac{7}{2} = \frac{77}{10}$ or $7\frac{7}{10}$

Estimate: $2\frac{1}{5}$ rounds to 2, $3\frac{1}{2}$ rounds to 4.

$2\cdot 4 = 8$ so the answer is reasonable.

17. $5\cdot 2\frac{1}{2} = \frac{5}{1}\cdot\frac{5}{2} = \frac{25}{2}$ or $12\frac{1}{2}$

21. $2\frac{2}{3}\div\frac{1}{7} = \frac{8}{3}\cdot\frac{7}{1} = \frac{56}{3}$ or $18\frac{2}{3}$

25. $8\frac{1}{3}$ rounds to 8.

$1\frac{1}{2}$ rounds to 2.

$8 + 2 = 10$

The best estimate is b.

29. Exact: $10\frac{3}{14} + 3\frac{4}{7} = 10\frac{3}{14} + 3\frac{8}{14} = 13\frac{11}{14}$

Estimate: $10\frac{3}{14}$ rounds to 10, $3\frac{4}{7}$ rounds to 4.

$10 + 4 = 14$ so the answer is reasonable.

33.

$\begin{array}{r} 12\frac{3}{14} \\ 10 \\ +\,25\frac{5}{12} \\ \hline \end{array}$ $\begin{array}{r} 12\frac{18}{84} \\ 10 \\ +\,25\frac{35}{84} \\ \hline 47\frac{53}{84} \end{array}$

37.

$\begin{array}{r} 3\frac{5}{8} \\ 2\frac{1}{6} \\ +\,7\frac{3}{4} \\ \hline \end{array}$ $\begin{array}{r} 3\frac{15}{24} \\ 2\frac{4}{24} \\ +\,7\frac{18}{24} \\ \hline 12\frac{37}{24} = 12 + 1\frac{13}{24} = 13\frac{13}{24} \end{array}$

41.

$\begin{array}{r} 10\frac{13}{14} \\ -\,3\frac{4}{7} \\ \hline \end{array}$ $\begin{array}{r} 10\frac{13}{14} \\ -\,3\frac{8}{14} \\ \hline \end{array}$

Exact: $7\frac{5}{14}$

Estimate: $10\frac{13}{14}$ rounds to 11, $3\frac{4}{7}$ rounds to 4.

$11 - 4 = 7$ so the answer is reasonable.

45.

$\begin{array}{r} 6 \\ -\,2\frac{4}{9} \\ \hline \end{array}$ $\begin{array}{r} 5\frac{9}{9} \\ -\,2\frac{4}{9} \\ \hline 3\frac{5}{9} \end{array}$

49.

$\begin{array}{r} 2\frac{3}{4} \\ +\,1\frac{1}{4} \\ \hline 3\frac{4}{4} = 3 + 1 = 4 \end{array}$

53. $3\frac{1}{9}\cdot 2 = \frac{28}{9}\cdot\frac{2}{1} = \frac{56}{9} = 6\frac{2}{9}$

57. $22\frac{4}{9} + 13\frac{5}{18} = 22\frac{8}{18} + 13\frac{5}{18} = 35\frac{13}{18}$

61.

$\begin{array}{r} 15\frac{1}{5} \\ 20\frac{3}{10} \\ +\,37\frac{2}{15} \\ \hline \end{array}$ $\begin{array}{r} 15\frac{6}{30} \\ 20\frac{9}{30} \\ +\,37\frac{4}{30} \\ \hline 72\frac{19}{30} \end{array}$

65. $4\frac{2}{7}\cdot 1\frac{3}{10} = \frac{30}{7}\cdot\frac{13}{10}$

$= \dfrac{30\cdot 13}{7\cdot 10}$

$= \dfrac{3\cdot 10\cdot 13}{7\cdot 10}$

$= \dfrac{39}{7}$

$= 5\frac{4}{7}$

69. $-5\frac{2}{7}$ decreased by a number translates as $-5\frac{2}{7} - x$.

73. Subtract the standard gauge in the United States from the standard gauge in Spain.

$$
\begin{array}{cc}
65\frac{9}{10} & 65\frac{9}{10} \\[6pt]
-56\frac{1}{2} & -56\frac{5}{10} \\[6pt]
& 9\frac{4}{10} = 9\frac{2}{5}
\end{array}
$$

Spain's standard gauge is $9\frac{2}{5}$ inches wider than the U.S. standard gauge.

77. $2 \cdot 1\frac{3}{4} = \frac{2}{1} \cdot \frac{7}{4} = \frac{2 \cdot 7}{1 \cdot 4} = \frac{2 \cdot 7}{1 \cdot 2 \cdot 2} = \frac{7}{2}$

The area is $\frac{7}{2}$ or $3\frac{1}{2}$ square yards.

81.

$$
\begin{array}{cc}
5\frac{1}{3} & 5\frac{8}{24} \\
5 & 5 \\
7\frac{7}{8} & 7\frac{21}{24} \\
+\ 3 & +\ 3 \\
& 20\frac{29}{24} = 20 + 1 + \frac{5}{24} = 21\frac{5}{24}
\end{array}
$$

The perimeter is $21\frac{5}{24}$ meters.

85. $12\frac{3}{4} \div 4 = \frac{51}{4} \div \frac{4}{1} = \frac{51}{4} \cdot \frac{1}{4} = \frac{51}{16}$

The patient walked $\frac{51}{16} = 3\frac{3}{16}$ miles per day.

89.

$$
\begin{array}{cc}
2\frac{9}{20} & 2\frac{27}{60} \\
2\frac{2}{3} & 2\frac{40}{60} \\
+\ 4\frac{7}{15} & +\ 4\frac{28}{60} \\
& 8\frac{95}{60} = 8\frac{19}{12} = 8 + 1 + \frac{7}{12} = 9\frac{7}{12}
\end{array}
$$

The total duration of the eclipses is $9\frac{7}{12}$ minutes.

93. $-4\frac{2}{5} \cdot 2\frac{3}{10} = -\frac{22}{5} \cdot \frac{23}{10}$

$\qquad\qquad = -\frac{2 \cdot 11 \cdot 23}{5 \cdot 2 \cdot 5}$

$\qquad\qquad = -\frac{253}{25}$

$\qquad\qquad = -10\frac{3}{25}$

97. $-31\frac{2}{15} + 17\frac{3}{20} = -31\frac{8}{60} + 17\frac{9}{60}$

$\qquad\qquad = -30\frac{68}{60} + 17\frac{9}{60}$

$\qquad\qquad = -13\frac{59}{60}$

101. $11\frac{7}{8} - 13\frac{5}{6} = 11\frac{21}{24} - 13\frac{20}{24}$

$\qquad\qquad = -\left(13\frac{20}{24} - 11\frac{21}{24}\right)$

$\qquad\qquad = -\left(12\frac{44}{24} - 11\frac{21}{24}\right)$

$\qquad\qquad = -1\frac{23}{24}$

105. $\frac{1}{3}(3x) = \left(\frac{1}{3} \cdot 3\right)x = 1 \cdot x = x$

109. a. $9\frac{5}{5} = 9 + 1 = 10$

    b. $9\frac{100}{100} = 9 + 1 = 10$

    c. $6\frac{44}{11} = 6 + 4 = 10$

    d. $8\frac{13}{13} = 8 + 1 = 9$

    a, b, and c are equivalent to 10.

113. answers may vary

117. answers may vary

Exercise Set 4.8

1. $\qquad x + \frac{1}{3} = -\frac{1}{3}$

$\qquad x + \frac{1}{3} - \frac{1}{3} = -\frac{1}{3} - \frac{1}{3}$

$\qquad\qquad x = -\frac{2}{3}$

Check: $\qquad x + \frac{1}{3} = -\frac{1}{3}$

$\qquad\qquad -\frac{2}{3} + \frac{1}{3} \overset{?}{=} -\frac{1}{3}$

$\qquad\qquad \frac{-2 + 1}{3} \overset{?}{=} -\frac{1}{3}$

$\qquad\qquad -\frac{1}{3} = -\frac{1}{3}$ True

The solution is $-\frac{2}{3}$.

5. $3x - \frac{1}{5} - 2x = \frac{1}{5} + \frac{2}{5}$

$\qquad x - \frac{1}{5} = \frac{3}{5}$

$\qquad x - \frac{1}{5} + \frac{1}{5} = \frac{3}{5} + \frac{1}{5}$

$\qquad\qquad x = \frac{4}{5}$

Check: $\quad 3x - \frac{1}{5} - 2x = \frac{1}{5} + \frac{2}{5}$

$\qquad 3 \cdot \frac{4}{5} - \frac{1}{5} - 2 \cdot \frac{4}{5} \overset{?}{=} \frac{1}{5} + \frac{2}{5}$

$\qquad \frac{12}{5} - \frac{1}{5} - \frac{8}{5} \overset{?}{=} \frac{1}{5} + \frac{2}{5}$

$\qquad\qquad \frac{3}{5} = \frac{3}{5}$ True

The solution is $\frac{4}{5}$.

9. $\qquad \frac{2}{5} + y = -\frac{3}{10}$

$\qquad \frac{2}{5} + y - \frac{2}{5} = -\frac{3}{10} - \frac{2}{5}$

$\qquad\qquad y = -\frac{3}{10} - \frac{2}{5} \cdot \frac{2}{2}$

$\qquad\qquad y = -\frac{3}{10} - \frac{4}{10}$

$\qquad\qquad y = -\frac{7}{10}$

Check:
$$\frac{2}{5} + y = -\frac{3}{10}$$
$$\frac{2}{5} + \left(-\frac{7}{10}\right) \overset{\underline{?}}{=} -\frac{3}{10}$$
$$\frac{2}{5} \cdot \frac{2}{2} + \left(-\frac{7}{10}\right) \overset{\underline{?}}{=} -\frac{3}{10}$$
$$\frac{4}{10} + \left(-\frac{7}{10}\right) \overset{\underline{?}}{=} -\frac{3}{10}$$
$$-\frac{3}{10} = -\frac{3}{10} \quad \text{True}$$

The solution is $-\frac{7}{10}$.

13.
$$-\frac{2}{9} = x - \frac{5}{6}$$
$$-\frac{2}{9} + \frac{5}{6} = x - \frac{5}{6} + \frac{5}{6}$$
$$-\frac{2}{9} \cdot \frac{2}{2} + \frac{5}{6} \cdot \frac{3}{3} = x$$
$$-\frac{4}{18} + \frac{15}{18} = x$$
$$\frac{11}{18} = x$$

Check: $\quad -\frac{2}{9} = x - \frac{5}{6}$
$$-\frac{2}{9} \overset{\underline{?}}{=} \frac{11}{18} - \frac{5}{6}$$
$$-\frac{2}{9} \cdot \frac{2}{2} \overset{\underline{?}}{=} \frac{11}{18} - \frac{5}{6} \cdot \frac{3}{3}$$
$$-\frac{4}{18} \overset{\underline{?}}{=} \frac{11}{18} - \frac{15}{18}$$
$$-\frac{4}{18} = -\frac{4}{18} \quad \text{True}$$

The solution is $\frac{11}{18}$.

17. $\quad \frac{1}{4}x = 3$
$$4 \cdot \frac{1}{4}x = 4 \cdot 3$$
$$x = 12$$

21. $\quad -\frac{4}{9}z = -\frac{3}{2}$
$$-\frac{9}{4} \cdot \frac{4}{9}z = -\frac{9}{4} \cdot -\frac{3}{2}$$
$$z = \frac{27}{8}$$

25. $\quad -3x = -\frac{6}{11}$
$$-\frac{1}{3} \cdot -3x = -\frac{1}{3} \cdot -\frac{6}{11}$$
$$x = \frac{6}{3 \cdot 11}$$
$$x = \frac{2}{11}$$

29.
$$\frac{x}{3} + 2 = \frac{7}{3}$$
$$3\left(\frac{x}{3} + 2\right) = 3 \cdot \frac{7}{3}$$
$$3 \cdot \frac{x}{3} + 3 \cdot 2 = 7$$
$$x + 6 = 7$$
$$x + 6 - 6 = 7 - 6$$
$$x = 1$$

33.
$$\frac{1}{2} - \frac{3}{5} = \frac{x}{10}$$
$$10\left(\frac{1}{2} - \frac{3}{5}\right) = 10 \cdot \frac{x}{10}$$
$$10 \cdot \frac{1}{2} - 10 \cdot \frac{3}{5} = x$$
$$5 - 6 = x$$
$$-1 = x$$

37. $\quad \frac{x}{7} - \frac{4}{3} = \frac{x}{7} \cdot \frac{3}{3} - \frac{4}{3} \cdot \frac{7}{7}$
$$= \frac{3x}{21} - \frac{28}{21}$$
$$= \frac{3x - 28}{21}$$

41. $\quad \frac{3x}{10} + \frac{x}{6} = \frac{3x}{10} \cdot \frac{3}{3} + \frac{x}{6} \cdot \frac{5}{5}$
$$= \frac{9x}{30} + \frac{5x}{30}$$
$$= \frac{14x}{30}$$
$$= \frac{2 \cdot 7 \cdot x}{2 \cdot 15}$$
$$= \frac{7x}{15}$$

45.
$$\frac{2}{3} - \frac{x}{5} = \frac{4}{15}$$
$$15\left(\frac{2}{3} - \frac{x}{5}\right) = 15 \cdot \frac{4}{15}$$
$$15 \cdot \frac{2}{3} - 15 \cdot \frac{x}{5} = 4$$
$$10 - 3x = 4$$
$$10 - 3x - 10 = 4 - 10$$
$$-3x = -6$$
$$\frac{-3x}{-3} = \frac{-6}{-3}$$
$$x = 2$$

49. $\quad -3m - 5m = \frac{4}{7}$
$$-8m = \frac{4}{7}$$
$$-\frac{1}{8} \cdot -8m = -\frac{1}{8} \cdot \frac{4}{7}$$
$$m = -\frac{1 \cdot 4}{2 \cdot 4 \cdot 7}$$
$$m = -\frac{1}{14}$$

53. $\dfrac{5}{9} - \dfrac{2}{3} = \dfrac{5}{9} - \dfrac{2}{3} \cdot \dfrac{3}{3} = \dfrac{5}{9} - \dfrac{6}{9} = -\dfrac{1}{9}$

57. $\dfrac{5}{7}y = -\dfrac{15}{49}$

$$\dfrac{7}{5} \cdot \dfrac{5}{7}y = \dfrac{7}{5} \cdot -\dfrac{15}{49}$$

$$y = -\dfrac{7 \cdot 15}{5 \cdot 49}$$

$$y = -\dfrac{7 \cdot 3 \cdot 5}{5 \cdot 7 \cdot 7}$$

$$y = -\dfrac{3}{7}$$

61. $-\dfrac{5}{8}y = \dfrac{3}{16} - \dfrac{9}{16}$

$$16\left(-\dfrac{5}{8}y\right) = 16\left(\dfrac{3}{16} - \dfrac{9}{16}\right)$$

$$-10y = 16 \cdot \dfrac{3}{16} - 16 \cdot \dfrac{9}{16}$$

$$-10y = 3 - 9$$

$$-10y = -6$$

$$\dfrac{-10y}{-10} = \dfrac{-6}{-10}$$

$$y = \dfrac{3}{5}$$

65. $\dfrac{7}{6}x = \dfrac{1}{4} - \dfrac{2}{3}$

$$12 \cdot \dfrac{7}{6}x = 12\left(\dfrac{1}{4} - \dfrac{2}{3}\right)$$

$$14x = 12 \cdot \dfrac{1}{4} - 12 \cdot \dfrac{2}{3}$$

$$14x = 3 - 8$$

$$14x = -5$$

$$\dfrac{14x}{14} = \dfrac{-5}{14}$$

$$x = -\dfrac{5}{14}$$

69. $\dfrac{x}{3} + 2 = \dfrac{x}{2} + 8$

$$6\left(\dfrac{x}{3} + 2\right) = 6\left(\dfrac{x}{2} + 8\right)$$

$$6 \cdot \dfrac{x}{3} + 6 \cdot 2 = 6 \cdot \dfrac{x}{2} + 6 \cdot 8$$

$$2x + 12 = 3x + 48$$

$$2x + 12 - 2x = 3x + 48 - 2x$$

$$12 = x + 48$$

$$12 - 48 = x + 48 - 48$$

$$-36 = x$$

73. 327 rounded to the nearest ten is 330.

77. $\dfrac{19}{53} = \dfrac{353x}{1431} + \dfrac{23}{27}$

$$1431 \cdot \dfrac{19}{53} = 1431\left(\dfrac{353x}{1431} + \dfrac{23}{27}\right)$$

$$27 \cdot 19 = 1431 \cdot \dfrac{353x}{1431} + 1431 \cdot \dfrac{23}{27}$$

$$513 = 353x + 1219$$

$$513 - 1219 = 353x + 1219 - 1219$$

$$-706 = 353x$$

$$\dfrac{-706}{353} = \dfrac{353x}{353}$$

$$-2 = x$$

Chapter Test

1. 7 of the 16 equal parts are shaded: $\dfrac{7}{16}$.

5. $-\dfrac{42x}{70} = -\dfrac{3 \cdot 14 \cdot x}{5 \cdot 14} = -\dfrac{3x}{5}$

9. $495 = 3 \cdot 165 = 3 \cdot 3 \cdot 55 = 3 \cdot 3 \cdot 5 \cdot 11 = 3^2 \cdot 5 \cdot 11$

13. The LCD of 7 and x is $7x$.

$$\dfrac{1}{7} - \dfrac{3}{x} = \dfrac{1}{7} \cdot \dfrac{x}{x} - \dfrac{3}{x} \cdot \dfrac{7}{7} = \dfrac{x}{7x} - \dfrac{21}{7x} = \dfrac{x - 21}{7x}$$

17. $-\dfrac{8}{15y} - \dfrac{2}{15y} = \dfrac{-8 - 2}{15y} = \dfrac{-10}{15y} = -\dfrac{2 \cdot 5}{3 \cdot 5 \cdot y} = -\dfrac{2}{3y}$

21. $\begin{array}{r} 19 \\ \underline{-2\tfrac{3}{11}} \\ \end{array}$ $\begin{array}{r} 18\tfrac{11}{11} \\ \underline{-2\tfrac{3}{11}} \\ 16\tfrac{8}{11} \end{array}$

25. $\dfrac{1}{2} \div \dfrac{2}{3} \cdot \dfrac{3}{4} = \dfrac{1}{2} \cdot \dfrac{3}{2} \cdot \dfrac{3}{4} = \dfrac{1 \cdot 3 \cdot 3}{2 \cdot 2 \cdot 4} = \dfrac{9}{16}$

29. $\dfrac{5 + \tfrac{3}{7}}{2 - \tfrac{1}{2}} = \dfrac{14\left(5 + \tfrac{3}{7}\right)}{14\left(2 - \tfrac{1}{2}\right)} = \dfrac{14 \cdot 5 + 14 \cdot \tfrac{3}{7}}{14 \cdot 2 - 14 \cdot \tfrac{1}{2}} = \dfrac{70 + 6}{28 - 7} = \dfrac{76}{21}$

33. $-5x = -5\left(-\dfrac{1}{2}\right) = \dfrac{5}{1} \cdot \dfrac{1}{2} = \dfrac{5 \cdot 1}{1 \cdot 2} = \dfrac{5}{2}$

37. Education: $\dfrac{1}{50}$

Transportation: $\dfrac{1}{5}$

Clothing: $\dfrac{1}{25}$

$$\dfrac{1}{50} + \dfrac{1}{5} + \dfrac{1}{25} = \dfrac{1}{50} + \dfrac{1}{5} \cdot \dfrac{10}{10} + \dfrac{1}{25} \cdot \dfrac{2}{2}$$

$$= \dfrac{1}{50} + \dfrac{10}{50} + \dfrac{2}{50}$$

$$= \dfrac{13}{50}$$

$\dfrac{13}{50}$ of spending goes for education, transportation, and clothing.

CHAPTER 5

Exercise Set 5.1

1. 5.62 in words is five and sixty-two hundredths.

5. -0.205 in words is negative two hundred five thousandths.

9. 3000.04 in words is three thousand and four hundredths.

13. 2.43 in words is two and forty-three hundredths.

17. The check should be paid to "Bell South," for the amount of "59.68," which is written in words as "Fifty-nine and $\dfrac{68}{100}$."

21. Nine and eight hundredths is 9.08.

25. Forty-six ten-thousandths is 0.0046.

29. $0.27 = \dfrac{27}{100}$

33. $5.4 = 5\dfrac{4}{10} = 5\dfrac{2}{5}$

37. $7.008 = 7\dfrac{8}{1000} = 7\dfrac{1}{125}$

41. $0.3005 = \dfrac{3005}{10,000} = \dfrac{601}{2000}$

45. In words, 0.077 is seventy-seven thousandths. As a fraction,

$0.077 = \dfrac{77}{1000}$.

49. 0.57   0.54

$\qquad$ ↑   ↑

$\qquad$ 7 > 4

so 0.57 > 0.54

Thus $-0.57 < -0.54$.

53. 0.54900   0.549

$\qquad$ ↑   ↑

$\qquad$ 9   =   9

so 0.54900 = 0.549

57. 1.062 1.07

$\qquad$ ↑   ↑

$\qquad$ 6 < 7

so 1.062 < 1.07

Thus, $-1.062 > -1.07$.

61. 0.023   0.024

$\qquad$ ↑   ↑

$\qquad$ 3 < 4

so 0.023 < 0.024

Thus, $-0.023 > -0.024$.

65. To round 98,207.23 to the nearest ten, observe that the digit in the ones place is 7. Since this digit is greater than 5, we add 1 to the digit in the tens place, change the digit in the ones place to 0, and drop all digits to the right of the decimal point. The number 98,207.23 rounded to the nearest ten is 98,210.

69. To round 0.5942 to the nearest thousandth, observe that the digit in the ten-thousandths place is 2. Since this digit is less than 5, we do not add 1 to the digit in the thousandths place. The number 0.5942 rounded to the nearest thousandth is 0.594.

73. To round $\pi \approx 3.14159265$ to the nearest thousandth, observe that the digit in the ten-thousandth place is 5. Since this digit is 5, we add 1 to the digit in the thousandths place. The number $\pi \approx 3.14159265$ rounded to the nearest thousandth is 3.142.

77. To round 0.1992 to the nearest hundredth, observe that the digit in the thousandths place is 9. Since this digit is at least 5, we add 1 to the digit in the hundredths place. The number 0.1992 rounded to the nearest hundredth is 0.2. The amount is $0.20.

81. To round 1.7283 to the nearest hundredth, observe that the digit in the thousandths place is 8. Since this digit is at least 5, we add 1 to the digit in the hundredths place. The number 1.7283 rounded to the nearest hundredth is 1.73. The time is 1.73 hours.

85. To round 24.6229 to the nearest thousandth, observe that the digit in the ten-thousandths place is 9. Since this digit is at least 5, we add 1 to the digit in the thousandths place. The number 24.6229 rounded to the nearest thousandth is 24.623. The day length is 24.623 hours.

89. $\quad$ 82

$\quad \underline{-\ 47}$

$\quad$ 35

93. To round 2849.1738 to the nearest hundredth, observe that the digit in the thousandths place is 3. Since this digit is less than 5, we do not add 1 to the digit in the hundredths place. 2849.1738 rounded to the nearest hundredth is 2849.17, which is choice a.

97. $7\dfrac{12}{100} = 7.12$

101. answers may vary

105. 0.0612 rounded to the nearest hundredth is 0.06. 0.066 rounds to 0.07. 0.0586 rounds to 0.06. 0.0506 rounds to 0.05. Thus, 0.0612 and 0.0586 round to 0.06.

109. 602.2 rounded to the nearest ten is 600. 33.8 rounded to the nearest ten is 30.

$\dfrac{600}{30} = 20$

An individual music video has a value of approximately $20.

Exercise Set 5.2

1. $\quad$ 5.6

$\quad \underline{+\ 2.1}$

$\quad$ 7.7

5. $\quad$ 24.6000

$\quad$ 2.3900

$\quad \underline{+\ \ 0.0678}$

$\quad$ 27.0578

9. $18.56 + (-8.23)$

Subtract the absolute values.

$\quad$ 18.56

$\quad \underline{-8.23}$

$\quad$ 10.33

Attach the sign of the larger absolute value.

$18.56 + (-8.23) = 10.33$

13. Exact:
$\qquad$ 1 11

$\quad$ 100.009

$\quad$ 6.080

$\quad \underline{+\ 9.034}$

$\quad$ 115.123

Estimate: $\quad$ 100

$\qquad\qquad$ 6

$\qquad \underline{+\ 9}$

$\qquad\qquad$ 115

17. $\quad$ 12.6 $\qquad\qquad$ *Check:* $\quad$ 4.4

$\quad \underline{-\ 8.2}$ $\qquad\qquad\qquad\qquad \underline{+\ 8.2}$

$\quad$ 4.4 $\qquad\qquad\qquad\qquad\quad$ 12.6

21. $\quad$ 654.90 $\qquad\qquad$ *Check:* $\qquad$ 11 1

$\qquad\qquad\qquad\qquad\qquad\qquad$ 598.23

$\quad \underline{-\ 56.67}$ $\qquad\qquad\qquad\qquad \underline{+\ 56.67}$

$\quad$ 598.23 $\qquad\qquad\qquad\qquad$ 654.90

25. Exact: $\quad$ 1000.0 $\qquad$ *Check:* $\qquad$ 11 11

$\qquad\qquad\qquad\qquad\qquad\qquad\qquad$ 876.6

$\qquad \underline{-\ \ 123.4}$ $\qquad\qquad\qquad\qquad \underline{+\ 123.4}$

$\qquad\quad$ 876.6 $\qquad\qquad\qquad\qquad\quad$ 1000.0

Estimate: $\quad$ 1000

$\qquad\qquad \underline{-\ 100}$

$\qquad\qquad$ 900

29. $-1.12 - 5.2 = -1.12 + (-5.2)$

Add the absolute values.

$$\begin{array}{r} 1.12 \\ + \ 5.20 \\ \hline 6.32 \end{array}$$

Attach the common sign.

-6.32

33. $-2.6 - (-5.7) = -2.6 + 5.7$

Subtract the absolute values.

$$\begin{array}{r} 5.7 \\ - \ 2.6 \\ \hline 3.1 \end{array}$$

Attach the sign of the larger absolute value.

3.1

37.
$$\begin{array}{r} 23.0 \\ - \ 6.7 \\ \hline 16.3 \end{array}$$

$$Check: \quad \begin{array}{r} \overset{11}{16.3} \\ + \ 6.7 \\ \hline 23.0 \end{array}$$

41. $-6.06 + 0.44$

Subtract the absolute values.

$$\begin{array}{r} 6.06 \\ - \ 0.44 \\ \hline 5.62 \end{array}$$

Attach the sign of the larger absolute value.

-5.62

45. $50.2 - 600 = 50.2 + (-600)$

Subtract the absolute values.

$$\begin{array}{r} 600.0 \\ - \ 50.2 \\ \hline 549.8 \end{array}$$

Attach the sign of the larger absolute value.

-549.8

49.
$$\begin{array}{r} 100.009 \\ 6.080 \\ + \ 9.034 \\ \hline 115.123 \end{array}$$

53. $-102.4 - 78.04 = -102.4 + (-78.04)$

Add the absolute values.

$$\begin{array}{r} 102.40 \\ + \ 78.04 \\ \hline 180.44 \end{array}$$

Attach the common sign.

-180.44

57. $x + z = 3.6 + 0.21 = 3.81$

61. $y - x + z = 5 - 3.6 + 0.21$

$\qquad = 5.00 - 3.60 + 0.21$

$\qquad = 1.40 + 0.21$

$\qquad = 1.61$

65. $\qquad 27.4 + y = 16$

$27.4 + (-11.4) \overset{?}{=} 16$

$\qquad\qquad\quad 16 = 16$ True

Yes, -11.4 is a solution.

69. $30.7x + 17.6 - 23.8x - 10.7$

$= 30.7x - 23.8x + 17.6 - 10.7$

$= (30.7 - 23.8)x + (17.6 - 10.7)$

$= 6.9x + 6.9$

73. Subtract the opening price from the closing price.

$$\begin{array}{r} 22.07 \\ - \ 21.90 \\ \hline 0.17 \end{array}$$

The price of each share increased by $0.17.

77. Perimeter $= 3.6 + 2.0 + 3.6 + 2.0 = 11.2$ in.

81. The phrase "How much faster" indicates that we should subtract the average wind speed from the record speed.

$$\begin{array}{r} 321.0 \\ - \ 35.2 \\ \hline 285.8 \end{array}$$

The highest wind speed is 285.8 miles per hour faster than the average wind speed.

85. To find the total, we add.

$$\begin{array}{r} \overset{1 \ \ 2 1}{600.78} \\ 460.99 \\ + \ 436.72 \\ \hline 1498.49 \end{array}$$

The total ticket sales were $1498.49 million.

89. Add the lengths of the sides to get the perimeter.

$$\begin{array}{r} \overset{11}{12.40} \\ 29.34 \\ + \ 25.70 \\ \hline 67.44 \end{array}$$

67.44 feet of border material is needed.

93. The tallest bar indicates the greatest chocolate consumption per person, so Switzerland has the greatest chocolate consumption per person.

97.

Country	Pounds of Chocolate per Person
Switzerland	22.0
Norway	16.0
Germany	15.8
United Kingdom	14.5
Belgium	13.9

101. $\left(\dfrac{1}{5}\right)^3 = \dfrac{1}{5} \cdot \dfrac{1}{5} \cdot \dfrac{1}{5} = \dfrac{1 \cdot 1 \cdot 1}{5 \cdot 5 \cdot 5} = \dfrac{1}{125}$

105. 3 nickels, 3 dimes, and 3 quarters:

$0.05 + 0.05 + 0.05 + 0.10 + 0.10 + 0.10 + 0.25$

$\qquad\qquad + \ 0.25 + 0.25 = 1.2$

The value of the coins shown is $1.20.

109. answers may vary

113. $14.271 - 8.968x + 1.333 - 201.815x + 101.239x$

$= -8.968x - 201.815x + 101.239x + 14.271 + 1.333$

$= (-8.968 - 201.815 + 101.239)x + (14.271 + 1.333)$

$= -109.544x + 15.604$

SOLUTIONS TO SELECTED EXERCISES **A55**

Exercise Set 5.3

1.
$$
\begin{array}{rl}
0.17 & \text{2 decimal places} \\
\underline{\times \quad 8} & \text{0 decimal places} \\
1.36 & 2 + 0 = \text{2 decimal places}
\end{array}
$$

5. The product $(-2.3)(7.65)$ is negative.

$$
\begin{array}{rl}
7.65 & \text{2 decimal places} \\
\underline{\times \quad 2.3} & \text{1 decimal place} \\
2\,295 & \\
\underline{15\,300} & \\
-17.595 & 2 + 1 = \text{3 decimal places and} \\
& \text{include the negative sign}
\end{array}
$$

9. Exact: $\quad 6.8 \qquad$ Estimate: $\quad 7$
$$
\begin{array}{rr}
\underline{\times\ 4.2} & \underline{\times\ 4} \\
1\,36 & 28 \\
\underline{27\,20} & \\
28.56 &
\end{array}
$$

13. Exact: $\ 1.0047 \quad$ Estimate: $\quad 1$
$$
\begin{array}{rr}
\underline{\times \quad 8.2} & \underline{\times\ 8} \\
20094 & 8 \\
\underline{8\,03760} & \\
8.23854 &
\end{array}
$$

17. $6.5 \times 10 = 65$

21. $(-7.093)(1000) = -7093$

25. $(-9.83)(-0.01) = 0.0983$

29.
$$
\begin{array}{r}
0.123 \\
\underline{\times \quad 0.4} \\
0.0492
\end{array}
$$

33.
$$
\begin{array}{r}
8.6 \\
\underline{\times\ 0.15} \\
430 \\
\underline{860} \\
1.290 \ \text{ or } 1.29
\end{array}
$$

37. $562.3 \times 0.001 = 0.5623$

41. 1.5 billion $= 1.5 \times 1$ billion
$$
\begin{aligned}
&= 1.5 \times 1{,}000{,}000{,}000 \\
&= 1{,}500{,}000{,}000
\end{aligned}
$$
The cost at launch was \$1,500,000,000.

45. $xy = 3(-0.2) = -0.6$

49.
$$
\begin{aligned}
0.6x &= 4.92 \\
0.6(14.2) &\overset{?}{=} 4.92 \\
8.52 &= 4.92 \quad \text{False}
\end{aligned}
$$
No, 14.2 is not a solution.

53. $C = \pi d$ is $\pi(10 \text{ cm}) = 10\pi$ cm
$$
C \approx 10(3.14) \text{ cm} = 31.4 \text{ cm}
$$

57. Multiply the number of ounces by the number of grams of saturated fat in 1 ounce.
$$
\begin{array}{r}
6.2 \\
\underline{\times\ 4} \\
24.8
\end{array}
$$
There are 24.8 grams of saturated fat in a 4-ounce serving of cream cheese.

61. Circumference $= \pi \cdot$ diameter
$$
C = \pi \cdot 250 = 250\pi
$$
$$
\begin{array}{r}
250 \\
\underline{\times\ 3.14} \\
10\,00 \\
25\,00 \\
\underline{750\,00} \\
785.00
\end{array}
$$
The circumference is 250π feet, which is approximately 785 feet.

65. Multiply her height in meters by the number of inches in 1 meter.
$$
\begin{array}{r}
39.37 \\
\underline{\times\ 1.65} \\
1\,9685 \\
23\,6220 \\
\underline{39\,3700} \\
64.9605
\end{array}
$$
She is approximately 64.9605 inches tall.

69. a. Circumference $= 2 \cdot \pi \cdot$ radius

Smaller circle:
$$
C = 2 \cdot \pi \cdot 10 = 20\pi
$$
$$
C \approx 20(3.14) = 62.8
$$
The circumference of the smaller circle is approximately 62.8 meters.

Larger circle:
$$
C = 2 \cdot \pi \cdot 20 = 40\pi
$$
$$
C \approx 40(3.14) = 125.6
$$
The circumference of the larger circle is approximately 125.6 meters.

b. Yes, the circumference gets doubled when the radius is doubled.

73.
$$
\begin{array}{r}
115.26 \\
\underline{\times \quad 675} \\
576\,30 \\
8\,068\,20 \\
\underline{69\,156\,00} \\
77{,}800.50
\end{array}
$$
675 U.S. dollars is equivalent to 77,800.5 Japanese yen.

77.
$$
\begin{array}{r}
8 \\
365 \overline{)\ 2920} \\
\underline{-2920} \\
0
\end{array}
$$

81.
$$
\begin{array}{r}
3.60 \\
\underline{+\ 0.04} \\
3.64
\end{array}
$$

85. The product of a negative number and a positive number is a negative number.
$$
\begin{array}{r}
0.221 \\
\underline{\times \quad 0.5} \\
0.1105
\end{array}
$$
The product is -0.1105.

89. answers may vary

Exercise Set 5.4

1. $6\overline{)\,27.6}$ = 4.6
$$\underline{-24}$$
$$3\,6$$
$$\underline{-3\,6}$$
$$0$$

5. $0.06\overline{)18}$ becomes $6\overline{)\,1800}$ = 300
$$\underline{-18}$$
$$000$$

9. Exact: $5.5\overline{)36.3}$ becomes $55\overline{)\,363.0}$ = 6.6
$$\underline{-330}$$
$$33\,0$$
$$\underline{-33\,0}$$
$$0$$

Estimate: $6\overline{)36}$ = 6

13. A positive number divided by a negative number is a negative number.

$0.06\overline{)36}$ becomes $6\overline{)\,3600}$ = 600
$$\underline{-36}$$
$$0$$

$36 \div (-0.06) = -600$

17. $0.27\overline{)1.296}$ becomes $27\overline{)\,129.6}$ = 4.8
$$\underline{-108}$$
$$21\,6$$
$$\underline{-21\,6}$$
$$0$$

21. $4.756 \div 0.82$

$0.82\overline{)4.756}$ Move the decimal points 2 places

$82\overline{)\,475.6}$ = 5.8
$$\underline{-410}$$
$$65\,6$$
$$\underline{-65\,6}$$
$$0$$

25. Exact: $7.2\overline{)70.56}$ becomes $72\overline{)\,705.6}$ = 9.8
$$\underline{-648}$$
$$57\,6$$
$$\underline{-57\,6}$$
$$0$$

Estimate: $7\overline{)70}$ = 10

29. $0.027\overline{)1.215}$ becomes $27\overline{)\,1215}$ = 45
$$\underline{-108}$$
$$135$$
$$\underline{-135}$$
$$0$$

$\dfrac{1.215}{0.027} = 45$

33. $3.78\overline{)0.02079}$ becomes $378\overline{)\,2.0790}$ = 0.0055
$$\underline{-1\,890}$$
$$1890$$
$$\underline{-1890}$$
$$0$$

37. $0.6\overline{)68.39}$ becomes $6\overline{)\,683.90}$ = 113.98 ≈ 114.0
$$\underline{-6}$$
$$08$$
$$\underline{-6}$$
$$23$$
$$\underline{-18}$$
$$5\,9$$
$$\underline{-5\,4}$$
$$50$$
$$\underline{-48}$$
$$2$$

41. $\dfrac{26.87}{10} = 2.687$

45. $7\overline{)\,88.2}$ = 12.6
$$\underline{-7}$$
$$18$$
$$\underline{-14}$$
$$4\,2$$
$$\underline{-4\,2}$$
$$0$$

49. $\dfrac{456.25}{10,000} = 0.045625$

53. $4.2 \div (-0.6)$

$0.6\overline{)4.2}$ becomes $6\overline{)\,42}$ = 7
$$\underline{-42}$$
$$0$$

$4.2 \div (-0.6) = -7$

57. $0.02\overline{)42}$ becomes $2\overline{)\,4200}$ = 2100
$$\underline{-4}$$
$$02$$
$$\underline{-2}$$
$$0$$

61. $87 \div (-0.0015)$

$$
0.0015\overline{)87} \text{ becomes } 15\overline{)\begin{array}{r} 58{,}000 \\ 870{,}000 \end{array}}
$$
$$
\begin{array}{r}
-75 \\ \hline
120 \\
-120 \\ \hline
0
\end{array}
$$

$87 \div (-0.0015) = -58{,}000$

65. $-2.4 \div (-100) = \dfrac{-2.4}{-100} = 0.024$

69. $z \div y = 4.52 \div (-0.8)$

$$
0.8\overline{)4.52} \text{ becomes } 8\overline{)\begin{array}{r} 5.65 \\ 45.20 \end{array}}
$$
$$
\begin{array}{r}
-40 \\ \hline
5\,2 \\
-4\,8 \\ \hline
40 \\
-40 \\ \hline
0
\end{array}
$$

$z \div y = 4.52 \div (-0.8) = -5.65$

73. $\dfrac{x}{4} = 3.04$

$\dfrac{12.16}{4} \stackrel{?}{=} 3.04$

$3.04 = 3.04$ True

Yes, 12.16 is a solution.

77.
$$
52\overline{)\begin{array}{r} 10.5 \approx 11 \\ 546.0 \end{array}}
$$
$$
\begin{array}{r}
-52 \\ \hline
26 \\
-0 \\ \hline
26\,0 \\
-26\,0
\end{array}
$$

Since he must buy whole quarts, 11 quarts are needed.

81. Divide the number of crayons by 64.

$$
64\overline{)\begin{array}{r} 11.40 \text{ rounded to the nearest tenth is 11.4 boxes.} \\ 730.00 \end{array}}
$$
$$
\begin{array}{r}
-64 \\ \hline
90 \\
-64 \\ \hline
26\,0 \\
-25\,6 \\ \hline
40
\end{array}
$$

85. From Exercise 83, we know that there are 24 teaspoons in 4 fluid ounces. Thus, there are 48 half teaspoons (0.5 tsp) or doses in 4 fluid ounces. To see how long the medicine will last, if a dose is taken every 4 hours, there are $24 \div 4 = 6$ doses taken per day. 48 (doses) $\div$ 6 (per day) = 8 days. The medicine will last 8 days.

89.
$$
\begin{array}{r}
52 \\
\times\ 40 \\ \hline
2080
\end{array}
$$

There are 2080 hours in 52 40-hour weeks.

$$
2080\overline{)\begin{array}{r} 5109.615 \approx 5109.62 \\ 10{,}628{,}000.000 \end{array}}
$$
$$
\begin{array}{r}
-10\,400 \\ \hline
228\,0 \\
-208\,0 \\ \hline
20\,00 \\
-\quad 0 \\ \hline
20\,000 \\
-18\,720 \\ \hline
1\,280\,0 \\
-1\,248\,0 \\ \hline
32\,00 \\
-20\,80 \\ \hline
11\,200 \\
-10\,400 \\ \hline
800
\end{array}
$$

His hourly wage would have been $5109.62.

93. $\dfrac{3}{5} - \dfrac{7}{10} = \dfrac{3}{5}\cdot\dfrac{2}{2} - \dfrac{7}{10} = \dfrac{6}{10} - \dfrac{7}{10} = -\dfrac{1}{10}$

97.
$$
\begin{array}{r}
1.278 \\
+\ 0.300 \\ \hline
1.578
\end{array}
$$

101.
$$
\begin{array}{r}
1000.00 \\
-\quad 95.71 \\ \hline
904.29
\end{array}
$$

105. $78.6 \div 97$ is approximately $78.6 \div 100 = 0.786$, which is choice b.

109. Area $=$ (length)(width)

$$
4.5\overline{)38.7} \text{ becomes } 45\overline{)\begin{array}{r} 8.6 \\ 387.0 \end{array}}
$$
$$
\begin{array}{r}
-360 \\ \hline
27\,0 \\
-27\,0 \\ \hline
0
\end{array}
$$

The length is 8.6 feet.

113.
$$
1.15\overline{)75} \text{ becomes } 115\overline{)\begin{array}{r} 65.21 \approx 65.2 \\ 7500.00 \end{array}}
$$
$$
\begin{array}{r}
-690 \\ \hline
600 \\
-575 \\ \hline
250 \\
-230 \\ \hline
200 \\
-115 \\ \hline
85
\end{array}
$$

$$
1.15\overline{)95} \text{ becomes } 115\overline{)\begin{array}{r} 82.60 \approx 82.6 \\ 9500.00 \end{array}}
$$
$$
\begin{array}{r}
-920 \\ \hline
300 \\
-230 \\ \hline
700 \\
-690 \\ \hline
100 \\
-0 \\ \hline
100
\end{array}
$$

The range of wind speeds is 65.2–82.6 knots.

Exercise Set 5.5

1.

$$5\overline{)\begin{array}{l}0.2 \\ 1.0\end{array}}$$
$$\underline{-1.0}$$
$$0$$

$\dfrac{1}{5} = 0.2$

5.

$$4\overline{)\begin{array}{l}0.75 \\ 3.00\end{array}}$$
$$\underline{-28}$$
$$20$$
$$\underline{-20}$$
$$0$$

$\dfrac{3}{4} = 0.75$

9.

$$4\overline{)\begin{array}{l}2.25 \\ 9.00\end{array}}$$
$$\underline{-8}$$
$$1\,0$$
$$\underline{-8}$$
$$20$$
$$\underline{-20}$$
$$0$$

$\dfrac{9}{4} = 2.25$

13.

$$40\overline{)\begin{array}{l}0.425 \\ 17.000\end{array}}$$
$$\underline{-16\,0}$$
$$1\,00$$
$$\underline{-80}$$
$$200$$
$$\underline{-200}$$
$$0$$

$\dfrac{17}{40} = 0.425$

17.

$$3\overline{)\begin{array}{l}0.333... \\ 1.000\end{array}}$$
$$\underline{-9}$$
$$10$$
$$\underline{-9}$$
$$10$$
$$\underline{-9}$$
$$1$$

$-\dfrac{1}{3} = -0.\overline{3}$

21.

$$11\overline{)\begin{array}{l}0.636363... \\ 7.000000\end{array}}$$
$$\underline{-6\,6}$$
$$40$$
$$\underline{-33}$$
$$70$$
$$\underline{-66}$$
$$40$$
$$\underline{-33}$$
$$70$$
$$\underline{-66}$$
$$40$$
$$\underline{-33}$$
$$7$$

$\dfrac{7}{11} = 0.\overline{63}$

25.

$$125\overline{)\begin{array}{l}0.624 \\ 78.000\end{array}}$$
$$\underline{-7\,50}$$
$$3\,00$$
$$\underline{-2\,50}$$
$$500$$
$$\underline{-500}$$
$$0$$

$\dfrac{78}{125} = 0.624$

29. $\dfrac{7}{16} = 0.4375 \approx 0.44$

33.

$$91\overline{)\begin{array}{l}0.615 \approx 0.62 \\ 56.000\end{array}}$$
$$\underline{-54\,6}$$
$$1\,40$$
$$\underline{-91}$$
$$490$$
$$\underline{-455}$$
$$35$$

37.

$$50\overline{)\begin{array}{l}0.02 \\ 1.00\end{array}}$$
$$\underline{-1\,00}$$
$$0$$

41.

$$200\overline{)\begin{array}{l}0.215 \\ 43.000\end{array}}$$
$$\underline{-40\,0}$$
$$3\,00$$
$$\underline{-2\,00}$$
$$1\,000$$
$$\underline{-1\,000}$$
$$0$$

$0.215 = \dfrac{43}{200}$

45.

$$6\overline{)\begin{array}{l}0.833... \\ 5.000\end{array}}$$
$$\underline{-4\,8}$$
$$20$$
$$\underline{-18}$$
$$20$$
$$\underline{-18}$$
$$2$$

$\dfrac{5}{6} = 0.8\overline{3}$ and $0.\overline{6} < 0.8\overline{3}$, or $0.\overline{6} < \dfrac{5}{6}$

49.

$$7\overline{)\begin{array}{l}0.571 \\ 4.000\end{array}}$$
$$\underline{-3\,5}$$
$$50$$
$$\underline{-49}$$
$$10$$
$$\underline{-7}$$
$$3$$

$\dfrac{4}{7} \approx 0.57$ and $0.57 > 0.14$, so $\dfrac{4}{7} > 0.14$.

53.

$$64\overline{)\begin{array}{l}7.125 \\ 456.000\end{array}}$$
$$\underline{-448}$$
$$8\,0$$
$$\underline{-6\,4}$$
$$1\,60$$
$$\underline{-1\,28}$$
$$320$$
$$\underline{-320}$$
$$0$$

$\dfrac{456}{64} = 7.125$ and $7.123 < 7.125$, so

$7.123 < \dfrac{456}{64}$.

57. $0.49 = 0.490$

$0.49, 0.491, 0.498$

61. $\dfrac{5}{8} = 0.625$

$0.612 = 0.612$

$0.649 = 0.649$

$0.612, \dfrac{5}{8}, 0.649$

65. $\dfrac{1 + 0.8}{-0.6} = \dfrac{1.8}{-0.6} = \dfrac{18}{-6} = -3$

69. $(5.6 - 2.3)(2.4 + 0.4) = (3.3)(2.8) = 9.24$

73. $\dfrac{7 + 0.74}{-6} = \dfrac{7.74}{-6} = -1.29$

77. $\dfrac{1}{4}(-9.6 - 5.2) = \dfrac{1}{4}(-14.8) = -3.7$

81. $\text{Area} = l \cdot w = (0.62)\left(\dfrac{2}{5}\right) = (0.62)(0.4) = 0.248$

85. $x - y$

$6 - (0.3) = 5.7$

89. $\dfrac{9}{10} + \dfrac{16}{25} = \dfrac{9}{10} \cdot \dfrac{5}{5} + \dfrac{16}{25} \cdot \dfrac{2}{2} = \dfrac{45}{50} + \dfrac{32}{50} = \dfrac{77}{50}$

93. $1.0 = 1$

97. $\dfrac{99}{100} = 0.99$

$\dfrac{99}{100} < 1$

101.

2019	rounds to	2000
1324	rounds to	1300
773	rounds to	800
684	rounds to	700
703	rounds to	700
502	rounds to	500
		6000

The total number of stations is estimated to be 6000.

Exercise Set 5.6

1. $\quad x + 1.2 = 7.1$

$x + 1.2 - 1.2 = 7.1 - 1.2$

$x = 5.9$

5. $\quad 6.2 = y - 4$

$6.2 + 4 = y - 4 + 4$

$10.2 = y$

9. $\quad -3.5x + 2.8 = -11.2$

$-3.5x + 2.8 - 2.8 = -11.2 - 2.8$

$-3.5x = -14$

$\dfrac{-3.5x}{-3.5} = \dfrac{-14}{-3.5}$

$x = 4$

13. $\quad 2(x - 1.3) = 5.8$

$2x - 2.6 = 5.8$

$2x - 2.6 + 2.6 = 5.8 + 2.6$

$2x = 8.4$

$\dfrac{2x}{2} = \dfrac{8.4}{2}$

$x = 4.2$

17. $\quad 7x - 10.8 = x$

$10(7x - 10.8) = 10x$

$10(7x) - 10(10.8) = 10x$

$70x - 108 = 10x$

$70x - 108 - 70x = 10x - 70x$

$-108 = -60x$

$\dfrac{-108}{-60} = \dfrac{-60x}{-60}$

$1.8 = x$

21. $\quad y - 3.6 = 4$

$y - 3.6 + 3.6 = 4 + 3.6$

$y = 7.6$

25. $\quad 6.5 = 10x + 7.2$

$6.5 - 7.2 = 10x + 7.2 - 7.2$

$-0.7 = 10x$

$\dfrac{-0.7}{10} = \dfrac{10x}{10}$

$-0.07 = x$

29. $\quad 200x - 0.67 = 100x + 0.81$

$200x - 0.67 + 0.67 = 100x + 0.81 + 0.67$

$200x = 100x + 1.48$

$200x - 100x = 100x - 100x + 1.48$

$100x = 1.48$

$x = 0.0148$

33. $\quad 8x - 5 = 10x - 8$

$8x - 5 + 8 = 10x - 8 + 8$

$8x + 3 = 10x$

$8x + 3 - 8x = 10x - 8x$

$3 = 2x$

$\dfrac{3}{2} = \dfrac{2x}{2}$

$1.5 = x$

37. $-0.9x + 2.65 = -0.5x + 5.45$

Multiply each term by 100.

$-90x + 265 = -50x + 545$

$-90x + 265 + 90x = -50x + 545 + 90x$

$265 = 40x + 545$

$265 - 545 = 40x + 545 - 545$

$-280 = 40x$

$\dfrac{-280}{40} = \dfrac{40x}{40}$

$-7 = x$

41. $0.7x + 13.8 = x - 2.16$

Multiply each term by 100.

$70x + 1380 = 100x - 216$

$70x + 1380 + 216 = 100x - 216 + 216$

$70x + 1596 = 100x$

$70x + 1596 - 70x = 100x - 70x$

$1596 = 30x$

$\dfrac{1596}{30} = \dfrac{30x}{30}$

$53.2 = x$

45. $\dfrac{6x}{5} \cdot \dfrac{1}{2x^2} = \dfrac{6x \cdot 1}{5 \cdot 2x^2} = \dfrac{2 \cdot 3 \cdot x}{5 \cdot 2 \cdot x \cdot x} = \dfrac{3}{5x}$

49.
$$b + 4.6 = 8.3$$
$$b + 4.6 - 4.6 = 8.3 - 4.6$$
$$b = 3.7$$

53.
$$5y - 1.2 - 7y + 8 = 5y - 7y - 1.2 + 8$$
$$= -2y + 6.8$$

57.
$$4.7x + 8.3 = -5.8$$
$$4.7x + 8.3 - 8.3 = -5.8 - 8.3$$
$$4.7x = -14.1$$
$$\frac{4.7x}{4.7} = \frac{-14.1}{4.7}$$
$$x = -3$$

61.
$$5(x - 3.14) = 4x$$
$$5 \cdot x - 5 \cdot 3.14 = 4x$$
$$5x - 15.7 = 4x$$
$$5x - 5x - 15.7 = 4x - 5x$$
$$-15.7 = -x$$
$$15.7 = x$$

65.
$$9.6z - 3.2 - 11.7z - 6.9 = 9.6z - 11.7z - 3.2 - 6.9$$
$$= -2.1z - 10.1$$

69.
$$-5.25x = -40.33575$$
$$\frac{-5.25x}{-5.25} = \frac{-40.33575}{-5.25}$$
$$x = 7.683$$

Exercise Set 5.7

1. Mean: $\dfrac{15 + 23 + 24 + 18 + 25}{5} = \dfrac{105}{5} = 21$

Median: Write the numbers in order: 15, 18, 23, 24, 25

The middle number is 23.

Mode: There is no mode, since each number occurs once.

5. Mean:

$$\frac{0.5 + 0.2 + 0.2 + 0.6 + 0.3 + 1.3 + 0.8 + 0.1 + 0.5}{9}$$
$$= \frac{4.5}{9}$$
$$= 0.5$$

Median: Write the numbers in order:

0.1, 0.2, 0.2, 0.3, 0.5, 0.5, 0.6, 0.8, 1.3

The middle number is 0.5.

Mode: Since 0.2 and 0.5 occur twice, there are two modes, 0.2 and 0.5.

9. Mean:

$$\frac{1679 + 1483 + 1483 + 1450 + 1380}{5} = \frac{7475}{5}$$
$$= 1495 \text{ feet}$$

13. answers may vary

17. $\text{gpa} = \dfrac{4 \cdot 3 + 4 \cdot 3 + 4 \cdot 4 + 3 \cdot 3 + 2 \cdot 1}{3 + 3 + 4 + 3 + 1}$
$$= \frac{51}{14}$$
$$\approx 3.64$$

21. Mode: 6.9 since this number appears twice.

25. Mean: $\dfrac{\text{sum of 15 pulse rates}}{15} = \dfrac{1095}{15} = 73$

29. There are 9 rates lower than the mean. They are 66, 68, 71, 64, 71, 70, 65, 70, and 72.

33. $\dfrac{18}{30y} = \dfrac{3 \cdot 6}{5 \cdot 6 \cdot y} = \dfrac{3}{5y}$

37. Since the mode is 35, 35 must occur at least twice in the set.

Since there are an odd number of numbers in the set, the median, 37 is in the set.

Let n be the remaining unknown number.

Mean: $\dfrac{35 + 35 + 37 + 40 + n}{5} = 38$

$$\frac{147 + n}{5} = 38$$
$$5 \cdot \frac{147 + n}{5} = 5 \cdot 38$$
$$147 + n = 190$$
$$147 - 147 + n = 190 - 147$$
$$n = 43$$

The missing numbers are 35, 35, 37, and 43.

Chapter Test

1. 45.092 in words is forty-five and ninety-two thousandths.

5. Subtract the absolute values.

```
 30.25
-9.83
-----
 20.42
```

Attach the sign of the larger absolute value.

$$9.83 - 30.25 = -20.42$$

9. 0.8623 rounded to the nearest thousandth is 0.862.

13. $-24.73 = -24\dfrac{73}{100}$

17. $\dfrac{0.23 + 1.63}{-0.3} = \dfrac{1.86}{-0.3} = -6.2$

21. Mean: $\dfrac{26 + 32 + 42 + 43 + 49}{5} = \dfrac{192}{5} = 38.4$

Median: The numbers are listed in order. The middle number is 42.

Mode: There is no mode since each number occurs once.

25. Area $= \dfrac{1}{2}(4.2 \text{ miles})(1.1 \text{ miles})$
$$= 0.5(4.2)(1.1) \text{ square miles}$$
$$= 2.31 \text{ square miles}$$

CHAPTER 6

Exercise Set 6.2

1. The ratio of 16 to 24 is $\dfrac{16}{24} = \dfrac{8 \cdot 2}{8 \cdot 3} = \dfrac{2}{3}$.

5. The ratio of 4.63 to 8.21 is

$$\dfrac{4.63}{8.21} = \dfrac{4.63 \cdot 100}{8.21 \cdot 100} = \dfrac{463}{821}.$$

9. The ratio of \$32 to \$100 is $\dfrac{32}{100} = \dfrac{4 \cdot 8}{4 \cdot 25} = \dfrac{8}{25}$.

13. The ratio of $3\frac{1}{2}$ to $12\frac{1}{4}$ is

$$\dfrac{3\frac{1}{2}}{12\frac{1}{4}} = 3\frac{1}{2} \div 12\frac{1}{4}$$

$$= \dfrac{7}{2} \div \dfrac{49}{4}$$

$$= \dfrac{7}{2} \cdot \dfrac{4}{49}$$

$$= \dfrac{7 \cdot 2 \cdot 2}{2 \cdot 7 \cdot 7}$$

$$= \dfrac{2}{7}$$

17. Perimeter $= 2 \cdot 125 + 2 \cdot 130 = 250 + 260 = 510$

The ratio of height to perimeter is

$$\dfrac{130 \text{ feet}}{510 \text{ feet}} = \dfrac{10 \cdot 13}{10 \cdot 51} = \dfrac{13}{51}$$

21. The ratio of women to men is

$$\dfrac{125 \text{ women}}{100 \text{ men}} = \dfrac{5 \cdot 25}{4 \cdot 25} = \dfrac{5}{4}.$$

25. The ratio of the base to the height is

$$\dfrac{6 \text{ feet}}{18 \text{ feet}} = \dfrac{1 \cdot 6}{3 \cdot 6} = \dfrac{1}{3}.$$

29. The ratio of average amount eaten by New Zealanders to the average amount eaten by

Americans is $\dfrac{56 \text{ pints}}{48 \text{ pints}} = \dfrac{7 \cdot 8}{6 \cdot 8} = \dfrac{7}{6}$.

33. The rate of 15 returns for 100 sales is

$$\dfrac{15 \text{ returns}}{100 \text{ sales}} = \dfrac{3 \text{ returns}}{20 \text{ sales}}.$$

37. The rate of 18 gallons of pesticide for 4 acres of

crops is $\dfrac{18 \text{ gallons}}{4 \text{ acres}} = \dfrac{9 \text{ gallons}}{2 \text{ acres}}$

41.
$$
\begin{array}{r}
75 \\
5\overline{)375} \\
-35 \\
\hline
25 \\
-25 \\
\hline
0
\end{array}
$$

375 riders in 5 subway cars is $\dfrac{75 \text{ riders}}{1 \text{ car}}$ or 75 riders/car.

45.
$$
\begin{array}{r}
50,000 \\
20\overline{)1,000,000} \\
-1\ 00 \\
\hline
0
\end{array}
$$

\$1,000,000 paid over 20 years is $\dfrac{\$50,000}{1 \text{ year}}$ or \$50,000/year.

49.
$$
\begin{array}{r}
300 \\
40\overline{)12,000} \\
-12\ 0 \\
\hline
0
\end{array}
$$

12,000 good products to 40 defective is $\dfrac{300 \text{ good}}{1 \text{ defective}}$ or 300 good/ defective.

53. a.
$$
\begin{array}{r}
31.25 \\
8\overline{)250.00} \\
-24 \\
\hline
10 \\
-8 \\
\hline
2\ 0 \\
-1\ 6 \\
\hline
40 \\
-40 \\
\hline
0
\end{array}
$$

The unit rate for Charlie is 31.25 boards/hour.

b.
$$
\begin{array}{r}
33.5 \\
12\overline{)402.0} \\
-36 \\
\hline
42 \\
-36 \\
\hline
6\ 0 \\
-6\ 0 \\
\hline
0
\end{array}
$$

The unit rate for Suellen is 33.5 boards/hour.

c. Suellen can assemble boards faster, since $33.5 > 31.25$.

57.
$$
\begin{array}{r}
11.50 \\
5\overline{)57.50} \\
-5 \\
\hline
07 \\
-5 \\
\hline
2\ 5 \\
-2\ 5 \\
\hline
0
\end{array}
$$

The unit price is \$11.50 per compact disc.

61.
$$
\begin{array}{r}
0.1487 \approx 0.149 \\
8\overline{)1.1900} \\
-8 \\
\hline
39 \\
-32 \\
\hline
70 \\
-64 \\
\hline
60 \\
-56 \\
\hline
4
\end{array}
$$

The 8-ounce size costs \$0.149 per ounce.

$$\frac{0.1325 \approx 0.133}{12)\overline{1.5900}}$$
$$\underline{-1\,2}$$
$$39$$
$$\underline{-36}$$
$$30$$
$$\underline{-24}$$
$$60$$
$$\underline{-60}$$
$$0$$

The 12-ounce size costs $0.133 per ounce.

The 12-ounce size is the better buy.

65.
$$\frac{0.1908}{12)\overline{2.2900}} \approx 0.191$$
$$\underline{-1\,2}$$
$$1\,09$$
$$\underline{-1\,08}$$
$$100$$
$$\underline{-96}$$
$$4$$

The 12-ounce size costs $0.191 per ounce.

$$\frac{0.1862 \approx 0.186}{8)\overline{1.4900}}$$
$$\underline{-8}$$
$$69$$
$$\underline{-64}$$
$$50$$
$$\underline{-48}$$
$$20$$
$$\underline{-16}$$
$$4$$

The 8-ounce size costs $0.186 per ounce.

The 8-ounce size is the better buy.

69.
$$\frac{2.3}{9)\overline{20.7}}$$
$$\underline{-18}$$
$$2\,7$$
$$\underline{-2\,7}$$
$$0$$

73. No; answers may vary

77. 10 defective to 200 good is $\dfrac{10\,\text{defective}}{200\,\text{good}} = \dfrac{1\,\text{defective}}{20\,\text{good}}$.

Yes, the machine should be repaired.

81.
$$\frac{543.6 \approx 544}{88300)\overline{48000000.0}}$$
$$\underline{-441500}$$
$$385000$$
$$\underline{-353200}$$
$$318000$$
$$\underline{-264900}$$
$$531000$$
$$\underline{-529800}$$
$$1200$$

There are 544 students per school.

85. No; answers may vary

Exercise Set 6.2

1. $\dfrac{10\,\text{diamonds}}{6\,\text{opals}} = \dfrac{5\,\text{diamonds}}{3\,\text{opals}}$

5. $\dfrac{6\,\text{eagles}}{58\,\text{sparrows}} = \dfrac{3\,\text{eagles}}{29\,\text{sparrows}}$

9. $\dfrac{22\,\text{vanilla wafers}}{1\,\text{cup cookie crumbs}} = \dfrac{55\,\text{vanilla wafers}}{2.5\,\text{cup cookie crumbs}}$

13. $\dfrac{5}{8} \stackrel{?}{=} \dfrac{4}{7}$

$5 \cdot 7 \stackrel{?}{=} 8 \cdot 4$

$35 \neq 32$

The proportion is false.

17. $\dfrac{5}{8} \stackrel{?}{=} \dfrac{625}{1000}$

$5 \cdot 1000 \stackrel{?}{=} 8 \cdot 625$

$5000 = 5000$

The proportion is true.

21. $\dfrac{4.2}{8.4} \stackrel{?}{=} \dfrac{5}{10}$

$4.2 \cdot 10 \stackrel{?}{=} 8.4 \cdot 5$

$42 = 42$

The proportion is true.

25. $\dfrac{2\frac{2}{5}}{\frac{2}{3}} \stackrel{?}{=} \dfrac{\frac{10}{9}}{\frac{1}{4}}$

$2\dfrac{2}{5} \cdot \dfrac{1}{4} \stackrel{?}{=} \dfrac{2}{3} \cdot \dfrac{10}{9}$

$\dfrac{12}{5} \cdot \dfrac{1}{4} \stackrel{?}{=} \dfrac{20}{27}$

$\dfrac{12}{20} \neq \dfrac{20}{27}$

The proportion is false.

29. $\dfrac{10}{15} \stackrel{?}{=} \dfrac{4}{6}$

$10 \cdot 6 \stackrel{?}{=} 15 \cdot 4$

$60 = 60$

The proportion $\dfrac{10}{15} = \dfrac{4}{6}$ is true.

33. $\dfrac{0.15}{3} \stackrel{?}{=} \dfrac{0.35}{7}$

$0.15 \cdot 7 \stackrel{?}{=} 3 \cdot 0.35$

$1.05 = 1.05$

The proportion $\dfrac{0.15}{3} = \dfrac{0.35}{7}$ is true.

37. $\dfrac{x}{5} = \dfrac{6}{10}$

$x \cdot 10 = 5 \cdot 6$

$10x = 30$

$\dfrac{10x}{10} = \dfrac{30}{10}$

$x = 3$

41. $\dfrac{30}{10} = \dfrac{15}{y}$

$30 \cdot y = 10 \cdot 15$

$30y = 150$

$\dfrac{30y}{30} = \dfrac{150}{30}$

$y = 5$

45. $\dfrac{24}{x} = \dfrac{60}{96}$

$24 \cdot 96 = x \cdot 60$

$2304 = 60x$

$\dfrac{2304}{60} = \dfrac{60x}{60}$

$38.4 = x$

49. $\dfrac{n}{0.6} = \dfrac{0.05}{12}$

$n \cdot 12 = 0.6 \cdot 0.05$

$12n = 0.03$

$\dfrac{12n}{12} = \dfrac{0.03}{12}$

$n = 0.0025$

53. $\dfrac{\frac{1}{3}}{\frac{3}{8}} = \dfrac{\frac{2}{5}}{n}$

$\dfrac{1}{3} \cdot n = \dfrac{3}{8} \cdot \dfrac{2}{5}$

$\dfrac{n}{3} = \dfrac{3}{20}$

$3 \cdot \dfrac{n}{3} = 3 \cdot \dfrac{3}{20}$

$n = \dfrac{9}{20}$

57. $\dfrac{n}{1\frac{1}{5}} = \dfrac{4\frac{1}{6}}{6\frac{2}{3}}$

$n \cdot 6\dfrac{2}{3} = 1\dfrac{1}{5} \cdot 4\dfrac{1}{6}$

$n \cdot \dfrac{20}{3} = \dfrac{6}{5} \cdot \dfrac{25}{6}$

$\dfrac{20}{3}n = 5$

$\dfrac{3}{20} \cdot n = \dfrac{3}{20} \cdot 5$

$n = \dfrac{3}{4}$

61. $\dfrac{3.2}{0.3} = \dfrac{x}{1.4}$

$3.2 \cdot 1.4 = 0.3 \cdot x$

$4.48 = 0.3x$

$\dfrac{4.48}{0.3} = \dfrac{0.3x}{0.3}$

$14.9 \approx x$

65. $\dfrac{7}{18} = \dfrac{x}{5}$

$7 \cdot 5 = 18 \cdot x$

$35 = 18x$

$\dfrac{35}{18} = \dfrac{18x}{18}$

$1.9 \approx x$

69. $8.01 \quad 8.1$

$\uparrow \qquad \uparrow$

$0 \;<\; 1$

$8.01 < 8.1$

73. $\dfrac{75}{125} = \dfrac{3 \cdot 25}{5 \cdot 25} = \dfrac{3}{5}$

77. $\dfrac{9}{15} = \dfrac{3}{5}$

$\dfrac{9}{3} = \dfrac{15}{5}$

$\dfrac{5}{15} = \dfrac{3}{9}$

$\dfrac{15}{9} = \dfrac{5}{3}$

81. $\dfrac{a}{b} = \dfrac{c}{d}$

$\dfrac{d}{b} = \dfrac{c}{a}$

$\dfrac{a}{c} = \dfrac{b}{d}$

$\dfrac{b}{a} = \dfrac{d}{c}$

85. $\dfrac{x}{7} = \dfrac{0}{8}$

$x \cdot 8 = 7 \cdot 0$

$8x = 0$

$\dfrac{8x}{8} = \dfrac{0}{8}$

$x = 0$

89. $\dfrac{222}{1515} = \dfrac{37}{y}$

$222 \cdot y = 1515 \cdot 37$

$222y = 56{,}055$

$\dfrac{222y}{222} = \dfrac{56{,}055}{222}$

$y = 252.5$

Exercise Set 6.3

1. Let x be the number of baskets made.

baskets $\rightarrow \dfrac{45}{100} = \dfrac{x}{800} \leftarrow$ baskets
attempts $\rightarrow$ $\qquad\qquad \leftarrow$ attempts

$45 \cdot 800 = 100 \cdot x$

$36{,}000 = 100x$

$\dfrac{36{,}000}{100} = \dfrac{100x}{100}$

$360 = x$

He made 360 baskets.

5. Let x be the number of applications received.

$$\text{accepted} \to \frac{2}{7} = \frac{180}{x} \leftarrow \text{accepted}$$
$$\text{applied} \to \frac{2}{7} = \frac{180}{x} \leftarrow \text{applied}$$
$$2 \cdot x = 7 \cdot 180$$
$$2x = 1260$$
$$\frac{2x}{2} = \frac{1260}{2}$$
$$x = 630$$

The school received 630 applications.

9. Let x be the number of square feet required.

$$\text{floor space} \to \frac{9}{1} = \frac{x}{30} \leftarrow \text{floor space}$$
$$\text{students} \to \frac{9}{1} = \frac{x}{30} \leftarrow \text{students}$$
$$9 \cdot 30 = 1 \cdot x$$
$$270 = x$$

30 students require 270 square feet of floor space.

13. Let x be the distance between Milan and Rome.

$$\text{kilometers} \to \frac{30}{1} = \frac{x}{15} \leftarrow \text{kilometers}$$
$$\text{cm on map} \to \frac{30}{1} = \frac{x}{15} \leftarrow \text{cm on map}$$
$$30 \cdot 15 = 1 \cdot x$$
$$450 = x$$

Milan and Rome are 450 kilometers apart.

17. Let x be the number of hits the player is expected to get.

$$\text{hits} \to \frac{3}{8} = \frac{x}{40} \leftarrow \text{hits}$$
$$\text{at bats} \to \frac{3}{8} = \frac{x}{40} \leftarrow \text{at bats}$$
$$3 \cdot 40 = 8 \cdot x$$
$$120 = 8x$$
$$\frac{120}{8} = \frac{8x}{8}$$
$$15 = x$$

The player would be expected to get 15 hits.

21. Let x be the number of applications she should expect.

$$\text{applications} \to \frac{4}{3} = \frac{x}{14} \leftarrow \text{applications}$$
$$\text{ounces} \to \frac{4}{3} = \frac{x}{14} \leftarrow \text{ounces}$$
$$4 \cdot 14 = 3 \cdot x$$
$$56 = 3x$$
$$\frac{56}{3} = \frac{3x}{3}$$
$$18\frac{2}{3} = x$$

She should expect 18 applications from the 14-ounce bottle.

25. Let x be the number of servings he can make.

$$\text{milk} \to \frac{1\frac{1}{2}}{4} = \frac{4}{x} \leftarrow \text{milk}$$
$$\text{servings} \to \frac{1\frac{1}{2}}{4} = \frac{4}{x} \leftarrow \text{servings}$$
$$1\frac{1}{2} \cdot x = 4 \cdot 4$$
$$\frac{3}{2}x = 16$$
$$\frac{2}{3} \cdot \frac{3}{2}x = \frac{2}{3} \cdot 16$$
$$x = \frac{32}{3} = 10\frac{2}{3}$$

He can make $10\frac{2}{3}$ servings.

29. a. Let x be the number of teaspoons of granules needed.

$$\text{water} \to \frac{25}{1} = \frac{450}{x} \leftarrow \text{water}$$
$$\text{granules} \to \frac{25}{1} = \frac{450}{x} \leftarrow \text{granules}$$
$$25 \cdot x = 1 \cdot 450$$
$$25x = 450$$
$$\frac{25x}{25} = \frac{450}{25}$$
$$x = 18$$

18 teaspoons of granules are needed.

b. Let x be the number of tablespoons of granules needed.

$$\text{tsp} \to \frac{3}{1} = \frac{18}{x} \leftarrow \text{tsp}$$
$$\text{tbsp} \to \frac{3}{1} = \frac{18}{x} \leftarrow \text{tbsp}$$
$$3 \cdot x = 1 \cdot 18$$
$$3x = 18$$
$$\frac{3x}{3} = \frac{18}{3}$$
$$x = 6$$

6 tablespoons of granules are needed.

33. Let x be the estimated head-to-toe height of the Statue of Liberty.

$$\text{height} \to \frac{x}{42} = \frac{5\frac{1}{3}}{2} \leftarrow \text{height}$$
$$\text{arm length} \to \frac{x}{42} = \frac{5\frac{1}{3}}{2} \leftarrow \text{arm length}$$
$$x \cdot 2 = 42 \cdot 5\frac{1}{3}$$
$$2x = 42 \cdot \frac{16}{3}$$
$$2x = 224$$
$$\frac{2x}{2} = \frac{224}{2}$$
$$x = 112$$

The estimated height is 112 feet.

$$112 - 111\frac{1}{12} = \frac{11}{12}$$

The difference is $\frac{11}{12}$ foot or 11 inches.

37. Let x be the estimated height of the Empire State Building.

$$\text{height} \to \frac{x}{102} = \frac{881}{72} \leftarrow \text{height}$$
$$\text{stories} \to \frac{x}{102} = \frac{881}{72} \leftarrow \text{stories}$$
$$x \cdot 72 = 102 \cdot 881$$
$$72x = 89,862$$
$$\frac{72x}{72} = \frac{89,862}{72}$$
$$x \approx 1248$$

The height of the Empire State Building is approximately 1248 feet.

41. Let x be the number expected to have worked in the restaurant industry.

$$\text{restaurant} \to \frac{x}{84} = \frac{1}{3} \leftarrow \text{restaurant}$$
$$\text{workers} \to \frac{x}{84} = \frac{1}{3} \leftarrow \text{workers}$$
$$x \cdot 3 = 84 \cdot 1$$
$$3x = 84$$
$$\frac{3x}{3} = \frac{84}{3}$$
$$x = 28$$

You would expect 28 of the workers to have worked in the restaurant industry.

45. a. Let x be the number of gallons of oil needed.

$$\text{oil} \to \frac{x}{5} = \frac{1}{50} \leftarrow \text{oil}$$
$$\text{gas} \to 5 \qquad 50 \leftarrow \text{gas}$$

$$x \cdot 50 = 5 \cdot 1$$

$$50x = 5$$

$$\frac{50x}{50} = \frac{5}{50}$$

$$x = \frac{1}{10} = 0.1$$

0.1 gallon of oil is needed.

b. $0.1\,\text{gallon} = 0.1(1\,\text{gallon})$

$$= 0.1(128\,\text{fluid ounces})$$

$$= 12.8\,\text{fluid ounces}$$

0.1 gallon is approximately 13 fluid ounces.

49. $200 = 2 \cdot 100$
$$\downarrow \quad \downarrow \quad \searrow$$
$$= 2 \cdot 4 \cdot 25$$
$$\downarrow \quad \swarrow\searrow \quad \swarrow\searrow$$
$$= 2 \cdot 2 \cdot 2 \cdot 5 \cdot 5$$
$$= 2^3 \cdot 5^2$$

53. Let x be the number of ml.

$$\text{mg} \to \frac{15}{1} = \frac{12}{x} \leftarrow \text{mg}$$
$$\text{ml} \to \quad\quad \leftarrow \text{ml}$$

$$15 \cdot x = 1 \cdot 12$$

$$15x = 12$$

$$\frac{15x}{15} = \frac{12}{15}$$

$$x = \frac{4}{5} = 0.8$$

0.8 ml of the medicine should be administered.

57. 11 muffins are approximately 1 dozen (12) muffins.

$$1.5 \cdot 8 = 12$$

Approximately 12 cups of milk will be needed.

61. answers may vary

Exercise Set 6.4

1. $\sqrt{4} = 2$ because $2^2 = 4$.

5. $\sqrt{\dfrac{1}{81}} = \dfrac{1}{9}$ because $\left(\dfrac{1}{9}\right)^2 = \dfrac{1}{9} \cdot \dfrac{1}{9} = \dfrac{1}{81}$.

9. $\sqrt{3} \approx 1.732$

13. $\sqrt{31} \approx 5.568$

17. $\sqrt{256} = 16$ because $16^2 = 256$.

21. $\sqrt{\dfrac{49}{144}} = \dfrac{7}{12}$ because $\left(\dfrac{7}{12}\right)^2 = \dfrac{7}{12} \cdot \dfrac{7}{12} = \dfrac{49}{144}$.

25. Let $a = 5$ and $b = 12$.

$$a^2 + b^2 = c^2$$

$$5^2 + 12^2 = c^2$$

$$25 + 144 = c^2$$

$$169 = c^2$$

$$13 = c$$

The missing length is 13 inches.

29. Let $a = 22$ and $b = 48$.

$$c^2 = a^2 + b^2$$

$$c^2 = 22^2 + 48^2$$

$$c^2 = 484 + 2304$$

$$c^2 = 2788$$

$$c = \sqrt{2788} \approx 52.802$$

The missing length is approximately 52.802 meters.

33.

$$\text{hypotenuse} = \sqrt{(\text{leg})^2 + (\text{other leg})^2}$$

$$= \sqrt{(3)^2 + (4)^2}$$

$$= \sqrt{9 + 16}$$

$$= \sqrt{25}$$

$$= 5$$

The hypotenuse has length 5 units.

37.

$$\text{hypotenuse} = \sqrt{(\text{leg})^2 + (\text{other leg})^2}$$

$$= \sqrt{(10)^2 + (14)^2}$$

$$= \sqrt{100 + 196}$$

$$= \sqrt{296}$$

$$\approx 17.205$$

The hypotenuse is about 17.205 units.

41.

$$\text{hypotenuse} = \sqrt{(\text{leg})^2 + (\text{other leg})^2}$$
$$= \sqrt{(30)^2 + (30)^2}$$
$$= \sqrt{900 + 900}$$
$$= \sqrt{1800}$$
$$\approx 42.426$$

The hypotenuse is about 42.426 units.

45.

$$\text{hypotenuse} = \sqrt{(\text{leg})^2 + (\text{other leg})^2}$$
$$= \sqrt{(7.5)^2 + (4)^2}$$
$$= \sqrt{56.25 + 16}$$
$$= \sqrt{72.25}$$
$$= 8.5$$

The hypotenuse has length 8.5 units.

49. $\text{leg} = \sqrt{(\text{hypotenuse})^2 - (\text{other leg})^2}$
$$= \sqrt{(32)^2 - (20)^2}$$
$$= \sqrt{1024 - 400}$$
$$= \sqrt{624}$$
$$\approx 25.0$$

The tree is about 25 feet tall.

53. $\dfrac{10}{12} = \dfrac{2 \cdot 5}{2 \cdot 6}$
$$= \dfrac{5}{6}$$

57. $\dfrac{3x}{9} - \dfrac{5}{9} = \dfrac{3x - 5}{9}$

61. Since 38 is between $36 = 6 \cdot 6$ and $49 = 7 \cdot 7$,
$\sqrt{38}$ is between 6 and 7; $\sqrt{38} \approx 6.16$.

65. answers may vary

Exercise Set 6.5

1. The triangles are congruent by Side-Side-Side.

5. The triangles are congruent by Angle-Side-Angle.

9. $\dfrac{12}{11} = \dfrac{14}{7} = \dfrac{12}{6} = \dfrac{2}{1}$
The ratio of corresponding sides is $\dfrac{2}{1}$.

13. $\dfrac{x}{3} = \dfrac{9}{6}$
$$6 \cdot x = 3 \cdot 9$$
$$6x = 27$$
$$\dfrac{6x}{6} = \dfrac{27}{6}$$
$$x = 4.5$$

17. $\dfrac{y}{3.75} = \dfrac{12}{9}$
$$9 \cdot y = 12 \cdot 3.75$$
$$9y = 45$$
$$\dfrac{9y}{9} = \dfrac{45}{9}$$
$$y = 5$$

21. $\dfrac{x}{3.25} = \dfrac{17.5}{3.25}$
$$3.25 \cdot x = 17.5 \cdot 3.25$$
$$3.25x = 56.875$$
$$\dfrac{3.25x}{3.25} = \dfrac{56.875}{3.25}$$
$$x = 17.5$$

25. $\dfrac{z}{60} = \dfrac{15}{32}$
$$32 \cdot z = 60 \cdot 15$$
$$32z = 900$$
$$\dfrac{32z}{32} = \dfrac{900}{32}$$
$$z = 28.125$$

29. $\dfrac{x}{25} = \dfrac{40}{2}$
$$2 \cdot x = 40 \cdot 25$$
$$2x = 1000$$
$$\dfrac{2x}{2} = \dfrac{1000}{2}$$
$$x = 500$$

The building is 500 feet tall.

33. $\dfrac{x}{18} = \dfrac{24}{30}$
$$30 \cdot x = 24 \cdot 18$$
$$30x = 432$$
$$\dfrac{30x}{30} = \dfrac{432}{30}$$
$$x = 14.4$$

The shadow of the tree is 14.4 feet long.

37. Let $a = 200$ and $c = 430$.

$$a^2 + b^2 = c^2$$
$$200^2 + b^2 = 430^2$$
$$40{,}000 + b^2 = 184{,}900$$
$$40{,}000 - 40{,}000 + b^2 = 184{,}900 - 40{,}000$$
$$b^2 = 144{,}900$$
$$b = \sqrt{144{,}900} \approx 381$$

The gantry is approximately 381 feet tall.

41. $(0.41)(-3) = -1.23$

45.
$$\frac{n}{58.7} = \frac{11.6}{20.8}$$
$$20.8 \cdot n = 11.6 \cdot 58.7$$
$$20.8n = 680.92$$
$$n = \frac{680.92}{20.8}$$
$$n \approx 32.7$$

Chapter Test

1. The ratio 4500 trees to 6500 trees is
$$\frac{4500 \text{ trees}}{6500 \text{ trees}} = \frac{9 \cdot 500}{13 \cdot 500} = \frac{9}{13}.$$

5. The ratio 414 feet to 231 feet is
$$\frac{414 \text{ feet}}{231 \text{ feet}} = \frac{3 \cdot 138}{3 \cdot 77} = \frac{138}{77}.$$

9.
$$\begin{array}{r} 0.1487 \approx 0.149 \\ 8\overline{)1.1900} \\ \underline{-8} \\ 39 \\ \underline{-32} \\ 70 \\ \underline{-64} \\ 60 \\ \underline{-56} \\ 4 \end{array}$$

The 8-ounce size costs \$0.149/ounce.

$$\begin{array}{r} 0.1575 \approx 0.158 \\ 12\overline{)\ 1.8900} \\ \underline{-1\ 2} \\ 69 \\ \underline{-60} \\ 90 \\ \underline{-84} \\ 60 \\ \underline{-60} \\ 0 \end{array}$$

The 12-ounce size costs \$0.158/ounce.

The 8-ounce size is the better buy.

13.
$$\frac{n}{3} = \frac{15}{9}$$
$$n \cdot 9 = 3 \cdot 15$$
$$9n = 45$$
$$\frac{9n}{9} = \frac{45}{9}$$
$$n = 5$$

17. Let x be the length of the home in feet.

$$\begin{array}{l} \text{feet} \rightarrow \\ \text{inches} \rightarrow \end{array} \frac{x}{11} = \frac{9}{2} \begin{array}{l} \leftarrow \text{feet} \\ \leftarrow \text{inches} \end{array}$$
$$x \cdot 2 = 11 \cdot 9$$
$$2x = 99$$
$$\frac{2x}{2} = \frac{99}{2}$$
$$x = 49\frac{1}{2}$$

The home is $49\frac{1}{2}$ feet long.

21. $\sqrt{157} \approx 12.530$

25. Let x be the height of the tower.

$$\frac{x}{5\frac{3}{4}} = \frac{48}{4}$$
$$x \cdot 4 = 5\frac{3}{4} \cdot 48$$
$$4x = \frac{23}{4} \cdot 48$$
$$4x = 276$$
$$\frac{4x}{4} = \frac{276}{4}$$
$$x = 69$$

The tower is approximately 69 feet tall.

CHAPTER 7

Exercise Set 7.1

1. $\dfrac{96}{100} = 96\%$

5. 37 of the adults preferred football, while 13 preferred soccer. Thus, $37 + 13 = 50$ preferred football or soccer.

$$\frac{50}{100} = 50\%$$

9. $6\% = 6(0.01) = 0.06$

13. $73.6\% = 73.6(0.01) = 0.736$

17. $0.6\% = 0.6(0.01) = 0.006$

21. $32.58\% = 32.58(0.01) = 0.3258$

25. $4\% = 4 \cdot \dfrac{1}{100} = \dfrac{4}{100} = \dfrac{1 \cdot 4}{4 \cdot 25} = \dfrac{1}{25}$

29. $175\% = 175 \cdot \dfrac{1}{100} = \dfrac{175}{100} = \dfrac{7 \cdot 25}{4 \cdot 25} = \dfrac{7}{4}$ or $1\dfrac{3}{4}$

33. $10\dfrac{1}{3}\% = 10\dfrac{1}{3} \cdot \dfrac{1}{100} = \dfrac{31}{3} \cdot \dfrac{1}{100} = \dfrac{31}{300}$

37. $0.006 = 0.006(100\%) = 0.6\%$

41. $5.3 = 5.3(100\%) = 530\%$

45. $0.2228 = 0.2228(100\%) = 22.28\%$

49. $0.7 = 0.7(100\%) = 70\%$

53. $\dfrac{4}{5} = \dfrac{4}{5} \cdot 100\% = \dfrac{400}{5}\% = 80\%$

57. $\dfrac{3}{8} = \dfrac{3}{8} \cdot 100\% = \dfrac{300}{8}\% = \dfrac{75}{2}\% = 37\dfrac{1}{2}\%$

61. $4\dfrac{1}{2} = 4\dfrac{1}{2} \cdot 100\% = \dfrac{9}{2} \cdot 100\% = \dfrac{900}{2}\% = 450\%$

65. $\dfrac{9}{11} = \dfrac{9}{11} \cdot 100\% = \dfrac{900}{11}\%$

$$
\begin{array}{r}
81.818 \approx 81.82 \\
11\overline{)\,900.000} \\
-88 \\
\hline
20 \\
-11 \\
\hline
9\,0 \\
-8\,8 \\
\hline
20 \\
-11 \\
\hline
90 \\
-88 \\
\hline
2
\end{array}
$$

$\dfrac{9}{11}$ is approximately 81.82%.

69.

Percent	Decimal	Fraction
60%	0.6	$\dfrac{3}{5}$
$23\dfrac{1}{2}\%$	0.235	$\dfrac{47}{200}$
80%	0.8	$\dfrac{4}{5}$
$33\dfrac{1}{3}\%$	$0.333\overline{3}$	$\dfrac{1}{3}$
87.5%	0.875	$\dfrac{7}{8}$
7.5%	0.075	$\dfrac{3}{40}$

73. $83\% = 38(0.01) = 0.38$

$38\% = 38 \cdot \dfrac{1}{100} = \dfrac{38}{100} = \dfrac{19}{50}$

77. $32.2\% = 32.2(0.01) = 0.322$

$$
\begin{aligned}
32.2\% &= 32.2 \cdot \frac{1}{100} \\
&= \frac{32.2}{100} \\
&= \frac{32.2}{100} \cdot \frac{10}{10} \\
&= \frac{322}{100} \\
&= \frac{161}{500}
\end{aligned}
$$

81. $12\% = 12(0.01) = 0.12$

$12\% = 12 \cdot \dfrac{1}{100} = \dfrac{12}{100} = \dfrac{3}{25}$

85. $\dfrac{7}{1000} = \dfrac{7}{1000}(100\%) = \dfrac{700}{1000}\% = \dfrac{7}{10}\% = 0.7\%$

89.
$$
\begin{aligned}
\frac{3}{4} - \frac{1}{2} \cdot \frac{8}{9} &= \frac{3}{4} - \frac{1 \cdot 8}{2 \cdot 9} \\
&= \frac{3}{4} - \frac{1 \cdot 2 \cdot 4}{2 \cdot 9} \\
&= \frac{3}{4} - \frac{4}{9} \\
&= \frac{3}{4} \cdot \frac{9}{9} - \frac{4}{9} \cdot \frac{4}{4} \\
&= \frac{27}{36} - \frac{16}{36} \\
&= \frac{27 - 16}{36} \\
&= \frac{11}{36}
\end{aligned}
$$

93. a. 52.8647% rounded to the nearest tenth of a percent is 52.9%.

    b. 52.8647% rounded to the nearest hundredth of a percent is 52.86%.

97. $45\% + 40\% + 11\% = 96\%$

$100\% - 96\% = 4\%$

4% of the U.S. population have AB blood type.

101. A fraction written as a percent is greater than 100% when the numerator is greater than the denominator.

105. The longest bar corresponds to network systems and data communication analysts, so that is predicted to be the fastest growing occupation.

109. answers may vary

Exercise Set 7.2

1. 18% of 81 is <u>what number</u>?

$\downarrow \quad \downarrow \quad \downarrow \quad \downarrow \qquad \downarrow$

18% $\cdot$ 81 = $\quad x$

5. 0.6 is 40% of <u>what number</u>?

$\downarrow \quad \downarrow \quad \downarrow \quad \downarrow \qquad \downarrow$

0.6 = 40% $\cdot$ $\quad x$

9. <u>What number</u> is 9% of 43?

$\qquad \downarrow \qquad \downarrow \; \downarrow \; \downarrow \; \downarrow$

$\quad x \qquad = 9\% \; \cdot \; 43$

13. $10\% \cdot 35 = x$

$0.10 \cdot 35 = x$

$3.5 = x$

10% of 35 is 3.5.

17. $1.2 = 12\% \cdot x$

$1.2 = 0.12 \cdot x$

$\dfrac{1.2}{0.12} = \dfrac{0.12x}{0.12}$

$10 = x$

1.2 is 12% of 10.

21. $x \cdot 80 = 88$

$\dfrac{x \cdot 80}{80} = \dfrac{88}{80}$

$x = 1.1$

$x = 110\%$

88 is 110% of 80.

25. $0.1 = 10\% \cdot x$

$0.1 = 0.10 \cdot x$

$\dfrac{0.1}{0.1} = \dfrac{0.1x}{0.1}$

$1 = x$

0.1 is 10% of 1.

29. $82.5 = 16\frac{1}{2}\% \cdot x$

$82.5 = 0.165 \cdot x$

$\dfrac{82.5}{0.165} = \dfrac{0.165x}{0.165}$

$500 = x$

82.5 is $16\frac{1}{2}\%$ of 500.

33. $x = 42\% \cdot 60$

$x = 0.42 \cdot 60$

$x = 25.2$

25.2 is 42% of 60.

37. $120\% \cdot x = 42$

$1.20 \cdot x = 42$

$\dfrac{1.2x}{1.2} = \dfrac{42}{1.2}$

$x = 35$

120% of 35 is 42.

41. $x \cdot 600 = 3$

$\dfrac{x \cdot 600}{600} = \dfrac{3}{600}$

$x = 0.005$

$x = 0.5\%$

0.5% of 600 is 3.

45. $1575 = x \cdot 2500$

$\dfrac{1575}{2500} = \dfrac{x \cdot 2500}{2500}$

$0.63 = x$

$63\% = x$

1575 is 63% of 2500.

49. $\dfrac{27}{x} = \dfrac{9}{10}$

$27 \cdot 10 = x \cdot 9$

$270 = 9x$

$\dfrac{270}{9} = \dfrac{9x}{9}$

$30 = x$

53. $\dfrac{17}{12} = \dfrac{x}{20}$

57. $5 \cdot n = 32$

$\dfrac{5 \cdot n}{5} = \dfrac{32}{5}$

$n = \dfrac{32}{5}$

Choice c is correct.

61. What number is thirty-three and one-third percent of twenty-four?

65. Since 55% is less than 100%, which is 1, 55% of 45 is less than 45; c.

69. Since 100% is 1, 100% of 45 is equal to 45; a.

73. $1.5\% \cdot 45,775 = x$

$0.015 \cdot 45,775 = x$

$686.625 = x$

1.5% of 45,775 is 686.625.

Exercise Set 7.3

1. 98% of 45 is <u>what number?</u>

  ↓   ↓   ↓

percent base   amount $= a$

$$\frac{a}{45} = \frac{98}{100}$$

5. 14.3 is 26% of <u>what number?</u>

  ↓   ↓   ↓

amount percent base $= b$

$$\frac{14.3}{b} = \frac{26}{100}$$

9. <u>What percent</u> of 400 is 70?

     ↓   ↓   ↓

percent $= p$ base amount

$$\frac{70}{400} = \frac{p}{100}$$

13. $\dfrac{a}{65} = \dfrac{40}{100}$

$a \cdot 100 = 65 \cdot 40$

$100a = 2600$

$\dfrac{100a}{100} = \dfrac{2600}{100}$

$a = 26$

40% of 65 is 26.

17. $\dfrac{90}{b} = \dfrac{15}{100}$

$90 \cdot 100 = b \cdot 15$

$9000 = 15b$

$\dfrac{9000}{15} = \dfrac{15b}{15}$

$600 = b$

15% of 600 is 90.

21. $\dfrac{42}{35} = \dfrac{p}{100}$

$42 \cdot 100 = 35 \cdot p$

$4200 = 35p$

$\dfrac{4200}{35} = \dfrac{35p}{35}$

$120 = p$

42 is 120% of 35.

25. $\dfrac{3.7}{b} = \dfrac{10}{100}$

$3.7 \cdot 100 = b \cdot 10$

$370 = 10b$

$\dfrac{370}{10} = \dfrac{10b}{10}$

$37 = b$

3.7 is 10% of 37.

29. $\dfrac{160}{b} = \dfrac{16}{100}$

$160 \cdot 100 = b \cdot 16$

$16,000 = 16b$

$\dfrac{16,000}{16} = \dfrac{16b}{16}$

$1000 = b$

160 is 16% of 1000.

33. $\dfrac{a}{62} = \dfrac{89}{100}$

$a \cdot 100 = 62 \cdot 89$

$100a = 5518$

$\dfrac{100a}{100} = \dfrac{5518}{100}$

$a = 55.18$

55.18 is 89% of 62.

37. $\dfrac{105}{b} = \dfrac{140}{100}$

$105 \cdot 100 = b \cdot 140$

$10,500 = 140b$

$\dfrac{10,500}{140} = \dfrac{140b}{140}$

$75 = b$

140% of 75 is 105.

41. $\dfrac{4}{800} = \dfrac{p}{100}$

$4 \cdot 100 = 800 \cdot p$

$400 = 800p$

$\dfrac{400}{800} = \dfrac{800p}{800}$

$0.5 = p$

0.5% of 800 is 4.

45. $\dfrac{a}{48} = \dfrac{20}{100}$

$a \cdot 100 = 48 \cdot 20$

$100a = 960$

$\dfrac{100a}{100} = \dfrac{960}{100}$

$a = 9.6$

20% of 48 is 9.6.

49. $-\dfrac{11}{16} - \left(-\dfrac{3}{16}\right) = \dfrac{-11-3}{16} = \dfrac{-14}{16} = -\dfrac{2 \cdot 7}{2 \cdot 8} = -\dfrac{7}{8}$

53. $\overset{1}{0.41}$

$\underline{+0.29}$

0.70

57. answers may vary

61. $\dfrac{36}{12} = \dfrac{p}{100}$

$\dfrac{36}{12} \overset{?}{=} \dfrac{50}{100}$

$3 \overset{?}{=} \dfrac{1}{2}$ False

The percent is not 50.

$\dfrac{36}{12} = \dfrac{p}{100}$

$36 \cdot 100 = 12 \cdot p$

$3600 = 12p$

$\dfrac{3600}{12} = \dfrac{12p}{12}$

$300 = p$

The percent is 300 (300%).

65. $\dfrac{88{,}542}{110{,}736} = \dfrac{p}{100}$

$88{,}542 \cdot 100 = 110{,}736 \cdot p$

$8{,}854{,}200 = 110{,}736p$

$\dfrac{8{,}854{,}200}{110{,}736} = \dfrac{110{,}736p}{110{,}736}$

$79.958 \approx p$

80.0% of 110,736 is 88,542.

Exercise Set 7.4

1. 4% of 220 is what number?

Method 1:

$4\% \cdot 220 = x$

$0.04 \cdot 220 = x$

$8.8 = x$

The minimum weight resistance is 8.8 pounds.

Method 2:

$\dfrac{a}{220} = \dfrac{4}{100}$

$a \cdot 100 = 220 \cdot 4$

$100a = 880$

$\dfrac{100a}{100} = \dfrac{880}{100}$

$a = 8.8$

The minimum weight resistance is 8.8 pounds.

5. 378 is what percent of 2700?

Method 1:

$378 = x \cdot 2700$

$\dfrac{378}{2700} = \dfrac{x \cdot 2700}{2700}$

$0.14 = x$

$14\% = x$

The student spent 14% of last semester's college cost on books.

Method 2:

$\dfrac{378}{2700} = \dfrac{p}{100}$

$378 \cdot 100 = 2700 \cdot p$

$37{,}800 = 2700p$

$\dfrac{37{,}800}{2700} = \dfrac{2700p}{2700}$

$14 = p$

The student spent 14% of last semester's college cost on books.

9. 35 is what percent of 535?

Method 1:

$35 = x \cdot 535$

$\dfrac{35}{535} = \dfrac{x \cdot 535}{535}$

$0.065 \approx x$

$6.5\% \approx x$

6.5% of the members of the 109th U.S. Congress attended a community college.

Method 2:

$\dfrac{35}{535} = \dfrac{p}{100}$

$35 \cdot 100 = 535 \cdot p$

$3500 = 535p$

$\dfrac{3500}{535} = \dfrac{535p}{535}$

$6.5 \approx p$

6.5% of the members of the 109th U.S. Congress attended a community college.

13. What number is 49% of 63,000?

Method 1:

$x = 49\% \cdot 63{,}000$

$x = 0.49 \cdot 63{,}000$

$x = 30{,}870$

The number of people employed as physician assistants is expected to be $63{,}000 + 30{,}870 = 93{,}870$.

Method 2:

$\dfrac{a}{63{,}000} = \dfrac{49}{100}$

$a \cdot 100 = 63{,}000 \cdot 49$

$100a = 3{,}087{,}000$

$\dfrac{100a}{100} = \dfrac{3{,}087{,}000}{100}$

$a = 30{,}870$

The number of people employed as physician assistants is expected to be $63{,}000 + 38{,}870 = 93{,}870$.

17. 10 is what percent of 80?

Method 1:

$$10 = x \cdot 80$$

$$\frac{10}{80} = \frac{x \cdot 80}{80}$$

$$0.125 = x$$

$$12.5\% = x$$

12.5% of the total calories come from fat.

Method 2:

$$\frac{10}{80} = \frac{p}{100}$$

$$10 \cdot 100 = 80 \cdot p$$

$$1000 = 80p$$

$$\frac{1000}{80} = \frac{80p}{80}$$

$$12.5 = p$$

12.5% of the total calories come from fat.

21. 26,250 is 15% of what number?

Method 1:

$$26{,}250 = 15\% \cdot x$$

$$26{,}250 = 0.15 \cdot x$$

$$\frac{26{,}250}{0.15} = \frac{0.15 \cdot x}{0.15}$$

$$175{,}000 = x$$

The price of the home was $175,000.

Method 2:

$$\frac{26{,}250}{b} = \frac{15}{100}$$

$$26{,}250 \cdot 100 = b \cdot 15$$

$$2{,}625{,}000 = 15b$$

$$\frac{2{,}625{,}000}{15} = \frac{15b}{15}$$

$$175{,}000 = b$$

The price of the home was $175,000.

25. What number is 4.5% of 19,286?

Method 1:

$$x = 4.5\% \cdot 19{,}286$$

$$x = 0.045 \cdot 19{,}286$$

$$x = 867.87$$

The price of the car will increase by $867.87. The new price of that model will be $19,286 + $867.87 = $20,153.87.

Method 2:

$$\frac{a}{19{,}286} = \frac{4.5}{100}$$

$$a \cdot 100 = 4.5 \cdot 19{,}286$$

$$100a = 86{,}787$$

$$a = \frac{86{,}787}{100}$$

$$a = 867.87$$

The price of the car will increase by $867.87. The new price of that model will be $19,286 + $867.87 = $20,153.87.

29. What number is 80% of 35?

Method 1:

$$x = 80\% \cdot 35$$

$$x = 0.80 \cdot 35$$

$$x = 28$$

The increase is projected to be 28 million. The population is projected to be (35 + 28) million = 63 million.

Method 2:

$$\frac{a}{35} = \frac{80}{100}$$

$$a \cdot 100 = 80 \cdot 35$$

$$100a = 2800$$

$$a = \frac{2800}{100}$$

$$a = 28$$

The increase is projected to be 28 million. The population is projected to be (35 + 28) million = 63 million.

33.

Original Amount	New Amount	Amount of Increase	Percent Increase
65	117	$117 - 65 = 52$	$\frac{52}{65} = 0.8 = 80\%$

37.

Original Amount	New Amount	Amount of Decrease	Percent Decrease
160	40	$160 - 40 = 120$	$\frac{120}{160} = 0.75 = 75\%$

41. percent decrease $= \dfrac{\text{amount of decrease}}{\text{original amount}}$

$$= \frac{10{,}845 - 10{,}700}{10{,}845}$$

$$= \frac{145}{10{,}845}$$

$$\approx 0.013$$

The decrease in cable TV systems was 1.3%.

45. percent increase $= \dfrac{\text{amount of increase}}{\text{original amount}}$

$$= \frac{464{,}000 - 455{,}000}{455{,}000}$$

$$= \frac{9000}{455{,}000}$$

$$\approx 0.0198$$

The increase in computer programmers is expected to be 2%.

49. percent decrease $= \dfrac{\text{amount of decrease}}{\text{original amount}}$

$$= \frac{7151 - 5713}{7151}$$

$$= \frac{1438}{7151}$$

$$\approx 0.201$$

The decrease in cinema sites was 20.1%.

53. percent increase $= \dfrac{\text{amount of increase}}{\text{original amount}}$

$$= \dfrac{212,000 - 110,000}{110,000}$$

$$= \dfrac{102,000}{110,000}$$

$$\approx 0.927$$

The increase in cellular subscribers was 92.7%.

57. $\begin{array}{r} {}^{1\,1}9.20 \\ +\ 1.98 \\ \hline 11.18 \end{array}$

61. $2\dfrac{4}{5} \div 3\dfrac{9}{10} = \dfrac{14}{5} \div \dfrac{39}{10}$

$$= \dfrac{14}{5} \cdot \dfrac{10}{39}$$

$$= \dfrac{14 \cdot 10}{5 \cdot 39}$$

$$= \dfrac{14 \cdot 2 \cdot 5}{5 \cdot 39}$$

$$= \dfrac{28}{39}$$

65. To find the percent increase, she should have divided by the original amount, which is 150.

$$\text{percent increase} = \dfrac{30}{150} = 0.2 = 20\%$$

Exercise Set 7.5

1. sales tax $= 5\% \cdot \$150 = 0.05 \cdot \$150 = \$7.50$

The sales tax is $7.50

5. $\$335.30 = r \cdot \$4790$

$$\dfrac{\$335.30}{\$4790} = r$$

$$0.07 = r$$

The sales tax rate is 7%.

9. sales tax $= 6.5\% \cdot \$1800 = 0.065 \cdot \$1800 = \$117$

total price $= \$1800 + \$117 = \$1917$

The total price of the bracelet is $1917.

13. $\$98.70 = r \cdot \$1645$

$$\dfrac{\$98.70}{\$1645} = r$$

$$0.06 = r$$

The sales tax rate is 6%.

17. commission $= 4\% \cdot \$1,329,401 = 0.04 \cdot \$1,329,401 = \$53,176.04$

Her commission was $53,176.04.

21. commission $= 1.5\% \cdot \$325,900 = 0.015 \cdot \$325,900 = \$4888.50$

His commission will be $4888.50.

	Original Price	Discount Rate	Amount of Discount	Sale Price
25.	$89	10%	10% · $89 = $8.90	$89 − $8.90 = $80.10
29.	$410	35%	35% · $410 = 143.50	$410 − $143.50 = $266.50

33. discount $= 15\% \cdot \$300 = 0.15 \cdot \$300 = \$45$

sale price $= \$300 - \$45 = \$255$

The discount is $45 and the sale price is $255.

	Purchase Price	Tax Rate	Sales Tax	Total Price
37.	$56	5.5%	5.5% · $56 = $3.08	$56 + $3.08 = $59.08

	Sale	Commission Rate	Commission
41.	$17,900	$\dfrac{\$1432}{\$17,900} = 0.08 = 8\%$	$1432

45. $400 \cdot \dfrac{3}{100} \cdot 11 = 12 \cdot 11 = 132$

49. Round $68 to $70 and 9.5% to 10%.

$10\% \cdot \$70 = 0.10 \cdot \$70 = \$7$

$\$70 + \$7 = \$77$

The best estimate of the total price is $77; d.

	Bill Amount	10%	15%	20%
53.	$72.17 ≈ $72.00	$7.20	$7.20 + $\dfrac{1}{2}$($7.20) = $7.20 + $3.60 = $10.80	2($7.20) = $14.40

57. $7.5\% \cdot \$24,966 = 0.075 \cdot \$24,966 = \$1872.45$

$\$24,966 + \$1872.45 = \$26,838.45$

The total price of the necklace is $26,838.45.

Exercise Set 7.6

1. simple interest $=$ principal $\cdot$ rate $\cdot$ time $= (\$200)(8\%)(2)$

$$= (\$200)(0.08)(2) = \$32$$

5. simple interest $=$ principal $\cdot$ rate $\cdot$ time $= (\$5000)(10\%)\left(1\dfrac{1}{2}\right)$

$$= (\$5000)(0.10)(1.5) = \$750$$

9. simple interest = principal · rate · time = $(\$2500)(16\%)\left(\dfrac{21}{12}\right)$

$\qquad = (\$2500)(0.16)(1.75) = \$700$

13. simple interest = principal · rate · time = $\$5000(9\%)\left(\dfrac{15}{12}\right)$

$\qquad = \$5000(0.09)(1.25) = \$562.50$

Total = $\$5000 + \$562.50 = \$5562.50$

17. $A = P\left(1 + \dfrac{r}{n}\right)^{n \cdot t}$

$\qquad = 6150\left(1 + \dfrac{0.14}{2}\right)^{2 \cdot 15}$

$\qquad = 6150(1.07)^{30}$

$\qquad \approx 46{,}815.37$

The total amount is $46,815.37.

21. $A = P\left(1 + \dfrac{r}{n}\right)^{n \cdot t}$

$\qquad = 10{,}000\left(1 + \dfrac{0.09}{2}\right)^{2 \cdot 20}$

$\qquad = 10{,}000(1.045)^{40}$

$\qquad \approx 58{,}163.65$

The total amount is $58,163.65.

25. $A = P\left(1 + \dfrac{r}{n}\right)^{n \cdot t}$

$\qquad = 2000\left(1 + \dfrac{0.08}{1}\right)^{1 \cdot 5}$

$\qquad = 2000(1.08)^{5}$

$\qquad \approx 2938.66$

The total amount is $2938.66.

29. perimeter = $10 + 6 + 10 + 6 = 32$

The perimeter is 32 yards.

33. $\dfrac{x}{4} + \dfrac{x}{5} = \dfrac{x}{4} \cdot \dfrac{5}{5} + \dfrac{x}{5} \cdot \dfrac{4}{4} = \dfrac{5x}{20} + \dfrac{4x}{20} = \dfrac{9x}{20}$

37. answers may vary

Chapter Test

1. $85\% = 85(0.01) = 0.85$

5. $6.1 = 6.1(100\%) = 610\%$

9. $0.2\% = \dfrac{0.2}{100} = \dfrac{2}{1000} = \dfrac{1}{500}$

13. $\dfrac{1}{5} = \dfrac{1}{5} \cdot \dfrac{100\%}{1} = \dfrac{100}{5}\% = 20\%$

17. *Method 1:*

$\qquad 567 = x \cdot 756$

$\qquad \dfrac{567}{756} = \dfrac{x \cdot 756}{756}$

$\qquad 0.75 = x$

$\qquad 75\% = x$

$\qquad$ 567 is 75% of 756.

Method 2:

$\qquad \dfrac{567}{756} = \dfrac{p}{100}$

$\qquad 567 \cdot 100 = 756 \cdot p$

$\qquad 56{,}700 = 756p$

$\qquad \dfrac{56{,}700}{756} = \dfrac{756p}{756}$

$\qquad 75 = p$

567 is 75% of 756.

21. percent increase $= \dfrac{\text{amount of increase}}{\text{original amount}}$

$\qquad = \dfrac{26{,}460 - 25{,}200}{25{,}200}$

$\qquad = \dfrac{1260}{25{,}200}$

$\qquad = 0.05$

The increase in population was 5%.

25. simple interest = principal · rate · time

$\qquad = (\$2000)(9.25\%)\left(3\dfrac{1}{2}\right)$

$\qquad = (\$2000)(0.0925)(3.5)$

$\qquad = \$647.5$

CHAPTER 8

Exercise Set 8.1

1. Washington has the greatest number of apples, so the greatest quantity of apples was produced in the state of Washington.

5. New York is represented by 2.5 apples. From the pictograph, California, with 1.5 apples, and Pennsylvania, with 1 apple, together produced as many apples as New York.

9. The year 2002 has 7.5 flames and each flame represents 12,000 wildfires, so there were approximately 7.5(12,000) = 90,000 wildfires in 2002.

13. 2004 has 6.5 flames and 2005 has 5.5 flames, which is one fewer. Thus, the decrease in the number of wildfires from 2004 to 2005 is 12,000.

17. The longest bar corresponds to September, so the month in which most hurricanes made landfall is September.

21. Two of the six hurricanes that made landfall in July did so in 2005. The fraction is $\dfrac{2}{6} = \dfrac{1}{3}$.

25. The longest bar corresponding to a city in the United States is the bar for New York. The population is approximately 21.9 million or 21,900,000.

29.

Fiber Content of Selected Foods

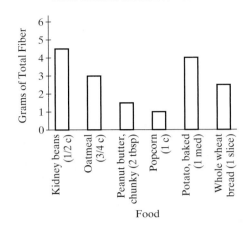

33. The height of the bar for 100–149 miles per week is 15, so 15 of the adults drive 100–149 miles per week.

37. 15 of the adults drive 100–149 miles per week and 9 of the adults drive 150–199 miles per week, so 15 + 9 = 24 of the adults drive 100–199 miles per week.

41. 9 of the 100 adults surveyed drive 150–199 miles per week, so the ratio is $\dfrac{9}{100}$.

45. According to the bar graph, approximately 21 million house-holders will be 55–64 years old.

49. There will be approximately 25 million householders 45–54 years old and 21 million householders 55–64 years old. There will be 25 − 21 = 4 million more householders 45–54 years old.

53.

Class Interval (Scores)	Tally	Class Frequency (Number of Games)
90–99	ⅢⅢ Ⅲ	8

57.

Class Interval (Account Balances)	Tally	Class Frequency (Number of People)
$200–$299	ⅢⅢ Ⅰ	6

61.

Golf Scores

65. The highest point on the graph corresponds to 1982, so the average number of goals per game was the greatest in 1982.

69. The dots for 1990 and 2006 are below the 2.5-level, so the average number of goals per game was less than 2.5 in 1990 and 2006.

73. 10% of 62 is 0.10 · 62 = 6.2

77. $\dfrac{17}{50} = \dfrac{17 \cdot 2}{50 \cdot 2} = \dfrac{34}{100} = 34\%$

81. The lowest point on the graph of low temperatures corresponds to Sunday and the low temperature on Sunday was 68°F.

85. answers may vary

Exercise Set 8.2

1. The largest sector corresponds to the category "parent or guardian's home," thus most of the students live in a parent or guardian's home.

5. 180 of the students live in campus housing while 320 live in a parent or guardian's home.

$\dfrac{180}{320} = \dfrac{9}{16}$

The ratio is $\dfrac{9}{16}$.

9. 30% + 7% = 37%

37% of the land on Earth is accounted for by Europe and Asia.

13. Australia accounts for 5% of the land on Earth.

5% of 57,000,000 = 0.05 · 57,000,000
= 2,850,000

Australia is 2,850,000 square miles.

17. The second-largest sector corresponds to nonfiction, so the second-largest category of books is nonfiction.

21. Children's fiction accounts for 22% of the books.

22% of 125,600 = 0.22 · 125,600
= 27,632

The library has 27,632 children's fiction books.

25.

Energy Source	Percent	Degrees in Sector
Coal	50.6%	50.6% of 360° = 0.506(360°) ≈ 182°
Natural gas	15.8%	15.8% of 360° = 0.158(360°) ≈ 57°
Hydroelectric	8.2%	8.2% of 360° = 0.082(360°) ≈ 30°
Nuclear	20.6%	20.6% of 360° = 0.206(360°) ≈ 74°
Petroleum	1.6%	1.6% of 360° = 0.016(360°) ≈ 6°
Other	3.2%	3.2% of 360° = 0.032(360°) ≈ 12°

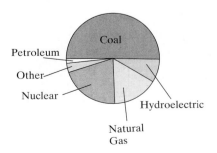

29. $20 = 2 \cdot 10 = 2 \cdot 2 \cdot 5 = 2^2 \cdot 5$

33. $85 = 5 \cdot 17$

37. Pacific Ocean: $49\% \cdot 264{,}489{,}800 = 129{,}600{,}002$ square kilometers

41. $21.5\% \cdot 2800 = 602$ respondents

45. $\dfrac{\text{number of respondents who spend \$0–\$15}}{\text{number of respondents who spend \$15–\$175}} = \dfrac{602}{1674}$

$$= \frac{2 \cdot 301}{2 \cdot 837} = \frac{301}{837}$$

Exercise Set 8.3

1.

5. Point A has coordinates $(0, 0)$.

Point B has coordinates $\left(3\frac{1}{2}, 0\right)$.

Point C has coordinates $(3, 2)$.

Point D has coordinates $(-1, 3)$.

Point E has coordinates $(-2, -2)$.

Point F has coordinates $(0, -1)$.

Point G has coordinates $(2, -1)$.

9. $x - y = 3$

$1 - 2 \overset{?}{=} 3$

$-1 = 3$ False

No, $(1, 2)$ is not a solution of $x - y = 3$.

13. $x = -3y$

$6 \overset{?}{=} -3(-2)$

$6 = 6$ True

Yes, $(6, -2)$ is a solution of $x = -3y$.

17. $x - 5y = -1$

$3 - 5(1) \overset{?}{=} -1$

$3 - 5 \overset{?}{=} -1$

$-2 = -1$ False

No, $(3, 1)$ is not a solution of $x - 5y = -1$.

21.

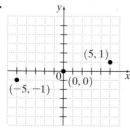

25. $y = -9x$

$y = -9(1)$

$y = -9$

The solution is $(1, -9)$.

$y = -9x$

$y = -9(0)$

$y = 0$

The solution is $(0, 0)$.

$y = -9x$

$-18 = -9x$

$\dfrac{-18}{-9} = \dfrac{-9x}{-9}$

$2 = x$

The solution is $(2, -18)$.

29. $x + y = -2$

$-2 + y = -2$

$-2 + 2 + y = -2 + 2$

$y = 0$

The solution is $(-2, 0)$.

$x + y = -2$

$1 + y = -2$

$1 - 1 + y = -2 - 1$

$y = -3$

The solution is $(1, -3)$.

$x + y = -2$

$x + 5 = -2$

$x + 5 - 5 = -2 - 5$

$x = -7$

The solution is $(-7, 5)$.

33. $y = 3x - 5$

$y = 3 \cdot 1 - 5$

$y = -2$

The solution is $(1, -2)$.

$y = 3x - 5$

$y = 3 \cdot 2 - 5$

$y = 1$

The solution is $(2, 1)$.

$y = 3x - 5$

$4 = 3x - 5$

$4 + 5 = 3x - 5 + 5$

$9 = 3x$

$\dfrac{9}{3} = \dfrac{3x}{3}$

$3 = x$

The solution is $(3, 4)$.

37.
$$4 + 2y = -8$$
$$4 - 4 + 2y = -8 - 4$$
$$2y = -12$$
$$\frac{2y}{2} = \frac{-12}{2}$$
$$y = -6$$

The solution is $(4, -6)$.

$$x + 2y = -8$$
$$x + 2(-3) = -8$$
$$x - 6 = -8$$
$$x - 6 + 6 = -8 + 6$$
$$x = -2$$

The solution is $(-2, -3)$.

$$x + 2y = -8$$
$$0 + 2y = -8$$
$$2y = -8$$
$$\frac{2y}{2} = \frac{-8}{2}$$
$$y = -4$$

The solution is $(0, -4)$.

41. $5.6 \cdot 3.9 = 21.84$

45. To plot (a, b), start at the origin and move a units to the right and b units up. The point will be in quadrant I. Thus, (a, b) is in quadrant I is a true statement.

49. To plot $(0, -b)$, start at the origin and move 0 units to the right and b units down. The point will be on the y-axis. Thus, $(0, -b)$ lies on the x-axis is a false statement.

53. To plot $(4, -3)$, start at the origin and move 4 units to the right, so the point $(4, -3)$ is plotted to the right of the y-axis.

57. $P = 2l + 2w$

Let $l = 6$ and $w = 4$.

$P = 2 \cdot 6 + 2 \cdot 4 = 12 + 8 = 20$

The perimeter is 20 units.

Exercise Set 8.4

1. $x + y = 4$

Find any three ordered-pair solutions.

Let $x = 0$.

$$x + y = 4$$
$$0 + y = 4$$
$$y = 4$$

$(0, 4)$

Let $x = 2$.

$$x + y = 4$$
$$2 + y = 4$$
$$2 - 2 + y = 4 - 2$$
$$y = 2$$

$(2, 2)$

Let $x = 4$.

$$x + y = 4$$
$$4 + y = 4$$
$$4 - 4 + y = 4 - 4$$
$$y = 0$$

$(4, 0)$

Plot $(0, 4), (2, 2),$ and $(4, 0)$. Then draw the line through them.

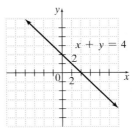

5. $y = 4x$

Find any three ordered-pair solutions.

Let $x = 0$.

$$y = 4x$$
$$y = 4(0)$$
$$y = 0$$

$(0, 0)$

Let $x = 1$.

$$y = 4x$$
$$y = 4(1)$$
$$y = 4$$

$(1, 4)$

Let $x = -1$.

$$y = 4x$$
$$y = 4(-1)$$
$$y = -4$$

$(-1, -4)$

Plot $(0, 0), (1, 4)$ and $(-1, -4)$. Then draw the line through them.

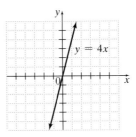

9. $x = -3$

No matter what y-value we choose, x is always -3.

x	y
-3	-4
-3	0
-3	4

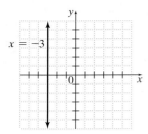

13. $x = 0$

No matter what y-value we choose, x is always 0.

x	y
0	-3
0	0
0	3

17. $y = -2$

No matter what x-value we choose, y is always -2.

x	y
-4	-2
0	-2
4	-2

21. $x = 6$

No matter what y-value we choose, x is always 6.

x	y
6	-5
6	0
6	5

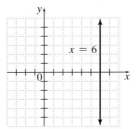

25. $x = y - 4$

Find any three ordered-pair solutions.

Let $y = 0$.
$x = y - 4$
$x = 0 - 4$
$x = -4$
$(-4, 0)$

Let $y = 4$.
$x = y - 4$
$x = 4 - 4$
$x = 0$
$(0, 4)$

Let $y = 6$.
$x = y - 4$
$x = 6 - 4$
$x = 2$
$(2, 6)$

Plot $(-4, 0)$, $(0, 4)$, and $(2, 6)$. Then draw the line through them.

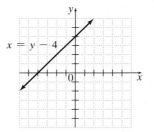

29. $y = -\frac{1}{4}x$

Find any three ordered-pair solutions.

Let $x = -4$.

$y = -\frac{1}{4}x$

$y = -\frac{1}{4}(-4)$

$y = 1$

$(-4, 1)$

Let $x = 0$.

$y = -\frac{1}{4}x$

$y = -\frac{1}{4} \cdot 0$

$y = 0$

$(0, 0)$

Let $x = 4$.

$y = -\frac{1}{4}x$

$y = -\frac{1}{4} \cdot 4$

$y = -1$

$(4, -1)$

Plot $(-4, 1), (0, 0)$, and $(4, -1)$. Then draw the line through them.

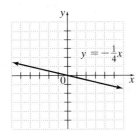

33. $y = 4x + 2$

Find any three ordered-pair solutions.

Let $x = -1$.

$y = 4x + 2$

$y = 4(-1) + 2$

$y = -4 + 2$

$y = -2$

$(-1, -2)$

Let $x = 0$.

$y = 4x + 2$

$y = 4(0) + 2$

$y = 0 + 2$

$y = 2$

$(0, 2)$

Let $x = 1$.

$y = 4x + 2$

$y = 4(1) + 2$

$y = 4 + 2$

$y = 6$

$(1, 6)$

Plot $(-1, -2), (0, 2)$ and $(1, 6)$. Then draw the line through them.

37. $x = -3.5$

No matter what y-value we choose, x is always -3.5.

x	y
-3.5	-3
-3.5	0
-3.5	3

41. $(0.5)\left(-\frac{1}{8}\right) = \left(\frac{1}{2}\right)\left(-\frac{1}{8}\right) = -\frac{1 \cdot 1}{2 \cdot 8} = -\frac{1}{16}$

45. $\frac{2}{11} - \frac{x}{11} = \frac{2 - x}{11}$

49. answers may vary

53. The DVD line is between 20 and 25 above 2005. Approximately $23.3 billion was spent on DVD home entertainment in 2005.

57. The drawn line crosses the $40 billion line between 2009 and 2010, so the amount spent on DVD home entertainment is expected to be $40 billion between 2009 and 2010.

Exercise Set 8.5

1.

12 outcomes

5.

16 outcomes

9.

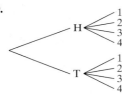

8 outcomes

13. A 1 or a 6 are two of the six possible outcomes. The probability is $\frac{2}{6} = \frac{1}{3}$.

17. Four of the six possible outcomes are numbers greater than 2. The probability is $\frac{4}{6} = \frac{2}{3}$.

21. A 1, a 2, or a 3 are three of three possible outcomes. The probability is $\frac{3}{3} = 1$.

25. One of the seven marbles is red. The probability is $\frac{1}{7}$.

29. Four of the seven marbles are either green or red. The probability is $\frac{4}{7}$.

33. The blood pressure did not change for 10 of the 200 people. The probability is $\frac{10}{200} = \frac{1}{20}$.

37. $\frac{1}{2} \cdot \frac{1}{3} = \frac{1 \cdot 1}{2 \cdot 3} = \frac{1}{6}$

41. One of the 52 cards is the king of hearts. The probability is $\frac{1}{52}$.

45. Thirteen of the 52 cards are hearts. The probability is $\frac{13}{52} = \frac{1}{4}$.

49.

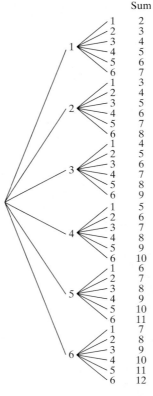

Sum

Five of the 36 sums are 6. The probability is $\frac{5}{36}$.

53. answers may vary

Chapter Test

1. There are $4\frac{1}{2}$ dollar symbols for the second week. Each dollar symbol corresponds to $50. $4\frac{1}{2} \cdot \$50 = \frac{9}{2} \cdot \$50 = \frac{\$450}{2} = \$225$
$225 was collected during the second week.

5. The shortest bar corresponds to February. The normal monthly rainfall in February in Chicago is 3 centimeters.

9. The line graph is above the 3 level for 1990, 1991, and 2000. Thus the inflation rate was greater than 3% in 1990, 1991, and 2000.

13. Services employed 31% of the labor force.

31% of 132,000,000 = 0.31 × 132,000,000
                   = 40,920,000

40,920,000 people were employed in the service industry.

17.

Class Interval (Scores)	Tally	Class Frequency (Number of Students)
40–49	|	1
50–59	|||	3
60–69	||||	4
70–79	|||||	5
80–89	|||| |||	8
90–99	||||	4

21. Point C has coordinates $(-3, 4)$.

CHAPTER 9

Exercise Set 9.1

1. The figure extends indefinitely in two directions.
It is line CD, line l or $\overleftrightarrow{CD}$.

5. The figure has two rays with a common endpoint. It is an angle, which can be named $\angle GHI$, $\angle IHG$, or $\angle H$.

9. Two other ways to name $\angle x$ are $\angle CPR$ and $\angle RPC$.

13. $\angle S$ is a straight angle.

17. $\angle Q$ measures between $90°$ and $180°$. It is an obtuse angle.

21. The complement of an angle that measures $23°$ is an angle that measures $90° - 23° = 67°$.

25. The complement of an angle that measures $58°$ is an angle that measures $90° - 58° = 32°$.

29. $52° + 38° = 90°$, so $\angle PNQ$ and $\angle QNR$ are complementary.
$60° + 30° = 90°$, so $\angle MNP$ and $\angle RNO$ are complementary.

33. $m\angle x = 74° - 47° = 27°$

37. $\angle x$ and the angle marked $150°$ are supplementary, so $m\angle x = 180° - 150° = 30°$. $\angle y$ and the angle marked $150°$ are vertical angles, so $m\angle y = 150°$. $\angle z$ and $\angle x$ are vertical angles so $m\angle z = m\angle x = 30°$.

41. $\angle x$ and the angle marked $80°$ are supplementary, so $m\angle x = 180° - 80° = 100°$. $\angle y$ and the angle marked $80°$ are alternate interior angles, so $m\angle y = 80°$. $\angle x$ and $\angle z$ are corresponding angles, so $m\angle z = m\angle x = 100°$.

45. $\angle x$ can also be named $\angle ABC$ or $\angle CBA$.

49. $m\angle ABC = 15°$

53. $m\angle DBA = m\angle DBC + m\angle CBA$
$\qquad\qquad = 50° + 15°$
$\qquad\qquad = 65°$

57. $\dfrac{7}{8} + \dfrac{1}{4} = \dfrac{7}{8} + \dfrac{2}{8} = \dfrac{9}{8}$ or $1\dfrac{1}{8}$

61. $3\dfrac{1}{3} - 2\dfrac{1}{2} = \dfrac{10}{3} - \dfrac{5}{2}$
$\qquad\qquad = \dfrac{20}{6} - \dfrac{15}{6}$
$\qquad\qquad = \dfrac{5}{6}$

65. The supplement of an angle that measures $125.2°$ is an angle with measure $180° - 125.2° = 54.8°$.

69. True; since $120°$ is less than $180°$, it is possible to find the supplement of a $120°$ angle.

73. No; answers may vary

Exercise Set 9.2

1. $P = 2 \cdot l + 2 \cdot w$
$\quad = 2 \cdot 17\text{ ft} + 2 \cdot 15\text{ ft}$
$\quad = 34\text{ ft} + 30\text{ ft}$
$\quad = 64\text{ ft}$
The perimeter is 64 feet.

5. $P = a + b + c$
$\quad = 5\text{ in.} + 7\text{ in.} + 9\text{ in.}$
$\quad = 21\text{ in.}$
The perimeter is 21 inches.

9. All sides of a regular triangle have the same length.
$P = a + b + c = 14\text{ in.} + 14\text{ in.} + 14\text{ in.} = 42\text{ in.}$
The perimeter is 42 inches.

13. Sum the lengths of the sides.
$P = 5\text{ ft} + 3\text{ ft} + 2\text{ ft} + 7\text{ ft} + 4\text{ ft}$
$\quad = 21\text{ ft}$
The perimeter is 21 feet.

17. $P = 2 \cdot l + 2 \cdot w$
$\quad = 2 \cdot 120\text{ yd} + 2 \cdot 53\text{ yd}$
$\quad = 240\text{ yd} + 106\text{ yd}$
$\quad = 346\text{ yd}$
The perimeter of the football field is 346 yards.

21. The amount of stripping needed is 22 feet.
22 feet $\cdot$ \$2.50 per foot $=$ \$55
The total cost of the stripping is \$55.

25. $P = 4 \cdot s = 4 \cdot 7\text{ in.} = 28\text{ in.}$
The perimeter is 28 inches.

29. The unmarked vertical side must have length
$28\text{ m} - 20\text{ m} = 8\text{ m}$.
The unmarked horizontal side must have length
$20\text{ m} - 17\text{ m} = 3\text{ m}$.
Sum the lengths of the sides.
$P = 17\text{ m} + 8\text{ m} + 3\text{ m} + 20\text{ m} + 20\text{ m} + 28\text{ m}$
$\quad = 96\text{ m}$
The perimeter is 96 meters.

33. The unmarked vertical side must have length
$5 \text{ cm} + 14 \text{ cm} = 19 \text{ cm}$.

The unmarked horizontal side must have length
$18 \text{ cm} - 9 \text{ cm} = 9 \text{ cm}$.

Sum the lengths of the sides.

$P = 18 \text{ cm} + 19 \text{ cm} + 9 \text{ cm} + 14 \text{ cm} + 9 \text{ cm} + 5 \text{ cm}$

$\quad = 74 \text{ cm}$

The perimeter is 74 centimeters.

37. $C = 2 \cdot \pi \cdot r$

$\quad = 2 \cdot \pi \cdot 8 \text{ mi}$

$\quad = 16\pi \text{ mi}$

$\quad \approx 50.24 \text{ mi}$

41. $\pi \cdot d = \pi \cdot 15 \text{ ft} = 15\pi \text{ ft} \approx 47.1$

He needs 15π feet of netting or 47.1 feet.

45. Sum the lengths of the sides.

$P = 9 \text{ mi} + 6 \text{ mi} + 11 \text{ mi} + 4.7 \text{ mi}$

$\quad = 30.7 \text{ mi}$

The perimeter is 30.7 miles.

49. The sides of a regular pentagon all have the same length. Sum the lengths of the sides.

$P = 8 \text{ mm} + 8 \text{ mm} + 8 \text{ mm} + 8 \text{ mm} + 8 \text{ mm}$

$\quad = 40 \text{ mm}$

The perimeter is 40 millimeters.

53. $5 + 6 \cdot 3 = 5 + 18 = 23$

57. $(18 + 8) - (12 + 4) = 26 - 16 = 10$

61. a. The first age category that 8-year-old children fit into is "Under 9," thus the minimum width is 30 yards and the minimum length is 40 yards.

b. $P = 2 \cdot l + 2 \cdot w$

$\quad = 2 \cdot 40 \text{ yd} + 2 \cdot 30 \text{ yd}$

$\quad = 80 \text{ yd} + 60 \text{ yd}$

$\quad = 140 \text{ yd}$

The perimeter of the field is 140 yards.

65. a. Smaller circle:

$C = 2 \cdot \pi \cdot r$

$\quad = 2 \cdot \pi \cdot 10 \text{ m}$

$\quad = 20\pi \text{ m}$

$\quad \approx 62.8 \text{ m}$

Larger circle:

$C = 2 \cdot \pi \cdot r$

$\quad = 2 \cdot \pi \cdot 20 \text{ m}$

$\quad = 40\pi \text{ m}$

$\quad \approx 125.6 \text{ m}$

b. Yes, when the radius of a circle is doubled, the circumference is also doubled.

69. The three linear sides have lengths of 7 feet, 5 feet, and 7 feet. The length of the curved side is half of the circumference of a circle with diameter 5 feet, or

$$\frac{1}{2}\pi \cdot d = \frac{1}{2}\pi \cdot 5 = 2.5\pi \approx 7.9 \text{ feet}$$

$7 \text{ ft} + 5 \text{ ft} + 7 \text{ ft} + 7.9 \text{ ft} = 26.9 \text{ ft}$

The perimeter of the window is 26.9 feet.

Exercise Set 9.3

1. $A = l \cdot w = 3.5 \text{ m} \cdot 2 \text{ m} = 7 \text{ sq m}$

5. $A = \dfrac{1}{2} \cdot b \cdot h = \dfrac{1}{2} \cdot 6 \text{ yd} \cdot 5 \text{ yd} = 15 \text{ sq yd}$

9. $A = b \cdot h = 7 \text{ ft} \cdot 5.25 \text{ ft} = 36.75 \text{ sq ft}$

13. $A = \dfrac{1}{2}(b + B) \cdot h$

$\quad = \dfrac{1}{2}(7 \text{ yd} + 4 \text{ yd}) \cdot 4 \text{ yd}$

$\quad = \dfrac{1}{2}(11 \text{ yd}) \cdot 4 \text{ yd}$

$\quad = 22 \text{ sq yd}$

17. $A = b \cdot h$

$\quad = 5 \text{ in.} \cdot 4\dfrac{1}{2} \text{ in.}$

$\quad = 5 \text{ in.} \cdot \dfrac{9}{2} \text{ in.}$

$\quad = \dfrac{45}{2} \text{ sq in.}$

$\quad = 22\dfrac{1}{2} \text{ sq in.}$

21.

Rectangle 1:

$A = l \cdot w = 10 \text{ mi} \cdot 5 \text{ mi} = 50 \text{ sq mi}$

Rectangle 2:

$A = l \cdot w = 12 \text{ mi} \cdot 3 \text{ mi} = 36 \text{ sq mi}$

The area of the figure is

$50 \text{ sq mi} + 36 \text{ sq mi} = 86 \text{ sq mi}.$

25. $A = \pi r^2 = \pi(6 \text{ in.})^2 = 36\pi$ sq in. $\approx 113\frac{1}{7}$ sq in.

29. $V = s^3 = (8 \text{ cm})^3 = 512$ cu cm

$SA = 6s^2 = 6(8 \text{ cm})^2 = 6 \cdot 64$ sq cm $= 384$ sq cm

33. $r = \frac{1}{2} \cdot d = \frac{1}{2} \cdot 10 \text{ in.} = 5 \text{ in.}$

$V = \frac{4}{3}\pi r^3$

$ = \frac{4}{3}\pi(5 \text{ in.})^3$

$ = \frac{4}{3}\pi \cdot 125$ cu in.

$ = \frac{500}{3}\pi$ cu in.

$ \approx \frac{500}{3} \cdot \frac{22}{7}$ cu in. $= 523\frac{17}{21}$ cu in.

$SA = 4\pi r^2$

$ = 4\pi(5 \text{ in.})^2$

$ = 4\pi \cdot 25$ sq in.

$ = 100\pi$ sq in.

$ \approx 100 \cdot \frac{22}{7}$ sq in. $= 314\frac{2}{7}$ sq in.

37. $V = \frac{1}{3}s^2 h$

$ = \frac{1}{3}(5 \text{ cm})^2(9 \text{ cm})$

$ = \frac{1}{3} \cdot 25$ sq cm $\cdot 9$ cm

$ = 75$ cu cm

41. $V = lwh = (2 \text{ ft})(1.4 \text{ ft})(3 \text{ ft}) = 8.4$ cu ft

$SA = 2lh + 2wh + 2lw$

$ = 2(2 \text{ ft})(3 \text{ ft}) + 2(1.4 \text{ ft})(3 \text{ ft}) + 2(2 \text{ ft})(1.4 \text{ ft})$

$ = 12$ sq ft $+ 8.4$ sq ft $+ 5.6$ sq ft

$ = 26$ sq ft

45. $A = l \cdot w = 7 \text{ ft} \cdot 6 \text{ ft} = 42$ sq ft

$4 \cdot 42$ sq ft $= 168$ sq ft

Four panels have an area of 168 square feet.

49. $A = \frac{1}{2}(b + B) \cdot h$

$ = \frac{1}{2}(140 \text{ ft} + 90 \text{ ft}) \cdot 80 \text{ ft}$

$ = \frac{1}{2} \cdot 230 \text{ ft} \cdot 80 \text{ ft}$

$ = 9200$ sq ft

53. $A = \frac{1}{2}(b + B)h$

$ = \frac{1}{2}(25 \text{ ft} + 36 \text{ ft}) \cdot 12\frac{1}{2} \text{ ft}$

$ = \frac{1}{2} \cdot 61 \text{ ft} \cdot 12\frac{1}{2} \text{ ft}$

$ = 381\frac{1}{4}$ sq ft

To the nearest square foot, the area is 381 square feet.

57. $r = \frac{1}{2} \cdot d = \frac{1}{2} \cdot 4 \text{ ft} = 2 \text{ ft}$

$A = \pi r^2 = \pi(2 \text{ ft})^2 = \pi \cdot 4$ sq ft

The area of the pizza is 4π square feet, or approximately $4 \cdot 3.14 = 12.56$ square feet.

61. $A = l \cdot w = 16 \text{ ft} \cdot 10\frac{1}{2} \text{ ft}$

$ = 16 \text{ ft} \cdot \frac{21}{2} \text{ ft}$

$ = 168$ sq ft

The area of the wall is 168 square feet.

65. $5^2 = 5 \cdot 5 = 25$

69. $1^2 + 2^2 = 1 \cdot 1 + 2 \cdot 2 = 1 + 4 = 5$

73. A fence goes around the edge of a yard, thus the situation involves perimeter.

77. Paint covers the surface of the wall, thus the situation involves area.

81. Note that the dimensions given are the diameters of the pizzas.

12-inch pizza:

$r = \frac{1}{2} \cdot d = \frac{1}{2} \cdot 12 \text{ in.} = 6 \text{ in.}$

$A = \pi \cdot r^2 = \pi(6 \text{ in.})^2 = 36\pi$ sq in.

Price per square inch $= \dfrac{\$10}{36\pi \text{ sq in.}} \approx \$0.0884$

8-inch pizzas:

$r = \frac{1}{2} \cdot d = \frac{1}{2} \cdot 8 \text{ in.} = 4 \text{ in.}$

$A = \pi \cdot r^2 = \pi(4 \text{ in.})^2 = 16\pi$ sq in.

$2 \cdot A = 2 \cdot 16\pi$ sq in. $= 32\pi$ sq in.

Price per square inch: $\dfrac{\$9}{32\pi \text{ sq in.}} \approx \$0.0895$

Since the price per square inch for the 12-inch pizza is less, the 12-inch pizza is the better deal.

85. no; answers may vary

89. The skating area is a rectangle with a half circle on each end.

Rectangle:

$A = l \cdot w = 22 \text{ m} \cdot 10 \text{ m} = 220 \text{ sq m}$

Half circles:

$A = 2 \cdot \dfrac{1}{2} \cdot \pi \cdot r^2$

$= \pi (5 \text{ m})^2$

$= 25\pi \text{ sq m} \approx 78.5 \text{ sq m}$

$220 \text{ sq m} + 78.5 \text{ sq m} = 298.5 \text{ sq m}$

The skating surface has area of 298.5 square meters.

Exercise Set 9.4

1. $60 \text{ in.} = \dfrac{60 \text{ in.}}{1} \cdot \dfrac{1 \text{ ft}}{12 \text{ in.}} = \dfrac{60}{12} \text{ ft} = 5 \text{ ft}$

5. $42{,}240 \text{ ft} = \dfrac{42{,}240 \text{ ft}}{1} \cdot \dfrac{1 \text{ mi}}{5280 \text{ ft}}$

$= \dfrac{42{,}240}{5280} \text{ mi}$

$= 8 \text{ mi}$

9. $10 \text{ ft} = \dfrac{10 \text{ ft}}{1} \cdot \dfrac{1 \text{ yd}}{3 \text{ ft}} = \dfrac{10}{3} \text{ yd} = 3\dfrac{1}{3} \text{ yd}$

13. $162 \text{ in.} = \dfrac{162 \text{ in.}}{1} \cdot \dfrac{1 \text{ ft}}{12 \text{ in.}} \cdot \dfrac{1 \text{ yd}}{3 \text{ ft}}$

$= \dfrac{162}{36} \text{ yd}$

$= 4.5 \text{ yd}$

17. $40 \text{ ft} = \dfrac{40 \text{ ft}}{1} \cdot \dfrac{1 \text{ yd}}{3 \text{ ft}} = \dfrac{40}{3} \text{ yd}$

```
     13 yd 1 ft
3) 40
   −3
   ───
    10
    −9
    ──
     1
```

21. $10{,}000 \text{ ft} = \dfrac{10{,}000 \text{ ft}}{1} \cdot \dfrac{1 \text{ mi}}{5280 \text{ ft}} = \dfrac{10{,}000}{5280} \text{ mi}$

```
         1 mi 4720 ft
5280) 10,000
      −5280
      ─────
       4720
```

25. $8 \text{ yd } 2 \text{ ft} = \dfrac{8 \text{ yd}}{1} \cdot \dfrac{3 \text{ ft}}{1 \text{ yd}} + 2 \text{ ft}$

$= 24 \text{ ft} + 2 \text{ ft}$

$= 26 \text{ ft}$

29. $3 \text{ ft } 10 \text{ in.} + 7 \text{ ft } 4 \text{ in.} = 10 \text{ ft } 14 \text{ in.}$

$= 10 \text{ ft } + 1 \text{ ft } 2 \text{ in.}$

$= 11 \text{ ft } 2 \text{ in.}$

33.
```
    22 ft 8 in.
  −16 ft 3 in.
  ───────────
    6 ft 5 in.
```

37. $28 \text{ ft } 8 \text{ in.} \div 2 = 14 \text{ ft } 4 \text{ in.}$

41. $60 \text{ m} = \dfrac{60 \text{ m}}{1} \cdot \dfrac{100 \text{ cm}}{1 \text{ m}} = 6000 \text{ cm}$

45. $500 \text{ m} = \dfrac{500 \text{ m}}{1} \cdot \dfrac{1 \text{ km}}{1000 \text{ cm}} = \dfrac{500}{1000} \text{ km} = 0.5 \text{ km}$

49. $1500 \text{ cm} = \dfrac{1500 \text{ cm}}{1} \cdot \dfrac{1 \text{ m}}{100 \text{ cm}} = \dfrac{1500}{100} \text{ m} = 15 \text{ m}$

53. $7 \text{ km} = \dfrac{7 \text{ km}}{1} \cdot \dfrac{1000 \text{ m}}{1 \text{ km}} = 7000 \text{ m}$

57. $20.1 \text{ mm} = \dfrac{20.1 \text{ mm}}{1} \cdot \dfrac{1 \text{ dm}}{100 \text{ mm}}$

$= \dfrac{20.1}{100} \text{ dm}$

$= 0.201 \text{ dm}$

61.
```
   8.6 m
 +0.34 m
 ───────
  8.94 m
```

65.
```
  24.8 mm        24.8 mm       2.48 cm
 −1.19 cm      −11.9 mm  or  −1.19 cm
 ────────      ────────      ────────
               12.9 mm        1.29 cm
```

69. $18.3 \text{ m} \times 3 = 54.9 \text{ m}$

73.

	Yards	Feet	Inches
Chrysler Building in New York City	$348\dfrac{2}{3}$	1046	12,552

	Meters	Millimeters	Kilometers	Centimeters
77. Length of elephant	5	5000	0.005	500
81. Distance from London to Paris	342,000	342,000,000	342	34,200,000

85.
```
  6000 ft
 − 900 ft
 ────────
  5100 ft
```

The Grand Canyon of the Colorado River is 5100 feet deeper than the Grand Canyon of the Yellowstone River.

89.
```
   80 mm        80.0 mm
 −5.33 cm      −53.3 mm
 ────────      ────────
               26.7 mm
```

The ice must be 26.7 mm thicker before skating is allowed.

93.
```
        3.35
   20) 67.00
      −60
      ───
       70
      −60
      ───
       100
      −100
      ────
         0
```

Each piece will be 3.35 meters long.

97. $0.21 = \dfrac{21}{100}$

101. $\dfrac{1}{4} = \dfrac{1}{4} \cdot \dfrac{25}{25} = \dfrac{25}{100} = 0.25$

105. Yes, glass for a drinking glass being 2 millimeters thick is reasonable.

109. 5 yd 2 in. is close to 5 yd. 7 yd 30 in. is close to 7 yd 36 in. = 8 yd. Estimate: 5 yd + 8 yd = 13 yd.

113. answers may vary

Exercise Set 9.5

1. $2 \text{ lb} = \dfrac{2 \text{ lb}}{1} \cdot \dfrac{16 \text{ oz}}{1 \text{ lb}} = 2 \cdot 16 \text{ oz} = 32 \text{ oz}$

5. $18{,}000 \text{ lb} = \dfrac{18{,}000 \text{ lb}}{1} \cdot \dfrac{1 \text{ ton}}{2000 \text{ lb}}$

$\qquad = \dfrac{18{,}000}{2000} \text{ tons}$

$\qquad = 9 \text{ tons}$

9. $3500 \text{ lb} = \dfrac{3500 \text{ lb}}{1} \cdot \dfrac{1 \text{ ton}}{2000 \text{ lb}}$

$\qquad = \dfrac{3500}{2000} \text{ tons}$

$\qquad = \dfrac{7}{4} \text{ tons}$

$\qquad = 1\dfrac{3}{4} \text{ tons}$

13. $4.9 \text{ tons} = \dfrac{4.9 \text{ tons}}{1} \cdot \dfrac{2000 \text{ lb}}{1 \text{ ton}}$

$\qquad = 4.9 \cdot 2000 \text{ lb}$

$\qquad = 9800 \text{ lb}$

17. $2950 \text{ lb} = \dfrac{2950 \text{ lb}}{1} \cdot \dfrac{1 \text{ ton}}{2000 \text{ lb}}$

$\qquad = \dfrac{2950}{2000} \text{ tons}$

$\qquad = \dfrac{59}{40} \text{ tons}$

$\qquad \approx 1.5 \text{ tons}$

21. $5\dfrac{3}{4} \text{ lb} = \dfrac{23}{4} \text{ lb} = \dfrac{\frac{23}{4} \text{ lb}}{1} \cdot \dfrac{16 \text{ oz}}{1 \text{ lb}}$

$\qquad = \dfrac{23}{4} \cdot 16 \text{ oz} = 23 \cdot 4 \text{ oz} = 92 \text{ oz}$

25. $89 \text{ oz} = \dfrac{89 \text{ oz}}{1} \cdot \dfrac{1 \text{ lb}}{16 \text{ oz}} = \dfrac{89}{16} \text{ lb}$

$$\begin{array}{r} 5 \text{ lb } 9 \text{ oz} \\ 16)\overline{89} \\ \underline{-80} \\ 9 \end{array}$$

$89 \text{ oz} = 5 \text{ lb } 9 \text{ oz}$

29. $3 \text{ tons } 1820 \text{ lb} + 4 \text{ tons } 930 \text{ lb}$

$\qquad = 7 \text{ tons } 2750 \text{ lb}$

$\qquad = 7 \text{ tons} + 1 \text{ ton } 750 \text{ lb}$

$\qquad = 8 \text{ tons } 750 \text{ lb}$

33.
$$\begin{array}{r} 12 \text{ lb } 4 \text{ oz} \\ -3 \text{ lb } 9 \text{ oz} \end{array} \qquad \begin{array}{r} 11 \text{ lb } 20 \text{ oz} \\ -3 \text{ lb }\ \ 9 \text{ oz} \\ \hline 8 \text{ lb } 11 \text{ oz} \end{array}$$

37. $6 \text{ tons } 1500 \text{ lb} \div 5 = \dfrac{6}{5} \text{ tons } 300 \text{ lb}$

$\qquad = 1\dfrac{1}{5} \text{ tons } 300 \text{ lb}$

$\qquad = 1 \text{ ton} + \dfrac{2000 \text{ lb}}{5} + 300 \text{ lb}$

$\qquad = 1 \text{ ton} + 400 \text{ lb} + 300 \text{ lb}$

$\qquad = 1 \text{ ton } 700 \text{ lb}$

41. $4 \text{ g} = \dfrac{4 \text{ g}}{1} \cdot \dfrac{1000 \text{ mg}}{1 \text{ g}} = 4000 \text{ mg}$

45. $48 \text{ mg} = \dfrac{48 \text{ mg}}{1} \cdot \dfrac{1 \text{ g}}{1000 \text{ mg}} = \dfrac{48}{1000} \text{ g} = 0.048 \text{ g}$

49. $15.14 \text{ g} = \dfrac{15.14 \text{ g}}{1} \cdot \dfrac{1000 \text{ mg}}{1 \text{ g}}$

$\qquad = 15.14 \cdot 1000 \text{ mg}$

$\qquad = 15{,}140 \text{ mg}$

53. $35 \text{ hg} = \dfrac{35 \text{ hg}}{1} \cdot \dfrac{10{,}000 \text{ cg}}{1 \text{ hg}}$

$\qquad = 35 \cdot 10{,}000 \text{ cg}$

$\qquad = 350{,}000 \text{ cg}$

57. $205 \text{ mg} + 5.61 \text{ g} = 0.205 \text{ g} + 5.61 \text{ g} = 5.815 \text{ g}$

or

$205 \text{ mg} + 5.61 \text{ g} = 205 \text{ mg} + 5610 \text{ mg}$

$\qquad\qquad\qquad\quad = 5815 \text{ mg}$

61. $1.61 \text{ kg} - 250 \text{ g} = 1.61 \text{ kg} - 0.250 \text{ kg} = 1.36 \text{ kg}$

or

$1.61 \text{ kg} - 250 \text{ g} = 1610 \text{ g} - 250 \text{ g} = 1360 \text{ g}$

65. $17 \text{ kg} \div 8 = \dfrac{17}{8} \text{ kg}$

$$\begin{array}{r} 2.125 \\ 8)\overline{17.000} \\ \underline{-16} \\ 10 \\ \underline{-8} \\ 20 \\ \underline{-16} \\ 40 \\ \underline{-40} \\ 0 \end{array}$$

$17 \text{ kg} \div 8 = 2.125 \text{ kg}$

69.

Object	Tons	Pounds	Ounces
A 12-inch cube of osmium	$\frac{269}{400}$ or 0.6725	1345	21,520

73.

Object	Grams	Kilograms	Milligrams	Centigrams
A six-year-old boy	21,000	21	21,000,000	2,100,000

77. $0.09 \text{ g} = \frac{0.09 \text{ g}}{1} \cdot \frac{1000 \text{ mg}}{1 \text{ g}} = 0.09 \cdot 1000 \text{ mg} = 90 \text{ mg}$

$90 \text{ mg} - 60 \text{ mg} = 30 \text{ mg}$

The extra-strength tablet contains 30 mg more medication.

81.

$\begin{array}{r} 64 \text{ lb } 8 \text{ oz} \\ -28 \text{ lb } 10 \text{ oz} \end{array}$ $\begin{array}{r} 63 \text{ lb } 24 \text{ oz} \\ -28 \text{ lb } 10 \text{ oz} \\ \hline 35 \text{ lb } 14 \text{ oz} \end{array}$

Carla's zucchini was 35 lb 14 oz lighter than the record weight.

85. $3 \times 16 = 48$

3 cartons contain 48 boxes of fruit.

$3 \text{ mg} \times 48 = 144 \text{ mg}$

3 cartons contain 144 mg of preservatives.

89. $3 \text{ lb } 4 \text{ oz} \times 10 = 30 \text{ lb } 40 \text{ oz}$

$= 30 \text{ lb} + 2 \text{ lb } 8 \text{ oz}$

$= 32 \text{ lb } 8 \text{ oz}$

Each box weighs 32 lb 8 oz.

$32 \text{ lb } 8 \text{ oz} \times 4 = 128 \text{ lb } 32 \text{ oz}$

$= 128 \text{ lb} + 2 \text{ lb}$

$= 130 \text{ lb}$

4 boxes of meat weigh 130 lb.

93. $\frac{4}{25} = \frac{4}{25} \cdot \frac{4}{4} = \frac{16}{100} = 0.16$

97. No, a pill containing 2 kg of medication is not reasonable.

101. No, a professor weighing less than 150 g is not reasonable.

105. True, a kilogram is 1000 grams.

Exercise Set 9.6

1. $32 \text{ fl oz} = \frac{32 \text{ fl oz}}{1} \cdot \frac{1 \text{ c}}{8 \text{ fl oz}} = \frac{32}{8} \text{ c} = 4 \text{ c}$

5. $14 \text{ qt} = \frac{14 \text{ qt}}{1} \cdot \frac{1 \text{ gal}}{4 \text{ qt}} = \frac{14}{4} \text{ gal} = 3\frac{1}{2} \text{ gal}$

9. $2 \text{ qt} = \frac{2 \text{ qt}}{1} \cdot \frac{2 \text{ pt}}{1 \text{ qt}} \cdot \frac{2 \text{ c}}{1 \text{ pt}} = 2 \cdot 2 \cdot 2 \cdot \text{c} = 8 \text{ c}$

13. $42 \text{ c} = \frac{42 \text{ c}}{1} \cdot \frac{1 \text{ qt}}{4 \text{ c}} = \frac{42}{4} \text{ qt} = 10\frac{1}{2} \text{ qt}$

17. $5 \text{ gal } 3 \text{ qt} = \frac{5 \text{ gal}}{1} \cdot \frac{4 \text{ qt}}{1 \text{ gal}} + 3 \text{ qt}$

$= 5 \cdot 4 \text{ qt} + 3 \text{ qt}$

$= 20 \text{ qt} + 3 \text{ qt}$

$= 23 \text{ qt}$

21. $58 \text{ qt} = 56 \text{ qt} + 2 \text{ qt}$

$= \frac{56 \text{ qt}}{1} \cdot \frac{1 \text{ gal}}{4 \text{ qt}} + 2 \text{ qt}$

$= \frac{56}{4} \text{ gal} + 2 \text{ qt}$

$= 14 \text{ gal } 2 \text{ qt}$

25. $2\frac{3}{4} \text{ gal} = \frac{11}{4} \text{ gal}$

$= \frac{\frac{11}{4} \text{ gal}}{1} \cdot \frac{4 \text{ qt}}{1 \text{ gal}} \cdot \frac{2 \text{ pt}}{1 \text{ qt}}$

$= \frac{11}{4} \cdot 4 \cdot 2 \text{ pt}$

$= 22 \text{ pt}$

29. $1 \text{ c } 5 \text{ fl oz} + 2 \text{ c } 7 \text{ fl oz} = 3 \text{ c } 12 \text{ fl oz}$

$= 3 \text{ c} + 1 \text{ c } 4 \text{ fl oz}$

$= 4 \text{ c } 4 \text{ fl oz}$

33.

$\begin{array}{r} 3 \text{ gal } 1 \text{ qt} \\ -1 \text{ qt } 1 \text{ pt} \end{array}$ $\begin{array}{r} 2 \text{ gal } 5 \text{ qt} \\ -1 \text{ qt } 1 \text{ pt} \end{array}$ $\begin{array}{r} 2 \text{ gal } 4 \text{ qt } 2 \text{ pt} \\ -1 \text{ qt } 2 \text{ pt} \\ \hline 2 \text{ gal } 3 \text{ qt } 1 \text{ pt} \end{array}$

37. $9 \text{ gal } 2 \text{ qt} \div 2 = (8 \text{ gal } 4 \text{ qt} + 2 \text{ qt}) \div 2$

$= 8 \text{ gal } 6 \text{ qt} \div 2$

$= 4 \text{ gal } 3 \text{ qt}$

41. $0.16 \text{ L} = \frac{0.16 \text{ L}}{1} \cdot \frac{1 \text{ kl}}{1000 \text{ L}} = 0.00016 \text{ kl}$

45. $3.2 \text{ L} = \frac{3.2 \text{ L}}{1} \cdot \frac{100 \text{ cl}}{1 \text{ L}} = 320 \text{ cl}$

49. $64 \text{ ml} = \frac{64 \text{ ml}}{1} \cdot \frac{1 \text{ L}}{1000 \text{ ml}} = \frac{64}{1000} \text{ L} = 0.064 \text{ L}$

53. $3.6 \text{ L} = \frac{3.6 \text{ L}}{1} \cdot \frac{1000 \text{ ml}}{1 \text{ L}} = 3600 \text{ ml}$

57. $2700 \text{ ml} + 1.8 \text{ L} = 2.7 \text{ L} + 1.8 \text{ L} = 4.5 \text{ L}$

or

$2700 \text{ ml} + 1.8 \text{ L} = 2700 \text{ ml} + 1800 \text{ ml} = 4500 \text{ ml}$

61. $17,500 \text{ ml} - 0.9 \text{ L} = 17,500 \text{ ml} - 900 \text{ ml}$

$= 16,600 \text{ ml}$

or

$17,500 \text{ ml} - 0.9 \text{ L} = 17.5 \text{ L} - 0.9 \text{ L} = 16.6 \text{ L}$

65. $81.2 \text{ L} \div 0.5 = 81.2 \text{ L} \div \frac{1}{2}$

$= 81.2 \text{ L} \cdot 2$

$= 162.4 \text{ L}$

69.

Capacity	Cups	Gallons	Quarts	Pints
Your kidneys filter about this amount of blood every minute	4	$\frac{1}{4}$	1	2

73. 354 ml + 18.6 L = 0.354 L + 18.6 L = 18.954 L

There were 18.954 liters of gasoline in her tank.

77. 5 pt 1 c + 2 pt 1 c = 7 pt 2 c

$$= 7 \text{ pt} + 1 \text{ pt}$$
$$= 8 \text{ pt}$$
$$= \frac{8 \text{ pt}}{1} \cdot \frac{1 \text{ qt}}{2 \text{ pt}}$$
$$= 4 \text{ qt}$$
$$= \frac{4 \text{ qt}}{1} \cdot \frac{1 \text{ gal}}{4 \text{ qt}}$$
$$= \frac{4}{4} \text{ gal}$$
$$= 1 \text{ gal}$$

Yes, the liquid can be poured into the container without causing it to overflow.

81. $\frac{20}{25} = \frac{4 \cdot 5}{5 \cdot 5} = \frac{4}{5}$

85. $\frac{72}{80} = \frac{8 \cdot 9}{8 \cdot 10} = \frac{9}{10}$

89. No, a tub filled with 3000 ml of hot water is not reasonable.

93. answers may vary

97. B indicates 1.5 cc.

101. B indicates 54 u or 0.54 cc.

Exercise Set 9.7

1. $756 \text{ ml} \approx \frac{756 \text{ ml}}{1} \cdot \frac{1 \text{ fl oz}}{29.57 \text{ ml}} \approx 25.57 \text{ fl oz}$

5. $1000 \text{ g} \approx \frac{1000 \text{ g}}{1} \cdot \frac{0.04 \text{ oz}}{1 \text{ g}} \approx 40 \text{ oz}$

9. $14.5 \text{ L} \approx \frac{14.5 \text{ L}}{1} \cdot \frac{0.26 \text{ gal}}{1 \text{ L}} \approx 3.77 \text{ gal}$

13.

	Meters	Yards	Centimeters	Feet	Inches
The height of a woman	1.5	$1\frac{2}{3}$	150	5	60

17. $10 \text{ cm} = \frac{10 \text{ cm}}{1} \cdot \frac{1 \text{ in.}}{2.54 \text{ cm}} \approx 3.94 \text{ in.}$

The balance beam is approximately 3.94 inches wide.

21. $200 \text{ mg} = 0.2 \text{ g} \approx \frac{0.2 \text{ g}}{1} \cdot \frac{1 \text{ oz}}{28.35 \text{ g}} \approx 0.007 \text{ oz}$

25. $4500 \text{ km} = \frac{4500 \text{ km}}{1} \cdot \frac{1 \text{ mi}}{1.61 \text{ km}} \approx 2795 \text{ mi}$

The trip is about 2795 miles.

29. 1.5 lb − 1.25 lb = 0.25 lb

$0.25 \text{ lb} \approx \frac{0.25 \text{ lb}}{1} \cdot \frac{0.45 \text{ kg}}{1 \text{ lb}} \cdot \frac{1000 \text{ g}}{1 \text{ kg}} \approx 112.5 \text{ g}$

The difference is approximately 112.5 g.

33. $8 \text{ m} \approx \frac{8 \text{ m}}{1} \cdot \frac{3.28 \text{ ft}}{1 \text{ m}} \approx 26.24 \text{ ft}$

The base diameter is approximately 26.24 ft.

37. One dose every 4 hours results in $\frac{24}{4} = 6$ doses per day and $6 \times 7 = 42$ doses per week.

$5 \text{ ml} \times 42 = 210 \text{ ml}$

$210 \text{ ml} \approx \frac{210 \text{ ml}}{1} \cdot \frac{1 \text{ fl oz}}{29.57 \text{ ml}} \approx 7.1 \text{ fl oz}$

8 fluid ounces of medicine should be purchased.

41. A liter has greater capacity than a quart; b.

45. An $8\frac{1}{2}$ ounce glass of water has a capacity of about 250 ml $\left(\frac{1}{4} \text{ L}\right)$; d.

49. $C = \frac{5}{9}(F - 32)$

$$= \frac{5}{9}(77 - 32)$$
$$= \frac{5}{9}(45)$$
$$= 25$$

77°F is 25°C

53. $F = \frac{9}{5}C + 32$

$$= \frac{9}{5}(50) + 32$$
$$= 90 + 32$$
$$= 122$$

50°C is 122°F

57. $C = \frac{5}{9}(F - 32)$

$$= \frac{5}{9}(20 - 32)$$
$$= \frac{5}{9}(-12)$$
$$\approx -6.7$$

20°F is approximately −6.7°C.

61. $F = 1.8C + 32$

$$= 1.8(92) + 32$$
$$= 165.6 + 32$$
$$= 197.6$$

92°C is 197.6

65. $C = \dfrac{5}{9}(F - 32)$

$= \dfrac{5}{9}(134 - 32)$

$= \dfrac{5}{9}(102)$

≈ 56.7

134°F is approximately 56.7°C.

69. $C = \dfrac{5}{9}(F - 32)$

$= \dfrac{5}{9}(70 - 32)$

$= \dfrac{5}{9}(38)$

≈ 21.1

70°F is approximately 21.1°C.

73. $F = 1.8C + 32$

$= 1.8(4000) + 32$

$= 7200 + 32$

$= 7232$

4000°C is 7232°F

77. $3[(1 + 5) \cdot (8 - 6)] = 3[(6) \cdot (2)] = 3[12] = 36$

81. No, a fever of 40°F is not reasonable.

85. Yes, a fever of 40°C is reasonable.

89. $40 \text{ in.} = \dfrac{40 \text{ in}}{1} \cdot \dfrac{2.54 \text{ cm}}{1 \text{ in}} = 101.6 \text{ cm}$

$\text{BSA} = \sqrt{\dfrac{50 \cdot 101.6}{3600}} \approx 1.19$

The BSA is approximately 1.19 square meters.

93. $C = \dfrac{5}{9}(F - 32)$

$= \dfrac{5}{9}(918{,}000{,}000 - 32)$

$= \dfrac{5}{9}(917{,}999{,}968)$

$\approx 510{,}000{,}000$

918,000,000°F is approximately 510,000,000°C

Chapter Test

1. The complement of an angle that measures 78° is an angle that measures 90° − 78° = 12°.

5. $\angle x$ and the angle marked 73° are vertical angles, so $m\angle x = 73°$.
$\angle x$ and $\angle y$ are alternate interior angles, so $m\angle y = m\angle x = 73°$.
$\angle x$ and $\angle z$ are corresponding angles, so $m\angle z = m\angle x = 73°$.

9. $P = 2 \cdot l + 2 \cdot w$

$= 2(7 \text{ yd}) + 2(5.3 \text{ yd})$

$= 14 \text{ yd} + 10.6 \text{ yd}$

$= 24.6 \text{ yd}$

$A = l \cdot w = 7 \text{ yd} \cdot 5.3 \text{ yd} = 37.1 \text{ sq yd}$

13. $P = 4 \cdot s = 4 \cdot 4 \text{ in.} = 16 \text{ in.}$

The perimeter of the frame is 16 inches.

17. $2\dfrac{1}{2} \text{ gal} = \dfrac{2\frac{1}{2} \text{ gal}}{1} \cdot \dfrac{4 \text{ qt}}{1 \text{ gal}} = 10 \text{ qt}$

21. $40 \text{ mg} = \dfrac{40 \text{ mg}}{1} \cdot \dfrac{1 \text{ g}}{1000 \text{ mg}} = 0.04 \text{ g}$

25. $0.83 \text{ L} = \dfrac{0.83 \text{ L}}{1} \cdot \dfrac{1000 \text{ ml}}{1 \text{ L}} = 830 \text{ ml}$

29. $5 \text{ gal } 2 \text{ qt} \div 2 = 4 \text{ gal } 6 \text{ qt} \div 2$

$= \dfrac{4}{2} \text{ gal } \dfrac{6}{2} \text{ qt}$

$= 2 \text{ gal } 3 \text{ qt}$

33. $F = \dfrac{9}{5}C + 32 = \dfrac{9}{5}(12.6) + 32 = 22.68 + 32 \approx 54.7$

12.6°C is about 54.7°F

37. $\begin{array}{r} 2 \text{ ft } 9 \text{ in.} \\ \times \qquad 6 \\ \hline \end{array}$

12 ft 54 in. = 12 ft + 4 ft 6 in. = 16 ft 6 in.

Thus, 16 ft 6 in. of material is needed.

CHAPTER 10

Exercise Set 10.1

1. $(2x + 3) + (-7x - 27) = (2x - 7x) + (3 - 27)$

$= -5x + (-24)$

$= -5x - 24$

5. $(12y - 20) + (9y^2 + 13y - 20)$

$= 9y^2 + (12y + 13y) + (-20 - 20)$

$= 9y^2 + 25y - 40$

9. $-(9x - 16) = -1(9x - 16)$

$= -1(9x) + (-1)(-16)$

$= -9x + 16$

13. $(8a - 5) - (3a + 8) = (8a - 5) + (-3a - 8)$
$$= 8a - 5 - 3a - 8$$
$$= 8a - 3a - 5 - 8$$
$$= 5a - 13$$

17. $(10y^2 - 7) - (20y^3 - 2y^2 - 3)$
$$= (10y^2 - 7) + (-20y^3 + 2y^2 + 3)$$
$$= (-20y^3) + (10y^2 + 2y^2) + (-7 + 3)$$
$$= -20y^3 + 12y^2 - 4$$

21.
$$\begin{array}{rr} 13y^2 - 6y - 14 & \quad 13y^2 - 6y - 14 \\ -(5y^2 + 4y - 6) & \quad + -5y^2 - 4y + 6 \\ \hline & \quad 8y^2 - 10y - 8 \end{array}$$

25. $(4y + 4) - (3y + 8) = (4y + 4) + (-3y - 8)$
$$= (4y - 3y) + (4 - 8)$$
$$= y - 4$$

29. $(5x + 4.5) + (-x - 8.6) = (5x - x) + (4.5 - 8.6)$
$$= 4x - 4.1$$

33. $(21y - 4.6) - (36y - 8.2)$
$$= (21y - 4.6) + (-36y + 8.2)$$
$$= (21y - 36y) + (-4.6 + 8.2)$$
$$= -15y + 3.6$$

37. $(2b^3 + 5b^2 - 5b - 8) + (8b^2 + 9b + 6)$
$$= 2b^3 + (5b^2 + 8b^2) + (-5b + 9b) + (-8 + 6)$$
$$= 2b^3 + 13b^2 + 4b - 2$$

41. $\left(3z + \dfrac{6}{7}\right) - \left(3z - \dfrac{3}{7}\right) = \left(3z + \dfrac{6}{7}\right) + \left(-3z + \dfrac{3}{7}\right)$
$$= (3z - 3z) + \left(\dfrac{6}{7} + \dfrac{3}{7}\right)$$
$$= \dfrac{9}{7}$$

45. $x^2 - 6x + 3$
Let $x = 2$.
$(2)^2 - 6(2) + 3 = 4 - 12 + 3 = -5$

49. $2x + 10$
Let $x = 5$.
$2(5) + 10 = 10 + 10 = 20$

53. $2x^2 + 4x - 20$
Let $x = 5$.
$2(5)^2 + 4(5) - 20 = 2(25) + 20 - 20 = 50$

57. $3000 + 20x$
Let $x = 10$.
$3000 + 20(10) = 3000 + 200 = 3200$
It costs \$3200 to manufacture 10 file cabinets.

61. 2007 corresponds to $x = 7$.
$17x^2 + 210x + 700$
Let $x = 7$.
$17(7)^2 + 210(7) + 700 = 17(49) + 1470 + 700$
$$= 3003$$
3003 million cellular subscribers are expected in 2007.

65. $(-5)^2 = (-5) \cdot (-5) = 25$

69. $2 \cdot 2 \cdot a \cdot a \cdot a \cdot a = 2^2 a^4$

73. $(7x - 10) - (3x + 5) = (7x - 10) + (-3x - 5)$
$$= (7x - 3x) + (-10 - 5)$$
$$= 4x - 15$$
The missing length is $(4x - 15)$ units.

77. $7a^4 - 6a^2 + 2a - 1$
Let $a = 1.2$.
$7(1.2)^4 - 6(1.2)^2 + 2(1.2) - 1 = 7.2752$

Exercise Set 10.2

1. $x^5 \cdot x^9 = x^{5+9} = x^{14}$

5. $3z^3 \cdot 5z^2 = (3 \cdot 5)(z^3 \cdot z^2) = 15z^5$

9. $2x \cdot 3x \cdot 7x = (2 \cdot 3 \cdot 7)(x \cdot x \cdot x) = 42x^3$

13. $(-5x^2y^3)(-5x^4y) = (-5)(-5)(x^2 \cdot x^4)(y^3 \cdot y)$
$$= 25x^6y^4$$

17. $(x^5)^3 = x^{5 \cdot 3} = x^{15}$

21. $(b^7)^6 \cdot (b^2)^{10} = b^{7 \cdot 6} \cdot b^{2 \cdot 10}$
$$= b^{42} \cdot b^{20}$$
$$= b^{42+20}$$
$$= b^{62}$$

25. $(a^{11}b^8)^3 = a^{11 \cdot 3}b^{8 \cdot 3} = a^{33}b^{24}$

29. $(-3y)(2y^7)^3 = (-3y) \cdot 2^3(y^7)^3$
$$= (-3y) \cdot 8y^{21}$$
$$= (-3)(8)(y^1 \cdot y^{21})$$
$$= -24y^{22}$$

33. $7(x - 3) = 7x - 21$

37. $9(x + 2y - 3) = 9x + 18y - 27$

41. Area $= \dfrac{1}{2}bh$
$$= \dfrac{1}{2} \cdot (6a^3b^4) \cdot (4ab)$$
$$= \left(\dfrac{1}{2} \cdot 6 \cdot 4\right)(a^3 \cdot a)(b^4 \cdot b)$$
$$= 12a^4b^5$$
The area is $12a^4b^5$ square meters.

45. $(8.1x^{10})^5 = 8.1^5(x^{10})^5 = 34{,}867.84401x^{50}$

49. $(a^{20}b^{10}c^5)^5 \cdot (a^9b^{12})^3 = a^{100}b^{50}c^{25} \cdot a^{27}b^{36}$

$$= a^{127}b^{86}c^{25}$$

Exercise Set 10.3

1. $3x(9x^2 - 3) = 3x \cdot 9x^2 + 3x \cdot (-3)$

$$= (3 \cdot 9)(x \cdot x^2) + (3)(-3)(x)$$

$$= 27x^3 + (-9x)$$

$$= 27x^3 - 9x$$

5. $7x^2(6x^2 - 5x + 7)$

$$= (7x^2)(6x^2) + (7x^2)(-5x) + (7x^2)(7)$$

$$= (7 \cdot 6)(x^2 \cdot x^2) + (7)(-5)(x^2 \cdot x) + (7 \cdot 7)x^2$$

$$= 42x^4 - 35x^3 + 49x^2$$

9. $(2x - 6)(x + 4)$

$$= 2x(x + 4) - 6(x + 4)$$

$$= 2x \cdot x + 2x \cdot 4 - 6 \cdot x - 6 \cdot 4$$

$$= 2x^2 + 8x - 6x - 24$$

$$= 2x^2 + 2x - 24$$

13. $(a + 6)(a^2 - 6a + 3)$

$$= a(a^2 - 6a + 3) + 6(a^2 - 6a + 3)$$

$$= a \cdot a^2 + a(-6a) + a \cdot 3 + 6 \cdot a^2 + 6(-6a) + 6 \cdot 3$$

$$= a^3 - 6a^2 + 3a + 6a^2 - 36a + 18$$

$$= a^3 - 33a + 18$$

17. $(x^3 + 2x + x^2)(3x + 1 + x^2)$

$$= x^3(3x + 1 + x^2) + 2x(3x + 1 + x^2) + x^2(3x + 1 + x^2)$$

$$= x^3 \cdot 3x + x^3 \cdot 1 + x^3 \cdot x^2 + 2x \cdot 3x + 2x \cdot 1 + 2x \cdot x^2 + x^2 \cdot 3x$$
$$+ x^2 \cdot 1 + x^2 \cdot x^2$$

$$= 3x^4 + x^3 + x^5 + 6x^2 + 2x + 2x^3 + 3x^3 + x^2 + x^4$$

$$= x^5 + 4x^4 + 6x^3 + 7x^2 + 2x$$

21. $-2y^2(3y + y^2 - 6)$

$$= -2y^2 \cdot 3y + (-2y^2) \cdot y^2 + (-2y^2)(-6)$$

$$= -6y^3 - 2y^4 + 12y^2$$

25. $(2a + 3)(2a - 3) = 2a(2a - 3) + 3(2a - 3)$

$$= 2a \cdot 2a + 2a(-3) + 3 \cdot 2a + 3(-3)$$

$$= 4a^2 - 6a + 6a - 9$$

$$= 4a^2 - 9$$

29. $\left(b + \dfrac{3}{5}\right)\left(b + \dfrac{4}{5}\right) = b\left(b + \dfrac{4}{5}\right) + \dfrac{3}{5}\left(b + \dfrac{4}{5}\right)$

$$= b^2 + \dfrac{4}{5}b + \dfrac{3}{5}b + \dfrac{3}{5} \cdot \dfrac{4}{5}$$

$$= b^2 + \dfrac{7}{5}b + \dfrac{12}{25}$$

33. $(7x + 5)^2 = (7x + 5)(7x + 5)$

$$= 7x(7x + 5) + 5(7x + 5)$$

$$= 49x^2 + 35x + 35x + 25$$

$$= 49x^2 + 70x + 25$$

37. $(2x^2 - 3)(4x^3 + 2x - 3)$

$$= 2x^2(4x^3 + 2x - 3) - 3(4x^3 + 2x - 3)$$

$$= 8x^5 + 4x^3 - 6x^2 - 12x^3 - 6x + 9$$

$$= 8x^5 - 8x^3 - 6x^2 - 6x + 9$$

41.

$$
\begin{array}{r}
2z^2 - z + 1 \\
\times\ 5z^2 + z - 2 \\
\hline
-4z^2 + 2z - 2 \\
2z^3 - z^2 + z \\
10z^4 - 5z^3 + 5z^2 \\
\hline
10z^4 - 3z^3\ \ \ \ \ \ \ + 3z - 2
\end{array}
$$

45. $72 = 2 \cdot 2 \cdot 2 \cdot 3 \cdot 3 = 2^3 \cdot 3^2$

49. $(y - 6)(y^2 + 3y + 2)$

$$= y(y^2 + 3y + 2) - 6(y^2 + 3y + 2)$$

$$= y^3 + 3y^2 + 2y - 6y^2 - 18y - 12$$

$$= y^3 - 3y^2 - 16y - 12$$

The area is $(y^3 - 3y^2 - 16y - 12)$ square feet.

53. answers may vary

Exercise Set 10.4

1. $48 = 2 \cdot 2 \cdot 2 \cdot 2 \cdot 3$

$15 = 3 \cdot 5$

GCF $= 3$

5. $12 = 2 \cdot 2 \cdot 3$

$20 = 2 \cdot 2 \cdot 5$

$36 = 2 \cdot 2 \cdot 3 \cdot 3$

GCF $= 2 \cdot 2 = 4$

9. $y^7 = y^2 \cdot y^5$

$y^2 = y^2$

$y^{10} = y^2 \cdot y^8$

GCF $= y^2$

13. $x^3y^2 = x \cdot x^2 \cdot y^2$

$xy^2 = x \cdot y^2$

$x^4y^2 = x \cdot x^3 \cdot y^2$

GCF $= x \cdot y^2 = xy^2$

17. $2 = 2$

$14 = 2 \cdot 7$

$18 = 2 \cdot 3 \cdot 3$

GCF $= 2$

$z^3 = z^3$

$z^5 = z^3 \cdot z^2$

$z^3 = z^3$

GCF $= z^3$

GCF $= 2z^3$

21. $10a^6 = 5a^6 \cdot 2$

$5a^8 = 5a^6 \cdot a^2$

GCF $= 5a^6$

$10a^6 - 5a^8 = 5a^6 \cdot 2 - 5a^6 \cdot a^2$

$\qquad\qquad = 5a^6(2 - a^2)$

25. $z^7 = z^5 \cdot z^2$

$6z^5 = z^5 \cdot 6$

GCF $= z^5$

$z^7 - 6z^5 = z^5 \cdot z^2 - z^5 \cdot 6 = z^5(z^2 - 6)$

29. $12a^5 = 12a^5$

$36a^6 = 12a^5 \cdot 3a$

GCF $= 12a^5$

$12a^5 - 36a^6 = 12a^5 \cdot 1 - 12a^5 \cdot 3a$

$\qquad\qquad = 12a^5(1 - 3a)$

33. $80\% = \dfrac{80}{100} = \dfrac{4}{5}$

37. area on the left: $x \cdot x = x^2$

area on the right: $2 \cdot x = 2x$

total area: $x^2 + 2x$

Notice that $x(x + 2) = x^2 + 2x$.

41. Let $x = 2$ and $z = 7$.

$(xy + z)^x = (2y + 7)^2$

$\qquad\qquad = (2y)^2 + 2y \cdot 7 + 7 \cdot 2y + 7^2$

$\qquad\qquad = (2y)^2 + 2 \cdot 2y \cdot 7 + 7^2$

$\qquad\qquad = 4y^2 + 28y + 49$

Chapter Test

1. $(11x - 3) + (4x - 1) = (11x + 4x) + (-3 - 1)$

$\qquad\qquad\qquad\qquad = 15x + (-4)$

$\qquad\qquad\qquad\qquad = 15x - 4$

5. $x^2 - 6x + 1 = (8)^2 - 6(8) + 1$

$\qquad\qquad\quad = 64 - 48 + 1$

$\qquad\qquad\quad = 17$

9. $(6a^3)(-2a^7) = (6)(-2)(a^3 \cdot a^7) = -12a^{10}$

13. $-2y(y^3 + 6y^2 - 4)$

$= -2y \cdot y^3 - 2y \cdot 6y^2 - 2y \cdot (-4)$

$= -2y^4 - 12y^3 + 8y$

17. Area:

$(x + 7)(5x - 2) = x(5x - 2) + 7(5x - 2)$

$\qquad\qquad\qquad\quad = 5x^2 - 2x + 35x - 14$

$\qquad\qquad\qquad\quad = 5x^2 + 33x - 14$

The area is $(5x^2 + 33x - 14)$ square inches.

Perimeter:

$2(2x) + 2(5x - 2) = 4x + 10x - 4 = 14x - 4$

The perimeter is $(14x - 4)$ inches.

21. $10a^2 = 2a \cdot 5a$

$12a = 2a \cdot 6$

GCF $= 2a$

$10a^2 + 12a = 2a \cdot 5a + 2a \cdot 6 = 2a(5a + 6)$

APPENDICES

Appendix C Exercise Set

1. $\dfrac{x^3}{x} = \dfrac{x^3}{x^1} = x^{3-1} = x^2$

5. $\dfrac{p^7 q^{20}}{pq^{15}} = \dfrac{p^7}{p^1} \cdot \dfrac{q^{20}}{q^{15}} = p^{7-1} \cdot q^{20-15} = p^6 q^5$

9. $7^0 = 1$

13. $-7^0 = -(7^0) = -1$

17. $4^{-3} = \dfrac{1}{4^3} = \dfrac{1}{64}$

21. $3^{-1} + 2^{-1} = \dfrac{1}{3} + \dfrac{1}{2} = \dfrac{1}{3} \cdot \dfrac{2}{2} + \dfrac{1}{2} \cdot \dfrac{3}{3} = \dfrac{2}{6} + \dfrac{3}{6} = \dfrac{5}{6}$

25. $\dfrac{x^{-2}}{x} = \dfrac{x^{-2}}{x^1} = x^{-2-1} = x^{-3} = \dfrac{1}{x^3}$

29. $3^{-2} + 3^{-1} = \dfrac{1}{3^2} + \dfrac{1}{3^1} = \dfrac{1}{9} + \dfrac{1}{3} = \dfrac{1}{9} + \dfrac{3}{9} = \dfrac{4}{9}$

33. $\dfrac{1}{p^{-4}} = p^4$

37. $(x^8 y^{-6})(x^{-2} y^{12}) = x^{8+(-2)} \cdot y^{-6+12}$

$\qquad\qquad\qquad\qquad = x^6 y^6$

41. $(5x^{-7})(3x^4) = 5 \cdot 3 \cdot x^{-7+4}$

$\qquad\qquad\qquad = 15 \cdot x^{-3}$

$\qquad\qquad\qquad = 15 \cdot \dfrac{1}{x^3}$

$\qquad\qquad\qquad = \dfrac{15}{x^3}$

45. $(8m^5 n^{-1})(7m^2 n^{-4}) = 8 \cdot 7 \cdot m^{5+2} n^{-1+(-4)}$

$\qquad\qquad\qquad\qquad = 56 \cdot m^7 \cdot n^{-5}$

$\qquad\qquad\qquad\qquad = 56 \cdot m^7 \cdot \dfrac{1}{n^5}$

$\qquad\qquad\qquad\qquad = \dfrac{56m^7}{n^5}$

49. $\dfrac{a^9 b^{14}}{ab} = \dfrac{a^9}{a^1} \cdot \dfrac{b^{14}}{b^1} = a^{9-1} \cdot b^{14-1} = a^8 b^{13}$

53. $3z^0 = 3 \cdot z^0 = 3 \cdot 1 = 3$

57. $8x^{-9} = 8 \cdot x^{-9} = 8 \cdot \dfrac{1}{x^9} = \dfrac{8}{x^9}$

61. $\dfrac{z^{-8}}{z^{-1}} = z^{-8-(-1)} = z^{-8+1} = z^{-7} = \dfrac{1}{z^7}$

65. $(a^{-2}b^3)(a^{10}b^{-11}) = a^{-2+10} \cdot b^{3+(-11)}$

$$= a^8 \cdot b^{-8}$$

$$= a^8 \cdot \dfrac{1}{b^8}$$

$$= \dfrac{a^8}{b^8}$$

Appendix D Exercise Set

1. $78{,}000 = 7.8 \times 10^4$

5. $0.00635 = 6.35 \times 10^{-3}$

9. $13{,}600 = 1.36 \times 10^4$

13. $3.3 \times 10^{-2} = 0.033$

17. $7.0 \times 10^8 = 700{,}000{,}000$

21. $1.23 \times 10^{12} = 1{,}230{,}000{,}000{,}000$

25. $(1.2 \times 10^{-3})(3 \times 10^{-2}) = 1.2 \cdot 3 \times 10^{-3+(-2)}$

$$= 3.6 \times 10^{-5}$$

$$= 0.000036$$

29. $\dfrac{8 \times 10^{-1}}{16 \times 10^5} = \dfrac{8}{16} \cdot \dfrac{10^{-1}}{10^5}$

$$= 0.5 \times 10^{-1-5}$$

$$= 0.5 \times 10^{-6}$$

$$= 0.0000005$$

33. $(7.5 \times 10^5)(3600) = (7.5 \times 10^5)(3.6 \times 10^3)$

$$= 7.5 \cdot 3.6 \times 10^{5+3}$$

$$= 27 \times 10^8$$

$$= 2.7 \times 10^1 \times 10^8$$

$$= 2.7 \times 10^9$$

On average, 2.7×10^9 gallons of water flow over Niagara Falls each hour.

INDEX

X

Y

Z

Photo Credits

CHAPTER 8 CO PARK STREET/PhotoEdit Inc.

CHAPTER 9 CO Tom Bean **Page 617** AP Wide World Photos; Will & Deni McIntyre/Photo Researchers, Inc. **Page 625** © Tom and Susan Bean, Inc./DRK PHOTO **Page 641** William West/AFP; Rich Iwasaki **Page 646** Theo Allofs/Image Bank/Getty Images **Page 681** Bill Bachmann/© Index Stock Imagery, Inc.; © Jon Feingersh/Corbis/Bettmann

CHAPTER 10 CO Simon Marcus/Corbis

ELAYN MARTIN-GAY: PREALGEBRA, 5E, CHAPTER TEST PREP VIDEO CD

0-13-157633-X
© 2008 Pearson Education, Inc.
Pearson Prentice Hall
Pearson Education, Inc.
Upper Saddle River, NJ 07458
Pearson Prentice Hall™ is a trademark of Pearson Education, Inc.

YOU SHOULD CAREFULLY READ THE TERMS AND CONDITIONS BEFORE USING THE CD-ROM PACKAGE. USING THIS CD-ROM PACKAGE INDICATES YOUR ACCEPTANCE OF THESE TERMS AND CONDITIONS.

Pearson Education, Inc. provides this program and licenses its use. You assume responsibility for the selection of the program to achieve your intended results, and for the installation, use, and results obtained from the program. This license extends only to use of the program in the United States or countries in which the program is marketed by authorized distributors.

LICENSE GRANT

You hereby accept a nonexclusive, nontransferable, permanent license to install and use the program ON A SINGLE COMPUTER at any given time. You may copy the program solely for backup or archival purposes in support of your use of the program on the single computer. You may not modify, translate, disassemble, decompile, or reverse engineer the program, in whole or in part.

TERM

The License is effective until terminated. Pearson Education, Inc. reserves the right to terminate this License automatically if any provision of the License is violated. You may terminate the License at any time. To terminate this License, you must return the program, including documentation, along with a written warranty stating that all copies in your possession have been returned or destroyed.

LIMITED WARRANTY

THE PROGRAM IS PROVIDED "AS IS" WITHOUT WARRANTY OF ANY KIND, EITHER EXPRESSED OR IMPLIED, INCLUDING, BUT NOT LIMITED TO, THE IMPLIED WARRANTIES OF MERCHANTABILITY AND FITNESS FOR A PARTICULAR PURPOSE. THE ENTIRE RISK AS TO THE QUALITY AND PERFORMANCE OF THE PROGRAM IS WITH YOU. SHOULD THE PROGRAM PROVE DEFECTIVE, YOU (AND NOT PEARSON EDUCATION, INC. OR ANY AUTHORIZED DEALER) ASSUME THE ENTIRE COST OF ALL NECESSARY SERVICING, REPAIR, OR CORRECTION. NO ORAL OR WRITTEN INFORMATION OR ADVICE GIVEN BY PEARSON EDUCATION, INC., ITS DEALERS, DISTRIBUTORS, OR AGENTS SHALL CREATE A WARRANTY OR INCREASE THE SCOPE OF THIS WARRANTY.

SOME STATES DO NOT ALLOW THE EXCLUSION OF IMPLIED WARRANTIES, SO THE ABOVE EXCLUSION MAY NOT APPLY TO YOU. THIS WARRANTY GIVES YOU SPECIFIC LEGAL RIGHTS AND YOU MAY ALSO HAVE OTHER LEGAL RIGHTS THAT VARY FROM STATE TO STATE.

Pearson Education, Inc. does not warrant that the functions contained in the program will meet your requirements or that the operation of the program will be uninterrupted or error-free. However, Pearson Education, Inc. warrants the CD-ROM(s) on which the program is furnished to be free from defects in material and workmanship under normal use for a period of ninety (90) days from the date of delivery to you as evidenced by a copy of your receipt. The program should not be relied on as the sole basis to solve a problem whose incorrect solution could result in injury to person or property. If the program is employed in such a manner, it is at the user's own risk and Pearson Education, Inc. explicitly disclaims all liability for such misuse.

LIMITATION OF REMEDIES

Pearson Education, Inc.'s entire liability and your exclusive remedy shall be:
1. the replacement of any CD-ROM not meeting Pearson Education, Inc.'s "LIMITED WARRANTY" and that is returned to Pearson Education, or
2. if Pearson Education is unable to deliver a replacement CD-ROM that is free of defects in materials or workmanship, you may terminate this agreement by returning the program.

IN NO EVENT WILL PEARSON EDUCATION, INC. BE LIABLE TO YOU FOR ANY DAMAGES, INCLUDING ANY LOST PROFITS, LOST SAVINGS, OR OTHER INCIDENTAL OR CONSEQUENTIAL DAMAGES ARISING OUT OF THE USE OR INABILITY TO USE SUCH PROGRAM EVEN IF PEARSON EDUCATION, INC. OR AN AUTHORIZED DISTRIBUTOR HAS BEEN ADVISED OF THE POSSIBILITY OF SUCH DAMAGES, OR FOR ANY CLAIM BY ANY OTHER PARTY.

SOME STATES DO NOT ALLOW FOR THE LIMITATION OR EXCLUSION OF LIABILITY FOR INCIDENTAL OR CONSEQUENTIAL DAMAGES, SO THE ABOVE LIMITATION OR EXCLUSION MAY NOT APPLY TO YOU.

GENERAL

You may not sublicense, assign, or transfer the license of the program. Any attempt to sublicense, assign or transfer any of the rights, duties, or obligations hereunder is void.

This Agreement will be governed by the laws of the State of New York.

Should you have any questions concerning this Agreement, you may contact Pearson Education, Inc. by writing to:
ESM Media Development
Higher Education Division
Pearson Education, Inc.
1 Lake Street
Upper Saddle River, NJ 07458

Should you have any questions concerning technical support, you may write to:
New Media Production
Higher Education Division
Pearson Education, Inc.
1 Lake Street
Upper Saddle River, NJ 07458

YOU ACKNOWLEDGE THAT YOU HAVE READ THIS AGREEMENT, UNDERSTAND IT, AND AGREE TO BE BOUND BY ITS TERMS AND CONDITIONS. YOU FURTHER AGREE THAT IT IS THE COMPLETE AND EXCLUSIVE STATEMENT OF THE AGREEMENT BETWEEN US THAT SUPERSEDES ANY PROPOSAL OR PRIOR AGREEMENT, ORAL OR WRITTEN, AND ANY OTHER COMMUNICATIONS BETWEEN US RELATING TO THE SUBJECT MATTER OF THIS AGREEMENT.

System Requirements

Windows
Pentium II 300-MHz processor-based computer
Windows 2000 (Service Pack 4) or XP
64 MB RAM
7.2 MB available hard drive space (optional—for minimum QuickTime installation)
800 x 600 resolution
8x CD drive
QuickTime 7.x
Sound card
Internet Explorer 6.0 or Netscape 6.23, 7.2

Macintosh
Power PC G3 233 MHz
Mac OS 10.x
64 MB RAM
19 MB on OS X (optional—if QuickTime installation is needed)
800 x 600 resolution monitor
8x CD drive
QuickTime 6 or 7
Internet Explorer 5.2, Safari 1.2, or Netscape 7.2

Support Information

If you are having problems with this software, please call our Media Support Line (800) 677-6337 Monday through Friday 8:00 a.m. to 8:00 p.m. and Sunday 5:00 p.m. to 12:00 a.m. EST. You can also get support by filling out the web form located at: http://247.prenhall.com/mediaform

Our technical staff will need to know certain things about your system in order to help us solve your problems more quickly and efficiently. If possible, please be at your computer when you call for support. You should have the following information ready:
• Textbook ISBN
• CD-ROM ISBN
• corresponding product and title
• computer make and model
• Operating System (Windows or Macintosh) and Version
• RAM available
• hard disk space available
• Sound card? Yes or No
• printer make and model
• network connection
• detailed description of the problem, including the exact wording of any error messages.

NOTE: Pearson does not support and/or assist with the following:
• third-party software (i.e. Microsoft including Microsoft Office suite, Apple, Borland, etc.)
• homework assistance
• Textbooks and CD-ROMs purchased used are not supported and are non-replaceable. To purchase a new CD-ROM, contact Pearson Individual Order Copies at 1-800-282-0693.